Biotechnology

Second Edition

Volume 12

Legal, Economic and Ethical Dimensions

Biotechnology

Second Edition

Fundamentals

Volume 1
Biological Fundamentals

Volume 2
Genetic Fundamentals and
Genetic Engineering

Volume 3
Bioprocessing

Volume 4
Measuring, Modelling, and Control

Products

Volume 5
Genetically Engineered Proteins and
Monoclonal Antibodies

Volume 6
Products of Primary Metabolism

Volume 7
Products of Secondary Metabolism

Volume 8
Biotransformations

Special Topics

Volume 9
Enzymes, Biomass, Food and Feed

Volume 10
Special Processes

Volume 11
Environmental Processes

Volume 12
Legal, Economic and
Ethical Dimensions

Distribution:

VCH, P. O. Box 101161, D-69451 Weinheim (Federal Republic of Germany)

Switzerland: VCH, P. O. Box, CH-4020 Basel (Switzerland)

United Kingdom and Ireland: VCH (UK) Ltd., 8 Wellington Court, Cambridge CB1 1HZ (England)

USA and Canada: VCH, 220 East 23rd Street, New York, NY 10010–4606 (USA)

Japan: VCH, Eikow Building, 10-9 Hongo 1-chome, Bunkyo-ku, Tokyo 113 (Japan)

ISBN 3-527-28322-6
Set ISBN 3-527-28310-2

A Multi-Volume Comprehensive Treatise

Biotechnology

Second, Completely Revised Edition

Edited by
H.-J. Rehm and G. Reed
in cooperation with
A. Pühler and P. Stadler

Volume 12

Legal, Economic and Ethical Dimensions

Edited by
D. Brauer

Weinheim · New York
Basel · Cambridge · Tokyo

Series Editors:

Prof. Dr. H.-J. Rehm
Institut für Mikrobiologie
Universität Münster
Corrensstraße 3
D-48149 Münster

Prof. Dr. A. Pühler
Biologie VI (Genetik)
Universität Bielefeld
P.O. Box 100131
D-33501 Bielefeld

Dr. G. Reed
2131 N. Summit Ave.
Apartment #304
Milwaukee, WI 53202-1347
USA

Dr. P. J. W. Stadler
Bayer AG
Verfahrensentwicklung Biochemie
Leitung
Friedrich-Ebert-Straße 217
D-42096 Wuppertal

Volume Editor:
Dr. Dieter Brauer
Hoechst AG
Head Office Corporate
Research and Development
D-65926 Frankfurt am Main

This book was carefully produced. Nevertheless, authors, editors and publisher do not warrant the information contained therein to be free of errors. Readers are advised to keep in mind that statements, data, illustrations, procedural details or other items may inadvertently be inaccurate.

Published jointly by
VCH Verlagsgesellschaft mbH, Weinheim (Federal Republic of Germany)
VCH Publishers Inc., New York, NY (USA)

Editorial Director: Dr. Hans-Joachim Kraus
Editorial Manager: Christa Maria Schultz
Copy Editor: Karin Dembowsky
Production Manager: Dipl. Wirt.-Ing. (FH) Hans-Jochen Schmitt

Library of Congress Card No.: applied for

British Library Cataloguing-in-Publication Data:
A catalogue record for this book is available from the British Library

Die Deutsche Bibliothek – CIP-Einheitsaufnahme
Biotechnology : a multi volume comprehensive treatise / ed. by
H.-J. Rehm and G. Reed. In cooperation with A. Pühler and P.
Stadler. – 2., completely rev. ed. – Weinheim; New York;
Basel; Cambridge; Tokyo: VCH.

ISBN 3-527-28310-2
NE: Rehm, Hans J. [Hrsg.]

2., completely rev. ed.
Vol. 12. Legal, Economic and Ethical Dimensions / ed. by D. Brauer
 – 1995
  ISBN 3-527-28322-6
NE: Brauer, Dieter [Hrsg.]

Composition and Printing: Zechnersche Buchdruckerei, D-67330 Speyer.
Bookbinding: Fikentscher Großbuchbinderei, D-64205 Darmstadt.
Printed in the Federal Republic of Germany

Preface

In recognition of the enormous advances in biotechnology in recent years, we are pleased to present this Second Edition of "Biotechnology" relatively soon after the introduction of the First Edition of this multi-volume comprehensive treatise. Since this series was extremely well accepted by the scientific community, we have maintained the overall goal of creating a number of volumes, each devoted to a certain topic, which provide scientists in academia, industry, and public institutions with a well-balanced and comprehensive overview of this growing field. We have fully revised the Second Edition and expanded it from ten to twelve volumes in order to take all recent developments into account.

These twelve volumes are organized into three sections. The first four volumes consider the fundamentals of biotechnology from biological, biochemical, molecular biological, and chemical engineering perspectives. The next four volumes are devoted to products of industrial relevance. Special attention is given here to products derived from genetically engineered microorganisms and mammalian cells. The last four volumes are dedicated to the description of special topics.

The new "Biotechnology" is a reference work, a comprehensive description of the state-of-the-art, and a guide to the original literature. It is specifically directed to microbiologists, biochemists, molecular biologists, bioengineers, chemical engineers, and food and pharmaceutical chemists working in industry, at universities or at public institutions.

A carefully selected and distinguished Scientific Advisory Board stands behind the series. Its members come from key institutions representing scientific input from about twenty countries.

The volume editors and the authors of the individual chapters have been chosen for their recognized expertise and their contributions to the various fields of biotechnology. Their willingness to impart this knowledge to their colleagues forms the basis of "Biotechnology" and is gratefully acknowledged. Moreover, this work could not have been brought to fruition without the foresight and the constant and diligent support of the publisher. We are grateful to VCH for publishing "Biotechnology" with their customary excellence. Special thanks are due to Dr. Hans-Joachim Kraus and Christa Schultz, without whose constant efforts the series could not be published. Finally, the editors wish to thank the members of the Scientific Advisory Board for their encouragement, their helpful suggestions, and their constructive criticism.

H.-J. Rehm
G. Reed
A. Pühler
P. Stadler

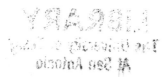

Scientific Advisory Board

Contents

Contributors

Dr. Dieter Brauer
Hoechst AG
Head Office
Corporate Research and Development
D-65926 Frankfurt am Main
Federal Republic of Germany
Chapters 3 and 8

Dr. Michael Bröker
Behringwerke AG
Emil-von-Behring-Straße 76
D-35041 Marburg
Federal Republic of Germany
Chapter 3

Dr. Mark F. Cantley
OECD (Organisation for Economic
Co-operation and Development)
Head of Biotechnology Unit
Directorate for Science, Technology and
Industry
2, rue André Pascal
F-75775 Paris Cedex 16
France
Chapter 18

Dr. Bong-Hyun Chung
Senior Research Scientist
Genetic Engineering Research Institute of
KIST
Taeduk Science Town
Taejon, South Korea
Chapter 13

Dr. Simone Christine Ehmig
Johannes Gutenberg-Universität
Institut für Publizistik
Jakob-Welder-Weg 20
D-55128 Mainz
Federal Republic of Germany
Chapter 17

Dr. Jens-Peter Gregersen
Behringwerke AG
Emil-von-Behring-Straße 76
D-35041 Marburg
Federal Republic of Germany
Chapters 7 and 10

Dr. Susanne L. Huttner
Director
UCLA Systemwide Biotechnology
Research and Education Program
Molecular Biology Institute
405 Hilgard Avenue
Los Angeles, CA 90024-1570
U.S.A.
Chapter 16

Dr. Alan J. Jones
Counsellor (Industry, Science and
Technology)
Australian Embassy
Godesberger Allee 106–107
D-53175 Bonn
Federal Republic of Germany
Chapter 13

Prof. Dr. Marion Leopold
Department of Sociology
Université du Quebec
P-8888 Succ A
Montreal, Quebec H3C 3P8
Canada
Chapter 12

Dr. Masao Kawai
Director
Japan Monkey Center
Chairman of the Committee on Biodiversity
Tokyo 105
Japan
Chapter 14

Dr. Cornelia Kellermann
Laboratorium für Molekulare Biologie
Genzentrum
Am Klopferspitz 18
D-82152 Martinsried
Federal Republic of Germany
Chapter 3

Prof. Dr. Hans Mathias Kepplinger
Johannes Gutenberg-Universität
Institut für Publizistik
Jakob-Welder-Weg 20
D-55128 Mainz
Federal Republic of Germany
Chapter 17

Prof. Dr. Darryl R. J. Macer
Institute of Biological Sciences
University of Tsukuba
Tsukuba Science City 305
Japan
and
Eubios Ethics Institute
31 Colwyn Street
Christchurch 5
New Zealand
Chapter 4

Dr. Sally L. McCammon
U.S. Department of Agriculture
APHIS Animal and Plant Health Inspection
Service
P.O. Box 96464
Washington, DC 20090-6464
U.S.A.
Chapter 6

Dr. Terry L. Medley
U.S. Department of Agriculture
APHIS Animal and Plant Health Inspection
Service
P.O. Box 96464
Washington, DC 20090-6464
U.S.A.
Chapter 6

Prof. Dr. Henry I. Miller
Stanford University
Hoover Institution and Institute for International Studies
Stanford, CA 94305-6010
U.S.A.
Chapter 2

Annette Millet
Biofutur
29, rue Buffon
F-75005 Paris
France
Chapter 15

Dr. Eleonore Poetzsch
FIZ Chemie Berlin GmbH
Fachinformationszentrum
Postfach 12 60 50
D-10593 Berlin
Federal Republic of Germany
Chapter 11

Dr. Susono Saono
Research and Development Centre for
Biotechnology
Jl. Ir. H. Juanda 18
P.O. Box 323
Bogor, 16122
Indonesia
Chapter 13

Prof. Dr. Horst Dieter Schlumberger
Bayer AG
Geschäftsbereich Pharma
Forschung und Entwicklung
Postfach 101709
D-42096 Wuppertal
Federal Republic of Germany
Chapter 8

Prof. Dr. Rolf D. Schmid
Institut für Technische Biochemie
Universität Stuttgart
Allmandring 31
D-70569 Stuttgart
Federal Republic of Germany
Chapter 13

Jeannie Scriven, B.A.
GBF Gesellschaft für Biotechnologische
Forschung mbH
Mascheroder Weg 1
D-38124 Braunschweig
Federal Republic of Germany
Chapter 13

Prof. Dr. Joseph Straus
Max-Planck-Institut für ausländisches und
internationales Patent-, Urheber- und
Wettbewerbsrecht
Siebertstraße 3
D-81675 München
Federal Republic of Germany
Chapter 9

Dr. Jane H. J. Tsai
Kurt-Schumacher-Weg 11
D-37073 Göttingen
Federal Republic of Germany
Chapter 13

Dr. George T. Tzotzos
UNIDO United Nations Industrial
Development Organization
ICGEB Science Coordinator
(International Centre for Genetic
Engineering and Biotechnology)
Vienna Office
P.O. Box 400
A-1400 Wien
Austria
Chapter 12

Prof. Dr. Guido Van Steendam
ifb International Forum for Biophilosophy
Director
Craenendonck 15
B-3000 Leuven
Belgium
Chapter 1

Prof. Dr. C. Theo Verrips
Unilever Research Laboratorium
P.O. Box 114
NL-3130 AC Vlaardingen
The Netherlands
Chapter 5

Prof. Dr. Ernst-Ludwig Winnacker
Laboratorium für Molekulare Biologie
Genzentrum
Am Klopferspitz 18
D-82152 Martinsried
Federal Republic of Germany
Chapter 3

Introduction

DIETER BRAUER

Frankfurt am Main, Federal Republic of Germany

"Biotechnology" undoubtedly is a fascinating field with far-reaching legal, economic and ethical implications for mankind. In this Volume 12 of the Second Edition of "Biotechnology" an experienced and diverse team of authors comprising scientists from academic research and industry, regulators and journalists examines the potential implications biotechnology may have from various angles. The perspective is thoroughly international and with many critical comparisons and views expressed. It is an indispensable reading for everyone seriously concerned with the potential risks and benefits of modern biotechnology and the more general problems associated with the introduction of a new technology into society.

From about 20 different technical and scientific disciplines including biochemistry, classical genetics, protein chemistry, microbiology etc., further biologically derived tools such as genetic engineering and hybridoma technology have been developed, which are often referred to as "modern biotechnology" and which represent the most exciting advance in the biological sciences this century. However, there is evidence that this new technology is not understood by the public, that there is a tendency to think of biotechnology as fundamentally different from other technologies. The speed of its introduction

seems to frighten some onlookers, and the safety of genetic modification is questioned. Hence the public demands regulation for its protection.

Not surprisingly therefore, since the pioneering phase of modern biotechnology we have seen the production of an enormous quantity of political, scientific, socio-economic and ethical literature evaluating the different aspects of biotechnology. There are also the questions of patenting "life" – needed for the protection of investments and rejected as "unmoral" by other interests and fears with regard to biotechnology have to be linked to ambiguous attitudes towards technology in general.

The contributions compiled in the five parts of this book are an attempt to cover the mainstream of thoughts and developments with respect to biotechnology from an international and multidisciplinary perspective:

Part I guides the reader into biotechnology and starts with an introduction into technology assessment in a fundamental way and not limited to biotechnology, only. Then the different concepts of risk assessment applied today are discussed and evaluated and are followed by a detailed introduction to biosafety issues in research and production. The introduction to biotechnology is concluded by combining the terms "biotechnology" and "bioethics" and the development of thoughts about what ethical biotechnology might be.

Part II focuses on the practical aspects of applying modern biotechnology in research and to the production of goods. It addresses the different expectations and needs of users of modern biotechnology, consumers and regulators who have to serve all interests appropriately. The topics range from structured risk assessment of rDNA products mainly in the food sector and the regulations applied to a safe development of transgenic plants to biomedicinal product development and a comparison of the different regulatory environments for modern biotechnology in the US, Japan and Europe.

Part III covers both intellectual property protection and bioinformatics. One chapter focuses on what "patent protection" actually is, and why lack of patent protection prevents investment into R & D confining activities to reengineering and copying. Only an effective use of intellectual property rights enables industry to recover the enormous costs involved in developing new products and processes. Because modern biotechnology is highly information-dependent, the availability of high-quality, up-to-date and comprehensive information is an important requirement on almost each level of R & D, production development and in intellectual property protection. Therefore, databases are an essential tool in biotechnology.

Part IV evaluates prospects for biotechnology in the developing world and specifically in the Asian-Pacific region. It is discussed in detail, to which extent the disjunction of priorities between the industrial and developing world influences the potential of biotechnology to solve health- and food supply-related problems in the developing world. Many countries in the Asian-Pacific region are in a state of rapid transformation from agricultural to industry- and service-based economies. The data available indicate that biotechnology has achieved a high priority status in the whole region which has become a strong contender to the development of biotechnology in North America and Europe.

Part V is an attempt to analyze the role of the public in the development of biotechnology and to compare the differing views in some key countries like France, USA and Germany. The focus is put on the activities of interest groups, the role of the media and the political responses. The concluding chapter in this book is a historical review of events and decisions in time relevant to the development of modern biotechnology with specific emphasis on the European perspective. However, this chapter is not limited to but goes well beyond biotechnology. In a narrative form it represents a case study in how societies cope with new knowledge in the last quarter of the twentieth century.

Everybody knows the fairy-tale of the innocent maiden and the ugly and horrible beast which turned out to be something honest and trustworthy after being treated with goodwill and trust rather than with fear and repulsion. But, unfortunately, there is no evidence for who is the beauty and who is the beast. Is it the innocent and trusting public which is confronted with a flourishing science threatening human life, the environment and the integrity of God's creation? Or is it a pure and beneficial science promising progress in many human problems which is rejected by an ignorant and distrustful public?

Hopefully, this comprehensive book will provide the interested reader with sufficient evidence to make him understand that biotechnology is neither a beauty nor a beast, and that science and technology are not a subsystem of our society in the sense that they could be regarded separately from other subsystems. Understanding this and the points made in this book may help the reader to find his own views and positions based on an educated choice not only towards biotechnology but also to "technology" in a broader sense.

Frankfurt am Main, November 1994

Dieter Brauer

I. Modern Biotechnology –
What is New About It?

1 The Evaluation of Technology as an Interactive Commitment-Building Process – The Failure of Technology Assessment

GUIDO VAN STEENDAM

Leuven, Belgium

1 Introduction

Is "Technology Assessment", or "TA", the magic brew which enables governments to guarantee the total quality and relevance of the technological developments they approve or stimulate? Can other initiators of technologies, such as biotech companies or academic research labs, make use of the same tool to convince the authorities of how well-prepared, beneficial and safe their plans are? Or does TA provide activists and environmentalists with an infallible instrument to collect irrefutable evidence of the danger of specific developments which they had since long suspected?

The answer to each of these questions is a rather disappointing but clear "No": TA is not the panacea it is still often believed to be. While being prepared by the American Congress in the sixties as a method with capital T and capital A, to understand and assess all aspects of technological developments, in reality it never worked that way, and it never will. Both the conception of "technology" and the ideas on "assessment" which underlie the TA approach appear to be exceedingly objective and therefore unrealistic. The TA experiment of the US Congress is an experiment that failed and was doomed to fail. This is the first point we want to make in this chapter.

The second point is that the TA ideals are a permanent threat to any serious attempt to monitor the relevance and harmlessness of technological developments. By holding up the prospect of providing a complete and objective analysis, TA plays along with our deeply human but utopian craving for making infallible decisions, and in this way diverts our attention from more realistic approaches to enhancing the quality of decision-making under uncertainty.

The most important point of this chapter is that proper management of technological developments requires a great diversity of activities of a large variety of institutions linked to all formal and informal social sectors. These institutions include research units, advisory bodies, discussion groups, international and interdisciplinary forums, and strategic groups – many of which already exist and function

well. They are important to the extent they succeed in developing activities that go beyond TA as designed by US Congress in the sixties, but contribute to realizing and sustaining a broader social process of informed, committed and interactive policy-making.

This chapter begins with a description of the ambiguous situation of the "golden sixties" during which the American Congress conceived the idea of Technology Assessment and coined the term. The sixties were dominated by technology which was first seen as the driving force behind wealth and abundance (Section 2) and later as a potential danger and a major source of an irrelevant and degenerating lifestyle (Section 3). Next follows a presentation of TA, as the attempt by US Congress to identify valuable technology and to separate the wheat from the chaff (Section 4). This outline is followed by an evaluation of this enterprise, explaining the untenability of the approach (Section 5) and paying attention to the positive lessons we can learn from the TA experiment (Section 6). The chapter concludes with an overview of a more realistic and workable, though fallible, approach to monitor technological developments. It explains that such endeavor should be seen as a further step in the burgeoning science and technology policy (Section 7) and that it takes account of our growing awareness and increasing understanding of the fallibility of human decision-making in a permanent situation of uncertainty (Section 8).

2 The Benefits of Technology

In the 1960s, few would have expected that anybody would ever seriously endorse the idea that new technological developments are in fact suspicious and in need of critical assessment before they be allowed to proceed. On the contrary, technological innovation was believed to be a good thing which should be stimulated by all means. Technology appeared to have revitalized all important aspects of social life.

2.1 Growing Social and Economic Welfare

The great affluence which so characterized the sixties was unrivalled by that of any other preceding period in the history of mankind. Never before had the people in Western industrialized countries experienced a similar wealth and comfort. Since World War II, the average productivity of human labor, the Gross National Product and the GNP per capita had steadily been increasing by about 3 to 4% per year. This trend of growing prosperity not only resulted in more dispensable income for the families and more money to invest for national states, but also in a decrease in the amount of working hours, which, in its turn, allowed for more time for leisure.

Technological innovation, based on strong scientific research, manifested itself to be at the heart of this new social order. It appeared that four hundred years of modern science and two hundred years of modern technology had finally reached their stage of maturity. And, as the academic and industrial world were only at the beginning of the exploration and exploitation of promising fields and powerful new technologies, much faith was placed in the fact that this process of increasing prosperity had been set into motion for good. Four fields were expected to lead to new revolutions with breathtaking new possibilities.

(1) The first one was the electronic revolution with the Integrated Circuit which at the time was still in its infancy. Despite its newness, however, its application had already drastically heightened economic productivity. It was expected that by the year 2000, the computer would enhance the productive capacity of human labor by another 300%, allowing every person to do the work – and to earn the income – of three.

(2) The second field revolved around the development of new and cheaper sources of energy. Here the promising new technology would allow the unleashing of enormous amounts of energy stored in the inner core of highly energetic material, by converting this hidden energy into electricity for everyday use. This movement is also called the nuclear revolution.

(3) Thirdly, there was a similar hope that certain traditional basic materials, like steel, whose elemental properties were difficult to engineer, could be replaced by more flexible man-made materials such as plastics. The properties of plastics present the advantage that they can be designed in a research facility and specifically adapted to their planned use, and, at a later stage, they can easily be manufactured in chemical plants by employing simple readily available raw materials.

(4) Finally, the sixties also announced a biological revolution. Genetics was believed to be the driving force behind this revolution. This discipline was seen as the study of the core elements of life, being genes and chromosomes, which encoded the characteristics of the different species and forms of life and which are used by living organisms to transmit hereditary traits to later generations. Research increasingly succeeded in unravelling the basic mechanisms of those mysterious units, and subsequently opened up prospects for powerful technological applications not only in the field of human health care, but also in agriculture and other fields of human activities. By the end of the sixties, science had already succeeded in isolating the first gene, but it would take until the seventies before the first technological uses of those genes would become possible.

These prospects, combined with the insights into the driving forces behind them not only stimulated an increased respect for academic science and technology, but went on to nurture the growth of industrial companies. Only large companies could afford to be involved in the basic research which was needed to guarantee ongoing economic growth. And it was only these same large companies that could afford to develop the mechanisms to turn these basic scientific discoveries and technological developments into real industrial products to be distributed in society.

In taking this beneficial effect of technology on industry into account, we cannot forget that life and the happiness of individuals and families remained at the center of those developments. New machines helped to reduce burdensome work and to organize and enrich daily life. Washing machines, dishwashers, refrigerators and freezers became available and accessible to each household. Color television opened a more vivid window on the real world and provided a private entertainment center for every living room. Communication was facilitated by a growing number of telephones and cars. These were the golden sixties, filled with new music, wealth and hope.

2.2 Maintaining Peace

Technology also left its mark on the widespread belief that there would never be another war like World War II which would be staged in the Western world. It was clear that war had not disappeared from the earth. But at the beginning of the sixties, massive war, with its atrocities and misery, did disappear from the daily life of Westerners. War as a concept with a capital W had become humiliated to the abstract notion of a mere diplomatic tension between two blocks – the Cold War between the East and the West – or to the theoretical knowledge of smaller, local conflicts in "hidden parts" of the world. There was a general faith that Western diplomats, politicians and military forces could master all existing threats because they could rely on the most sophisticated arsenal of high-tech weapons then known to man, which would easily be able to stop all attacks from the outside. Furthermore, it was clear to many that these weapons, including a whole battery of nuclear weapons, were so powerful that no other power would ever dare try to attack the West, because they realized that the West could immediately reply to any attack with the utter annihilation of the attacking force and indeed its entire nation by means of a series of atomic bombs many times more powerful than the bomb which destroyed Hiroshima. In the same way that large stone walls could protect the inhabitants of a medieval town or fortress, the shield of technology was believed to be protecting the modern Western world. Sophisticated weapons replaced bow and arrow while radars and satellites replaced the watchers' eye.

2.3 Prestige

In the sixties, technology was not just appreciated as the powerful engine of economic wealth and peace. It was also a deep source of human pride and a symbol of the unlimited possibilities of human rationality and of strong Western nations. Even the most unbelievable adventures appeared to be possible now. The exploration of space was the showpiece of this evolution. The manned landing on the moon was the first dream which became true. On May 26, 1961 President KENNEDY earned a long and thunderous applause while explaining his projects to the Congress:

"I believe that this nation should commit itself to achieving the goal, before this decade is out, of landing a man on the moon and returning him safely to the earth. No single space project in this period will be more impressive to mankind, or more important for the long-range exploration of space; none will be so difficult or expensive to accomplish." (JOHN F. KENNEDY, 1961).

As we all know, this dream became reality. On July 21, 1969, NEIL ARMSTRONG and BUZZ ALDRIN spent two hours walking on the moon. The landing on the moon was transmitted live by television all around the world. This technical feat was experienced as the clear symbol of the unlimited power and possibilities of mankind. For many, the conviction prevailed that man had entered a new phase in his existence, or, as NEIL ARMSTRONG expressed it, the very moment after putting the first human foot on the lunar soil: "That's one small step for man, one giant leap for mankind."

3 Technology under Attack

Being a major symbol of the successes of science and technology in the sixties, the lunar landing and space travel in general soon became a focal point in a growing movement of mistrust and critical reappraisal. Doubt was cast on the social relevance of many splendid technological and industrial developments. Furthermore, it became clear that science and technology, functioning as the engine of our increasing prosperity were accompanied by a host of social and environmental costs to which we had been blind for many years. Science and technology came under a growing attack in each of the domains in which they had proven so successful in the sixties.

3.1 Doubts on the Social Relevance

3.1.1 Prestige

The landing on the moon was soon followed by a postnatal depression. For ten years there had been a concerted effort to put an American on the moon, and several scientific, technological and industrial projects derived their *raison d'être* from this adventure. Now that the goal had been achieved, questions began to surface concerning the true nature of the project. What was the real intention behind the space program? Was the natural next step to stage a similar landing on the planet Mars? What was the point of the American public authorities' incorporating space travel into their main technological priorities? What had the nation – or mankind – actually gained from its lunar expedition?

Leading scientists had already considered these problems earlier. ALVIN WEINBERG, for example, a former member of the American President's Science Advisory Committee, wrote as early as 1966: "A crisis of mission will face our space program during the middle seventies. The present scale and pace of our space enterprise is determined by our commitment to send a man to the moon early in 1970. Now that this date is approaching, grave questions arise as to the next step. Should we go on to Mars; or should the space program revert to the scale it enjoyed before we decided to send a manned expedition to the moon? The existence of a massive space establishment, including many fine laboratories, surely will tend to weigh the balance strongly toward going to Mars after we have reached the moon. Yet I would hope that when the

time for this far-reaching decision comes, we shall have ready for this great scientific technical resource many other technical challenges that bear more directly on our human well-being than does a manned trip to Mars" (ALVIN M. WEINBERG, 1967).

3.1.2 Maintaining Peace

The domain of military research, another of Americans technological priorities came under attack by the same school of criticism. After all, had not the Vietnam experience proved that the "American war machine" had not succeeded, even with the most sophisticated weapons, in bringing the war to an end? This growing realization was linked to the general question as to why the Vietnam war had actually been necessary. In addition, more and more members of the American Congress felt that the nation would be better served by a policy different from the maintenance of a high-tech arms race. They also felt that research funds could be put to better use.

3.1.3 Growing Social and Economic Welfare

Economic and industrial growth had the same ambiguous history. While, in the sixties, few would deny that this growth had led to an increasing prosperity and general well-being, it also became gradually clear that it did not make sense to perceive unlimited growth as a goal in itself. These doubts dampened the evident relevance of the large investments in scientific and technological developments, designed to stimulate economic growth.

3.2 The Hidden Costs

There was another source of unrest and mistrust towards science and technology. The sixties marked the beginning of a growing awareness that some technologies, developed and applied with only the best of intentions, threatened our natural or cultural resources, long taken for granted.

3.2.1 Cost to the Natural Environment

In 1962 RACHEL CARSON published "Silent Spring". In this dramatic and alarming report, she convincingly demonstrates how the increased use of pesticides is slowly but surely poisoning our biological environment. It is the first influential and well-documented admonition to society that, by undermining the environment, our technological civilization could ultimately turn against humanity instead of promoting it. Her claim was followed by many negative reports about the effects of motorways, jet aircrafts and airports, nuclear energy and many other phenomena of our modern culture. Until the beginning of the 1960s it had always been naively presumed that the sea, rivers, groundwater, air, earth and the plant, animal and human kingdoms could withstand any conceivable shock that humans could deliver to them. Until that time, it had scarcely been thinkable that the increasing industrialization would ultimately lead to the widespread extinction of aquatic life, to heavily polluted coasts and beaches, or, indeed, to sickening and unhealthy air pollution in the large cities. In the seventies, when biotechnological methodologies were revolutionized, the very scientists and technologists behind these breakthroughs had already expressed their deep concern for the potential hazards of the use and laboratory study of recombinant DNA molecules – hazards to the human health as well as to the health of our environment. Leading molecular biologists even proclaimed a moratorium for all further research. This gesture of caution attracted the attention of the press and policy-makers and fuelled the widespread belief that the potential dangers associated with biotechnology are of a very special nature.

3.2.2 Cost to the Social Environment

Potential dangers, increasingly attributed to technological developments, were not limited to our natural environment. A great diversity of threats to the social environment were also increasingly considered. Unemployment was a manifest example of these social dangers. Less tangible perhaps was the danger that the ever-changing technologies in the workplace might adversely influence the motivation of workers, as it could be feared that they would not want to constantly adapt to new procedures or to be in need of permanent education.

Aside from the social environment in the workplace, the cultural and family-centered social environments were not spared from the technological implications of this new age. For example, it had been argued that television had signified the death of social and cultural life and even of family life. Also, the enormously increased mobility of the jet-age generation was described as a potential threat to family life.

Questions also arose concerning the necessity to adapt our legal and social mechanisms to the many new and powerful technologies, such as the broadened potentials of medicine in the fields of prenatal life and care of the elderly. Not everyone was convinced that the new social and legal order would ultimately bring out social optimality in a better and more humane way than the old order did. Again the oncoming biotechnology and its multiple uses in agriculture, industry and health care stirred the imagination. What should we think e.g. about a society in which specific, newly designed, living beings obtain an economic value and are patented? And how can our society survive in a situation where man can predefine the traits of their own children or design, "produce" and patent a genetically engineered worker or soldier?

4 The Rise of Technology Assessment

After years of economic growth and social euphoria, the 1960s ended with a deep awareness of crisis. The engine of our increasing prosperity, namely technological and industrial development, had proved to be accompanied by a host of social and environmental costs to which we had been blind for many years. National governments, which were themselves systematically involved in large-scale technological development programs, felt compelled to seriously examine and re-consider their earlier carefree approach and their priorities.

It was within the context of this reappraisal that the American Congress launched its own initiative, prepared by the subcommittee for "Science, Research and Development" of the "Science and Astronautics" Committee of the House of Representatives. The first report of this subcommittee, published on October 17, 1966, described how Congress planned to deal with the growing conviction that it is ab-solutely necessary to "keep [...] tab on the potential dangers, as well as the benefits in-herent in new technology". Two major con-cerns appear to form the basis of these Con-gressional efforts.

Their first objective was the ensuring of the attainability of in-depth knowledge into the potentially dangerous consequences of tech-nologies like "technological unemployment, toxic pesticides, pollution, exhaustion of re-sources, the disposal of radioactive wastes, and invasions of personal liberty by electronic snooping and computer data banks". They proposed reaching this objective by means of the development of a new, highly specific, scientific methodology to provide an all-en-compassing assessment procedure of all that could be described as "technology". For this new discipline, the subcommittee coined the name "Technology Assessment" (TA).

The second objective of the Congressional subcommittee was to make sure that such TA studies were actually carried out, and this in such a way that they would prevent the US Government from promoting and funding ir-relevant or dangerous technological develop-ments. This concern resulted in the firm belief that one had to establish some sort of institu-tional mechanism which would organize the necessary "TA studies" on a systematic basis. Congress believed that this institution should be established within Congress itself. This clarifies why, in 1973, a new Congressional Office was established: the "Office for Tech-nology Assessment" (OTA).

The following sections will expose in greater detail the actual birth of both the TA methodology and the TA Office (see e.g., Sect. 6.1.3).

4.1 The TA Methodology

Unexpected troubles like the environmen-tal crises of the 1960s had made Congress acutely aware of the importance of the ability of foresight. Only being able to recognize the destructive consequences of technological projects after they have been developed and widely used, was an embarrassing handicap Congress wanted to rid itself of. The alarming DDT experience should never happen again. Congressmen also staunchly opposed invest-ing in the development of technological solu-tions which afterwards would prove to be use-less or socially irrelevant. They realized that they were in need of a methodology that could provide them with the accurate infor-mation about future consequences of techno-logical developments.

The mechanism of foresight that was needed would have to be characterized by the following four elements:

(a) Firstly, the mechanism should function as a "first alert" system which would sound the alarm bell before the hazardous technolo-gy had gotten past the drawing board, or at least before the potential negative conse-quences had the opportunity to take place. In other words, the information about future consequences would have to be provided when it was still possible to stop the planned technology from being implemented, or at least to redirect or restrain it.

(b) Secondly, the early warning system should be systematic and exhaustive. Every consequence of the planned technologies should be explored including the "real and potential impacts of technology on social, economic, environmental, and political sys-tems and processes" (VARY T. COATES, 1972). However, this analysis was not to con-tent itself with the pure assessment of that trend or scenario, which at first glance ap-pears to be the only or most evident one. Many technological developments can in fact follow several scenarios depending on the impact of broad social or economic developments, specific human decisions or other external factors. In this case the consequences of each of these alternative scenarios should be studied. This included the special case of the consequences of the *zero option*, where it was decided not to initiate the technology at all. Finally, Congress wished that the scope of TA would be of such dimensions that all technological domains, harmless though they may appear to be, would be scrutinized.

(c) Thirdly, the methodology should not just name the consequences, but also indicate to what extent and in what sense they are hazardous. The methodology procedures should appraise the nature, significance and merits of the consequences, simultaneously providing a balanced appraisal of the technology under examination. Based on the results of a TA analysis, the policy-makers should be able to understand what is really at stake when new technological programs are introduced. TA studies should become a powerful tool in policy-making.

(d) Fourthly, the prognostications should not be the mere guesses of uninformed people, or the work of lobbyists trying to stop or promote a specific technological development. The forecast was to be reliable and objective for 100% and constitute an incontestable basis for sound and justified policy-making. In other words, the methodology of the investigations should follow some sort of "scientific method" and the result should be reliably scientific material. Congress was aware that no single existing scientific discipline was able to deliver the required assessment of all future consequences. So it believed that a great variety of existing analytical methods would have to be modified and combined to form a spanking brand-new discipline.

4.2 The TA Office

It was clear that the US Congress would already feel much more comfortable when a scientific methodology would be available for making complete and reliable predictions of the consequences of technological and industrial developments. But the mere availability

of a TA methodology could not completely reassure its members. Congress also wanted to regulate the institutional context of TA studies in such a way that they would be taken seriously. According to Congress, this would imply, first of all, that a new structure for TA studies should be created as an institutionalized element of the Congressional work unto itself, and secondly, that this new structure should be given the necessary staff and funding to organize TA studies in a professional, systematic and consistent way.

4.2.1 A Congressional Institution

There are at least three reasons why Congress wanted to develop the new TA institution within Congress itself.

(1) The basic reason is that, in the US, it is the President and his administration who are the major initiators and promoters of national technological projects, while Congress asserts that it is the only institution that can effectively put pressure on the President to stop or modify his technological projects, in the event the results of a TA study would have unmasked their irrelevance or hazards. The most direct tool of Congress consisted in its power to control the budget of the President. In the United States, the White House can do relatively little without the consent of Congress, which must approve the budget needed to carry out its plans. Thus the famous speech about man landing on the moon, delivered to the American Congress by President KENNEDY in 1961, was in fact not simply an expression of an undertaking by the President to his people, but was really a request to Congress to give financial support to his plans and to approve the budget. As long as science and technology were generally considered to be beyond any suspicion, Congress did not feel the need for intensive and systematic assessment studies. However, when the conviction grew that technological developments, and therefore also the technological policy of the government, might sometimes be hairy or even problematic, Congress wanted to give a clear sign to the American public that it was ready to shoulder its responsibility and to base its approval on a preliminary careful as-

sessment of the technologies proposed. Congress was naturally quite aware of the extent of the power of persuasion of public pressures (such as activists and lobbyists) in the White House. The tendencies towards a more direct democracy were understandably perceived as threatening by many members of Congress, who preferred to come up with a strong and potent plan to reassure the constituency that they were capable of and effective in fulfilling their mandate to control the executive power.

(2) A second reason for Congress' desire to keep TA under its direct control, is that Congress dwelt in the conviction that the impartiality of the studies could only be guaranteed by relying on inhouse technical and scientific expertise. Congress did not go as far as to request that this inhouse Congressional office for TA studies be not allowed to subcontract minor or larger parts of the research needed. It felt, however, that working with external groups always carried the risk that such teams would include biased and partial experts, who might have an interest in giving an incomplete or distorted picture of the effects of a planned technology. In this sense it was felt that only the Congressional office itself could bear the responsibility for the final presentation of the results.

(3) As a third important reason for having inhouse expertise available, Congress stressed that this method was the only means to avoid that certain studies which were relevant for policy-making would not be carried out because of lack of an interested external research partner.

4.2.2 A Strong Institution

There are also several reasons which help clarify why Congress was convinced that a proper functioning of TA required a strong, well-staffed and well-financed institute.

The obvious reason is that the founding of such a well-established institution presents the only guarantee that technological initiatives funded or controlled by the US government be assessed in a professional way.

More specifically, Congress wanted to take advantage of the growing belief in the need of

TA, to strengthen its internal power and to counter the power of the White House. In fact, Congress had not (until then) ever been really involved in decision-making in the field of new technological developments. Setting the agenda for Government-funded technology had always been the exclusive task of the President. The President had, of course, always been dependent on Congress for budget approvals, but Congress never commanded any serious means of discussing the proposals made by the President. Since the time of President ROOSEVELT, the White House had had the services of a special advisory body at its disposal. This body was led by the "Presidential Science Advisor", and was utilized to determine the technological initiatives of the US Government. As a result, Presidential dossiers were usually well-prepared and convincing. They were filled with tables, lists and figures, which authoritatively explained the supporting scientific evidence of the Presidential initiatives and which illustrated the many benefits these new programs would bring to the American nation. Consequently, Congress did not see any necessity for engaging in detailed discussions on the specific aspects of the plans of the President. Nor did it feel the need to, in doing so, prepare a tool of its own which would allow it to contest the conclusions of the Presidential experts. The result is that Congress did not dispose of the experience or expertise it required, when it gradually discovered the need to make its approval of the President's technology budget dependent upon the results of a serious assessment of the long-term relevance, benefits and dangers of the new technological developments proposed by the President. On the other hand, the growing scepticism did, of course, also change the attitude of the seasoned Presidential advisors. They increased the strength of the scientific evidence endorsing their plans, they multiplied the number of pages of their reports, and they used ever more sophisticated strategies to convince Congress of the necessity of the technological developments they proposed. ROBERT McNAMARA, President JOHNSON's Defense Secretary, was notorious for his use of the computer in his attempts to silence members of Congress in debates and hearings. In this context, Congress felt the need for a strong counter-attack, which would lead to the establishment of a strong in-house office for the study of technology.

This is why, in 1967, representative EMILIO Q. DADDARIO, chairman of the subcommittee which explored the need for TA, introduced bill H.R. 6698, which – being the result of several hearings and internal deliberations and external studies – finally led in 1973 to the establishment of a new Congressional Office, the "Office of Technology Assessment", or "OTA", directed by EMILIO Q. DADDARIO. It is clear that the White House was not inclined to favor this development. On the contrary, the Republican President NIXON, during whose Presidential term OTA was finally set up, did his utmost to combat this internal strengthening of the then predominantly Democratic Congress in the area of science policy. But when President NIXON finally signed the bill, OTA became established as an autonomous Congressional body. It was founded on the idea that it should not have any other master than the objective requirements of its neutral, scientific TA methodology, which is able to monitor the technological initiatives of the US Government. Consequently, it was accorded all kinds of rights to guarantee its autonomy, also towards the power and influence of the President and his administration. It was given the right to publish the results of its studies, whenever these appeared to make sense, albeit in the face of explicit disapproval by the executive powers. This is what happened in 1988 when, under strong protest of the Pentagon and the Department of Defense, OTA published its assessment of the Strategic Defense Initiative under the title "SDI: Technology Survivability and Software". OTA was also given the right to call witnesses and to demand all documents necessary for the success of its work. In this sense it had been given the power of a strong, political, parliamentary committee with the mandate to track and monitor the White House in its actions.

5 The End of Technology Assessment

The TA experiment of the US Congress is more than 20 years underway by now. OTA has become a well-known and internationally respected body. It functions as a paradigm for many well-inspired people all over the world, who are convinced that technological developments require strict control and guidance, but who fail to see how this can be done. OTA gives them hope. For them, OTA is a symbol of success. It is the living evidence that it is not just an unattainable dream to create an institution which can keep technological developments on the right track. OTA gives them the courage to fight and lobby in order to introduce a similar institution in their own country, allowing the local politicians to differentiate between technologies which are irrelevant and hazardous and developments which are relevant and safe. In short, the TA initiative of the US Congress gained wide acceptance as the appropriate way to deal with social problems linked to technology. Many people, when thinking about handling technological developments properly, do not feel the need to go any further than studying and implementing the same methodology which we described in the previous section and for which the US Congress coined the specific name "TA" or "Technology Assessment."

5.1 A Recognized Failure

Despite its initial success and despite the fact that the TA methodology appears to be unquestionably convincing and self-evident, everybody who has a closer look at what happened in OTA can only come to the conclusion that TA, as created by the US Congress, did not work and never will. By the end of the seventies, there existed a fairly general consent that OTA was not capable of achieving the goal for which it had been founded. These insights were shared by scholars like VARY T. COATES, who at the beginning of the 1970s had worked enthusiastically on the development of the theoretical structure of a scientific "TA discipline". In 1979 she writes:

"Technology Assessment ... had its day and has faded into the background along with planning by objectives, and Program-Planning and Budgetting Systems (PPBS), and other nonce phrases cum slogans recorded in the history of public administration" (VARY T. COATES, 1979).

Reports prepared by OTA virtually never achieved the level of scientific indisputability which had been envisaged. On the contrary, serious doubts arose at fairly regular intervals concerning the completeness or the "scientific" nature of the studies carried out. Nor did OTA succeed in establishing an early warning system, coming up with wide-ranging and timely forecasts of potential social problems and environmental risks associated with the continued development of science and technology. Finally, OTA reports hardly ever affected actual policy-making. The influence of the studies on policy can in fact be described as varying between minimal and non-existent.

For many analysts of technological developments, these rather negative observations did not come as a surprise. Even before OTA was established, as early as 1968, the National Academy of Sciences (NAS) in Washington, DC was asked by Congress to give it advice on how to organize an effective assessment of technological developments. This advice turned out to be a strong warning against any attempt to guarantee the relevance of technology by trying to install some type of agency which would function as a neutral judge who is able to find out what is good for "society" or for the "environment" and who is empowered to indicate which technological developments should be permitted and which prohibited. The NAS report clearly demonstrates that this is not the way technology can develop or can be controlled. It concludes that it is simply an illusion to hope that a so-called neutral and objective agency would be able to solve the problems linked to technology. Proper decisions about technology require assessment procedures that differ completely from the creating of a centralized office of TA, and completely different studies than the all-encompassing TA studies OTA was

dreaming of. We will expand on this later. But first we want to analyze what exactly went wrong in the implementation of OTA. The problems appear to be very fundamental. It turns out that the concept of "Technology Assessment" as developed by the US Congress was built on a complete misunderstanding of what both "technology" and "assessment" really are.

5.2 An Inadequate View on Technology

It is a common mistake to imagine science and technology as a totality of processes whose internal progress is determined by an implacable, strict, scientific method. In this view, technological developments can be compared to the movement of clocks which run according to their own dynamics. Once the process is started and its momentum kept up, its continued internal progress is regarded as being objectively fixed. According to this view, "external" institutions, such as public authorities, must refrain from any attempt to intervene in the "internal" dynamics of science and technology. Should they try to do this, the scientific quality of the processes can no longer be guaranteed. There is, however, another way for "external institutions" to influence the course of science and technology without endangering its internal logic. The basis of this type of interaction is the fact that many external conditions and bodies "from outside" are or can be involved in decisions about which scientific and technological processes will be permitted, started, encouraged, held back or stopped. In all these "external interactions", the internal logic of the technological process is left untouched. This possibility formed the opportunity the US Congress wanted to seize in order to prevent irrelevant or dangerous technological developments. This possibility was also the context for developing TA, which was believed to provide a solid basis for such "Go" or "No-go" decisions by presenting a reliable and comprehensive analysis of what would really happen when a new technological process would be allowed to start.

In real life technology, such solid *TA predictions* appear to be unrealistic, just as much as the belief that simple *Go/No-go decisions* by external actors can determine whether or not specific technological developments will or will not be realized.

5.2.1 Predictability

Technological developments appear to have little in common with the widespread idea that autonomous, homogeneous, predictable processes are completely determined by some internal scientific logic. Instead, they appear to be complex social processes with many possible side-branches and alternatives. In designing and developing new technologies, scientists arc continually faced by choices and are constantly forced to take creative decisions concerning the direction they will follow in the future. Many of these "internal" decisions are co-determined by a multitude of "external" influences to which they are exposed, which feed and monitor their work and which are taken on board. It has been widely demonstrated by now that the concrete form which a given scientific or technological development finally takes and the way it is implemented in society, is not just the result of the internal logic of some clear methodology, but also of a multitude of individual and collective decisions and interests. This concrete form is anything but predictable, due to the great number of variables and the high measure of uncertainty inherent to technological developments. A comprehensive analysis of the many types of consequences of this technology for social life or the environment proves even less feasible.

A proposed new way of avoiding this problem would be to ensure that an assessment study does not simply entail an analysis of the consequences of a technology, but would also entail a whole decision-making and information process which will accompany the development of that technology. A good assessment study then does not try to come up with a clear "prognosis" of the future technology and its consequences. Instead, it aims at presenting the total picture of the various possible future developments (scenarios). The

study does not attempt to forecast which of these scenarios will actually present itself at some time in the future. This, after all, will depend on the unpredictable way in which the scientists, the decision-makers and the public will react at crucial moments in the future. It is believed, however, that such a broader assessment of all scenarios, while being of a less rigorously predictive nature, is still able to provide valuable insights in the possible future of a particular technology.

The great problem with this "scenario thinking" is that it still clings to a classical view in which scientific and technological developments are regarded as predictable processes into which it is possible to have a continuous insight. Although scenario analysts show themselves ready to take account of various unpredictable influences and all kinds of social factors, they ultimately think it must be possible in principle to chart these influences with reasonable clarity. This view, however, makes no allowance for the individual nature of the many conscious and unconscious controlling factors which are constantly at work in scientific and technological development. This means that the scenarios described are those which the client and the research group, possibly after wide consultation and study, are willing and able to think of. The selection of scenarios is therefore subject to their conscious and unconscious preferences, and is still far removed from the dreamed-of "objective overview" of the future of a given technology, which a TA analysis promised to be.

5.2.2 Controllability

We have seen that the multifaceted complexity of real-life technological developments precludes the possibility of providing a full and reliable picture of their future consequences. Another aspect of this complexity is the impossibility for outsiders, like TA analysts of the members of the US Congress, to supervise the implementation of specific technological products by the simple Go/No-go decision giving specific technological processes the green light, while nipping others in the bud. Could Congress really prevent some dangerous or undesirable future consequences of a specific technological evolution by using all its power and authority to stop its development ?

We have more reason to be pessimistic than optimistic concerning the power and the potential of such clear "No-go decisions". Obviously, it is possible for a financier to call a halt to a good many concrete research projects by turning off the flow of cash. An authoritative government body can achieve the same effect in many other ways as well. And yet the influence of such decisions, even if they may be immediately successful, is always fairly limited and short-lived. Is it not sufficient for the research group simply to find another financial backer? Cannot the power of certain authorities be circumvented by continuing the same research in a location where the prohibiting authority has no say? There is a good chance that this could happen. After all, a decision to halt a scientific process always ends up in the midst of a complex force field in which a great number of other interests are already at work. The fact that one of the interested parties withdraws because he no longer finds the process interesting, or even feels it to be dangerous, does not necessarily mean that the others will follow him and that the process will come to a standstill. It is more likely that the other partners will go in search of ways to continue the process and that they will if necessary seek out new interested partners.

These are the sorts of considerations which regularly lead to the strange situation whereby persons or institutions are genuinely convinced that it would be better on the grounds of "higher motives" to halt an interesting development, while they in reality simply carry on. The reason then given is that it is never possible to prevent the same work being done by others in another place, so that it is ultimately felt to be better – if the work is to go ahead anyway – that those already involved should be the ones to reap the rewards.

Moreover, scientific and technological developments do not simply follow a single straight line towards a certain goal, but rather seek their way through a multitude of processes which in turn constantly branch out in various directions and cross each other again at various levels. In such a tangled web, an "undesirable" effect which was revealed as a potential consequence of one specific development may also come out of the blue as being the quasi-coincidental result of a whole series of completely unrelated technological developments.

5.3 An Inadequate View of Assessment

We mentioned that "Technology Assessment", as created by the US Congress, is not only built on a completely inadequate view on technology, but also on a completely unworkable view on assessment. In the previous section we explained that the sort of technology the US Congress wanted to assess does not exist in real life. The manifest conclusion was that the TA methodology is totally unsuitable for clarifying the possibilities, challenges and potential problems related to technological developments as they actually exist.

We will now explain the problems of the TA methodology which are due to inadequacies and ambiguities in its underlying conception of the general process of political assessment. In fact, even if real-life technology proved to be the simple, straightforward and predictable process as TA assumes it to be, the TA methodology of the US Congress would still be doomed to failure. We will explain first the inadequacy of its theoretical view on assessment, and later the ambiguity of its practical view.

5.3.1 Objective Overall Assessment

The whole expected purpose of TA was to provide the members of Congress with an evaluation, an *assessment* of a specific technological project under discussion. This means that a TA report should contain the scientific evidence required to demonstrate to what extent the different facets, results and consequences of a specific technological project will be dangerous or harmless, irrelevant or needed. EMILIO Q. DADDARIO, the driving force behind the establishment of OTA and its first director, put it as follows:

"Technology assessment is a form of policy research which provides *a balanced appraisal* to the policymaker" (EMILIO Q. DADDARIO, 1968).

This balanced evaluation is in fact why Congress originally felt the need for creating a mechanism for objective Technology Assessment. "Congress created OTA in 1972 to meet its need for a source of technical information that is nonpartisan, objective, and anticipatory ... Members of Congress began to wonder whom to believe as an expert because it appeared that there were experts on all sides of an issue. It became obvious that Congress needed its own source of nonpartisan, in-house technical analytic help" (CHRISTINE LAWRENCE, 1985). Congress wanted to make scientific evidence its only adviser. It was distrustful of industrial lobbyists or Presidential advisors who dwelled solely on the beneficial consequences of the technology they wanted to promote while downplaying potential risks. Congress also looked with distrust on activists and members of the anti-technology lobby who come up with endless lists of risky, hazardous, and bad consequences which are invariably presented as dangers to such important goods as human health, the environment, the air, social life, and many other elements. OTA was established as an independent resource which was capable of making an objective appraisal and allowing the members of Congress to make decisions based on facts, rather than on empty rhetoric. This is why Congress wanted the reports of its TA office to be explicit and exhaustive about anticipated problems and dangers as well as about expected benefits and promises.

In nurturing these beliefs the US Congress is a child of its time. It shares the expectations which have gradually gained widespread credibility since the eighteenth century, the age of the Enlightenment. Keywords here are notions like "harmony", "certainty", "objective knowledge". In fact, the ideas of a homogeneous and predictable science and technology are part of a particular embodiment of the Enlightenment ideals, which we will call the "certainty-based" or "completely-objectivistic" Enlightenment view. This Enlighten-

ment view embraces the ideas that a balanced appraisal of complex human situations can be achieved through the meticulous compiling, deducing and comparing of objective facts and that responsible decision-making is unequivocally determined by these objective and incontestable insights. It is clear that these types of Enlightenment beliefs have left their mark on every layer of our modern culture and our modern body of thought. The idea of a purely logical and controllable science and technology, which as we have seen is an unrealistic myth, is in fact only one of the many traces of this specific Enlightenment view. While the certainty-based Enlightenment project has never been without individual critics and while its failure already gave rise to widespread scepticism and relativism, only in the sixties did a powerful and sustained process of critique really break through.

It is not difficult to see that its ideals pertaining to strong and certain assessment and decision-making are as unrealistic as its conception of science and technology. Decisions on responsible action and a global assessment of a complex human situation are simply not as straightforward as we would hope and as the certainty-based Enlightenment vision wants us to believe. We do not need a lot of theoretical analysis to realize that there is something askew with this purely objectivistic view on appraisal and decision-making. It is somewhat more difficult to grasp how a more realistic understanding of rational decision-making can be developed. We will have to come back to this later. For the moment we can limit ourselves to unfolding the idea that "pure facts", and "an objective analysis of a situation" (should they exist altogether) can in no way present a balanced appraisal which could function as the unequivocable guide to responsible action.

We can illustrate this idea with an example from clinical genetics, where future parents sometimes seek to discover the risks they run of having a child with a particular disorder. No-one will doubt that ignorance on such questions is a great disadvantage. But the conscious or unconscious apprehensions which people nurture concerning precise figures or a precise description of the potential disorder are frequently exaggerated. Many secretly hope that they will ultimately know what to do when the objective figures are available, but these people tend to be cheated in their expectations. For example, there is absolutely no such objective guideline encompassed in the knowledge that there is a 25% risk of having a child with a given heart disorder, if there is no means of tracing or preventing this disorder before the birth. Should a couple give up a planned marriage in such a case? Should all plans to have children be abandoned? Should the couple use one of the technologically aided forms of reproduction? Knowledge of objective figures and data gives no answer to these and other questions. People then usually want to hear more from the counsellor than statistics and objective prognoses. They hope the counsellor can also give them objective information about the best course of action left open to them at that point. They try to find out what the counsellor would do in their situation. Counsellors are well-trained, however, not to answer these questions. They realize that any answer that they would give would go beyond the limits of the solid information they can provide and would simply be an expression of their subjective evaluation of the situation. They even know that their evaluation would be somewhat unrealistic and beside the point, because they are not in the exact same situation as their counsellees. The counsellors' subjective evaluation cannot take into account the complex mix of feelings they cannot possibly share with the counsellees. Counsellors realize that even exact figures, trends and facts do not give us a complete picture of their acceptability or desirability. Having a balanced assessment of a situation is always something more than knowing and accepting the generally recognized facts about the situation. Facilities for genetic counseling have developed strategies for dealing with those uncertainties. These strategies reflect the belief of these facilities that the information they can provide is of utmost importance as well as their understanding that this information has only a limited value. They know that they cannot provide a balanced appraisal of the situation of the counsellee and they are not in the position to tell them what their responsible decision should be.

The situation of policy-makers who deal with technological projects, is quite similar to the situation of parents. Both have to make appropriate choices among different alternative actions with far-reaching consequences. Both try to gain a better understanding of their situation by obtaining scientific information. In neither case, however, can a detailed report full of figures and data replace the balanced appraisal of the decision-makers involved. A major difference between the parents and the policy-makers is that the institutions which are created to help parents are well-aware of the difficult and complex nature of the decisions to be made, while the official ambition of OTA, as created by the US Congress, was to produce scientific information which provides a *balanced appraisal* of the situation. It should have become obvious by now that "assessment" of, and "responsible decision-making" in a complex situation simply does not work this way.

5.3.2 Subjective Overall Assessment

While absolute obedience to scientific objectivity is given lipservice in the official description of TA, experienced policy-makers like the members of the US Congress have always known better. In reality, they never believed that someone else could come up with a balanced appraisal of a situation where political decisions have to be made. It would have been consistent with the general approach of scientific neutrality to appoint a small group of authoritative, skilled researchers as staff members and to give them the responsibility for setting up the Congressional TA activity. This scientific group should then assume the responsibility of identifying the assessment studies to be done, for selecting, hiring and firing the scientific collaborators, for carrying out the research and for publishing the policy-preparing TA reports which would contain the overall balanced appraisal.

In practice, the members of Congress tried to be more closely involved in pulling the strings of OTA. In each and every step they embodied the idea that policy-making transcends scientific analysis.

(a) First of all, Congress undeniably considered the task of choosing the issues to be studied as a political and not a scientific matter. Members of Congress wanted to retain the exclusive right to initiate TA studies. OTA could only carry out those studies which were explicitly requested by the Congressional committees and subcommittees. In theory, the director and the Management Board of OTA (officially named the TAB or Technology Assessment Board), could also initiate studies. In practice, OTA made little or no use of this possibility. Besides, it would not really make much of a difference if it did. EMILIO DADDARIO, OTA's first director, was a former Congressman and could hardly be considered an independent scientist. Furthermore, the charter of OTA required that the twelve members of the Management Board were all elected members of Congress, six from each political party.

(b) Secondly, Congress did not only want to decide what was to be studied, it also wanted to monitor the work of the scientists who were actually carrying out the research. So, it proved to be usual in DADDARIO's time for each of the twelve Congressmen who were members of the TAB to be able to appoint a supporter as a key figure on the scientific staff. Political loyalty was more important in these appointments than scientific expertise. It was RUSSELL PETERSON who, as second director from January 1978 to April 1979, joined the battle against this practice and resolutely built up a scientific staff whom he could appoint and dismiss on his own authority and according to scientific criteria. PETERSON too, however, was more a politician than a scientist. Only with the arrival of JOHN GIBBONS in 1979 did OTA have an academic scientist at its head for the first time.

(c) Thirdly, once a study was completed, members of Congress have always reserved to themselves the right to interpret "the facts" revealed by the study, to draw the appropriate conclusions, and to make right political decisions accordingly. JOHN GIBBONS, while recognizing the value of policy-preparing research, is quite explicit about its limits.

"Although expert scientific advice had become increasingly important to good decision-making, Congress properly reserved to itself the responsibility to make policy" (JOHN H. GIBBONS, OTA director, 1979-1992, and HOLLY L. GWIN, 1985).

It should be understood that in practice neither the members of Congress nor the OTA really expected that an overall assessment of technology could be provided by scientific analysis. In reality, the overall assessment of a technological project was seen as an element of the broader process of political decision-making. In this context, a clear distinction is made between the realm of science, which is rigorously governed by strict unambiguous logic and objective facts, and the realm of policy-making, which is characterized by elements like options, choices, interests, negotiations and bargaining, and which are far more vague and subjective.

This realistic practical note does not, however, refute our earlier conclusion that the TA experiment, as designed and developed by the US Congress, is built on a concept of assessment which is not practicable. Even when we take our practical qualifications into account, the least we can say is that the US Congress holds a very ambiguous view on the relationship between the assessment of technology, analysis of technology, and responsible decisions related to technology. Whereas in some contexts the relationship between those elements is seen as strictly objective, in other contexts it is seen as strictly subjective. In the objective approach, the key position is held by the scientific analysis of technology, the TA study. This analysis is believed to include the overall assessment and to determine the content of responsible decision-making. What is expected from members of Congress is that they use their political power and mandate to give a legal force to the conclusion which was obtained in a scientific TA study. In the subjective approach, on the other hand, everything revolves around the complex and multifaceted situation of political decision-making. In this view a scientific TA report on the consequences of a specific decision is only one aspect which the decision-makers have to take into account. While they will readily accept that a TA report is important, they are convinced that such a study does not entail the overall assessment of a situation, which can only be attained as the result of complicated subjective processes of political negotiation. Our short description of these two approaches ought to have been sufficient to illustrate why US Congress makes a proper mess of assessing technological projects instead of providing clear and workable guidelines. The situation is even worse. The problem with the subjective approach is not only that Congress is not consistent enough and is mixing this subjective approach with the remaining elements of an exaggerated objective approach; no, even taken by itself, the subjective approach does not provide a clarifying model of how a specific TA methodology can contribute to keep the quality of technology in check. In a purely subjective context we do not get the slightest hint of how to distinguish a responsible decision about technological developments from mere arbitrariness, and the potential role of any policy-preparing research is not clear at all.

6 Lessons from TA

Without downplaying the fact that the TA methodology, as designed by US Congress, is completely unworkable, unrealistic, naive, and doomed to fail, it would be too simple and unfair to blame the members of the US Congress for the weakness of their attempts to steer technological developments. Congress failed to the extent its TA initiative is built on inadequate basic concepts which were widely accepted in the sixties, at least in Western culture. It did not invent the unworkable concepts of technology and assessment. Nor was it directly responsible for the ambiguous mixture of objectivity and subjectivity which is so characteristic of contemporary thought. All these are the common heritage of the endlessly objectifying Enlightenment culture and the corresponding endlessly subjectifying reaction against it. It is wise not to underestimate the merits and strengths of the Congressional initiative. US Congress has long since been a privileged witness of the

creation and development of many large technological projects. Over the years, members of Congress also gained an accurate and realistic insight in the way they could probably influence the quality of these technological developments. A deeper understanding of these backgrounds can help us to better grasp the real value of the Congressional TA initiative. Despite the complete failure of the TA approach in itself, the basic inspiration behind it is sound and extremely valuable. This school of thought rests upon the fundamental conviction that the monitoring of the quality and relevance of technological projects requires the reinforcement of an appropriate interaction among all parties who have the power to influence the design of the project in question. Five characteristics of this interaction appear to be of utmost importance. We will list them, and explain how an objectivist conception of technology and assessment could deprive this sound approach of its strength.

6.1 The Importance of Sound Interaction

6.1.1 Interaction with the Initiator

Congress knew very well that real-life technological projects, especially the mega-projects like the development of an atomic bomb or sending a man to the moon, are rarely if ever unassessed. Every time Congress is asked to approve the budget of specific technological projects, its members know that those projects have already been evaluated and approved by the President. Congressmen know that Presidents are developing a structured and targeted science and technology policy. What they have to assess is not just an uninterpreted, merely logical scheme or plan, but a specific down-to-earth technological enterprise designed in a predefined context to achieve specific goals which are considered to be interesting or needed for political, military, economic or other reasons. Each of these large-scale technological undertakings has its own internal assessment procedures. The real lesson Congress learned in the sixties was not that there is an urgent need to *start* assessing unassessed technologies, but that it makes sense to *improve existing* assessment procedures by making them interactive. Members of Congress did not claim that the idea of evaluating and streamlining technological developments was their brainchild. They knew they did not have to invent that idea. What they really wanted was to improve and strengthen the quality of their interaction with the President's science and technology policy. Up to the 1960s, Congress had tended to view its contribution to American science and technology policy as minimal. It was especially the White House administration that initiated the national technological projects and that prepared their assessment. More and more, Congress wanted to be involved in this process of assessing and directing technological developments. Assessing technology was seen as an interactive process. A first fundamental characteristic of this interaction is that the initiators and sponsors of technology, the original assessors, should be involved. It does not make sense to establish a second circuit to design an alternative science and technology policy. Such a second circuit will scarcely understand what is really at stake and will not have the power to prevent or stimulate specific technological developments. The only way to effectively contribute to the quality of an existing science and technology policy is to start an interaction with those who are already in charge and to enrich their internal assessment procedures. Assessing technology is and remains primarily the task of the initiators of technology. Any contribution to this assessment requires an interaction with those initiators.

6.1.2 A Powerful and Committed Interaction

We have described the link with the initiator of a technological project as a first and central feature of any workable interaction required to assess and monitor the total quality of this project. Congress was very much aware of the fact that the perspectives and assessments which move the initiator constitute a fundamental element of any global appraisal of the potentials of the project. Congress also realized that its own input into the interaction should be more than a mere token interest. Its input should be just as committed and powerful as the President's input. Confronted with the problems which came about in the sixties, members of Congress felt the strong need to provide a firm counter-voice to the President's technological plans. They realized that the President had his own priorities and objectives and that he might not be critical or suspicious enough for a technological development which appears to be crucial for realizing his goals. Confronted with the hidden impact of technologies on the environment or on social life, members of Congress developed their own ideas on which technologies or which aspects of those technologies might require an explicit and careful consideration and further study before declaring the projects desirable or acceptable. They knew that it was their task to challenge the evaluations made by the Presidential advisors by opening the political discussion to include those aspects which the President would never have placed on the agenda. They were convinced that they should take this task seriously, and develop a proper vision of the way a specific technology could promote or endanger the quality of social life. They also realized that they had to strengthen their cohesion and their internal consensus across party lines. All too often the achievement of a consensus within Congress was a very difficult matter, given that political power was divided between the two general assemblies (House and Senate) and the many committees and subcommittees, were dominated by either of the two political parties. The fact that Congress was often internally divided meant that the President rarely came up against a strong unified opposition from Congress. In its attempts to answer this situation the keywords were "access to specialized information" and "the creation of new structures".

6.1.3 Informed Interaction

For years, Congress had looked on with envy at the way the President and his cabinet were able to surround themselves with a group of specialized science advisors, who have always been well-staffed and well-equipped. Members of Congress knew that they would never be powerful partners in the interaction with the President without being properly informed themselves about all aspects of a technology which were relevant to them. They realized that more was needed than general, informal or incidental information. In the words of NANCY CARSON, an OTA program manager: "The tradition of high intelligence and informal decision-making that characterized Congressional staff no longer seemed sufficient to deal with the complexity of technology problems" and to reply to the well-prepared and cleverly presented "print-outs, calculations, and convincing projections" of the "executive branch witnesses" (NANCY CARSON, 1989). What Congress needed now, was the possibility to initiate careful studies of its own.

6.1.4 Properly Institutionalized Interaction

Many elements of the interaction between Congress and the President are institutionalized. Their occurrence does not depend on a lucky coincidence, on the accidental initiative of one of the partners or the goodwill of both. In this broad context, there has always been a structural link between the Presidential science and technology policy and Congress. But, as we have seen, the way this interaction was institutionalized did not guarantee Congress that it could remain as informed as the President, and, furthermore, due to its internal segmentation, Congress rarely came up with one, powerful voice in important matters like technology policy. When members of Congress realized that similar weaknesses were also present in several other domains of their interaction with the President, they decided to combat the growing Presidential power by establishing appropriately institutionalized structures. This resulted in the establishment in the 1970s of a number of supporting "offices" which contained members of both parties and which were charged with preparing the way for a consensus in Con-

gress. In this manner, for example, a "Congressional Budget Office" (CBO) was set up to perform economic analyses, and a "General Accounting Office" (GAO) was established to carry out careful financial analyses of public programs. In addition, a well-equipped parliamentary documentation center, the Congressional Research Service (CRS), was set up in the Library of Congress - the largest book collection in the world. This center maintains summaries and publishes short reports on various complex political questions of the day.

The Congressional Office of Technology Assessment (OTA) is another of the offices which was designed to strengthen the internal power of Congress and to counter the power of the White House. OTA is directed by a Technology Assessment Board (TAB), which consists of twelve members, six from each political party. Six members are appointed by the Chairman of the Senate and six by the Chairman of the House of Representatives. The TAB chooses its own Chairman and Vice-Chairman from among its members, with each political party taking at least one of these posts. The appointments of these posts alternate between the two parties every two years. Initially OTA could hire about 20 scientific collaborators to do the daily work. After some years this number had surpassed 150.

6.1.5 Interaction Among All Partners Involved

In the US, the official Science and Technology Policy is designed by the Presidential administration and in an increasing degree also by the Congress. So far we have focused on the interaction between these two political bodies and on the concrete measures taken by Congress to strengthen its role in this interaction. By doing so, Congress not only wanted to exercise its democratic mandate of controlling the President. As we have seen, it also wanted "to provide a countervailing force to public pressures for more direct democracy" (K. GUILD NICHOLS, 1978). This concern for its own position does not imply, however, that Congress wanted to prevent other social partners from being involved in the discussion and evaluation of technological projects. On the contrary, as a publicly accountable body, Congress believed in the importance of informing the public at large about new technological plans and their backgrounds. This is why Congress decided that all TA reports should be made available to the public at large. Congress believed in the necessity to inform each relevant social actor, so that each was given the possibility to react using the appropriate channels at their disposal.

6.2 The Allurement of Objective Certainty

Even though until the dawn of the 1960s, members of Congress had to content themselves with little more than a formal, symbolic and marginal input into the Governmental Science and Technology Policy, they were close enough to the decision-making process to be able to fully fathom its workings and how they could play a role in it. We can learn a lot from their sound intuition of the importance of an informed and committed interaction among all partners involved.

In spite of this solid point of departure, things went amiss during the process of further elaboration of their basic vision. We have already seen how this failure was caused by unworkable and unrealistic conceptions of the technological processes themselves and on an ambiguous, at the same time over-objective and over-subjective, conception of assessment processes. These weaknesses were identified as persistent remainders of a certainty-based Enlightenment view. A second lesson that we can derive from the TA experiment is that we need be wary of the "fatal attraction" of this objectivist view – its disarming seductiveness to even fool one's better judgement but also its power to kill viable and inspiring insights and beliefs.

Congress often assumed that it could obtain objective information about and an objective overall appraisal of technological projects and their consequences. This under-standing undermined any of the prerequisites of a serious and balanced interaction among all parties involved. To the extent that Congress believed to be the neutral monitor and judge based on the "objective appraisal" provided by its TA Office, there was no longer any need to engage in interactive consultations with the Presidential Administration, which was then considered to be the merely subjective, politicized, "unreliable instigator" who cannot add anything to the insights of Congress (first characteristic). Secondly, this objective basis eroded the role of Congress itself, which was restricted now to be the serving-hatch of objective information – a role which in practice appeared to be ambiguous, because on top of its objective pretensions Congress still wanted to maintain the right of subjective interpretation of the so-called objective facts (second characteristic). Thirdly, a real contribution of other partners was seen as completely superfluous. The objective truth revealed by OTA was meant to serve Congress as well as President or any other social group (fifth characteristic). In this objective, positivistic approach, the central responsibility for assessing technologies was passed on to one centralized, objective, impartial arbiter that was supposed to do the assessment job at the service of all parties involved. In this way, in the process of implementing OTA, an unworkable and objective method was adopted at the expense of a more promising, realistic and sound interactive approach.

Fortunately, Congress and OTA soon unearthed these true underlying problems and, after some years of wandering and rambling in an impossible objective world, returned to their basic aspiration of reinforcing the interactive decision-making process. Congress was wise and judicious enough to learn in practice what it hadn't learned in theory. Even before the foundation of the OTA, as early as 1968, the American National Academy of Sciences, in its very first recommendation written at the request of US Congress, had given an expressed and well-argued warning against an attitude where the responsibility to assess technological projects was to be placed in the hands of a single, so-called objective bureau which could directly influence policy. We quote from the report of the committee of the

NAS that dealt with the problem, and split up the quotation in five separate ones, because each of them is still capable of giving food for thought right up to the present day.

"The panel would emphatically oppose any scheme that would empower an agency to decide on behalf of something called 'society' or 'the environment', which technological developments will be permitted and which prohibited."

"Selections among alternative technologies require that choices be made among *competing and conflicting interests."*

"To the extent that those choices are made and enforced collectively rather than individually, they are essentially political in character and must therefore be the responsibility of the politically responsive branches of government and of those publicly accountable bodies that are specifically entrusted with regulatory responsibilities in narrowly circumscribed areas."

"The making of such choices is, in principle, indistinguishable from the resolution of the many other conflicts that beset society."

"To entrust the resolution of all those conflicts to a single, all-encompassing authority would be incompatible with representative government" (NATIONAL ACADEMY OF SCIENCES, 1968).

7 The Origins of Science and Technology Policy

While the sixties undeniably showed that there is something wrong with the way technologies were being planned and evaluated, it would be naive to infer from the growing problems and doubts concerning technology that no mechanisms for planning and evaluation existed and that there was an urgent need to create them. In fact, most of the problems with new technologies occurred precisely because of the priorities forwarded by a developing science and technology policy. We have to admit that science policy was still relatively young and inexperienced. Even the US and the Soviet Union, the two superpowers of the sixties which can be regarded as forerunners in this area, had to wait until World War II before well-structured mechanisms for science policy were established. The sixties marked a period of growing pains and teething troubles. Dealing with the problems of the sixties implies studying the existing mechanisms for science and technology policy and exploring how they can be improved.

7.1 Industrial Science and Technology Policy

Nowadays everyone finds it self-evident that fundamental scientific and technological research will sooner or later be translated into the goods which can be mass-produced and are of use to society. Thus, for example, research in electronics becomes visible after a time in the form of improved computers, more easily programmable dish-washing machines or more sophisticated hifi systems. In the same way, fundamental biological research is expected to lead to improved health care and increased food production. Two hundred years ago, even the most learned scientist or the most skilled craftsman could not have imagined that his work would ever be relevant for industrial production. The scientists of the 18th century were not concerned with enhancing the productivity of the manufacturers, which were the industrialists of the day. They identified their professional activities as the quest for and the discovery of the truth, and not as a contribution to change the world. Neither did artisans, craftsmen or technicians expect that their skills and techniques could ever be of any service to the activity of large-scale production. Their arts and crafts were not intended to be practised by untrained workers. Manufacturers, in their turn, did not feel this situation as a great lack. They took care of the easy production of consumables for the inhabitants of their town. What they needed was people, time, raw materials, good eyes, good hands, and care for their work. Their work did not require sophisticated skills or scientific insights. The movement which completely renewed the methods of industrial production and their relationship with science and technology is called the "industrial revolution", which is usually subdivided in three or four waves (see Fig. 1).

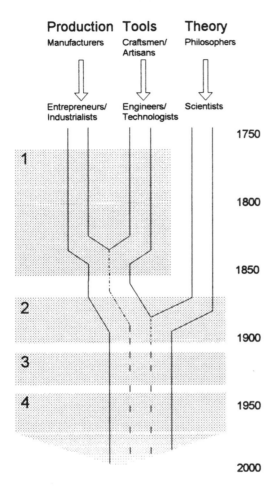

Fig. 1. The Industrial Revolution.

7.1.1 The First Industrial Revolution

The first wave was put in motion by a complex cluster of interacting social, demographic and technical processes, which started by the middle of the 18th century. The introduction of the steam engine in the mining industry is its showpiece. Further important innovations were the textile machine, the locomotive (STEPHENSON), and improvements in procedures in iron and steel production. But these are only the most conspicuous examples of a growing interaction between skilful craftsmen and manufacturers, which even intensified around 1800 and slowed down again towards the end of the first half of the 19th century. When this first industrial revolution came to an end, the first modern "industries" were developed. Sophisticated machinery and techniques, capable of repeated renewal and improvement, replaced the traditional production methods. An increasing productivity, a concentration of workers in the neighborhood of factories and growing towns were some of the consequences. This does not mean, though, that traditional manufactures ceased to exist as units of "industrial" production. Near the middle of the 19th century, the industrialization was limited to England and Scotland, where the revolution began, and France and Belgium.

7.1.2 The Second Industrial Revolution

The world of science had not yet been swept into this revolution. The primary innovations on which the revolution was based, like the steam engine or the locomotive were the work of inventive craftsmen and experimenting entrepreneurs. But while their creative intuition, inventiveness and practical understanding sufficed to get the revolution off the ground, they were not sufficient to keep the inertial momentum going. More and more, technicians began to realize that the further improvement of their machineries and tools required a real understanding of underlying scientific laws. They also experienced that several domains of industrial production would persistently resist their efforts to design appropriate methodologies and instruments, unless they could base their attempts on deeper scientific insights. It took several years before the scientific world was ready to show some interest in helping to solve these technical and industrial difficulties. But when this finally happened, after the

1870s, the Industrial Revolution gained speed again. A second wave began, resulting in the industrialization of most Western-European countries and the US.

From this moment on, industrial companies were quite active in searching and creating possible links between industrial processes, the technical infrastructure needed and "basic science". The German industry which specialized in the production of organic dyes was one of the earliest examples of a chemical industry founded on scientific research. It erected in the very first years of the second industrial revolution (the last quarter of the 19th century). Similarly, at the very beginning of the present century, scientific research was adopted as the basis for the electrical engineering industry in America. To guarantee a maximum of basic science which would explicitly serve the industrial objectives of the company, both branches of industry founded their own research laboratories from the start.

7.1.3 Later Revolutions

At the beginning of the 20th century, the second wave of the Industrial Revolution came to an end. It was succeeded by what is often called the Third Revolution, which transformed and revitalized industry by replacing steam with electricity and gas as the new sources of energy, and the fourth revolution, which introduced the automation and later the computerization of the production process and its administration. During all those years, industrial companies continued to manage their own Research and Development budgets, which were far from inconsiderable. In most countries industrial companies spent more on R&D than the public authorities. Only late in the sixties or even in the seventies did governmental science and technology policy become of equal importance.

7.2 Early Governmental Science and Technology Policy

7.2.1 Prestige and Maintaining Peace

JEAN-JACQUES SALOMON, former head of the Science Policy Division of the OECD, points out how contemporary governmental science policy found its roots in World War II. It is well known that wars can be a catalyst for governmental interests in science. Ever since the ancient times, generals and statesmen were eager to listen to the learned thinkers and skilled craftsmen hoping that they would be able to invent a superb machinery that would help them capture a city, break through the enemy lines or simply win the war. This was true in Greek times, with the Trojan Horse and the inventions by ARCHIMEDES. It was still true during the First World War, when institutions were created whose aim it was to promote research into military and strategic applications of science. It is also true, however, that the state's interest for science soon vanished once the war was over. Special war-oriented scientific enterprises were scrapped, following the conclusion of peace. It is exactly in this respect that World War II was different, according to JEAN-JACQUES SALOMON. We know that during the war the familiar scenario reoccurred. The Manhattan project instituted by the American public authorities made history in this respect with the development of the atom bomb. The harsh reality of the bombs lay at the basis of many historical developments, of which the emergence of a systematic science policy is one example. There are two major underlying reasons. First of all, the bomb and the release of nearly unlimited amounts of atomic energy became a symbol of the great, nearly unlimited powers hidden in the practice of science. The bomb made it abundantly clear that modern science was capable of a great deal more than the theoretical exploration of reality. The beginning of the Cold War was a second reason why, contrary to what normally happens at the end of a war, states did not simply send their scientists back to their academic labs and libraries. World War II did not really end. The tension which remained between the great powers, and the emerging Cold War, served as the great stimulating force behind the superpowers to keep their scientists "mobilized" and to institute large-scale coordinated scientific programs to serve the national interest. The public authorities considered their primary aim here to be the continued assurance of national security and the strengthening of national prestige.

"Science policy developed ... as a consequence of the impossibility of establishing real peace at the end of World War II" (J.J. SALOMON, 1971).

The American public authorities were forerunners in developing their own science policy, and the working model was set up in 1945 at the request of President ROOSEVELT by VANNEVAR BUSH, who during the War had been director of the United States Office for Scientific Research and Development. The Soviet Union, the other superpower, took similar initiatives. Other industrialized countries followed suit.

In this light, priority in the United States was given to three sectors: space travel research, nuclear research and military research. The development of radar, manned rockets, jet aircraft, atomic energy, DDT and computers are all consequences of this policy. The launch of the Sputnik, the first manned space flight, and the landing of the first man on the moon by the Americans are clear symptoms of a fierce competitive race between the superpowers, a development which led to the enormous size and rapid growth of the scientific disciplines concerned.

7.2.2 Growing Social and Economic Welfare

Once the public authorities saw how the injection of massive amounts of capital and the setting up of coordinated scientific programs had succeeded in creating technologies of national importance, a second phase in science policy was born, in which an economic edge was added to the old national goals. Enormous research budgets were made available for scientific research which was thought to increase the competitiveness of national industries.

In this way, the "golden sixties" brought not only great prosperity to the economies of the industrialized nations, but also to science and technology. Scientists and technologists, from whatever research area, had quickly understood that public authorities had the potential to become good customers if they could clearly see that the research could contribute to a military or economic breakthrough.

7.3 Spread of Governmental Science and Technology Policy

The OECD (Organization for Economic Co-operation and Development) played an important role in the further development of science and technology policy. The OECD was founded in December 1960 as the heir to the "Organization for European Economic Cooperation" (OEEC), which was founded in 1948 in the context of the post-war Marshall plan. The deed of foundation was signed by 24 Western industrialized countries. The stimulation of economic growth forms one of the primary objectives with which the organization was entrusted.

"The OECD is not a supranational organisation but a place where policy makers can meet and discuss their problems, where governments can compare their points of view and their experience. The secretariat is there to find and point out the way to go, to act as a catalyst. Its role is not academic; nor does it have the authority to impose its ideas. Its power lies in its capacity for intellectual persuasion" (JEAN-CLAUDE PAYE, Secretary General of the OECD).

It is clear that the OECD could not ignore the growing potentials of science and technology to stimulate economic growth and expansion on a global level. Within its secretariat it therefore created a Science Policy Division to study and monitor the possibilities of science policy. The OECD secretariat was convinced of the importance of a well-developed science policy. It wanted to discuss the issues with its member countries. In 1963 it convened a ministerial meeting between all the ministers responsible for science. In the majority of European countries this was the Minister of Cultural Affairs. "When in 1963 the first Ministerial Meeting on science took place at the OECD, the Ministers specifically in charge of scientific affairs could still be counted on the fingers of one hand" (E. G. MESTHENE, 1965). During this meeting the representatives from the various countries agreed that science is not merely a "cultural asset" to be protected and nurtured, but also an efficient provider of extremely valuable services for society. They concluded (a) that it was necessary for the national states to develop a governmental science policy and (b) that the states would have to create special organisms, advisory councils, secretariats and other institutional facilities in order to achieve this.

7.3.1 Science for Policy

The core of contemporary involvement by public authorities in science was described at the OECD meeting as "Science for Policy", meaning "Science to serve Public Policy". The authorities had understood that science and technology could prove valuable aids in the achievement of the goals which the authorities had set themselves. In fact, this comes down to the discovery that the scientific world is the perfect provider of the high-quality products which the authorities need. The only problem is that most of the products in which the public authorities show an interest are so grandiose and special that they are not in stock as standard in the normal research world, and are far from simple to manufacture. If the authorities wish to enjoy the end products, then they must finance not only the marginal cost of the product itself, but also the costs of the scientific and technological production mechanism. In this way, the authorities ended up as managers of extensive scientific and technological projects, in which they supported both "applied" and "fundamental" research.

7.3.2 Policy for Science

This development enables us to understand why the OECD claims that involvement in science by the public authorities also continues to demonstrate a component which it names "Policy for Science" ("Public Policy at the Service of Science"). The authorities see themselves as a large, wealthy and patient customer of science. As such, the authorities are aware that it is necessary to allow the scientific tree to blossom in an ideal growing climate if they wish, sooner or later, to be able to pluck the technological fruits of that tree. In this sense, the authorities are ready to furnish an environment in which various research activities are able to flourish undisturbed.

8 Making Science and Technology Policy Interactive

A major point we have made is that the whole mechanism of TA, while self-evident, convincing and attractive at first sight, becomes unworkable and inadequate on further consideration. We explained that the TA approach, with its perceived objective methodology and institutions, does not provide an appropriate way to deal with the growing and persisting doubts about the relevance and safety of technological developments. Both the conception of "technology" and the ideas on "assessment" which serve as the foundations of the TA approach were described as exceedingly objective and therefore unrealistic. In this concluding section we will illustrate how a workable alternative can be developed. We will expand on those ideas which have already been mentioned as sound elements of the basic inspiration of the members of the US Congress. The crucial point will be to develop these ideas without allowing unrealistic conceptions of "technology" and "assessment" to undermine their strength. Such an approach will be unworkable without explicit attention to the following five points.

8.1 Making Existing Policy Interactive

When technology was gradually losing its aura of being an incontrovertible blessing for mankind, man had the opportunity to react in two different ways. The TA experiment is the most well-known example of the first type of reaction. This response builds on the awareness that current technology is not governed by an absolute rationality which is perfectly reliable in guaranteeing its total quality. It concludes from this that it is time to distance oneself from such a lack of critical rationality and to actively develop an objective mechanism which is able to bring technological developments within the realm of reason. We have seen, however, why this first type of reaction must be rejected. Its supposedly objective solution is only an illusion. This leads us to the second type of reaction, which, like the first, recognizes the lack of absolute rationality in the planning of science and technology but which, unlike the first, accepts that this restrained rationality, with its limited value, is all we have. The only thing left to do is to discover and maximize its resources. This is also what this second reaction plans to do with the way technological developments were drawn up in the sixties. It would be wrong to deny the limitation and shortsightedness of these early science and technology policies. But it would be just as wrong to deny their existence and to think we have to reinvent the wheel. Instead of starting a second circuit in dream-TA land, it is better to remain down-to-earth and to see how existing science and technology policy can be improved.

What is in fact the problem of the science and technology policy as it existed in the sixties? What can be done to improve it? Did governments, industrial companies or academic research institutions make mistakes when they identified the proper technologies to be developed? Did the initiators of new technologies act against the firmly established and universally recognized standards of rationality? The fundamental problem appears to be much deeper and has to do with the standards of rationality themselves. According to traditional standards, which were evident in the sixties, some action or plan can be considered "right" once you can provide *one* single proof that it is right, just like *one* proof of guilt is enough to come to the conclusion that somebody is guilty, or just like you can rightly say that you have solved a mathematical problem once you have successfully followed *one* possible strategy leading to the solution. The events of the sixties, much more than theoretical works, were able to convincingly argue that such "one-dimensional" rationality was not always enough. We learned that an action which was rightly labeled as "good" was not necessarily good, when seen from another point of view. Even when an action is proven to be of utmost importance for the best goals – like maintaining peace, providing employment, destroying harmful insects or promoting the wealth and happiness of many – no guarantee can be given that sooner or later completely other considerations, approaching the same action from different angles, will not strongly recommend us to raise doubts about its good quality or even to reject it. This situation might resemble the old wisdom that even the most desirable thing may have its drawbacks, as expressed in the proverb "No rose without a thorn". But the lesson of the sixties goes much further than that. The negative aspects of technology which were revealed were much more than some minor drawbacks of something good and blessed. What we learned in the sixties is that the drawbacks may largely surpass the positive elements. What appeared to be a simple rose might after all be more a thorn. Furthermore, few would have expected to find those negative aspects in such unsuspicious reality as scientific and technological progress, which was the sacred cow of the sixties. We learned now that good, even when said of the good beyond suspicion, is no longer simply good.

The lesson of the sixties dictates that our one-dimensional, absolute conception of rationality does not hold water. The corresponding challenge was to feel the pressing need to develop a more appropriate conception of rational and wise decision-making; in other words, a conception of a rationality which allows room for uncertainty and temporariness and which accounts for the possibility of different and sometimes conflicting approaches and evaluations – an "interactive rationality" which is not locked in its single-minded logic and approach but which breaks out of the points of view it already acquired and is continually open to new evaluations which transcend its own logic. Much has been written about this new challenge to rationality. It is clear, however, that this new situation is in the first place a challenge for those who are involved in the practice of decision-making. Developing ways of interactive rational decision-making are especially important for the founding of a sound science and technology policy. The problems linked to large-scale technological projects were a major catalyst for opening the theoretical and practical debate on interactive rationality.

The OECD, in stimulating the development of science and technology policies, has always underlined that this requires an intense awareness of the complex interaction between all social actors involved. The need for a careful study of the interaction between science, technology and society was already one of the conclusions of the first meeting on science and technology policy in 1963. The OECD has held a leading position in preparing high-level studies about this relationship. These were closely linked with the general philosophical and sociological literature which was becoming increasingly critical of a one-dimensional understanding of scientific and technological rationality. In a series of shades and forms, the idea was developed that only an interactive decision-making process could guarantee that science and technology would genuinely evolve to serve society. According to the OECD, countries must work and experiment with various forms of active participation by the public in the decision-making process of the public authorities. From the very beginning, the OECD has also taken a critical stance towards the usefulness of the TA approach, as developed in the US. OECD studies have always pointed out that this American approach is too theoretical and too narrow and leaves no room for a real broadening of the decision-making process. The OECD has always refused to set up a formal "TA program". It is convinced that the

problems linked to technology require an ongoing integration of policy-making on technology in the more general mechanisms of democratic decision-making. Dealing with the problems linked to technology is just one element in the broader democratic experiment of the last centuries.

We have already seen that the National Academy of Sciences in Washington shares this view, and states that solving problems linked to technology "is, in principle, indistinguishable from the resolution of the many other conflicts that beset society." Decisions about technologies are closely related to other social decisions, in that they do not require any more sophisticated treatment. Most governments who were wise enough to ask for advice on how to deal with problems linked to technology received the similar advice not to narrow their decision-making basis by means of the establishment of some supposedly neutral institution but, on the contrary, to broaden the basis of their decision-making by stimulating all forms of social interaction. In most countries, experiments for realizing this are still in their infancy. Even in very active countries like the US or the Netherlands, there is still the strong temptation to turn back the clock and to reduce all democratic attempts at keeping up one, central, specialized research institution, linked to the parliament.

Sweden has probably made the most decisive advancements in developing workable experiments. It has a long tradition of "consensus democracy". This "ensured that even in the mid-1960s much attention was devoted to the societal consequences of technology" (RUUD SMITS and JOS LEYTEN, 1988). Sweden also digested much of the talk about specialized TA, but it was not impressed by the so-called need for new methodologies and special institutions. It was evident for the Swedish that social evaluation of technology can only make sense when it "is considered to be an integral part of [...] societal decision-making processes in which technology plays some role". It was also evident for them that the only way to enhance the quality of this evaluation consisted in "broadening and improving public discussion" on all relevant factors. From this perspective "TA in Sweden was never institutionalized in an independent organization". For this could do nothing better than averting the attention from the real things to be done.

8.2 Using All Possible Social Resources

It is good to warn against a potential misunderstanding of our plea for an interactive technology policy. What we said is not that governments, industrial companies and academic engineers are professionally blind and helpless, and that the rationality of their actions and plans can only be warranted by forcing them to listen to "society" which is then taken to be the source and guard of true wisdom. Interpreting the value of alternative technological developments is as difficult for "society at large" as for the small group of institutional decision-makers. Problems linked to technological developments and inadequacies in science and technology policy are caused by a lack of vision and consideration by society as a whole, and not just by some alleged shortsightedness of technologists, scientists, industrialists or administrators.

"Science policy is in disarray because society itself is in disarray, partly for the very reason that the power of modern science has enabled society to reach goals that were formerly only vague aspirations, but whose achievement has revealed their shallowness ..." (OECD, *The Brooks Report*, 1971).

Many, if not all, problems which are commonly believed to be "evoked" by modern science and technology are, in fact, embodiments of the underlying weaknesses in our cultural resources. By realizing our hollow and uncritical projects, science and technology often function as a magnifying-glass of preexistent questionable ambitions or lack of interest. While it is true that many technological developments need a broader vision of what is really relevant, meaningful or acceptable for society, it would be naive to believe that the basic intuition of non-scientists or non-technologists could serve as a source of reference for this vision. The spontaneous aspirations of different social groups are not better off than those of institutional decision-makers. The power of interactive technology policy does not reside in the fact that the alleged shortsightedness of policy-makers is managed by the supposedly authentic aspirations of society. It does not make sense to make science and technology even more dependent upon the uncritical tendencies in society. The basic ideas behind interactive rationality are, in fact, primo, that no cultural or social sector can claim to possess absolute reason, and secundo, that the best we can do is to nurture and enrich our authentic cultural resources – independent of the social sectors or social practices where they flourish.

8.3 An Information Process

The availability of appropriate information is essential to high-quality interaction. Especially when dealing with complex technologies with far-reaching impact, critical reflection, intensive discussion and accurate research are of utmost importance.

The information needed is, however, much richer and far more diverse but also more vulnerable than the knowledge US Congress was looking for and TA research was believed to produce. The information should be appropriate to help us in answering the interrelated questions about the interesting social activities and plans for us to realize and the interesting technologies for us to develop in this context. These questions are neither universal, eternal nor general but very specific and concrete. We are not inquiring about the eternal good practice or social life, but about what would make sense for us, giving our past, our context and our possibilities.

A great variety of information, about society in general as well as about specific technological developments, can help us in this endeavor. More specifically it is relevant to gain a deeper practical and theoretical understanding of the original context and inspiration behind the different traditions, institutions, and practices which constitute our social life. There is of course no justification in using these elements as the absolute yardstick for our current or future plans. Those institutions and practices only embody what was believed to be meaningful in the past. But to this extent, they at least offer a concrete point of departure for talking about relevant social projects. Furthermore, when we talk about social relevance, about improving or endangering social structures, this is the only realistic reference point we have. The only way to answer the question about our future, is to explore the best future for our past. The study of scientific, technological and industrial achievements and their impact on society and nature is simply part of this general attempt to deepen our understanding of our cultural resources.

We should not underestimate the degree of complexity of this intellectual enterprise. Contemporary society is a complicated web of a multitude of practices and traditions which in part conflict, overlap or evolve independently. Additionally, several of these practices are strongly colored by science and technology and have intensified their private impact on the social and natural environment of all. But, as indicated by analysts, the greatest difficulty of understanding the cultural assets of our time might well be that for the last three centuries Western culture has neglected to update the understanding of its resources. This did not simply happen due to forgetfulness, laziness or lack of time but as a result of the basic belief in an absolutely objective reason, which we already mentioned above. When people believe that this universal rationality can and should guide all things in life, a careful understanding of the limited, contingent, present-day achievements and of their equally limited historical backgrounds becomes completely irrelevant and obsolete. This lack of interest and the absence of evaluative discussion not only resulted in a gap in our understanding of several centuries of cultural development, it also caused a strongly diminished integration of more recent achievements of Western culture, resulting in an increased number of internal conflicts and incompatibilities. Only a sustained effort of multidisciplinary research and multisectorial and cross-cultural discussion, proportional to the complexity of three centuries of intensified cultural developments and analytical standstill, can ever restore the basis which is needed in order to seriously assess technolo-

gical enterprises or other social projects. The success of this undertaking demands the collaboration of researchers like physicists, engineers, biologists and physicians as well as thinkers like historians, psychologists, sociologists and philosophers. Furthermore, a close link with all social sectors is strictly necessary.

We are far away from the objectivist and narrow aims of the TA methodology US Congress was dreaming about. The urgent need for this broad social research does not, however, mean that the detailed study of new technological projects is not equally important. It remains a necessary component of our rational attitude towards technologies to construct a realistic image, conveying as lucidly as possible the way we think the newly planned technology will function, how it will influence those activities we and others value, who will be able to influence the development of the technology, and how further decisions will be organized. But even this type of research is much broader than and different from the narrow predictive attempts of TA. It does not aspire to provide certain knowledge about an autonomous technological development or to be a neutral and infallible guide for those who plan to become more closely involved with this technology. It only makes sense in the context of a broader and more permanent monitoring process of the technology and not just as a red or green light at the so-called beginning of its development.

8.4 A Process of Shared Commitment-Building

Rational assessment of a technology, like any other social assessment, is not just an intellectual exercise. It is much more than a process of information gathering, even though insight and knowledge play an important role. It is also a social process of commitment-building, of commonly accepting the responsibility for a social decision to be made. When we evaluate a specific development, we as a society decide whether and under what conditions this development can be seen as a further enrichment of our resources. We decide whether and sometimes also under which conditions we prefer to live with it or without it.

This decision is never irreversible. New insights may come up. The technology itself or other aspects of our society may have evolved in such a way that we believe it is better to retrace our steps and to reconsider earlier evaluations. We might decide to give a central place to a technology we earlier rejected, or we may become more hesitant towards a technology we used to endorse.

Properly functioning social assessment procedures will, however, exclude that we blame let us say governments for not starting earlier with a specific type of new technology, or industries for producing materials or tools with side-effects we dislike. To the extent that appropriate assessment procedures are functioning, we cannot be critical of every change and progress and blame our country for not being far-seeing or our health care for being old-fashioned. Just like we cannot simply applaud the fruits of economic development when everything appears to be all right and blame the initiators in case unforeseen problems crop up. In short, assessing is more a well-understood political process, which builds shared subjective commitment, rather than a research methodology, which yields objective knowledge. When a decision needs to be modified or even turns out to be completely wrong, nobody can disclaim his share in the responsibility.

The crucial importance of the social process, inherent in each rational assessment, does not diminish the significance of the role of information. A proper assessment is more than the accidental result of a discussion or political bargaining. Justified assessment is not just a shared commitment, but an "informed shared commitment". Decisions which are not founded upon the fullest possible practical and theoretical understanding of our cultural resources and of the technology to be evaluated, can hardly be called assessments, whether they are made by dictatorial decree or by unanimous vote. This also implies that even a complete consensus cannot make up for the necessarily incomplete and uncertain character of the available information. Assessment remains a human and falli-

ble process. The many difficulties and weaknesses of the social process of shared commitment-building still adds to the precarious nature of assessment. Consensus will not always be achieved. In this case one has to take refuge in systems like deciding by majority vote, which drastically reduces the level of commitment of those involved. The fragmented character of our culture, and the fact that many decisions have to be made on a global scale still adds to this difficulty. Furthermore, it is impossible for anyone to spend time and attention to each of the social problems in which he is directly or indirectly involved. In practice many decisions are taken on behalf of us, by people we give some sort of direct of indirect mandate. In short, the quality of our assessment depends on the general quality of the democratic and participatory structures we are developing. These structures are far from being perfect, but they are all we have. We can dream of a universal, theoretical, absolute procedure which makes unequivocal, clear and infallible decisions. No doubt that for many decisions we would be better off than now. But sheer dreaming does not make it reality.

8.5 Decentralized and Multi-Form Institutionalization

Scientific and technological developments are occupying an ever more important place in our social and economic life. They have an increasing impact on society which is often welcome, sometimes regarded as suspicious, but occasionally hazardous or just irrelevant. There is a broad consensus that it would be unwise to leave their planning and follow-up to chance or to casual decisions.

The conclusion we have to draw from this is that we need to institutionalize the planning mechanisms of technologies. Only in this way can we make sure that all relevant elements be included systematically in each planning process. Only in this way can we guarantee that the necessary intellectual and practical resources are made available to those who need them. We are warned, however, that no institutionalization will be workable unless it is implemented in an interactive way. This is exactly what the OECD encouraged its member countries to do in the sixties: to institutionalize their own science and technology policy and to take the interaction with other social actors seriously. These recommendations have laid the foundations, in the sixties, of the fledgling science policies of several national governments, which are still in the process of further maturation. But what is needed in order to guarantee the total quality of technology is much more. The national governments, members of the OECD, are not the only actors involved in the planning and designing of technologies. Other social actors who are directly or indirectly involved include banks, consumer organizations, patient support groups or just plain users of technology. The total quality of technology can only be guaranteed to the extent that the impact on technology of each of these other actors is also properly institutionalized. As there is a great variety in the type of impact each of these actors may have on technological developments, there will also be a large diversity with respect to the proper way their involvement should be properly institutionalized.

This plea for far-going institutionalization of our concern for the relevance of technology may appear strange. Didn't we strongly reject the attempts of US Congress to create OTA, which is in fact nothing more than a highly specialized institution established for doing exactly what we wanted to institutionalize: to guarantee the total quality of technology. It is not difficult to show that, despite these superficial similarities, US Congress was dreaming of something completely different than the type of institutionalized interaction we advocate. Let us therefore recall what exactly OTA wanted to institutionalize. This will give us a last opportunity to reformulate the points we made in this text and to bring out the sharp contrast with the TA approach. The basic goal of OTA has been well summarized by JOHN H. GIBBONS, director of OTA from 1979 to 1992:

"If the national legislature were to act as *'all wise rulers must: cope not only with present troubles, but also with ones likely to arise in the future, and assiduously forestall them'*, it would need an institutional mechanism for foresight, one that would make analysis of the consequences of technology a methodical enterprise" (JOHN H. GIBBONS and HOLLY, L. GWIN, 1985).

The institutionalized mechanism which can realize this goal has three main characteristics. (1) First of all: what is needed is a research institution. What US Congress wanted was a methodological, scientific, objective analysis of the consequences of technology. At the institutional level this results in the establishment of a neutral research institution which can provide the necessary knowledge. (2) This first characteristic quite simply results in the second one: what is needed is one single institution. Once one institute has come up with the appropriate and objective results, a second institute can only reinvent the wheel. (3) This again results in the third characteristic: what is needed is a centralized institution. Only when the TA research is done under the direct control of the elected representatives of the people can its objectivity and neutrality be guaranteed.

In contrast with this, we are talking about (a) multiform (b) many and (c) decentralized institutional elements. In an interactive approach the result of the evaluation process is not a report, but an informed shared commitment. What is needed to realize and support this is much more than research but a variety of activities. It is impossible to indicate a general structure and function for the various actors who are involved as formal and informal "decision centers" in the decision-making process with respect to technology. In practice, the actors display an enormous variety of forms and degrees of involvement, as well as an enormous variety of organizational forms, or the lack thereof. At an institutional level, promoting the total quality and relevance of technology implies the need for establishing a great variety of offices or bodies which provide support for enhancing the interactive quality of its own involvement in the social process of technology planning.

Such "commitment and interaction-promoting units" will differ for each type of actor. Their functioning and structure can only be determined as the result of a detailed analysis of the historical and sociological position of the actor involved, of his views on the future, and of his ongoing or expected interactions with other actors. This requires specific case studies and the planning, performing and thorough analyzing of social experiments. It is clear though, that the dimensions of such units will be proportional to the intensity of the involvement in technology planning. In the case of large and important decision centers, such as governments, parliaments, universities, large employers' associations, such a "commitment and interaction unit" might develop into a small or large group of full-time employees. It could also result in a standing contract with a properly equipped and motivated consultancy or research organization. At the other extreme, that of a committed individual, the "institutional unit" to be established will be closer to making a structural arrangement for the allocation of time for reflection on a regular basis about one's own private and social life, one's achievements, interests and planned future. It might also include the establishment of some limited structural link with more operational institutions, including becoming an active or passive member of an association or club for reading or discussion. Whatever their organizational framework and intensity should prove to be, the basic functions of each "commitment and interaction promoting unit" are clear. They should help the actor (a) to maintain a continually updated understanding and inspired vision of its own past, possible futures, and cultural resources, (b) to remain informed about the views and resources of other practices, (c) to inform other relevant actors about its own backgrounds and standpoints and (d) to participate in an appropriate way in the social processes of shared decision-making and commitment-building.

Even though the number of assessment-assisting structures has been multiplied, all these institutions together cannot guarantee that new technological developments will be relevant and harmless. They do not provide society with an infallible insight into its present state or with a mechanism for foresight of the impact of its future activities, including the development of technological projects. These institutions are much more modest. They are attached to one specific actor or type of actors and try to clarify his situation by providing him with the multidisciplinary, scientific and organizational support which might help him to put his own resources to their best possible use and to strengthen his input and contribution into the broader social reality. They cannot take away the fallible and uncertain nature of all human activities, including technological projects. They can only contribute to make the human plans and choices related to technology as responsible as possible.

Any attempt to create a centralized checkpoint in the name of infallible objectivity, means a declaration of war to all serious and committed attempts to enhance the total quality of our private and social life and of the technology to make this possible.

9 TA after TA

The end of TA is not the end of the need to plan and evaluate technologies. The failure of TA does not necessarily mean that mankind should give up its attempts to monitor scientific, technological and industrial developments. What we learned is that such assessment cannot be achieved through a single, objective, cool methodology or performed in some centralized, state-controlled and objective institution. We know that the total assessment of technological developments is in the first place a broad, decentralized social process, and that this process can be endorsed but never replaced by a great variety of research projects. Those intellectual efforts include "forecasting techniques" which extrapolate current scientific and social trends in order to construct a hypothetical picture of society within 20 or 50 years. Such a picture can never be "proven" and is probably never true, but it helps to understand and evaluate current trends and to see them in context. Other pertinent research projects include the systematic mapping of known or expected relevant consequences and aspects of ongoing or planned technologies, such as their use of energy, their expected conflicts with existing social or legal institutions, and their impact on environmental elements which our society wants to protect. Many of these intellectual experiments existed before TA was born. The fact that it has proven to be impossible to use them as building blocks for some unified and infallible TA methodology, is no reason to cast doubt on their value as a resource in the process of assessing technological developments. Most pioneering TA institutions and thinkers have been wise enough to later make it quite explicit that they had abandoned the original TA inspiration. Some of them explain that they are working at implementing "TA in a broadened sense" which includes the development of high-level expertise for promoting social interaction and for performing some of the relevant types of research. Inadvertently, these attempts to give a second life to the concept of "Technology Assessment" have fueled the naive illusion that "the old and powerful TA", which in fact was never more than a dream, is still alive. Many decision-makers, inspired by the praiseworthy conviction that technology demands appropriate assessment, have been attracted and mislead by the false belief that the old "Technology Assessment" still exists and have established TA initiatives in the old and completely inviable style. Much hope has been raised by and much credit been given to numerous completely unrealistic initiatives, simply because they were believed to perform powerful, traditional TA studies. Much confusion would have been avoided if those high-quality institutions which developed TA in the sixties, only to abandon it in the seventies, would have circumnavigated the term TA entirely, while persuing their mission of continuing the development of interesting programs.

10 References

CARSON, N. (1989), *Process, Prescience, and Pragmatism: The Office of Technology Assessment* (Internal Paper, prepared for the 11th Annual Research Conference of the Association of Public Policy Analysis and Management, November 2–4, 1989, Arlington, Virginia), OTA, Washington, DC.

COATES, V. T. (1979), *Technology Assessment in Federal Agencies. 1971–1976*, George Washington University, Washington, DC.

DADDARIO, E. Q. (1968), *Technology Assessment. Statement of the Subcommittee on Science, Research and Development* (90th Congress, 1st Session), US Government Printing Office, Washington, DC.

GIBBONS, J. H., GWIN, H. L. (1985), Technology and Governance, in: *Technology and Society*, Vol. 7, pp. 333–352.

LAWRENCE, C. (1985), OTA: Technical Guidance for Congressional Decisions, in: *EPRI J.*, Jan./Feb., 34–38.

MESTHENE, E. G. (1965), *Ministers Talk about Science*, OECD, Paris.

NATIONAL ACADEMY OF SCIENCES (1968), *Technology: Process of Assessment and Choice*, Government Printing Office, Washington DC.

NICHOLS, K. G. (1978), *Technology on Trial*, OECD, Paris.

SALOMON, J.-J. (1971), Science Policy Studies and the Development of Science Policy, in: SPIEGEL-RÖSING, I. (1977): *Science, Technology and Society. A Cross-Disciplinary Perspective*, pp. 43–70, Sage, London.

SMITS, R., LEYTEN, J. (1988), Key Issues in the Institutionalization of Technology Assessment, *Futures*, February, 19–36.

WEINBERG, A. M. (1967), *Reflections on Big Science*, Cambridge, Massachusetts – London.

2 Concepts of Risk Assessment: The "Process versus Product" Controversy Put to Rest

HENRY I. MILLER

Stanford, CA 94305-6010, USA

1 Introduction

The last two decades of regulation of products and processes of the "new biotechnology" has seen much to praise and also much to criticize. On the positive side, oversight by the United States National Institutes of Health (NIH) has been relatively enlightened, with progressively decreasing stringency of their guidelines for laboratory research, and the *de facto* elimination of NIH approvals of field trials. The approach of the United States Food and Drug Administration (FDA) to the new products, overseeing them in the same way as similar products made in other ways, has been validated for a large number of products, with more than 800 (mostly diagnostics, plus approximately 20 therapeutics) approved for marketing and more than 1500 clinical trials of drugs and biologics under way.

On the negative side, good science and good sense have not always been well represented in the formulation of regulatory policy, and too often the bureaucracy builders have instituted burdensome, gratuitous regulation or retained it long after its time. Planned introductions (field trials or "deliberate releases") in particular have frequently been overregulated. Several European countries and Japan have adopted a process-based approach to oversight and have been unnecessarily restrictive, and they have paid the price in diminished research and development activity. In the United States, an intermittently inhospitable regulatory climate and a lack of public understanding have discouraged research and development in certain areas. In particular, academic and industrial researchers have been leery of field trials of recombinant organisms – even in areas such as bioremediation and microbial biocontrol agents that were once highly touted – and investor interest in sectors that require field trials has diminished.

The emergence and perseverance of equivocal or hostile regulatory climates for the new biotechnology can be traced in part to the propagation of various myths that have not been balanced by vigorous efforts at public education. This disparity is unfortunate, because the vast majority of applications of the new biotechnology are fundamentally similar, on both theoretical and empirical grounds, to well-accepted, ubiquitous applications of the "old biotechnology", and they are demonstrably safe. Treating "new biotechnology" products as a discrete category that is hazardous – or even potentially hazardous – is unjustified and leads to a number of undesirable consequences, from holding superfluous conferences and doing the wrong risk assessment experiments to implementing flawed regulatory policies and creating disincentives to using the newest techniques. These outcomes will, in turn, cause the public ultimately to suffer from a failure or a delay in the development of new products.

A pivotal issue related to the destructive mythology attending new biotechnology is whether the trigger for regulatory jurisdiction or for a certain degree of regulatory scrutiny should be the use of a certain *process* of genetic manipulation; or whether the *characteristics of the product* (either a living organism or inert product) or its intended use should constitute the trigger. This conundrum has been referred to as the "process versus product" debate, which has dragged on long after the issue should have been put to rest by international scientific consensus.

The goal of the United States government's coordinated approach to the regulation of biotechnology products has been to limit potential risks while encouraging the innovation, development, and availability of new products (*Federal Register*, 1986). However, not only must products be safe, but the public must have confidence in their safety. At the same time, public misapprehensions cannot be permitted to stimulate regressive policies. Thus, the advent of the new biotechnology continues to pose a challenge that involves perception as much as reality, and public education as much as the presence of an effective and coherent regulatory framework. Only if we can correct the misconceptions and eliminate the harmful myths that have grown up around the buzzword "biotechnology" can its potential be realized.

2 The Myths of Biotechnology

2.1 Myth 1: Biotechnology is a Discrete New Technology

One myth is that biotechnology is something discrete or homogeneous, a corollary of which is that there exists *a* "biotechnology industry" that can or should be rigidly controlled. This view is facile and inaccurate. "Biotechnology" is merely a catch-all term for a broad group of useful, enabling technologies with wide and diverse applications in industry and commerce. This spectrum of tools has become so integrated into the work of practitioners such as plant breeders and microbiologists that it is difficult, and also pointless, to attempt to distinguish between "old" and "new". This theme is enlarged upon in Sect. 2.2.

A useful working definition of biotechnology used by several U.S. government agencies is "the application of biological systems and organisms to technical and industrial processes". This definition encompasses processes as different as fish farming, forestry, the development of disease-resistant crop plants, the production of enzymes for laundry detergents, and the genetic engineering of bacteria to clean up oil spills, kill insect larvae, or produce insulin.

Biotechnology thus encompasses myriad dissimilar processes, producing even greater numbers of dissimilar products for vastly dissimilar applications. These processes and products are so diverse and have so little in common with one another that it is difficult to construct valid generalizations about them, for whatever purpose (see also Sect. 5.2.5). Because of biotechnology's lack of systematic, uniform characteristics, it cannot be effectively legislated or overseen in the uniform, all-inclusive manner that is possible, for example, for the underground coal mining industry.

Biotechnology's diversity dictates that the regulation of so many end uses must be accomplished by several government agencies.

As the uses vary, so do agencies' jurisdiction and the nature of their evaluation of the products. For example, a review by the U.S. Environmental Protection Agency (EPA) of an enzyme used as a drain cleaner will be different from a review by the FDA of the same enzyme injected into patients to dissolve blood clots (*Federal Register*, 1984). The diversity of products and their applications argues against the usefulness of legislation or regulations that attempt to encompass unnatural groupings that would occur under general terms such as "biotechnology", "genetically manipulated organisms", or "planned introductions" (MILLER, 1986).

2.2 Myth 2: Biotechnology and Genetic Engineering are New

A second myth is that biotechnology is new. On the contrary, many forms of biotechnology have been widely used for millennia. Earlier than 6000 B.C. the Sumerians and the Babylonians exploited the ability of yeast to make alcohol by brewing beer. A "picture" of the ancients preparing and fermenting grain and storing the brew has actually been retained for posterity in a hieroglyphic (DEMAIN and SOLOMON, 1981). An even older subset of biotechnology is genetic engineering, the indirect manipulation of an organism's genes by the guided mating of animals or crop plants so as to enhance desired characteristics. In this process of domestication, changes are made at the level of the whole organism; that is, selection is made for desired phenotypes, and the genetic changes, often poorly characterized, occur concomitantly. The new technologies, in contrast, enable genetic material to be modified at the cellular and molecular levels. These new techniques yield more precise and deliberate variants of genetic engineering, and they consequently produce better characterized and more predictable results; however, they retain the aims of classical domestication.

Building on the domestication of microorganisms for the modification of foods, industry has used classical genetic methods to yield many valuable organisms. One example is the genetic improvement of *Penicillium chrysogenum*, the mold that produces penicillin: by several methods including the screening of thousands of isolates and the use of mutagens to accelerate variation, penicillin yields have been increased more than a hundred-fold during the past several decades. There are many similar examples; microbial fermentation is employed throughout the world to produce a variety of other important substances, including industrial detergents, antibiotics, organic solvents, vitamins, amino acids, polysaccharides, steroids, and vaccines (WHO, 1985). The value of these and similar products of conventional biotechnology is in excess of $ 100 billion annually. Today, with the identification of the genes responsible for the biosynthesis of various antibiotics, the empirical search for improved products or yields is being supplanted by the introduction of specific, directed genetic changes.

In addition to such "contained" industrial applications, there have been many beneficial planned introductions of other organisms into the environment, including insects, bacteria, and viruses. Insect release was used successfully to control troublesome weeds in Hawaii in the early 20th century. Other examples are the successful program for the biological control of St. Johnswort ("Klamath weed") in California by insects in the 1940s and 1950s and the more recent use of an introduced rust pathogen to control rush skeletonweed in Australia. There have been hundreds of planned introductions of natural predators of insect pests since the first such release of the Australian Vedalia beetle in California in 1988. This area of research, the biological control of weeds, nematodes, insects, and diseases, is actively promoted by the U.S. Departments of Interior and Agriculture (KLINGMAN and COULSON, 1982).

Currently, more than a dozen microbial biocontrol agents are approved and registered with the EPA, and these organisms are marketed in hundreds of different products for use in agriculture, forestry, and home gardening (BETZ et al., 1983). In another major area – growth promoters for plants – preparations containing the bacterium *Rhizobium*, which enhances the growth of leguminous plants such as soybeans, alfalfa, and beans, have been sold in this country since the late nineteenth century. These products induce the development on plant roots of structures in which bacteria can convert nitrogen into compounds that can be used by plants and thereby decrease the need for chemical fertilizers.

Some of the most ubiquitous planned introductions of genetically modified organisms have involved the vaccination of human and animal populations with live, attenuated viruses. Live viruses modified by various techniques and licensed in the United States comprise vaccines against mumps, measles, rubella, poliomyelitis, and yellow fever. Inoculation of a live viral vaccine entails not only "infection" of the recipient (with minimal adverse impact on health) but sometimes the possibility of further transmission of the virus in the community – with the presumed risk of serial propagation. Nevertheless, no vaccine virus has become established in the environment, despite the continual presence of vaccine viruses there. For example, studies in the United States and the United Kingdom have shown that vaccine strains of poliovirus are present in sewage, but this reflects the continuing administration and excretion of live virus vaccine rather than its serial propagation through human or animal hosts in the community (KILBOURNE, 1985).

Viral vaccines produced with older genetic techniques for attenuating virulence have been remarkably effective throughout the world and have completely eradicated the dread smallpox virus. Despite rare adverse reactions, the beneficial impact of vaccines has been, to this point, superseded in importance as a promoter of human longevity and quality of life only by the agricultural "green revolution", which was mediated in part by genetically modified plants.

Building on the successes of the past, the newest biotechnological techniques, including recombinant DNA, are already providing still more precise, better understood, and more predictable methods for manipulating the genetic material of viruses and other microorganisms for use as vaccines.

The modification of crop plants has been performed ever since ancient agriculturists selected for cultivation plants possessing desirable traits from domesticated relatives of wild species. The rediscovery in 1900 of MENDEL's concepts of inheritance ushered in the scientific application of genetic principles to crop improvement. Since then, each scientific advance has increased the ability to improve predictably the genotype (and more important for the consumer of the product, the phenotype); and now, a combination of several techniques is often used to improve plants. For example, an existing plant might have been modified by many generations of classical breeding and selection, and more recently by somaclonal variation and wide crosses with embryo rescue (see below). These plants are being further improved by the newer molecular techniques (such as recombinant DNA) and can then be reintroduced into classical breeding programs from which its descendants will be released into commerce.

In this century, plant breeders have increasingly used interspecific hybridization to transfer genes from certain non-cultivated plant species to a variety of different (but closely related) species. These interspecific transfers of traits from wild species to domesticated relatives in the same genus stimulated attempts at even wider crosses, including those between members of different genera. These "wide crosses", transcending natural barriers to mating, have been facilitated by "embryo rescue" – culture techniques in which a sexual cross yielding a viable embryo but abnormal endosperm is "rescued" by culturing the embryo. This is accomplished by providing the hybrid embryo with the life support normally supplied in the early stages of development by maternal tissue and the endosperm. A number of plants resulting from wide crosses have been used in further breeding, extensively field tested, and marketed in the United States and elsewhere. These plants include commonly available varieties of tomatoes, potatoes, corn, oats, sugar beets, rice, and bread and durum wheat.

The molecular techniques for the genetic manipulation of plants enable scientists to move specific and useful segments of genetic material readily between unrelated organisms. These techniques offer several advantages and complement existing breeding efforts by increasing the diversity of genes and

germ plasm available for incorporation into crops. The numerous molecular techniques for genetic manipulation of plants can be divided into two main types – vectored and non-vectored. Vectored modifications rely on the use of biologically active agents, such as plasmids and viruses, to facilitate the entry of the foreign gene into the plant cell. Non-vectored modifications rely on the foreign genes being physically inserted into the plant cell by such methods as electroporation, microinjection, or particle guns. Although in some respects the use of these molecular techniques is more complex than traditional plant breeding practices, it is likely to be at least as safe. In both kinds of approaches, new DNA enters the plant's genome and is stably maintained and expressed.

2.3 Myth 3: The Unknowns Outweigh the Knowns

It is excessively negative – and even misleading – to argue as some have that "as far as the ecological properties of microbes are concerned, the unknowns far outweigh the knowns (SHARPLES, 1987). In fact, it is not what is *unknown, per se*, that is material, but what is unknown that relates to risk that is important. In the quotation above, one could just as easily substitute "the functions of mutations in live poliovirus vaccines" for "the ecological properties of microbes", but for four decades these uncharacterized genetic changes have not prevented our using empirically developed, live, attenuated poliovirus vaccines. The statement is particularly dubious for many microorganisms of commercial interest, including several *Pseudomonas* species (*syringae, aureofaciens, fluorescens*), *Thiobacillus* species, *Bacillus thuringiensis, B. subtilis, Rhizobium*, and baculoviruses, as well as the microbes used in fermentation to produce bread, cheese, wine, beer, and yoghurt. In fact, the majority of microbes are essential to ecosystem processes or otherwise beneficial to man, and only a minuscule fraction are pathogenic or otherwise harmful.

Until the recent furor over "engineered" (meaning rDNA-manipulated) organisms, small-scale field trials of native or non-pathogenic microbes modified by classical techniques were long exempt in the United States from both the pesticide and toxic substances regulations, and the safety record extending over more than half a century is admirable. The scientific method and prior experience, when applied logically to risk assessment, do enable us to make useful predictions.

2.4 Myth 4: Novel and Dangerous Organisms will be Created

The degree of novelty of microorganisms or macroorganisms created by the new genetic engineering techniques has been wildly exaggerated. A corn plant that has had newly incorporated into its genome the gene that synthesizes a *Bacillus thuringiensis* toxin is still, after all, a corn plant. *Escherichia coli* K-12 which has been programmed to synthesize human alpha interferon by means of recombinant DNA techniques differs very little from unmanipulated siblings that manufacture only bacterial molecules. Moreover, in nature, innumerable recombinations via several mechanisms between even very distantly related organisms have likely occurred (DAVIS, 1976). Bacteria can take up naked DNA, though inefficiently, and in nature they have long been exposed to DNA from disintegrating mammalian cells, as in the gut, in decomposing corpses, and in infected wounds. The human population alone excretes on the order of 10^{22} bacteria per day; hence, over the past 10^6 years innumerable mammalian–bacterial hybrids are likely to have appeared and been tested and discarded by natural selection. An analogous argument can be made for recombination among fungi, bacteria, viruses, and plants.

Both genetic and ecological constraints operate to prevent the emergence of exceedingly pathogenic viral variants, though even a single point mutation can alter virulence (FIELDS and SPRIGGS, 1982). And while the variations that continuously arise on an enormous scale in nature do occasionally produce a modified pathogen, such as an influenza virus with increased virulence or the AIDS virus, it is hardly likely that they would produce

in one fell swoop a serious pathogen from a *non*pathogen. Furthermore, the chances of such an event arising from the small-scale changes made by man do not compare with the tremendous background "noise" of recombination and selection in nature.

It is pertinent that certain kinds of gene transfers thought until recently to be impossible in nature because of the phylogenetic distance between donor and recipient have now been shown to occur in the laboratory and may occur in nature as well. For example, BRISSON-NOEL et al. (1988) have demonstrated that a gene or genes for erythromycin resistance was transferred between the Gram-negative bacterium *Campylobacter* and unrelated Gram-positive bacteria. In recent laboratory experiments, it was demonstrated that gene transfer can occur between *E. coli* and *Streptomyces* or yeast, and that the crown gall disease in plants results from a natural transfer of DNA from a bacterial species (*Agrobacterium*) to plant cells (HEINEMANN and SPRAGUE, 1989; MAZODIER et al., 1989). (Knowledge of the nature of this mechanism has already led to development of a highly effective means of developing transgenic plants without danger of disease induction, using DNA fragments from *Agrobacterium*.)

Since evolution continually creates novelty, the distinction between "natural" and "unnatural" (or "novel") is not a clear one and is, arguably, irrelevant. Novel is not synonymous with dangerous, nor need it even imply uncertainty.

2.5 Myth 5: Nonpathogens will be Transformed into Pathogens

A frequently voiced concern is that genetic manipulation may inadvertently transform a nonpathogen into a pathogen. However, this view ignores the complexity and the multifactorial nature of pathogenicity. Pathogenicity is not a *trait* produced by some single omnipotent gene; rather, it requires the evolution of a special set of properties that involve a number of genes whose functions control factors such as fitness, virulence, and adhesion.

A pathogen must possess three general characteristics, which are themselves multifactorial. First, it must be able to metabolize and multiply in or upon host tissues; that is, the oxygen tension and pH must be satisfactory, the temperature (and, for plant pathogens, the tissue water potential) suitable, and the nutritional milieu favorable. Second, assuming an acceptable range for all of the many conditions necessary for metabolism and multiplication, the pathogen must be able to resist or avoid the host's defense mechanisms for a period of time sufficient to reach the numbers required actually to produce disease. Third, a successful pathogen must be able to survive and be disseminated to new host organisms.

The organism must be meticulously adapted to its pathogenic lifestyle. Even a gene coding for a potent toxin will not convert a harmless bacterium into an effective source of epidemic, or even localized, disease, unless many other required traits are present. For human and animal pathogens, these include, at the least, resistance to enzymes, antibodies, and phagocytic cells in the host; the ability to adhere to specific surfaces; and the ability to thrive on available nutrients provided by the host. No one of these traits *confers* pathogenicity, though a mutation that affects any essential one can *eliminate* pathogenicity.

Another important consideration is that severe pathogenicity is more demanding of favorable conditions and therefore, much rarer in nature than are mild degrees of pathogenicity. Hence, the probability of inadvertently creating an organism capable of producing a medical or agricultural catastrophe must be vanishingly small. (This point is important because of the claims of some biotechnology critics that while the likelihood of a mishap may be small, when one occurs, it will be cataclysmic.) The record of the laboratory use of rDNA techniques supports this prediction. Despite the release of the order of 10^7 recombinant microorganisms per investigator per year from the standard BL1 biosafety level laboratory (according to a 1983 analysis by LINCOLN et al. funded by the U.S. Environmental Protection Agency under Grant No. R-808317-01) for more than 15 years, not a single adverse reaction has been observed in humans, animals, or the environment.

2.6 Myth 6: All Technology is Intrinsically Dangerous

Another myth is that the application of all new technology is dangerous. The basis of this view may be the atavistic fear of disturbing the natural order and of breaking primitive taboos, combined with unfamiliarity with the statistical aspects of risk. Promulgators of this myth cite the hazards of toxic chemical waste dumps and the technical problems of the nuclear industry (which are of dubious relevance to biotechnology in any case), but they conveniently ignore the overwhelming successes of telephonic communication; vaccination; blood transfusions; microchip circuitry; and the improved productivity of plants, animals, and microbes wrought by genetic improvement. It is worth remembering that critics predicted electrocution from the first telephones, the creation of human monsters by JENNER's early attempts at smallpox vaccination, the impossibility of matching blood for transfusions, and epidemics of uncontrollable "Andromeda strains" resulting from rDNA manipulation of microbes. They said, in effect, "the costs will be too high; there is no such thing as a free lunch".

No responsible person would deny that certain of the practices, processes, or products of biotechnology are potentially hazardous in some way. Some of these hazards are already well known: workers purifying antibotics have experienced allergic reactions; laboratory workers have been infected inadvertently by pathogenic bacteria; and vaccines have occasionally elicited serious adverse reactions.

In addition, introductions of exotic species, such as English sparrows, gypsy moths, and kudzu have had serious economic consequences. An analogy is sometimes made between the introduction of these non-native (alien or "exotic") organisms and the possible consequences of field trials with organisms modified by the new techniques. However, the comparison is largely specious, relying on the assumption that recombinant DNA manipulation can alter the properties of an organism in a wholly unpredictable way that will cause it to affect the environment adversely. Both theory and experience make this outcome very unlikely. Generally, introductions will be of indigenous organisms changed minimally – often by only the insertion or deletion of a single structural gene – from organisms that are already present in the environment, and that will not enjoy a selective advantage over their wild-type cohorts. They will be subject to the same physical and biological limitations of their environments as their unmodified parents. Moreover, in the past, exotic pest organisms – whether introduced intentionally or inadvertently – arrived in a new environment ecologically fit or adapted and with their undesirable properties present. In contrast, there is no evidence that any organism has been *made* a pest by breeders' genetic manipulation of plants, animals, or microorganisms. Thus, for organisms modified by the techniques of the new biotechnology, the older system of selective breeding, testing, and use of domesticated plants, animals and microbes, with its long history and admirable safety record, is a more applicable model than the introduction of exotic organisms into environments in which a favorable niche existed.

In any case, complex and comprehensive regulatory apparatuses based in numerous governmental agencies in the United States, European countries, Japan, and elsewhere, have long overseen the safety of food plants and animals, pharmaceuticals, pesticides, and other products that can be produced by biotechnology. There is every reason to expect that for the products of new biotechnology these regulatory mechanisms will continue to be equal to the task and to perform in a way that does not stifle innovation. There may never be a "free lunch", but often we can make it an excellent value.

3 The Basis for Risk Assessment for the New Biotechnology: Product, Not Process

Among those scientifically knowledgeable about the new methods of genetic manipula-

tion, there is wide consensus that existing risk assessment methods are applicable and suitable for environmental applications of the new products; this is based on both the large body of knowledge about the biological world and evolutionary biology and from extensive experimentation using the newest techniques. Several kinds of procedures are available that make possible the assessment of risk in different ways. And while it is true that risk assessment for new and old biotechnology is not an exact, predictive discipline (any more than it is for other technologies or for making personal decisions about what automobile to buy or what foods to eat), the U.S. NATIONAL SCIENCE FOUNDATION (1985) has concluded that available methods provide both a useful foundation and "a systematic means of organizing a variety of relevant knowledge". In assessing the potential risks of the new biotechnology, the need is not for new methods but rather for ways of ascertaining the correct underlying assumptions. For risk assessment, as for many other aspects of the new biotechnology, the new products manufactured with the new processes do not necessarily require new regulatory or scientific paradigms (see Sects. 3.1 and 3.2).

3.1 Broad Scientific Consensus

International organizations and professional groups have explored repeatedly the question of what are the correct assumptions and whether the uses of the techniques of the new biotechnology are sufficient to elicit a new regulatory paradigm. Their conclusions have been remarkably congruent.

A joint statement from the International Council of Scientific Unions' (ICSU) Scientific Committee on Problems of the Environment (SCOPE) and the Committee on Genetic Experimentation (COGENE) (Bellagio, Italy 1987), concluded:

> "[t]he properties of the introduced organisms and its target environment are the key features in the assessment of risk. Such factors as the demographic characterization of the introduced organisms; genetic stability, including the potential for horizontal trans-

fer or outcrossing with weedy species; and the fit of the species to the physical and biological environment ... These considerations apply equally to both modified or unmodified organisms; and, in the case of modified organisms, they apply independently of the techniques used to achieve modification. That is, it is the organism itself, and not how it was constructed, that is important".

The report of a NATO ADVANCED RESEARCH WORKSHOP (1987) came to similar conclusions:

"In principle, the outcomes associated with the introduction into the environment of organisms modified by rDNA techniques are likely to be the same in kind as those associated with introduction of organisms modified by other methods. Therefore, identification and assessment of the risk of possible adverse outcomes should be based on the nature of the organism and of the environment into which it is introduced, and not on the method (if any) of genetic modification."

Echoing these contemporaneous view points, the UNIDO/WHO/UNEP Working Group on Biotechnology Safety concluded (report of the third meeting of the Working Group, Paris 1987) that "[t]he level of risk assessment selected for particular organisms should be based on the nature of the organism and the environment into which it is introduced".

Various national groups have both reflected and extended these conclusions. The United States National Academy of Sciences (NAS) published in 1987 a policy statement on the planned introduction of organisms into the environment, and it has had wide-ranging impacts in the United States and internationally (NAS, 1987). Several of its most significant conclusions and recommendations are:

- Recombinant DNA techniques constitute a powerful and safe new means for the modification of organisms
- Genetically modified organisms will contribute substantially to improved health care, agricultural efficiency, and the amelioration of many pressing environmental problems that have resulted from the extensive reliance on chemicals in both agriculture and industry

- There is no evidence of the existence of unique hazards either in the use of rDNA techniques or in the movement of genes between unrelated organisms
- The risks associated with the introduction of rDNA-engineered organisms are the same in kind as those associated with the introduction of unmodified organisms and organisms modified by other methods
- The assessment of risks associated with introducing recombinant DNA organisms into the environment should be based on the nature of the organism and of the environment into which the organism is to be introduced, and independent of the method of engineering *per se.*

In an expansion of this policy, the United States National Research Council (NRC) concluded in a landmark report that "no conceptual distinction exists between genetic modification of plants and microorganisms by classical methods or by molecular techniques that modify DNA and transfer genes", whether in the laboratory, in the field, or in large-scale environmental introductions. In order to leave no room for equivocation on this point, the NRC's report went on to reinforce it with related observations:

- "The committees [of experts commissioned by the NRC] were guided by the conclusion (NAS, 1987) that the *product* of genetic modification and selection should be the primary focus for making decisions about the environmental introduction of a plant or microorganism and not the *process* by which the products were obtained." (p. 14, bottom)
- "Information about the process used to produce a genetically modified organism is important in understanding the characteristics of the product. However, the nature of the process is not a useful criterion for determining whether the product requires less or more oversight." (p. 14, bottom)
- "The same physical and biological laws govern the response of organisms mod-

ified by modern molecular and cellular methods and those produced by classical methods." (p. 15, middle).

- "Recombinant DNA methodology makes it possible to introduce pieces of DNA, consisting of either single or multiple genes, that can be defined in function and even in nucleotide sequence. With classical techniques of gene transfer, a variable number of genes can be transferred, the number depending on the mechanism of transfer; but predicting the precise number of the traits that have been transferred is difficult, and we cannot always predict the phenotypic expression that will result. With organisms modified by molecular methods, we are in a better, if not perfect, position to predict the phenotypic expression." (p. 13, bottom)

- "With classical methods of mutagenesis, chemical mutagens such as alkylating agents modify DNA in essentially random ways; it is not possible to direct a mutation to specific genes, much less to specific sites within a gene. Indeed, one common alkylating agent alters a number of different genes simultaneously. These mutations can go unnoticed unless they produce phenotypic changes that make them detectable in their environments. Many mutations go undetected until the organisms are grown under conditions that support expression of the mutation." (p. 14, top)

- "Crops modified by molecular and cellular methods should pose risks no different from those modified by classical genetic methods for similar traits. As the molecular methods are more specific, users of these methods will be more certain about the traits they introduce into the plants." (p. 3, bottom)

- "The types of modifications that have been seen or anticipated with molecular techniques are similar to those that have been produced with classical techniques. No new or inherently different hazards are associated with the molecular techniques. Therefore, any oversight of field tests should be based on the plant's phenotype and genotype and

not on how it was produced." (p. 70, middle)

- "Established confinement options are as applicable to field introductions of plants modified by molecular and cellular methods as they are for plants modified by classical genetic methods." (p. 69, middle)

The NRC (1989) proposed that the evaluation of experimental field testing be based on three considerations: familiarity, i.e., the sum total of knowledge about the traits of the organism and the test environment; the ability to confine or control the spread of the organism; and the likelihood of harmful effects if the organism should escape control or confinement.

The same principles were emphasized in an excellent and wide-ranging report by the U.S. NATIONAL BIOTECHNOLOGY POLICY BOARD (1992), which was established by the Congress and is comprised of representatives from the public and private sector. The report concluded that

> "[t]he risks associated with biotechnology are not unique, and tend to be associated with particular products and their applications, not with the production process or the technology *per se*. In fact biotechnology processes tend to reduce risks because they are more precise and predictable. The health and environmental risks of not pursuing biotechnology-based solutions to the nation's problems are likely to be greater than the risks of going forward."

These conclusions were bolstered by the findings and recommendations of the United Kingdom's House of Lords Select Committee on Regulation of the U.K. Biotechnology Industry and Global Competitiveness (WARD, 1993), which are discussed below.

The Paris-based Organization for Economic Cooperation and Development (OECD), approaching this question from a different direction – the safety of new varieties of foods – came to similar conclusions. OECD's 1993 report "Concepts and Principles Underpinning Safety Evaluation of Foods Derived by Modern Biotechnology" described several concepts related to food safety:

- "In principle, food has been presumed to be safe unless a significant hazard was identified."
- "Modern biotechnology broadens the scope of the genetic changes that can be made in food organisms and broadens the scope of possible sources of foods. This does not inherently lead to foods that are less safe than those developed by conventional techniques. Therefore, evaluation of foods and food components obtained from organisms developed by the application of the newer techniques does not necessitate a fundamental change in established principles, nor does it require a different standard of safety."
- "For foods and food components from organisms developed by the application of modern biotechnology, the most practical approach to the determination of safety is to consider whether they are *substantially equivalent* to analogous conventional food product(s), if such exist."

3.2 A Useful Syllogism

A paradigm for regulation of products of the new genetic engineering may be summarized in a syllogism. Industry, government, and the public already possess considerable experience with the planned introduction of traditional genetically modified organisms: plants for agriculture and microorganisms for live attenuated vaccines and for other uses such as sewage treatment and mining. Existing regulatory mechanisms have generally protected human health and the environment effectively without stifling industrial innovation. There is no evidence that unique hazards exist either in the use of recombinant DNA techniques or in the movement of genes between unrelated organisms. *Therefore*, for recombinant DNA-manipulated organisms, there is no need for additional regulatory mechanisms to be superimposed on existing regulation. (In fact, as a former regulator who was involved for more than a decade with the evaluation of the organisms and oth-

er products resulting from new genetic engineering techniques, I continue to be impressed with the extraordinary predictability and safety of the newer techniques compared to the older, "conventional" ones; arguably, any disparity of regulatory treatment would logically propose *lesser* scrutiny of organisms crafted with the most precise and predictable techniques, other factors being equal.)

The syllogism assumes, of course, that leaving aside the new genetic engineering, there exists adequate governmental control over the testing and use of living organisms and their products. This assumption is certainly open to question, particularly in those countries where known dangerous pathogens, chemicals, and similar products are largely unregulated or where regulations are widely ignored. Nevertheless, throughout our pre-recombinant-DNA history, in the United States and elsewhere, scientists have had a high degree of unencumbered freedom of experimentation, with pathogens as well as nonpathogens, indoors and out. The resulting harmful incidents have been few, and the benefits, both intellectual and commercial, have far exceeded any detrimental effects.

The syllogism can be illustrated graphically. In Fig. 1a, the large triangle represents the entire universe of field trials. The horizontal lines divide the universe into classes according to the safety category of the experimental organism (with examples on the right side of the figure). These categories can take into account the effect of genetic changes, whether they are a consequence of spontaneous mutation or of the use of conventional or new techniques of genetic manipulation. Such genetic changes can cause the organism to be shifted from one safety category to another. For example, if one were to grow mutagenized cultures of *Neisseria gonorrhea* or *Legionella pneumophila* in the presence of increasing concentrations of antibiotics to select for antibiotic-resistant mutants, the classification could change from, say, Class III to Class IV. Conversely, deletion of the entire botulin toxin gene from *Clostridium botulinum* could move the organism from Class III to Class II or even Class I. The oblique lines divide the universe according to the use of various techniques (with techniques becoming newer,

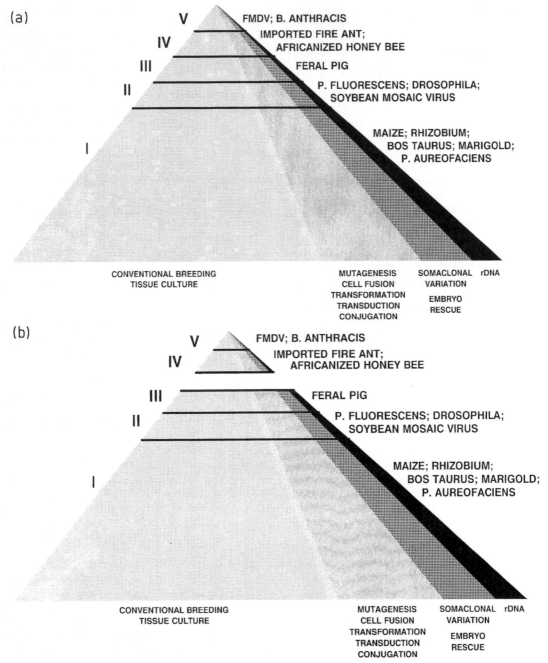

Fig. 1. Planned introductions into the environment. The large triangles represent the entire universe of field trials of plants, animals, and microbes. The horizontal lines divide the universe into classes according to the safety category of the experimental organism and take into account the nature of a genetic mutation or genetic change. Examples of organisms within the categories are listed on the right. The oblique lines divide the universe according to the use of various techniques (with techniques becoming more recent as one moves from left to right).

moving from left to right). Consideration of the figure reveals that the use of various techniques does not itself confer safety or risk (except insofar as a genetic change wrought with rDNA techniques is likely to be more precise, better characterized, and more predictable). Rather, risk is primarily a function of the phenotype of an organism (whether the wild type or with traits newly introduced or enhanced in some way) which, in turn, is determined by its genomic information: some organisms are destined primarily to exist symbiotically and innocuously on the roots of legumes (like the bacterium *Rhizobium*) and others to infect and kill mammals at low inoculum concentration (Lassa fever virus).

For an oversight scheme to be "risk-based", the use of genetic manipulation techniques generally or of certain techniques in particular should not, *per se*, dictate the degree of oversight required; rather, the trigger for oversight or for enhanced scrutiny should be a function of the characteristics of an organism – either as wild type or in any variants. Therefore, there is no coherent scientific rationale for defining the scope of a regulatory net as, say, the "rDNA" slice in Fig. 1, though that approach has often been proposed.

Fig. 1b differs from that in 1a by having Classes IV and V lifted off to illustrate that this, rather than the "rDNA slice", might be an appropriate "scope" for a regulatory net, particularly if one were designing a regulatory scheme *de novo*. This suggestion is merely illustrative; depending upon various considerations such as the amount of scrutiny judged to be appropriate and the regulatory burden on researchers and the government, what falls into the "net" could be stratified. For example, Class IV and V organisms might be circumscribed for "case by case every case" reviews by governmental agencies prior to field trials, Class III organisms for only a notification, with Class I and II organisms exempt. Such a conceptual approach moves us away from discredited technique-based oversight toward more fruitful consideration of factors relevant to cost-effective protection from risks.

Why have I belabored these principles – stated in almost the same words again and again by distinguished national and international scientific bodies and depicted in Fig. 1? There are two reasons. First, the principles are sound and their application can help to avoid false or shaky assumptions in formulating public policy; they can help to avoid substituting value judgements – especially those of new biotechnology's professional critics – for science, in crafting regulatory policy. Second, despite broad scientific consensus, the principles have not been universally applied. The flawed assumptions that have been adopted in their stead have given rise to various kinds of mischief, from holding the wrong conferences and performing superfluous, wasteful risk assessment experiments, to implementing regressive public policies (see below).

4 The Status of Regulation in the United States

The overwhelming scientific consensus on the process versus product issue has ineluctably influenced public policy in the United States. A watershed event for biotechnology regulatory policy was the release by the PRESIDENT'S COUNCIL ON COMPETITIVENESS (1991) of the "Report on National Biotechnology Policy". The report described the basic tenets of official policy, including seeking "to eliminate unneeded regulatory burdens for all phases of developing new biotechnology products – laboratory and field experiments, product development, and eventual sale and use". At least in theory, U.S. policy adheres to the principle that the extent of governmental oversight should be only that which is necessary and sufficient; or putting it another way, the object of regulation is not to regulate whenever possible, but only when there are unaddressed *and* unreasonable risks.

The report also described principles of regulatory review that underpin regulatory policy in the United States, including:

1. "Federal government regulatory oversight should focus on the characteristics and risks of the biotechnology product – not the process by which it is created." This principle echoes the major conclusions of U.S. and international scientific panels (see above) and should be reflected in future refinements of U.S. agencies' policies.
2. "For biotechnology products that require review, regulatory review should be designed to minimize regulatory burden while assuring protection of public health and welfare."
3. "Regulatory programs should be designed to accommodate the rapid advances in biotechnology. Performance-based standards are, therefore, generally preferred over design standards."

Another important milestone was the announcement in early 1992 of the U.S. policy, "Exercise of Federal Oversight Within Scope of Statutory Authority: Planned Introductions of Biotechnology Products into the Environment", the so-called Scope Announcement. It described

> "a risk-based, scientifically sound approach to the oversight of planned introductions of biotechnology products into the environment that focuses on the characteristics of the bio-technology product and the environment into which it is being introduced, not the process by which the product is created. Exercise of oversight in the scope of discretion afforded by statute should be based on the risk posed by the introduction and should not turn on the fact that an organism has been modified by a particular process or technique" (*Federal Register*, 1992b).

All federal agencies' regulations and guidelines were then expected to move toward conformance with these principles. However, the requirements of U.S. regulatory agencies are currently in various stages of readiness – and conformance. The last major coordinated statement of policy by the federal agencies was published in 1986 (*Federal Register*, 1986). The presence of a new administration makes future developments and directions

difficult to predict, but there is little reason to expect progressive policies.

In 1987, USDA's APHIS (Animal and Plant Health Inspection Service) promulgated Plant Pest Act regulations specifically for organisms that have been manipulated with recombinant DNA (rDNA) techniques and that contain any amount of DNA from a plant pest. Organisms were covered even in the face of compelling evidence of the extremely low probability of transforming a benign organism into a pest (MILLER, 1991; HUTTNER et al., 1992; HUTTNER, 1993), and field trials or even the *transport* (say, from one laboratory building to another) of organisms were subjected to a federal permitting process that included a comprehensive (and expensive) environmental assessment. In fact, *all of the more than 1400 permits for field trials or transport* issued under these regulations have been for organisms of negligible risk. In late 1992, with this realization (and also in order to implement the intent of the "scope" principles), the Bush administration sought to rationalize the APHIS rules. Accordingly, APHIS drafted a proposal to substitute notification to USDA to replace a permit under certain limited circumstances. The circumstances were so narrowly defined, however, that only a very few companies and almost no academic researchers would have benefitted. The effects of this proposal would have been merely the illusion of regulatory reform, the plight of researchers largely unchanged, and the vast regulatory bureaucracy undiminished.

White House officials despaired early on of removing completely the rDNA focus of the regulations and eventually settled for small but significant changes that made the proposal somewhat more risk-based and scientifically defensible, applicable to more research trials with more different species of plants, and that would have better encouraged innovative research. This proposal was published on November 6, 1992 (*Federal Register*, 1992a).

After the comments on the proposal were received and analyzed, the Clinton administration had its chance to correct the half-measures of the November 6 notice. But not only did they not correct the deficiencies in the proposal, they removed the improvements that had been wrought. On March 30, 1993, the final rule was published (*Federal Register*, 1993a).

Regulation by the EPA has been in a muddle for almost a decade. Manifestations of these problems include the decision to classify as a *pesticide* the ubiquitous and innocuous "ice-minus" *Pseudomonas syringae*, the onerous review to which it was subjected (KEYSTONE NATIONAL BIOTECHNOLOGY FORUM, 1989), and also the agency's proposed policies.

The EPA has put forth proposal after proposal in which the scope of regulation – that is, whether or not a product fell into the regulatory net – has turned on the use of the newest and most precise genetic manipulation techniques (HOYLE, 1993). This has occurred notwithstanding the wide scientific consensus and the official U.S. government "scope" policy to the contrary, and in the face of severe criticism of EPA's policies in separate reports by the U.S. NATIONAL BIOTECHNOLOGY POLICY BOARD (1992) and by the agency's own blue-ribbon panel (EPA, 1992). EPA has seemed determined to eschew *bona fide* risk considerations in favor of focusing on *techniques*. Eventually, forced to a compromise by the Bush administration's Office of Management and Budget, EPA did offer one explicitly risk-based alternative (of three) in its January 22, 1993, proposed FIFRA (Federal Insecticide, Fungicide, and Rodenticide Act) rule (*Federal Register*, 1993b). Predictably, EPA's preferred approach was not the explicitly risk-based alternative but one that focuses on organisms whose pesticidal properties have been imparted or enhanced *"by the introduction of genetic material that has been deliberately modified"* (emphasis added). In practice, this latter element is virtually synonymous with "manipulated with recombinant DNA techniques" because organisms created with chemical or other mutagenesis would not be captured (genetic material is not considered to be *introduced*) and other classical techniques such as conjugation or transformation are now seldom used. As pointed out by DUDLEY (1993) there are also important procedural differences between the different alternatives, and it should come as no surprise that EPA's preference is for a centralized ap-

proach; that is, for essentially all rDNA-manipulated biocontrol agents, *the EPA* would perform a review and decide on the applicability of the burdensome experimental use permit (EUP) regulations.

In contrast, the risk-based alternative is at the same time more scientific and more decentralized: the need for involvement of the EPA before a field trial depends on a variety of factors that are actually related to the risk of the product in the environment in which it will be tested. *The researcher*, after considering EPA's risk guidance, is assumed to be in an appropriate position to judge whether the experiment meets the criteria that cause it to require a permit from the EPA (DUDLEY, 1993).

The FIFRA rule represents another opportunity for scientific principles to determine public policy, but it seems a foregone conclusion that in the final rule, the EPA will be permitted to adopt their preferred, centralized, technique-based approach.

The research side of the Department of Agriculture appears to have abandoned its earlier intention to implement guidelines for field research (*Federal Register*, 1989). This leaves intact a "vertical" mechanism for oversight of research, with field trials subject to the jurisdiction of various agencies, according to the characteristics of the organism or its intended use. For example, veterinary vaccines and plant pests would be subject to USDA/APHIS, human vaccines to FDA, microbial pesticides to EPA, and so forth.

The FDA has imposed no new procedures or requirements for products made with the new biotechnology (MILLER and YOUNG, 1988). For more than a decade, FDA has regarded the techniques of new biotechnology as an extension, or refinement, of older genetic manipulation techniques; and the products of new biotechnology, therefore, have been subject to the same regulatory requirements and procedures as other products. In May 1992, FDA published a statement of policy on the oversight of new varieties of food plants. FDA's regulatory approach identifies scientific and regulatory issues for which the developers of the new variety may need to consult with the agency. These issues are related to characteristics of foods that raise safety questions and that elicit a higher level of FDA review. Such characteristics include the presence in the new variety of a substance that is completely new to the food supply, the presence of an allergen in an unusual or unexpected milieu, changes in levels of a major nutrient, or increased levels of toxins that are normally found in foods. New food varieties without such potentially worrisome characteristics are subject to an appropriately lower level of governmental scrutiny. The use of one or another technique of genetic manipulation does not in itself determine the need for or the level of governmental review.

5 Implications for Public Policy

5.1 Important Non-Scientific Issues

Of concern to many practitioners, regulators, and observers of biotechnology are four non-scientific issues that could play a pivotal role in the future of the new biotechnology's development and use in the United States and elsewhere. First, the regulatory climate could, if irrational or inappropriately risk-averse, impede industry's eventual introduction of products now being developed. This obstacle could lead to future withdrawal or diminution of new product development, with continued reliance on less sophisticated, and often more hazardous, alternative technologies.

Second, concerns within the financial community about the long-term stability and success of companies doing business in environmentally (over)regulated fields could depress the values of those companies, drying up capital for the continued development and testing necessary to satisfy regulatory requirements (KINGSBURY, 1986).

Third, the genetic approach to crop improvement has resulted in tens of thousands of individual improvements in plants in order to overcome locally important constraints on crop production. These are typically "deliv-

ered" into practice in the form of locally adapted crop varieties, an approach that will be not available to new biotechnology if the newer molecular methods of manipulation are subjected to an assortment of "case by case every case" regulatory disincentives.

Fourth, the continued antagonism of a handful of well-organized and vocal opponents of biotechnology (a term that seems an oxymoron, like "opponents of antibiotics" or "opponents of tangelos") has had some effect. Their efforts have confused and misled the public, and diverted government agencies from their more constructive bureaucratic pursuits by forcing them to contend with nuisance lawsuits, petitions, and the like.

Those who are responsible for regulatory policy toward biotechnology have a critical responsibility to act quickly and definitively to weigh the arguments and evidence – and to make those policy decisions that remain. With the accumulated wisdom of the analyses of the NAS, NRC, House of Lords Select Committee, and others (see Sect. 3.1), the broad scientific questions have been resolved and the correct assumptions identified. The notion that the products derived from the new biotechnology defy useful, accurate risk assessment or are too dangerous to introduce into the environment is without merit: both theory and experience repudiate this assertion. Policy must be guided in part by an understanding that there are genuine costs of overly risk-averse regulatory policies that prevent the testing and approval of new products. Real-world examples of these costs include the destruction of crops by frost while field trials of the obviously innocuous "ice-minus" bacteria were debated endlessly; the continued application of chemical pesticides while new biological control agents languished untested for years (and some were abandoned completely in the face of regulatory uncertainty or hostility); and the sluggish pace of development of rDNA-modified oil-degrading organisms, largely because of its overregulation. BENGT JÖNSSON is correct in stating that safety can only be bought at a cost (JÖNSSON, 1986), but one must wonder at the appropriateness of any "cost" – that is, whether we buy *any* true increment in "safety" – when regulators' resources are expended on such things as the environmental effects of the ice-minus *Pseudomonas syringae*, extended shelf-life tomatoes and petunias of a new color.

5.2 Real-World Effects of Irrational or Excessive Regulation

Concerns about overregulation are more than theoretical. Internationally, regulation has been a hodge-podge, and often excessive. Although Japan's regulatory approach has been illogical and process-based, its restrictions on pharmaceutical products made with the newest techniques have seemed merely annoying to industry rather than debilitating, and the new biotechnology applied to pharmaceuticals there has sustained significant progress. But some areas have undoubtedly been impeded by the government's conviction that the use of rDNA techniques, *per se*, raises new safety issues. For example, despite a medical and scientific infrastructure that could support clinical trials of human gene therapy, no Japanese group is close to moving into the clinic, and not a single company has been created there with gene therapy as its goal. By contrast, gene therapy trials are already under way in the United States, Italy, France, the Netherlands, and China, with more than 200 patients having been treated and the numbers rising exponentially.

5.2.1 Japan

Japan's stigmatization of the new biotechnology is similarly reflected in agricultural biotechnology. Only three field trials in Japan of recombinant DNA-manipulated plants (but none of microorganisms) have been carried out, and Japanese research and development in this area is far behind what one would expect. The Japanese government has provided little encouragement in the form of clear, predictable, risk-based regulation to those contemplating field trials. Moreover, the Japanese MINISTRY OF HEALTH AND WELFARE (1992) has imposed a strict regulatory regime specific to foods and food additives manufactured with rDNA techniques.

5.2.2 Europe

Denmark, the Federal Republic of Germany, and the European Union (EU), formerly the European Community (EC), are widely perceived as having created potent regulatory disincentives to the use of new biotechnology (*European BioNews*, 1992). Denmark's Novo/Nordisk Industri, the world's largest producer of enzymes, has threatened to move its research and development to Japan, and Germany has experienced a veritable hemorrhage of research and development (RauCon, 1989). Bayer AG has continued to establish research and development facilities abroad, especially in the United States and recently also in Japan (Bayer announced in February 1993 its intention to establish a Center for Allergy Research in Kansai Science City that would create 400–450 jobs). BASF has shelved investment plans for a $ 70 million research and development facility in Ludwigshafen and has erected it near Boston instead. Hoechst has discontinued investments in Germany and expanded their facilities in Australia, Japan and in the United States. A mid-sized company, Wiemer Pharmazeutisch-Chemische Fabrik, has abandoned its biotechnology production and liquidated its equipment; Abbott Diagnostics in Wiesbaden has postponed indefinitely its biotechnology plans (RauCon, 1989).

The EU has begun the implementation of extremely regressive directives (the equivalent of regulations in the United States) for both contained uses and planned introductions of genetically modified organisms into the environment (*Nature*, 1990). Strongly process-based, the directives' underlying premises are scientifically flawed; they are expected to hinder research and development activities throughout the scientifically and economically advanced EU countries; and the directives' provisions could potentially be used to erect non-tariff trade barriers to foreign products (WARD 1993; YOUNG and MILLER, 1989). The United States Congress' Office of Technology Assessment recently published this analysis:

"In enacting directives that specifically regulate genetically modified organisms, the EC has established a regulatory procedure that is significantly different from that of the United States. In the EC, regulation is explicitly based on the method by which the organism has been produced, rather than on the intended use of the product. This implies that the products of biotechnology are inherently risky, a view that has been rejected by regulatory authorities in the United States. In addition, manufacturers are concerned that their new biotechnology-derived products may face additional barriers before they can be marketed, for the product may also be subject to further regulations based on its intended use" (U.S. CONGRESS, 1991).

For evidence of the wide-ranging, real-world impact of the EU's regulatory philosophy and directives, it is instructive to consider three news articles on a *single page* (page 684) of the 21 October 1993 issue of *Nature*. One article, "German changes face opposition", describes the futile struggle of Germany, with one of the most restrictive rDNA regulatory regimes in Europe, to liberalize its law even in small ways. (However, many observers believe that even if this attempt were successful, it would be too little, too late for German biotechnology.) A second article, "Biotechnology rules: Spain falls into line," describes Spain's attempts to discharge its responsibilities to the EU directives by crafting a new law focused on rDNA. These attempts seem to have alienated virtually every part of the government and society that is involved. Finally, "UK report criticizes EC directives" describes the recommendation of the House of Lords Select Committee on Science and Technology for essential and fundamental changes in the "excessive regulation" that emanates from the EU directives (see below).

Responding to widespread criticism of the biotechnology directives from both outside and within Europe, the EU established in March 1991 a new high-level "interservice group", the Biotechnology Coordination Committee (BCC), to oversee EU biotechnology policy. According to an EU press release (number 1P(91) 277, March 1991), the main tasks of the committee will be three-fold: to examine new initiatives made by the Commission's services and to prepare the Commission's final decisions; to convene, as necessary, discussions among the Commission, industry, and other interested parties; and to *"evaluate the existing Community policy on biotechnology"* (emphasis added). The actions of the BCC may mitigate to some degree the full potential negative impact of the directives, but there has been little evidence of its influence.

Another spur to needed changes in the EU directives may be the October 1993 report of the UK House of Lords Select Committee on Science and Technology, "Regulation of the United Kingdom Biotechnology Industry and Global Competitiveness". Echoing other assessments of the EU's approach to biotechnology, the report excoriated the EU's regulatory approach to both contained uses and field trials of organisms and recommended reduced and rationalized regulation (WARD, 1993):

> "As a matter of principle, GMO-derived products should be regulated according to the same criteria as any other product ... U.K. regulation of the new biotechnology of genetic modification is excessively precautionary, obsolescent, and unscientific. The resulting bureaucracy, cost, and delay impose an unnecessary burden to academic researchers and industry alike."

5.2.3 The Developing World: Especially at Risk

The prospect of gratuitous or excessive regulation of the new biotechnology is especially ominous for the developing world. As destructive as overregulation and specious approaches to oversight have been to technologically advanced countries (see above), lesser developed countries, which often lack adequate resources for the regulation even of products that are known to be hazardous, are less able to absorb such excesses. If they are compelled (or feel compelled, which is much the same thing) to establish a regulatory apparatus specifically for "GMOs" (genetically modified organisms, i.e., defined as those manipulated with rDNA techniques), such countries are likely to have greatly reduced access to such products. A reduced array of technologies or products for local producers and consumers spells more limited environmental and social options. Proposals such as the UNIDO "code of conduct" (*Biotech Forum Europe*, 1992) for field trials of GMOs and the "safety provisions" in Article 19 of the Convention on Biological diversity – if they are focused on GMOs – are a lose-lose-lose proposition, particularly for those whose resources are meagre: they waste resources directly on gratuitous bureaucracy; provide disincentives to developing countries' performing R & D themselves; prevent developing countries from becoming real partners in the commercial enterprise; and send inaccurate messages to their public and their politicians about risk and, most especially, about relative or alternative risks. In view of the scientific consensus that "GMOs" are not conceptually distinct from other organisms, whether wildtype or manipulated with other techniques (NRC, 1989), it is difficult to defend the establishment of elaborate regulatory mechanisms to evaluate "case by case every case" the field trials of GMOs in countries that lack

adequate oversight of vaccines, pesticides, toxic chemicals, plant pests and other pathogens. For example, one difficult-to-reconcile outcome of a process-based regulatory approach could well be a regulatory net that requires governmental risk assessments of every rDNA-manipulated plant but omits scrutiny of field trials with "natural" organisms such as *Bacillus anthracis*, foot and mouth disease virus (FMDV), or an invasive plant such as bamboo or kudzu, even if these have been genetically manipulated with older, cruder techniques.

5.2.4 The United States

Investor confidence in two of the sectors of U.S. industry that use the new biotechnology, namely biopharmaceuticals and agriculture, provides a measure of the effect of excessive regulation (see Fig. 2). Over the period 1983–1992 investor interest in the biopharmaceutical sector was strong, outperforming the market by about 200%. By contrast, the agricul-

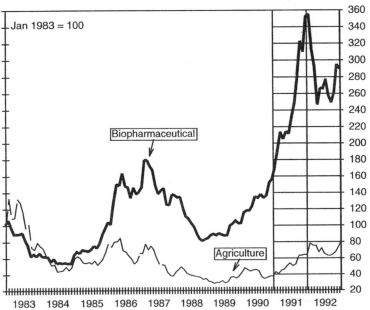

**Biotechnology
Biopharmaceutical & Agriculture Segments
Relative Performance vs. S&P 500 Index**

Fig. 2. Relative performance of two sectors of U.S. industry using the new biotechnology. The share-weighted stock performance of companies normalized to Standard and Poor's 500 Index is grouped by *biopharmaceuticals* (98 companies), and *agriculture* (11 companies). The indices were set at 100 for January 1983. (Data from L. I. MILLER, PaineWebber, Inc.).

tural biotechnology sector performed dismally and, despite stunning breakthroughs in the relevant basic science, actually underperformed the marked by about 20%.

One reason for this disparity is certainly the marked difference in the regulatory milieus. The FDA adopted a scientifically reasoned policy that regulates new biotechnology-derived products no differently from other products, approved approximately 20 drugs and biologics in significantly less than the usual approval times, and permitted more than 1500 clinical trials of such products. By contrast, field testing of new biocontrol and other agricultural products have faced additional regulatory requirements because rDNA techniques were employed, as well as much-publicized regulatory and judicial delays. Not surprisingly, private sector investors have shown reluctance to gamble funds on research and development in an area where the climate is inhospitable – where it could take years as well as millions of dollars simply to perform a single field test, let alone meet the requirements for marketing approval down the road, and where virtually every experiment requires government intervention and unnecessary delays.

Another U.S. example of the problems of overregulation is to be found in the state of Wisconsin, where by 1988 the prospect of months of regulatory paperwork to obtain approval for a single field trial was causing investigators, especially young untenured academic researchers, to shy away from research areas that require field testing (*The Capital Times*, 1988). A 1989 survey of agricultural research in the public and private sectors uncovered a similar problem. Approximately 10 percent of the researchers in the public sector (nonprofit institutions, including universities) and 20 percent in the private sector had actually developed organisms with the new biotechnology but had elected not to proceed with field trials; three-quarters of these researchers cited reasons related to governmental regulation as the reason for not proceeding (RATNER, 1990).

Thus, whole areas of research that could ultimately provide useful products, ranging from growth promotants to microbial biocontrol agents and biological treatments for toxic wastes, are being systematically avoided. In addition, some companies have felt compelled to eschew the newest and most precise techniques of molecular genetic engineering in favor of scientifically inferior but bureaucratically favored – in other words, less stringently regulated – techniques for constructing organisms for agricultural applications (U.S. NATIONAL BIOTECHNOLOGY POLICY BOARD, 1992; *Wall Street Journal*, 1989). These are the fruits of *dis*incentives to the application of the new biotechnology.

5.2.5 Serious Flaws in a "Horizontal" Approach to Biotechnology Risk

Irrational regulatory policies are not the only unfortunate legacy of what has been termed a "horizontal" approach to biotechnology risk – that is, "a fallacy based on scanning across experiments whose only common element is the use of the same genetic modification technique" (MILLER and GUNARY, 1993). This misconception has, in recent years, dictated the theme of major conferences and a survey of field trials of rDNA-modified organisms commissioned by the Organization for Economic Cooperation and Development (OECD). More such dubious exercises are planned. However, given the kinds of organisms that have been modified and the traits introduced, one might as well survey all experiments that were performed using plastic, as opposed to glass, pipettes; or ones that were begun on certain days of the week.

In addition to stimulating irrational public policies and superfluous conferences and surveys, there is evidence that the horizontal approach to biotechnology risk also encourages the wrong kinds of risk assessment experiments. Consider, for example, the elaborate risk assessment experiment performed by CRAWLEY et al. (1993), "to find out how ecological performance is affected by genetic engineering".

Conducted in three climatically distinct sites and four habitats, the experiment compared the invasiveness of an "unmodified" variety of oilseed rape with two variants of the plant manipulated with rDNA techniques to confer herbicide or antibiotic resistance. Some of the significant limitations of this well executed but poorly designed experiment were described in a commentary in the same issue of *Nature* (KAREIVA, 1993). Indeed, there is little reason to anticipate that, in the absence of the herbicide or antibiotic to which the modified plants were made resistant, selection pressure would favor the rDNA-modified forms. And one observed difference, reductions in seed survival in the modified plants, cannot be attributed unequivocally to the genetic modification, because of the presence of maternal effects and other variables.

The absence of unexpected persistence, invasiveness, or gene transfer in field trials with rDNA-modified plants that have been protected against interpollination with related crop or wild plants does not indicate that an incorporated trait may not under specific circumstances be hazardous. A rigorous demonstration that an rDNA-modified plant presents a different degree of hazard from a new non-rDNA-modified variety of that species (or plants of that species in regular use) must entail experiments that provide reasonable opportunities for the manifestation of identified hazards.

The performance of such a well executed but poorly designed risk assessment experiment, "one of the most comprehensive population studies ever undertaken in plant ecology" (KAREIVA, 1993), makes one wonder whether the real purpose of the study was more ecological public relations than scientific enquiry. In any case, it seems unlikely that an experiment of this design and magnitude could have been performed in the absence of confusion over whether to focus on the traits newly introduced into the plant or the technique used for genetic manipulation.

Additional, second-order effects of inaccurate and overly risk-averse messages extend more broadly to science and technology. DONALD KENNEDY (1987) has warned that vital research is already curtailed in research institutions in parts of the United States. He pointed out that the anti-research movement, whether from concern for laboratory animals

or from fear of scientific experimentation, had delayed the opening of Stanford's completed $ 13 million biomedical research facility and had declared war on others planned at the University of California and elsewhere. KENNEDY asked rhetorically where this opposition expects society, without such facilities, to find cures for AIDS and other scourges.

5.2.6 A *Caveat* for Policy Makers

FREEMAN DYSON (1975) has observed astutely that resisting a new technology is generally safer for regulators than embracing it, but that this course often has larger costs for society. Ultimately there is no real alternative but to try to channel and control new technologies, with only that amount of governmental intervention that is necessary and sufficient. One cannot attempt to avoid problems with new technology by resisting it, by trying to stuff the genie back into the bottle.

For these reasons, the professional practitioners and regulators of the new biotechnology must strive to demystify it and to provide an accurate perspective for the public, because it is the public who will benefit most. The stakes are high in terms of both economic and social benefit. In recent years the FDA has approved several products of new biotechnology that are medical milestones. These include alpha-interferons for the treatment of a lethal leukemia, beta-interferon for multiple sclerosis, a monoclonal antibody preparation for preventing the rejection of kidney transplants, a new-generation vaccine for the prevention of hepatitis, a hormone that reduces the blood transfusion requirements for kidney dialysis patients, and a growth factor that is an important adjunct to cancer chemotherapy. Among myriad other applications, biotechnology promises improved diagnostic devices for detection of harmful chemicals and pathogens; vaccines against scourges such as malaria, schistosomiasis, and AIDS; as well as new therapies that could ameliorate or cure for the first time such genetic diseases as cystic fibrosis and certain inherited immune deficiencies. Used for new generations of medicines and food plants and animals, biotechnology could provide partial solutions to the trinity of despair – hunger, disease, and the growing mismatch between material resources and population.

Finally, far from constituting an imminent threat to the environment, application of the new biotechnology may well provide solutions to existing environmental problems. For example, more efficient crop plants require a smaller commitment of land to agriculture, freeing it for recreation or for a return to wilderness; plants can be engineered that diminish the overall need for chemical pesticides and fertilizers; herbicide-resistant plants can offer farmers more choices, favoring the use of herbicides that are less suspect than those used now with some crops; microorganisms can be engineered to degrade toxic compounds; and biological control agents are likely to reduce costs and to be more predictable and safer than many chemicals.

6 Conclusions

Biotechnology has been widely applied for millennia, including many successful and beneficial uses that require introductions into the environment. In recent decades, the precision and power of the genetic manipulation of both macroorganisms and microorganisms have greatly increased with the development of molecular genetics. The techniques of the new biotechnology are best viewed as extensions, or refinements, of older techniques for genetic manipulation and domestication. For these reasons, there exists a broad scientific consensus that current risk assessment methods are suitable and applicable for the foreseeable applications of the new biotechnology, and that new products manufactured with the new processes do not necessarily require new regulatory paradigms. There is no scientific rationale for fear or avoidance of the newest techniques of genetic engineering or, indeed, of nonpathogenic, nontoxigenic organisms that have been modified by any technique. Likewise, there is no rationale for public policies that create *dis*incentives to research and development using the newest,

most precise techniques; or that send inaccurate messages to the public via discriminatory regulation, product labeling or other means.

Moreover, there are genuine costs of overly risk-averse regulatory policies that impede or prevent the testing and approval of new products. Such policies are anti-innovative, and they delay the benefits of new products to the public. These policies both emanate from and feed into a pernicious anti-science movement that threatens basic as well as applied scientific research. Much of the anti-science and anti-technology movement has less to do with concerns about safety than about attempting to arrogate control over research and development strategies and over consumers' choices.

The long-term antidote to both the public misapprehensions about biotechnology and to the (modest) successes of the antagonists of science and technology has two parts: first, improved education of consumers and, especially, public opinion leaders, about science, technology, and the basic aspects of risk analysis; and second, the establishment of rational public policies that are consistent with sound science.

MAX PLANCK said more than half a century ago that an important scientific innovation rarely makes its way by gradually winning over and converting its opponents; rather, its opponents gradually die out and the succeeding generation is familiarized with the idea from the beginning. It would be tragic if this were the fate of the new biotechnology, for it would result in prolonged delays in the widespread testing and use of valuable new products.

Acknowledgements

This chapter is dedicated to the memory of Professor BERNARD D. DAVIS. The author would like to thank Professors JAMES COOK and ARTHUR KELMAN for their comments on this manuscript.

7 References

BETZ, F., LEVIN, M., ROGUL, M. (1983), Safety aspects of genetically engineered microbial pesticides, *Recomb. DNA Tech. Bull.* **6**, 135–140.

Biotech Forum Europe **9**, 218–221 (1992), Release of organisms into the environment: Voluntary code of conduct.

BRISSON-NOEL, A., ARTHUR, M., COURVALIN, P. (1988), Evidence of natural gene transfer from Gram-positive cocci to *Escherichia coli*, *J. Bacteriol.* **170**, 1739.

The Capital Times (1988), UW researchers stymied by genetic test limits, March 16, p. 31, Madison, WI.

CRAWLEY, M. J., et al. (1993), Ecology of transgenic oilseed rape in natural habitats, *Nature* **363**, 620–623.

DAVIS, B. D. (1976), Evaluation, epidemiology and recombinant DNA, *Science* **193**, 442.

DEMAIN, A. L., SOLOMON, N. A. (1981), Industrial microbiology, *Sci. Am.* **245**, 66–75.

DUDLEY, S. E. (1993), *Bio/Technology* **11**, 524.

DYSON, F. J. (1975), On the hidden costs of saying 'No', *Bull. At. Sci.* **31**, 23.

EPA (Environmental Protection Agency) (1992), Safeguarding the future: Credible science, credible decisions, The Report of the Expert Panel on the Role of Science at EPA, *EPA Document* 600/9-91/050, March.

European BioNews (1992), **1**, October, p. 1.

Federal Register (1984), **49**, 50856.

Federal Register (1986), **51**, 23302–23393.

Federal Register (1989), **54**, 7027.

Federal Register (1992a), **57**, 53036–53043.

Federal Register (1992b), **57**, 6753–6762.

Federal Register (1993a), **58**, 17044–17059.

Federal Register (1993b), **58**, 5878–5902.

FIELDS, B. N., SPRIGGS, R. (1982), Attenuated reovirus type 3 strains generated by selection of haemagglutinin antigenic variants, *Nature* **297**, 68.

HEINEMANN, J. A., SPRAGUE, JR., G. G. (1989), Bacterial conjugative plasmids mobilize DNA transfer between bacteria and yeast, *Nature* **340**, 205.

HOYLE, R. (1993), Off to a fast start – and a promising, if uncertain, future, *Bio/Technology* **11**, 460–463.

HUTTNER, S. L. (1993), Risk and reason: An assessment of APHIS, in: *U.S. Agricultural Research: Strategic Challenges and Options* (WEAVER, R. D., Ed.), Bethesda, MD: Agricultural Research Institute.

HUTTNER, S. L., et al. (1992), Revising oversight of genetically modified plants, *Bio/Technology* **10**, 967–971.

JÖNSSON, B. (1986), The economic viewpoint, in: *IFPMA Symposium, Scientific Innovation in Drug Development: The Impact on Registration,* Geneva: IFPMA.

KAREIVA, P. (1993), Transgenic plants on trial, *Nature* **363**, 580–581.

KENNEDY, D. (1987), The antiscientific method, *The Wall Street Journal*, October 29, p. 11, New York.

KEYSTONE NATIONAL BIOTECHNOLOGY FORUM (1989), *An Analysis of the Federal Framework for Regulating Planned Introductions of Engineered Organisms,* Interim Summary Report, The Keystone Center, Keystone, CO, February.

KILBOURNE, E. D. (1985), Epidemiology of viruses genetically altered by man – predictive principles, in: *Genetically Altered Viruses and the Environment* (FIELDS, B., MARTIN, M. A., KAMELY, D., Eds.), *Banbury Report* **22**, Cold Spring Harbor, NY: Cold Spring Harbor Laboratories.

KINGSBURY, D. T. (1986), *Bio/Technology* **4**, 1071–1074.

KLINGMAN, D. L., COULSON, J. R. (1982), Guidelines for introducing foreign organisms into the USA for biological control of weeds, *Plant Dis.* **66**(12), 1205–1209.

MAZODIER, P., PETTER, R., THOMPSON, C. (1989), Intergeneric conjugation between *Escherichia coli* and *Streptomyces* species, *J. Bacteriol.* **171**, 3583.

MILLER, H. I. (1986), FDA and the growth of biotechnology, *Pharm. Eng.* **6**, 28.

MILLER, H. I. (1991), Regulation, in: *The Genetic Revolution* (DAVIS, B. D., Ed.), Baltimore: Johns Hopkins University Press.

MILLER, H. I., GUNARY, D. (1993), Serious flaws in the horizontal approach to biotechnology risk, *Science* **262**, 1500–1501.

MILLER, H. I., YOUNG, F. E. (1988), The FDA and biotechnology: Update 1989, *Bio/Technology* **6**, 1385–1392.

MINISTRY OF HEALTH AND WELFARE (1992), *Guidelines for Foods and Food Additives Produced by Recombinant DNA Techniques*, Ministry of Health and Welfare, Government of Japan.

NAS (U.S. National Academy of Sciences) (1987), *Introduction of Recombinant DNA-Engineered Organisms into the Environment: Key Issues*, Washington, DC: National Academy Press.

NATO ADVANCED RESEARCH WORKSHOP (1988), Summary, in: *Safety Assurance for Environmental Introductions of Genetically-Engineered Organisms* (FIKSEL, J., COVELLO, V. T., Eds.), Rome 1987, *NATO ASI Series*, Berlin – Heidelberg – New York: Springer-Verlag.

Nature (1990), New European release rules ratified (news item), *Nature* **233**, 371.

NRC (U.S. National Research Council) (1989), *Field Testing Genetically Modified Organisms: Framework for Decisions*, Washington, DC: National Academy Press.

OECD (Organization for Economic Cooperation and Development) (1993), *Safety Evaluation of Foods Derived by Modern Biotechnology: Concepts and Principles*, Paris: OECD.

THE PRESIDENT'S COUNCIL ON COMPETITIVENESS, (1990), *Report on National Biotechnology Policy*, Washington, DC: Executive Office of the President.

RATNER, M. (1990), *Bio/Technology* **8**, 196–198.

RauCon Biotechnology Consultants GmbH (1989), *Impact of Genetic Engineering Regulation on the West German Biotechnology Industry*, Dielheim, Federal Republic of Germany.

SHARPLES, F. E. (1987), *Science* **235**, 1329.

U.S. CONGRESS, Office of Technology Assessment (1991), *Biotechnology in a Global Economy*, OTA-BA-494, Washington, DC: U.S. Government Printing Office, October.

U.S. NATIONAL BIOTECHNOLOGY POLICY BOARD (1992), *Report*, Bethesda, MD: National Institutes of Health, Office of the Director.

U.S. NATIONAL SCIENCE FOUNDATION (1985), The Suitability and Applicability of Risk Assessment Methods for Environmental Applications of Biotechnology (COVELLO, V. T., FIKSEL, J. R., Eds.), Washington, DC: U.S. National Science Foundation.

The Wall Street Journal (1989), Clouds gather over the Biotech Industry, January 30, New York.

WARD M. (1993), Do U.K. regulations of GMOs hamper industry? *Bio/Technology* **11**, 1213.

WHO WORKING GROUP (1985), Health impact of biotechnology: Report, *Swiss Biotechnol.* **2**, 7–16.

YOUNG, F. E., MILLER, H. I. (1989), Deliberate releases in Europe: Over-regulation may be the biggest threat of all, *Gene* **75**, 1–2.

3 Biosafety in rDNA Research and Production

Dieter Brauer

Frankfurt am Main, Federal Republic of Germany

Michael Bröker

Marburg, Federal Republic of Germany

Cornelia Kellermann

Martinsried, Federal Republic of Germany

Ernst-Ludwig Winnacker

Martinsried, Federal Republic of Germany

1 Introduction

Modern biotechnology represents the most exciting advance in the biological sciences this century. Particularly genetic engineering opened up completely new horizons for general basic research in biological systems which subsequently are bound to be followed by new prospects for medical, agricultural and industrial applications. However, biotechnology is not in itself a science, an industry nor a product but a powerful set of tools which is increasingly used to develop processes and to manufacture many products for every-day use or consumption as well as for industrial uses.

Much of the current confusion and lack of policy coherence arises from the tendency to think of biotechnology as fundamentally different from other technologies and therefore requiring special rules particularly with respect to biosafety issues. This equally concerns the handling of recombinant organisms in the laboratory and production facility, consumption or use of products from such processes and the deliberate release of organisms into the environment with the objective of their survival therein. While all regulations of modern biotechnology worldwide differentiate and thereby exclude applications of non-recombinant methods like diagnosis, "*in vitro*" fertilization or embryo-splitting correctly, the interest groups profoundly hostile to biotechnology continue to scramble up all applications and construct horror scenarios well received by parts of the media.

This chapter will focus on biosafety issues which are relevant to the handling of microorganisms in "contained use" and thereby review the experience accumulated in both microbiology and molecular biology which is the basis for the assessment of biological risks of organisms containing recombinant DNA molecules (the widely used term "recombinant organism" is actually misleading and should only be used for organisms produced by cell-fusion technology). While for "naturally" occurring organisms, and particularly pathogenic microorganisms, much scientific knowledge has accumulated since the end of last century, the "safety issues" of the debate on "recombinant DNA technology" (to be differentiated from the "socio-economic" issues) was and still is largely based on the presumption that recombinant organisms are "unique". While the scientific community in vain tried to falsify this view, it was taken up by the media and particularly by the European environmental agencies and the European political establishment resulting in regulations specific for the application of recombinant DNA technology instead of more appropriately introducing regulations for operations with pathogenic organisms and biological entities potentially harmful to parts of the environment as well as product approval systems evaluating the safety of biological products primarily by their properties and not by the methods they are manufactured. Consequently, assessment procedures were introduced which allow the comparison of a recombinant organism with its non-recombinant wild type or comparable non-recombinant organisms. This leads to a classification of the recombinant organism either to be harmless or to be classified into one out of the three risk groups (three is the highest risk group) which are used for the classification of naturally occurring pathogenic microorganisms.

Safety issues related to food and pharmaceutical products and the deliberate release of GMOs are treated in different chapters in this volume.

2 The Development of Molecular Biology

Biotechnology is best characterized as an interdisciplinary issue and can be defined as follows (SAGB, 1990):

> "Biotechnology" refers to the application of living organisms and their cellular, subcellular or molecular components to create products and processes.

Modern biotechnology, and in particular genetic engineering, stems from knowledge and methods developed in different areas of research like microbiology (bacteriology and virology), cell biology, biochemistry and other specialist fields.

Since GREGOR MENDEL discovered the basic principles of the laws of heredity in 1866 (published under the title "Experiments with Plant Hybrids") and DARWIN's discovery of the evolution of organisms, the prime goal of basic research in biological science has been to find out what these hereditary characteristics and their relation to the evolution of organisms actually are and how they work. Ultimately, this scientific quest aims to understand life; at least its biological basis.

Although deoxyribonucleic acid (DNA) was discovered by MIESCHER long ago in 1869, it still took another 75 years for detecting the functional relationship between DNA and the genetic properties of living organisms by AVERY (1944). Once it had been demonstrated that the chromosomes located in the cell nuclei are the carriers of hereditary information, WATSON and CRICK succeeded in 1953 in making the pioneering discovery that DNA consists of two molecular chains in the form of strands which are arranged in a helical structure as a double strand. Since JACKSON, SYMONS and BERG (1972) first constructed a DNA molecule which was capable of multiplication ("recombinant" plasmid), genetic engineering is the major contributor to the dramatic progress in understanding biological systems in molecular terms and for unique applications from industrial processes to medical therapy.

As the genetic engineering methods can be rapidly understood, learnt and put to practical application using excellent manuals (AUSUBEL et al., 1989), they will not be presented here.

3 Determinants of Pathogenicity

Before organisms and the work to be conducted with them are allocated to a safety level, a risk evaluation must be conducted. This implies that it is necessary to define exactly what actually a pathogenic microorganism is: namely an organisms which is known to be able to cause a disease (or diseases) in humans, animals or plants or for which there is a well-founded suspicion that this may be the case. An important subject in this context is the measurement of pathogenicity because, in a strict sense, the term pathogenicity has only a comparative and no absolute meaning. Therefore, because one microbial population can be more pathogenic than another and standard species or strains are not defined (nor seriously can), pathogenicity often can be compared only by inoculating animals with graded doses of the respective strains until a disease is detected. However, species differences, location of (primary) infection and the general health "status" of the animals are difficult to compare. Not surprisingly therefore, reliable methods for growing and quantifying the pathogens or valid animal models often are not available. For the classification of pathogens, some general aspects need to be considered for the evaluation and in this context are:

- The quantification of the inoculum (infectious dose) which is relatively simple to determine for many bacteria but less satisfactory for viruses and fungi. Additionally, the route of inoculation (e.g., by aerosol or intraperitonial) often affects both the necessary dosage and virulence itself.
- The comparison of pathogenicity between species is almost impossible in quantitative and qualitative terms, when, e.g., the influenza virus causes a respiratory disease in ferrets and the lassa virus produces a hemorrhagic fever in mice or rabbits (these are the standard animal models).
- The ideal test animal is a natural host, which essentially only applies to veterinary diseases. The choice of experimental animals with similar disease syndromes is often difficult to achieve, and primates are usually not available for many studies: while there is no good animal model available for human measles, the aerosol infection of guinea pigs with *Legionella pneumophilia* is a good model of the respective human pneumonia (regarded as zoonosis), and the

oral infection of primates with *Shigella* spp. mimics the human bacillary dysentery (shigellosis).

- The in-depth analysis of the multifactorial nature of pathogenicity, e.g., with respect to mucosal infection and invasion, multiplication in or outside cells, evasion of host defenses like phagocytosis or complement action, immunopathology etc. has progressed much further with bacteria than with viruses, fungi or parasites.
- Pathogenic determinants can be identified by either chemical mutation and biochemical purification or by genetic modification. However, with respect to the isolation of surface markers it must be kept in mind that such markers are not necessarily identical with the determinant(s) of pathogenicity. Only the use of the recently developed genetic engineering methods allows the analysis of individual pathogenic determinants, e.g., by cloning a determinant from a pathogenic strain into *Escherichia coli* and subsequently reintroducing the determinant into a deficient strain thereby eventually reestablishing "natural" pathogenicity. In addition, for the demonstration of its biological significance, the cloned determinant can be easily mutated *in vitro* with a concomitant loss of its biological properties.

During the last 20 to 30 years much knowledge for many pathogenic species has accumulated on the multifactorial nature of particularly bacterial pathogenicity, and numerous components of pathogenicity have been identified and probed. A good example may be the following collection of pathogenic determinants of *Vibrio cholerae*, which may not even be comprehensive (MÜLLER, 1992):

- one heat-labile and one heat-stable toxin
- one hemolysine
- one N-acetylhexose aminase
- one neuraminidase
- one nicotinamide-adenine dinucleotidase
- several proteases

- fucose- and mannose-sensible hemagglutinins
- "outer-membrane" proteins and
- motile flagellae.

Since these pathogenic determinants will not have developed to have pathogenic effects on the human intestine, their natural and ecological functions have to be interpreted with respect to the natural niche of *Vibrio cholerae*, which is the colon and exoskeleton of crab species. This means that this human pathogen may be a harmless facultative or even obligate parasite in lower crustacea (HUQ et al., 1986) and underlines the introductory remark that pathogenicity is not a strict but relative term.

It remains to be seen what the functions of such determinants in their normal niches really are. For legionellas it was found that the normal ecology as well as pathogenicity coincide: Namely the ability to multiply intracellularly within free-living amoebas in the same manner as in human leukocytes within the human lung causing pneumonia. Insofar, listerias and "atypical" mycobacteria may exhibit similar patterns.

With respect to an understanding of pathogenicity it is certainly helpful to distinguish "obligate" pathogens such as *Vibrio cholerae*, *Shigella* spp. or *Salmonella typhi* (each carrying specific pathogenic determinants) from "facultative" pathogenic organisms such as *Pseudomonas aeruginosa*, certain *streptococci* or *E. coli*. While the presence of the classical pathogenic agents in water, food or even in feces may present a high risk for the community, the latter "intestinal" organisms belong to the natural flora of the human body or to those microorganisms which are typically found in the environment. Furthermore, it must be kept in mind that particularly feces may contain microorganisms which can be highly pathogenic to the host if introduced parentally into the host (e.g., *Clostridium tetani* via injury). These ubiquitous "facultative" pathogenic organisms only rarely develop strains with pathogenic determinants and mostly may cause mild infections only under certain unusual conditions.

Well known are also "opportunistic" enteropathogenic bacterial strains which usually do not cause infections except in, e.g., immunologically compromised HIV patients (RUF and POHLE, 1989). Ubiquitous microorganisms, such as some "atypical" strains of *Mycobacterium*, *Salmonella*, *Shigella*, *Campylobacter*, *Clostridium*, and some fungi, which were reported to use the opportunity to colonize and propagate in the immunocompromised body what they cannot in healthy patients. While these infections usually are accessible to, e.g., antibiotic therapy, infections recur frequently. However, it must be stressed that while such "secondary" infections with opportunistic pathogens can be observed, the often lethal infections in HIV patients are not caused by these opportunistic pathogens but by simultaneous infections with pathogenic organisms (e.g., causing pneumonia or influenza). It is noteworthy to mention, that *E. coli*, which may cause gastrointestinal infections in healthy patients, are not significant in such opportunistic infections, and that microorganisms such as *Saccharomyces cerevisiae*, *Bacillus subtilis* and others, which are classified as non-pathogens, have not been identified as opportunists in these immunocompromised patients.

Last but not least the infectious dose, i.e., the number of pathogenic organisms necessary to establish a disease, is relevant for the classification of pathogenicity, which ranges from about 3 cells for *Rickettsia tsutsugamushi* or 180 cells for *Shigella flexneri* to 10^8 cells for *Vibrio cholerae*.

For further details concerning issues relevant to pathogenicity, the interested reader is referred to the literature (SMITH, 1988; FINLAY and FALKOW, 1989).

In conclusion, the feasible but difficult task is to classify pathogenicity via the observation of prevalence, duration and severity of the disease produced by the different species in man and comparing this with the infectious dose determined in animal models. With respect to risk assessment, beyond the classification of pathogenic organisms additional criteria like the availability of cures (e.g., antibiotics) or preventive measures (e.g., vaccination) need to be inspected and valued.

While the criteria and aspects outlined above apply to pathogens only, more than 80% of the research work conducted worldwide uses non-pathogenic organisms such as *E. coli* K12, *S. cerevisiae*, *B. subtilis* or *Streptomyces* spp. For production purposes there is a clear priority to use non-pathogenic organisms because of the safety for workers considerations, investment costs and product safety validation. However, there exists ample experience with "classical" productions using pathogenic microorganisms for toxin production (e.g., pertussis toxoid produced by *Bordetella pertussis*) or the production of "inactivated" vaccines (e.g., rabies) with excellent safety records for both workers and the population (Tab. 1).

In July 1974, a committee of prominent scientists involved in rDNA research called for a temporary worldwide moratorium on certain types of recombinant experiments and also called for an international conference on potential biohazards (for review see OTA REPORT, 1981). In response to this initiative of the involved scientists, the NIH established the rDNA Advisory Committee (RAC) on 7 October 1974, and an international conference was held at the Asilomar Conference Center, Pacific Grove, California, in February 1975. As a consequence of these activities, the NIH GUIDELINES FOR RESEARCH INVOLVING RECOMBINANT DNA MOLECULES were established in 1976, revised in 1986 and expanded for large-scale uses in 1991. It should be noted that the NIH Guidelines regulate applications with "recombinant molecules" and "organisms containing recombinant DNA molecules" instead of using the rather inaccurate or vague term "genetically modified organism" used by the EU DIRECTIVES (1990).

As time passed, it became clear that the initial worries about certain risks of rDNA research were greatly overstated, and in particular no evidence has been brought forward until today which would support the early risk scenarios. While this assessment is based on general experience, this view was further supported by experiments designed to explore risks of rDNA experiments (ISRAEL et al., 1979; CHAN et al., 1979): In a series of experiments polyoma virus DNA was cloned into bacterial vectors. Neither the recombinant DNA molecules ("naked DNA") nor *E. coli* K12 cells carrying these vectors caused viral

Tab. 1. Vaccines for Infectious Diseases in Humans and Fermenter Volumes Used in Industrial Fermentations

Bacterial Diseases		Fermenter Volume (L)
Diphtheria	Toxoid	300
Tetanus	Toxoid	700
Pertussis	Killed *Bordetella pertussis*	600
Typhoid fever	Killed *Salmonella typhi*	4000
Paratyphoid fever	Killed *Salmonella paratyphi*	4000
Cholera	Killed/extracts *Vibrio cholerae*	300
Plague	Killed/extracts *Yersinia pestis*	<300
Tuberculosis	Attenuated *Mycobacterium tuberculosis*	<300
Meningitis	Polysaccharide from *Neisseria meningitidis*	1000
	Polysaccharide from *Haemophilus influenzae*	300
Bacterial pneumonia	Polysaccharide from *Streptococcus pneumoniae*	100
Typhus fever	Killed *Rickettsia prowazeki*	egg
Leprae	Killed *Mycobacterium leprae*	armadillo
Viral Diseases		
Smallpox	Attenuated virus	calf
Yellow fever	same	egg
Measles	same	cell culture
Mumps	same	cell culture
Rubella	same	cell culture
Polio	Attenuated (Sabin) or inactivated virus	cell culture
Influenza	Attenuated or inactivated virus (Salk)	egg
Rabies	Inactivated virus	cell culture

infections nor tumors in mice. Most significantly, while it has long been known that mammalian cells can take up DNA (GRONENBERG et al., 1975), biological effects depend on the nature of the transmitted DNA: baculovirus DNA is not transcribed and rapidly eliminated from mammalian cells (CARBONELL and MILLER, 1987), while the injection of v-onc DNA into animals (BURNS et al., 1991) may induce tumors similar to naturally occurring infections with oncogenic viruses. These observations have led to recommendations for the protection of laboratory personnel to apply certain laboratory practices while handling potentially oncogenic DNA (ADVISORY COMMITTEE ON GENETIC MODIFICATION, United Kingdom, 1990). However, the accumulated knowledge clearly indicates that cloned pathogenic viral DNA bears no increased hazards in comparison to the uncloned viral DNA itself.

Any number of scientific experiments may prove that no general or unique biological risks are associated with rDNA experiments, nevertheless, by definition a final risk in individual cases cannot be excluded. With respect to the classification of rDNA work into Good Laboratory Practice (GLP) or one out of three risk categories, the so-called "additive model" has proved to be a useful and reliable tool for assessing potential risks. This model is applied worldwide and basically assumes that recombinant DNA experiments can be classified by inspecting the properties of the host organism, the organism the DNA fragment is derived from and the vector to be used. In addition to this it is assumed that the introduction of DNA segments from a non-pathogenic organism into another non-pathogenic organism (host) does not result in a pathogenic organism. With respect to the determinants responsible for pathogenicity, as outlined before, this assumption appears both reasonable and practical. Moreover, even the introduction of pathogenic determinants (like the ones from *Vibrio cholerae*) into a non-pathogenic host such as *E. coli* K12 or yeast is not sufficient to convert the hosts into pathogenic organisms. On the other hand, it is obvious that the cloning of pathogenic determinants from, e.g., *Yersinia pestis* into *Vibrio cholerae* needs the appropriate safety precautions, firstly because the host is pathogenic, and secondly such a modification may (or may not) increase the pathogenicity of the recombinant host in comparison to the wild-type organism. However, if such a modification would be beneficial to the host, it is highly probable that such a modified organism would have established itself by means of a "natural" gene flow which is known to exist among microorganisms (LORENZ and WACKERNAGEL, 1994).

In practical evaluations, the different biological backgrounds into which a certain gene (or set of genes) is cloned and potential "synergistic effects" originating thereof are not fully predictable in any case but not necessarily new or ignored. At the genetic level, an example for such effects is the activation of recombinational events leading to genomic rearrangements in developing lymphocytes by two adjacent genes (OETTINGER et al., 1990). Similar effects are known from a variety of biological systems, e.g., *Drosophila*, the activation of oncogenes (positioning effects) or recombinational events during embryogenesis. Altogether, while such effects may amplify or weaken certain properties of an organism or result in modified properties or substances being synthesized, it is hard to imagine that such effects could convert a non-pathogenic organism into one with unique properties and risks. The vast experience accumulated in rDNA R&D and particularly its excellent safety record support this estimation and champion the additive assessment model as an adequate and pragmatic tool.

In conclusion, the knowledge accumulated in biological and genetic sciences is a sound basis for the evaluation of potential risks associated with the genetic modification of microorganisms. This applies to both classical and recombinant methods used for modifying microorganisms genetically. Beyond the safe handling of non-pathogenic microorganisms, the risk evaluation concerning the use of bacteria, viruses, fungi and parasites which are pathogenic to man principally has to take into account the following criteria:

- Virulence and/or pathogenicity of the organism, i.e., whether in fact an infection is caused and the severity of its course
- Route of infection: aerosols, direct or indirect contact, injuries, vectors, etc.
- Epidemiological behavior: incidence and distribution of a pathogenic organism, immune state of the host organism, role of vectors and possible reservoirs
- Survival conditions in the laboratory
- Ability to survive outside the laboratory/production facility

If a microorganism is classified pathogenic according to the above criteria, several factors should be considered which could be relevant to any kind of work which might change the pathogenic profile and especially increase the pathogenicity of a microorganism. This includes possible effects of the vector(s) being used:

- Production of one or more toxins
- Transfer of genetic elements which have an invasive effect (this may maintain the infection and could be necessary for multiplication)
- Infectiousness
- Pathogenic mechanisms
- Infection dose to establish a disease
- Invasiveness
- Change in host specificity
- Ability to survive outside the host
- Mechanism for the transfer of genetic elements
- Biological stability
- Antibiotic resistance
- Toxicology
- Antiphagocytic factors
- Organotropisms
- Triggering of allergies
- Expression rate of the product (protein)
- Biological properties of the product.

Because this list does not claim to be complete, and particularly with respect to the administrative biosafety regulations, the reader is referred to the specific literature (SIMON and FROMMER, 1992) or the requirements according to, e.g., the EU Directives 90/219/

EEC ("contained use") and 90/220/EEC ("deliberate release") or the NIH Guidelines.

The "contained use" of microorganisms in classified facilities means that the technical measures taken are a lower or higher hurdle for the microorganisms to escape a given containment. Therefore, a 100% containment is both not possible and not necessary. Instead, the efficiency of containment must be both validated and appropriate for the microorganisms to be contained.

While most of the contained use applications deal with microorganisms which do not survive in the environment, the deliberate release of organisms into the environment with the purpose of their survival therein, at least for a certain time period, raises different questions. Evaluations have to consider the fact that possible adverse consequences as a result of the release of a naturally occurring exotic organism into a foreign and perhaps favorable environment (for review see OTA REPORT, 1993) is to be expected much greater than the release of an organism which has been modified by genetic engineering and which originates from a known, harmless organism with respect to the environment it is released into. This equally applies to plants, animals and microorganisms which may be harmful to the environment or not. Therefore, in regard to both the contained use and the deliberate release of organisms, the overall properties of the organism in question should be assessed when evaluating the risk and not the methods used to obtain the "modification".

A wide-ranging discussion of issues which arose in connection with the deliberate release of genetically modified organisms into the environment is not subject of this chapter, but is reviewed and well documented in the scientific literature (DOMSCH et al., 1987; TIEDJE et al., 1989) and is treated by MEDLEY and McCAMMON in this volume.

4 Pathogenic Microorganisms

Microorganisms such as bacteria, viruses, parasites and fungi are available in large amounts (DAVIS et al., 1985) in soil, water and air, and only very few of them are pathogens. In fact, the great majority of microorganisms is harmless, and many are beneficial for animals as well as plants. Microorganisms pathogenic for man usually occur as a few types in species which mostly contain many non-pathogenic members (BERUFSGENOSSENSCHAFT DER CHEMISCHEN INDUSTRIE, 1992) and can often be further divided into virulent and avirulent strains.

Since the end of the last century, the "Golden Era" of medical bacteriology, non-pathogenic and particularly highly pathogenic microorganisms have been investigated in basic research – safely for the researchers and workers concerned, the general population and the environment. With respect to industrial applications it must be stressed that particularly the fermentation of pathogenic bacteria, e.g., for the production of vaccines has proved to be very safe (KÜENZI et al., 1985; TUIJNENBURG, 1987).

Regarding legal aspects, it must be noted that international cooperation in the control of epidemics has had a legal basis since the beginning of the century (International Agreement on Sanitation, 21 June, 1926). Since the introduction of the Imperial Epidemics Law and, e.g., the German Epidemics Law of 1962, infectious diseases caused by more than 50 pathogenic microorganisms have to be reported and are therefore subject to administrative control. Furthermore, the German Epidemics Law controls not only the handling of sick patients or animals, but also rules that an authorization is required for the handling of pathogenic microorganisms, e.g., for research purposes, in clinical and biochemical laboratories. Similar regulations are in place in most OECD countries.

Important contributions to the understanding of biological processes stem from the analysis of the Darwinian process of genetic adaptation of organisms to various ecological niches. Such evolution occurs rapidly with microorganisms. A well known practical example is the introduction of the myxomatose virus to Australia in order to reduce the crop destructions caused by rabbits. Initially, this lethal and easily communicable disease was very effective and reduced the rabbit population dramatically. However, within a few years the rabbit population returned to approximately the number before the introduction of the virus because of the survival of resistant rabbit strains and the outgrowth of less virulent myxomatose strains which evidently had better opportunities to spread. This example addresses various factors which are involved in infectious diseases, namely pathogenicity or virulence, infectivity and communicability of a pathogen, while the immunological defense mechanisms are necessary for survival or resistance of the host. Pathogenicity and virulence, which is a synonymous term, is the capacity of some microorganisms to produce a disease. It is convenient to distinguish pathogenicity from virulence: for example, *Bacillus anthracis* is more pathogenic than *E. coli*, but the Vollum strain of *B. anthracis* is more virulent than the H M strain. Microorganisms vary enormously in their pathogenicity: from those which may infect humans without being pathogenic (indigenous bacteria of the pharynx, lower intestine or genitals) through those which coexist with the host in a "truce" (like staphylococci and herpes simplex virus) that is only occasionally broken up to those microorganisms which are usually eradicated after causing an acute disease (like thyphoid bacilli or the measles virus).

To trigger off a disease, a microorganism must be both infectious and pathogenic. This implies that the microorganism may have to overcome several hurdles, namely:

- be communicated to the host
- attach itself in or on the host,
- resist host defenses,
- penetrate into its tissue via the natural entry points and through mucous surfaces or via tissue injuries,
- multiply within the tissue and
- cause damage.

The molecular basis of the biological requirements for pathogenicity are the determinants of pathogenicity. Any microbial factor and specifically surface components and extracellular products such as bacterial capsular materials, toxins and viral envelope components can be the respective molecules. *In vivo*, host factors like nutrients or temperature are also important, because they contribute to the growth of the pathogens and play a major role in the differing susceptibility of animal species to various diseases. Each of the requirements mentioned above is complex involving many facets of pathogen–host interaction, and many different determinants are needed for a pathogen to successfully accomplish survival and multiplication. Pathogenicity therefore is a multifactorial property. Depending on the pathogen in question, the loss of only one of the required properties listed above may result in a complete loss of pathogenicity or may lead to reduced pathogenic potential (FRETER, 1978).

For the investigation of the genetics of microorganisms, their properties usually have been changed by mutagenesis, i.e., modification of the hereditary material by irradiation or with mutagenic substances. The pathogenic potential of the mutants obtained was often lower or higher than that of the respective wild types. Nevertheless, the safety record demonstrates that it was possible to work very safely with these mutants simply by using hygienic procedures and technical measures (COLLINS, 1990). However, laboratory infections may occur caused by injuries with, e.g., syringes irrespective of the nature of a pathogen – mutated or not. Thus, numerous safe working methods are available today, which are based on about a centuries' experience of working with pathogens and which permit the safe handling of all kinds of different organisms if applied appropriately, including those that have been modified by genetic engineering. Principally, the technical measures taken to contain the different pathogenic organisms are designed to reduce the number of bacteria escaping the containment to a level which does not pose a hazard to workers or the protective goals in general. Such "primary containment" measures are the use of sterile work benches (lamina flows) and technical measures to prevent aerosols escaping from, e.g., centrifuges. It should be

Tab. 2. Comparison of the Major Containment Measures

Classification	GLP/GILSP[a]	Risk Groups		
		C1	C2	C3
Separated area	+	+	+ (shielded)	+ (separate building)
Doors, windows	closed	closed	closed, sealed, locked	
Air filters	–	–	+	+ (double filters)
Reduced air pressure	–	–	+	+
Safety hood	–	+	+ (class I or II)	+ (class III)
Waste inactivation	(–/+)	+	+	+
Waste water inactivation	(–/+)	+	+	+
Lab coats	+	+	+	+ (extra clothing)
Emergency power	–	–	(–/+)	+

[a] GLP, Good Laboratory Practice; GILSP, Good Industrial Large Scale Practice

understood that the key technical design differences between safety level 1 and safety levels 2 and 3 are additional measures like reduced pressure aiming to prevent pathogens from escaping the "secondary" containment (Tab. 2). Such technical measures do not add "safety" to those working inside the "secondary" containment area. However, such technical methods are neither applicable in keeping and treating infected patients nor in routine work in medical laboratories – not surprisingly infections occur accidentally (HUNTER, 1993).

The following description of the different risk groups of bacteria, viruses, fungi and parasites is mainly based on the assessments published by the German BERUFSGENOSSEN-SCHAFT DER CHEMISCHEN INDUSTRIE (1991–1992). However, in the following the classification system recommended by the OECD and upon which the EU Directive 90/219 EEC "contained use" is based, will be used. This system distinguishes harmless microorganisms (group I organisms in the EU Directive) to be handled following the rules of "Good Laboratory Practice (GLP)" and "Good Industrial Large Scale Practice (GILSP)", respectively, from harmful pathogenic organisms classified into the risk groups 1, 2 and 3 (group II organisms in the EU Directive) which may be only worked with in the respective containment level C1, C2 and C3 facilities. Unfortunately, other systems have been implemented like in the German Gene Technology Act, which classifies organisms into four risk groups and therefore does not really distinguish harmless from pathogenic organisms.

4.1 Bacteria

In contrast to many other microorganisms, bacteria are morphologically differentiated into only a few forms. These are spherical or straight, bent or spiral forms. At present, more than 3000 bacterial species are known, and about 150 new species are discovered each year. The ratio of the large surface area to the small volume of bacteria leads to a very efficient metabolic rate. Bacteria can metabolize up to 10000 times their own weight in sugar per hour. The human body would require about 30 years in order to achieve the same turnover.

Bacteria with a generation time of 30 minutes, could grow from one cell to 10^{14} cells weighing about 100 g within 24 hours and could theoretically multiply to 10^{28} cells the next 24 hours and attain a weight of 100000000000 tons. In reality, bacterial "growth" is soon halted by nutritional and other environmental restrictions.

Up to 10^9 bacteria are contained in 1 g of soil, and the approximately 400 bacterial species colonizing the human body on the skin and in the intestinal tract amount to about 10^{14} cells, that is more than the number of cells of a human body, which is approximately 10^{13} cells. While environmental bacteria usually are saprophytes, i.e., they grow on "non-living" organic material and often are also capable of using nutrients directly from living organisms like plants or animals, most bacteria are not capable of piercing the mechanical, chemical, cellular and immunological defense mechanisms of other organism. Overall, there is a great variety of bacteria ranging from saprophytes to obligate parasitic bacteria being classified into three risk groups (or classes) according to their pathogenic properties initially only to men but meanwhile also to animals and plants (e.g., German Gene Technology Act). Comprehensive lists have been compiled in many countries worldwide, and the experience reflected in these lists is the basis for the risk assessment introduced for genetic engineering.

Group I bacteria compromise those organisms where "no" risk is known to exist for man and vertebrates (therefore, there is no need to label such organisms with the prefix "risk"). There are several groups of bacteria which have no relevance as pathogens because of their specific properties:

- psychrophilic bacteria multiply only below 20 °C
- thermophilic bacteria multiply only above 50 °C
- acidophilic bacteria grow only below pH 4
- alkaliphilic bacteria grow only above pH 8.5
- obligate phototrophic bacteria
- obligate chemolithotrophic bacteria
- bacteria with a safe history of industrial use, e.g., *Bacillus subtilis*
- saprophytes
- bacterial strains which have completely lost their virulence and therefore are used as live vaccines (e.g., *Salmonella typhi* 21a).

Group II, Risk Group 1 bacteria comprise those organisms which may cause a disease in workers, but with respect to infectivity, pathogenicity and the availability of prophylactic and/or therapeutic measures (for both workers concerned and the general population) only low risks are to be expected. This group also contains organisms which are pathogenic only to vertebrates (but not to man). Typical organisms are:

- *Streptococcus mutans*, colonizes the mucous surface of the mouth and is involved in the development of caries. It is a facultative pathogen and only rarely causes endocarditis.
- *Clostridium tetani*, develops spores ubiquitous particularly in rural areas. Even in the absence of a vaccination, the regular contact with these spores is not sufficient to cause a disease.
- *Vibrio cholerae*, needs a high infection dosis. The infection occurs almost exclusively with "drinking" water, and infections do not occur via air or body contacts.

Group II Risk Group 2 comprises organisms which may cause a serious disease but pose a moderate risk only considering the criteria described for Risk Group 2. Typical examples are:

- *Mycobacterium tuberculosis*, causes a serious disease with a lengthy therapeutic treatment. Vaccination is possible.
- *Yersinia pestis*, can be transmitted by flea-bites and via air, then causing a very serious infection. Vaccination is possible.

There are no bacteria known which need to be classified into Group II Risk Group 3. This high risk group contains viruses only.

4.2 Viruses

Viruses display unique features not found in other microorganisms. These primarily concern the inability of viruses for self-multiplication, i.e., they infect a host first and then use the replication machinery provided by the host for amplification. Hence it is not surprising that no chemotherapy has been developed so far comparable to the efficient treatment of bacterial infections with antibiotics. The genome of viruses consists of either DNA or RNA with the latter usually being reverse transcribed into DNA, before replication of the viral genome can occur. Viruses are specialized to certain "permissive" cell types and can be found as cell parasites and/or pathogens for almost all kinds of cells including bacteria. Viruses can be transmitted horizontally between hosts and vertically between successive generations of permissive hosts. Similar to bacterial infections, the horizontal transmission of viruses may occur via:

- certain blood cell types transmitted by injections, transfusion or sexual intercourse (hepatitis B and C virus, HIV 1 and 2, etc.)
- bites (rabies)
- vectors in general (injuries, injection needles, food, infects, ticks, etc.)
- respiratory tract by aerosols, dust (influenza- and rhinoviruses)
- digestive tract (poliovirus)
- water (enteroviruses)
- skin (papilloma viruses)
- direct contacts (herpes viruses or measles)

Vertical transmissions may occur via:

- germline (retroviruses)
- hematogen-diaplacental (German measles, parvovirus-B-19)
- genital infections (herpes viruses)
- perinatal transmissions (HIV 1 and 2)

In general, viral infections vary from generalized, local, persisting, chronic, latent to "slow" forms with very different symptoms. Moreover, even serious viral infections with, e.g., hepatitis B virus may take a chronic form with almost no clinical symptoms. Nearly all herpes viruses tend to cause latent diseases with reactivations of the viruses after, e.g., stress or partial immunodeficiencies. At the stage of latency it is often extremely difficult to detect viral particles.

Various criteria need to be considered for the classification of viruses into the four risk groups which are mainly:

- virulence of the pathogen
- seriousness of the infection and its frequency
- transmission pathway
- epidemiological situation (immunity, reservoirs, etc.)
- infection dosis
- availability of vaccines

The classification of viruses resembles the classification of bacteria and ranges from "no" to low risk (Group I), low to moderate (Group II, Risk Group 1), moderate to unknown (Group II, Risk Group 2) to high risk (Group II, Risk Group 3). These variations within one risk group relate to the slightly different properties of human pathogens, animal pathogens and dual human/animal pathogens. The term "unknown" risk for Risk Group 2 viruses refers to "exotic" animal and dual human/animal viruses, which are not found in Europe and for which therefore the epizootic behavior is not known.

Group I comprises viruses which are completely non-pathogenic to humans (and most vertebrates), such as bacterial phages, plant viruses and live vaccines (e.g., poliomyelitis type 1, 2 and 3), even though the latter may cause diseases (under certain conditions, i.e., only if reverse mutations occur, and very rarely). It must be noted that live vaccines which multiply in, e.g., humans can be transmitted and may propagate into revertants.

Group II Risk Group 1 comprises a very broad spectrum of viruses ranging from spuma viruses (human diseases are not documented), hepatitis B virus to human herpes viruses like HSV 1 and cytomegalo virus. For these human herpes viruses in more than 80% of the adults tested antibodies are found, i.e., one or the other form of primary infection (mostly inapparent) must have occurred. Even in the presence of high titers of neutralizing antibodies a substantial fraction of the population is subject to recurrent herpes.

Group II Risk Group 2 contains viruses which are characterized by diseases with up to 100% both manifestation and lethality and for which often neither therapies nor prophylactic vaccination are available. While the human immunodeficiency viruses (HIV 1 and 2) are transmitted only via the transfer of body fluids, yellow fever presents a complex ecological situation with two distinct epidemiological types and is only transmitted by insects which do not exist in Europe (vaccination is possible). The herpes B virus of monkeys may infect man only accidentally but can be transmitted via air (laboratory infections have been reported).

Group II Risk Group 3 comprises those viruses which are highly contagious with almost 100% manifestation and lethality of the disease. These viruses have a high probability to cause pandemics and can be transmitted even by vaccinated individuals. Examples are lassa virus, variola major/minor and marburg virus, which need specific precautions to be taken if they are subject to diagnostic or research operations.

With respect to their genetic contents, viruses vary considerably in size from small RNA tumor viruses with only 4 genes (e.g., Rous sarcoma virus with the *gag, pol, env* and *src* gene) to herpes viruses, pox viruses or insect viruses with up to more than hundreds of genes and genome sizes up to 300000 nucleotides. By now, several thousands of animal virus isolates have been analyzed which could be classified into a reasonably small number of virus families. The accumulated genetic knowledge clearly indicates that the replication mechanisms of viruses rely on complex interactions with the host, which leads to the conclusion, that the evolution of a successful virus type must be a very rare event.

For more than 100 years viruses have been the object of intensive research. Following the introduction of genetic engineering, major discoveries about the genetics of higher cells came from first studying their viruses. If the necessary technical precautions are taken for containment, even the viruses of the highest risk group can be handled safely for both research and industrial operations as the practical experience proves.

4.3 Fungi and Parasites

Fungi are abundant in soil and on plants and are found in all climate zones. They have traditionally been regarded as plantlike organisms, grow either as single cells (yeasts) or as multicellular filamentous colonies (molds and mushrooms) and produce ubiquitous airborne spores. It is difficult to define fungi comprehensively and unambiguously because there are at least 100000 fungi with an enormous diversity (HAWKSWORTH et al., 1983). While only few fungi cause diseases in man, other pathogenic fungi have serious impacts on human welfare as causes of plant diseases destroying harvests or by toxic or carcinogenic metabolic byproducts such as aflatoxins and ochratoxins from aspergilli. Fungi reproduce by both sexual and asexual cycles as well as by a parasexual process, and most of the fungi that are pathogenic for man lack sexuality. Fungi pathogenic to man usually are classified with respect to healthy workers and therefore do not include serious immunodeficiencies or other diseases like tuberculosis, AIDS or diabetes with respect to fungal infections.

Parasites are organisms which live either on or in the body of other organisms (ectoparasites and endoparasites, respectively) and may cause parasitoses. The most frequent disease is caused by *Malaria tropica* with currently more than 300 millions of new infections annually with more than one million lethal cases (mostly children). Beyond the individual health problems caused by parasites, the economic costs for health care in both human and production losses with domestic animals are considerable. Most parasites have complicated life cycles with usually more than one host being involved. Differing from the classification schemes used for bacteria, for viruses and fungi a different and more individual classification approach is necessary. With respect to parasites it is important, whether vectors and/or intermediate hosts are used in the respective work. The examples given are restricted to fungi.

Generally, fungi and parasites are classified by definitions similar to those used for viruses. There are no fungi and parasites known, which need to be classified into Group II Risk Group 3.

Group I fungi comprise those organisms where no dangers for man are associated with (*Saccharomyces cerevisiae*) and those which may be opportunistic pathogens in severely immunocompromised patients only (e.g., *Aspergillus niger* ubiquitous in nature, and *Malassezia furfur*, which colonizes the healthy human skin but may cause systemic infections in severely immunocompromised children).

Group II Risk Group 1 contains fungi which may cause infections in case of minor disturbances of the immune system and for which effective therapy is available. Examples are *Candida albicans* (which is found in the mouth and the intestinal tract and only in cases like diabetes or chemotherapy may cause systemic candidosis), and *Trichophyton mentagrophytes* which is the agent causing common athlete's foot under certain conditions.

Group II Risk Group 2 may cause severe mycotic diseases as is the case with *Coccidioides immitis* which is found in certain areas in southwest USA and which may, via an infection of the lung, cause further infections of other organs.

Several fungi are non-pathogens by definition but may cause damage by the production of mycotoxins which may even pass the skin like trichothecene from *Fusarium* sp. With respect to the inactivation of fungi including their spores, toxins may survive the treatment and therefore sometimes afford extra attention.

5 General Assessment of Biosafety

With about 10^7 to 10^8 cells per gram of soil and thousands of different species/subspecies, microorganisms are in a fierce competitive struggle. They have to establish themselves in the face of numerous stress factors and competitors (ALEXANDER, 1981). Countless investigations with laboratory or wild strains which had been released into the environ- ment (e.g., for agricultural uses or research purposes) have shown that the number of live microorganisms usually decreases very rapidly by several powers of ten (ARMSTRONG et al., 1987) which mainly depends on the nutrients available (DEVANAS et al., 1986).

The classification of microorganisms into three risk groups and accordingly the use of different technical safety measures in three "safety" levels (beyond GLP) have a sound scientific basis and depend on the properties of the relevant organisms. Thus, for example, *E. coli* K12 laboratory strains and baker's yeast are completely harmless to humans, animals, plants and the environment (Group I; GLP and GILSP, respectively). Polio viruses (wild strains) are harmless to animals and the environment but constitute a serious risk for individuals who have not been vaccinated (Risk Group 2; Safety level 2). Similarly, smallpox viruses represent a lethal risk only for humans (but do not infect animals), and the foot-and-mouth disease virus constitutes a serious risk only to cattle (both are classified into Risk Group 3; Safety level 3), but carries no risks to humans and plants.

Based on the organism, vector and genetic material used for genetic engineering work, such work too is classified into one out of the four categories and allocated to either GLP/ GILSP or the respective containment level with the appropriate technical safety measures (SINGER and SOLL, 1973). This system has a proven safety record and permits risk-minimized handling of pathogenic organisms under appropriate containment. In practical terms, technical measures extend, for example, from the use of gloves (GLP) to lamina flow work benches, autoclaves and laboratory exhaust air filters to special facilities in the containment level 3, which must be kept completely under reduced air pressure in order to prevent the escape of microorganisms from the controlled area.

These technical safety measures do not apply solely to genetic engineering work, however. The conventional genetic and biochemical investigation of all kinds of pathogenic microorganisms are of course also carried out under similarly graded conditions.

Only with the introduction of genetic engineering techniques the technical safety measures applied to the safe handling of microorganisms could be supplemented and extended by biological safety measures. Using certain combinations of safe microorganisms and vectors, dual host/vector systems were developed which allow to classify certain genetic engineering work into the lowest safety level or even exempt them from regulatory overview. The exempt category, as defined in the NIH GUIDELINES (1984), covers approximately 80–90% of all rDNA experiments, whereby most work is conducted with *E. coli* K12, *S. cerevisiae* and asporogenic *B. subtilis* host–vector systems.

Theoretically, an additional safety system could be established if microorganisms could be constructed which, if they escape from the laboratory, would produce a factor that resulted in immediate death of the respective cell (suicide system). Naturally, tests would have to be carried out in order to establish how reliable such a system would be.

What does the introduction of biological safety measures additionally to technical safety measures mean in practical terms?

First, a certain combination of technical safety measures must be applied to the multiplication and subsequent research conducted on infectious pathogenic organisms such as HIV, hepatitis C or *Bacillus anthracis* (which are all classified into Risk Group 2). Such bacteriological or virological research, recombinant or not, is subject to safety provisions as mandatory according to the rDNA rules for containment level 2 for rDNA research.

Second, the analysis of individual genes and the respective gene products of such pathogenic organisms, once they are introduced into safe host/vector systems like *E. coli* K12 in conjunction with certain conjugation-negative plasmids, usually can be carried out under GLP/GILSP conditions. The background is that a "safe" non-pathogenic organism like *E. coli* K12 cannot be transformed into a pathogenic one by the transfer of any combination of genes, because pathogenicity is a too highly specific and multifactorial phenomenon.

Additionally, it should be kept in mind, that DNA from eukaryotic cells (including viruses) usually cannot be expressed in prokaryotic cells, e.g., *E. coli*, and *vice versa*. Conse-

quently, eukaryotic proteins will not be synthesized, and hence no biological risks develop.

In this context, it must be borne in mind that patients infected with the naturally occurring microorganisms quoted above, of course are not subject to the technical safety measures applicable to containment level 2 as applied in the laboratory, even though the count of infectious particles released by patients may be significant.

Using the dual system of technical and biological safety measures described above, the pathogenicity factors of virtually all (micro)organisms can be easily and safely investigated in the low safety levels. The introduction of genetic engineering methods into medical, bacteriological, virological and ecological research and for biotechnical production processes therefore has led, in many respects, to a marked increase in safety at the workplace for the researchers and other personnel concerned and for the environment.

6 Species-Specificity of Genes

The genes (DNA) transferred from one species to another (for example, between mammals, microorganisms, and plants) cannot, with a few exceptions, be directly used by other species. This is due to the great specificity of the regulatory structures (promoters) and the specific mechanisms of protein biosynthesis at the ribosomes which control the transformation of the information stored in the genes into the relevant gene products (i.e., protein molecules). The relevant processes, named "transcription" and "translation", are highly specific and can utilize DNA only from organisms which are genetically compatible. As a consequence of the genetic differences mentioned above, a large part of the application-oriented genetic engineering research and process development work deals with the exchange and improvement of such regulatory structures in order to achieve maximum expression rates of the cloned coding sequences in prokaryotic or eukaryotic cells.

In addition to the non-compatibility of the promoters, chromosomal eukaryotic genes usually are structured into intron and exon regions, which is not the case for bacterial genes (only the fused exon regions are finally used in eukaryotic protein biosynthesis). Not surprisingly, bacteria are not able to remove the intron regions from the original "eukaryotic" mRNA transcript (which is an exact copy of the coding DNA strand) into the "final mRNA", which in eukaryotic cells finally is translated by the ribosomes into the respective protein(s).

Therefore, prokaryotes usually are not capable of expressing chromosomal eukaryotic genes into biologically relevant amounts of functional proteins (and *vice versa*) because:

• the regulatory structures are not compatible and
• because of the intron/exon structures.

This inability also applies to the few known cases of eukaryotic gene sequences which do not contain introns. For details on the molecular mechanisms of gene expression and other related issues the reader is referred to the literature, e.g., WATSON et al. (1987).

As, for example, the human proinsulin gene is associated with the typical properties described for eukaryotic cells, it cannot be directly used for production purposes in bacterial cells. Therefore, instead of using the chromosomal insulin gene, firstly the "final mRNA" is isolated from tissue and reverse transcribed into the corresponding DNA structure using the appropriate experimental procedures. Secondly, a suitable bacterial promoter is fused to this DNA structure resulting in an artificial gene now composed of both a promoter and coding region. Thirdly, this gene is integrated into an appropriate vector and transformed into, e.g., *E. coli* K12 cells which only under appropriate nutritional conditions can synthesize the proinsulin molecule.

This hybrid gene constructed *in vitro* from the bacterial promoter and the section of DNA determining the proinsulin protein structure can, however, only be used in bacte-ria after the modification. It can no longer be activated in mammalian cells because these conversely cannot use the bacterial promoter.

The transfer of DNA segments between organisms using similar DNA transcription and message-processing mechanisms automatically results in the transfer of genes, because these can principally be expressed by the recipient cell. This usually applies to the cloning of bacterial DNA, e.g., into *E. coli* cells and is similar to natural processes which are known to occur in nature (TREVORS et al., 1987). Different to this, the DNA transferred between organisms using different DNA transcription and message-processing mechanisms usually cannot be processed by the recipient cell and therefore cannot be considered as transfer of genes but of non-functional DNA only. This applies to, e.g., the construction of eukaryotic gene libraries in *E. coli* cells.

The basis of this distinction is, that a gene in fact only exists, when a cell is able to express it into a biological function – otherwise the DNA sequence in question must be considered as non-coding DNA only. Whether or not such a sequence can be activated by cellular mechanisms like recombination (or others), would be an event which is not unique to genetic engineering and comparable to effects caused by, e.g., transposons, retroviruses and, more generally, by repair mechanisms.

For the reasons described, it is impossible to construct a "super-gene" (and hence a super-enzyme) that could simultaneously be:

• expressed in all species,
• biologically active in all species with a multitude of biological properties.

The fundamental differences between prokaryotic and eukaryotic life forms with respect to gene expression have been identified at the molecular level (hence molecular biology) and constitute a highly relevant and intrinsic aspect of biosafety.

After successful completion of the "construction" work at the DNA level, as described, no further genetic or genetically engineered changes need to be carried out for the production of proteins for research and commercial purposes (e.g., for crystallization experiments and as pharmaceuticals, respectively). Genetic engineering in the narrower sense is therefore restricted to the laboratory.

All further measures are identical to the conventional fermentation methods and have to be approached, e.g., with respect to the production of substances for clinical testing, in terms of process and facility validation as usual.

7 Gene Transfer Mechanisms

In genetic engineering DNA is usually transferred by means of vectors, i.e., plasmids (WINNACKER, 1987) or viruses, and multiplied together with these. Naturally occurring vectors can be transmitted between different cells via three different mechanisms: conjugation (and mobilization), transduction and transformation (ARBER, 1987; STRAUSS et al., 1986; ROSZAK and COLWELL, 1987; STEWART and CARLSON, 1986; MORRISON et al., 1978).

For "conjugation" a conjugative plasmid is required which must first multiply in the donor cell so that it can be transferred to a recipient cell. For mobilization, a helper plasmid must make the necessary enzymes available. "Transduction" is based on viral DNA molecules which are integrated into the genome of a cell and multiplied by cell division. Such proviruses are activated by a wide range of influences and finally are packaged into infectious viral particles. During this process host genes located next to the provirus can be "inadvertently included".

In contrast to the two "biological" mechanisms mentioned, "transformation" is a passive process in which the nucleic acid is transported into the cell through the cell wall/membrane by unspecific means. Subsequently, transformed DNA is multiplied like plasmids or can be integrated into the cells' chromosome by recombination or repair mechanisms.

E. coli cells require chemical pretreatment to make them "competent" for a transformation as described.

Unlike the naturally occurring bacterial plasmids (COUTURIER et al., 1988), most vectors (i.e., safety plasmids or phages) used in genetic engineering for the multiplication of DNA fragments cannot be transferred to other bacteria by conjugation or transduction because, particularly in the case of plasmids, the genetic elements necessary for the transfer have been removed. In the case of phages frequently used in addition to plasmids, the host specificity is restricted almost exclusively to the *E. coli* K12 laboratory strains (BLATTNER et al., 1977; DENHARDT et al., 1978).

The transfer of plasmids between different bacteria follows certain biochemical rules and requires the plasmids to carry special structures from which the "transfer mechanisms" can proceed (BASTIA, 1978). The transfer genes (at least 12 *tra* genes; WILLETS, 1988) needed for transfer, the genes of the respective enzymes (*mob*; BOYD et al., 1989) and the mobilization site (*nic/bom*; FINNEGAN and SHERRATT, 1982) are not present on the safety plasmids employed today. For this reason, safety plasmids of this type cannot be transferred with helper functions from other plasmids (TWIGG and SHERRATT, 1980).

Even naturally occurring broad-host-range plasmids (such as RP4 and others) which can multiply in many bacterial species from *E. coli* to *Alcaligenes* strains and *Agrobacterium tumefaciens* (SAYLER and STACEY, 1986; ELSAS et al., 1990) do *not* multiply uncontrolled in all possible bacteria, as is sometimes claimed even of the safety plasmids.

If this were the case, it would *never* have been possible to develop antibiotic therapy for bacterial infections with compounds such as penicillin, tetracycline, cephalosporin or streptomycin, because antibiotic resistance is known to be passed on by plasmids, and consequently all existing bacteria would be bound to be resistant (KELCH and LEE, 1978).

The molecular genetic principles for the widely varying and often very limited ability of naturally occurring plasmids to multiply have been investigated in depth (e.g., for ColE1 derivatives) and are based on numerous genetic properties:

– For their stabilization in bacteria plasmids require a section located in the vicinity of the "origin of replication" which is known as *par* (from partition) (MEACOCK and COHEN, 1980). This section interacts with certain cellular components which bring about uniform distribution of the plasmids on the daughter cells during division. If this section is removed from the plasmid or if the section is biologically inactive (as a result of mutations), the plasmid is slowly lost in a growing bacterial culture (JONES et al., 1980). If, however, a plasmid of this type contains an antibiotic resistance gene, it can be stabilized in a growing bacterial culture as a result of the selection pressure exerted by, e.g., adding the appropriate antibiotic. Only bacteria containing the plasmid can survive under these selection conditions. Therefore, care must be taken in R & D in order to ensure that the plasmids are stabilized by selection pressure in the host bacteria used and that they are not lost as a result of carelessness.

– Naturally occurring plasmids are divided into more than 30 different "incompatibility groups" (NOVICK, 1987; THOMAS and SMITH, 1987). This classification is necessary because genetically similar plasmids cannot multiply at the same time in the same bacterium. One type of plasmid is rapidly eliminated. This genetic property also contributes to the fact that plasmids do not multiply in all bacteria occurring in nature.

– Restriction enzymes are isolated from microorganisms and used to break down DNA into defined fragments. Today, more than 200 different enzymes from differing bacteria are known (ROBERTS and MACELIS, 1992). Many of the commercially available restriction enzymes are obtained by genetic engineering techniques. The restriction coenzymes are one part of the "restriction-modification system" of bacteria whose effect can be interpreted as an "immune system". Its aim is to protect bacteria from penetrating "foreign DNA". This is achieved by the foreign DNA being digested by the cell-specific restriction enzymes, while the host DNA first is changed chemically by the cells' modification enzyme (methylation) to such an extent that it is protected against the cell-specific restriction enzymes. Naturally occurring bacteria possess further effective mechanisms which can restrict the invasion and spread of DNA (RAYSSIGUIER et al., 1989).

– As mentioned above, the vast majority of plasmids that are used for genetic engineering are derived from the naturally occurring plasmid ColE1 (HERSHFIELD et al., 1974). This plasmid is relatively small and carries the genetic information for a protein molecule described as "colicine E1". (Colicines are proteins which help bacteria to destroy competitors; KONISKY, 1978.) Because of the properties of its very thoroughly investigated "origin of replication" (KÜES and STAHL, 1989), the ColE1 plasmid has an extremely narrow host spectrum and can only be fermented in coliform enterobacteria, but not, for example, in bacilli or pseudomonads. The reasons for this specificity are the sequences and enzymes needed for multiplication which are largely host-specific.

Since the naturally occurring ColE1 plasmid does not multiply in non-enteric bacteria, its origin of replication proved eminently suitable for the construction of the various plasmids in use today.

The DNA transfers between Gram-negative and Gram-positive microorganisms (MAZODIER et al., 1989), bacteria and yeast cells (HEINEMANN and SPRAGUE, 1989) and between bacteria and plants (BUCHANAN-WOLLASTON et al., 1987) described in the literature show that a slight horizontal gene transfer can be detected under artificial conditions. Usually special selection procedures precede the detection, and the following conditions have to be considered:

- no safety plasmids should be used,
- the recipient cells must be made competent or
- experiments must be carried out with high cell densities,
- sterilized soils or water samples must be used,
- special buffers, plasmid concentrations and temperatures are required.

There are so far no reports of conjugation, transduction or mobilization processes which did not employ one or the other of the above experimental conditions.

Horizontal gene transfer (or the supposedly uncontrollable spread of genes) is an often discussed topic and has already been addressed to the definition of genes and the question when they are merely to be considered non-coding DNA. In respect of safety to humans and to the environment, such reflections can be based on broad practical experience and on the biological mechanisms that are widely encountered in nature.

Many organisms are bound to take in, with their food, large amounts of DNA from the cells of plants or meat. This means, that DNA is an important part of the diet and that not only the human body has learned to break down and utilize orally ingested DNA, irrespective of its origin. This applies equally to DNA which is present in the microorganisms ingested every day with the food and from the environment. These microorganisms may have been capable of causing even fatal diseases in individual cases over thousands of years, but they have not been "transmitted" to mammals and have therefore never caused hereditary genetic problems.

A prerequisite for any stable uptake of DNA would be that certain sections of the DNA, which can exert specific enzymatic functions in the recipient organism, must be incorporated into the organism's chromosome (for microorganisms) or the germline (e.g., for mammals). Thereby the new genetic trait would be made available to the hereditary and evolutionary processes the individuals in question underly. Irrespective of the fact that this is not bound to be disadvantageous *per se* to the species concerned, such events and mechanisms today can only be the sub-ject of speculation, because there is too little knowledge at the molecular level and particularly no well-founded suspicion that this might be relevant.

Depending on the genetic relationship between the donor and the recipient, the DNA naturally transferred between species may be immediately usable by the recipient. In such a case, phenotypical properties may be added and could be detected immediately by biochemical methods – irrespective of whether the DNA becomes integrated into a chromosome or is maintained in extrachromosomal elements like plasmids or viruses. Otherwise, if the DNA is not usable because of genetic incompatibility between donor and recipient, the DNA will have no generic biologic effect except for possible indirect effects on neighboring genetic elements. Therefore, biochemical effects are not detectable in such cases, and the DNA segment in question may have effects only after accumulating mutational events. Since evolution is not target-oriented, i.e., does not aim for a certain property of a DNA segment or even the properties of a gene product, the long-term effect cannot be estimated. It is questionable whether this scenario has any relevance at all, as a simple comparison (ZINKE, 1990) of the first 90 nucleotides of

- human prepro*insulin* (BELL et al., 1980) with
- *β-lact*amase from *E. coli* (pBR 322; SUTCLIFFE, 1978)
- *glyc*ogen synthetase from *E. coli* (KUMAR et al., 1986)
- *tetanus* toxin from *Clostridium tetani* (FAIRWEATHER and LYNESS, 1986) and
- a statistically compiled *random sequence*

indicates, which show a similarity between 20 and 37% of absolute identity:

		Similarity in %		
β-lact.	22.5			
glyc. s.	20	36.6		
random s.	27	18	22	
tetanus t.	21	30	33	22
	insulin	*β-lact.*	*glyc. s.*	*random s.*

This simple comparison shows that a "new" random sequence or any other sequence can principally be derived by a comparable number of successive mutations from many other DNA sequences. Because of the properties of the genetic code (the first nucleotide in a codon is more "specific" than the third nucleotide) even less mutations will be necessary for the development of the respective biological activity (e.g., peptide or protein). Irrespective of whether this comparison has a minor or major significance, it can be concluded that by introducing eukaryotic DNA (visible and non-visible genes) into prokaryotic hosts, no "new type" of potential risk is being formed.

Since there are no biochemical differences between coding DNA, non-coding DNA and recombinant DNA, the same biological laws apply. Therefore, both theoretically and according to the empirical experience with classical mutagenesis and selection procedures, short- and long-term effects both positive and negative (how to define?) by "modified" organisms only can occur, if:

- the collective properties of an organism allow its long-term survival *and*
- positive or negative effects actually develop.

However, these two essential properties, namely, the capacity of surviving and a certain potential for measurable effects, are difficult to obtain with non-environmental organisms, such as *E. coli* K12, *S. cerevisiae* and eukaryotic cells, and have not been observed with recombinant DNA technology since its invention. This is due to the fact that the typical properties of a "safe" organism (e.g., non-pathogenic, no environmental germ, dependence on certain nutrients and temperatures, non-competitive in its species' ecological niche) cannot be changed by introducing a couple of foreign genes or by mutations of a recombinant gene. Therefore, even the introduction of recombinant genes used in production processes into other bacteria can be assessed as not relevant in biological terms.

Thus, in more general terms, there is no unique risk potential associated with the transfer of DNA sequences by rDNA beyond what happens with microbes anyway. Instead, a potential risk arises solely from the function of a DNA sequence or, more specifically, of the corresponding gene product (RAMAGHA-DRAN, 1987). This applies, for instance, to the manufacturing of toxins in, e.g., *E. coli* K12, which does not raise problems of bacterial pathogenicity but those of product toxicity.

Numerous vaccines, especially live vaccines, contain nucleic acids, i.e., DNA and/or RNA. Since the last century, these nucleic acids have been injected as constituents of the actual vaccines or administered orally, without causing any fundamental problems. Even the cases described (WHO, 1987) of vaccines inadvertently contaminated with viruses that are pathogenic to animals (yellow fever vaccine contaminated with avian leucosis virus and polio vaccine contaminated with SV 40 virus) have had no adverse effect on the vaccinated persons. The side effects of vaccines that can be observed from time to time are not due to the nucleic acids (DNA/RNA) but to other causes, mostly to failures of the immune system, and chimeras or superorganisms that might jeopardize mankind have of course not occurred.

The above evaluations are consistent with the observations that DNA is constantly released from carcasses of mammals and associated microorganisms, from plants, plankton (PAUL et al., 1987) and from bacteria (MUTO and GOTO, 1987) and the therefore permanent availability of DNA ("gene pool") for exchange within the microflora of the soil (LORENZ and WACKERNAGEL, 1994).

Last but not least, since innumerous species of microorganisms, often in very similar but distinguishable subforms, have developed and stabilized, there is no reason to pursue the view that a unique role should be assigned to the transfer of recombinant DNA between different species different from the role the transfer of non-recombinant DNA plays in nature (VIDAVER, 1985).

8 The *Escherichia coli* Bacteria

Since the turn of the century, and particularly from the work of THEODOR ESCHERICH, it is has been well established that the "bacterium coli communalis" is part of the natural intestinal microbial flora of humans and animals, and that it may be a causative agent for intestinal and extraintestinal infections. These pathogenic *Escherichia coli* (*E. coli*) bacteria are characterized by carrying pathogenic determinants which are not found in the laboratory strain *E. coli* K12. "Normal" intestinal bacteria of the genus *E. coli* represent up to 5% of the over 400 different species of bacteria (in addition to fungi) comprising the normal flora in the human and animal intestine (LINZENMEIER, 1989; DANCYGIER, 1989) where they feed on organic substrates (LEVINE, 1984; LEE, 1985). As the comprehensive literature proves (GLANTZ and JACKS, 1967), they are harmless and in temperate latitudes do not survive outside their host. As *E. coli* bacteria are easily detectable, they are often determined as a "fecal indicator" representative for cholera and typhoid pathogens. Whereas tropical warm waters usually contain little oxygen and therefore are low in nutrients (e.g., contain small amounts of plankton), the determination of coliform bacteria may temporarily be necessary in other waters which are heavily charged with organic material (LOPEZ-TORRES et al., 1988; BERMUDEZ and HAZEN, 1988).

The family of *E. coli* K12 bacteria differs markedly from the naturally occurring *E. coli* wild strains and has been widely used since 1922. Meanwhile, several hundred variants have been developed from the original K12 bacteria. *E. coli* K12 has been very closely investigated as model organism for a wide variety of genetic, medical and biochemical investigations, and at present it is the best studied organism in biochemical and particularly in genetic terms (BACHMANN, 1990).

By international agreement *E. coli* K12 is regarded, both by the scientific community and by the authorities (e.g., see *Recombinant DNA Technical Bulletin 10*, 1987), as harmless to humans, animals, plants and the environment. Therefore, in genetic engineering *E. coli* K12 serves as host for cloning and expression of both prokaryotic and eukaryotic genes. This is one of the reasons, why the early biosafety studies in the mid seventies were conducted with this bacterial strain.

8.1 Pathogenicity of *Escherichia coli*

E. coli bacteria belong to the group of facultative pathogenic microorganisms with more than 90% of the variants characterized as completely non-pathogenic. The following pathogenic variants may cause intestinal infections, i.e., different forms of diarrhea:

- enterotoxic *E. coli* (with cholera-like symptoms)
- enteropathogenic variants (infections in newborns)
- enterohemorrhagic *E. coli* (hemorrhagic colitis)
- enteroinvasive *E. coli* (dysenteric symptoms).

Extraintestinal infections are mainly those of the

- urinary tract (uropathogenic variants)
- very rarely meningitis of newborns or septic infections,
- bovine mastitis and newborn chicken sepsis or
- enterotoxemic infections in pigs.

In contrast to the rare *E. coli* strains that are pathogenic to humans and which can cause the infections listed above, the laboratory strain K12 is not pathogenic (HACKER et al., 1991). Moreover, because of numerous genetic differences, it is even less harmful than the vital non-pathogenic wild strains of the genus *E. coli* which occur naturally in the human or animal intestine.

In particular, *E. coli* K12 cannot form colonies on the skin (ROMERO-STEINER et al., 1990), on the nasal mucous membrane (WINKLER and HOCKSTRA, 1985) nor in the human or animal intestine (ANDERSON, 1975), since it has lost the properties necessary to adhere to the intestinal cells. K12 has lost the adhesive factors (fimbriae-adhesins) for the recognition of oligosaccharide receptors at the eukaryotic cell surfaces, it misses the K antigens necessary for pathogenicity, and the normal "smooth" lipopolysaccharide with the O antigen side-chains (ORSKOV et al., 1977) is reduced. No *E. coli* K12 bacteria have been detected in the feces of volunteers later than 6 days after oral administration (LEVY et al., 1980), and the lethal dose (LD_{50}) in mice is 10^2 to 10^3-fold compared to pathogenic wild strains.

Absent, too, are virulence determinants such as the capsule antigens which are located at several sites of the bacterial chromosome (KLEMM, 1985) as well as the toxins necessary for pathogenicity (LUPSKI and FEIGIN, 1988):

- enterotoxic (heat-labile and/or heat-stable) toxins,
- enteroinvasive toxins,
- enteropathogenic toxins.

E. coli K12 is sensitive to the acidic conditions prevailing in the stomach, and some laboratory strains are extremely sensitive to bile acids because of numerous mutations which cause structural changes in the outer cell membrane (WALLACE et al., 1989). When present in the tissue or in the blood stream, this laboratory strain is very sensitive to defense mechanisms (HAHN and BRANDEIS, 1988) such as lysozyme (TAYLOR and KROLL, 1983) and the immune response by antibodies or the very effective bactericidal principles which are mediated by T-lymphocytes, granulocytes and macrophages (phagocytosis) as well as complement (TAYLOR, 1983). There are also no indications that *E. coli* K12 would cause autoimmune diseases, and even in the case of immunocompromised HIV patients no specific gastrointestinal infections have been observed (RUF and POHLE, 1989).

As the pathogenicity of microorganisms depends on a variety of factors, the specific transfer of other virulence factors such as colicine V or hemolysin is not sufficient to transform *E. coli* K12 into a virulent strain (MINSHEW et al., 1978). Even the transfer of virulence factors from *Shigella flexneri* 2a (SANSONETTI et al., 1983) or from *Yersinia* strains (ISBERG, 1989) into *E. coli* K12 did not restore *in vivo* pathogenicity.

The many functions described above which are missing in *E. coli* K12 are typical for wild type *E. coli* and are independent of each other. The many genes involved in the pathogenicity of the facultative pathogen *E. coli* are distributed over the entire bacterial genome, and there appears to exist no means in nature to transfer these genes to *E. coli* K12, which misses them. Hence the assessment that the introduction of foreign DNA into *E. coli* K12 is safe as well.

This classification of *E. coli* K12 has been practically proven by a study which was carried out for a period of 12 months in a production plant for recombinant interferon alpha-2a (WEIBEL and SEIFFERT, 1993), that asked the questions:

1. Can we observe changes in the antibiotic resistance of the gut flora of the staff?
2. Can we find the production organism or rDNA in the gut flora of the personnel?
3. Can we efficiently destroy the rDNA and resistance genes in the process?
4. Can we find anti-product antibodies in the serum of the employees?

Using various analytical methods, including the polymerase chain reaction, no change of the antibiotic resistance pattern nor rDNA was found in the gut flora, no anti-product antibodies were found, and the rDNA was completely destroyed in the process. These findings are in accordance with the data collected and published in a report by the FONDS DER CHEMISCHEN INDUSTRIE (1993).

In summary, *E. coli* K12 represents a completely non-pathogenic strain of the *E. coli* species, and the properties described above are the basis for the classification of *E. coli* K12 as a biological safety measure in addition to technical safety measures as included in most regulations for genetic engineering.

8.2 Escherichia coli in the Environment

E. coli bacteria disappear from the environment within a very short time, as they are sensitive to dryness and react with UV light (E. coli K12 to an even greater extent) and cannot demonstrably survive neither by a cryptic (non-discoverable) state of dormancy (DEVANAS and STOTZKY, 1986) nor by drifting into deeper layers of soil (COKER, 1983). No cases of transfer to birds (GLANTZ and JACKS, 1967) or long-term survival in sterile soils, in which neither competing microorganisms nor predatory protozoa or nematodes are present (SCHILF and KLINGMÜLLER, 1983), have been observed.

Moreover, it has long been known that "the environment" is not "toxic" nor "lethal" to normal intestinal bacteria such as E. coli, but they cannot compete with "environmental" organisms and hence do not survive. E. coli bacteria are no environmental bacteria, but instead are adapted to the intestines of some mammals. As one gram of human feces contains about 10^6 E. coli bacteria out of a total of about 10^{11} microorganisms (HOBSON and POOLE, 1988), the approximately two million people living in the Rhein/Main district in Germany alone produce 6–8 tons E. coli bacteria per day (other pathogenic microorganisms from hospital effluent and the bacteria liberated by domestic animals not taken into account). The intestinal microorganisms are efficiently removed by competing microorganisms in sewage treatment plants, or they are removed from the environment within a very short time after soil fertilizing with manure or sewage. The environmental problems sometimes associated with the latter examples resulted from the introduction of too large amounts of organic compounds like urea (over-fertilizing) but were not related to the release of enteric microorganisms. Thus "normal" E. coli bacteria, even in large quantities, do not represent any hazard to humans or the environment. This applies to an even greater extent to the comparatively less viable laboratory strain K12 whose chances of survival in nature are still further reduced, as has been demonstrated by many experiments and a

wealth of data. E. coli K12 therefore is classified as a safety strain and has a wide range of applications throughout the world.

9 Risk Assessment of Recombinant Saccharomyces cerevisiae Strains

The generic name Saccharomyces cerevisiae was introduced by MEYEN in 1838 for brewer's yeast. In 1870, REESS distinguished between the beer yeast S. cerevisiae and S. ellipsoideus, which also fermented fruit juices. Later on, S. cerevisiae was described by HANSEN as forming large spheroidal cells measuring 5–11 μm and as to be found in the sediment in hopped wort. VAN DER WALT (1970) classified more than 41 species within the genus Saccharomyces, the most important of which is S. cerevisiae for wine, beer and bread production, production of vitamins, enzymes, nucleic acids and their components, coenzymes, lipids as well as a live therapeutic agent itself.

At present, only ten species within the genus Saccharomyces are accepted (BARNETT, 1992). Many well-known and long established yeast names (specific epithets), though belonging to the genus Saccharomyces, are no longer accepted as those in current taxonomy. Background of this is that no "correct" classification system for Saccharomyces species exists and the rules vary with the techniques used and the prejudices of taxonomists.

Apart from E. coli, S. cerevisiae is one of the best studied organisms in the fields of cell biology, physiology, biochemistry and classical and molecular genetics. S. cerevisiae does not produce any toxins and is widely accepted for human consumption by both the consumers and the authorities. With the advent of recombinant DNA technology, it was soon realized that the new technique offered an enormous potential for producing proteins of high commercial value. E. coli is not the ideal host, since this prokaryote lacks many posttranslational modification systems, does not exhibit the cellular environment to express any eukaryotic protein in the correct folding and generally does not secrete proteins into the culture fluid.

The yeast *S. cerevisiae* has unequivocally been shown to be an ideal host for modern biotechnology. One of the first recombinant proteins used in human medicine, the hepatitis B virus surface antigen (HBsAg), was expressed in *S. cerevisiae*. This recombinant antigen has been proven to be highly immunogenic, and the subunit vaccine is safe and in use worldwide (STEPHENNE, 1990). A number of heterologous proteins have also been expressed in *S. cerevisiae*, including lymphokines, hormones and enzymes. Many of them can now be produced in sufficient quantities with appropriate biological activities. Additionally, an area of major interest is the strain improvement of *S. cerevisiae* by genetic engineering for its use in the food industry, e.g., for the manufacturing of food ingredients especially for baking, distilling, beer- and wine making (LANG-HINRICHS and HINRICHS, 1992).

9.1 Survival in the Environment

The natural habitats of *S. cerevisiae* are ripe fruit and exudates of tree bark. *S. cerevisiae* is not found in the atmosphere nor in aquatic habitats, marine and river sediments, and was only occasionally and temporally detected in soil (DiMENNA, 1957). The yeast is transferred from the fruit to insects – mainly bees, wasps and flies like *Drosophila*. A number of investigations (BENDA, 1964) indicate that *S. cerevisiae* does not survive in the soil during the winter, and that bees rather than wind may play a dominant role in the survival of yeasts and their distribution.

This ecological niche is due to the following biological features of *S. cerevisiae*:

- *S. cerevisiae* mainly metabolizes sugars such as glucose and galactose, but not pentoses like ribose and xylose nor polysaccharides like starch, many fruit acids and sugar alcohols

- *S. cerevisiae* cannot assimilate nitrate
- *S. cerevisiae* tolerates high concentrations of sugars and alcohol
- *S. cerevisiae* tolerates low pH values.

Higher eukaryotic cells like animal tissue culture cells are highly restricted to specific media and growth conditions, which can only be provided in laboratories. Without any doubt, these cells die immediately upon being set free into the environment. This is generally not the case with microorganisms, since these in general exhibit a cell wall which protects the cells against physical stress. Furthermore, they are less demanding as regards nutrition and culture conditions (e.g., temperature and concentration of gas), and some of them can develop cell forms like spores (bacilli), which enable the cell to survive under disadvantageous conditions.

So far, many experiments have been carried out on the survival of bacteria, especially *E. coli*, in their natural environment as well as under laboratory conditions resembling the natural ones (DEVANAS and STOTZKY, 1986). For *S. cerevisiae*, only a limited number of such investigations have been made. LIANG et al. (1982) studied the survival of *S. cerevisiae* laboratory strain GRF104 (not transformed with any plasmid) and observed a decrease in living cells under various, non-sterile conditions. In a comparative study (BRÖKER, 1990), the survival of a prototrophic isolate of *S. cerevisiae* was compared with an auxotrophic laboratory strain and the laboratory strain transformed with a plasmid which confers the expression of a human protein under sterile and non-sterile conditions. The number of yeast cells declined slowly and continuously in sterile water and sterile soil/water suspensions, whereas the yeasts died rapidly under non-sterile conditions – independent of the strain and whether the cells carried a plasmid or not. In a similar study, JENSEN (1986) reported that recombinant *S. cerevisiae* cells decay even faster than the wild-type strain.

These studies indicate that prototrophic as well as the auxotrophic laboratory strains of *S. cerevisiae* can survive in sterile medium for a rather long time (several weeks), but that they are eliminated quickly (within some days or less) upon introduction into the natural environment, because they are susceptible to predation, parasitism, lytic enzymes and toxins produced by the indigenous microbial community. Up to now, no invasion of recombinant *S. cerevisiae* into the indigenous microflora is expected in the case of an accidental release of yeast cells from a production plant, and no or only a limited interference with the existing community of organisms is expected.

Since the essential factors such as the sugar-containing environment of grapes and berries would be lacking in the case of an accidental release of recombinant yeast and due to the specialized eco-niche of *S. cerevisiae*, a heterologous gene expression with respect to risk assessment has no relevance.

9.2 *Saccharomyces cerevisiae* – A Non-Pathogen

9.2.1 Colonization of the Human Gut

The human gastrointestinal tract is colonized with about 10 times more microorganisms (10^{14}) than the human body consists of cells (10^{13}). The habitats of the microorganisms are the ileum, the caecum, the colon ascendens and colon transversum. In addition, bacteria can be found in the duodenum and the jejunum in healthy humans, although they are normally not found in these regions. About 90 to 99% of all microbial flora are bacteria belonging to the *Bifidus* group, the *Bacteroides* group, streptococci, lactobacilli and the *coli* group. Only about 1% or less of the microorganisms include aerobic spore-forming bacteria, clostridia, staphylococci, pseudomonads and yeasts. *S. cerevisiae* does not belong to the normal flora of the gut in contrast to some other yeast species, e.g., the genus *Candida* (*C. albicans, C. glabrata, C. parapsilosis, C. tropicalis, C. krusei, C. pseudotropicalis*). These species belong to the commensal flora of the gastrointestinal tract, and more than 75% of the central-European people carry these yeast species. *Candida* cell numbers of 10^4 per gram of feces seem to be

without any clinical relevance, whereas higher numbers indicate immunological or vascular disorders. Mycological studies of feces isolated from cats and dogs have shown that yeast can frequently be isolated (WEBER and SCHÄFER, 1993). In most cases, *C. albicans* was the yeast species found, but *S. cerevisiae* was also isolated. Nevertheless, there is no indication that cats and dogs serve as a reservoir for infections of humans with yeast.

For the treatment of functional disorders of the intestine such as colitis or diarrhea, commercial preparations (Perenterol®) of live *S. cerevisiae* Hansen CBS 5926 (synonym: *S. boulardii*) are available. More than 60 years ago it was shown that yeast can passage the human stomach without relevant loss in viability (MONTGOMERY et al., 1930). This has been confirmed for *S. cerevisiae* by experiments on the survival of the yeast in artificial stomach juice (GEDEK and HAGENHOFF, 1988) and by several *in vivo* pharmacokinetic experiments in healthy volunteers (KAPPE and MÜLLER, 1987). When baker's yeast cells are introduced into the gastrointestinal tract of mice, nude mice, rats, cynomolgus monkeys or man, they do not multiply, but are eliminated with the feces either live or dead within a few days (PACQUET et al., 1991). The authors could not find any evidence for the passage from the intestinal lumen to any other part of the body. However, in the hypothetic case that *S. cerevisiae* cells would pass from the intestine to the lymphatic system, the majority of the cells would probably be persorbed in the liver and would therefore never reach the peripheral sites, due to mannose receptors of the liver macrophages clearing the yeast cells effectively from the circulation. In the early stage of infection, the mannan-binding protein (MBP) found in the serum may play a role prior to the generation of the specific humoral and cellular defense system directed against yeasts. The MBP is a serum lectin, which can act as an activator of the classical pathway of complement after binding to the suitable carbohydrate ligands (THIEL, 1992).

9.2.2 Possible Signs of Pathogenicity

Many mycelial fungi and yeast, which are non-pathogenic for healthy humans or which even belong to the normal commensal flora of the human skin or the gut, such as *Candida* or *Torulopsis*, emerge as major nosocomial pathogens in seriously weakened patients. These opportunistic microorganisms include several species of the genus *Candida*, *Fusarium* and *Trichosporon* as well as *Malassezia furfur* (MARCON and POWELL, 1992), *Cryptococcus neoformans* and *Torulopsis glabrata*. High-risk patients include those with leukemia, solid tumors and leukopenia, bone marrow transplant recipients, patients with gastrointestinal diseases, burn patients, HIV patients and premature infants (PFALLER and WENZEL, 1992).

Despite considerable human oral uptake of foods produced by means of adding live *S. cerevisiae* and the occasional presence of this yeast in normal oral and gastrointestinal tract flora, *S. cerevisiae* is generally regarded as non-pathogenic. Furthermore, live *S. cerevisiae* CBS 5926 (Perenterol®) has been orally administered as a prophylaxis and therapy for diarrhea, enteritidis and colitis for many years to both adults and children with good results (RIECKHOF, 1993).

There are only very few case reports on *S. cerevisiae* as a human pathogen. The first report concerned a septicamia probably caused by *S. cerevisiae* (ESCHETE and BURTON, 1980) nosocomially in a hyperalimented burned man. Fungemias caused by *S. cerevisiae* have been described in immunocompromised patients with chronic renal failure (CIMOLAI et al., 1987), mitral prosthetic valve (RUBINSTEIN et al., 1974), postoperative peritonitis (DOUGHERTY and SIMMONS, 1982), leukemia (ENG et al., 1984) and in a patient with AIDS (SETHIA and MANDELL, 1988). Vulvovaginal infections have also been described (NICOLI et al., 1974). This list is of course not complete, and the number of cases will increase with the number of immunocompromised humans studied, especially tumor and AIDS patients. However, it must be emphasized that there is no indication classifying *S. cerevisiae* as a potential pathogen for healthy persons.

The interconversion of a yeast form to a filamentous form is typical of many pathogenic fungi, e.g., *Candida albicans* and *Malassezia furfur*. GIMENO et al. (1992) reported on a dimorphic transition of *S. cerevisiae* which involved changes in cell shape and resulted in invasive filamentous growth under nitrogen limiting conditions. Thus, the genetic potential of *S. cerevisiae* in its filamentous state has yet to be established.

MCCUSKER and DAVIS (1991) have characterized an *S. cerevisiae* strain isolated from an AIDS patient. This clinical isolate has some unusual properties, as it can grow at 43°C, a temperature at which laboratory strains are completely unable to grow. Furthermore, the isolate is able to secrete a protease as determined by liquefaction of gelatin. In this context it is interesting to note that protease secretion is thought to be a virulence factor in pathogenic fungi. In addition, the authors found that the cell and colony morphology can switch between multiple colony- and cell-type forms. So far, there is no evidence of *S. cerevisiae* causing disease, and no death has been observed, when clinical *S. cerevisiae* isolates were tested in experimental animal infections. These results were observed despite the fact that large numbers of cells (10^6–10^7 colony-forming units) were injected intravenously into mice.

It is very unlikely that clinical strains could cause problems for healthy persons. The "virulence" of the clinical *S. cerevisiae* isolates appears to be a quantitative trait polygenic with complex additive and epistatic effects. Realistically, a person who is so debilitated as to harbor *S. cerevisiae*, is at a far greater risk of infection from commensal yeast (e.g., *Candida albicans*) or other opportunistic or pathogenic microorganisms found in the environment.

There have been reports of patients suffering from repeated attacks of alcoholic intoxication without any prior intake of alcohol. This is called "intragastrointestinal alcohol fermentation syndrome" or "autobrewery syndrome" (KAJI et al., 1984). This syndrome can be caused by yeast, mainly those species belonging to the *Candida* group, but also by *S. cerevisiae*, if they have proliferated abnormally in the gastrointestinal tract. This rare syndrome consists of the following essential factors:

- parasitism and abnormal proliferation of the causative agent (yeast), which is capable of alcohol fermentation in the alimentary tract,
- abnormal stagnation of foods caused by organic or functional disorders of the alimentary tract,
- intake of a carbohydrate diet as the substrate for the alcohol fermentation,
- a low threshold of the host patient to alcohol.

High ethanol concentrations of up to 19.7 mg/dL blood in a patient have been reported among a group of patients with a variety of disorders and clinically suspected yeast fermentation who had been orally loaded with large amounts of glucose after an overnight fast (HUNNISETT et al., 1990).

The autobrewery syndrome has also been discussed as playing a role in the etiology of "sudden infant death". BININ and HEINEN (1985) studied the ethanol production of some yeast species including *S. cerevisiae*, when grown on infant food formulas and supplements or Coca Cola. The highest alcohol production was determined, when *S. cerevisiae* was cultivated in infant food preparations, but additional studies are needed to determine if there is a relationship between ethanol production in the gastrointestinal system and of "sudden infant death".

9.2.3 Allergy

Allergies to fungi are an increasing problem and are correlated to the increase of technically produced hydrolytic enzymes from fungi for industrial purposes or as components of food. Furthermore, hydrolases are added to creams, perfumes, toothpastes, etc. and are used in the production of textiles. Several hundred tons of glucoamylase from *Aspergillus* or *Rhizopus* are produced every year, and hydrolases, beta-amylase from *Aspergillus*, cellulase from *Penicillium* or invertase from *Saccharomyces* are produced and

used in high amounts in various industrial productions (SCHATA and JORDE, 1988).

In contrast to many other mycelial fungi and yeasts (e.g., *Candida*) (GUMOWSKI et al., 1991), *S. cerevisiae* does not seem to be a causative agent of allergic reactions. Even after more than 35 years of Perenterol® production, no allergies have been observed in the workers of the respective production facilities (GEDEK and AMSELGRUBER, 1990). Whenever allergic reactions to beer have been accounted to *S. cerevisiae*, thorough investigations led to the conclusion that the allergens originated from the *Fusarium* species which enters the beer via the hop (JORDE, personal communication). Besides this assessment, the following two observations need to be discussed:

BALDO and BAKER (1988) reported on allergic reactions to enolase from yeast and showed a high cross-reactivity between the enolases from *S. cerevisiae* and from *Candida albicans*. Recently, the cDNA of the enolase from *C. albicans* has been cloned (SUNDSTROM and ALIAGA, 1992). The analysis of the predicted amino acid sequence of the *C. albicans* enolase showed strong conservation structure with the enzyme from *S. cerevisiae*. The enolase is a major circulating antigen in patients with disseminated candidiasis, although the enzyme is not present on the surface of *C. albicans*. Regardless of the true source of the fungal hypersensitivity, the authors conclude that *S. cerevisiae* enolase may prove to be useful for diagnosis of fungal hypersensitivity in at least a population of subjects showing symptoms of inhalation of molds.

Raised levels of antibodies to *S. cerevisiae* have been found in patients with Crohn's disease, but not in patients with ulcerative colitis when compared with normal controls (MCKENZIE et al., 1990; LINDBERG et al., 1992). The patient sera reacted preferentially to a cell wall-associated, water-soluble and heat-stable glycoprotein of *S. cerevisiae* (HEELAN et al., 1991). It has been suggested that the immunological sensitization can occur in man through ingestion of products from brewing and baking. Recently, GIAFFER et al. (1992) concluded from their findings that the presence of IgG antibodies to *S. cere*-

visiae is characteristic but not specific to Crohn's disease. Nonetheless, the immunological basis for the association of an immune response to this *S. cerevisiae* antigen in Crohn's disease and the pathogenic importance of raised IgA antibody levels in Crohn's disease remain to be explained. Recently, SAVOLAINEN et al. (1992) and KORTEKANGAS-SAVOLAINEN et al. (1993) reported that the severity of atopic dermatitis, studied in a population of more than 200 patients, correlated with the skin prick test response to *S. cerevisiae*. The correlation in turn correlated with serum total IgE and skin prick test response to other yeast species studied, e.g., *C. albicans, Crypotococcus albidus* and *Pityrosporum ovale*. Because there was a skin response to all yeast species tested, one may conclude that the causative agent was not *S. cerevisiae* but another yeast species with cross-reacting antigen(s), e.g., the enolase.

One of the most commonly used recombinant protein isolated from *S. cerevisiae* may be the surface antigen of the hepatitis B virus (HBsAg). Intensive studies on the immunogenicity and the reactogenicity of the yeast-derived hepatitis B vaccine have shown that there were no significant differences in the immunogenicity and reactogenicity when comparing the recombinant and the plasma-derived vaccine. In addition, no signs of vaccine-induced hypersensitivity were observed (WIEDERMANN et al., 1987).

No significant changes in the levels of anti-yeast IgG or IgE and no new species of anti-yeast antibodies were observed after vaccination with the recombinant hepatitis B vaccine (WIEDERMANN et al., 1988; ANDRÉ, 1987; DENTICO et al., 1992). This indicates that putative yeast contaminants apparently play no major role after immunization with the recombinant hepatitis B vaccines. Comparatively good toleration should be expected when other recombinant proteins isolated from *S. cerevisiae* come into clinical use, which have been highly purified.

9.3 Plasmids and Gene Transfer

9.3.1 Expression Plasmids

The plasmid vectors which are used to transform *S. cerevisiae* are normally *E. coli/S. cerevisiae* shuttle vectors. The prokaryotic part usually originates from pBR322 or derivatives thereof and therefore contains the ColE1 type of origin of replication, the ampicillin resistance marker and/or the tetracycline resistance marker. This arrangement allows *E. coli* to be used as a primary host cell for various manipulations associated with DNA. The eukaryotic part of the shuttle vectors is based on two types of replication systems. The first is dependent on chromosomal DNA fragments isolated from *S. cerevisiae* and contains autonomously replicating sequences (ARSs). The second type of shuttle vectors is based on elements derived from the endogenous yeast 2 μm plasmid.

The ARS-based vectors can transform *S. cerevisiae* efficiently and can replicate to a high copy number within the nucleus of the cells, but they are mitotically very unstable under non-selective and even under selective conditions. Therefore, ARS-based vectors are currently of limited use for biotechnological purposes. The instability of the ARS vectors can be overcome by the addition of a segment of DNA originating from the yeast centromere. These *CEN* vectors also transform *S. cerevisiae* with high efficacy, are almost entirely stable and are maintained at a copy number of 1–2 per cell. This low copy number is of disadvantage in most biotechnical processes, because the cloned gene of interest consequently has the same low copy number. Hence the production yields are relatively low.

Most of the expression vectors are based on the 2 μm plasmid. This plasmid is present in most laboratory strains of *S. cerevisiae* at a copy number ranging from 20 to 200 per cell. It is a circular molecule of 6318 bp containing two inverted repeats of 599 bp, and the replication of this plasmid and its maintenance in the cell are dependent on two *cis*-acting elements (origin of replication and REP3 or STB) and two *trans*-acting elements (REP1 and REP2). In addition, there is a major coding sequence, the *FLP* gene, encoding a recombinase which mediates site-specific recombination between the inverted repeats (MEACOCK et al., 1989). Vectors based on the essential function of the natural 2 μm plasmid can be used to transform *S. cerevisiae* at high frequencies and are maintained at copy numbers ranging from 20 to 400 per cell.

The expression plasmids are generally transformed into laboratory strains which are auxotrophic for amino acids (e.g., leucine, tryptophan) or bases (e.g., uracil, adenine), and the selection is carried out by the respective functional genes cloned into the plasmid. That means, a yeast strain which has the genetic markers leu2 or ura3 can be transformed with a plasmid which harbors the *LEU2* gene coding β-isopropyl malate dehydrogenase or which carries the *URA3* gene coding for orotidine 5-decarboxylase. Strains with double or triple markers (e.g., leu2, ura3, trp1) and which are transformed with an expression plasmid that carries only one selection marker (e.g., *LEU2*) are still auxotrophic for the other marker(s); in this case: ura3 and trp1. Therefore, the transformants have to be supplemented with the amino acids or bases which they cannot synthesize when grown in minimal media; but these transformants usually have no markedly retarded growth rate or other disadvantages under laboratory conditions.

Besides the selection of transformants by means of complementation of defects in enzymes of the biosynthetic pathway, vectors have been developed with dominant selection markers. These plasmids confer, e.g., resistance against antibiotics like G418 or heavy metals like copper. Nevertheless, such selection systems are only rarely used, and such plasmids are mainly introduced when prototrophic strains have to be transformed where no selection based on complementation of basal enzymatic activities is possible. When yeast strains are transformed with plasmid to express heterologous proteins either on laboratory or industrial scale, this is normally not the case.

Besides the essential sequences which are needed for the selection and propagation in *E. coli* and *S. cerevisiae*, these shuttle vectors contain yeast-specific promoters (transcription initiation elements) and transcription-termination elements. These DNA elements are in nearly all cases derived from genes of *S. cerevisiae* to fulfill optimal expression. It is unusual to observe transcription of a heterologous gene in *S. cerevisiae* because baker's yeast is usually unable to accurately initiate transcription on a heterologous promoter – even when it originates from a eukaryotic cell. In most cases, either no transcription takes place or an aberrant transcription, e.g., from the promoter of the *Dictyostelium* discoidin I gene (JELLINGHAUS et al., 1982) and the bean phaseolin gene (CRAMER et al., 1985), can be obtained.

Although promoters of *S. cerevisiae* are highly efficient only in this species, there are examples of *S. cerevisiae* genes being expressed in other yeast species. These include genes from the biosynthetic pathway, e.g., the *LEU2* gene and the *URA3* gene which are active in the fission yeast *Schizosaccharomyces pombe* and can complement leu1 and ura4 auxotrophies in this species. Constitutive promoters from the *S. cerevisiae ADC I* gene (alcohol dehydrogenase I) and *CYC I* gene (iso-1-cytochrome c) have even been used to allow the expression of heterologous proteins in *S. pombe* under the control of these transcription initiation elements (BRÖKER et al., 1987). Although *S. cerevisiae* promoters are normally not functional in prokaryotes, it cannot be totally excluded that certain yeast DNA sequences may have promoter activity in bacteria. KORMANEC (1991), e.g., identified a DNA fragment from *S. cerevisiae* by chance which acts as a promoter in *E. coli*. This DNA sequence, part of the *S. cerevisiae POP1* gene, exhibits the optimal promoter consensus structure (-35)TTGACA-17 ± 1 bp$-(-10)$TATAAT. This may be interesting in evolutionary terms.

Yeast promoters contain many of the canonical sequences common to higher eukaryotic promoters, although the locations of these sequences, namely the "TATA" box and the "CAAT" box, are more heterogeneous than observed in higher eukaryotes with respect to the mRNA start site(s). Yeast genes also contain upstream *cis*-acting elements (upstream activator sequences; *UAS*), which modulate transcription.

The *GAL1* and *GAL10* divergent promoters contain a *UAS* sequence at which the GAL4-encoded protein binds for optimal transcription. In *S. cerevisiae*, the *GAL1* and *GAL10* promoters are repressed in the presence of glucose, but are induced when glucose is replaced by galactose. This is regulated by the binding of the GAL4-encoded protein at the *UAS* of the *GAL1/GAL10* promoter region (JOHNSTON, 1987). Whenever this positive regulator protein interferes with the GAL80 encoded regulated protein in the presence of glucose or when the GAL4 protein is absent, no transcription of the *GAL1* or *GAL10* promoter-controlled expression can be observed. This explains why promoters like the *GAL1/GAL10* are much more species-specific than constitutive promoters from glycolytic genes. In fact, the *GAL1/GAL10* promoter is not even active in the fission yeast *S. pombe* (which does not synthesize a protein functionally related to the *S. cerevisiae* GAL4 encoded protein), unless the *S. cerevisiae GAL4* gene is cloned in addition into the *GAL1/GAL10* promoter containing plasmid (BRÖKER, 1990).

9.3.2 Gene Transfer

Gene transfer between microorganisms has been investigated intensively since the discovery of the phenomena of bacterial transformation, conjugation and transduction in the 1950s and 1960s. The frequency of gene transfer and its implication for risk assessment, especially in respect of gene transfer between bacteria, has been discussed in Sect. 7.

Among the possible gene transfer mechanisms within yeast, no transduction has been detected so far, and no yeast viruses have been identified. Retrotransposons (or viruslike particles, VPL) are well known in yeast and have been intensively characterized, but these VPLs are non-infectious, and no horizontal transfer has been detected (WICKNER, 1992).

Transformation of *S. cerevisiae* with plasmid DNA is routinely carried out in molecular biology laboratories. *S. cerevisiae* cells are not competent to pick up DNA during certain stages of their life cycles as, e.g., *Bacillus subtilis* is. On the contrary, tedious procedures have to be carried out in order to transform *S. cerevisiae* by plasmid DNA. To make *S. cerevisiae* cells competent, they have to be grown under special conditions, the cells are then treated in special buffers, containing lithium salts and/or polyethyleneglycol and are heat-shocked. By improvements of this transformation protocol, up to 10^6 transformants/µg plasmid DNA can be obtained. Alternative transformation procedures are either the electroporation (DELORME, 1989) or the transfer of DNA into cells which have been converted into protoplasts upon digestion of the cell wall with lytic enzymes (HUBBERSTEY and WIDEMAN, 1991). An inspection of the transformation protocols published by the above mentioned authors makes evident that the transformation of *S. cerevisiae* cells by the mechanism of transformation needs special requirements. These can only be provided under artificial laboratory conditions. No DNA transformation of *S. cerevisiae* cells has been reported so far where the cells have not been made competent.

Conjugations between bacteria and yeast have been observed recently under experimental conditions, between *E. coli* and *S. cerevisiae* (HEINEMANN and SPRAGUE, 1989) and *Schizosaccharomyces pombe* (SIKORSKY et al., 1990). The DNA transmission mechanism from *E. coli* to yeast is physically and genetically comparable to a conjugal DNA transfer between bacteria. The frequency at which DNA is experimentally transmitted to yeast by bacterial conjugation is similar to the frequency of transmission observed in crosses between bacteria. During the conjugational event, the thick yeast cell wall must be bypassed to allow productive cell to cell contact. This might occur during a discrete stage of the cell cycle. The transfer and the establishment of the plasmid in *S. cerevisiae* does not need yeast-specific *ars* sequences for stable replication of the incoming DNA. NISHIKAWA et al. (1992) have shown that foreign plasmid DNA can integrate into the recipient host chromosome by recombination events. However, integration is rare, probably because, like *E. coli*, *S. cerevisiae* demands extensive sequence homology between molecules before it permits their recombination.

Bacteria–yeast conjugation is a laboratory phenomenon of horizontal gene transfer. At present, this process is of uncertain biological significance. In the natural environments of *S. cerevisiae*, e.g., in ripening fruit, many bacteria are found which carry conjugal plasmids. In the case of such a conjugal DNA transfer from prokaryotes to yeast, the foreign DNA must enter the nucleus and must be replicated in order to be passed onto the yeast daughter cell(s). There are some examples known from the literature that prokaryotic sequences can be replicated in *S. cerevisiae*, e.g., broad host plasmids like RP4 and staphylococcal plasmids (GOURSOT et al., 1982). However, such DNA sequences are unstable, replicate only transiently and are not established in the new host cell.

Conjugational transfer of genes from yeast to bacteria or from one yeast to another have not been reported so far. By means of plasmids with the right choice of replicons and selection markers such an event should be easily detectable under laboratory conditions if it occurred.

10 Biosafety Aspects of Animal Cell Cultures

The use of animal cells in tissue culture for the production of pharmaceuticals and as a tool of biomedical research has a history which dates back at least five decades. Issues of biological safety pertaining to the use of tissue culture are therefore not new (PETRICCIANI, 1987). The recent introduction of recombinant DNA techniques, however, has broadened and widened the scope of tissue culture applications to such an extent that safety questions must be reconsidered in the light of these developments.

Mammalian cells in culture are currently employed for the following purposes:

- propagation of viruses for vaccines production,
- tools for biomedical research, i.e., for the study of transcription and replication patterns as well as mechanisms of growth control,
- sources of nucleic acids for the preparation of genomic and cDNA libraries,
- sources of proteins,
- sources of recombinant eukaryotic proteins which cannot be expressed in bacteria, i.e., proteins of high molecular weight and with extensive posttranslational modifications.

The safe handling of mammalian cells critically depends on the origin and the status of the cells in question. In tissue culture we distinguish between primary cell culture and established permanent cell lines. Primary cultures are obtained from explants of tissue or from organs. They can be propagated only for a very limited number of passages and have been used to produce the early polio vaccines (summarized by HILLEMAN, 1990). The use of such cells may be quite hazardous because, if prepared from free-living animals, they may be contaminated with pathogenic viruses, bacteria, mycoplasms, fungi, and other parasites of the donor organism. There are only very few instances of laboratory infections, however, which could be traced back to contaminated cell cultures. Among these are the outbreaks of Marburg virus infections in the sixties as well as the contamination of polio vaccines with the oncogenic Simian virus SV40. Such infections can arise in animals even after long periods of quarantine and are particularly disturbing in view of the constant danger of unknown and therefore non-detectable viruses.

Primary cells, if kept in culture for extended periods of time, undergo a "crisis" from which they may recover as permanent cultures that can be cultured almost indefinitely. This is true for cell lines derived from the Syrian hamster, the Chinese hamster, and the mouse. Cells, however, derived from humans and chicken, invariably fail to survive cell culture conditions and thus do not give rise to non-malignant continuous cell lines. In these cases, primary cell cultures can be immortalized and thus developed into continuous cell lines by transformation with tumor viruses or oncogenes. Immortalized cells can be obtained also directly from tumors and often develop into permanent malignant cell lines. Hybridomas, which are hybrids of normal and transformed lymphocytes, also belong to this category. Most permanent and malignant cell lines are aneuploid, i.e., they do not contain the proper diploid set of chromosomes. Malignant as well as non-malignant continuous cell lines possess features that are particularly relevant to the issue of biological safety:

- the transformed character of some of these cells,
- the possible contamination with viruses and
- the risks associated with the use of viral vectors.

10.1 Risks Associated with the Transformed Character of Tissue Culture Cells

Protooncogenes are normal constituents of tissue culture cells. They are part of endogenous retroviral genomes present in birds and mammals, i.e., in many, if not all, tissue culture lines derived thereof. Germ line DNA from all inbred strains of mice contains about 50 copies of murine leukemia virus-related sequences (STOYE and COFFIN, 1988). Among the non-defective proviruses we distinguish ecotropic, xenotropic and polytropic viruses. Ecotropic viruses can replicate only within their host cells, xenotropic viruses replicate only on non-host cells, while polytropic viruses in general have a very broad host range. It is thought that most, if not all, inbred strains of mice share a single female ancestor (FERRIS et al., 1982). The observed differences in the proviral content of different inbred strains which have consistently been observed and documented thus must arise by (a) the mobility of some of the proviruses and (b) by the differential fixation of the proviruses during different inbreeding strategies. There is no doubt that the horizontal transfer of xenotropic proviruses can cause insertional mutagenesis and that these viruses can become partners in recombinational processes

leading to novel types of retroviruses, i.e., variants with a different host range.

The human genome contains thousands of endogenous retroviral copies. As far as we know they are all defective, i.e., they contain only pseudogenes or lack parts of the elements required for proviral activation and replication. Their ancestry has been studied in considerable detail (MARIANI-CONSTANTINI et al., 1989). Some are related to rodent intracisternal A-particle genes (ONO, 1986), some to simian sarcoma virus sequences. Recently it was shown that some even are related to human immunodeficiency virus type 1 (HORWITZ et al., 1992). Two of these endogenous sequences, named EHS-1 and EHS-2, display sequence similarity with the domain of the envelope protease cleavage site of HIV-1 and with the overlapping reading frame for Rev and gp41.

A variety of reports have demonstrated that endogenous proviruses can be expressed in both normal and malignant human cells (reviewed in LARSSON et al., 1989). Expression of endogenous retrovirus genes can also be induced by X-ray irradiation and certain mutagens, e.g., 5-azacytidine (HSIAO et al., 1986), although in human cells many if not most of the proviral genomes are unavailable for expression. Infectious retroviruses corresponding to endogenous proviral genomes thus are probably quite rare and therefore do not pose health hazards to anyone working with human tissue culture cells. The situation is quite different in the rodent system. Here endogenous proviruses can be readily induced and activated (STOYLE and COFFIN, 1987). Using RNA fingerprinting techniques and PCR methodologies, they can be easily identified and followed analytically (SHIH et al., 1991).

The transformed character of tissue culture cells can also be induced by DNA tumor viruses, e.g., papilloma and certain herpes viruses. Papilloma viruses cannot be propagated in most tissue culture cells (ZUR HAUSEN, 1981, 1991) except in keratinocytes (MUNGAL et al., 1992). Bovine papilloma virus, however, is able to transform bovine and murine cells in culture. The transformed cells generally carry the viral genome in an episomal form. In humans, common skin warts do not convert into malignant tumors while some papillomas, e.g., certain types of genital warts, have strong tendencies to undergo malignant conversion. Tissue culture cells derived from these tumors, for example, HeLa and their derivatives, but also primary cells derived from cervical carcinomas, contain human papilloma virus genomes in both integrated and episomal forms (SCHWARZ et al., 1985; BEAUDENON et al., 1986).

Some human tumors, for example, Burkitts lymphoma and human nasopharyngeal carcinoma, are thought to be connected with a human herpes virus, the Epstein-Barr virus (EBV) (MARX, 1989). Cancerous cells derived from these tumors, for example, B- and T-lymphocytes, harbor the EBV genome in a circular form. EBV infection is often used to immortalize human B-cells and to establish continuous B-cell lines. A well known example of such a cell line is the Namalva line which is widely used in vaccine production. It is transformed by EBV and carries an EBV genome which, however, is thought to be defective. Since EBV is also the cause of infectious mononucleosis in man, cell lines derived from EBV-derived tumors must be treated under the biosafety level conditions required for experiments with EBV virus (biosafety level 2).

There is little doubt then that genomic DNA from tumor cell lines carrying non-defective endogenous retroviruses and/or oncogenic papilloma viruses or EBV genomes must be treated with considerable caution. An immediate skin contact certainly must be avoided. Several reports in the literature attest to these difficulties. DNA derived from the cloned viral gene v-src, if infected into chicken, can induce tumors at the site of injection (FUNG et al., 1983), and human T24 H-ras gene preparations induce tumors when rubbed onto the surface of the skin of mice (BURNS et al., 1991). In these cases, however, homogeneous oncogene preparations were used, while genomic DNA would contain eight orders of magnitude less of a particular oncogene at a given DNA concentration (LÖWER, 1990). In addition, oncogenesis is a multistep process both in vitro and in vivo. The risk of developing tumors by working with genomic DNA even from malignant cell lines must be regarded as remote.

Nevertheless, the "Pasteur incident" is not forgotten. It relates to the observation of five fatal rare cancers in 1986 at the Institut Pasteur, Paris, among workers younger than 50. The matter has never been fully elucidated, although a mortality study among 3765 people who had worked for at least 6 months in the Institut Pasteur between 1971 and 1986 was initiated following these incidences (CORDIER, 1990). The study showed that the total mortality was less than expected compared with national rates, but that certain groups of people working mostly in bacteriology laboratories had an excess risk of death from bone, brain and pancreatic cancer. No connection was found as to the suspicion that some of the observed cases might be linked to chemical carcinogenesis. Nevertheless, both British and German authorities have recently given a strong advisory to avoid the simultaneous use of oncogenic nucleic acids and chemical or physical carcinogenic agents. In this context it may be reassuring to know that DNA from tumor samples of two molecular biologists who had died from malignant lymphoma and adenocarcinoma of the colon, respectively, and who had worked with SV40 virus as well as SV40 transformed cells did not contain SV40 DNA, even when assayed with PCR techniques. Since transformation with DNA tumor viruses requires the viral genome to be present in the transformed cells, SV40 cannot have played a part in these tragic incidences (HOWLEY et al., 1991).

Direct contact should also be avoided with intact, full-length proviral DNA from retroviruses. In a recent study, recombinant bacteriophage lambda DNA containing SIV-proviral DNA was injected intramuscularly into M. fascicularis monkeys (LETVIN et al., 1991). All animals not only seroconverted within several weeks of inoculation, but also intact SIV particles could be isolated from peripheral blood lymphocytes of most of the transfected animals. The study suggests that viral nucleic acids are not only infectious in tissue culture but also in vivo in their permissive hosts.

10.2 Contamination of Cell Lines with Viruses

Contaminations of tissue culture cells with viruses may have various origins. They may be the results of acute or latent infections of the donor, and hence cells and organs taken from such donors are most likely also infected with these agents. Other potential sources for contaminations are medium additives, for example, serum, hormones, growth factors. Cross-contamination from other sources within the laboratory may also be a factor to be considered.

The potential hazards of primary human cell lines are mainly associated with human immunodeficiency virus (HIV), hepatitis B (HBV) and hepatitis C virus. PCR technology and sensitive immunoassays can be employed to readily ascertain whether such infections are to be expected in material derived from particular donors.

Primary cells from non-human primates can be contaminated by simian foamy virus (SFV) (TEICH, 1984) and simian immunodeficiency virus (SIV) (DANIEL et al., 1985). Most African green monkeys kept in captivity are infected with these viruses (KANKI et al., 1985), although they do not show any symptoms of disease. When transferred to macaques, an Asian old world primate, the lentivirus SIV, however, induces HIV-like symptoms in these animals. While foamy viruses are not known to induce diseases in natural herds as well as in laboratory animals (COFFIN, 1990), SIV-related lentiviruses have been isolated from humans in West Africa. In order to minimize hazards associated with the use of primary cultures derived from non-human primates, these cells should be kept in biosafety level 2 containment facilities.

Among the risks of tissue culture associated with exogenous sources looms the specter of contamination from serum. In most cases its source is from cattle, as fetal calf or calf serum. While many of the more common bovine virus disease agents can be assayed readily and therefore avoided, a major concern remains with the potential risk associated with bovine spongiform encephalopathy (BSE), one of four prion diseases recognized in animals. The disease was first identified in Britain in November 1986 and is thought to be caused by cattle feed containing ground offal from scrapie-infected sheep carcasses. The incubation time of BSE ranges from 2 to 8 years. Since the British government banned the feeding of meat and bone to ruminants in July 1988, it may take until the turn of the century until it disappears, unless it becomes endemic, as scrapie in sheep. The origin of the prion diseases is controversial (PABLOS-MÉNDEZ et al., 1993). In the absence of diagnostic tests, contamination of tissue culture can only be avoided by relying on sera from herds with no known history of BSE or on serum-free media (KIMBERLIN, 1990; HODGSON, 1993).

A final source of concern is cross-contamination with viruses within the laboratory. The most recent and conspicuous examples concern the relationship between different HIV isolates, some of which having been identified as laboratory contaminations (MULDER, 1988; ALDHOUS, 1991). Such contaminations are difficult to avoid, especially when handling new isolates, which can readily be outgrown by contaminating viruses present in the laboratory and already adapted to tissue culture conditions. Viruses can survive in aerosol droplets as well as in glass-ware. Rigorous rules and schedules thus have to be established in laboratories working with novel viral isolates.

Finally, the issue of accidental infection of laboratory workers exposed to contagious viruses remains enigmatic. It is not new but has recently been revived due to the expanding intensity of laboratory work with HIV. A recent study assessing such risks concludes that strict observance of biosafety level 3 containment as well as constant refinement of procedures should be able to contain the problem (US DEPARTMENT OF HEALTH AND HUMAN SERVICES, 1988; WEISS et al., 1988; MILMAN and D'SOUZA, 1990).

10.3 Safety Aspects of Viral Vectors Used in Eukaryotic Tissue Culture

Vectors are vehicles used in recombinant DNA technology to transfer genes into recipient cells. They are thus necessary prerequisites for the establishment of cell lines producing recombinant proteins including vaccines, as well as for somatic gene therapy (MILLER, 1992). In mammalian systems these vectors are mainly based on viral genomes. These include DNA viruses such as adeno-associated viruses, papova- and herpes viruses, but also poxviruses and retroviruses. In most cases these viral genomes are used in a form which lacks possible virulence factors and thus makes them considerably safer to use than the corresponding viruses themselves. Due to the different strategies these viruses use to invade their host cells, these virulence factors may be oncogenes, enzymes of nucleotide metabolism, DNA sequences serving as targets for homologous recombination with endogenous host cell sequences, and others.

More often than not, vectors contain antibiotic resistance genes. Such genes when present, for example, as DNA impurities in proteins prepared from recombinant cells, could potentially be transferred to a patient and/or his/her bacterial flora. This chain of events not only has to be considered as highly improbable, it can and will be efficiently avoided by adhering to WHO or even better standards as to the amount of DNA impurities permitted in drugs. Nevertheless, the problem is not negligible because of the current medical crisis spawned by the spread of drug resistance following an uncontrolled use of antibiotics in the past decades (NEU, 1992). It would certainly be advisable not to aggravate the problem and to avoid the use of resistance markers whenever possible.

10.3.1 Vectors Derived from Poxvirus

Poxviruses, in particular vaccinia virus, have recently enjoyed a growing interest as efficient vectors to express foreign genes for the production of functional and posttranslationally modified proteins, often from other viruses (SPEHNER et al., 1991; MORRISON et al., 1991). Recombinant poxviruses are generated by *in vivo* homologous recombination of the insertion vector with an appropriate virus strain, e.g., the WR (Western Reserve) strain of vaccinia virus. The general enthusiasm about the success of these vectors, however, has been tempered by safety questions regarding vaccinia virus. Although the chances of accidental infections with recombinant vaccinia virus under biosafety level 2 conditions can be regarded as low, such infections can occur and have occurred in the past. As in smallpox vaccination such infections may result in adverse reactions, e.g., encephalopathy or exanthematous disease. Efforts have thus been undertaken recently to develop highly attenuated vaccinia strains which have lost their virulence. This was achieved by deleting 18 genes from the viral genome, including some genes involved in nucleotide metabolism, the thymidine kinase gene and the gene for ribonucleotide reductase (TARTAGLIA et al., 1992). This will prevent replication of the virus in the quiescent cells of the nervous system which do not express the corresponding cellular enzymes. It has also been proposed to use non-replicating poxviruses as vectors, e.g., canary poxvirus (CADOZ et al., 1992). Since, however, these grow quite inefficiently on their permissive host cells, chicken embryo fibroblasts, it remains to be seen whether this particular approach will eventually prevail.

In summary, it can be stated that the use of the highly attenuated vaccinia strains as expression vectors will reduce the hazards associated with the use of vaccinia virus and thus should be strongly recommended and encouraged. This is particularly true because many countries do not preserve registered smallpox vaccines anymore since the WHO has declared smallpox as eradicated from the globe in 1980.

10.3.2 Vectors Derived from Retroviruses

Recombinant retroviruses are widely used as vectors in human somatic gene therapy.

The current strategies employ defective genomes from which all viral genes have been removed. They cannot replicate in their respective target cells and simply serve as one-way vehicles to transfer a foreign gene into a desired genome. Potential risks associated with these vectors are of two kinds: insertional mutagenesis and helper-virus production. Retroviruses integrate at random chromosomal sites. The integration event therefore may interrupt important genes or may lead to the undesired activation of neighboring genes, i.e., cellular oncogenes (CULLEN et al., 1984). This has been observed in mice in the course of infection with replication-competent retrovirus, but never during work with replication-deficient genomes. Nevertheless, the problem cannot be avoided as a matter of principle and will only be eliminated if retroviruses were to be designed which integrate preferentially at known and specific chromosomal sites.

The problem of helper virus production relates to the difficulty of having to propagate and package defective viral genomes which themselves lack the viral genes and their products required for packing into viral particles and for their propagation. In principle, these functions can be provided by coinfection with appropriate helper viruses. Since this is undesirable, the problem has been circumvented by using "packaging" cell lines. These produce all of the viral protein but not infectious virus, because the corresponding infected viral genome lacks a packaging signal required for packaging of viral RNA into virions. Transfer of a retroviral vector into such a cell line thus produces vector RNA containing virions which can "infect" an appropriate target cell.

This system unfortunately carries with itself the potential of helper-virus formation through homologous recombination between the vector genome and the endogenous genome in the packaging cell. Homologous recombination in retroviruses is an efficient process. It has been well documented (LUSSO et al., 1990) and studied in detail (STUHL-MANN and BERG, 1992). In this particular situation it can be minimized by the use of vectors with little or no overlap with the viral sequences in the packaging cell line.

By employing the most advanced vectors currently available and approved by the FDA as well as the appropriate cell lines (MILLER and BUTTIMORE, 1986) recombinated helper-virus production has been virtually eliminated even by the most stringent assay standards (MILLER and ROSMAN, 1989; LYNCH and MILLER, 1991). Nevertheless, a "break-through" event, i.e., the generation of a replication-competent helper virus has recently been described in a commercial production facility for retroviral vectors (FOX, 1993). Whether this rare recombination event happened with the most advanced vector technology or with less advanced systems is not known. Nevertheless, the incident reinforces the need for both improvements in vector technology and possibly improvements in the tests for the absence of helper virus.

10.3.3 Vectors Derived from Adenoviruses

Adenovirus-derived vectors – unlike retroviral vectors – have the advantage of being capable of delivering recombinant genes into non-proliferating cells (BERKNER, 1988). Since they are tropic for the respiratory epithelium, they have been used to transfer the human α1-antitrypsin gene as well as the human cystic fibrosis transmembrane conductance regulator gene to the respiratory epithelium of experimental animals (ROSEN-FELD et al., 1991, 1992). Adenovirus vectors are even considered as recombinant virus vaccines. A construct expressing herpes simplex virus (HSV) glycoproteins, for example, has been used to protect mice from lethal challenges with HSV (JOHNSON, 1991). With regard to the safety of such vectors, it must be realized that while human adenoviruses are widely spread in the general production, some serotypes are known to be (a) involved in respiratory diseases and (b) to be oncogenic in newborn rodents. The serotypes 2 and 5, however, which are currently used as the basis for vector constructions, are non-oncogenic. In addition, they are constructed to be replication-defective by removal of at least two early region genes, the *E1a* and *E1b* structural genes. A major negative manifestation of an adenoviral infection, the lytic infection resulting eventually in cell death, thus can be avoided. Potentially it could be envisaged that certain host cells could overcome this replication defect by providing the missing *E1a* function in *trans* or that a wild-type adenovirus infection, latent or acute, could induce replication of the recombinant virus by recombination or complementation (ROSENFELD et al., 1992). Whether such events turn into significant risk factors, may depend on the immune status of the patients.

Many adenovirus vectors not only contain deletion of the *E1a/E1b* regions but also of early region *E3*. This region is non-essential in tissue culture but codes for a number of proteins which modulate and influence immunosurveillance in the infected host (WOLD and GOODING, 1991). One of the E3-proteins down-regulates the level of class I MHC polypeptide heavy chains on the plasma membrane and thus reduces the number of targets for recognition of infected cell antigens by cytotoxic T-lymphocytes. Another protein prevents cytolysis by tumor necrosis factor. Deletions in this gene markedly increase the inflammatory responses in experimental animals (GINSBERG et al., 1989). It thus has to be kept in mind that even the avirulent serotypes 2, 4, 5 and 7 may become virulent strains when used as vectors with E3 deletions.

10.3.4 Adeno-Associated Viruses

Adenovirus-associated virus (AAV) is a non-pathogenic parvovirus which has potential for certain application in human gene therapy. It requires a human adenovirus as helper virus for its propagation. In the absence of helper virus the parvovirus genome integrates into a single chromosomal site in the long arm of human chromosome 19 (SAMULSKI et al., 1991). While an integrated viral genome is rescued by adenovirus superinfection, vector DNAs remain stably integrated and do not replicate under such conditions (MURO-CACHO et al., 1992). AAV-derived transducing thus appears to be both a highly efficient and safe delivering system for human

gene therapy and is being considered for the treatment of hemoglobinopathies (WALSH et al., 1992) as well as in gene transfer strategies directed towards the respiratory tract (FLOTTE et al., 1992).

10.3.5 Vectors Derived from Herpes Viruses

Herpes simplex virus (HSV-1)-derived vectors are only beginning to be considered as gene transfer vehicles. In particular, due to the known neurotropism of this virus they are envisaged to provide useful tools for gene transfer to cells of the mammalian nervous system. Nevertheless, vector development is only in an early state. So far, the replication-defective vectors which lack the essential regulatory gene *IE3* are toxic to cultured CNS neurons and glias (JOHNSON et al., 1992; HUANG et al., 1992). This problem eventually will have to be solved. In addition, herpes viruses may be explored for their tropism to other cell types, e.g., *H. saimiri* for the gene transfer into human T-cells (BIESINGER et al., 1992).

10.3.6 Vectors Derived from Bovine Papilloma Virus

BPV-1 viral DNA can transform rodent cells in culture. Although there are reports which describe the identification of BPV-1 DNA integrated into host chromosomal sequences (ALLSHIRE and BOSTOCK, 1986), it generally persists in an unintegrated episomal form in the transformed cells. Since it could be shown that the latter is also true for BPV-1 DNA cloned into the *E. coli* plasmid pBR322 (HOWLEY et al., 1980), the potential of BPV-1 DNA as a shuttle vector for the expression of foreign proteins in transformed mouse cells has been recognized and even commercially exploited (WURM and ZETTLMEISSL, 1989). Hazards associated with such vector constructs relate either to the malignant potential of the transforming DNA or the potential spread of shuttle vectors based on BPV-1 and

E. coli plasmid sequences into the environment. If this should be a problem, it can be eliminated by avoiding the use of shuttle vectors.

As to the malignant potential of BPV-1-derived vector DNA, it can be argued that BPV-1 only induces benign tumors in its natural host, the cattle. Nevertheless, there are indications that BPV-derived malignant tumors can arise in cattle under circumstances of co-carcinogenesis with environmental carcinogens (JARRETT et al., 1978). BPV-1 DNA has been isolated from equine sarcoids (AMTMANN et al., 1980). Although a clearcut cause/effect relationship has not yet been established, PCR-studies with a BPV-1- and BPV-2-derived set of primers indicate that BPV play an important role in the development of equine sarcoids (TEIFKE and WEISS, 1991). There is no evidence that human tumors are causcd by BPV-1 or have ever been found to contain BPV-1 DNA. Humans working with BPV-1 infected cattle can develop benign skin warts but not tumors. Extensive studies on warts in butchers and fish-merchants have convincingly demonstrated that they contain only human papilloma virus DNA, i.e., HPV-7 (JABLONSKA et al., 1988; RUDLINGER et al., 1989). There is no doubt, however, that the issue of tumorigenicity of BPV-1 DNA containing cell lines and protein products derived thereof has to be addressed carefully prior to the licensing of such processes and products.

11 Risks of Genetic Engineering: Facts and Fiction

For more than 40 years enzymes, amino acids, antibiotics, vaccines and other substances have been produced by fermentation with numerous microorganisms (including pathogenic organisms), often on a very large scale (up to 250 m^3). Compared with the wild strains isolated from nature, the high-performance strains obtained by mutagenesis and selection carry numerous unidentified modifications in their genome and have to be handled under safety conditions. As a result of all the genetic modifications successively collected, the penicillin yield from culture medium, for example, has been increased from $7 \mu\text{g}$ per liter with the wild strain *Penicillium chrysogenum* to over 30 g per liter with the production strains used today.

In the initial phase of genetic engineering and the development of recombinant organisms, the scientists asked themselves the important question, whether conjectual risks could arise with the new technique. The focal point of consideration was, whether the transfer of sections of DNA of foreign origin, which themselves contain no pathogenicity genes, into non-pathogenic organisms could possibly create pathogenic organisms by "synergistic" effects. If this were possible, "new types" of risks could occur which would have to be assessed differently or more severely than the hazards already known from the work with pathogenic (and hence hazardous) bacteria and viruses (and their hereditary material).

This question (BERG et al., 1974) was posed immediately after the first genetic engineering experiments between 1972 and 1974 and led to the congress in Asilomar (USA) in 1975 (BERG et al., 1975; GROBSTEIN, 1977) at which the first guidelines were drafted and experiments to answer this question were planned. In retrospective, it is fair to state that the issues and motives for the biosafety debate initiated by the inventors of genetic engineering, i.e., PAUL BERG and colleagues, never have been understood by the public and instead have triggered the regulations specific for genetic engineering.

Public debate still speculates on the "chances and risks" of modern biotechnology, particularly of genetic engineering, which are said to be associated with these new methods of modifying the genetic composition of organisms. In fact, such discussions are found with respect to almost any issue, albeit in differing magnitudes. These debates usually mix the terms "risk" and "hazard" rather unspecifically, which therefore need some reflection. This applies also to the term "chances" often used in conjunction with "risk" and "hazard".

A "hazard" exists, if a microorganism threatens the host's health and thereby causes damage. The extent of the damage caused, and therefore the severity of the disease, are directly related to the hazard posed by the microorganism. With respect to "risk", in addition to the magnitude of the damage caused, the dimension of "probability" needs to be considered (risk equals the product of probability × damage caused).

For practical evaluations, this means that very different scenarios can be drawn, and that the term "risk" needs to be used rather carefully because it can easily be misinterpreted, misunderstood or misused:

- The probability of microorganisms to escape from facilities applying GLP or GILSP is rather high, but the probability of these apathogenic microorganism to cause any damage is extremely low.
- The probability of a microorganism to escape from a containment level 2 or 3 facility is rather low, but the probability to cause damage is rather high for a certified host – e.g., hepatitis viruses for humans and foot and mouth disease virus for cattle – not *vice versa* and both not for plants or insects nor other species.
- The probability of a recombinant cow, producing a pharmaceutically relevant protein in the milk, to escape from the stable, to disappear in the environment and to cause a damage therein, is rather low. But there is a high probability that this animal might be a competitive economic threat to classical industrial facilities worldwide which produce the same protein by conventional cell culture techniques in a less efficient way.
- A transgenic crop plant with higher nutritional value is no threat to the environment but probably to the balance of the EU budget from which 60% are spent for agricultural subsidies. In other world areas such a plant will not be considered a "risk" and be welcomed instead.

There should be no illusions: as with any innovative technology, biotechnology will change economic and competitive conditions on the market. Indeed, economic renewal through innovation is the motor force of democratic societies. In respect of (not only) genetic engineering it is therefore necessary to identify issues and not to mix terms which do not belong together. Therefore, the assessment of whether a microorganism is harmful or not, cannot be made by sociological, legal or ethical arguments, and social or ethical questions which arise in prenatal diagnostics (e.g., using DNA probes) cannot be answered by biosafety assessments (modified organisms are not created nor involved). Unfortunately, the scientific contexts usually are not interpretable by the general public, and it seems not possible to comprehensively present the scientific issues in a simple fashion – hence uneasiness remains in the public, which can be exploited for various purposes not necessarily related to the science in question.

Since about 1976 it has been repeatedly established and confirmed (PROCEEDINGS FROM A WORKSHOP HELD AT FALMOUTH, MA, 1978; UNIDO, 1989) that no "new types of risk" are involved and no specific accidents or damages have been reported. For this reason, the safety requirements of the safety guidelines which had been introduced in many countries since about 1978, have been continuously relaxed (LEVIN, 1984). This also applies to the "Guidelines for the protection against risks constituted by *in-vitro* recombinant nucleic acids" (BMFT, 1986) which were valid in Germany in its fifth amended version until the Gene Technology Act came into force on 1 July 1990. As in the US, the German chemical industry had voluntarily agreed to abide by these guidelines which were mandatory for government-sponsored projects only (PHARMA-KODEX ZUR GENTECHNOLOGIE, 1988). In addition, since 1988 the Accident Prevention Regulations for Biotechnology have been in force which control the handling of biological material to protect workers and environment (BERUFSGENOSSENSCHAFT DER CHEMISCHEN INDUSTRIE, 1988).

In contrast to the different classical methods used in animal and plant breeding which employ the entire genetic information (the DNA of all chromosomes) of the organisms cultivated, genetic engineering work is restricted to the handling of relatively small sections of DNA compared to the size of chromosomal DNA. This is because of the physicochemical properties of DNA and is basically limited by the fragility of DNA. The efficiency of the genetic engineering methods must therefore not be unrealistically overestimated as it often is by the general public.

An assessment corresponding to the international evaluation of genetic engineering was also made by the "Enquiry Committee of the German Parliament: Chances and Risks of Genetic Engineering" in their report to the Parliament (CATENHUSEN and NEUMEISTER, 1987):

> "Ten years' intensive fundamental research with these systems in which countless gene transfer experiments were carried out worldwide showed no indication of hypothetical new risks. This research, which was conducted with special precautions, had implicitly the function of safety research. It is moreover likely that the risks would have been identified if there were any."

A wholesale "no" to genetic engineering research or products manufactured by recombinant methods as stated by the German B.U.N.D. (Association for Environment and Protection of Nature) is basically a nonsensical and irrational assessment (BUND, 1988):

> "The BUND demands a prohibition of any genetic manipulation of life forms and viruses. This prohibition includes any research, production and use in this area."

Interestingly, this much publicized declaration has partly been overruled by an almost unpublicized declaration (BUND, 1990):

> "The necessary medical/pharmaceutical research and applications without reasonable alternatives can – case by case and including a risk assessment – be excluded from this prohibition."

The fundamental misconception in such statements is that a technology can never be dangerous in itself. Basically, only the ingredients used in a certain production process or a certain product manufactured by means of certain methods, alone or in combination with other substances or factors, can be ad-

vantageous or dangerous. Applied to genetic engineering, this means that it is not the method of genetic engineering itself or the handling of nucleic acids (DNA or RNA) that involves risks, but that in individual cases a recombinant organism, protein molecule or a secondary metabolite may be associated with a certain hazard to man, animals or local environments. As outlined above, it might be advantageous to address "hazards" instead of "risks" in order to separate scientific issues from sociopolitical scenarios. Above and beyond the known risks involved in the handling of non-pathogenic, pathogenic and other harmful organisms, no "unique risk" has been observed with organisms modified by genetic engineering.

This assessment was already filed in 1987, e.g., by the U.S. National Academy of Science (KELMAN et al., 1987):

- rDNA techniques constitute a powerful and safe means for the modification of organisms.
- There is no evidence that unique hazards exist either in the use of rDNA techniques or in the movement of genes between unrelated species.
- The risks associated with the introduction of rDNA-engineered organisms are the same in kind as those associated with the introduction of unmodified organisms and organisms modified by other methods.
- The assessment of risks associated with introducing rDNA organisms into the environment should be based on the nature of the organism; based on the environment into which the organism is to be introduced; and independently of the method of engineering *per se.*

Meanwhile, this view repeatedly was confirmed, e.g., by a review of the drugs and diagnostic products approved by the FDA and at a public hearing in the German Parliament on experiences with the Gene Technology Act (CATENHUSEN et al., 1992), where the chairman of the German ZKBS (rDNA Advisory Committee) and the representative of the Berufsgenossenschaft der chemischen Industrie (professional cooperative of the chemical industry) reported that they were not aware of any single event that would indicate specific hazards or safety concerns with recombinant DNA technology.

12 Conclusions

Genetic engineering has become a normal and often essential feature of research, technical application and student training on both international (BAP, 1982–1986; BRIDGE, 1990–1994) and national scale (IN DEVELOPMENT: BIOTECHNOLOGY MEDICINE, 1990).

Many diseases and in particular epidemics have posed a threat to mankind since primeval times. These risks have initially been countered by the introduction of hygienic measures or quarantine, although the scientific principles on which these initiatives were based had not been researched or understood. This was followed by vaccination and by antibiotic treatment.

Nowadays many details are known of the wide variety of organisms that are used with good effect in traditional biotechnology, both in research and in industrial applications, including, of course, organisms obtained by mutagenesis or cell fusion.

The new genetic engineering methods are being used intensively in more than 20000 laboratories worldwide, and in almost 20 years innumerous different microorganisms have been modified and constructed by genetic engineering, with their behavior being monitored and their properties being analyzed.

Based on these findings and the microbiological knowledge accumulated, a reliable risk evaluation can be carried out. Based on this, further studies pointing the way forward are envisaged which will give further insights into biological processes. These are then bound to give rise to new questions in a never ending process.

Reviewing the available data and the practical experience gained from different branches of biological research, the conclusion is that microorganisms – modified without and with recombinant methods – have been handled extremely safely as a result of technical measures alone. This applies to both the people working with these organisms and to the plants, animals and the remainder of the environment indirectly affected. Only with the new methods of molecular biology an additional safety level has become available which was described as a biological safety measure.

The risk which an "organism" can constitute for scientists, the general population, animals, plants or the environment is always interrelated with its specific properties and independent of the kind of its manufacture. The views of biological safety discussed here have carefully been developed and investigated by competent scientists in worldwide collaboration. As knowledge increased, this gave rise to the international agreement that no "new" risks are associated with genetic engineering methods, that biological risks are clearly recognizable and that they can be particularly well controlled when the dual concept of technical and biological safety measures is adhered to in both research and production.

13 References

ADVISORY COMMITTEE ON GENETIC MODIFICATION (1990), United Kingdom, Note 1: *Guidance on Construction of Recombinants Containing Potentially Oncogenic Nucleic Acids.*

ALDHOUS, P. (1991), Spectre of contamination, *Nature* **349**, 359.

ALEXANDER, M. (1981), Why predators and parasites do not eliminate their prey and hosts, *Annu. Rev. Microbiol.* **35**, 113–133.

ALLSHIRE, R. C., BOSTOCK, C. J. (1986), Structure of bovine papillomavirus type 1 DNA in a transformed mouse cell line, *J. Mol. Biol.* **188**, 1–13.

AMTMANN, E., MÜLLER, H., SAUER, G. (1980), Equine connective tissue tumors contain unintegrated bovine papilloma virus DNA, *J. Virol.* **35**, 962–964.

ANDERSON, E. S. (1975), Viability and transfer of a plasmid from *E. coli* K12 in the human intestine, *Nature* **255**, 502–504.

ANDRÉ, F. E. (1989), Summary of safety and efficacy data on a yeast derived hepatitis B vaccine, *Am. J. Med.* **87**, Suppl. 3A, 14–20.

ARBER, W. (1987), Natural mechanisms of interspecific gene transfer, *Swiss Biotech.* **5-2a**, 11–12.

ARMSTRONG, J. L., KNUDSEN, G. R., SEIDLER, R. J. (1987), Microcosm method to assess survival of recombinant bacteria associated with plants and herbivorous insects, *Curr. Microbiol.* **15**, 229–232.

AUSUBEL, F. M., et al. (1989), *Current Protocols in Molecular Biology*, Vol. I and II, Harvard Medical School/Massachusetts General Hospital, New York: Green Publishing Associates.

BACHMANN, B. (1990), Linkage map of *Escherichia coli* K12, Edition 8, *Microbiol. Rev.* **54** (2), 130–197.

BALDO, B. A., BAKER, R. S. (1988), Inhalant allergies to fungi: reactions to bakers' yeast (*Saccharomyces cerevisiae*) and identification of bakers' yeast enolase as an important allergen, *Int. Arch. Allergy Appl. Immunol.* **86**, 201–208.

BAP (1982–1986), *Biomolecular Engineering Programme*, Programme zur Förderung und Entwicklung der Gentechnik (Programs for Promotion and Development of Gene Technology), Brussels: Commission of the European Community (GD XII).

BARNETT, J. A. (1992), The taxonomy of the genus *Saccharomyces cerevisiae* Meyen ex Reess: a short review for non-taxonomists, *Yeast* **8**, 1–23.

BASTIA, D. (1978), Determination of restriction sites and the nucleotide sequence surrounding the relaxation site of ColE1, *J. Mol. Biol.* **124**, 601–639.

BEAUDENON, S., KREMSDORF, D., CROISSANT, O., JABLONSKA, S., WAIN-HOBSON, S., ORTH, G. (1986), A novel type of human papillomavirus associated with genital neoplasias, *Nature* **321**, 246–249.

BELL, G. I., et al. (1980), Sequence of the human insulin gene, *Nature* **284**, 26–32.

BENDA, I. (1964), Die Hefeflora des fränkischen Weinbaugebietes, *Weinbau u. Keller* **11**, 67–80.

BERG, P., BALTIMORE, D., BOYER, H. W., COHEN, S. N., DAVIS, R. W., HOGNES, D. S., NATHANS, D., ROBLIN, R., WATSON, J. D., WEISSMAN, S., ZINDER, N. D. (1974), Potential biohazards of recombinant DNA molecules, *Proc. Natl. Acad. Sci. USA* **71**, 2593–2594.

BERG, P., BALTIMORE, D., BRENNER, S., ROBLIN, R. O., SINGER, M. F. (1975), Asilomar conference on recombinant DNA molecules, *Science* **188**, 991–994.

BERKNER, K. L. (1988), Development of adenovirus vectors for the expression of heterologous genes, *BioTechniques* **5**, 616–629.

BERMUDEZ, M., HAZEN, T. C. (1988), Phenotypic and genotypic comparison of *Escherichia coli* from pristine tropical waters, *Appl. Environ. Microbiol.* **54**, 979–983.

BERUFSGENOSSENSCHAFT DER CHEMISCHEN INDUSTRIE (professional cooperative of the chemical industry, Germany) (1988), *Unfallverhütungsvorschrift* (regulations for accident prevention) Abschnitt 31: Biotechnologie. Heidelberg: Jedermann-Verlag Dr. Pfeffer oHG.

BERUFSGENOSSENSCHAFT DER CHEMISCHEN INDUSTRIE (professional cooperative of the chemical industry, Germany) (1991–1992), 1/92, *Eingruppierung biologischer Agenzien: Bakterien, Viren, Pilze, Parasiten* (classification of biological substances: bacteria, viruses, fungi, parasites). Merkblätter B 006, B 004, B 007, B 005.

BIESINGER, B., MÜLLER-FLECKENSTEIN, I., SIMMER, B., LANG, G., WITTMANN, S., PLATZER, E., DESROSIERS, R., FLECKENSTEIN, B. (1992), Stable growth transformation of human T-lymphocytes by *herpesvirus saimiri*, *Proc. Natl. Acad. Sci. USA* **89**, 3116–3119.

BININ, W. S., HEINEN, B. N. (1985), Production of ethanol from infant food formulas by common yeasts, *J. Appl. Bacteriol.* **58**, 355–357.

BLATTNER, F. R., WILLIAMS, B. G., BLECHL, A. E., DENISTON-THOMPSON, K., FABER, H. E., FURLONG, L. A., GRUNWALD, D. J., KIEFER, D. O., MOORE, D. D., SCHUMM, J. W., SHELDON, E. L., SMITHIES, O. (1977), Charon phages: Safer derivatives of bacteriophage lambda for DNA cloning, *Science* **196**, 161–169.

BMFT (Bundesminister für Forschung und Technologie) (1986), *Richtlinien zum Schutz vor Gefahren durch in-vitro neukombinierte Nukleinsäuren* (Guidelines for Protection against Risks Caused by *in vitro* Recombinant Nucleic Acids), 5th revised Ed., 5/1986, Bonn: BMFT.

BOYD, A. C., ARCHER, J. A. K., SHERRATT, D. J. (1989), Characterization of the ColE1 mobilization region and its protein products, *Mol. Gen. Genet.* **217**, 488–498.

BRIDGE (1990–1994), *Biotechnology Research for Innovation, Development and Growth in Europe*, Programme zur Förderung und Entwicklung der Gentechnik (Programs for Promotion and Development of Gene Technology), Brussels: Commission of the European Community (GD XII).

BRÖKER, M. (1990), A study on the survival of wild-type, laboratory and recombinant strains of the baker yeast *Saccharomyces cerevisiae* under sterile and nonsterile conditions, *Zbl. Hyg.* **190**, 547–557.

BRÖKER, M., RAGG, H., KARGES, H. E. (1987), Expression of human antithrombin III in *Saccharomyces cerevisiae* and *Schizosaccharomyces pombe*, *Biochim. Biophys. Acta* **908**, 203–213.

BUND (Bund für Umwelt und Naturschutz Deutschland e.V.) (1988), *Decision of the delegates at the annual assembly in Lüneburg*.

BUND (Bund für Umwelt und Naturschutz Deutschland e.V.) (1990), *Decision of the delegates at the annual assembly in Hagen.*

BUCHANAN-WOLLASTON, V., PASSIATORE, J. E., CANNON, F. (1987), The mob and oriT mobilization functions of a bacterial plasmid promote its transfer to plants, *Nature* **328**, 172–175.

BURNS, P. A., JACK, A., NEILSON, F., HADDOW, S., BALMAIN, A. (1991), Transformation of mouse skin endothelial cells *in vivo* by direct application of plasmid DNA encoding the human T24 H-ras oncogene, *Oncogene* **6**, 1973–1978.

CADOZ, M., STRADY, A., MEIGNIER, B., TAYLOR, J., TARTAGLIA, J., PAOLETTI, E., PLOTKIN, S. (1992), Immunisation with canary pox virus expressing rabies glycoprotein, *Lancet* **339**, 1429–1432.

CARBONELL, L. F., MILLER, L. K. (1987), Baculovirus interaction with nontarget organisms: a virus-borne reporter gene is not expressed in two mammalian cell lines, *Appl. Environ. Microbiol.* **53**, 1412–1417.

CATENHUSEN, W. M., NEUMEISTER, H. (1987), *Chancen und Risiken der Gentechnologie* (Enquiry Commission of the German Parliament), Dokumentation des Berichts an den Deutschen Bundestag, München: J. Schweitzer Verlag.

CATENHUSEN, W. M., THOMAE, D., ALTHERR, W. (1992), *Experiences with the Law for the Regulation of Aspects concerning Gene Technology*, Hearing of the Committees of the German Parliament, Bonn, 12 February, 1992.

CHAN, H. W., et al. (1979), Molecular cloning of polyoma virus DNA in *E. coli*: Lambda phage vector system, *Science* **203**, 887–892.

CIMOLAI, N., GILL, M. J., CHURCH, D. (1987), *Saccharomyces cerevisiae* fungemia: Case report and review of literature, *Diagn. Microbiol. Infect. Dis.* **8**, 113–117.

COFFIN, J. M. (1990), Retroviruses and their replication, in: *Virology* (FIELDS, B. N., KNIPE, D. M., eds.), p. 1437, New York: Raven Press.

COKER, E. G. (1983), Biological aspects of the disposal utilization of sewage sludge on land, *Adv. Appl. Biol.* **9**, 257–322.

COLLINS, C. H. (1990), Safety in industrial microbiology and biotechnology: UK and European classifications of microorganisms and laboratories, *TIBTECH* **8**, 345–348.

CORDIER, S. (1990), Risk of cancer among laboratory workers, *Lancet* **335**, 1097.

COUTURIER, M., BEX, F., BERGQUIST, P. L., MAAS, W. K. (1988), Identification and classification of bacterial plasmids, *Microbiol. Rev.* **52** (3), 375–395.

CRAMER, J. H., LEA, K., SLIGHTOM, J. L. (1985), Expression of phaseolonic DNA genes in yeast under control of natural plant DNA sequences, *Proc. Natl. Acad. Sci. USA* **82**, 334–338.

CULLEN, B. R., LOMEDICO, P. T., JU, G. (1984), Transcriptional interference in avian retroviruses – implications for the promoter insertion model of leukaemogenesis, *Nature* **307**, 241–245.

DANCYGIER, H. (1989), Bakterien und intestinales Immunsystem, *Internist* **30**, 370–381.

DANIEL, M. D., LETVIN, N. L., KING, N. W., KANNAGI, M., SEHGAL, P. K., HUNT, R. D., KANKI, P. J., ESSEX, M., DESROSIERS, R. C. (1985), Isolation of T-cell tropic HTLV-III-like retrovirus from macaques, *Science* **228**, 2101–2104.

DAVIS, B. D., DULBECCO, R., EISEN, H. N., GINSBERG, H. S. (1985), *Microbiology*, 3rd Ed., New York: Harper & Row.

DELORME, E. (1989), Transformation of *Saccharomyces cerevisiae* by electroporation, *Appl. Environ. Microbiol.* **55**, 2242–2246.

DENHARDT, D. T., DRESSLER, P., RAY, C. G. (1978), *The Single-Stranded DNA Phages*, Cold Spring Harbor, NY: Cold Spring Harbor Laboratory Press.

DENTICO, P., BUONGIORNO, R., VOLPE, A., ZAVOIANNI, A., PASTORE, G., SCHISALDI, O. (1992), Long-term immunogenicity safety and efficacy of a recombinant hepatitis B vaccine in healthy adults, *Eur. J. Epidemiol.* **8**, 650–655.

DEVANAS, M. A., STOTZKY, G. (1986), Fate in soil of a recombinant plasmid carrying a *Drosophila* gene, *Curr. Microbiol.* **13**, 279–283.

DEVANAS, M. A., RAFAELI-ESHKOL, R., STOTZKY, K. (1986), Survival of plasmid-containing strains of *Escherichia coli* in soil: effect of plasmid size and nutrients on survival of hosts and maintenance of plasmids, *Curr. Microbiol.* **13**, 269–277.

DIMENNA, M. E. (1957), The isolation of yeasts from soil, *J. Gen. Microbiol.* **17**, 678–688.

DOMSCH, K. H., DRIESEL, A. J., GOEBEL, W., ANDERSCH, A., LINDENMAIER, W., LOTZ, W., REBER, H., SCHMIDT, F. (1987), Überlegungen zur Freisetzung gentechnisch veränderter Mikroorganismen in der Umwelt, *Forum Mikrobiol.* **10**, 475–483.

DOUGHERTY, S. H., SIMMONS, R. L. (1982), Postoperative peritonitis caused by *Saccharomyces cerevisiae*, *Arch. Surg.* **117**, 248–249.

ELSAS, J. D., TREVORS, J. T., STARODUB, M. E., OVERBEECK, L. S. (1990), Transfer of plasmid RP4 between pseudomonads after introduction into soil; influence of spatial and temporal aspects of inoculation, *FEMS Microbiol. Ecol.* **73**, 1–12.

ENG, R. H. K., DREHMEL, R., SMITH, S. M., GOLDSTEIN, E. J. C. (1984), *Saccharomyces cerevisiae* infection in man, *Saboraudia* **22**, 403–407.

ESCHETE, M. L., BURTON, C. W. (1980), *Saccharomyces cerevisiae* septicemia, *Arch. Intern. Med.* **140**, 1539.

EU Directives (1990),
Council Directive of 23 April 1990 on the contained use of genetically modified microorganisms (90/219/EEC), *Off. J. EC* **L 117/1**, May 8, 1990.
Council Directive of 23 April 1990 on the deliberate release into the environment of genetically modified organisms (90/220/EEC), *Off. J. EC* **L 117/15**, May 8, 1990.

FAIRWEATHER, N. F., LYNESS, V. A. (1986), The complete nucleotide sequence of tetanus toxin, *Nucleic Acids Res.* **14**, 7809–7812.

FERRIS, S. D., SAGE, R. D., WILSON, A. C. (1982), Evidence from mtDNA sequences that common laboratory strains of inbred mice are descended from a single female, *Nature* **295**, 163–165.

FINLAY, B. B., FALKOW, S. (1989), Common themes in microbial pathogenicity, *Microbiol. Rev.* **53**, 510–530.

FINNEGAN, J., SHERRATT, D. (1982), Plasmid ColE1 conjugal mobility: the nature of *bom*, a region required in *cis* for transfer, *Mol. Gen. Genet.* **185**, 344–351.

FLOTTE, T. R., SOLOW, R., OWENS, R. A., AFIONE, S., ZEITLIN, P. L., CARTER, B. J. (1992), Gene expression from adeno-associated virus vectors in airway epithelial cells, *Am. J. Resp. Cell. Mol. Biol.* **7**, 349–356.

FONDS DER CHEMISCHEN INDUSTRIE (1993), Informationsband *Sichere Biotechnologie*, Heft 32, Frankfurt am Main, ISSN 0174-2728.

FOX, J. L. (1993), NIHRAC and FDA ponder gene-therapy risks, *Biotechnology* **11**, 28–29.

FRETER, R. (1978), Possible effects of foreign DNA on pathogenic potential and intestinal proliferation of *Escherichia coli*, *J. Infect. Dis.* **137**, 624–629.

FUNG, Y. K. T., CRITTENDEN, L. B., FADLY, A., KUNG, H. J. (1983), Tumor induction by direct infection of cloned v-src DNA into chickens, *Proc. Natl. Acad. Sci. USA* **80**, 353–357.

GEDEK, B. R., AMSELGRUBER, W. (1990), Mikrobieller Antagonismus: Zur Eliminierung von enteropathogenen *E. coli*-Keimen und Salmonellen aus dem Darm durch *Saccharomyces boulardii*, in: *Ökosystem Darm II* (OTTENJANN, R., MÜLLER, J., SEIFERT, J., Eds.), pp. 180–188, Heidelberg: Springer-Verlag.

GEDEK, B., HAGENHOFF, G. (1988), Orale Verabreichung von lebensfähigen Zellen des Hefestammes *Saccharomyces cerevisiae* Hansen CBS 5926 und deren Schicksal während der Magen-Darm-Passage, *Therapiewoche* **38**, 33–40.

GIAFFER, M. H., CLARK, A., HOLDSWORTH, C. D. (1992), Antibodies to *Saccharomyces cerevisiae* in patients with Crohn's disease and their possible pathogenic importance, *Gut* **33**, 1071–1075.

GIETZ, D., ST. JEAN, A., WOODS, R. A., SCHIESTL, R. H. (1992), Improved method for high efficiency transformation of intact yeast cells, *Nucleic Acids Res.* **20**, 1425.

GIMENO, C. J., LJUNNGDAHL, P. O., STYLES, C. A., FINK, G. R. (1992), Unipolar cell divisions in the yeast *S. cerevisiae* lead to filamentous growth: regulation by starvation and RAS, *Cell* **68**, 1077–1090.

GINSBERG, H. S., LUNDHOLM-BEAUCHAMP, U., HORSWOOD, R. L., PERNIS, B., WOLD, W. S., CHANOCK, R. M., PRINCE, G. A. (1989), Role of early region 3 (E3) in pathogenesis of adenovirus disease, *Proc. Natl. Acad. Sci. USA* **86**, 3823–3827.

GLANTZ, P. J., JACKS, T. M. (1967), Significance of *Escherichia coli* serotypes in wastewater effluent, *J. Water Pollut. Control Fed.* **39**, 1918–1921.

GOURSOT, R., GOZE, A., NIAUDET, B., EHRLICH, S. D. (1982), Plasmids from *Staphylococcus aureus* replicate in yeast *Saccharomyces cerevisiae*, *Nature* **298**, 488–490.

GROBSTEIN, C. (1977), The recombinant-DNA debate, *Sci. Am.* **237**, 22–23.

GRONENBERG, J., BROWN, D. T., DOERFLER, W. (1975), Uptake and fate of DNA of adenovirus type 2 in KB cells, *Virology* **64**, 115–131.

GUMOWSKI, P. I., LATGÉ, J.-P., PARIS, S. (1991), Fungal allergy, in: *Handbook of Applied Mycology*, Vol. 2: *Humans, Animals, and Insects* (ARORA, D. K., AJELLO, L., MUKERJI, K. G., Eds.), pp. 163–204, New York: Marcel Dekker.

GYLES, C. L. (1992), *Escherichia coli* cytotoxins and enterotoxins, *Can. J. Microbiol.* **38**, 734–746.

HACKER, J., OTT, M., TSCHÄPE, H. (1991), The problem of *Escherichia coli* pathogenicity and its consequences for the recombinant DNA-technology, *BIOforum* **14**, 150–157.

HAHN, H., BRANDEIS, H. (1988), Infektion und Infektabwehr, in: *Lehrbuch der Medizinischen Mikrobiologie* (BRANDIS, H., PULVERER, G., Eds.), pp. 65–92, Stuttgart: Gustav Fischer Verlag.

HAWKSWORTH, D. K., SUTTON, B. C., AINSWORTH, G. C. (1983), *Ainsworth & Bisby's Dictionary of the Fungi* (including the lichens), Kew, Surrey: Commonwealth Mycological Institute.

HEELAN, B. T., ALLAN, S., BARNES, R. M. R. (1991), Identification of a 200-kDa glycoprotein antigen of *Saccharomyces cerevisiae*, *Immunol. Lett.* **28**, 181–186.

108 *3 Biosafety in rDNA Research and Production*

HEINEMANN, J. A., SPRAGUE, G. F. (1989), Bacterial conjugative plasmids mobilize DNA transfer between bacteria and yeast, *Nature* **340**, 205–209.

HERSHFIELD, V., BOYER, H. W., YANOFSKI, C., LOVETT, M. A., HELINSKI, D. R. (1974), Plasmid ColE1 as a molecular vehicle for cloning and amplification of DNA, *Proc. Natl. Acad. Sci. USA* **71**, 3455–3459.

HILLEMAN, M. R. (1990), History, precedent and progress in the development of mammalian cell culture systems for preparing vaccines: Safety considerations revisited, *J. Med. Virol.* **31**, 5–12.

HOBSON, P. N., POOLE, N. J. (1988), Water pollution and its prevention, in: *Micro-organisms in Action: Concepts and Applications in Microbial Ecology* (LYNCH, J. M., HOBBLE, J. E., Eds.), pp. 207–237, Boston: Blackwell Scientific Publications.

HODGSON, J. (1993), Fetal bovine serum revisited, *Biotechnology* **11**, 49–53.

HORWITZ, M. S., BOYCE-JACINO, M. T., FARAS, A. J. (1992), Novel human endogenous sequences related to human immunodeficiency virus type 1, *J. Virol.* **66**, 2170–2179.

HOWLEY, P. M., LAW, M. F., HEILMAN, C. A., ENGEL, L., ALONSO, M. C., ISRAEL, M. A., LOWY, D. R., LANCASTER, W. D. (1980), Molecular characterization of papilloma virus genomes, *Cold Spring Harbor Conf. Cell Prolif.* **7**, 233–247.

HOWLEY, P. M., LEVINE, A. J., LI, F. P., LIVINGSTON, D. M., RABSON, A. S. (1991), Lack of SV40 DNA in tumors from scientists working with SV40 virus, *N. Engl. J. Med.* **324**, 494.

HSIAO, W. L., GATTONI-CELLI, S., WEINSTEIN, I. B. (1986), Effects of 5-azacytidine on expression of endogenous retrovirus-related sequences in C3H 10T 1/2 cells, *J. Virol.* **57**, 1119–1126.

HUANG, Q., VONSATTEL, J. P., SCHAFFER, P. A., MARTUZA, R. L., BREAKEFIELD, X. O., DI FIGLIA, M. (1992), Introduction of a foreign gene into rat neostratial neurons using herpes simplex virus mutants: a light and electronmicroscopic study, *Exp. Neurol.* **115**, 303–316.

HUBBERSTEY, A. V., WIDEMAN, A. G. (1991), Transformation of *Saccharomyces cerevisiae* by use of frozen spheroblasts, *Trends Genet.* **7**, 41.

HUNNISETT, A., HONORD, J., DAVIES, S. (1990), Gut fermentation (or the "auto-brewery") syndrome: a new clinical test with initial observations and discussions of clinical and biochemical implications, *J. Nutrit. Med.* **1**, 33–38.

HUNTER, P. R. (1993), Occupational infections of workers in medical laboratories, *Microbiol. Eur.* **1**(2), 8–12.

HUQ, A., et al. (1986), Colonization of the gut of the blue crab (*Callinectes sapidus*) by *Vibrio cholerae*, *Appl. Environ. Microbiol.* **52**, 586–588.

IN DEVELOPMENT: BIOTECHNOLOGY MEDICINE (1990), 1990 Annual Survey: *2 New Medicines Approved; 104 Drugs, Vaccines in Human Testing*, Washington, DC: Pharmaceutical Manufacturers Association.

ISBERG, R. R. (1989), Mammalian cell adhesion functions and cellular penetration of enteropathogenic *Yersinia* species, *Mol. Microbiol.* **3**, 1449–1453.

ISRAEL, M. A., et al. (1979), Molecular cloning of polyoma virus DNA in *E. coli*: Plasmid vector system, *Science* **203**, 883–887.

JABLONSKA, S., OBALEK, S., GOLEBIOWSKA, A., FAVRE, M., ORTH, G. (1988), Epidemiology of butchers' warts, *Arch. Dermatol. Res.* **280**, Suppl. S24–28.

JACKSON, D. A., SYMONS, R. H., BERG, P. (1972), Biochemical methods for inserting new genetic information into DNA of Simian virus 40: Circular SV 40 DNA molecules containing lambda phage genes and the galactose operon of *E. coli*, *Proc. Natl. Acad. Sci. USA* **69**, 2904–2909.

JARRETT, W. F. H., McNEIL, P. E., GRIMSHAW, W. T. R., SELMAN, I. E., McINTIYRE, W. I. M. (1978), High incidence area of cattle cancer with a possible interaction between an environmental carcinogen and a papilloma virus, *Nature* **274**, 215–217.

JELLINGHAUS, V., SCHATZLE, V., SCHMID, W., ROEWEKAMP, W. (1982), Transcription of a *Dictyostelium* discoidin-I gene in yeast: Alternative promoter sites used in two different eukaryotic cells, *J. Mol. Biol.* **159**, 623–636.

JENSEN, T. S. (1986), Environmental experiments with genetically engineered yeast cells, *Novo Annual Report 1986*, pp. 37–39, Bagsvaerd, Denmark: Novo Industri A/S.

JOHNSON, D. C. (1991), Adenovirus vectors as potential vaccines against herpes simplex virus, *Rev. Infect. Dis.* **13**, Suppl. 11, S912–916.

JOHNSON, P. A., MIYANOHARA, A., LEVINE, F., CAHILL, T., FRIEDMANN, T. (1992), Cytotoxicity of a replication-defective mutant of herpes simplex virus type 1, *J. Virol.* **66**, 2952–2965.

JOHNSTON, M. (1987), A model fungal gene regulatory mechanism: the *GAL* genes of *Saccharomyces cerevisiae*, *Microbiol. Rev.* **51**, 458–476.

JONES, I. M., PRIMROSE, S. B., ROBINSON, A., ELLWOOD, D. C. (1980), Maintenance of some ColE1-type plasmids in chemostat culture, *Mol. Gen. Genet.* **180**, 579–584.

KAJI, H., ASANUMA, Y., YAHARA, O., SHIBUE, H., HISAMURA, M., SAITO, N., KENAKAMI, Y., MURAO, M. (1984), Intragastrointestinal alcohol fermentation syndrome, report of two cases and review of the literature, *J. Forensic Sci. Soc.* **24**, 461–471.

KANKI, P. J., KURTH, R., BECKER, W., DREESMAN, G., McLANE, M. F., ESSEX, M. (1985), Antibodies to simian T-lymphotropic retrovirus type III in African green monkeys and recognition of STLV-III viral proteins by AIDS and related sera, *Lancet* **1**, 1330–1332.

KAPPE, R., MÜLLER, J. (1987), Cultural and serological follow-up of two oral administrations of baker's yeast to a human volunteer, *Mykosen* **30**, 357–368.

KELCH, W. J., LEE, J. S. (1978), Antibiotic resistance patterns of gram-negative bacteria isolated from environmental sources, *Appl. Environ. Microbiol.* **36**, 450–456.

KELMAN, A., ANDERSON, W., FALKOW, S., FEDAROFF, N. V., LEVIN, S. (1987), *Introduction of Recombinant DNA-engineered Organisms into the Environment: Key Issues*, Washington, DC: National Academy of Science Press.

KIMBERLIN, R. H. (1990), Taking stock of the issue, *Nature* **345**, 763–764.

KINGSMAN, S. M., KINGSMAN, A. J., MELLOR, J. (1987), The production of mammalian proteins in *Saccharomyces cerevisiae*, *TIBTECH* **5**, 54–57.

KLEMM, P. (1985), Fimbrial adhesions of *Escherichia coli*, *Rev. Infect. Dis.* **7**, 321–340.

KONISKY, J. (1978), The bacteriocins, in: *The Bacteria – A Treatise on Structure and Function, Vol. VI: Bacterial Diversity* (ORNSTON, L. N., SOKATCH, J. R., Eds.), New York: Academic Press.

KORMANEC, J. (1991), A yeast chromosomal fragment having strong promoter activity in *Escherichia coli*, *Gene* **106**, 139–140.

KORTEKANGAS-SAVOLAINEN, O., LAMMINTANSTA, K., KALIMO, K. (1993), Skin prick test reactions to brewer's yeast (*Saccharomyces cerevisiae*) in adult atopic dermatitis patients, *Allergy* **48**, 147–150.

KÜENZI, M., et al. (1985), Safe biotechnology – General considerations. A report prepared by the Safety in Biotechnology Working Party of the European Federation of Biotechnology, *Appl. Microbiol. Biotechnol.* **21**, 1–6.

KÜES, U., STAHL, U. (1989), Replication of plasmids in gram-negative bacteria, *Microbiol. Rev.* **53**, 491–516.

KUMAR, A., LARSEN, C. E., PREISS, J. (1986), Biosynthesis of bacterial glycogen: Primary structure of *E. coli* ADP-glucose: alpha-1,4-glucan-4-glucosyltransferase as deduced from the nucleotide sequence of the *glgA* gene, *J. Biol. Chem.* **261**, 16256–16259.

LANG-HINRICHS, C., HINRICHS, J. (1992), Recombinant yeasts in food and food manufacture – possibilities and perspectives, *AGRO-Industry Hi-Tech*, in press.

LARSSON, E., KATO, N., COHEN, M. (1989), Human endogenous proviruses, *Curr. Top. Microbiol. Immunol.* **148**, 115–132.

LEE, A. (1985), Neglected niches: the microbial ecology of the gastrointestinal tract, *Adv. Microb. Ecol.* **8**, 115–162.

LETVIN, N. L., LORD, C. I., KING, N. W., WYAND, M. S., MYRICK, K., HASELTINE, W. A. (1991), Risks of handling HIV, *Nature* **349**, 573.

LEVIN, B. (1984), Changing views of the hazards of recombinant DNA manipulation and the regulation of these procedures, *Recomb. DNA Tech. Bull. USA* **7**, 107–114.

LEVINE, M. M. (1984), *Escherichia coli Infections. Bacterial Vaccines*, pp. 187–235, London: Academic Press.

LEVY, S. B., MARSHALL, B., ROWSE-EAGLE, D., ONDERDONK, A. (1980), Survival of *Escherichia coli* host-vector systems in the mammalian intestine, *Science* **209**, 391–394.

LIANG, L. N., SINCLAIR, J. L., MALLORY, L. M., ALEXANDER, M. (1992), Fate in model ecosystems of microbial species of potential use in genetic engineering, *Appl. Environ. Microbiol.* **44**(3), 708–714.

LINDBERG, E., MAGNUSSON, K.-E., TYSK, C., JARNEROT, G. (1992), Antibody (IgG, IgA and IgM) to baker's yeast (*Saccharomyces cerevisiae*), yeast mannan, gliadin, ovalbumin and betalactoglobulin in monozygotic twins with inflammatory bowel disease, *Gut* **33**, 909–913.

LINZENMEIER, G. (1989), Darmflora und Chemotherapie, *Internist* **30**, 362–366.

LÖWER, J. (1990), Risk of tumor induction *in vivo* by residual cellular DNA: Quantitative considerations, *J. Med. Virol.* **31**, 50–53.

LOPEZ-TORRES, A. J., PRIETO, L., HAZEN, T. C. (1988), Comparison of the *in situ* survival and activity of *Klebsiella pneumoniae* and *Escherichia coli* in tropical marine environments, *Microb. Ecol.* **15**, 41–57.

LORENZ, M. G., WACKERNAGEL, W. (1994), Bacterial gene transfer by free DNA in the environment, *Microbiological Reviews*, submitted.

LUPSKI, J. R., FEIGIN, R. D. (1988), Molecular evolution of pathogenic *Escherichia coli*, *J. Infect. Dis.* **157**, 1120–1123.

LUSSO, P., DI MARZO VERONESE, F., ENSOLI, B., FRANCHINI, G., JEMMA, C., DEROCCO, S. E., LAKYANARAMAN, V. S., GALLO, R. C. (1990), Expanded HIV-1 cellular tropism by phenotypic mixing with murine endogenous retroviruses, *Science* **247**, 848–852.

LYNCH, C. M., MILLER, A. D. (1991), Production of high-titer helper virus-free retroviral vectors by co-cultivation of packaging cells with different host ranges, *J. Virol.* **65**, 3887–3890.

MARCON, M. J., POWELL, D. A. (1992), Human infections due to *Malassezia* spp., *Clin. Microbiol. Rev.* **5**, 101–119.

MARIANI-CONSTANTINI, R., HORN, T. M., CALLAHAN, R. (1989), Ancestry of human endogenous retrovirus family, *J. Virol.* **63**, 4982–4985.

MARX, J. L. (1989), How DNA viruses may cause cancer, *Science* **243**, 1012–1013.

MAZODIER, P., PETTER, R., THOMPSON, C. (1989), Intergeneric conjugation between *Escherichia coli* and *Streptomyces* species, *J. Bacteriol.* **171**, 3583–3585.

MCCUSKER, J., DAVIS, R. (1991), A novel cell type switching system, high temperature growth and protease secretion in a clinical *S. cerevisiae* isolate, *Yeast Genetics and Molecular Biology Meeting, San Francisco*, Poster Abstract, p. 778.

MCKENZIE, H., MAIN, J., PENNINGON, C. R., PARRATT, D. (1990), Antibody to selected strains of *Saccharomyces cerevisiae* (baker's and brewer's yeast) and *Candida albicans* in Crohn's disease, *Gut* **31**, 536–538.

MEACOCK, P. A., COHEN, S. N. (1980), Partitioning of bacterial plasmids during cell division: a *cis*-acting locus that accomplishes stable plasmid inheritance, *Cell* **20**, 529–542.

MEACOCK, P. A., BRIEDEN, K. W., CASHMORE, A. M. (1989), The two-micron circle: module replicon and yeast vector, in: *Molecular and Cell Biology of Yeast* (WALTON, E. F., YARRANTON, G. T., Eds.), pp. 330–359, Glasgow: Blackie & Son Ltd.

MIDDLEBROOK, J. L., DORLAND, R. B. (1984), Bacterial toxins in cellular mechanisms of action, *Microbiol. Rev.* **48** (3), 199–221.

MILLER, D. A. (1992), Human gene therapy comes of age, *Nature* **357**, 455–460.

MILLER, D. A., BUTTIMORE, C. (1986), Redesign of retrovirus packaging cell lines to avoid recombination leading to helper virus production, *Mol. Cell. Biol.* **6**, 2895–2902.

MILLER, D. A., ROSMAN, G. J. (1989), Improved retroviral vectors for gene transfer and expression, *BioTechniques* **7**, 980–990.

MILMAN, G., D'SOUZA, P. (1990), HIV infections in SCID mice: Safety considerations, *ASM News* **56**, 639–642.

MINSHEW, B. H., JORGENSEN, J., SWANSTRUM, M., GROOTES-REUVENCAMP, G. A., FALKOW, S. (1978), Some characteristics of *Escherichia coli* strains isolated from extraintestinal infections of humans, *J. Infect. Dis.* **137**, 648–654.

MONTGOMERY, B. E., BOOR, A. K., ARNOLD, L.,

BERGEIM, O. (1930), Destruction of yeast in the normal human stomach, *Proc. Soc. Exp. Biol. Med.* **28**, 385.

MORRISON, W. D., MILLER, R. V., SAYLER, G. S. (1978), Frequency of F116-mediated transduction of *Pseudomonas aeruginosa* in a freshwater environment, *Appl. Environ. Microbiol.* **36**, 724–730.

MORRISON, H. G., GOLDSMITH, C. S., REGNERY, H. L., AUPERIN, D. D. (1991), Simultaneous expression of the Lassa virus N and GPC genes from a single recombinant vaccinia virus, *Virus Res.* **18**, 231–241.

MULDER, C. (1988), A case of mistaken non-identity, *Nature* **331**, 562–563.

MÜLLER, H. E. (1992), Relationship between ecology and pathogenicity of microorganisms, *BIOforum* **1–2**/92.

MÜLLER, J., OTTENJANN, R., SEIFERT, J. (1989), *Ökosystem Darm*, Heidelberg: Springer-Verlag.

MUNGAL, S., STEINBERG, B. M., TAICHMAN, L. B. (1992), Replication of plasmid-derived human papillomavirus type 12 DNA in cultured keratinocytes, *J. Virol.* **66**, 3220–3224.

MURO-CACHO, C. A., SAMULSKI, R. J., KAPLAN, D. (1992), Gene transfer in human lymphocytes using a vector based on adeno-associated virus, *J. Immunother.* **11**, 231–237.

MUTO, Y., GOTO, S. (1987), Transformation by cellular DNA produced by *Pseudomonas aeruginosa*, *Microbiol. Immunol.* **30**, 621.

NEU, H. C. (1992), The crisis in antibiotic resistance, *Science* **257**, 1064–1073.

NICOLI, R.-M., SEMPE, M., RUSSO, M. S. (1974), Un nouveau peuplement vaginal inosilite: une levurose conjugale à la levure de panification, *Ann. Parasitol.* **49**, 369–370.

NIH GUIDELINES FOR RESEARCH INVOLVING RECOMBINANT DNA MOLECULES (1984), *Fed. Reg.* **46**, 266–291.

NIH GUIDELINES FOR RESEARCH INVOLVING RECOMBINANT DNA MOLECULES (1986), *Fed. Reg.* **27**, 902 (June 1976); *Fed. Reg.* **51**, 16358–16985, No. 88 (May 7, 1986), Physical Containment for Large-Scale Uses of Organisms Containing Recombinant DNA Molecules.

NIH GUIDELINES FOR RESEARCH INVOLVING RECOMBINANT DNA MOLECULES (1991), *Fed. Reg.* **58**, 33178–33182, No. 138 (July 18, 1991).

NISHIKAWA, M., SUZUKI, K., YOSHIDA, K. (1990), Structural and functional stability of IncP plasmids during stepwise transmission by transkingdom mating: Prosmiscuous conjugation of *Escherichia coli* and *Saccharomyces cerevisiae*, *Jpn. J. Genet.* **65**, 323–334.

NISHIKAWA, M., SUZUKI, K., YOSHIDA, K. (1992), DNA integration into recipient yeast chromosomes by trans-kingdom conjugation between *Escherichia coli* and *Saccharomyces cerevisiae*, *Curr. Genet.* **21**, 101–108.

NOVICK, R. P. (1987), Plasmid incompatibility, *Microbiol. Rev.* **51** (4), 381–395.

OETTINGER, M. A., SCHATZ, D. G., GORKA, C., BALTIMORE, D. (1990), RAG-1 and RAG-2, Adjacent genes that synergistically activate V(D)J recombination, *Science* **248**, 1517–1523.

ONO, M. (1986), Molecular cloning and long terminal repeat sequences of human endogenous retrovirus genes related to types A and B retrovirus genes, *J. Virol.* **58**, 937–944.

ORSKOV, F., ORSKOV, I., JANN, B., JANN, K. (1977), Serology, chemistry and genetics of O and K antigens of *Escherichia coli*, *Bacteriol. Rev.* **41**, 667–710.

OTA Report (Office of Technology Assessment) (1981), *Impacts of Applied Genetics: Micro-Organisms, Plants and Animals* (April 1981), Washington, DC: U.S. Congress.

OTA Report (Office of Technology Assessment) (1993), *Harmful Non-indigenous Species in the U.S.* (September 1993), Washington, DC: U.S. Congress.

PABLOS-MÉNDEZ, A., NETTO, E. M., DEFENDINI, R. (1993), Infectious prions or cytotoxic metabolites? *Lancet* **341**, 159–161.

PACQUET, S., GUILLAUMIN, D., TANCREDE, C., ANDREMONT, A. (1991), Kinetics of *Saccharomyces cerevisiae* elimination from the intestine of human volunteers and effect of this yeast on resistance to microbial colonization in gnotobiotic mice, *Appl. Environ. Microbiol.* **57**, 3049–3051.

PAUL, J. H., JEFFREY, W. H., DE FLAUN, M. F. (1987), Dynamics of extracellular DNA in the marine environment, *Appl. Environ. Microbiol.* **53**, 170–179.

PETRICCIANI, J. C. (1987), Should continuous cell lines be used as substrates for biological products? *Dev. Biol. Standard.* **66**, 3–11.

PFALLER, M. A., WENZEL, R. P. (1992), Impact of the changing epidemiology of fungal infections in the 1990s, *Eur. J. Clin. Microbiol. Infect. Dis.* **11**, 287–291.

PHARMA-KODEX ZUR GENTECHNOLOGIE (1988), *Eur. Chem.* **4**, 45.

PROCEEDINGS FROM A WORKSHOP HELD AT FALMOUTH, MA (1978), (June 20 and 21, 1977), Risk assessment of recombinant-DNA experimentation with *Escherichia coli* K12, *J. Infect. Dis.* **137**.

RAMAGHADRAN, T. V. (1987), Products from genetically engineered mammalian cells: benefits and risk factors, *Trends Biotechnol.* **5**, 175–179.

RAYSSIGUIER, C., THALER, D. S., RADMAN, M. (1989), The barrier to recombination between *Escherichia coli* and *Salmonella typhimurium* is disrupted in mismatch-repair mutants, *Nature* **342**, 396–401.

Recombinant DNA Technical Bulletin **10**, 107–110 (1987), and *Federal Register* **52**, No. 163, August 24 (1987).

RIECKHOF, B. (1993), Antimykotische Therapie mit *Saccharomyces boulardii*, *Aerztl. Nachr.* **34**, 657.

ROBERTS, R. J., MACELIS, D. (1992), Restriction enzymes and their isoschizomers, *Nucleic Acids Res.* **20**, 2167–2180.

ROMERO-STEINER, S., WITEK, T., BALSIH, E. (1990), Adherence of skin bacteria to human epithelial cells, *J. Clin. Microbiol.* **28**, 27–31.

ROSENFELD, M. A., SIEGFRIED, W., YOSHIMURA, K., YONEYAMA, K., FUKAYAMA, M., STIER, L. E., PÄÄKÖ, P. K., GILARDI, P., STRATFORD-PERRICAUDET, L. D., PERRICAUDET, M., JALLET, S., PAVIRANI, A., LECOCQ, J. P., CRYSTAL, R. G. (1991), Adenovirus-mediated transfer of a recombinant alpha-1-antitrypsin gene to the lung epithelium, *Science* **252**, 431–434.

ROSENFELD, M. A., YOSHIMURA, K., TRAPNELL, B.-C., YONEYAMA, K., ROSENTHAL, E. R., DALEMANS, W., FUKAYAMA, M., BARGON, J., STIER, L. E., STRAATFORD-PERRICAUDET, L., PERRICAUDET, M., GUGGINO, W. B., PAVIRANI, A., LECOCQ, J. P., CRYSTAL, R. G. (1992), *In vivo* transfer of the human cystic fibrosis transmembrane conductance regulator gene to the airway epithelium, *Cell* **68**, 143–155.

ROSZAK, R., COLWELL, R. (1987), Survival strategies of bacteria in the natural environment, *Microbiol. Rev.* **51**, 365–379.

RUBINSTEIN, E., NOREIGA, E. R., SIMBERKOFF, M. S., HOLZMAN, R., RAHAL, J. J. Jr. (1974), Fungal endocarditis: analysis of 24 cases and review of the literature, *Medicine* **54**, 331–344.

RUDLINGER, R., BUNNEY, M. H., GROB, R., HUNTER, J. A. (1989), Warts in fish handlers, *Br. J. Dermatol.* **120**, 375–381.

RUF, B., POHLE, H. D. (1989), Bakterien als Erreger opportunistischer Infektionen des Gastrointestinaltraktes bei HIV-Infektionen, *Internist* **30**, 352–357.

SAGB (Senior Advisory Group Biotechnology) (1990), *Community Policy for Biotechnology, Priorities and Actions*, Brussels: CEFIC.

SAMULSKI, R. J., ZHU, X., XIAO, X., BROOK, J. D., HOUSMAN, D. E., EPSTEIN, N., HUNTER, L. A. (1991), Targeted integration of adeno-associated virus (AAV) into human chromosome 19, *EMBO J.* **10**, 3941–3950.

SANSONETTI, P. J., HALE, T. L., DAMMIN, G. J., KAPFER, C., COLLINS JR., H. H., FORMAL, S. B. (1983), Alterations in the pathogenicity of *Escherichia coli* K12 after transfer of plasmid and chromosomal genes from *Shigella flexneri*, *Infect. Immun.* **39**, 1392–1402.

SAVOLAINEN, O., LAMMINTAUSTA, K., KALIMO, K. (1992), Baker's yeast and atopic dermatitis, *Acta Derm. Venereol. Suppl.* **176**, 140.

SAYLER, G., STACEY, G. (1986), Methods for evaluation of microorganisms properties, in: *Biotechnology Risk Assessment: Issues and Methods for Environmental Introductions* (FISKEL, J., COVELLO, V. T., Eds.), pp. 35–55, Oxford: Pergamon Press.

SCHATA, M., JORDE, W. (eds.) (1988), *Klinische Objektivierung von Schimmelpilzallergien*, Mönchengladbacher Allergie-Seminare, Band 2, Deisenhofen, FRG: Dustri-Verlag.

SCHILF, W., KLINGMÜLLER, W. (1983), Experiments with *Escherichia coli* on the dispersal of plasmids in environmental samples, *Recomb. DNA Tech. Bull.* **6**, S 101.

SCHWARZ, E., FREESE, U. K., GISSMANN, L., MAYER, W., ROGGENBUCK, B., STREMLAU, A., ZUR HAUSEN, H. (1985), Structure and transcription of human papillomavirus sequences in cervical carcinoma cells, *Nature* **314**, 111–114.

SETHIA, N., MANDELL, W. (1988), *Saccharomyces* fungemia in a patient with AIDS, *N. Y. State J. Med.* **88**, 278–279.

SHIH, A., COUTAVAS, E. E., RUSH, M. G. (1991), Evolutionary implications of primate endogenous retroviruses, *Virology* **152**, 495–502.

SIKORSKI, R. S., MICHAND, W., LEVIN, H. L., BOCKE, J. D., HIETER, P. (1990), *Nature* **345**, 582–583.

SIMON, R., FROMMER, W. (1993), Safety aspects in biotechnology, in: *Biotechnology 2nd Ed.*, Vol. 2 *Genetic Fundamentals and Genetic Engineering* (REHM, H.-J., REED, G., PÜHLER, A., STADLER, P., Eds.), pp. 835–853, Weinheim: VCH.

SINGER, M., SOLL, D. (1973), Guidelines for DNA hybrid molecules, *Science* **181**, 1114.

SMITH, H. W. (1988), The state and future of studies on bacterial pathogenicity, in: *Virulence Mechanisms of Bacterial Pathogens* (ROTH, J., Ed.), pp. 365–382, Washington, DC: American Society for Microbiology.

SPEHNER, D., KIM, A., DRILLIEN, R. (1991), Assembly of nucleocapsid structures in animal cells infected with a vaccinia virus recombinant encoding the measles virus nucleoprotein, *J. Virol.* **65**, 6296–6300.

STEWART, G. J., CARLSON, C. A. (1986), The biology of natural transformation, *Annu. Rev. Microbiol.* **40**, 211–235.

STEPHENNE, J. (1990), Production in yeast versus mammalian cells of the first recombinant DNA human vaccine and its proved safety, efficacy, and economy: Hepatitis B vaccine, *Adv. Biotechnol. Processes* **14**, 279–299.

STOYE, J. P., COFFIN, J. M. (1987), The four classes of endogenous murine leukemia virus: Structural relationships and potential for recombination, *J. Virol.* **61**, 2659–2669.

STOYE, J., COFFIN, J. M. (1988), Polymorphism of murine endogenous proviruses revealed by using virus class-specific oligonucleotide probes, *J. Virol.* **62**, 168–175.

STRAUSS, H. S., HATTIS, D., PAGE, G., HARRISON, K., VOGEL, S., CALDART, C. (1986), Genetically engineered microorganisms: II. Survival, multiplication and gene transfer, *Rec. DNA Tech. Bull.* **9**, 69–88.

STUHLMANN, H., BERG, P. (1992), Homologous recombination of copackaged retrovirus RNAs during reverse transcription, *J. Virol.* **66**, 2378–2388.

SUNDSTROM, P., ALIAGA, G. (1992), Molecular cloning and analysis of protein secondary structure of *Candida albicans* enolase, an abundant, immunodominant glycolytic enzyme, *J. Bacteriol.* **174**, 6789–6799.

SUTCLIFFE, J. G. (1978), Complete nucleotide sequence of the *E. coli* plasmid pBR 322, *Cold Spring Harbor Symp. Quant. Biol.* **43**, 77–90.

TARTAGLIA, J., PERKUS, M. E., TAYLOR, J., VORTON, E. K., AUDONNET, A. C., COX, W. I., DAVIS, S. W., VAN DER HOEVEN, J., MEIGNIER, B., RIVIERE, LANGUET, B., PAOLETTI, E. (1992), NYVAC: A highly attenuated strain of vaccinia virus, *Virology* **188**, 217–232.

TAYLOR, P. W. (1983), Bactericidal and bacteriolytic activity of serum against gram-negative bacteria, *Microbiol. Rev.* **47**, 46–83.

TAYLOR, P. W., KROLL, H. P. (1983), Killing of an encapsulated strain of *Escherichia coli* by human serum, *Infect. Immun.* **39**, 122–131.

TEICH, N. (1984), Taxonomy of retroviruses, in: *RNA Tumor Viruses* (WEISS, R. A., TEICH, N., VARMUS, H., COFFIN, J. M., Eds.), p. 25, Cold Spring Harbor, NY: Cold Spring Harbor Laboratory Press.

TEIFKE, J. P., WEISS, E. (1991), Nachweis boviner Papillomavirus-DNA in Sarkoiden des Pferdes mittels der Polymerase-Ketten-Reaktion (PCR), *Berl. Münch. Tierärztl. Wschr.* **104**, 185–187.

THIEL, S. (1992), Mannan-binding protein, a complement activating animal lectin, *Immunopharmacology* **24**, 91–99.

THOMAS, C. M., SMITH, C. A. (1987), Incompatibility group P plasmids: Genetics, evolution and use in genetic manipulation, *Annu. Rev. Microbiol.* **41,** 77–101.

TIEDJE, J. M., COLWELL, R. K., GROSSMAN, Y. L., HODSON, R. E., LENSKI, R. E., MACK, R. N., REGAL, P. J. (1989), The planned introduction of genetically engineered organisms: ecological considerations and recommendations, Report prepared for The Ecological Society of America, *Ecology* **70** (2), 97–315.

TREVORS, J., BARKAY, T., BOURGUIN, A. (1987), Gene transfer among bacteria in soil and aquatic environments: a review, *Can. J. Microbiol.* **33,** 191–197.

TUIJNENBURG, C. (1987), Air and surface contamination during microbial processing, *Swiss Biotech.* **5** (2a), 43–49.

TWIGG, A. J., SHERRATT, D. (1980), Trans-complementable copy-number mutants of plasmid ColE1, *Nature* **283,** 216–218.

UNIDO (1989), *An International Approach to Biotechnology Safety*, Vienna: UNIDO.

U.S. DEPARTMENT OF HEALTH AND HUMAN SERVICES (1988), *Biosafety in Microbiological and Biomedical Laboratories*, 2nd Ed., HHS publication # NIH 88-8395, Washington, DC: US Government Printing Office.

VAN DER WALT, J. P. (1970), Genus 16. *Saccharomyces* Meyen emend. Reess, in: *The Yeasts, a Taxonomic Study* (LODDER, J., Ed.), pp. 555–718, Amsterdam: Elsevier Biomedical Press.

VIDAVER, A. K. (1985), Plant-associated agricultural applications of genetically engineered microorganisms: projection and constraints, *Rec. DNA Tech. Bull.* **8,** 97–102.

WALLACE, R. J., FALCONER, M. L., BHARGAVA, P. K. (1989), Toxicity of volatile fatty acids at rumen pH prevents enrichment of *Escherichia coli* by sorbitol in rumen contents, *Curr. Microbiol.* **19,** 277–281.

WALSH, C. E., LIU, J. M., XIAO, X., YOUNG, N. S., NIENHUIS, A. W., SAMULSKI, R. J. (1992), Regulated high level expression of a human gamma-globin gene introduced into erythroid cells by an adeno-associated virus vector, *Proc. Natl. Acad. Sci. USA* **89,** 7257–7261.

WATSON, J. D., CRICK, F. H. (1953), Molecular structures of nucleic acids: a structure for deoxyribose nucleic acid, *Nature* **171,** 737–738.

WATSON, J. D., HOPKINS, N. H., ROBERTS, J. W., STEITZ, J. A., WEINER, A. M. (1987), *Molecular Biology of the Gene*, 4th Ed., Menlo Park, CA: Benjamins/Cummings Publishing Company, Inc.

WEBER, A., SCHÄFER, R. (1993), Vorkommen von Hefen in Kotproben von Hunden und Katzen, *Hautnah. Mykol.* **3,** 130–131.

WEIBEL, E. K., SEIFFERT, B. D. (1993), Biosafety investigations in an r-DNA production plant, *Appl. Microbiol. Biotechnol.* **39,** 227–234.

WEISS, R. A., TEICH, N., VARMUS, H., COFFIN, J. M. (1985), *RNA Tumor Viruses*, Cold Spring Harbor, NY: Cold Spring Harbor Laboratory Press.

WEISS, H. W., GOEDERT, J. J., GARTNER, S., POPOVIC, M., WATERS, D., MARKHAM, P., DI MARZO VERONESE, F., GAIL, M. H., BARKLEY, E., GIBBONS, J., GILL, F. A., LEUTHER, M., SHAW, G. M., GALLO, R. C., BLATTNER, W. (1988), Risk of human immunodeficiency virus (HIV-1) infection among laboratory workers, *Science* **239,** 68–71.

WHO (World Health Organization) (1987), Acceptability of cell substrates for production of biologicals, *Technical Report Series* **747,** Geneva: WHO.

WICKNER, R. B. (1992), Double-stranded and single-stranded RNA viruses of *Saccharomyces cerevisiae*, *Annu. Rev. Microbiol.* **46,** 347–375.

WIEDERMANN, G., AMBROSCH, F., KREMSNER, P., ANDRÉ, F., SAFARY, A. (1987), Reactogenicity and immunogenicity of different lots of a yeast-derived hepatitis B vaccine, *Postgrad. Med. J.* **63** (Suppl. 2), 109–112.

WIEDERMANN, G., SCHREINER, O., AMBROSCH, F., KRAFT, D., KOLLARITSCH, H., KREMSNER, P., HAUSER, P., SIMOEN, E., ANDRÉ, F. E., SAFARY, A. (1988), Lack of induction of IgE and IgG antibodies to yeast in humans immunized with recombinant hepatitis B vaccines, *Int. Arch. Allergy Appl. Immunol.* **85,** 130–132.

WILLETS, N. (1988), Conjugation, *Methods Microbiol.* **212,** 50–77.

WINKLER, E. L., HOCKSTRA (1985), *Safe Microbiological Technique*, p. 29.

WINNACKER, E.-L. (1987), *From Genes to Clones. Introduction to Gene Technology*, Weinheim: VCH.

WOLD, W. S. M., GOODING, L. R. (1991), Region E3 of adenovirus: A cassette of genes involved in host immunosurveillance and virus-cell interactions, *Virology* **184,** 1–8.

WURM, F., ZETTLMEISSL, G. (1989), Biotechnologische Zellkultur: Herstellung von Biologika für die Medizin, *BioEngineering* **5,** 43–50.

ZINKE, H. (1990), Institut für Biochemie, TH Darmstadt, personal communication.

ZUR HAUSEN, H. (1981), Papilloma viruses, in: *DNA Tumor Viruses* (TOOZE, J., Ed.), 2nd Ed., pp. 371–382d, Cold Spring Harbor, NY: Cold Spring Harbor Laboratory Press.

ZUR HAUSEN, H. (1991), Human papilloma viruses in the pathogenesis of anogenital cancer, *Virology* **184,** 9–13.

4 Biotechnology and Bioethics: What is Ethical Biotechnology?

DARRYL R. J. MACER

Tsukuba Science City, Japan,
Christchurch, New Zealand

1 Biotechnology and Bioethics

As has been described in other volumes of this series, modern biotechnology has had a great impact on medicine and agriculture. It can only be expected to have an even more dominating impact on future science and technology. It's impact is not limited to the technical impact that these advances have upon industry, medicine and agriculture, any technology influences society, and one can expect that life science technology potentially has the greatest impact.

Biotechnology has also influenced the thinking of society, as will be discussed in this chapter, and we can expect further paradigm shifts to occur. These paradigm shifts include the switch to biodegradable products, industrial pressures to restructure scientific information sharing, the paradigm of sustainable and limited economic growth, and the paradigm of intervention in nature rather than observation and participation in it. Biotechnology has also been a catalyst to the consideration of bioethical issues (MACER, 1990), and the two words, biotechnology and bioethics, have coevolved.

Before extending discussion it is essential to define what is meant by the words, biotechnology and bioethics. This in itself is no easy task because different people with different interests can broaden or narrow these concepts. In this chapter a broad meaning of biotechnology is taken; the use or development of techniques using organisms (or parts of organisms) to provide or improve goods or services. Bioethics is the study of ethical issues associated with life, including medical and environmental ethics. Both of these words are recent terms, but both topics are found throughout human history.

2 Bioethics

There are large and small problems in ethics; there are global, regional, national, community and individual issues. We can think of ethical issues raised by biotechnology that involve the whole world, and issues which involve a single person. A global problem such as global warming may be aided by global applications of biotechnology, for example, to reduce net atmospheric carbon dioxide increase by reducing emissions or increasing biomass, however, excess consumption and energy use can only be solved by individual action, to reduce energy use. A regional issue is the risk presented by the introduction of new organisms or of an unstable genetically modified organism (GMO) into the environment, but it also involves individual responsibility to ensure that sufficient care and monitoring of the release is made. Other ethical issues arising from biotechnology that are thought of as individual issues such as genetic testing, or use of gene therapy, also have societal implications.

We hardly need to ask why we need ethics, rather we need to ask what principles and factors are crucial for guiding decision-making, especially over such a diverse spectrum of issues. Medical ethics involves decision-making on a personal level, it concerns the patient and the health care professional, especially the physician. At a further level away may be many others who will be indirectly affected by such questions as the cost of very expensive treatment that takes funds away from other patients. At this level higher policy-making is required, as in the case of issues such as environmental risk, or intellectual property protection policy.

Some key principles of ethics are outlined below, with brief discussion of their relevance to biotechnology issues. We should balance the implications that arise from each principle to arrive at more ethical decisions. We may need to develop further principles, and bioethics is still being developed (MACER, 1990, 1994).

2.1 Autonomy

All people are different. This is easy to see, if we look at our faces, sizes and the clothes that we chose to wear. This is also true of the choices that we make. We may decide to play tennis, or golf, or chess, read a book, or watch television. These are all personal choices. In a democratic society we recognize that we have a duty to let people make their own choices. Above the challenges of new technologies, and increasing knowledge, the challenge of respecting people as equal persons with their own set of values is a challenge for all. This is also expressed in the language of rights, by recognizing the right of individuals to make choices.

2.2 Rights

Legal rights are claims that would be currently backed by the law if the case went to court, while human rights are critical to maintaining human dignity but may not have yet attained legal recognition. The recognition of human rights has changed the situation in many countries, and many countries in the world have signed the U.N. Declaration of Human Rights (SIEGHART, 1985), or one of the regional versions of this. This can be applied to many situations, for example, we all have a right to be involved in decisions about our country, the freedom of religion, or speech, to raise a family, to share in the benefits arising from scientific advances, and a right to a reasonable future. Some of these rights are difficult to define, as human "rights" have always been. Respect for personal rights should change the nature of relationships between people in power and people without power from being characterized by authoritarianism or paternalism to becoming a partnership.

Ethics is not the same as law. Ethics is a higher pursuit, doing more than the law requires. The law is needed to protect people and to set a minimum standard, but you cannot determine good moral behavior by settling cases in a court of law. We only need to think of medical litigation or environmental damage penalties, which can lead to huge sums of money being paid for accidents (or negligence) which cannot really by compensated by monetary reimbursement. The solution is to have more careful and moral physicians, companies, lawyers, and politicians, and the replacement of monetary balance

sheets by ethical values, as the primary motive of decision-making.

2.3 Beneficence

One of the underlying philosophical ideas of society is to pursue progress. The most cited justification for this is the pursuit of improved medicines and health. It has often been assumed that it is better to attempt to do good than to try not to do harm. A failure to attempt to do good, working for people's best interests, is taken to be a sin of omission. Beneficence is the impetus for further research into ways of improving health and agriculture, and for protecting the environment. Beneficence supports the concept of experimentation, if it is performed to lead to possible benefits.

The term beneficence suggests more than actions of mercy, for which charity would be a better term. The principle of beneficence asserts an obligation to help others further their important and legitimate interests. It means that if you see someone drowning, providing you can swim, you have to try to help by jumping in the water with him/her. It also includes the weighing of risks, to avoid doing harm.

Governments have a duty to offer their citizens the opportunity to use new technology, providing it does not violate other fundamental ethical principles. Just the definition of what fundamental ethical principles are, may be culturally and religiously dependent, especially in the way that they are balanced when opposing principles conflict (see Sect. 3). Although different cultures vary, they all share some concepts of beneficence and do no harm. People should be offered the option of using new technology in medicine and agriculture, and such applications should be made, providing internationally accepted ethical and safety standards are applied.

Beneficence also asserts an obligation upon those who possess life-saving technology, in medicine or agriculture, to share their technology with others who need it. This is relevant to biotechnology companies also, who may hold patent rights on particular processes, beneficence would assert that they must share it with others, even if they cannot pay for it. This may mean that companies share developments with developing countries, or give new drugs to individuals too poor to purchase them, and may conflict with so-called business ethics.

2.4 Do No Harm

The laws of society generally attempt to penalize people who do harm, even if the motive was to do good. There needs to be a balance between these two principles, and it is very relevant to areas of science and technology where we can expect both benefits and risks. Importantly, we must balance risks versus benefits of different and often alternative technologies, then apply these comparisons to our own behavior, as well as in determining government policy.

Do no harm is a very broad term, but is the basis for the principles of justice and confidentiality, and philantropy. It can also be expressed as respect for human life and integrity. This feature is found in the Hippocratic tradition and all other traditions of medical and general ethics. To do no harm is expressed more at an individual level, whereas justice is the expression of this concept at a societal level. Do no harm has been called the principle of non-maleficence.

Biotechnology and genetic engineering are providing many benefits, but there are also many risks. It is also unclear who will really benefit the most. It is important to see these benefits and risks in an international way because the world is becoming smaller and ever more interdependent. Biotechnology affects the lives of people throughout the world (WALGATE, 1990). All people of the world can benefit if it is used well, through medicines, and more environmentally sustainable agriculture. However, biotechnological inventions that allow industrialized countries to become self-sufficient in many products will change the international trade balances and prosperity of people in developing and industrialized countries. If developing countries cannot export products because of product substitution, the result may be political instability and war. This may in the end become

the biggest risk. For example, the use of enzymic conversion of corn starch into high fructose corn syrup causes serious damage to the economies of sugar exporting nations (SASSON, 1988), and may already have caused political instability there. We need to remember national and international issues.

Although we will continue to enjoy the many benefits to humanity, and we may hope for environmental benefits, the price of the new technology is that it may make us think about our decisions more than in the past. This is long overdue! International food safety and environmental standards should be speedily developed to ensure that all people of the world share their protection, and no country becomes a testing ground for new applications.

2.5 Justice

Those who claim that individual autonomy comes above societal interests need to remember that the reason for protecting society is because it involves many human lives, which must all be respected. Individual freedom is limited by respect for the autonomy of all other individuals in society and the world. People's well-being should be promoted, and their values and choices respected, but equally, which places limits on the pursuit of individual autonomy. We also need to consider interests of future generations which places limits on this generation's autonomy. We also need to apply this principle globally, as discussed above; no single country should pursue policies which harm people of any country.

The key principle arising from the high value of human life is respect for autonomy of each individual human being. This means they should have the freedom to decide major issues regarding their life, and is behind the idea of human rights. This idea is found in many religions also. Part of autonomy is some freedom to decide what to do, as long as it does not harm others, also called individual liberty or privacy. Well-being includes the principle of "do no harm" to people, and to work for people's best interests.

Internationally, the area of biotechnology patent policy should be examined in light of public opinion and the principle of justice. Naturally occurring genetic resources should not be able to be owned by any one individual or company, rather they should be common. At the same time, some patent protection for specific applications involving biotechnology need to be protected to encourage further research, and to make the results of such research immediately open for further scientific research (see Sect. 7).

2.6 Confidentiality

The emphasis on confidentiality is very important. Personal information should be private. There may be some exceptions when criminal activity is involved or when third parties are at direct risk of avoidable harm. It is very difficult to develop good criteria for exceptions, and they will remain rare. We must be careful when using computer databanks that contain personal information, and if they cannot be kept confidential, the information should not be entered into such a bank.

A feature of the ethical use of new genetics is the privacy of genetic information. This is one of the residual features of the existing medical tradition that needs to be reinforced. It is not only because of respect or people's autonomy, but it is also needed to retain trust with people. If we break a person's confidences, then we cannot be trusted. If medical insurance companies try to take only low risk clients by prescreening the applicants, there should be the right to refuse such questions (HOLTZMAN, 1989). The only way to ensure proper and just health care is to enforce this on employers and insurance companies, or what is a better solution, a national health care system allowing all access to free and equal necessary medical treatment. We need to protect individuals from discrimination that may come in an imperfect world, one that does not hold justice as its pinnacle.

2.7 Animal Rights

These above principles apply to human interactions with other humans. However, we also interact with animals, and the environment.

The moral status of animals, and decisions about whether it is ethical for humans to use them, depends on several key internal attributes of animals; the ability to think, the ability to be aware of family members, the ability to feel pain (at different levels), and the state of being alive. All will recognize, inflicting pain is bad, so if we do use animals we should avoid pain (SINGER, 1976). If we believe that we evolved from animals we should think that some of the attributes that we believe humans have, which confer moral value on humans, may also be present in some animals (RACHELS, 1990). Although we cannot draw black and white lines, we could say that because some primates or whales and dolphins appear to possess similar brain features, similar family behavior and grief over the loss of family members to humans, they possess higher moral status than animals that do not exhibit these. Therefore, if we can achieve the same end by using animals that are more "primitive" than these, such as other mammals, or animals more primitive than mammals, then we should use the animals at the lowest evolutionary level suitable for such an experiment, or for food production (which is by far the greatest use of animals). If we take this line of reasoning further, we conclude that we should use animal cells rather than whole animals, or use plants or microorganisms for experiments, or for testing the safety of food.

Animals are being used for genetic engineering, for use as models of human disease, for use in the production of useful substances such as proteins for medical use, and in the more traditional uses in agriculture. Some of these uses, such as the production of mutations in strains of animals to study human disease will have human benefit, but are more ethically challenging because some of these strains may be deliberately made sick and will therefore probably feel pain (MACER, 1989, 1991a).

2.8 Environmental Ethics

Humans also have interactions with the environment, and in fact depend upon the health of the environment for life. The easiest way to argue for the protection of the environment is to appeal to the human dependence upon it. There are also human benefits that come from products we find in nature, from a variety of species we obtain food, clothing, housing, fuel and medicine. The variety of uses also supports the preservation of the diversity of living organisms, biodiversity. As we have learnt, the ecosystem is delicately balanced, and the danger of introducing new organisms into the environment is that this may upset that balance. This is another key issue raised by genetic engineering. However, we have been using agricultural selection for 10000 years, so the introduction and selection of improved and useful microorganisms, plants and animals is nothing new, and we should learn from mistakes of the past.

The above arguments should convince people of the value of the environment, and that is a first stage. However, it appeals to our sense of values based on human utility. There is a further way to argue for the protection of nature and the environment, and it is a more worthy paradigm. It is that nature has value in itself because, it is there. We should not damage other species, unless it is absolutely necessary for the survival of human beings (not the luxury of human life). Nature has life, thus it has some value. Another paradigm for looking at the world is a religious view, that God made the world so the world has value, and we are stewards of the planet, not owners. This paradigm can make people live in a better way than if they look at the world only with the paradigm of human benefit.

There needs to be examination of the view of nature that different people have, so that we can find what the commonly acceptable limits to modification of nature, plant and animal varieties, and human beings are. In the modern world any new science can easily spread, so researchers are accountable to all peoples of the world. There will be future possible applications of technology which are against "common morality". We need to know what these perceived limits of changing nature are, before we grossly change the characters of individual organisms, or make irreversible changes to the ecosystem and human society. A study on the images revealed in open questions about "life" and "nature" in different countries of an International Bioethics Survey showed universality of thinking about these concepts (MACER, 1994).

Microorganisms are generally placed at the lowest end of the "scale" of ethical status, because the only internal character they have is the state of being alive. External factors from a human esthetic viewpoint mean that the only argument usually applied to them is human utility.

Biodiversity may have some value in itself, though it is yet to be defined in non-religious terms. If we want to preserve biodiversity, it is essential that we separate parts of nature on land and ocean as nature reserves or parks, away from the parts of nature which are agricultural areas. However, while we separate these areas physically we should not separate them psychologically as areas which we can abuse and areas which we protect. This applies both in terms of sustainable environment protection and animal rights. In fact, agricultural biodiversity is of direct human utility, and we should attempt to stop its continued loss (FOWLER and MOONEY, 1990).

2.9 Decision-Making

To anyone who starts to try to apply these principles in his/her daily life or to decisions concerning biotechnology, it will very soon be apparent that there needs to be a balancing of conflicting principles of ethics. Different interests will conflict, so, for example, there are exceptions to the maintenance of privacy and confidentiality if many people or large environmental damage, are threatened. How do we balance protecting one person's autonomy with the principle of justice, that is protecting all people's autonomy. Many medical and scientific procedures are challenging because they involve technology with which both benefits and risks are associated, and will always be associated.

Human beings are challenged to make ethical decisions, and to balance the benefits and risks of alternatives, they have to. The benefits are great, but there are many possible risks. In this regard utilitarianism, that we should attempt to produce the most happiness and benefit, will always have some place, though it is very difficult to assign values to different interests and to the degree of "happiness" or "harm". Although our life may become easier due to technological advances, so that it may appear that we do not need to make so many decisions, we are challenged to make more decisions than in the past. The more possibilities we have, the more decisions we make (MACER, 1990).

Standards of education are increasing, but it is another thing whether people are educated for decision-making. People need to be taught more about how to make decisions, and the education system should accommodate this need of modern life. Even if they are, this may still be no guarantee that the right decisions will be made.

3 Cross-Cultural Bioethics

Any attempt to develop international bioethical approaches must involve consideration of the values of all peoples. We could call this cross-cultural bioethics. This means something different from universalism – attempts to define an international ethical code of what is ethical and what is not, or a table of acceptable and unacceptable risks based on consideration of ethical principles.

Universal ethics (MACER, 1994) may be possible to a significant degree, however, we even have difficulty in universal recognition of basic laws such as those respecting human rights. The existence of international environmental laws, e.g., The Law of the Sea, and charters of human rights (SIEGHART, 1985), is some encouragement for the future progress of limited universalism. We also see attempts within regions, such as by the Council of Europe, to devise a European Convention on Bioethics (EP, 1991; MUNDELL, 1992; HOLM, 1992).

Cross-cultural bioethics involves mutual understanding of various cultural, religious, political and individual views that people have. The diversity of individual viewpoints in any one culture appears to exceed the differences between any two. For example, in any culture one can find people fervently opposed to induced abortion and those who support it as a "right" for women's choice. The opinions expressed in the responses to an International Bioethics Survey, conducted in Australia, Hong Kong, India, Israel, Japan, New Zealand, The Philippines, Russia, Singapore and Thailand in 1993, suggest that people in these diverse countries have a similar variety of reasoning (MACER, 1992a, 1994). This type of research provides to better understand the reasoning of people. People in many countries do share the same hopes and fears, which strengthens the call for international standards.

If we look at declarations of ethical codes made by different religious groups, professional groups, and among different nations, we can see the principles of bioethics that were outlined in the above section in most. A key question in cross-cultural bioethics is how the concept of do no harm should be applied, and to what beings it applies. For example: At what stage of development should human embryos be legally protected, for *in vitro* research or abortion decisions? Which animals should be protected from which research or use? How do we balance justice within national boundaries with global distributive justice, and justice to future generations? How much individual liberty do we allow when individual choices affect society values and options for other people or beings? What is necessity and what is human desire or luxury? What is the level of acceptable risk of harm?

These are wide questions, and this chapter will discuss some of them. For the purposes of this volume the discussion will be focused around the question of what ethical biotechnology is, and developing approaches that may allow us to better answer this question for policy development.

4 Perceptions of Ethical Biotechnology

4.1 "Moral" is Not the Same as Ethical

What we call "ethical biotechnology" cannot be decided just by public opinion. However, something which is morally offensive to the majority of people in a country, or region, or world-wide, is judged to be immoral and is likely to be outlawed. What is seen as immoral is often also unethical, though unethical practices are often tolerated by a society and thus our definition of moral would say that they are "morally acceptable", because it is "common morality". For example, people living in industrialized countries enjoy the fruits of an economic system that is disruptive to people living in developing countries and the environment. By use of basic ethical principles of distributive justice, and justice to future generations who will have to live in a polluted and changed world, we would say it is unethical. However, this situation is "common morality" to a majority of the people living in the rich countries, though the proportion may be falling, and it is morally unacceptable to the poor of developing countries. If we drew our definition of morality at national or regional borders, we would see this mixed morality standard, but if we drew our morality from a global majority we would see it as immoral.

Decisions may be made democratically in a country if a consensus supports them, if the rights of minority groups are respected, and if it makes sense in the long term, both nationally and internationally. However, not all decisions made this way will be ethical, society can make unethical majority decisions and will continue to do so. In the area of biology and genetics, we should never forget the unethical compulsory eugenics that swept the world in the first half of this century, when more than 40 countries made laws to enforce mandatory sterilization and selective immigration policies (KELVES, 1985), nor should we forget the environmental destruction that still continues today. We cannot say that these abuses are always based on ignorance, rather they are sustained by groups of people pursuing their own interests who can lead the public into following the pattern of living that will sustain the people in power in those positions. Usually appeals are made to the selfish side of human personality, that we all possess. Rather, we should be concerned with global sustainability and protection of the rights of all people.

We must remember this distinction between ethical and moral when we look at public opinion. Law is often based on the so-called common morality of a country, and in the area of biotechnology we can see varying laws established by different countries, and even within Europe there are conflicting laws, for example, in the area of assisted reproduction and the use of human embryo experiments for research. Germany prohibits research as a criminal offense (DEUTSCH, 1992), and Britain permits approved research until the embryo is 14 days old (BOLTON et al., 1992). The laws on the contained use or release of genetically engineered microorganisms vary between different countries, due partly to different public perceptions of risks.

Whenever we consider the results of opinion surveys, we need to remember the axiom "Lies, damn lies and statistics". Nevertheless, they are an important gauge of public opinion, and when combined with the results of methods that allow the thinking behind such results to be determined, they are important in sociological study. Governments and companies involved in biotechnology research have become careful in their monitoring of public opinion, for in the case of governments it can mean they are not reelected, and public opposition to companies can be expensive in terms of time delays and lost sales.

Most people receive information via the mass media, especially the newspaper and television. The media have a large responsibility to communicate science issues well, and scientists should also inform people about science. The media have a responsibility to present balanced information, on the benefits and risks of alternative technologies and to do this independently of commercial interests. Public opinion can be influenced by groups who have a special interest, such as political groups, and other groups, whose members spend time to publicize their opinions, and who can get media coverage of their views. The media may also have their own special interests.

4.2 Mixed Perception of Benefit and Risk

There have been some opinion polls conducted on the topics of biotechnology, and these are the subject of other chapters. Many of these opinion polls have limited meaning because they ask set questions with set responses, allowing little room for free response. The responses are often suggestive, and cannot give us the real picture of what the public is thinking. The extra time spent in analysis of free response questions may be well worth the investment if the underlying reasoning is to be determined. Even the use of questions looking at the balance of benefit and risk are more useful than asking single questions (MACER, 1992a, 1993, 1994).

In August–October 1991, a series of public opinion surveys were conducted in Japan (MACER, 1992a). Mailed nationwide opinion surveys on attitudes of biotechnology were conducted in Japan, among randomly selected samples of the public ($N=551$), high school biology teachers ($N=228$) and scientists ($N=555$). The results of several of the 20 questions are summarized in this chapter, as they are useful in examining what are seen as "bioethical" conflicts of biotechnology, that will be examined in Sect. 5. The results were compared with the results of the same questions used in New Zealand in May 1990 (COUCHMAN and FINK-JENSEN, 1990), among samples of the public ($N=2034$), high school biology teachers ($N=277$) and scientists ($N=258$).

People were first asked about their awareness of eight developments of science and technology (Q5a), then asked whether they thought each development would have a benefit for Japan or not (Q5b). Q5c and Q5d examined their perceptions about the risks of

technology by asking them how worried they were about each development. The questions were:

Q5. We will ask you about some particular scientific discoveries and developments. Can you tell me how much you have heard or read about each of these. Please answer from this scale.
1 I have not heard of this.
2 I have heard of this, but know very little/nothing about it.
3 I have heard of this to the point I could explain it to a friend.
How much have you heard or read about?
Biological pest control   Silicon chips
Biotechnology   Fiber optics
Agricultural pesticides   *in vitro* Fertilization
Superconductors   (IVF)
                           Genetic engineering

Q5b. For each of these developments that you have heard of, do you personally believe (DEVELOPMENT) would be a worthwhile area for scientific research in Japan (NZ)?
1 Yes   2 No   3 Don't know

Q5c. In the area of (DEVELOPMENT) do you have any worries about the impact of research or its applications?
1 Yes   2 No   3 Don't know

Q5d. For each development where you are worried, could you please tell me how worried you are, using this scale ... about the impacts of (DEVELOPMENT)?
1 I am slightly worried about this.
2 I am somewhat worried about this.
3 I am very worried about this.
4 I am extremely worried about this.

Japanese have a very high awareness of biotechnology, 97% saying that they had heard of the word (Tab. 1). They also have a high level of awareness of *in vitro* fertilization (IVF) and genetic engineering. In New Zealand only 57% said they had heard of biotechnology. These results were confirmed with other surveys in Japan, and in the 1993 International Bioethics survey (MACER, 1994). In a 1988 survey of 2000 public in the U.K. only 38% of respondents said they had heard of biotechnology (RSGB, 1988), considerably less than in New Zealand, and compared to 97% in Japan in this survey in 1991. The result of Q5 suggests that the Japanese public is

Tab. 1. Attitudes to Developments in Science and Technology

Sample	Public		High School Biology Teachers		Scientists	
	Japan	NZ	Japan	NZ	Japan	NZ
Biological Pest Control						
Heard of (*N*)	403	1668	211	276	511	256
Worthwhile	84.1	85.8	94.8	99.3	96.3	96.9
Not worried	43.7	49.7	45.4	34.1	60.3	32.4
Slightly worried	14.1	6.0	12.8	47.8	7.2	48.4
Somewhat worried	16.9	11.3	20.9	9.4	11.4	12.5
Very worried	10.9	22.2	12.3	3.6	8.4	3.9
Extremely worried	4.2	9.6	4.7	0.7	2.2	0.8
Don't know	26.0	0.7	8.2	0	3.7	2.0
Biotechnology						
Heard of (*N*)	516	1151	217	253	542	225
Worthwhile	84.9	71.7	93.5	84.6	97.4	81.3
Not worried	38.2	68.4	35.0	42.3	53.7	46.2
Slightly worried	14.1	2.5	8.3	39.5	11.4	39.1
Somewhat worried	16.9	6.0	22.6	6.7	16.4	6.2
Very worried	13.0	14.3	20.7	2.0	11.6	0.4
Extremely worried	8.5	7.2	8.3	0.4	6.1	0
Don't know	21.5	1.6	6.9	9.1	4.4	8.0
Pesticides						
Heard of (*N*)	509	1861	217	259	533	241
Worthwhile	89.2	84.5	87.6	79.2	94.7	82.2
Not worried	27.1	38.9	24.0	7.7	43.9	15.4
Slightly worried	7.9	14.2	6.0	22.4	7.1	30.7
Somewhat worried	14.9	18.1	13.4	32.4	16.7	28.6
Very worried	25.1	22.1	29.0	18.9	20.6	13.3
Extremely worried	18.1	6.3	22.6	13.1	11.1	8.3
Don't know	16.1	0.4	7.3	0	2.6	3.7
Genetic Engineering						
Heard of (*N*)	495	1492	214	273	540	247
Worthwhile	76.2	57.4	91.6	85.7	94.4	79.8
Not worried	19.4	43.8	16.4	10.6	43.3	14.2
Slightly worried	11.1	14.6	7.9	37.4	11.3	39.3
Somewhat worried	18.2	13.8	14.5	31.9	16.5	22.3
Very worried	21.4	18.6	31.3	10.6	17.2	13.8
Extremely worried	19.8	8.2	25.2	8.4	12.8	8.9
Don't know	19.8	0.9	7.9	1.1	3.2	1.6

The values are expressed as % of the number of respondents to Q5 who had heard of, or could explain, each development (*N*). Results from Japan are from MACER (1992a), and New Zealand results are from the survey of COUCHMAN and FINK-JENSEN (1990)

comparatively very well exposed to the word 'biotechnology', with 33% (29% in 1993) saying they could explain it, compared to only 8% (15% in 1993) in New Zealand.

The responses revealed that there are mixed perceptions of benefit and risk from the use of these technologies (Fig. 1). Biotechnology was seen to be worthwhile by 85% of the public, while 40% were worried about research. Genetic engineering was said to be a worthwhile research area for Japan by 76%, while 58% perceived research on IVF

Public perception of science developments in Japan

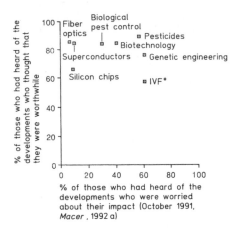

% of those who had heard of the
developments who were worried
about their impact (October 1991,
Macer, 1992 a)

Public perceptions of science developments in New Zealand

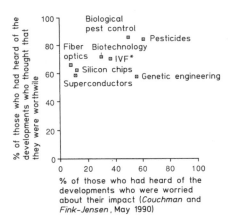

% of those who had heard of the
developments who were worried
about their impact (*Couchman* and
Fink-Jensen, May 1990)

Fig. 1. Comparative perceptions of science developments between Japan and New Zealand. The results are based on the number of respondents who said that they had heard of each development (Q5a), and are presented as scattergrams, with number of respondents who thought each development was worthwhile for their country versus the number of respondents who were worried about the impact of the developments.

as being worthwhile; however, 61% were worried about research on IVF or genetic engineering (Tab. 1). Japanese expressed more concern about IVF and genetic engineering than New Zealanders. People of all groups expressed a relatively high degree of worry about biotechnology, IVF and genetic engi-

neering when compared to other developments of science and technology (Fig. 2). These results were also confirmed in 1993, using extra research area of "computers", and "nuclear power" (MACER, 1994). Nuclear power gathered significantly more and greater concern than other areas, in all countries. The later survey also included open questions on the reasoning for both the plus and minus side of each area which is reported fully elsewhere (MACER, 1994).

From the results of this question it appears that people had a mixed view of the benefits of science. The attitudes to biotechnology vary even within Europe (EB, 1993). A following question in these surveys asked them "Overall do you think science and technology do more harm than good, more good than harm, or about the same of each?" This looked at their general attitude to the benefit and/or harm perceived to be done from science in general. The Japanese high school biology teachers and the public gave similar responses in 1991 (MACER, 1992a) with 58% and 56%, respectively, thinking that they did more good than harm. Only 6% of the public and 3% of the teachers thought that science and technology did more harm. Scientists had a more optimistic picture, with 78% saying science and technology did more good and only 2% saying it did more harm than good.

In a 1989 public survey in the U.K., ($N=1020$) for the same question, only 44% answered "more good", 37% said "about the same", 9% said "more harm", and 10% "didn't know" (KENWARD, 1989). These results were similar to the U.K. in 1985. In Australia, in 1989, 757 public were asked the same question, and 56% said "more good", 26% said "about the same", and 10% said "more harm", with 2% saying they "didn't know" (ANDERSON, 1989). In 1993 in Australia 66% were found who said "more good", 27% said "about the same", and 4% said "more harm", with 3% saying "don't know", but in Japan reduced support with 42% saying "more good", 45% "about the same", 5% "more harm", and 8% "don't known" (MACER, 1994). Australians were the most optimistic of all respondents in the International Bioethics Survey for the broad question, but not for specific questions about different

The degree of worry about the impact of science developments in Japan

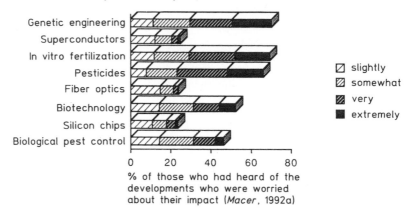

% of those who had heard of the
developments who were worried
about their impact (*Macer*, 1992a)

The degree of worry about the impact of science developments in New Zealand

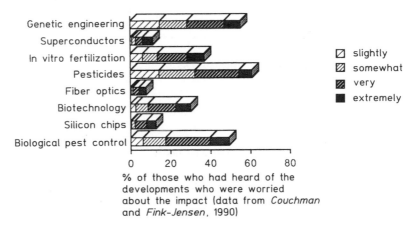

% of those who had heard of the
developments who were worried
about the impact (data from *Couchman*
and *Fink-Jensen*, 1990)

Fig. 2. Comparative public concern about the impact of developments in science and technology between Japan and New Zealand. The results are based on the number of respondents who said that they had heard of each development (Q5a), and the degree of worry about the impact of the developments (Q5d) is plotted. * IVF, *in vitro* fertilization.

areas of technology. Internationally, the most optimistic respondents to this question were respondents to a 1989 survey in Beijing, China ($N=4911$), where 82% said more good, 2% said more harm, 12% said the same and 5% said "don't know" (ZHANG, 1991).

In most countries science and technology have been promoted as being of benefit to people, by government and industry. Awareness of the developments of science and technology in Q5 was not directly correlated with the perception of these technologies. The responses are more complex than Q6 may indicate, for example, New Zealand scientists and high school teachers expressed more general concern about the impact of all developments in Q5 than their peers in Japan, whereas the New Zealand public expressed less concern about all these developments than the Japanese public! Bioethically, a more mature group may express most selectivity in concern between different applications of technology.

4.3 Reasoning Behind Acceptance or Rejection of Genetic Manipulation

It is apparent that people do, on balance, see more benefits coming from science than harm. However, at the same time people also perceive risks, such as human misuse of technology. We need to determine more about these perceptions. More specific questions than those asked in Q5, were used in Q7 (MACER, 1992a). Rather than testing concerns about the techniques included by the broad term "genetic engineering", the views of genetic manipulation on four types of organisms were examined: humans, animals, plants and microbes, with room for free response to list reasons for acceptance, benefits and risks perceived. The questions were:

Q7a. Can you tell me how much you have heard or read about ...?
Manipulating genetic material in human cells
Manipulating genetic material in microbes
Manipulating genetic material in plants
Manipulating genetic material in animals
Use this scale ...
1 I have not heard of this.
2 I have heard the words but no more.
3 I have heard the words and have some understanding of the idea behind it.

Q7b. Please answer the questions below: Which, if any, of those biological methods you've heard of are acceptable to you for any reasons?
1 Acceptable
2 Unacceptable (If unacceptable write why each one is not acceptable to you)

Q7c. Which of those biological methods, if any, of those you've heard of could provide benefits for Japan?
1 No benefit
2 Benefit (If a benefit, what benefits do you believe each one could produce?)

Q7d. Which, if any, of those biological methods could present serious risks or hazards in Japan?
1 Risk
2 No risk (If a risk, what serious risks or hazards do you believe each one could present in Japan?)

In both Japan and New Zealand, genetic manipulation of plants is the most acceptable type, followed by genetic manipulation of microbes, animals and humans, in order of decreasing acceptability (Tab. 2). About half of the teacher and scientist samples think that human genetic manipulation (human cells) is acceptable, in both countries. This preference order is the same as that obtained in the USA in November 1986 (OTA, 1987). This could represent two major thoughts, a scale of biological complexity associated with increasing ethical 'status' from microbes to plants to animals to humans. This is complicated by the higher perceived danger of genetic manipulation using microbes, which are associated with disease and which are environmentally more mobile, and animals which are mobile. The results of questions on the benefits and risks (Q7c, d) found most people perceived both benefit and risk, as in Q5. The results for public and scientists are schematically represented in Fig. 3.

The reasons for unacceptability, benefit or risk perception were asked, and they are perhaps the most interesting result. For different organisms they were different, reflecting reality. This is particularly useful for those working on an organism such as microorganisms, the focus of this book series, for the perceptions will vary greatly from those associated with human genetic manipulation. The method used to analyze the reasoning was to assign the comments to categories. A total of 38 different categories were used in the computer data analysis, some of which were later combined in data presentation. Although a variety of comments were written, generally they could easily be assigned to categories. For each distinct reason given in the comment, a count of 1 was scored in one of the categories of the data sheet in the computer. The most reasons given for a single comment were 3, but generally there were only 1 or 2 reasons. Also, a high proportion did not write any comment. A summary of the reasoning is presented in Tabs. 2, 3 and 4.

University students and academics in social sciences were also surveyed in Japan; for the results and many examples of the actual comments, see MACER (1992a). Interestingly, more public wrote comments for reasons for acceptability than scientists.

Tab. 2. Reasons Given for Unacceptability of Genetic Manipulation

Organism	Group	Japan				New Zealand			
		H	P	M	A	H	P	M	A
% who said it was acceptable	P	26.0	80.9	72.8	54.2	42.5	85.4	71.1	56.4
	T	46.6	87.8	82.4	75.6	48.7	87.3	72.2	81.6
	S	54.6	92.5	89.9	24.4	53.8	82.7	75.2	77.4
Unacceptable for the following reasons (% total respondents):									
Interfering with nature, unnatural	P	12.3	4.7	4.1	9.5	16.1	5.1	6.4	9.6
	T	3.6	2.2	2.2	3.6	5.1	1.4	1.0	3.0
	S	2.8	1.1	1.3	2.4	3.7	2.6	2.0	3.2
Playing God	P	10.8	2.3	2.0	4.9	—	—	—	—
	T	5.0	0.9	0.9	1.3	—	—	—	—
	S	4.5	0	0.4	1.5	—	—	—	—
Unethical	P	4.9	0.2	0.1	3.0	9.2	—	—	15.2
	T	9.4	0.4	0.5	2.4	22.1	0	0	6.1
	S	6.0	0.2	0	2.2	3.7	—	—	2.7
Disaster, out of control	P	5.1	0.6	1.2	2.6	9.2	1.7	3.5	3.9
	T	2.2	1.8	4.5	2.3	6.2	1.1	5.4	3.3
	S	1.9	0.6	1.7	1.5	6.5	3.6	8.7	4.5
Fear of unknown	P	6.1	1.8	1.5	4.7	4.6	1.6	4.6	3.5
	T	7.2	0.9	1.8	2.7	5.6	1.2	1.5	3.3
	S	6.5	1.1	1.7	2.4	4.6	2.2	4.2	4.1
Ecological effects	P	5.7	5.1	3.8	6.5	—	1.7	—	5.2
	T	1.4	2.2	1.8	3.2	0	0.9	0.5	0
	S	0.8	0.9	1.3	1.9	0	3.6	1.5	0.5
Feeling	P	4.9	1.8	1.8	3.1	—	—	—	—
	T	0.5	0	0	0.5	—	—	—	—
	S	1.9	0.2	0.4	1.5	—	—	—	—
Humanity changed	P	3.1	0	0.1	0	—	—	—	—
	T	1.8	0	0	0	3.1	0	0.5	0.4
	S	1.9	0	0	0.2	3.2	—	1.0	0.5
Insufficient controls	P	2.8	0.2	0.1	1.2	4.0	1.2	2.9	3.0
	T	5.0	2.2	3.1	3.2	9.8	1.4	2.6	4.4
	S	9.0	0.4	0.2	2.2	9.7	4.8	7.7	6.1
Danger of human misuse	P	3.0	0.6	0.4	1.7	5.2	1.2	3.8	2.6
	T	2.2	1.7	2.3	2.3	6.2	0.2	1.8	1.3
	S	4.1	0.7	0.6	1.5	5.1	1.7	4.7	2.7
Eugenics, cloning	P	3.7	0	0	0.2	2.9	—	—	—
	T	4.5	0	0	0	5.1	0	0	0
	S	2.2	0	0	0.6	2.3	—	—	—
Deformities, mutations	P	1.7	0.2	0	0.6	1.7	—	—	0.9
	T	1.4	0.4	0	0.9	1.1	—	—	0.4
	S	1.2	0	0	0.2	0.9	0.5	0	1.4
Human health effect	P	0.8	0	0.8	0.2	2.9	—	0.9	—
	T	0.9	0.4	1.4	0.5	1.5	0	2.6	0.4
	S	0.8	0	0.6	0.4	—	0.5	0.5	0.9
Not stated	P	19.5	4.9	7.3	13.6	5.2	3.6	9.5	4.4
	T	0.9	2.7	3.6	6.8	—	—	—	—
	S	12.7	3.7	3.9	7.7	1.4	0.5	1.0	0.9

The values are expressed as % of the total respondents who answered Q7; in Japan, public $N=509$, teachers $N=222$, scientists $N=535$ (MACER, 1992a); and New Zealand, public $N=2034$, teachers $N=277$, scientists $N=258$ (COUCHMAN and FINK-JENSEN, 1990). Organism: H, human cells; P, plants; M, microbes; A, animals. Group: P, public; T, high school biology teacher; S, scientist. The absence of data is indicated by '—'

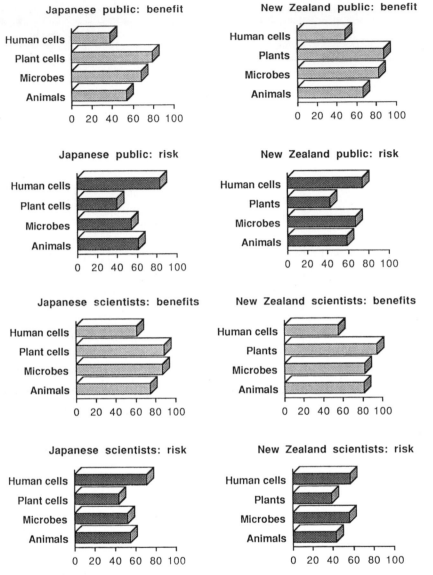

Fig. 3. Comparative perceptions of the benefits and risks of genetic manipulation in Japan and New Zealand by public and scientists. Results from Japan are from Q7 of MACER (1992a), and New Zealand results are from the survey of COUCHMAN and FINK-JENSEN (1990).

The major reasons cited for the unacceptability of genetic manipulation (Tab. 2) can be grouped into categories:

1. Unnatural, playing God, unethical, feeling
2. Disaster, fear of unknown, ecological and environmental effects
3. Human misuse, insufficient controls, eugenics, cloning, humanity changed
4. Health effects, mutations
5. Not stated.

Group 1 concerns may persist with development of the technology, but group 2 and 4

Tab. 3. Benefits of Genetic Manipulation Cited by Respondents

Organism	Group	Japan				New Zealand			
		H	P	M	A	H	P	M	A
% who saw a benefit	P	37.7	78.9	68.5	53.1	48.4	87.5	62.7	66.4
	T	53.5	86.8	80.5	71.0	59.8	96.6	81.2	81.6
	S	60.8	88.0	86.5	74.3	55.0	94.2	81.9	81.6
Reasons cited as benefits (% total respondents):									
Cure or prevent genetic disease	P	8.3	0.4	0.2	0.2	10.6	—	—	—
	T	24.4	0	0	0.5	29.9	0	1.6	0
	S	24.1	0	0	0	20.9	0	0	0
Disease control	P	5.4	2.7	2.3	1.2	15.0	1.7	16.9	10.6
	T	10.0	5.2	1.8	3.6	7.2	42.5	13.8	26.9
	S	11.8	6.4	1.9	2.7	11.0	35.8	7.4	22.0
Medical advance, cancer cure	P	5.4	1.9	10.1	2.7	8.7	1.7	8.2	2.0
	T	9.1	2.2	10.9	5.9	16.7	19.3	3.2	6.5
	S	7.0	1.7	7.1	4.4	31.9	2.8	18.8	4.9
Make medicines	P	0	0.2	4.4	0	—	—	—	—
	T	0.5	0	23.1	0.5	—	—	—	—
	S	0.2	0.6	9.3	1.9	—	—	—	—
Make useful substances, industry	P	0.2	1.3	2.3	0.8	—	—	—	2.0
	T	0	2.2	14.8	1.8	—	—	47.1	—
	S	0.6	4.3	20.5	5.2	—	—	37.7	—
Scientific knowledge	P	0.6	0.2	1.4	1.8	3.4	—	6.3	4.0
	T	3.2	3.0	4.5	6.8	0.6	3.9	0.8	1.6
	S	3.1	3.3	4.6	8.1	1.1	1.9	6.6	2.4
Agricultural advance	P	0	2.0	1.0	0.4	—	19.2	8.2	6.6
	T	0	6.7	3.1	4.5	—	23.2	20.3	11.4
	S	0	3.8	1.6	2.7	—	20.7	24.6	8.2
Increased yield, to make more food	P	0.4	16.9	4.7	8.0	—	20.1	—	10.6
	T	0	18.8	7.2	11.8	—	40.6	0	31.8
	S	0	23.0	9.1	13.5	—	44.3	—	31.0
Different varieties	P	0	11.5	1.9	6.8	—	17.5	—	—
	T	0	27.6	3.6	19.9	—	29.0	0	12.2
	S	0	22.4	5.8	16.4	—	22.6	—	10.6
Increased quality	P	1.9	9.7	2.5	5.0	—	33.2	—	32.5
	T	0	3.8	0	6.7	—	21.3	—	32.6
	S	0.8	3.5	1.7	3.5	—	24.5	—	36.7
Exports increase, economics	P	2.3	3.5	3.7	3.5	1.9	7.9	1.3	7.3
	T	0.5	0.4	0.5	0.5	0	2.9	7.3	7.3
	S	0.9	3.8	4.6	2.9	1.1	10.4	4.1	11.4
Environmental advantage	P	0	3.6	4.9	1.4	—	—	6.9	—
	T	0	1.5	1.8	0.5	—	—	—	—
	S	0	5.3	5.4	1.0	—	—	—	—
Humanity benefits, whole world benefits	P	5.1	8.0	7.4	6.6	10.6	2.6	5.6	4.0
	T	1.8	4.2	6.8	7.2	2.4	0	0	0
	S	5.4	6.1	6.9	6.0	6.0	0.9	4.9	1.6
Doubtful benefit	P	0.4	0.6	0.4	0	—	—	—	—
	T	0.9	0.7	0.9	0.5	—	—	—	—
	S	0.8	0.8	0.8	0.6	—	—	—	—
Benefit not stated	P	12.8	30.1	25.8	19.4	9.7	7.0	18.8	9.3
	T	8.6	15.0	17.2	20.5	—	—	—	—
	S	14.1	26.3	26.1	22.0	1.6	3.8	4.9	2.4

The values are expressed as % of the total respondents who answered Q7; in Japan, public $N=485$, teachers $N=221$, scientists $N=518$ (MACER, 1992a); and New Zealand, public $N=2034$, teachers $N=277$, scientists $N=258$ (COUCHMAN and FINK-JENSEN, 1990). Organism: H, human cells; P, plants; M, microbes; A, animals. Group: P, public; T, high school biology teacher; S, scientist. The absence of data is indicated by '—'

Tab. 4. Risks of Genetic Manipulation Cited by Respondents

Organism	Group	Japan				New Zealand			
		H	P	M	A	H	P	M	A
% total who saw risk	P	83.3	39.5	53.6	61.3	74.4	42.4	67.4	58.4
	T	85.6	54.7	69.6	68.7	43.9	25.6	57.5	25.0
	S	70.8	42.9	51.7	54.3	56.7	38.9	56.1	43.3
Reasons cited as risks (% total respondents):									
Unethical, ethical abuse	P	3.9	0	0	1.0	3.7	—	—	3.5
	T	10.4	0	0	1.4	12.3	0	0	4.5
	S	7.1	—	0.4	2.3	13.6	0.5	0	3.5
Playing God, unnatural	P	7.2	2.5	3.1	4.9	6.0	5.5	4.7	5.3
	T	4.1	0.9	0.9	2.2	3.5	4.1	2.3	3.8
	S	2.7	1.7	1.5	2.7	1.1	5.1	1.7	4.3
Disaster, out of control	P	5.3	2.3	3.7	3.7	18.6	10.6	16.8	12.3
	T	2.7	5.9	7.7	5.0	6.6	4.6	25.3	5.0
	S	4.2	2.5	3.3	3.9	12.5	11.3	26.9	11.7
Fear of unknown	P	9.7	6.8	7.8	8.5	9.7	5.1	7.4	7.6
	T	10.4	8.2	9.0	9.5	4.4	3.1	4.6	2.0
	S	10.2	7.5	7.9	7.9	10.2	9.3	8.0	9.5
Ecological effects	P	7.4	9.1	8.9	12.1	—	5.1	3.4	6.4
	T	5.4	14.5	14.0	15.4	0	10.2	6.3	4.5
	S	3.4	10.0	7.3	10.6	0	5.1	7.4	6.9
Biohazard, spread of genes	P	0.6	1.9	2.9	1.5	—	5.1	—	3.5
	T	1.8	9.5	5.9	12.7	1.8	0.5	18.4	1.3
	S	1.7	3.3	3.7	2.5	—	4.6	5.1	—
Danger of human misuse, biowarfare	P	8.4	4.1	6.6	4.9	8.2	3.4	6.7	5.3
	T	11.7	5.5	9.5	6.4	7.5	2.3	2.9	3.0
	S	15.8	9.9	11.8	11.0	9.1	5.7	7.4	4.8
Eugenics	P	3.5	0	0	0.4	—	—	—	—
	T	8.2	0	0	0.9	—	—	—	—
	S	5.0	0	0	0.6	—	—	—	—
Cloning, human reproduction abused	P	2.2	0	0	0	—	—	—	—
	T	3.2	0	0	0	—	—	—	—
	S	1.9	0	0	0	—	—	—	—
Humanity changed	P	5.2	0	0	0.4	7.0	—	—	—
	T	2.7	0	0	0	3.5	0	0.6	0.8
	S	6.8	0.8	0.6	1.2	2.8	0	0	0.9
Deformities, mutations	P	4.9	1.5	0.4	2.1	17.1	—	—	6.4
	T	10.0	2.3	2.7	5.0	4.8	0	0	2.0
	S	4.2	1.7	1.5	2.1	1.1	0	0	0.4
Insufficient controls, need public discussion	P	4.5	2.2	2.2	3.3	8.2	6.4	15.5	7.0
	T	3.2	3.2	3.6	3.1	4.4	3.8	4.6	4.0
	S	7.0	5.6	5.6	6.2	10.8	10.8	9.1	8.7
Economic corruption of safety standards	P	1.0	0.8	0.8	1.0	—	—	—	—
	T	0.4	0.5	0	0.5	—	—	—	—
	S	1.1	1.2	1.2	1.2	—	—	—	—
Not stated	P	33.2	13.8	19.8	24.7	10.4	—	14.8	10.5
	T	23.6	13.6	15.8	16.7	—	—	—	—
	S	22.0	12.1	15.2	17.4	5.1	3.1	2.3	2.6

The values are expressed as % of the total respondents who answered Q7; in Japan, public $N=485$, teachers $N=221$, scientists $N=518$ (MACER, 1992a); and New Zealand, public $N=2034$, teachers $N=277$, scientists $N=258$ (COUCHMAN and FINK-JENSEN, 1990). Organism: H, human cells; P, plants; M, microbes; A, animals. Group: P, public; T, high school biology teacher; S, scientist. The absence of data is indicated by '—'

concerns may be lessened by development of technology. Group 3 concerns can be lessened by regulations, as will discussed in Sect. 8. People who did not cite a reason may feel less strongly about the issue, but there is no real indication of what concerns they had. We should also note that many people expressed reasoning across several of these types of concern.

The most common response in both countries for a benefit from genetic manipulation of human cells were medical reasons, as from microbes where the benefit of making useful substances was also often cited (Tab. 3). Economic benefits were not cited much, with more respondents in New Zealand listing these benefits, perhaps because the economy is so dependent upon biotechnology, in terms of agriculture, and the economic recession has been much harder there. In the reasons cited for genetic manipulation of animals, many more New Zealanders cited disease control of animals, as a reason. In both countries similar proportions cited "new varieties" or "increased production and food" as the main benefits of genetic manipulation of plants and animals, with a trend for more New Zealanders to cite the latter.

There was also a wide diversity of responses to the reasons why people perceived risks from genetic manipulation (Tab. 4). The frequency of the common responses to Q7d did not differ greatly from those given to Q7b, though many respondents listed different reasons in response to these two questions. The risks were in general more involving human misuse, and activity, rather than abstract concerns such as "interfering with nature". They were also more specific, so that more respondents listed deformities and mutations as a problem. In addition to ecological and environmental concerns, there were also substantial numbers who cited a risk connected with the spread of genes, viruses, and genetically modified organisms (GMOs), generally labelled "biohazard" in the categories in Tab. 4. A few said that science was always associated with danger.

In a recent European public opinion poll in the U.K., France, Italy and Germany (performed in 1990 by Gallup for Eli Lily, $N=3156$, DIXON, 1991), the respondents were asked to choose the largest benefit that they saw coming from biotechnology, between one of four possible benefits from biotechnology. Over half rated cures for serious diseases as the most important benefit. Another option was reducing our dependence upon pesticides and chemical fertilizers, which 26% of Italians, 24% of French, 22% of British and 16% of Germans, chose as the largest benefit.

The European respondents were asked a similar question about their largest concern. Potential health hazards from laboratory genetic research were named by 29% in Italy, 17% in France, 11% in Britain and 10% in Germany. 40% of French, 35% of Germans, and 25% of British and Italian respondents chose eugenics, and slightly lower proportions overall chose environmental harm, 34% in Britain, 33% in France, 22% in Italy and 21% in Germany.

In the European survey, overall one third of respondents feel that biotechnology is ethical, and one third feel that it is unethical, and one third think it is in between, "neither". This compares to a more favorable acceptability of genetic manipulation in Japan and New Zealand, especially for non-human applications (Tab. 2). In the USA when people ($N=1273$) were asked whether they thought that human gene therapy was morally wrong, 42% said it was, and 52% said it was not, with 6% unsure (OTA, 1987). However, only 24% of the USA sample said that creating hybrid plants or animals by genetic manipulation was morally wrong, but 68% said it was not, 4% said it depends, and 4% said they were unsure. The people who found it morally unacceptable were asked for reasons, and these reasons can be compared to those given in Tab. 2 from Japan and New Zealand. The US results, expressed as % of the total sample, were that 3.1% said that it was interfering with nature, 2.8% said it was playing God, and about 1% said they were afraid of unknown results, 0.6% said it was morally wrong and did not cite a reason, and another 2–3% had other reasons. It does reinforce the idea that abstract reasons are a major concern about genetic manipulation, but further data are needed.

In the European survey, eugenics was the

major concern (DIXON, 1991). However, in the Japanese survey and in the New Zealand survey, the proportion of people who cited eugenic concerns from genetic manipulation of humans was equivalent to about 4% of the total respondents, half of the proportion who expressed concern because of environmental reasons, and much lower than the number of respondents who cited reasons related to perceived interference with nature, playing God, ethics, or fear of the unknown (Tabs. 2 and 4). In 1993 surveys in response to the above question Q5d (Sect. 4.2) people were asked to give their reason for concern about "genetic engineering". In New Zealand and Australia 7% of the total expressed eugenic concerns, while only 0.6% of Japanese did (MACER, 1994). Because free response questionnaire data are unavailable from Europe, we cannot directly compare the apparently higher concern about eugenics in Europe as opposed to New Zealand or Japan. One could speculate that it may be related to self-acknowledgement of the past eugenic abuses in Europe, and due to media coverage of the eugenic concerns raised by organized feminist and Green groups in Europe. Free response questions may provide a better estimate of people's opinions and provide a better picture of actual perceptions than agreement with suggestive concerns.

Another feature of the free response surveys was the low proportion of respondents who cited environmental benefits (Tab. 3). In the European survey, discussed above, the choice of the benefit of reduced pesticide use and environmental benefits was popular. In Japan and New Zealand further questions concerning science were included in Q16.

Q16. To what extent do you agree or disagree with the following statements that other people have made?
1 strongly disagree   2 disagree
3 neither agree   4 agree   .5 agree strongly
  nor disagree

Q16f. Genetically modified plants and animals will help Japanese agriculture become less dependent on chemical pesticides.

The response to Q16f, which asked if people thought GMOs would have an environmental advantage, was quite supportive (Fig. 4). The 1993 survey found the same values in New Zealand and Japanese public as 1991, with similar values in Australia (MACER, 1994). The free response questions (Q7c, Tab. 3), however, suggest that it may not actually be a common feeling. Environmental benefits of biotechnology may be very unfamiliar, despite the high level of concern expressed in Q5 about pesticides (Figs. 1 and 2). In both New Zealand and Japan, there should be more publicity associated with this environmental benefit, though the chemical companies who make pesticides may have different

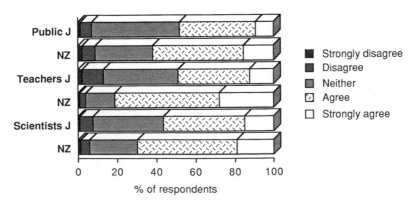

Fig. 4. Perception of environmental benefits from applications of genetic engineering in agriculture in Japan (J) and New Zealand (NZ). Results from Japan are from Q16f of MACER (1992a), and New Zealand results are from the survey of COUCHMAN and FINK-JENSEN (1990).

priorities (see Sect. 7). As in the risk of eugenics discussed above, we see clearly the lack of reliable data about public perceptions, at least, among what has been published.

4.4 Concerns about Consuming Products of GMOs

Products produced by genetically engineered microorganisms, such as human insulin or growth hormone, have been approved for medical use for over a decade. The first food products for human consumption have been approved for consumption, and it is expected that vegetable products derived from GMOs will be approved in 1994. We can expect many products to be approved as safe for human consumption in the next few years, so a pressing question is what concerns the public has that could result in conflict.

The views on the safety of products made by genetic manipulation were examined by Q8b (MACER, 1992a), as had also been used by COUCHMAN and FINK-JENSEN (1990). 75% of the Japanese public said that they were aware that GMOs could be used to produce food and medicines, similar to 73% of the public in New Zealand, and in both countries, 97% of scientists said that they were aware of this. Free response was requested of the concerns people had:

Q8b. If any of the following were to be produced from genetically modified organisms, would you have any concerns about using them?
1 No concern   2 Concern
For each product that you are concerned about, what concerns would you have about using it?
Dairy products   Vegetables
Meat   Medicines

The results are listed in Tab. 5. Vegetables were of less concern, especially among the public, and meat was the product with the highest concern. Dairy products were of intermediate concern. Medicines were still of considerable concern. New Zealanders appear to be somewhat less concerned about consuming products containing GMOs or made from them, than are the Japanese. The biggest difference is in the opinions of high school bio-

logy teachers, which are very concerned in Japan. Scientists in Japan are also more concerned than their peers in New Zealand, though many Japanese company scientists showed less concern than government scientists (MACER, 1992a). The concerns included significant numbers who saw the products as unnatural (see Sect. 5.1), and who had health concerns. Similar reasons were cited in the 1993 International Bioethics Survey (MACER, 1994).

There appears to be joint perception of benefits and risks as for genetic manipulation. In a European-wide survey ($N = 12800$, MACKENZIE, 1991), 65% of people approved of genetic engineering to improve food and drink quality, but 72% said that it was "risky". In an earlier US survey (OTA, 1987), 80% had not heard of risks associated with genetically engineered products while only 19% had. However, since the time of that survey, we can expect that many people have heard of concerns about consuming products of genetic engineering. A related issue has been continued controversy about the use of genetically engineered bovine somatotropin, despite evidence that it is safe for human consumption. These concerns will be considered in Sect. 7.

5 Past and Present "Bioethical" Conflicts in Biotechnology

From the results of the above public opinion surveys, and other data, we can see which issues among the variety that have been expressed by academics and protest groups in the decades of debate on biotechnology, are common and which are not. In this section we will consider the major reasons cited for rejection of genetic manipulation research. The emotions concerning these technologies are complex, and we should avoid using simplistic public opinion data as measures of public perceptions, rather we need to address the expressed concerns and apply policy measures

Tab. 5. Concerns about Consuming Products Made from GMOs

Product	Group	Japan				New Zealand			
		Dairy	Veget.	Meat	Med.	Dairy	Veget.	Meat	Med.
% total with concern	P	51.6	41.0	55.4	50.5	42.8	38.4	48.3	34.1
	T	34.9	31.5	36.1	35.5	13.0	9.7	13.7	9.7
	S	36.1	32.3	38.0	28.8	24.0	21.7	24.4	19.8
Concerns cited about consuming such products (% total):									
Unnatural, will taste bad	P	6.8	6.1	8.5	5.7	11.1	12.3	7.8	7.2
	T	5.0	4.5	4.5	3.1	1.4	1.5	1.1	1.1
	S	4.1	3.7	4.2	3.0	2.4	2.8	2.7	2.0
Don't know what we are consuming	P	0.4	0.6	0.7	0.4	6.0	4.6	7.2	3.4
	T	1.4	1.3	1.4	1.4	1.0	0.7	0.7	0.4
	S	0	0	0	0	0.7	0.9	0.7	0.4
Unknown health effect	P	8.9	7.4	9.5	7.8	9.4	7.7	10.1	7.8
	T	7.2	6.3	7.7	7.7	4.0	2.9	4.7	0.4
	S	5.5	5.5	5.7	5.2	4.3	3.5	5.1	4.0
Long-term risk	P	2.3	2.4	2.6	1.9	—	—	—	—
	T	0.9	0.4	1.4	0.9	—	—	—	—
	S	2.0	2.2	2.2	1.8	—	—	—	—
New disease	P	1.5	0.6	1.8	2.1	1.3	0.8	1.4	0.3
	T	1.8	1.3	1.4	2.3	—	—	—	0.4
	S	3.3	2.2	3.3	1.7	—	—	—	—
Side effects	P	2.1	0.6	1.9	5.9	3.0	1.9	2.4	4.1
	T	1.8	1.3	3.2	5.5	3.2	2.5	3.4	3.6
	S	0.7	0.6	0.8	1.3	3.6	3.0	3.4	3.2
Safety doubts, need to test properly	P	6.1	5.3	5.9	5.5	5.1	3.8	4.3	5.1
	T	9.5	9.1	9.0	9.5	1.8	1.9	1.8	1.5
	S	9.4	9.6	9.9	8.5	8.7	7.8	7.1	8.1
Unknown research area	P	0.7	0.6	0.7	1.0	1.7	1.5	1.9	1.7
	T	1.8	0.9	0.9	0.9	0.8	0.7	0.7	0.7
	S	1.1	0.9	1.1	1.0	1.2	1.1	1.2	0.4
No guarantee of purity or quality	P	1.3	1.3	1.1	1.3	1.3	1.5	1.9	0.3
	T	1.8	2.2	2.3	2.3	—	—	—	—
	S	4.6	3.3	4.6	3.3	—	—	—	—
Because it is food, daily use	P	1.5	1.5	1.9	1.1	—	—	—	—
	T	0	0.4	0.4	0	—	—	—	—
	S	0.2	0.2	0.2	0	—	—	—	—
Environmental effects	P	0.7	0.7	0.7	0.8	—	—	—	—
	T	1.4	1.8	2.3	0.9	—	—	—	—
	S	0.7	0.7	0.7	0.6	—	—	—	—
Other reasons: Medicines – patients are weak; dairy – given to children	P	0.9	0	0	0.4	—	—	—	—
	T	0.9	0	0	0.5	—	—	—	—
	S	0	0	0	0.1	—	—	—	—
Lack information, information is hidden	P	3.2	3.0	3.4	2.9	4.7	3.8	4.3	4.1
	T	1.8	1.8	1.8	1.4	0.8	0.7	1.1	0.7
	S	1.8	1.8	1.9	1.7	0.7	0.9	0.7	0.4
Economic corruption of safety standards, ethical concerns	P	0.7	0.7	0.9	1.1	—	—	—	—
	T	1.4	1.3	1.4	1.4	—	—	—	—
	S	0.7	0.7	0.7	0.7	—	—	—	—
Not stated	P	17.7	12.1	19.6	15.9	4.7	4.2	5.8	4.8
	T	8.6	7.7	9.0	8.1	0.8	0.4	0.4	0.4
	S	8.8	9.6	9.7	7.2	2.6	2.8	3.2	1.6

The values are expressed as % of the total respondents who answered Q8; in Japan, public $N=527$, teachers $N=221$, scientists $N=543$ (MACER, 1992a); and New Zealand, public $N=2034$, teachers $N=277$, scientists $N=258$ (COUCHMAN and FINK-JENSEN, 1990). Group: P. public; T, high school biology teacher; S, scientist. Absence of data is indicated by '—'

to lessen the conflict that people find with biotechnology. Because beneficence is a basic ethical principle, we can assume that there are important grounds for pursuing research and applying technology, providing we are consistent in respecting the other ethical principles, such as to do no harm.

5.1 Interference with Nature or "Playing God"

There were also significant proportions of respondents who thought that genetic manipulation was interfering with nature, or that it was profanity to God, or said that they had a bad feeling about it. Also many saw genetic manipulation, especially of humans and animals, as unethical (Tab. 2, Sect. 4.3). These respondents may see these techniques as unacceptable, regardless of the state of technology and regulation. In the US survey, 46% said that we have no business meddling with nature, while 52% disagreed (OTA, 1987). Although many scientists react to people with these views as irrational, it is noteworthy that about 16% of the scientists and teachers in New Zealand and Japan who found these techniques unacceptable also shared these views, and these reasons were also cited regarding genetic manipulation of microbes.

The questions about food also illustrate this concern. In Japan 12–16% of the public who were concerned about consuming products made from GMOs, said that such foodstuffs or medicines would be unnatural, while in New Zealand the values were much higher (Tab. 5). These results were confirmed in the 1993 International Bioethics Survey (MACER, 1994). Australians were similar to New Zealanders, with other Asians expressing this idea less, like Japanese. While rationally we can say such foods are just as natural as foods made from any modern crop or animal breed, 10–12% of scientists also said this. In a 1988 public opinion survey in Britain, 70% agreed that "natural vitamins are better for us than laboratory-made ones", while only 18% disagreed (DURANT et al., 1989). Other studies of attitudes to consuming products of genetic engineering revealed similar fears in The Netherlands (HAMSTRA, 1993). In fact the use of varieties bred using genetic engineering should allow the avoidance of chemical pesticides and preservatives during crop growth, food storage and processing, which could actually make such foods "more natural".

We have yet to understand what people believe nature really is. It is a changing concept and varies between individuals, religions and cultures. As societies become urbanized they lose touch with nature. However, there is also a recent trend to buy products from "organic farming", or preferences for "free-range eggs" over eggs from battery farmed chickens. All people have some limit in the extent to which they support changing nature, or the application of technology. Bioethics still needs to be developed in order to approach this abstract area of thinking. In the meantime, scientists as well as the public perceive limits to what is acceptable, or "ethical", biotechnology, and further research is needed to determine what these limits are.

By reducing the use of chemicals in agriculture, food processing, and medicine, biotechnology may actually be able to make these areas more "natural". Also, if efficiency of agriculture is increased, and genetic diversity increased, biotechnology may allow some agricultural land to revert to more random natural vegetation. The potential is there, if society demands it. However, increased use of microorganisms for industrial and environmental processes may lessen the use of chemicals in these applications, would this also lessen people's concern – or raise it?

5.2 Fear of Unknown

In general the other frequently cited comments in all sample groups for all organisms were connected with the unknown nature or danger of the results of genetic manipulation. Some people saw this in terms of a disaster, while others have less dramatic concerns. There is also fear associated with unknown research fields, which is true of any area. We could subdivide this concern into health and ecological concerns.

5.2.1 Health Concerns

Fear of the unknown was found to be a common concern that people had about genetic engineering (Tabs. 2 and 4). Fear of health effects are related to this, and were also common. When asked a later question concerning food safety, the principal concern was regarding health effects, including side effects (Tab. 5). Food safety represents the principle, to do no harm. It is a responsibility of developers and marketers of new varieties of GMOs and products from GMOs. Sufficient regulatory procedures should be in place already as for existing products, and further regulations have been added in some countries (like Japan or the European Community) but not in others (like the USA) (see Sects. 5.3 and 8.1).

5.2.2 Environmental and Ecological Risks

There is growing concern about environmental issues, which must be welcome to all who are concerned about environmental ethics and justice to current and future generations. Part of the reason for the increased concern has been recognition of global pollution that human activity has caused, and a fear for the survival of the planet (WRI, 1992). The phrase "sustainable technology" has been added as an ethical criterion of technology. Clearly, with the changed UV light and predicted increased temperature and climate change, the current conditions are not sustainable – rather we need to look at the steady state that may be possible to stabilize in another century.

Nevertheless, the earlier we act, the less different the future world will be, and sustainability has become recognized as an ethical criterion of technology. Any biotechnology which is in conflict with this principle will be called "unethical". As discussed in Sect. 5.1, biotechnology has potential for ecological improvement, if applications are targeted with that objective. The ecological concern that many people have, refers to the introduction of new organisms into the environment, in-

cluding GMOs. Ecological studies are needed, and monitoring of releases of GMOs, to gain data to allow prediction of the ecological impact of such introductions. Some data are already available, after over 1000 field tests of GMOs already performed, but until safety has been demonstrated for each type and species of GMO in question we should not have commercial introduction on a large scale. We should support programs such as the PROSAMO study in the U.K. which are designed to provide methods to monitor the release and survival of GMOs (KILLHAM, 1992). In the cases where safety has been shown, and in cases where the GMO presents less ecological risk than the current varieties, we should use the GMO. It is basic risk management. It is the topic of other chapters of this volume.

5.3 Regulatory Concerns

There was also much concern expressed in Japan and New Zealand about insufficient controls, especially by teachers and scientists. If what are seen to be safe and adequate controls are established, the people who had these reasons for objecting to genetic manipulation, may accept it. It is up to the researchers to prove that the results represent an acceptable level of risk, and to adjust regulatory procedures to those that are seen to be adequate. A discussion of the regulation of biotechnology is in Sect. 8, and biosafety is discussed in Chapter 3 of this volume.

There was also qualified acceptability by some respondents, depending on the introduction of appropriate control measures. About 7% of scientists and teachers, and 2% of the public, wrote such comments for plants and microbe applications, and 19–25% of all respondents in Japan who said that these techniques were acceptable wrote such comments for genetic manipulation of human cells (MACER, 1992a). The actual number of respondents who were concerned about controls should include these respondents in addition to those who said that the area was unacceptable because of insufficient controls. It may be significant that such a high proportion of respondents who said that the techniques were acceptable, did spontaneously write down some qualification to their response choice.

Field releases of GMOs are regulated in all countries of the world, officially, as they should be (MACER, 1990; Chapter 6). The procedures vary, as does the public satisfaction with such procedures. They are also subject to political climate, and bureaucratic regulations conflict with industry and with the principle of beneficence, whereas inadequate regulations risk harm, as discussed above. Most countries are shifting to a product-based system, which is more scientific than a system of exclusion or inclusion based on production method. However, the exclusion categories of organisms from regulations may be broadened so much as to conflict with sufficient protection of human and ecological safety.

In 1992 the US FDA said that they would not impose any special regulations on GMO food products (KESSLER et al., 1992). This has pleased industry (FOX, 1992), but it may lead to more people having fear of what they perceive are unknown food products. In the European Community and Japan, a committee will examine each case, to determine whether extra safety tests are required (MACER, 1992a). Here we see different balancing of ethical principles, beneficence versus do no harm. It remains to be seen how much conflict there is in the acceptance of food made from varieties bred using genetic engineering, it may depend on possible future cases if harm is related to such products.

5.4 Human Misuse

There was also concern about human misuse of these techniques, which again, could be eased by further guarantees over who uses these techniques. For human beings, another major response was concerns about eugenics, and cloning. These fears may be eased by the introduction of laws, but we should note that in Europe where there are some laws to prevent such abuses, there is still much concern with eugenics (see Sects. 4.3 and 8.2). It may be good to maintain a high level of such fears in society as the most effective method to prevent future abuses of biotechnology from be-

ing made. However, it is important that people learn to distinguish medical uses of genetics from racial applications that are associated with the word eugenics. They can already distinguish between applications involving different organisms.

A related issue, and one more relevant to microbial biotechnology, is the use of biotechnology in biowarfare research. Biotechnology is being applied to military research, to develop biosensors to detect poisons, and to improve immunity (NRC, 1992). If the motive is defensive, and the results are distributed to the general public, globally, we could find ethical justification for such research. However, such research would conflict with ethics if it is not shared with all people, and if there is any possible escalation of offensive biological warfare research. We cannot eliminate the possibility that biotechnology will be applied to biowarfare development by individual terrorists, so such defensive research could have benefit. Additionally, any research to improve immunity will be useful as we face infectious diseases, and in the future as the increased incidence of UV light may decrease our immunity. We can only attempt to ensure that biotechnology is not misused.

6 Future "Bioethical" Conflicts in Biotechnology

6.1 Changing Perceptions of Nature

As discussed in Sect. 5.1, a significant proportion of people, including scientists, see genetic manipulation as interfering with nature. As will be discussed in Sect. 8.2, people who said genetic manipulation of human cells is playing God, may still support medical use of human gene therapy, or environmental applications of GMOs (MACER, 1992a, c). This represents the most common reason why people may overcome their fear of playing God, to treat disease. Disease is natural in the sense that it may not involve human action, but it can be perceived as unnatural when it is curable. There are of course limits to this change, as we will all die – death is natural. What is thus important is not whether it is natural or not, but whether we should treat it or not. In medicine, there are other words which distinguish these concepts, ordinary treatment and extraordinary treatment.

This already presents conflicts to the unlimited use of biotechnology, such as life support technology used in intensive care. We can expect these conflicts to become more familiar and more common, in questions such as which genes do we consider important to "fix" in gene therapy?, which genes do we perform genetic screening for?, what is disease?, how far do we genetically engineer animals as transgenic production systems for pharmaceuticals?, and how far do we transform our environment with genetically modified plants for agriculture and ornaments?, and should we introduce genetically engineered microorganisms into the wide environment in efforts to clean up pollution, by removing or degrading toxic chemicals or heavy metals?

The list of questions is enormous. Biotechnology may not have a fundamental conflict with nature, and many would see new technology as a gift from God; however, we must not confuse ourselves with God. We are likely to make mistakes in the future, and apply technology too far. One basic criterion for examination of the extent to which we should proceed and what is ethical application of technology is that we should not use technology which will mean that future generations lose the possibility of reverting back to social and environmental conditions that existed in our generation. This may mean that we limit the introduction of genetically engineered microorganisms into the wider environment, though many may arise via the selective pressures present in polluted environments. Are these selected organisms different from specifically designed ones?, are they any safer?, we would say in general no.

However, elimination of disease is ethical biotechnology; there should be no objection to the elimination of human diseases, such as smallpox. There may be more debate about

the elimination of recessive disease-causing alleles, that may have some unknown advantage, like the sickle cell anemia trait which offers improved protection against malaria, but no baby should be denied the possibility of therapy for the disease, be it by safe gene therapy or other options. However, carriers may want to eliminate their risk of transmitting a harmful allele to their offspring – what is called germ-line gene therapy. This is another future conflict that faces us, even if some countries enact regulations that prohibit germ-line gene therapy for the present, we cannot escape the question.

6.2 Pursuit of Perfection – a Social Goal

The paradigm of beneficence argues that we should pursue benefit. This is sometimes confused with pursuit of perfection. As any human being carries 10–20 lethal recessive alleles, no one is perfect, and we can never expect to become "perfect" genetically. Because social standards also differ, no one will be socially "perfect" to all. Thus the pursuit of perfection is impossible. There is also no perfect environment, or food, we all have different preferences, and we must also weigh the ethical interests of other organisms.

However, ideals of phenotype are not impossible. A recent example illustrates this, the case of using recombinant human growth hormone to make short children, without a growth hormone gene defect, tall (DIEKEMA, 1990). Such a study was supported by the NIH in the USA, and is under some ethical review (STONE, 1992). Is this ethical biotechnology? Support for it only comes from the recognition of parents autonomy – however, we already limit this when it may damage the child or society. Is it a waste of resources? It is a waste of health care resources, but is it any worse than spending large sums of money on luxury pursuits?

The principal objection must come from the social effect of allowing parents to make their children taller by extraordinary methods (though good food may have an equal effect), which is consistent with a social ideal that tall children are better – which is inconsistent with human equality and could have harmful social consequences. It also has harmful environmental consequences, as big people use more resources. Above all, such treatment recognizes a failure of society to tolerate differences, and if it is approved, suggests society has given-up. If this is so, we can expect further discrimination against those who are different from a narrowing social norm.

This argument applies to all cases of treating abnormality – however, the crucial ethical factor that means that it is ethical to treat someone with a serious genetic disease and it is not ethical to perform cosmetic therapy like that example, is that the principle of beneficence demands us to assist those who suffer from disease, be it mental or physical, but the principle of do no harm to society asserts that we do not develop a society less tolerant of abnormality. We balance these two principles in favor of individual beneficence when the disease is serious.

Another example of this is the pursuit of efficiency in agriculture. Battery farming of chickens may have been somewhat cheaper than free-range farming, however, in Switzerland and Sweden it is being rejected due to ethical reasons. As agricultural systems to more efficiently feed animals are developed, people may say that these are unethical, and prefer to pay a little extra price for having free-range animals. Perhaps the economists may even redo their equations and find that the tourist income from grazing sheep scenes more than compensates for a possible higher production cost compared to factory farming. If we have sufficient production capacity, and some would say even if not, there is no ethical excuse for animal cruelty. Biotechnology has the potential to increase production efficiency so that more free-range animal systems can be used. The question of whether this will happen, is addressed in Sect. 7.

6.3 Limitation of Individual Autonomy

The above example about growth hormone illustrates the conflict between individual autonomy and society. Autonomy must be limited by society in order to preserve the autonomy of all individuals to an equal degree. Ethically, we should apply this globally, based on justice and equality of human rights, so that individual autonomy should be limited.

A clear contradiction to this comes in the policies on environmental use. Currently, industrialized countries produce much more pollution, and use much more environmental resources than developing countries. Should any person use more than his/her share of communal environmental resources or produce too much pollution? Yet, societies who like to claim high value on liberty, claim that individuals can use their money to satisfy their desires, without reference to these factors. The only way to combat this selfish greed, that most of us possess, would be to introduce fair environmental quotas.

A different issue is whether we should prevent individuals from exposing themselves to risky behavior. We allow individuals to play dangerous sports, but we impose stricter limits on occupational health. Already there are some genetic risk factors identified that could be used to screen out workers at high risk in particular environments (DRAPER, 1991). Should society be paternalistic, or should it let individuals have freedom. As further genes are identified for genetic predispositions, we will have more issues at stake.

6.4 Human Genetic Engineering

In the future when techniques for targeted gene insertion become safe to use on humans, treatment of non-disease conditions will be called for. It is a further issue whether it is ethical to attempt to improve the genes of humans as we may do with agricultural organisms. Although such genetic engineering would be considered unethical by many people today, it is likely that this conflict will arise when increasing numbers of people want to improve characters such as immunity to disease, improve the efficiency of digestion of foodstuffs, or numerous psychological or social traits. Immunization by gene therapy entered a clinical trial in 1992 (ANDERSON, 1992b), so it may be a more immediate prospect. These applications will cause future conflict, and it is also related to the above question of when individual autonomy should be restricted.

7 Bioethics versus Business: A Conflict?

In short, the answer is yes, the reason is that ethical concerns rely on principles such as just distribution of wealth and equality, and on factors such as beneficence. However, the goal of business is to make profit, and many businesses aim at economic growth and high profits. Biotechnology may allow production of consumer goods from renewable biomass sources, however, energy is still required to transform raw materials into finished products, thus economic growth requires continual energy input. The economic policies, based on Schumpter Dynamics, are not compatible with sustainable development (KRUPP, 1992), therefore if biotechnology aims to be ethical it must use a different economic theory. When businesses consider raw materials, they may attempt to use the lowest cost materials, which may mean that international common assets such as environmental resources are used, the so-called problem of the commons. They may also ignore the future costs of pollution caused by the production and use of technology.

Much of the new wave in biotechnology research is being performed by private companies. These companies are being encouraged to perform research in their countries' national interests, including the hope of more export earnings from the sale of products and/or technology (OTA, 1991c). Some of the conflicts relevant to ethical biotechnology are discussed below.

7.1 Intellectual Property Protection

There are several forms of intellectual property protection, and they are outlined in Chapter 10 of this volume. There continues to be much controversy regarding the patenting of plants and animals, and of genetic material from living organisms, especially humans. There is less controversy regarding patenting of microorganisms.

Patents and variety rights are supported to act as incentives for technology development (LESSER, 1991), consistent with beneficence. However, should there be subject matter which is exempted from patent protection, such as plant and animal varieties are exempted in the clause 53(b) of the European Patent Convention? There is also a question of what is novel, when gene sequences consist of information that is already existing in nature – even though this information can be shuffled into new vectors. Another question that is important for the future of biotechnology patents and for gene sequencing patents is whether the application of robotic sequencing methods is non-obvious. The policy should be made considering all the economic, environmental, ethical and social implications, and it should be internationally consistent.

In 1991 a controversy arose when a single patent application for 337 human genes was made in the USA, and in February 1992 a further application for patents on another 2375 genes was also made by the NIH (MACER, 1992a). Modern technology has the ability to sequence all of the 100000 human genes within several years, and the company The Institute for Genomic Research has promised to publish most human gene (cDNA) sequences by early 1995. There is no demonstrated utility, so this type of broad application is expected to fail, regardless of ethical or policy issues. The patents were applied for on behalf of the US National Institutes of Health, though many inside the NIH are against it (ROBERTS, 1992). This government body wanted to sublicense particular US companies to pursue research on these genes in an attempt to "protect" the US biotechnology industry from international competition.

Researchers in Britain, France and Japan are also obtaining many gene sequences (including some of the same genes and sequences), so a patent war was threatened which would damage international scientific

Tab. 6. Attitudes Towards Patenting by Company, Government and University Scientists in Japan

Item	Company			Government			University			Total		
	Yes	No	DN	Yes	No	DN	Yes	No	DN	Yes	No	DN
Inventions	97	2	1	93	3	4	92	4	4	94	3	3
Books	80	13	7	83	9	8	83	11	6	82	11	7
Plants	88	5	7	76	8	14	71	11	18	78	8	14
Animals	84	7	9	73	11	16	67	12	21	74	10	16
Plant and animal genetic material	62	21	17	35	30	35	45	31	24	46	28	26
Human genetic material	52	32	16	24	41	35	34	37	29	35	37	28

Respondents were asked the question, "In your opinion, for which of the following should people be able to obtain patents and copyright?", Responses: Yes = approve, No = disapprove, DN = don't know (from MACER, 1992a)

cooperation in the human genome project. The US Patent Office made a relatively quick decision rejecting the initial applications. Government action to prevent such patents on random cDNA fragments has been widely called for (KILEY, 1992). The French government, and Japanese genome researchers (SWINBANKS, 1992), have announced that they will not apply for similar patents because of ethical reasons. England's Medical Research Council (MRC) applied for a similar patent on more than 1000 genes, but in 1993 withdrew the application. In the end, the rapid progress in sequencing the human genes, as referred to just above, has defused the issue.

The human genome is common property of all human beings, and no one should be able to patent it (MACER, 1991b). Public opinion could force a policy change regarding the patenting of genetic material, even if it is judged to be legally valid. People in Japan and New Zealand were asked if they agreed whether patents should be obtainable for different subject matter (MACER, 1992a, b). 90–94% of all groups agreed with the patenting of inventions in general, such as consumer products. There was less consensus on the patenting of other items, though the same relative order of items was followed in all groups in Japan and New Zealand. There was less acceptance of patenting new plant or animal varieties than of inventions in general. Only 51% of the public agreed with patenting of "genetic material extracted from plants and animals" in New Zealand, but even less, 38%, in Japan. There was even lower acceptance of patent-

ing "genetic material extracted from humans", in Japan only 29% of the public agreed, while 34% disagreed. In all groups more people disagreed with the patenting of genetic material extracted from humans than those who agreed with it. Among scientists, however, company scientists were much more supportive of patents than government scientists, as can be seen in Tab. 6. The rejection of patents on human genetic material was also seen in the International Bioethics Survey (MACER, 1994).

7.2 Global Issues of Technology Transfer

Most research is conducted in "rich" industrialized countries because of their resources and company base. Genes can be derived from one country, which may be a developing country, and the research conducted in another country. Therefore, a patent may be applied for by the researchers who by virtue of their money could invest time into sequencing a gene associated with a useful property. In 1992 a Biodiversity Treaty was signed by most countries of the world at the World Environment and Development Summit in Rio de Janeiro, which would offer some protection to developing countries against such opportunistic investment. They may argue that already the rich countries are gaining much advantage from free transfer of agricultural crops that originated in developing countries, developed by old biotechnology, so the developing countries should be able to at least share equally, or even gain some compensation, for future uses of genetic resources by rich countries who apply new biotechnology. The question is whether old biotechnology is so different from new biotechnology, and what is justice. We can only expect this conflict to continue, as the Biodiversity Treaty is applied to patent policy and law.

There is still no reward given to the farmers who for millenia have established crop varieties, which plant breeders use as starting materials. It is ironic that small farmers continue to lose their farms in the development of commercial biotechnology. In 1983, at a UN Food and Agriculture Organization conference, representatives from 156 countries recognized that "plant resources were part of the common heritage of mankind an should be respected without any restriction" (JUMA, 1989). Since then an international network of gene banks has begun to be established, which will provide genetic material worldwide. These also preserve genetic material from species that are becoming extinct because of environmental destruction.

All people should share in the benefits of biotechnology. There is ethical and religious support for this, such as "love thy neighbour as thyself", and the utilitarian ideal that we should try to benefit as many people as possible, and from the ethical/legal principle of justice. In the United Nations Declaration of Human Rights, Article 27(1), is a basic commitment that many countries in the world have agreed to observe (in their regional versions of this declaration, SIEGHART, 1985). These are "(1) *Everyone* has the right *freely* to participate in the cultural life of the community, to enjoy the arts and *to share in scientific advancement and its benefits*" (italics added for emphasis). The common claims to share in the benefits of technology should be considered in all aspects of biotechnology, also including the questions of who should make decisions concerning its applications. This question is important for the sharing of technology.

7.3 Short-Term versus Long-Term Perspectives

Most businesses, and governments, are run on short-term windows, rather than long-term perspectives. Until long-term perspectives are adopted, sustainable technology is likely to be considered too expensive to compete with short-term resources. We only need to think of the price of petroleum for energy use and transport to see a key example. Biomass-derived energy sources, which are based on renewable solar energy, may not be adopted until they can compete economically. Because pollution damage from petroleum emissions is not included into their cost analysis, they are seen as cheaper energy sources. There is thus an economic conflict, which prevents the rapid introduction of biotechnology-derived energy sources.

There has been much controversy regarding the use and FDA approval for sale of recombinant bovine somatotropin (BST) to increase milk production in the dairy industry. The major concern is not the safety of the product (KESSLER et al., 1992), but the socioeconomic impact on farming communities (OTA, 1991a). Although BST increases milk yield, it may accentuate the current trend for small dairy farmers to go out of business, and for larger farms, which are more economically efficient, to succeed. Here we have a short-term economic view versus a long-term perspective related to social issues or urbanization, and secondary economic issues about the creation of new employment. This is not unique to agricultural biotechnology, and other factors such as nutrition of feed can alter milk production, but it is a question that requires a long-term perspective to answer in an ethical way.

Societal change as a result of international trade and changing agricultural practice has occurred increasingly in recent decades. This challenges the lack of forward planning in free market economics, however, the economic failure of most centrally planned communist economies illustrates the dangers of unrealistic plans. Although GATT was agreed at the end of 1993, a number of ethical, social and environmental questions remain with the accepted economic freedom to pursue new technology adopted by short-term economic and social views.

7.4 Safety versus Costs

Such a heading may be provocative, but it is a real conflict. More time spent in testing the safety of a new drug or foodstuffs, or the environmental safety of a new organism, means more money is invested. Ethically, we may say do no harm has priority, and require long periods for testing of new products. However, this means that the average costs for development of new drugs are so large that only large companies can take a product through to the market, after safety approval.

Society allowing private industry appears to succeed better, as shown from the recent experiments with communism in the 20th century, so society wants to promote industry as a means to competitive and more efficient production of drugs or foods. It is to the benefit of society to support some industry, though they must be careful when giant monopolies are formed. Therefore, there is a conflict between safety and development costs. This conflict led to the use of special measures such as the Orphan Drug Act in the USA, which allows the earlier use of new drugs if there is a strong medical need, encouraging industry to spend time developing them.

Nevertheless, society does impose safety standards to protect human and environmental health. Another method of attempting to ensure safety is to allow liability suits in courts, which is an additional protection. However, there also need to be limits on liability claims, otherwise research into such areas as contraceptives, or vaccines, may be inhibited, due to company fears of future litigation for unrealistic monetary sums in such sensitive areas.

In early 1991, the US government attempted to restrict regulations on biotechnology products (PCC, 1991), such as foodstuffs, as an incentive to encourage further industrial investment. We will not know whether this compromises human or environmental health until the future if mishaps occur. Large industry may be cautious about liability suits, and better ensure safety of products, but it has been suggested that allowing industry the option of not asking for independent review of product safety, risks exposing the public to untested products marketed by small companies trying to make a quick profit.

7.5 Is New Technology Better?

Companies naturally have a desire to recover development costs of new products, so they will attempt to market their products once they are approved. There has been recent controversy over the use of several products of genetic engineering in microorganisms that are examples of the type of ethical conflict that can arise if these products may not be the most clinically suitable or cheapest.

Recombinant human insulin has been associated with unawareness of hypoglycemia, unlike porcine insulin, and there are growing calls to examine its effects (EGGER et al., 1992). Human insulin was speedily adopted in most countries of the world as a replacement for porcine insulin, a decade ago; when new data come to hand regarding adverse effects, we need to reexamine its use. An example involving much more economic interest regards the reported lower effectiveness of tissue plasminogen activator (TPA) in some medical applications compared to streptokinase (ISIS-3, 1992). TPA is much more expensive than streptokinase, so that even if their effectiveness is equal, we should use the cheaper product, according to just distribution of health care money. However, commercial interests want TPA to succeed. One can even see the infamous case of the apparent French delay in use of HIV-tests until a French-produced kit was available, to protect national industrial interests (ANDERSON, 1992a). Another area of economic benefit to pharmaceutical companies is development of new broad-spectrum antibiotics, yet these can be linked to increasing health problems caused by antibiotic-resistant microorganisms (NEU, 1992).

There are clearly many conflicts raised by commercial interests and marketing. Ethical biotechnology is often under challenge, and will be subordinate to commercial biotechnology interests. There may even be unethical calls for patronage of research areas made by scientists who want to develop new technology, when older techniques are available, as in the case of calls for a malaria vaccine versus use of the money in vector control (GAJDU-SEK, 1992). The public may sustain this by belief that new technology is better – it is not always.

8 Resolution of Conflicts

8.1 Who Can be Trusted?

Scientists will win more public support for research by involving the public in decision-making, and being open. If it is ethical biotechnology, then the public may, by a large majority, support such an application. The public has a high level of suspicion of safety statements made by scientists, especially those involving commercial decisions. In surveys conducted in Japan (MACER, 1992a) and New Zealand (COUCHMAN and FINK-JENSEN, 1990), high school biology teachers and government scientists were even more suspicious than the public, when asked the following questions:

Q16. To what extent do you agree or disagree with the following statements that other people have made?
1 strongly disagree   2 disagree
3 neither agree   4 agree   5 agree strongly
  nor disagree

Q16b. If a scientist working in a government department made a statement about the safety of a research project, I would believe it.

Q16c. I would usually believe statements made by a company about the safety of a new product it had released.

Q16d. The activities of scientists in Japan should be more closely regulated to protect public safety.

In Q16b, 35% of the public said that they would believe a statement made by a scientist working in a government department about product safety, and 21% said that they would not, and 44% said that they would not be sure (Fig. 5). Q16c asked about the credibility of company statements about product safety, and there was less trust of such statements, with 17% of the public saying that they would believe such statements and a greater proportion, 36%, saying that they would not believe it, and half, 46%, said that they would neither agree nor disagree.

The responses to Q16c clearly show how company safety statements are not trusted by people, even by company scientists. Only 12% of scientists said that they would usually believe a statement made by a company about the safety of a product it had released. Only 6% of government scientists would usually believe company safety statements, but 24% of company scientists said they would believe them, whereas 26% of government scientists, and 35% of company scientists, said they would believe safety statements by government scientists (MACER, 1992a). In the USA, people were also asked whether they would believe statements made about the risks of products by different groups (OTA, 1987), and companies were less trusted than government agencies, with university scientists being the most trusted. Only 6% said that they would definitely believe a statement made by a company about its product, and 15% said they would not believe, with 37% inclined not to believe and 39% inclined to believe. Similar distrust of companies was found in the International Bioethics Survey (MACER, 1994), and in Europe (EB, 1993).

Committee meetings involved in the regulation of biotechnology and genetic engineering should be open to the public. Such open decision-making would gain more public support than closed meetings, and openness would improve public confidence in regulators. It may also result in better safety than regulations which put industry on the defensive and result in closed-door discussions. Moreover, an open approach may be better at winning public support than the current approach of spending money on advertising campaigns that could be seen as pro-biotechnology "propaganda" campaigns. Most peo-

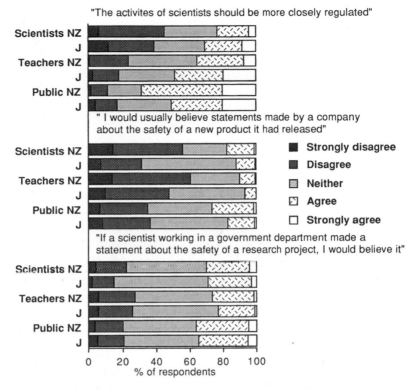

Fig. 5. Comparative attitudes towards scientists in Japan (J) and New Zealand (NZ). Results from Japan are from Q16b, c, d of MACER (1992a), and New Zealand results are from the survey of COUCHMAN and FINK-JENSEN (1990).

ple were already aware of the benefits of biotechnology, but they will remain concerned about decision-making that is hidden.

8.2 Provision of Information May Obtain Higher Public Approval

There was higher support for specific applications of genetic engineering than there was for general research, suggesting that the public will better support worthy applications of technology if they are told the details of them. In the results of the opinion surveys conducted in many countries (MACER, 1992a, 1994), and the USA (OTA, 1987), higher acceptance with biotechnological techniques was found when specific information was given.

When people were asked whether they would use gene therapy to cure serious genetic diseases, the majority in all countries overwhelmingly accept (80–90%) the use of human genetic manipulation for curing serious genetic diseases (MACER, 1992c, 1994; OTA, 1987). Q7 was a general question and was expected to show lower approval of genetic manipulation on humans than the specific questions (Tab. 2).

A similar effect is seen regarding the approval for environmental release of GMOs (MACER, 1992a, 1994; OTA, 1987; HOBAN and KENDALL, 1992).

Q19. If there was no direct risk to humans and only very remote risks to the environment, would you approve or disapprove of the environmental use of genetically engineered organisms designed to produce …?

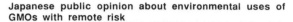

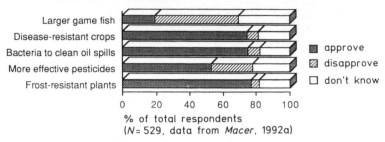

Japanese public opinion about environmental uses of GMOs with remote risk

% of total respondents
(*N* = 529, data from *Macer*, 1992a)

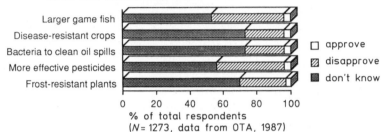

U.S. opinions about environmental uses of GMOs with remote risks

% of total respondents
(*N* = 1273, data from OTA, 1987)

Fig. 6. Attitudes to environmental release of GMOs (genetically modified organisms) in Japan and the USA.

1 Approve   2 Disapprove   3 Don't know
Frost-resistant crops
More effective pesticides
Bacteria to clean up oil spills
Disease-resistant crops
Larger game fish

The results from the 1991 Japanese and 1986 US surveys are shown in Fig. 6. There was clear approval for environmental release of disease or frost-resistant crops, with less approval for bacteria to help fight oil spills. There was lower acceptance of developing better pesticides, though still a majority of all groups supported this. There was rejection of the idea of making big game fish in all countries, with least rejection in the USA. A survey in the U.K. (MARTIN and TAIT, 1992) found a similar differentiation by British people. These survey results are a clear mandate for further research to develop some products involving GMOs, and to have further field releases of new varieties of plants to test their performance, prior to farmer's use. They also clearly illustrate that the public can differentiate their support depending on the perceived benefits or risks, and the information they receive.

8.3 Public Education Not Propaganda

There is a significant public policy decision to be made regarding public education programs. There has been an information campaign underway for a decade in Japan supported by members of the Japan Bioindustry Association, involving government and industry, to promote biotechnology. It appears to have resulted in high awareness of biotechnology, with mixed perceptions. There are also calls in Europe by industry groups to promote biotechnology, with the goal of reducing what is seen as a high level of concern about the technology. Such campaigns can include publication of books, which can also be useful to promote discussion of ethical issues (BMA, 1992). Some US companies have also had large public relations campaigns, such as Monsanto. Recently, following a survey of scientists in the USA engaged in recombinant DNA research, which found that more saw public attention on genetic engineering research as beneficial than harmful to their research, public education programs to stress the benefits of biotechnology have been called for (RABINO, 1991), though some are underway already in the USA. A similar survey and conclusion was made among European scientists (RABINO, 1992).

The results of the surveys described in MACER (1992a, 1994) question the effectiveness of such programs, and also whether their goal is desirable. Rather than attempting to dismiss feelings of concern, society should value and debate these concerns to improve the bioethical maturity of society. However, media responsibility is crucial (see Chapter 17).

8.4 Sufficient Regulations

8.4.1 Regulation of Environmental Risk

Regulations for the environmental release of GMOs have been enacted in many countries of the world, some being statutory and others guidelines. The European Parliament set minimum legal standards for European Community countries, though regulations vary between strict, as in Germany, and non-existent in other countries – which rely on the default European regulations. In the USA a variety of agencies have taken responsibility for different areas of biotechnology applications, and the guidelines are evolving into product-based regulations with exemptions (OTA, 1991c). The USDA has approved many releases of GMOs (see Chapter 6). In Japan, each of the major ministries have their own regulations, and there has been only one field release of a GMO (MACER, 1992a).

Islands may develop particularly different regulations and enforce them, but regions, such as Europe, need common minimum regulations, as neighboring countries are at risk. Conversely, any country which imposes extra regulations must suffer the lower industrial development of their neighbors, without a significant reduction in risk.

We must also gather information from past releases of new organisms and their ecological consequences. We can hope that the information is shared globally, to avoid others making the same mistakes, and to ensure that all countries have a similar minimum standard of protection. It is clear that the authorities and committees that have the most experience with releases should have developed the most skill in assessing the ecological risk. Review should of course be independent, to avoid conflict of interests.

There may be beneficial environmental consequences of biotechnology, as discussed in Sect. 4.3. There may also be less risk associated with using living organisms to produce raw materials for industries, such as the chemical industry (OTA, 1991b). We only need to think of accidents like Bhopal in India, where a disaster was caused by dangerous chemical intermediates. Biotechnology using safer intermediates and raw materials may not only be more sustainable but also safer.

8.4.2 Product Safety

Independent clinical review of drug safety is already standard in most countries, and to be ethical, we must ensure that all people of the world share its protection. Such protection should be standardized, but it is a more difficult question when a country wants to impose stricter standards. A government has a duty to allow beneficial products and technologies to be used by its citizens, including unconventional treatments such as somatic cell gene therapy.

We should ensure that all people of the world enjoy the protection of similarly high safety standards, and that they are kept informed of the content of their food. We may not need to apply any additional regulations to food, unless novel components are introduced to the food (WHO, 1991). The policy has recently been formulated in several countries. In a rapidly moving and new area, an independent committee approach to regulation is the only way to efficiently and safely examine food safety. The Japanese Ministry of Health and Welfare has published bilingual guidelines for foods and food additives produced by recombinant DNA techniques (MHW, 1992). They exclude organisms that have gene deletions from these guidelines, only including organisms which contain "recombinant DNA" sequences or parts of vectors. The "expert committee" of the Ministry will review all cases "to ensure and sustain reasonable criteria", and they can decide whether to insist on additional data from safety tests or not. The data presented must be published in peer reviewed journals. The key question is whether they decide the foods are novel or not, because if they are novel, extensive safety tests must be performed. The guidelines state "the novelty depends upon comparison of identity and promotion of product with existing foods or food additives". The test of the guidelines will be how the committee works, whether they make their proceedings public, and where it draws the line of novelty. It is certainly safer to force a committee examination, than to exempt that as has been adopted in the USA.

8.5 Public Involvement in Regulatory Processes

In some democracies the public has a clear role in the process of regulation, and clear opportunities to voice concerns. This opportunity to voice concerns is important to gain public trust, especially considering the lack of trust (see Sect. 8.1). In some countries hearings are conducted in public, as in the RAC committee hearings on human gene therapy in the USA. This may have lowered public concern in a controversial area, though companies do not require RAC approval, only FDA review which is private (ANDERSON, 1992b). In other countries there is almost no openness to the public, as in Japan.

In the above mentioned survey responses (MACER, 1992a), the public, high school biology teachers and academics gave very similar depth of responses to many questions, suggesting that the public can make well reasoned arguments concerning biotechnology risk and benefit. The public should be involved more in committees making science policy and regulating applications of science. This requires more public willingness to be involved, and the scientists and bureaucrats should allow third party and public entry to committees. As a minimum standard for en-

suring ethical biotechnology, decisions should be made in forums open to public knowledge.

9 Ethical Limits of Biotechnology

9.1 Absolute or Relative?

We can ask whether there should be an absolute ban on certain applications, such as sex selection, or germ-line gene therapy. Such bans are only justified when there is a clear risk of social harm. There are other areas of biotechnology where there clearly should not be an absolute ban, as in the use of genetic engineering on any organism. There are some conditional bans imposed, such as the guidelines to protect human subjects from unconsented experimentation or dangerous experimentation. Some would call for a ban on experimentation using animals, and the proportion supporting such a ban on primates as significant. Laws may be introduced to impose absolute or conditional bans on some research, and this is needed, at least in some areas. Everyone is subject to laws, and scientists should not expect to be exempt. Laws, as said above may not always be ethical, they may be relative, as public opinion is not always reflecting an ethical view – and there may not even be an absolute ethically correct answer.

Is there a clear distinction between disease and non-disease, which may determine whether prenatal genetic screening and selective abortion is ethical or not? At the extremes there is, but there are gray areas in the middle, and we must ask how to regulate this – do we impose absolute limits, or do we take up the principle of autonomy and allow women to decide what is a disease or not. The ideal would be to allow women a completely free choice in the adoption of the disease markers they want to use, and that they would responsibly do this. However, abuses may occur, sometimes due to social forces upon the women – such as a medial insurance companies refusal to provide medical care to children suffering from disease who had a "preventable birth". Such a discussion will be shocking to some people, but in other cultures a majority may be more pragmatic, and it would not be seen as immoral. Can we have absolute limits?

Another example is whether we should have absolute exclusions on subject matter for patents. Should we make exceptions in areas of important benefit, or to encourage beneficial research that would not otherwise be performed. The answer is there may be exceptions, and a response to the demand for flexibility is the use of independent committees for decision-making.

9.2 Timeless or Transient?

As discussed above in Sect. 6, there will be future conflicts in determining what is ethical biotechnology. Our concepts will change, and there is no guarantee that unethical applications will be made, and even supported, by future public majorities. We need to remember history, and also may need to introduce some international laws which make it more difficult for future unethical uses to occur. However, we need to be flexible, as we gather experience we may need less stringent regulations, as in the case of release of GMOs or gene therapy.

Biotechnology has had a positive effect on the debate or bioethics, and we must ensure that the debate matures and is applied more widely to other technologies, old and new. The introduction of new techniques may even be necessary to change general patterns of ethical and social thinking, for example, the introduction of nondirective genetic counseling that is associated with genetic screening could hasten the introduction of the concept of informed consent into general medical practice.

9.3 Scientific Responsibility for Ethical Applications

Scientists are called upon to take responsibility for the social consequences of their research. Recently we can see the growth of ELSI (ethical, legal and social impact) grants from human genome research programs, so that the NIH in the USA is spending over 5% of the human genome project grants on ELSI issue research. We can also see the emergence of movements such as the Universal Movement for Scientific Responsibility (MURS). MURS is attempting to introduce articles into the UN Charter of Human Rights (MELANCON, 1992), and such moves represent important steps in the growing maturity of scientists. These may illustrate a paradigm shift among scientists to concentrate more attention on the social impacts of their research, especially in areas such as biotechnology and genetics.

10 Criteria to Assess Whether Biotechnology Research is Ethical

We can think of some summary criteria which may be useful in determining whether any given application of biotechnology is ethical. Although it is possible to develop useful numerical scoring systems, as has been attempted for animal experiments (PORTER, 1992), they are still only guides and may be more useful in directing attention to the better and more ethical design of experiments. Therefore, only some criteria important to assess the ethicity of an experiment or application are given below:

1. What is the benefit? To whom? Is it life-saving? Human benefit is greater than monetary benefit.
2. Do no harm to humans. What is an acceptable level of risk? Follow ethical codes on free and informed consent for human experimentation.
3. Do not cause pain. Protect animal rights as much as possible, use less sentient animals for research, and develop non-animal alternatives.

4. Do no harm to the environment. Use the technology that is most environmentally sustainable over the long term. Minimize consumption, may need to introduce environmental quotas to do this according to just distribution of global assets and introduce maximum levels for individual production of pollution.
5. Protect biodiversity. Protect endangered species. Allow farmers affordable or free access to breeding stock, and encourage planting of diverse crops.
6. Justice to all people, and future generations. Share benefits and risks.
7. Independent open decision-making on safety questions, consider ethical and social impact.
8. Inform and educate the public and scientists about all dimensions of the projects, scientific, social, economic and ethical, using third-party media.

11 Conclusion

We need to understand that perceptions of the impacts of technology are more complex than simple perception of benefit or risk, as they should be. The capacity to balance benefit and risk of alternative technologies, while respecting human autonomy and justice and the environment, while simultaneously being under the continual influence of commercial advertisements and media stories of varying quality and persuasion, may prove to be an important indicator of the social and bioethical maturity of a society. In addition, to develop the bioethical maturity of society, global human rights need to be increasingly respected so that we get social progress as well as scientific progress. All people should equally share both the benefits of new technology and the risks of its development.

12 References

ANDERSON, I. (1989), A first look at Australian attitudes toward science, *New Sci.* (16 Sept.), 42–43.

ANDERSON, C. (1992a), AIDS scandal indicts French government, *Nature* **353**, 197.

ANDERSON, C. (1992b), US company's gene therapy trial is first to bypass RAC for approval, *Nature* **357**, 615.

BMA (British Medical Association) (1992), *Our Genetic Future. The Science and Ethics of Genetic Technology*, Oxford: Oxford University Press.

BOLTON, V., OSBORN, J., SERVANTE, D. (1992), The Human Fertilisation and Embryology Act 1990 – A British case history for legislation on bioethical issues, *Int. J. Bioethics* **3**, 95–101.

COUCHMAN, P. K., FINK-JENSEN, K. (1990), *Public Attitudes to Genetic Engineering in New Zealand, DSIR Crop Research Report* **138**, Christchurch: Department of Scientific and Industrial Research.

DEUTSCH, E. (1992), Fetus in Germany. The Fetus Protection Law of 12.13.1990, *Int. J. Bioethics* **3**, 85–93.

DIEKEMA, D. S. (1990), Is taller really better? Growth hormone therapy in short children, *Perspect. Biol. Med.* **34**, 109–123.

DIXON, B. (1991), Biotech a plus according to European poll, *Biotechnology* **9**, 16.

DRAPER, E. (1991), *Risky Business. Genetic Testing and Exclusionary Practices in the Hazardous Workplace*, Cambridge: Cambridge University Press.

DURANT, J. R., EVANS, G. A., THOMAS, G. P. (1989), The public understanding of science, *Nature* **340**, 11–14.

EB (1993), *Eurobarometer 39.1, Biotechnology and Genetic Engineering, What Europeans Think About It in 1993,* Brussels: European Commission.

EGGER, M., SMITH, G. D., TEUSCHER, A. (1992), Human insulin and unawareness of hypoglycaemia: need for a large randomised trial, *Br. Med. J.* **305**, 351–355.

EP (European Parliamentary Assembly) (1991), *Recommendation* **1160**.

FOWLER, C., MOONEY, P. (1990), *The Threatened Gene. Food, Politics, and the Loss of Genetic Diversity*, Cambridge: Lutterworth Press.

FOX, J. L. (1992), FDA dishes up food policy, *Biotechnology* **10**, 740–741.

GAJDUSEK, D. C. (1992), Scientific responsibility, in: *Human Genome Research and Society* (FUJIKI, N., MACER, D. R. J., Eds.), pp. 205–210, Christchurch: Eubios Ethics Institute.

HAMSTRA, A. M. (1993), *Consumer Acceptance of Food Biotechnology,* Den Haag: SWOKA.

HOBAN, T. J., KENDALL, P. A. (1992), *Consumer Attitudes About the Use of Biotechnology in Agriculture and Food Production,* Raleigh, NC: North Carolina State University.

HOLM, S. (1992), A common ethics for a common market?, *Br. Med. J.* **304**, 434–436.

HOLTZMAN, N. A. (1989), *Proceed with Caution. Predicting Genetic Risk in the Recombinant DNA Era,* Baltimore: Johns Hopkins University Press.

ISIS-3 (Third International Study on Infarct Survival Collaborative Group) (1992), ISIS-3: a randomised comparison of streptokinase vs tissue plasminogen activator vs anistreplase and of aspirin plus heparin vs aspirin alone among 41 299 cases of suspected acute myocardial infarction, *Lancet* **339**, 753–770.

JUMA, C. (1989), *The Gene Hunters. Biotechnology and the Scramble for Seeds*, Princeton, NJ: Princeton University Press.

KELVES, D. J. (1985), *In the Name of Eugenics*, New York: Knopf.

KENWARD, M. (1989), Science stays up the poll, *New Sci.* (16 Sept.), 39–43.

KESSLER, D. A., TAYLOR, M. R., MARYANSKI, J. H., FLAMM, E. L., KAHL, L. S. (1992), The safety of foods developed by biotechnology, *Science* **256**, 1747–1749, 1832.

KILEY, T. D. (1992), Patents on random complementary DNA fragments?, *Science* **257**, 915–918.

KILLHAM, K. (1992), Positively luminescent, *Biotechnology* **10**, 830–831.

KRUPP, H. (1992), *Energy Politics and Schumpeter Dynamics. Japan's Policy Between Short-Term Wealth and Long-Term Global Welfare*, Tokyo: Springer-Verlag.

LESSER, W. H. (1991), *Equitable Patent Protection in the Developing World: Issues and Approaches*, Christchurch: Eubios Ethics Institute.

MACER, D. (1989), Uncertainties about 'painless' animals, *Bioethics* **3**, 226–235.

MACER, D. R. J. (1990), *Shaping Genes: Ethics, Law and Science of Using Genetic Technology in Medicine and Agriculture*, Christchurch: Eubios Ethics Institute.

MACER, D. R. J. (1991a), New creations?, *Hastings Center Report* (Jan./Feb.), 32–33.

MACER, D.R. J. (1991b), Whose genome project?, *Bioethics* **5**, 183–211.

MACER, D. R. J. (1992a), *Attitudes to Genetic Engineering: Japanese and International Comparisons*, Christchurch: Eubios Ethics Institute.

MACER, D. R. J. (1992b), Public opinion on gene patents, *Nature* **358**, 272.

MACER, D. R. J. (1992c), Public acceptance of human gene therapy and perceptions of human genetic manipulation, *Hum. Gene Ther.* **3**, 511–518.

MACER, D. R. J. (1993), Perception of risks and benefits of *in vitro* fertilization, genetic engineering and biotechnology, *Soc. Sci. Med.* **38**, 23–33.

MACER, D. R. J (1994), *Bioethics for the People by the People*, Christchurch: Eubios Ethics Institute.

MACKENZIE, D. (1991), *New Sci.* (13 July), 14.

MARTIN, S., TAIT, J. (1992), Attitudes of selected public groups in the UK to biotechnology, in: *Biotechnology in Public. A Review of Recent Research* (DURANT, J., Ed.), pp. 28–41, London: Science Museum.

MELANCON, M. J. (1992), Scientist's responsibilities in acquiring knowledge and developing intervention technologies with regard to the human genome, in: *Human Genome Research and Society* (FUJIKI, N., MACER, D. R. J., Eds.), pp. 198–204, Christchurch: Eubios Ethics Institute.

MHW (Japanese Ministry of Health and Welfare) (1992), *Guidelines for Foods and Food Additives Produced by Recombinant DNA Techniques*, Tokyo: Chuou Houki, Shuppann.

MUNDELL, I. (1992), Europe drafts a conversion, *Nature* **356**, 368.

NEU, H. C. (1992), The crisis in antibiotics research, *Science* **257**, 1064–1073.

NRC (National Research Council) (1992), *Strategic Technologies for the Army*, Washington, DC: National Academy Press.

OTA (US Congress Office of Technology Assessment) (1987), *New Developments in Biotechnology, 2: Public Perceptions of Biotechnology* – Background Paper, Washington, DC: USGPO.

OTA (US Congress Office of Technology Assessment) (1991a), *Agricultural Commodities as Industrial Raw Materials*, Washington, DC: USGPO.

OTA (US Congress Office of Technology Assessment) (1991b), *US Dairy Industry at a Crossroad: Biotechnology and Policy Choices*, Washington, DC: USGPO.

OTA (US Congress Office of Technology Assessment) (1991c), *Biotechnology in a Global Economy*, Washington, DC: USGPO.

PORTER, D. G. (1992), Ethical scores for animal experiments, *Nature* **356**, 101–102.

PCC (President's Council on Competitiveness) (1991), *Report on National Biotechnology Policy*, Washington, DC.

RABINO, I. (1991), The impact of activist pressures on recombinant DNA research, *Sci. Technol. Hum. Val.* **16**, 70–87.

RABINO, I. (1992), A study of attitudes and concerns of genetic engineering scientists in Western Europe, *Biotech Forum Europe* (Oct.), 636–640.

RACHELS, J. (1990), *Created from Animals*, Oxford: Oxford University Press.

ROBERTS, L. (1992), NIH gene patents, round two, *Science* **251**, 912–913.

RSGB (Research Surveys of Great Britain) (1988), *Public Perceptions of Biotechnology: Interpretative Report*, UK: RSGB Ref. 4780.

SASSON, A. (1988), *Biotechnologies and Development*, Paris: UNESCO.

SIEGHART, P. (1985), *The Lawful Rights of Mankind*, Oxford: Oxford University Press.

SINGER, P. (1976), *Animal Liberation*, London: Jonathan Cape.

STONE, R. (1992), NIH to size up growth hormone trials, *Science* **257**, 739.

SWINBANKS, D. (1992), Japanese researchers rule out gene patents, *Nature* **356**, 181.

WALGATE, R. (1990), *Miracle or Menace? Biotechnology and the Third World*, London: Panos Institute.

WHO (World Health Organization) (1991), *Report of a Joint FAO/WHO Consultation, Strategies for Assessing the Safety of Foods Produced by Biotechnology*, Geneva: WHO.

WRI (World Resources Institute) (1992), *World Resources 1992–93*, Oxford: Oxford University Press.

ZHANG, Z. (1991), People and science: Public attitudes in China toward science and technology, *Sci. Publ. Pol.* **18**, 311–317.

II. Product Development and Legal Requirements

5 Structured Risk Assessment of rDNA Products and Consumer Acceptance of These Products

C. Theo Verrips

Vlaardingen, The Netherlands

1 General Introduction

Biotechnological products based on recombinant DNA have been a subject of debate at many levels since the Asilomar Conference in 1975. It started as a debate among conscientious scientists who wanted to think about the consequences of new techniques in molecular biology before applying these in the construction of new combinations of DNA. Already during the Asilomar Conference this debate became clouded by a number of unscientific arguments and personal interests (CAMPBELL, 1990). Some journalists started to describe this technology in the media as a modern version of Cerberus or the molecular version of the monster of Frankenstein. Unfortunately, the scientific community was not very well prepared for this debate and did not react uniformly. Moreover, the scientific community started to compare the risk related to rDNA technology to other risks, e.g., the risk of spontaneous or deliberate mutations or classical breeding resulting in dangerous combinations of DNA. These types of comparisons were scientifically incorrect and led to further confusion among the public.

Most unfortunately, rDNA technology has often been compared with nuclear energy technology. This comparison is extremely misleading, as for nuclear energy the intrinsic hazard is certain (in the language of this chapter "1.00"), and its effects may be very large and will affect all living species. Therefore, the probability of such a hazard occurring should be extremely small, and the debate on nuclear energy always focuses on the aspect whether physical containment can ensure this extreme low probability. For rDNA technology the intrinsic hazard was often unknown. In general this hazard will depend on the nature of the foreign gene, the host cell and various other factors, and it will range from zero to one (e.g., in case of the transfer of a gene coding for a toxic protein to human cells). During the initial discussions it was assumed that in general the hazard of rDNA products was not zero, and therefore, the debate concentrated on the physical and biological containment of the experiments to reduce the risk.

Soon after the Asilomar Conference in most countries committees were formed to set up rules to ensure that the intrinsic hazard of the foreign genetic property was low and that biological and physical containment would guarantee that the probability that the hazard would become manifest outside the laboratory would be extremely small. The arguments of certain individual manufacturers that the rules proposed to evaluate hazards and risks related to rDNA could jeopardize their future, were not correct and contributed to a further polarization of views. This all resulted in a delay of the acceptance or even in a rejection of rDNA technology outside the area of health care by the public.

It is beyond any doubt that the application of rDNA technology to elucidate fundamental questions in biology has contributed enormously to our understanding of the basic biological processes. After just 20 years it can be concluded that PAUL BERG (1974) and others who organized the Asilomar Conference were absolutely right in that rDNA technology would have an enormous impact on mankind and in their plea for a careful evaluation of this technology, since so little was known about it. A large number of discoveries that have been made since 1975 such as the existence of introns, the role of transposons in many species, RNA as enzyme, the reprogramming of transcriptional competence, a number of translation processes that deviate from the general rules for translation, some preliminary rules for protein folding and various processes resulting in horizontal fluxes of genes DNA between species, the influence of ecological pressure on the selection of species created by the horizontal and vertical fluxes of genes, demonstrate that in fact proper risk assessment of rDNA processes or products was not possible at that time.

Although at present still much more has to be learned, the results of millions of rDNA experiments show that indeed there is unity in biology and therefore rDNA technology is as such not intrinsically dangerous. Moreover, there is now sufficient understanding of basic processes in biology to allow a rational approach of the evaluation of hazards and risks related to rDNA technology. This knowledge of molecular biology together with a structured approach to evaluate the various aspects of rDNA technology separately can contribute to end the confusion about real or perceived hazards and risks related to rDNA technology.

Whether after careful evaluation of these factors rDNA products other than pharmaceuticals will be accepted by society is a completely different story. However, as will be discussed later, a risk assessment showing that the risk is very low will increase the acceptance of the product by the public (see Sect. 5, Tab. 15).

It is the aim of this chapter to discuss the various technical aspects of hazard analysis and risk assessment of rDNA products and to indicate briefly the complex but extremely important aspects such as ethics, labeling and consumer acceptance. As in most countries the area of hazard analysis and risk assessment and the legislation of rDNA products are far from established, no definitive view can be given. However, a decision scheme to analyze hazards and risks containing the most useful elements of a number of national guidelines will be presented. Implementation of a decision scheme in line with the proposed one might facilitate the approval of rDNA products in many countries.

To structure the discussion it may be worthwhile to put these aspects in a framework comprising all aspects related to rDNA technology:

1. Technical aspects
   a) Intrinsic hazards of recombinant DNA
   b) Risk of creation of intrinsic hazard in the recombinant DNA
   c) Risk of accidental release of rDNA modified microorganisms during processing
   d) Risks associated with cell-free rDNA products
   e) Risks associated with the deliberate release of microorganisms
2. Structured methods to control intrinsic hazard and risk of DNA processes and products
3. Benefits versus real and perceived risks
4. Ethical aspects
5. Consumer acceptance.

The first two items concern matters between regulatory bodies and scientists in academia and industry. The aspects 3 and 4 have much more players, like consumers and consumer organizations, retailers, religious groups, politicians, pressure or action groups, health authorities, regulatory bodies and manufacturers. Moreover, the risk/benefit analysis cannot be generalized, as the risk/benefit ratio for rDNA products that inhibit processes related to AIDS, like inhibition of reverse transcriptase, RNase H, protease or the regulatory protein TAT of the HIV virus, will differ substantially from that of frost-resistant strawberries or from a microorganism capable of degrading halogenated benzene. Moreover, it should be borne in mind that to cure certain diseases highly specific inhibitors will be required and that only rDNA technology will be able to produce molecules having the required specificity.

Also consumer acceptance cannot be simply defined, as it will be a function of the religious, cultural and educational background of the consumers as well as of their economic situation. This was demonstrated clearly during a discussion on this subject at the Third International Congress of Plant Molecular Biology in Tucson in 1991. After listening to several speakers from the Western societies, a representative of China stated that he considered the whole discussion on the application of rDNA technology to improve yields of crops as a luxury problem of Western society and that his government and fellow scientists considered it their duty to use rDNA technologies, e.g., to develop new rice varieties that can be cultivated on land that cannot be used at present.

2 Introduction into Some Technical Aspects of rDNA

If one wishes to carry out a hazard analysis and risk assessment of an rDNA process or product, it is essential to take into account all aspects that may influence these analyses.

This seems to be an obvious statement, but unfortunately, too often the terms hazard analysis and risk assessment have been used for improper analysis. Therefore, in this chapter the terms hazard and risk will be defined and the biological processes that may influence these analyses will be discussed briefly. In Sect. 4 the hazards and risks related to the construction of rDNA molecules in research and development stages will be analyzed taking into account the following biological processes:

1. Errors introduced during DNA duplication and other events on DNA level
2. Errors in transcription and translation
3. Stability of episomal vectors and integrated foreign DNA
4. Reprogramming of metabolic pathways in the host cell
5. Transfer of the foreign genes from selected hosts to unknown recipient cells.

2.1 Definition of Hazard and Risk

Before starting any discussion on rDNA technology it is essential to define the words hazard and risk. In this chapter *hazard* is defined as the potential (toxicological or ecotoxicological) harmful intrinsic property of the product encoded by the newly constructed genetic material or by the host carrying this genetic material or its translation product. *Risk* is the probability of the hazard occurring.

rDNA technology can be used to construct genetic material that is potentially harmful, and in fact this possibility was one of the major concerns in the early days of rDNA technology. Potentially harmful means that when this genetic material is transcribed and translated into a protein, that protein is either by itself harmful or the protein (enzyme) is able to catalyze (a) reaction(s) that directly or indirectly produce a harmful compound. As the rDNA technology has been developed using *Escherichia coli* as host for the foreign DNA, in the discussion on the hazards of this technology the intrinsic harmful properties of *E. coli*, e.g., their production of endotoxins and the potential of certain *E. coli* strains to pro-

duce enterotoxins have played an important role. Therefore, at the start of rDNA technology it was advocated to use only certain strains of *E. coli* K12 or to use other microorganisms, e.g., *Bacillus* or *Saccharomyces cerevisiae* (baker's yeast) as hosts for genes transferred by rDNA technology. Unfortunately, compared to *E. coli*, insufficient understanding of the molecular biology, especially regulation of transcription and translation processes of *Bacillus* and *S. cerevisiae* existed in the mid seventies.

For the transformation of plants an intrinsic hazardous plasmid was used, and for the expression in certain hosts promoters were used originating from viruses (e.g., CaMV for plants and SV40 for mammalian cells). Together with the lack of understanding of the basic principles of biology by the public, the use of biological material that is intrinsically harmful or was perceived to be harmful has contributed largely to the confusion about the hazards and risks related to rDNA technology.

If vectors are used that do not contain unknown sequences or known sequences that can code (directly or indirectly) for harmful products or genes that can give an ecological advantage to a second generation of recipient cells, one may define such vectors as safe. If one knows that the gene transferred by rDNA technology has been sequenced and proved to contain only the intended information, this then guarantees the absence of the hazard that the foreign gene will code for a protein different from the intended protein. However, all this does not guarantee the absence of any intrinsic hazard. Firstly, the foreign gene will be subject to normal vertical flux of genes (see Sect. 2.2.1). Secondly, there is a small, but not negligible probability that the foreign genetic material will be transcribed or translated differently from the normal transcription and translation processes (see Sects. 2.3 and 2.4). Finally a gene not coding for a potential harmful property in a certain host cell may in the second generation of recipients of the new genetic combination code for a protein that contributes indirectly to a hazard. The probability of such events occurring should be assessed (compare Sect. 4.5).

Often opponents of rDNA technology use "unforeseeable synergistic effects" as an argument against this technology. Although enormous progress has been made in our understanding of cell biology, it is certainly true that we do not understand cell biology in all its details. One aspect, feedback control mechanisms, has been analyzed in detail in various systems, and from these studies it can be concluded that it is extremely unlikely that transfer into a new host cell of a gene not coding for proteins involved in signal transduction or transcription, will result in "unforeseeable synergistic effects". Moreover, if one requires that new food products can only be introduced into the market if one can guarantee that no "unforeseeable synergistic effects" will occur, this would mean that we have to block the introduction of any new food product. In fact, no existing food product will comply with such a requirement. Moreover, food products derived from recombinant organisms will be analyzed in much more detail than any other food product, and not just for the "known effects", but also for unexpected effects, e.g., by the application of the technique recently developed by the group of ARTHUR PARDEE that can detect differences in the expression of "unknown" genes (LIANG and PARDEE, 1992). Also the "90 days feeding trial" required in most countries will reduce the probability of "unforeseeable synergistic effects".

2.2 Events Resulting in Changes in DNA

2.2.1 Vertical Flux of Genes

Even before 1975 it was known that recombination on the DNA level was a common process in *E. coli*. In fact, recombination is an important factor in generating genetic diversity, and recombination is involved in the repair of chromosome breakage and other DNA damage, and according to SMITH (1988) multiple pathways of homologous recombination for multiple reasons exist. This clearly indicates that DNA is not the inert genetic blueprint of the living cell as was thought before 1970.

Although the reliability of DNA duplication is very high and the few mistakes are often corrected by ingenious mechanisms and counterbalanced by back-mutations, DNA duplication is not an errorless process. Foreign DNA will, therefore, also be a subject of these errors and constitute a factor that should be included in the hazard analysis and risk assessment of rDNA products. However, the mutations are not directed towards a certain property or function of the gene product, but occur just by chance. This means that for genes introduced by rDNA technology the mutation frequency will be equal to or less than that for endogenous genes. Recently W. ARBER (1990) has described the events that may change DNA:

1. Fidelity of DNA replication
2. Mutagenesis induced by external factors
3. Recombination processes
   a) General genetic recombination
   b) Transposition
   c) Site-specific recombination
   d) Illegitimate recombination.

The probability of a mutation in a certain gene in cells, growing under conditions that do not stimulate mutations, is low (about 10^{-5} per bacterial genome per generation) but it cannot be excluded. So also mutation in foreign genes can occur. The probability that a mutation may render the gene product from an intrinsically innocent into a hazardous protein is very small. Moreover, as outlined in some more detail in Sects. 2.4 and 2.9, techniques are available which can be incorporated into the normal QA/QC procedure of manufacturers of food enzymes or starter cultures (these procedures should be followed also for endogenous gene products). These procedures will eliminate the possibility that products encoded by mutated foreign genes will pose a hazard.

2.2.2 Horizontal Flux of Genes

It is difficult to define which discovery initiated the development of modern biotechnology. The work by AVERY et al. (1944), using transformation experiments with S and R type pneumococci resulted in the conclusion "that a nucleic acid of the deoxyribose type (DNA) is the fundamental unit of the transforming principle of *Pneumococcus*" could certainly be considered as a landmark in biotechnology. Transformation requires that the recipient cells are competent (i.e., able to take up DNA from the environment). Competence is part of the normal physiology of many species and can also be induced artificially by chemical and physical treatments. In fact, transformation proved to be a biological process that has played an important role in biological evolution (Fig. 1).

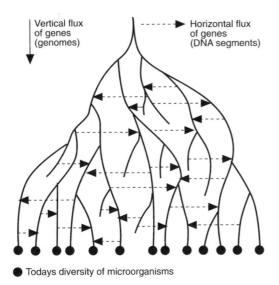

Fig. 1. Flux of genes in biological evolution.

The frequency of transformations in Nature is not very well known, but it has been reported that rDNA microorganisms introduced in soil could transform *Bacillus subtilis* present in that soil (GRAHAM and ISTOCK, 1978). Therefore, transformation of rDNA from the selected host into new recipients in the environment cannot be excluded and should be considered in the risk assessment. Numerous experiments have been performed to determine the probability that genetic material will be transferred from the host strain to a recipient cell. One of the most convincing

experiments was done by ISRAEL et al. (1979). The complete genome of Polyoma virus was transferred into *E. coli* K12 and $5 \cdot 10^9$ transformed cells were injected into mice. In control experiments the virus itself, naked viral DNA either in the single-strand or double-strand forms, integrated in a bacterial plasmid as such or treated with a restriction enzyme were used. The outcome of these experiments was that the rDNA construct was more than 10000 times safer than the naked genome and more than 10^{10} safer than the virus itself (note: no tumor formation with the rDNA construct could be detected). In another experiment the guts of 100 mice were colonized by transformed *E. coli* for 6 weeks. No tumor formation was found which means that the probability of transfer must be less than 1 in $2 \cdot 10^{13}$ per day. In fact these extremely low values mean that in a risk assessment this probability can be considered as zero.

The above results may be explained by the lack of competence of epithelial cells to take up DNA or an extremely well developed system of the epithelial cells to destroy foreign DNA or that DNA in the hostile condition of the gut is not very stable. Another matter of concern was the transfer of rDNA from the host cell to cells of the gut flora. LEVINE et al. (1983) carried out extensive studies with *E. coli* that hosted a self-transmissible plasmid and a poorly transmissible plasmid pBR325 containing a gene resulting in tetracycline resistance. LEVINE was unable to detect pBR325 in the *E. coli* cells belonging to the gut flora. From his experiments it can be estimated that transfer of DNA between related species in high-density populations as exist in the gut, is extremely rare. Bacilli used in common fermentation processes, *Saccharomyces* sp., *Kluyveromyces* sp., or *Aspergillus niger* or *A. awamori* are unable to colonize the gut and, therefore, the probability that these microorganisms will transfer rDNA to microbial cells of the gut flora is even lower than for *E. coli* and can therefore be considered as zero in risk assessments.

Transduction (the exchange of genetic material mediated by bacteriophages or viruses) is another method to transfer DNA from one species into another. This is of direct relevance, if the foreign DNA is located on a phage or virus, however, this is an unusual procedure and should not be applied on an industrial scale. However, as a risk factor transduction cannot be excluded completely, if plasmids or integration vectors are used as vehicles for rDNA. In aquatic and terrestrial ecosystems quite high titers of temperate and lytic bacteriophages have been determined. If an rDNA containing bacterium enters this environment, it is possible that one of these phages recognizes this bacterium and occasionally plasmid or even chromosomal DNA can be packed into the capsids producing a transducing particle that may contain rDNA (SAYE et al., 1987). However, the probability of such events is very low.

Conjugation (the parasexual transfer of plasmids and/or chromosomal DNA) of bacteria or mating of lower eukaryotes is the most important mechanism to transfer genetic material between related species. There is overwhelming evidence that if sufficiently high numbers of antibiotic-resistant cells are present in nature and sufficient selection

pressure exists, transfer of genetic material, even between very distinct species occurs, as is proven by the widespread antibiotic resistance among enteric bacteria. Tab. 1 shows some figures for the horizontal flux between related microorganisms, however, a considerable amount of evidence exists that also between unrelated species like *E. coli* and *S. cerevisiae* conjugational transfer of genetic information can occur (HEINEMANN and SPRAGUE, 1989).

Mating between lower eukaryotes seems to be restricted to the members of the same genus due to the role of specific pheromones and receptors for the mating factors in this process (SPRAGUE, 1991). In production processes the probability of conjugation or mating can be eliminated by proper strain selection, and therefore it is not necessary to include these aspects in the risk assessment. However, conjugation and mating should be included in the risk assessment of emitted rDNA organisms from production processes. Conjugation of deliberately released microorganisms with microorganisms in the environ-

Tab. 1. Horizontal Flux of Genes Between Bacteria

Plasmid	Donor → Recipient	Rate[a]
pBC16 (Tra$^+$)	*Bacillus subtilis* → *B. subtilis*	10^{-5}–10^{-2}
	B. subtilis → *B. licheniformis*	10^{-3}
	B. subtilis → *B. megaterium*	10^{-6}
RP4 (Tra$^+$)	*Escherichia coli* → *Thiobacillus novellus*	10^{-5}
R68.45 (Tra$^+$)	*Pseudomonas aeruginosa* → *P. aeruginosa*	10^{-4}
		10^{-5}
		10^{-2}
		10^{-8}
PMQ1 (Tra$^+$)	*P. aeruginosa* → *P. aeruginosa*	10^{-1} (R)
		10^{-2} (R)
		10^{-4} (R)
		10^{-7} (R)
pBR325 (Tra$^-$Mob$^+$)	*E. coli* → *E. coli*	10^{-5}
		10^{-7}
pFT30 (Tra$^+$)	*B. cereus* → *B. subtilis*	10^{-6}
R Plasmids (Tra$^+$)	*Klebsiella* sp. → *Klebsiella* sp.	10^{-6}
pBC16 (Tra$^+$)	*B. thuringiensis* → *B. thuringiensis*	10^{-3}
	B. thuringiensis → Soil bacteria	10^{-6}–10^{-5}
pDN705 (Tra$^-$Mob$^+$	*E coli* → *Alcaligenes eutrophus*	10^{-4}–10^{0}
		10^{-6}–10^{0}
pIJ673 (Tra$^+$)	*Streptomyces lividans* → *S. lividans*	10^{-2}–10^{-1}
	S. violaceolatus → *S. lividans*	10^{-3}

[a] Transconjugant per donor cell, or transconjugant per recipient cell (R)

ment cannot be excluded and, therefore, should form part of the risk assessment. Mating in the environment outside the production facilities can be excluded, because in the environment the concentrations of wild-type and rDNA microorganisms of the opposite mating type will be very low.

Besides the horizontal flow of genetic material between microorganisms and in rare cases between microorganisms and plants, this flow also exists between plants by pollen exchange. Whereas new traits of commercial crops introduced via classical breeding are often not of any use for their wild relatives, transgenic plants will often be equipped with new properties like salt and drought tolerance and herbicide or insecticide resistance, and these properties may be beneficial to the wild relatives as well. ELLSTRAND (1988) has summarized the probability of horizontal flow of genes via either wind or insect transferred pollen as a function of the distance between the donor and recipient plant (see Tab. 2).

These data clearly show that a horizontal flow of genes (also rDNA) can occur with a rather high probability when wild relatives are present within 1000 meters, and therefore this should be included in risk assessment of transgenic plants, in particular if the foreign gene codes for properties that provide the recipient with a clear ecological advantage.

Tab. 2. Gene Flow by Pollen

Species	Isolation	Gene Flow (%)
Herbs		
Raphanus sativus	100–1000 m	4.5–18.0
Agrostis tenuis	8 km	>1
Phlox drummondii	50–100 m	8
Trees		
Gleditsia triacanthos	200 m	5.8
Pinus taeda	122 m	36
Pseudotsuga menziesii	161 m	20.4–26.6

The available data demonstrate that pollen can act as a vehicle for the transfer of engineered genes from crops to their wild relatives. The spread of these genes into natural populations will probably be rapid even if the wild relatives occur 1000 m from the crop and if the engineered gene confers an advantage to the wild species.

In the event that by one of the above described mechanisms for horizontal flow of genes, genes have been transferred, this does not mean that the gene will be expressed in the recipient cell. Even between relatively closely related species as *E. coli* and *Pseudomonas aeruginosa* many genes are not transcribed because of inefficient promoters. It is important to keep this in mind when reading publications on horizontal flows of genes based on the PCR methodology, as this methodology does not give any indication of gene expression.

Fig. 2 summarizes experimental evidence of transfer of genetic information between unrelated species. The barrier between various species is clearly not as absolute as was thought in the early seventies, and the argument that rDNA technology is hazardous only because it surpasses natural barriers is at least disputable.

2.3 RNA Splicing and mRNAs Derived from Synthetic Genes

RNA splicing can be considered as a common process in eukaryotes and soon after the discovery of this phenomenon, DARNELL and DOOLITTLE (1986) proposed that RNA splicing has played an important role in evolution. The nucleotide sequences of 5′ and 3′ splice sites are well known and reasonably conserved, although differences in nucleotide are permitted. Even taking into account that the sequences are not conserved absolutely, it is not too difficult to scan foreign genes for the potential that they may contain RNA splice sites that may result in a completely differently processed mRNA and consequently in a completely different protein. Especially if chimeric genes are constructed and these constructs are transferred to eukaryotic hosts, it is essential to check for the creation of splice sites to eliminate unintended splicing as a hazard. For prokaryotes splicing is not a factor of the hazard analysis.

To improve the translation rate of foreign genes they are quite often synthesized chemically in the codons preferred by the new host (ANDERSON and KURLAND, 1990). Also

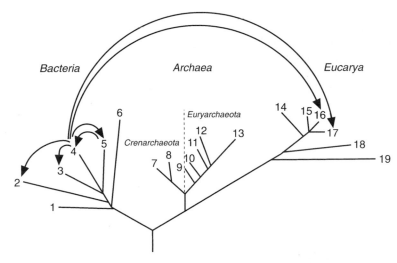

Fig. 2. Transfer of genetic information between unrelated species. **Bacteria: 1**, Thermotogales; **2**, flavobacteria and relatives; **3**, cyanobacteria; **4**, purple bacteria; **5**, Gram-positive bacteria; **6**, green nonsulfur bacteria. **Archea:** kingdom Crenarchaeota: **7**, genus *Pyrodictium*; **8**, genus *Thermoproteus*; and kingdom Euryarchaeota: **9**, Thermococcales; **10**, Methanococcales; **11**, Methanobacteriales; **12**, Methanomicrobiales; **13**, extreme halophiles. **Eucarya**: **14**, animals; **15**, ciliates; **16**, green plants; **17**, fungi; **18**, flagellates; **19**, microsporidia (from DAVIES, 1990; WOESE et al., 1990).

genes coding for enzymes that have to be modified to improve their usefulness in certain application (e.g., detergents) are often synthesized as a modular system containing different DNA cassettes. Normally each cassette is flanked by a unique restriction site enabling the rapid exchange of such a cassette by another carrying the modified genetic code. The introduction of restriction sites quite often requires silent mutations. Finally genes are often (partly) synthesized chemically to improve the stability of the mRNA or their rate of translation, although general models that describe the stability of mRNA hardly exist (BAIM and SHERMAN, 1988). Synthetic genes that have new nucleotide sequences should be screened for frameshift mutations, splicing and premature termination as all these processes may result in the synthesis of unknown proteins. Consequently they can contribute to the intrinsic hazard of rDNA products. Careful analysis of the transcription products of the foreign (synthetic) gene can reduce largely the above mentioned phenomena as a source of hazard related to the rDNA technology.

2.4 Reliability of Translation Processes and Post-Translational Modifications

Although gene expression is mainly regulated at the transcriptional level, there are a number of ingenious regulation mechanisms that have been developed on a translational level. Some of these mechanisms involve frameshifting or initiation of the translation at a site different from the normal start codon. These frameshifts result in proteins that differ from the protein derived from the nucleotide sequence, and therefore this phenomenon is important in the discussion on the intrinsic hazard of foreign genes. Various mechanisms of recoding the message of an mRNA have been discovered:

1. *Plus-one frameshifts.* One of the best studied examples of +1 frameshifts is the release factor 2 (RF2). In RF2 a stop codon (UGA) is located in position 26 (CRAIGEN et al., 1985). However, with a probability of about 50% a +1 frameshift

(UGA → GAC) occurs resulting in the biosynthesis of RS2 (CRAIGEN and CASKEY, 1986). This type of frameshifts is only possible, if a certain sequence in the mRNA, called stimulator, is present to which the 16S RNA of the ribosome can bind. Recently also for the eukaryotic chromosomal gene involved in the post-translational modification of ornithine decarboxylase, a +1 frameshifting has been discovered (GESTELAND et al., 1992).

2. *Minus-one frameshifts* occur quite frequently in the translation of retroviruses. The stimulator is a stem-loop structure located downstream of the place of frameshift (BRIERLEY et al., 1989). Minus-one frameshifts are not restricted to retroviruses, but occur also in *E. coli* (TSUCHIHASHI and KORNBERG, 1990).

3. *Hopping* is another translation phenomenon that can cause the synthesis of a protein different from the protein expected on the basis of the nucleotide sequence of the gene. This phenomenon is quite well documented for bacteriophage T4. At a certain sequence the ribosome leaves the mRNA molecule and rejoins it about 50 bases later (HUANG et al., 1988).

4. In the genetic information of certain retroviruses and some plant and bacterial viruses, codons are encoding amino acids differently from the codon table as established by NIRENBERG and others. For instance, in certain mRNAs that contain particular sequences (ZINONI et al., 1990), the stop codon UGA encodes selenocysteine (HILL et al., 1991). Although in higher eukaryotes also UGA is used to incorporate selenocysteine in proteins, the mechanism in these cells is less well known, but it is likely that again downstream sequences play an important role (BERRY et al., 1991).

5. Normally translation starts at the triplet ATG. However, at a low probability translation starts at GUG (8%) and even more rarely at UUG or CUG (KOZAK, 1983). This may occur also in the translation of an rDNA gene encoded product.

Although recoding of the genetic information is a rare event and mainly restricted to viruses and bacteriophages, it has been detected in normal mRNAs of bacteria, lower and higher eukaryotes and, therefore, it is essential that for each gene transferred into a new host at least a theoretical evaluation is made whether recoding can occur. This is of particular relevance, if the transferred gene is chemically synthesized to adapt the codons or to design a gene that is very suitable for protein engineering, as there is a small probability that a stimulator sequence has been introduced. On top of theoretical analysis, careful inspection of the proteins produced by the rDNA strain by 2D electrophoresis and Western blotting and in cases of any doubt determination of the C- and N-terminal amino acids can eliminate recoding as a hazard.

Besides recoding of the mRNA, foreign genes may be translated into a protein different from the protein in the original host by differences in post-translational modification. The best studied difference in post-translational modification is glycosylation. A eukaryotic gene carrying the amino acid sequences Asn X Thr or Ser that serve as a recognition site for glycosylation in the endoplasmic reticulum and Golgi apparatus, will when transferred into a prokaryotic host not be glycosylated. Alternatively, if the gene is transferred from one eukaryote to another, e.g., from a plant to a yeast or mold, the N-glycosylation site will be recognized by the new host, but the type of glycosylation will be different. Besides N-glycosylation also O-glycosylation exists, but the exact mechanism by which O-glycosylation takes place is not fully known and is consequently an uncertainty in rDNA technology. Moreover, in addition to glycosylation, other hardly known post-translational modifications, such as N- or C-terminal processing resulting in smaller proteins, phosphorylation, farnesylation, etc. can occur. All these changes will be noticed already in the initial research phase and when observed, the choice of host should be reconsidered (with the exception of differences in N-glycosylation) to eliminate the biosynthesis of a protein of which the intrinsic property is unknown. With respect to N-glycosylation the situation is different. Firstly, the degree of glycosylation is not constant but seems to vary with cultivation conditions. Secondly, quite a number of proteins have been made via rDNA technology that have the same amino acid sequence but different glycosylation. The enzymic properties of these enzymes are very similar as, e.g., has been proven for the α-galactosidase from *Cyamopsis tetragonoloba* when expressed in various hosts (OVER-BEEKE et al., 1990). Although any change in N-glycosylation may in principle create an intrinsic hazardous protein, the absence of any indication for this until now proves that it is extremely unlikely that this will really occur. However, the allergic properties of an enzyme produced in a new host may differ from those of the original enzyme. In most decision schemes for the approval of rDNA products for food applications, the protein will be tested in feeding trials. If the results of these tests are similar to those for the natural protein, no hazard has been introduced by the transfer of the gene from the original source to the new host.

2.5 Random and Site-Directed Mutated Genes

One of the great promises of rDNA technology is that enzymes can be obtained that can catalyze reactions under non-physiological conditions. This is of particular relevance to the chemical industry with respect to the development of processes in which enzymes are used as catalysts. These processes are more environmentally friendly than most of the presently used processes in the chemical industry. Also for the development of enzymes for cleaning products random or site-directed mutagenesis is very important (WELLS and ESTELL, 1988). Both random and site-directed mutagenesis will create enzymes that most probably do not exist in nature. Arguments that nature has had four billion years to try out all possible combinations are not valid because firstly, it has been calculated that by far not all possible combinations have been made (MANFRED EIGEN, personal communication) and secondly, there has not been any environmental pressure in nature to select those enzymes the chemical or detergent industry is interested in.

In most of the decision trees for approval of new food products no attention is paid to genes that have been modified in a well known way and on applying the schemes very strictly, modified genes have to be rejected or will at least require much more testing as they are not originating from, or are not (yet) found in a natural, approved source. Provided that all the test procedures for the wild-type gene product have been carried out and the gene product has been approved, a simple scheme to cope with this problem for enzymes used in food or chemical processes, or in foods or in personal care or detergent products is proposed (Fig. 3, Tab. 3). This scheme takes into account the possibility that the modified gene product will have different immunological or enzymic properties. If the modified gene product deviates in significant aspects from the wild-type gene product, then a completely new approval procedure has to be followed. On the other hand, if in all but a few, from a safety point of view secondary aspects, the modified gene product is equal to the wild-type gene product, a more simple approval procedure is proposed.

2.6 Protein Folding

In most cases the transcription and translation of the rDNA gene will be performed without any failure. Still this does not guaran-

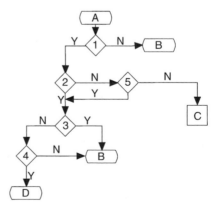

Fig. 3. Decision scheme to evaluate the hazard of proteins derived from site-directed or randomly mutagenized genes.

Tab. 3. Proposed Approval Scheme for Mutagenized Proteins

Entry, Actions and Results
A.   Mutagenized gene
B.   Start complete approval procedure
C.   Determine nucleotide sequence of the gene(s)
D.   Gene(s) or gene product(s) do (does) not pose a hazard

Questions
Q1.   Has the non-modified gene product been approved by an official body?
Q2.   Has the modification been done using site-directed mutagenesis?
Q3.   Has (have) the gene(s) due to the modification any new sequence that may result in recoding phenomena or has (have) the modified gene product(s) a new stretch of at least 6 amino acids identical to another class of proteins?
Q4.   Has (have) the gene product(s) encoded by the modified gene(s) essentially the same immunological, physical and enzymic properties?
Q5.   Has (have) the nucleotide sequence(s) of the modified gene(s) been determined?

tee that the rDNA product will be folded correctly. Intrinsically protein folding is a very complex phenomenon, moreover, it depends among other things on the translocation processes and the redox potential in various cell compartments or organelles (HWANG et al., 1992) and with a relative low probability this may result in (partially) improperly folded proteins. Sometimes very specific helper proteins are required for the correct folding. If this helper protein is absent, improper folding occurs (e.g., FRENKEN et al., 1993a, b). Although our understanding of protein folding is rapidly increasing, it is still impossible to predict folding of rDNA proteins in new hosts. For enzymes the specific activity (k_{cat} per mass, determined by enzymic and immunological methods) is a very accurate measurement of proper folding. For non-enzymic proteins it is more difficult to determine whether the protein is folded correctly, but a combination of physical and immunological measurements can reduce this probability of an unexpected hazard due to improper folding considerably.

2.7 Choice of Host Strain/Variety for Foreign DNA

Directly after the discovery that between evolutionary unrelated species genetic material could be transferred, the discussion about the safety of the host strain started. This was mainly due to the fact that the gut bacterium *E. coli* had become the work horse of the molecular biologists. To decrease the probability of a pathogenic strain being used as host in rDNA technology, the strain *E. coli* K12 has been selected. This strain, originally isolated in 1922, has lost during the many generations of cultivation on rich laboratory media four of the five essential properties to be a pathogenic bacterium, notably the property of adhesion in the gut, the production of enterotoxins, its resistance against phagocytosis. Moreover, *E. coli* K12 became sensitive to the human immune system and lost its property to penetrate through the epithelial wall of the gut. The production of endotoxin was reduced largely but not completely lost (BERGMANS et al., 1992).

An important lesson can be learned from this. Microorganisms maintained for a large number of generations on rich laboratory media will with a high probability lose some of the capabilities they have acquired during evolution. This is not only valid for *E. coli* K12, but for many other strains selected on their performance in this type of media. Consequently this means that strains, optimized to perform under standard conditions in a fermentation process will have a largely reduced capability to survive under the severe conditions outside the fermentation process.

The other lesson *E. coli* K12 taught us is that pathogenicity is quite a complex property. Experiments of GUINEE (1977) to reconstruct the pathogenicity of *E. coli* K12 by transferring the genetic material coding for the adhesion factor and the enterotoxin production failed. Similar results were obtained by ISBERG and FALKOW (1985) who transferred the adhesion factors back into *E. coli* K12.

When microorganisms used as host for rDNA technology are functional microorganisms in fermented food products (Tab. 4) and

Tab. 4. Main Functional Microorganisms in European Fermented Foods

Foods	Microorganisms					
	Bacteria		Yeasts		Molds	
Baked Goods	*Lactobacillus*	*farciminus* *plantarum* *acidophilus* *delbrueckii* *brevis* *buchneri* *fermentum* *san francisco*	*Saccharomyces* *Pichia*	*cerevisiae* *exigueris* *inusitus* *saitoi*		
Wine and Brandy	*Leuconostoc* *Pediococcus*	*gracile* *venos* *plantarum* *casei* *fructiocrans* *desidiosus* *hilgardii* *brevis* *cerevisiae*	*Candida* *Hanseniaspora* *Kluyveromyces* *Saccharomyces*	*vini* *uvarum* *apiculata* *cerevisiae* *rosei* *uvarum* *oviformis*		
Beer			*Saccharomyces*	*cerevisiae* *uvarum*		
Cheese and Dairy Products	*Brevibacterium* *Lactococcus* *Leuconostoc* *Pediococcus* *Streptococcus*	*linens* *lactis* *cremoris* *casei* *helveticus* *bulgaricus* *reuteri* *plantarum* *cremoris* *acidilactici* *pentosaceum* *thermophilus* *faecum*	*Kluyveromyces*	*lactis* *fragilis*	*Penicillum*	*camemberti* *caseicolum* *roquefortii*
Cabbage and Cucumbers	*Lactobacillus* *Leuconostoc* *Pediococcus*	*brevis* *plantarum* *meretoroides* *cerevisiae*				
Olives	*Lactobacillus* *Streptococcus* sp. *Pediococcus* sp. *Leuconostoc* sp.	*plantarum* *brevis* *delbrueckii*	*Saccharomyces* sp. *Kluyveromyces* sp. *Hansenula* sp. *Debaryomyces* sp.			
Meat	*Lactobacillus* *Pediococcus* *Micrococcus*	*plantarum* *lactis* *acidilactici* *pentosaceum* *caseolyticus* *varians*	*Debaryomyces* *Saccharomyces*	*hansenii* *cannosus*		

Deducted from *Biotechnology* 1st Ed. Vol. 5 Chapters 1–8

have a long safety record, the probability that such a strain becomes pathogenic through the transfer of one or a few well defined genes that code(s) for intrinsically safe proteins is extremely small.

To put it in perspective, this risk is considerably lower than the famous 12 D concept (the probability that less than one tin can in 10^{12} will be contaminated with *Clostridium botulinum* and that this will result in the formation of a detectable amount of toxin) which is used successfully in the food industry to ensure the safety of canned foods (SMELT, 1980).

Moreover, the probability that strains selected for their performance in a fermentation process will survive in the environment will be very small.

2.8 Plasmid Stability and Stability of Integrated Foreign DNA

Stability of the plasmids or transposons carrying the foreign gene was a matter of great concern in the early days of rDNA technology, and this concern is depicted in Fig. 4 (SAYRE and MILLER, 1991).

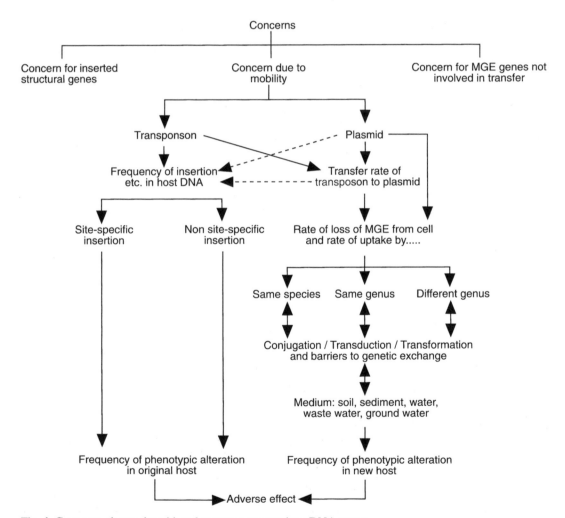

Fig. 4. Concerns about plasmid and transposon carrying rDNA genes.

The stability of plasmids derived from pBR322 in *E. coli* or ARS- or 2 μm derived vectors in *Saccharomyces cerevisiae* has been described in numerous papers. However, the reason for concern in relation to risk assessment of rDNA was not very clear. It can be argued that if a plasmid is unstable, many pedigrees of the transformed cell will not carry the foreign gene(s) and, therefore, these pedigrees are of no concern in the risk assessment.

Nearly always selection pressure is required for good stability of vectors. Often antibiotics are used for the selection pressure. Antibiotics will not be present in the environment in sufficient quantities, and consequently the fate of the plasmid in the transformed host in the environment will be even worse than the fate of plasmids under selection pressure, which already is not very good (e.g., see for yeast vectors MURRAY and SZOSTAK, 1983). This diminishes the probability of transfer of the plasmid to unknown recipients considerably. More recently, so-called food grade vectors (fully characterized vectors that contain, with the exception of the foreign gene, only homologous sequences and these sequences should only code for normal anabolic functions in the host cell) have been developed both for prokaryotes and eukaryotes. These vectors are maintained in the host, because they contain a gene encoding an essential enzyme of the anabolism of the host, e.g., an enzyme of the biosynthesis of an amino acid. In principle this selection pressure will be maintained in the environment, however, these host cells are in general quite fastidious. Therefore, the host cells will not survive very well in the environment, and this will reduce the risk of horizontal flow of gene to unknown recipients. Moreover, the recipient will have its own gene for that anabolic function which most likely is more effective in the recipient than in the foreign gene.

In fermentation processes stability of vectors containing the gene of interest is of major importance, as the yield of the gene product is often directly related to the number of gene copies. Even vectors that are maintained by selection pressure are not completely stable, and therefore vectors have been developed which integrate into the chromosome of the host. These integration vectors are very stable (stability most often identical to that of the chromosome), but the drawbacks are that the vector is not always integrated at the selected site (illegitimate integration) and the number of gene copies is quite low. Recently these disadvantages have been overcome in an elegant way by constructing an integration vector that contains upstream of a gene (encoding an enzyme essential for the anabolism of the host) a defective promoter and part of the genes coding for ribosomal RNA (UNILEVER, patent application, 1989). In this way the vector is integrated in a multimeric form precisely at a predefined locus in the chromosome of the host. The advantage of integration vectors is that the probability of horizontal flow is extremely low and may be neglected in risk assessments.

Vectors used to transform plants are nearly all based on the tumor-inducing plasmid of *Agrobacterium tumefaciens* (SCHELL, 1987). From these vectors only that part is maintained which is essential for integration into the chromosome of the host plant. The disadvantage of this procedure is that the site of integration is not known beforehand and is difficult to determine. This means that a certain metabolic function can be damaged in the host cell and that is a hazard. The attractiveness of using Ti-plasmid to transform plants is that the foreign gene is stably integrated in the chromosome of the host and that the probability of horizontal flux of the foreign gene is extremely low. More recently other vectors and methods to transform plants have been developed, but all result in stable, site-unspecific integration, although very recently indications have been obtained that site-specific integration in plants is possible (P. HOOIJKAAS, personal communication).

Transformation of plants always means integration of rDNA into the chromosome of the plant. This is also the case in the often used anti-sense technology to repress the translation of mRNA. During the anti-sense studies quite often effects of the position of integration are observed as well as the rather surprising result that also sense genes can repress translation (GRIERSON et al., 1991). These unexpected results show once more that it is important to determine the place of integration and the effects on the metabolism of the host cell that integration can have.

2.9 Reprogramming of Metabolic Pathways in Cells Hosting Foreign DNA

One of the targets for rDNA technology will be the reprogramming of the metabolic pathway of the host cell. Intended reprogramming of metabolic fluxes can be illustrated with four examples:
(1) Anti-sense polygalacturonase (PG) in tomatoes. In these tomatoes a gene coding for the endogenous PG gene is introduced but in the anti-sense direction (SMITH et al., 1988). Although the anti-sense mechanism is not completely understood (GRIERSON et al., 1991), one explanation is that the sense and anti-sense mRNAs "coding" for PG recognize each other and form a complex that cannot be transcribed. In this way the amount of PG is reduced, and this will decrease the cell wall degradation, and consequently the shelf life of the tomato is extended.
(2) Using the same principles Calgene Inc. has developed rapeseeds in which the conversion of stearic acid into oleic acid by a desaturase is blocked by anti-sense δ-C9-stearoyl-desaturase. This results in a higher level of stearic acid in rapeseeds, which is of advantage for certain industrial applications (CALGENE, 1991).
(3) To improve the gassing power of baker's yeasts, GIST-BROCADES has developed a yeast strain in which maltose permease and maltase are expressed constitutively, thereby avoiding glucose repression of the endogenous genes. Indeed this resulted in an increase in the gassing power of about 25% (GIST-BROCADES, 1987).
(4) To improve the ethanol production rate of yeasts all genes of the glycolysis have been cloned and over-expressed. However, due to extensive feedback control mechanisms the ethanol production rate increased only by about 5% (SCHAAFF et al., 1989).
Whereas in the first example only a minor pathway in the metabolism was reprogram-

med, the rapeseed and baker's yeast examples are examples of a change of one particular step in a main metabolic route. Both approaches were very successful. However, the reprogramming of a whole main metabolic route in brewer's yeast was unsuccessful, because cells have many feedback inhibitions and other control mechanisms of main metabolic routes. The above examples deal with intended reprogramming of the metabolism, thereby creating cells that may differ considerably from the original host and should therefore be tested according to the full safety scheme.

In most rDNA projects metabolic reprogramming is not planned, but may be caused by intended or illegitimate integration (Sect. 2.8). Moreover, change in metabolism can be caused by the intrinsic property of the introduced foreign gene or by a helper gene, if the maintenance of the vector is based on the anabolic function of the helper gene product. Therefore, in the hazard analysis the possibility that the normal fluxes of metabolites in the host strain will be changed by the introduction of foreign genes encoding enzymes should be considered. Using molecular biological techniques the occurrence of metabolic reprogramming could be determined. Using Southern blotting coupled with RFLP/RAPID analysis the site of integration could be determined, and if this integration resulted in disruption of a structural gene, the function of this gene should be determined. Alternatively, by very sensitive analysis differences in the mRNA, profiles between wild-type and transformed cells can be determined (LIANG and PARDEE, 1992). Also with enzymic or immunological methods the translation products of these mRNAs can be determined. Knowing or calculating the levels of enzymes it is possible to estimate the difference in fluxes of substrate conversions due to these changed levels (POSTMA, 1990) and therefore the importance of unexpected aspects of metabolic reprogramming.

3 Various Stages in the Development of an rDNA Process or Product

Soon after the Asilomar Conference in a number of countries committees were established to develop guidelines to evaluate the safety of rDNA experiments in the laboratories. These committees focused their effort on creating frameworks for the categorization of rDNA experiments on the basis of intrinsic hazard of the foreign gene, nature of the vector used in the transformation procedures, perceived hazard based on the origin of the foreign gene, possible errors in the cloning procedures and the intrinsic hazard of the host cell.

Initially there were many differences in the views of these committees, but after a number of years in nearly all Western countries and Japan more or less the same rules for the evaluation of the safety of laboratory experiments were applied.

At the end of the seventies the first guidelines for experiments exceeding the 10 liter scale were established, and again these rules were quite similar in various countries. Early 1980 rules for pilot-plant and large-scale production were formulated. In most countries for each containment level general approval for the facilities can be obtained, however, for each new type of process or product approval of the used strain, equipment and processing are necessary.

The volume of 10 liters for which an additional approval procedure must be started is quite arbitrary. Additional approval for pilot-plant or large-scale production does in general not pose a problem to manufacturers, as transfer from lab-scale to pilot plant and from pilot plant to large scale is always a point of careful evaluation of all aspects of any new process/product. Although often the approval of pilot and production facilities and the process has been rather time-consuming with a few exceptions, e.g., of the rDNA production of human insulin by Hoechst in Germany, approval was obtained. However, approval to produce an rDNA product does not mean that a new product can be marketed. For pharmaceutical products the procedures for approval of new products, including rDNA products are quite clear. For rDNA products in food processing or for their use in food or detergent products or in other consumer products, unclear procedures varying from country to country exist or even no procedures are in place at all. The absence of clear and uniform guidelines in Western Europe (EEC), USA and Japan has clearly contributed to a considerable delay in the introduction of rDNA products and resulted in certain companies reducing or even losing their interest in rDNA technology.

For companies it is extremely important that it is clear already at the start of a new research and development program what the approval procedures for the end product will be. If a company knows the approval procedures, a scheme for the development of an rDNA process/product, including the approval procedures and an estimated time scale, can be set up. A typical example of such a scheme is given in Fig. 5. In spite of the many approval procedures, the development of rDNA processes and products by a company that uses such an integrated development/approval scheme will in general not be slowed down by these procedures.

Many of the questions of the decision schemes can already be answered at the moment the project reaches the state of transfer from shaking flask to fermentation or small-scale glass house experiments. Approval of the facilities can be obtained, even if the project is still in its research phase, although in most countries approval for pilot-plant and large-scale production per project is necessary. However, this approval is normally not a problem. If necessary, the feeding and toxicological studies can start as soon as the outline of the process on pilot-plant scale has been established. The extensive trials to get approval for pharmaceutical products will take more time; however, also for new pharmaceuticals produced via conventional techniques these trials determine largely the time of introduction.

Such an approach also implies that during the actual research stage of the development of a new rDNA product the various steps in

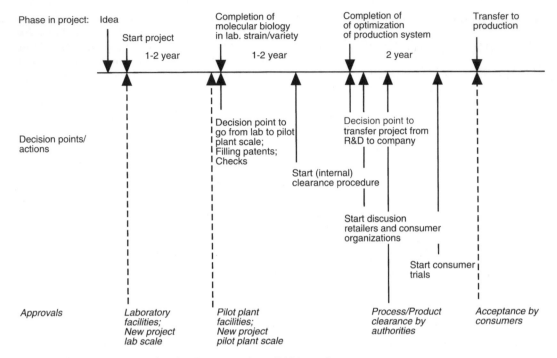

Fig. 5. Different phases in the development of an rDNA product.

the construction of the new gene and its expression are carried out in such a way that the probability that approval will be obtained is optimal. In Sect. 2 the possible sources or errors in the rDNA technology that can render an intrinsically safe gene or gene product present in an intrinsically safe host organism into an intrinsically hazardous product have been discussed. Most of these sources can be detected during the research stage, and their absence should be checked by independent external experts. Spontaneous mutations can occur after completion of the laboratory experiments and even during actual use of the transgenic organism in production processes or in agriculture. By applying common quality assurance/control protocols spontaneous mutations of the gene resulting in a change of catalytic or antigenic properties or molecular mass or physical characteristics or in a different restriction pattern of the foreign gene and its flanking regions can be detected. Spontaneous mutations that are silent in all these aspects can be considered as safe (as the expres-

sion products are identical to the original products). Larger rearrangements at DNA level will be detected by Southern blotting of microorganisms or RFLP or RAPID analysis of plant DNA. Both sets of techniques are (should be) standard procedures in fermentation or seed breeding industries, respectively, as such rearrangements may have an influence on other quality traits of the organisms as well. Consequently there is no technical reason not to know for certain whether a transgenic organism carries a hazard. An important feature of the proposed scheme is (Fig. 5) that discussions with consumer organizations and others start soon after filing patents but certainly long before the new process is transferred to a production facility. In this way sufficient time is created to discuss the risk/benefit ratio of the new rDNA product with consumer organizations and others, and this may increase significantly the probability of acceptance of the rDNA product by consumers.

4 Decision Schemes for Various Types of rDNA Products and Various Applications of These Products

4.1 General

A number of committees (have) develop-(ed) decision schemes for the approval of rDNA products that will be applied in various types of products, e.g., pharmaceutical, personal care products, cleaning products, and in the food and process industry. This section will focus on decision schemes for food products. As no uniform guidelines for these decision schemes exist worldwide (KOK et al., 1993), it was necessary to start with different schemes developed by universities or governmental institutes, several national or international guidelines and to develop these schemes. In Tab. 5 a number of guidelines and their main characteristics are summarized.

Tab. 5. Comparison Between Major Characteristics of Various Guidelines for Food Products Made by, or Containing rDNA Material (KOK et al., 1993)

Reference	A	B	C	D	E	F
IFBC (1990)	+	+	−	−	−	+
NNT (1991)	−	+	?	+	+	−
ACNFP (1991)	+	+	−	+/−	+	+
WHO/FAO (1991)	−	+	−	−	+	?
Voedingsraad (1993)	+	+	−	+	−	−
EEC (1990 a, b)	?	+	+	?	?	−
OECD (1991)	−	+	−	−	−	?
FDA (1992)	+	+	−	−	+	+

A, uses decision trees
B, case-by-case approach
C, new regulation necessary
D, risk assessment involves animal feeding trials
E, allergy is considered as an important factor
F, manufacturer is fully responsible
+/− in certain but not all cases required

In this section the guidelines proposed by the Department of Health of the U.K. for novel foods serve as a starting point. This choice was made because the scheme includes all types of novel foods including foods derived from biological species obtained via classical breeding and chemically or enzymatically modified foods or food ingredients. Moreover, the proposed decision scheme could be connected easily to the schemes proposed in the Statement of Policy of the FDA on "Foods Derived From New Plant Varieties" and to recent proposals of the GEZOND-HEIDSRAAD (1992) and the VOEDINGSRAAD (1993) to the Dutch government which fit quite well within the EEC directives and may together with the U.K. guidelines serve as a framework for EEC decision schemes.

4.2 Decision Tree Derived from "Guidelines on the Assessment of Novel Foods and Processes"

In 1991 the Department of Health in the U.K. issued a report in which the Advisory Committee on Novel Foods and Processes gives "Guidelines on the Assessment of Novel Foods and Processes" (ACNFP, 1991). In this report a decision tree for all novel foods is given, and this tree includes also foods derived from genetically modified organisms. Although it is called a decision tree, it does not end in qualifications like "approved" or "not allowed", but in a large number of actions.

Unfortunately, the necessary actions to obtain approval are not always clearly defined in this report. However, as the decision tree includes all foods, it is a good scheme to start with. Fig. 6 and Tab. 6 show that part of the decision tree for novel foods derived from genetically modified organisms which is relevant in the framework of this chapter.

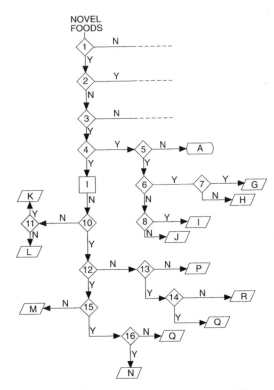

Fig. 6. Guidelines on the assessment of novel foods and processes. Truncated decision tree.

Tab. 6. Guidelines on the Assessment of Novel Foods and Processes (Truncated Decision Tree)

Entry, Actions and Results

A.   This is not a novel food
     Exit point: Information requirements:
G.   I, II, III, V, VI, VII
H.   I, II, III, IV, V, VI, VII
I.   I, III, V, VI, VII, VIII, IX
J.   I, III, IV, V, VI, VII, VIII, IX
K.   I, II, III, IV, V, VI, VII, VIII, X, XII
L.   I, III, IV, V, VI, VII, VIII, IX, X, XII
M.   I, III, V, VI, VII, VIII, X, XI, XII, XIII
N.   I, III, V, VI, VII, VIII, X, XI, XII, XIII, XIV
O.   I, III, VI, VII, VIII, X, XI, XII, XIII, XIV, XV
P.   I, III, V, VI, VII, VIII, IX, X, XI, XII, XIII
Q.   I, III, V, VI, VII, VIII, IX, X, XI, XII, XIII, XIV
R.   I, III, VI, VII, VIII, IX, X, XI, XII, XIII, XIV, XV
Z.   The food material must be genetically modified

Tab. 6. (Continued)

Questions
1.   Novel material?
2.   Naturally occurring strain?
3.   Genetic change?
4.   Classical breeding?
5.   Significant change in genetic material?
6.   Human exposure?
7.   GMO?
8.   GMO?
10. Does the food product contain genetic material?
11. Human exposure?
12. Human exposure?
13. GMO?
14. GMO a seed?
15. GMO?
16. GMO a seed?

The numbers I–XV above refer to the information requirements listed below

I.   Instructions for use
II.   Evidence of previous human exposure
III.   Intake/extent of use
IV.   Technical details of processing and product specifications
V.   Nutritional studies
VI.   History of organism
VII.   Characterization of derived strain
VIII. Toxicological assessment
IX.   Human studies
X.   Assessment of a genetic modification procedure
XI.   Effect of a genetic modification procedure on the known properties of the parent organism
XII.   Genetic stability of a modified organism
XIII. Site of expression of any novel genetic material
XIV.   Transfer of the novel genetic material
XV.   Assessment of a modified organism for survivability, colonization and replication/amplification in the human gut

4.3 Decision Scheme for "Foods Derived from New Plant Varieties" as Applied by the FDA

The FDA issued a Statement of Policy on foods derived from new plant varieties (FDA, 1992). The various decision schemes given in this statement have been integrated into one

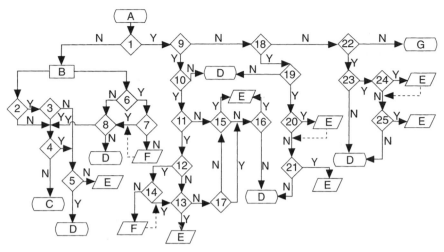

Fig. 7. FDA Statement of Policy on "Foods derived from new plant varieties". Integration of their decision schemes into one decision tree.

scheme (Fig. 7, Tab. 7). This scheme can quite easily be connected to the general scheme that covers all types of novel foods. Compared with the scheme proposed in the U.K. (see Sect. 4.2) and the Dutch schemes (see Sect. 4.4) it is remarkable that the FDA schemes include a number of questions on the (expected) allergic properties of the new varieties. This is in contrast with the two European schemes. On the other hand, the FDA schemes do not ask many details on DNA constructions. This is consistent with their view that it is not the way by which the intrinsic properties of a product are obtained, but the intrinsic properties as such that matter.

4.4 Decision Schemes Developed for the Dutch Government for Foods Containing rDNA-Encoded Materials

The guidelines produced in The Netherlands are largely based on the rules proposed initially by PARIZA and FOSTER (1983). Different decision schemes have been developed for four food product categories (GEZOND-HEIDSRAAD, 1992; VOEDINGSRAAD, 1993):

a) Single chemicals or well defined mixtures of chemicals produced by biological systems modified by rDNA technology (decision scheme given in Fig. 8, questions in Tab. 8)
b) Foods derived from transgenic plants (Fig. 9, Tab. 9)
c) Foods derived from transgenic animals (Fig. 10, Tab. 10)
d) Foods or food ingredients made by rDNA-modified microorganisms (Fig. 11, Tab. 11).

The Dutch schemes are the outcome of a number of discussions between consumer organizations, governmental and academic experts on rDNA technology, toxicology and communication and industry and cover the whole range of food products. During the discussions it became clear that the traditional 90 days feeding trials are not (always) appropriate for the evaluation of food products made by rDNA technology. Therefore, much attention has been paid to find molecular biological approaches to guarantee the safety of these products. As not very much experience in this area exists, it was decided to use the traditional feeding trials and the molecular approaches simultaneously and to decide after a period of at least three years whether in-

Tab. 7. Foods Derived from New Plant Varieties
Decision Tree Based on FDA Statement of Policy, May 1992

Entry, Actions and Results
A.   Food derived from a new (via rDNA technology modified) plant variety?
B.   Assess the safety of the host plant and the donor of the DNA (gene)
C.   New variety is *not* acceptable
D.   No concern
E.   Consult FDA
F.   Consult FDA for additional or alternative testing
G.   The product does not fit in this decision scheme

Questions
Q1.   Will the modification result in nutritional effects?
Q2.   Does the host species have a history of safe use?
Q3.   Do characteristics of the host species, related species, or progenitor lines warrent analytical or toxicological tests?
Q4.   Do test result provide evidence that toxicant levels in the new plant variety do not present a safety concern?
Q5.   Is the concentration and bio-availability of important nutrients in the new variety within the range ordinarily observed in the host species?
Q6.   Is food from the donor commonly allergenic?
Q7.   Can it be demonstrated that the allergenic determinant has not been transferred to the new variety of host?
Q8.   Do characteristics of the donor species, related species, or progenitor lines warrent analytical or toxicological tests?
Q9.   Will the modification result in the expression of protein(s) in the new variety?
Q10. Is the newly introduced protein present in food derived from the plant?
Q11. Is the protein derived from a food source substantially, or similar to an edible protein?
Q12. Is food from the donor commonly allergenic?
Q13. Is the introduced protein reported to be toxic?
Q14. Can it be demonstrated that the allergenic determinant has not been transferred to the new variety of host?
Q15. Does the biological function of the introduced protein raise any safety concern, or is the introduced protein reported to be toxic?
Q16. Is the introduced protein likely to be a macroconstituent of the human or animal diet?
Q17. Will the intake of the donor protein in the new variety be generally comparable to the intake of the same or similar protein in the donor or other food?
Q18. Will the modification result in the biosynthesis of new or modified carbohydrates in the new variety?
Q19. Has there been an intentional alteration in the structure, composition, or level of carbohydrates in the new variety?
Q20. Have any structural features or functional groups been introduced into the carbohydrate that do not normally occur in food carbohydrates?
Q21. Have there been any alterations that could affect digestibility or nutritional quality in a carbohydrate that is likely to be a macroconstituent of the diet?
Q22. Will the modification result in the biosynthesis of new or modified fats or oils in the new variety?
Q23. Has there been an intentional alteration in the identity, structure, or composition of fats or oils in the new variety?
Q24. Have the intentional alterations been in a fat or oil that will be a macroconstituent of the diet?
Q25. Are any unusual or toxic fatty acids produced in the new variety?

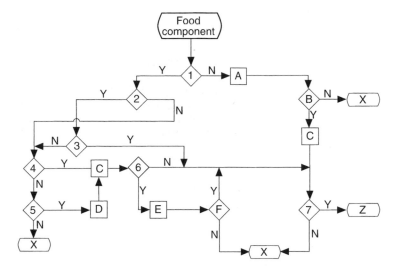

Fig. 8. Decision tree for well-defined single food components.

Tab. 8. Decision Tree for Well Defined Single Food Components

---

Entry, Actions and Results
A. Carry out evaluation studies to determine the safety of the product and make specifications
C. Make new specifications
D. Apply a process to reduce the level of undesirable components
E. Carry out a 90-day feeding trial
X. It is not allowed to bring the component on the market
Z. The component is approved for use in foods

Questions
1. Is the use of the *component* in foods allowed at this moment?
2. Does the *component* comply with existing specifications on identity and purity?
3. Are the existing specifications sufficient to control the presence of undesirable site components or too high levels of the intended *component*?
4. Are the levels of known components within the safety specifications?
5. Is it possible to reduce the level of undesirable components during processing in order to comply with the existing specifications?
6. Is it possible that the product contains unknown components?
7. If the intended or assumed consumption of the *component* results in a change in eating habits will the new habit still be considered as safe?
B. Does the evaluation show that the *component* is safe?
F. Does the 90-day feeding trial show that the *component* is safe?

Note: Questions B and F are not (yet) included in the Dutch decision trees as separate questions but form part of *action A* and *question 5*, respectively

---

deed the molecular biological methods are better than the traditional methods (see Fig. 8). Chymosin is most probably the first and most widely applied rDNA enzyme in the foods area. The gene coding for preprochymosin has been cloned (UNILEVER, 1981) and expressed in the GRAS yeast *Kluyveromyces lactis* (GIST-BROCADES, 1982). Many publications on the properties, safety evaluation and production of chymosin have been issued, and therefore chymosin can serve as a model.

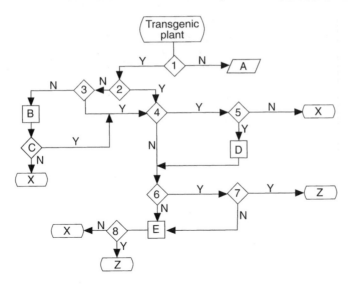

Fig. 9. Decision tree for foods derived from transgenic plants.

Tab. 9. Foods Derived from Transgenic Plants

Entry, Actions and Results
A. Start a procedure for novel foods
B. Carry out evaluation studies and determine the safety of the product
C. Do these evaluation studies show that the expression product(s) is (are) safe?
D. Apply a process to reduce the endogenous or new components to an acceptable level
E. Carry out a 90-day feeding trial
X. It is not allowed to bring products derived from this plant on the market
Z. Products derived from this plant are approved for use in *foods*

Questions
1. Is there sufficient know-how on host organism and donor DNA to start the approval procedure?
2. Are the components of the food only of an endogenous nature?
3. (Are) is the expression product(s) of the donor DNA an endogenous expression product(s) of a/ another food?
4. Is there a possibility that consumption of the new component in the intended or assumed quantities will affect the health of the consumer negatively?
5. Is it possible to reduce the (active) components to the allowed level during processing?
6. Is (are) the site(s) of integration of the donor DNA on the plant chromosome(s) exactly known?
7. Is that knowledge sufficient to conclude that the metabolism of the plant is not changed or are the changes within the naturally occurring variations?
8. Does the 90-day feeding trial show that the component is safe?

Fig. 12 shows a flowsheet of the rDNA chymosin production process developed by GIST-BROCADES. The flowsheet demonstrates clearly that the probability that genetically modified organisms (GMOs) that have produced chymosin will be present in the end product is practically zero. Firstly, the fermentation liquid is filtered twice to separate the GMOs from the enzyme, secondly, the liquid containing the enzyme is acidified with a solution containing about 200 mg/kg benzoic acid (pH 2). The latter is necessary to transform prochymosin into the active enzyme chymosin and provides an additional guarantee that the end product is free of GMOs. Finally, in a later stage of the process an ultrafiltration step

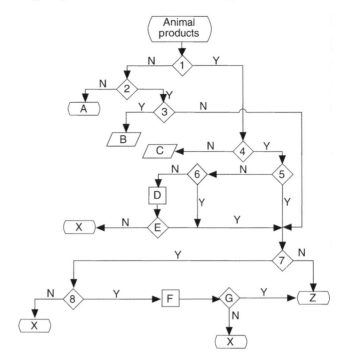

Fig. 10. Decision tree for foods derived from animals treated with products made by modern biotechnology or obtained from transgenic animals.

Tab. 10. Foods Derived from Animals Treated with Products Obtained by Means of Modern Biotechnology or from Transgenic Animals

---

Actions and Results
A. Start a procedure for novel foods
B. Evaluate the product in accordance with the national/international legislation for veterinary products
C. This product should be analyzed as a *novel food*
D. Evaluate the safety of the new components
E. Does the evaluation of the new component show its safety?
F. Develop a procedure/process to reduce the level of the endogenous or new component
G. Does the evaluation of the procedure or process to reduce the level of the endogenous or new component show that the level is acceptable?
X. It is not allowed to bring products derived from this animal on the market
Z. Products derived from this animal are approved for use in foods

Questions
1. Is the food or food product derived from a transgenic animal?
2. Is the food or food product derived from an animal treated or fed with a product made via rDNA technology?
3. Has the animal been treated with a veterinary product?
4. Is there sufficient knowledge and documentation about the host animal and the genetic donor material to start the approval procedures?
5. Are the components of the food or food product derived from the transgenic animal endogenous?
6. Is (are) the expression product(s) encoded by the genetic donor material endogenous in other foods?
7. Is there a possibility that consumption of the endogenous or new components in the intended or assumed quantities will affect the health of the consumer negatively?
8. Is it possible to reduce the endogenous or new components to an acceptable level?

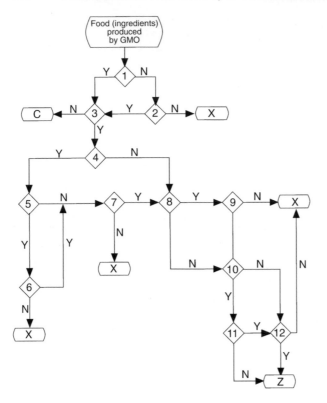

Fig. 11. Decision tree for foods or food ingredients derived from GMO.

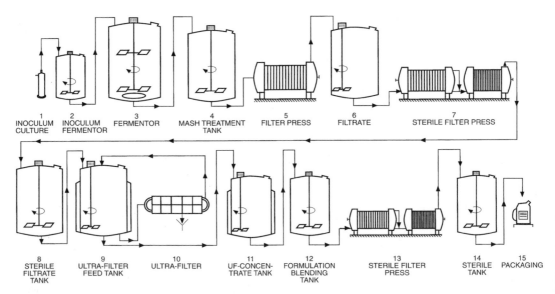

Fig. 12. Production of chymosin by rDNA *Kluyveromyces.*

Tab. 11. Foods or Food Ingredients Produced with the Aid of Genetically Modified Microorganisms (see also Fig. 11)

Actions and Results
C. This product should be analyzed as a novel food
X. It is not allowed to bring food products containing ingredients produced with this GMO on the market
Z. Food products containing ingredients produced with this GMO are approved

Questions
1. Does the unmodified microorganism have a record of safe use in food products?
2. Can on the basis of feeding and/or toxicological studies the unmodified microorganism be considered as safe in food products?
3. Is there sufficient knowledge and documentation that the new genetic material codes for (a) product(s) that is (are) acceptable in food products?
4. Does the GMO or an inherent part of it or the product(s) encoded by the new genetic material remain in the food product?
5. It is intended that the modified microorganism fulfills a functional role in the gastrointestinal tract of the consumer?
6. Has the intended functionality been demonstrated?
7. Is the modified microorganism free of genes encoding antibiotic resistance?
8. May the consumption of the food, in particular the GMO or an inherent part of it, or the product(s) encoded by the new genetic material in the intended or expected consumed quantities result in any negative aspect on the health of the consumer?
9. Is it possible to reduce the quantity of the GMO or an inherent part of it or the product(s) encoded by the new genetic material to an acceptable level?
10. Is the physical state or the integration into the chromosome of the host of the new genetic material fully known?
11. Does the integration of the new genetic material disturb the metabolism of the host in such a way that hazardous products may be formed?
12. Does a 90-day feeding trial with the food product containing the GMO or an inherent part of it show that the introduction of the new genetic material into the host does not have an effect on the metabolism of the host cell resulting in (a) hazardous compound(s)?

---

is included to concentrate the chymosin. The probability that a GMO will escape from the process upstream of the first filtration unit is low, but certainly not zero. Naturally the fermentation unit has been designed to exclude the possibility that microorganisms from outside will penetrate the fermentation unit.

However, the equipment is not aseptic and due to the high number of yeast cells in the fermentation unit, there is a probability that from time to time the rDNA yeasts will escape and enter the environment. However, as explained in Sect. 2.7, the probability that the production microorganism, in this case *Kluyveromyces lactis*, will survive the conditions present in the environment is extremely low. TEUBER (1990) described the safety evaluation of chymosin process and product. In this paper the chymosin gene is still assumed to be

located on a plasmid containing an antibiotic resistance marker. When applying the scheme of Fig. 8 for approval of this product, this is not allowed. At present, the construct carrying the chymosin gene is integrated into the chromosome and does not carry an antibiotic resistance marker (GIST-BROCADES, personal communication).

4.5 Decision Scheme for the Approval of the Release of GMOs

It is likely that foods containing plants or microorganisms derived from genetically modified organisms will be the first GMOs that will enter the market. Normally foods derived from plant material will not be used to (re)generate plants; however, food products like bread or yoghurt may contain rDNA microorganisms that are still alive and are able to grow out. It is the release of this category of rDNA organisms that is still a matter of much debate. Although certainly more complex than dead material derived from GMOs there is no fundamental difference between the approaches to assess the risk of these products. In Figs. 13 and 14 and in Tab. 12, a rational approach to assess the risk of living lactic acid bacteria in a dairy product is given.

An important question in this type of risk assessment is whether one should stop with the transfer of genetic material from the host

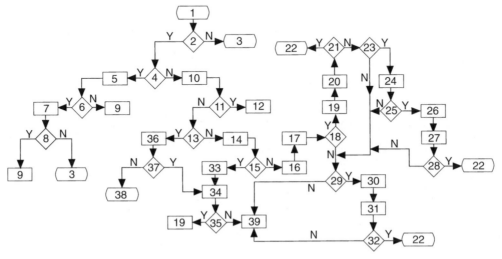

Fig. 13. Decision tree for the risk assessment of the release of genetically modified microorganisms present in food products for human consumption.

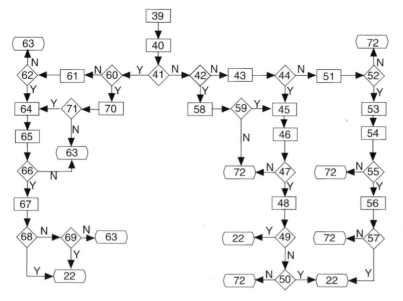

Fig. 14. Decision tree for the risk assessment of the release of genetically modified microorganisms present in food products into the environment.

Tab. 12. A Proposal for a Structured Assessment of the Risk Related to the Introduction of Genetically Modified Microorganisms (GMOs) in (or as) Food Products

Entry, Questions, Actions and Results

Food Products

1 (E). Genetically modified microorganism (GMO)

2 (Q). Has the product containing the GMO formal clearance based on animal feeding trials?

3 (R). The release of this GMO cannot be evaluated before it has this formal clearance

4 (Q). Is this GMO produced as a mixed culture?

5 (A). Determine the probability $P(a)$ that the genetic information will be transferred to (one of) the other microorganisms present in the mixed culture

6 (Q). Is $P(a) > a$?
(Note: in the research phase attention should have been paid to minimizing the probability of such a transfer)

7 (A). This secondary transformed microorganism (STO) should be cleared in the same way as the GMO (compare *question 2*)

8 (Q). Official clearance obtained?

9 (A). Go to *action 10*

10 (A). Determine the amount of product that comes into the environment (into the sewage system either from factories or households or directly in the soil) before consumption (= spilled product)

11 (Q). Is V (spilled) $> bV$ (produced)?

12 (A). Go to *action 40* to evaluate the behavior of GMO (and if appropriate the STO) in spilled products and continue with *question 13* for consumed product

13 (Q). Will the product containing the GMO receive a physical/chemical treatment before consumption and will that treatment result in an inactivation of GMO by a factor $> c$?

14 (A). Determine the distribution of the residence times of GMO in the gastro/intestinal (g/i) tract of the consumers. Take the time corresponding with 95% of this distribution curve as $t(r)$

15 (Q). Will the GMO lyse with a probability $> d$ in the g/i tract?

16 (A). Although the correct procedure will be the determination where lysis will occur and the determination of the distribution of the time of intact cells in the g/i tract, a worst-case scenario is used assuming the concentration of intact cells is not changed by the lysis and that the intact cells can transfer their genetic information during $t(r)$ to other microorganisms in the g/i tract

17 (A). Determine the probability $P(e)$ that intact cells of the GMO transfer genetic information to normal inhabitants of the g/i tract. Use in these studies $t(r)$ as contact time and the conditions of the g/i tract. Determine also the probability $P(f)$ that the GMO will be transformed by genetic material originating from the common g/i tract microorganisms (= modified GMO)

18 (Q). Is $P(e) > e$?

19 (A). Determine whether the transformed inhabitants (or STO) obtain an advantage over untransformed inhabitants in the g/i tract: $A(i)$
Define $A(i)$ in either faster growth rates $t'(g)$; better adhesion h'; or higher production of certain metabolites $\{p(x)', x = 1, ...\}$

20 (A). Determine the probability $P(g)$ that any of the events described under *action 19* will result in the formation of a hazardous microorganism (producing toxins) or that the STO will replace beneficial microorganisms in the g/i tract) microorganisms

21 (Q). Is $\{P(e) + P(l)\} \cdot P(g) > g$ or $P(m) \cdot P(g) > g$?

22 (A). This risk is unacceptable and the GMO should not be released

23 (Q). Is the number of STOs larger than h per volume unit g/i tract (compare *action 19*)?

24 (A). Determine the probability $P(i)$, that the STO can transfer that part of its genetic content that contains the genetic information from the GMO to another microorganism of the g/i tract resulting in a tertiary transformed microorganism (TTO)

25 (Q). Is $\{P(e) + P(i)\} \cdot P(i) > i$ or $P(m) \cdot P(i) > i$?

26 (A). Determine whether the TTOs have an advantage over untransformed cells in the g/i tract (compare *action 19*)

27 (A). Determine the probability $P(j)$ that any event described in *action 26* will result in a hazardous microorganism

28 (Q). Is $P(e) \cdot P(i) \cdot P(j) > j$ or $P(m) \cdot P(i) \cdot P(i) > j$?

29 (Q). Is $P(f) > f$?

Tab. 12. (Continued)

30 (A). Determine whether the modified GMO gains an advantage over untransformed GMOs: $B(i)$. Define $B(i)$ in either faster growth rates $t''(g)$; better adhesion h''; or higher production of certain metabolites $\{p(x)'', x=1, \ldots\}$

31 (A). Determine the probability $P(k)$ that any of the events described under *action 30* will result in the formation of an hazardous GMO

32 (Q). Is $P(f) \cdot P(k) > k$?

33 (A). As described in *action 16* a worst-case scenario is used to determine the probability that DNA of lysed GMO cells transforms other microorganisms of the g/i tract. (Use for these studies $t(r)$ as contact time and the g/i tract conditions)

34 (A). Determine the probability $P(l)$ that DNA originating from lysed GMO transforms normal inhabitants of the g/i tract (resulting in STO)

35 (Q). Is $P(l) > l$?

36 (A). Determine the probability $P(m)$ that DNA originating from the inactivated GMOs transforms normal inhabitants of the g/i tract (resulting in STO). (Use for these studies $t(r)$ as contact time and the g/i tract conditions)

37 (Q). Is $P(m) > m$?

38 (R). The GMO can be released

39.   This is the end of the risk assessment in the g/i tract. The next phase will be the risk assessment of GMOs, modified GMOs and STOs in the environment ("Sewage System").

"Sewage System"
   (This scheme can also be used for the release of microorganisms in the environment for non-foods products)
   This decision tree deals with genetically modified microorganisms (GMOs), or modified GMOs and secondary transformed microorganisms (STOs)

40 (A). Determine the average residence times of the GMOs or modified GMOs and STOs in sewage $(t(r2), t(r3)$ and $t(r4)$, respectively)

41 (Q). Is the microorganism considered STO?

42 (Q). Will the GMO or modified GMO lyse with a probability of $>n$ in the "sewage system" (s-system)?

43 (A). Determine the probability $P(o)$ that the GMO or modified GMO will transfer its genetic information to other inhabitants of the s-system. Use in these studies $t(r2)$ and $t(r3)$, respectively, as contact time and the various s-system conditions to determine $P(o)$

44 (Q). Is $P(o) > o$?

45 (A). Determine whether the transformed inhabitants gain an advantage (or new property) over untransformed inhabitants of the s-system: $C(i)$
   Define $C(i)$ in either faster growth rates $t'''(g)$; better survival s''' or higher production of certain metabolites $\{p(x)''', x=1, \ldots\}$

46 (A). Determine the probability $P(p)$ that any of the events described under *action 45* will result in the formation of a hazardous microorganism

47 (Q). Is $\{P(o) + P(u)\} \cdot P(p) > p$?

48 (A). Determine the probability $P(q)$ that a transformed hazardous microorganism in the s-system (STO) will enter the food chain

49 (Q). Is $\{P(o) + P(u)\} \cdot P(p) \cdot P(q) > q$?

50 (Q). Will the STO change the ecosystem considerably, e.g., replace the natural microflora?

51 (A). Determine the probability $P(r)$ that the GMO will be transformed by DNA of the normal inhabitants of the s-system

52 (Q). Is $P(r) > r$?

53 (A). Determine whether the transformed GMO gains an advantage or new property over untransformed GMO in the s-system: $D(i)$
   Define $D(i)$ in either faster growth rates: $t''''(g)$, better survival: s'''', or higher production of certain metabolites $\{p(x)'''', x=1, \ldots\}$

54 (A). Determine the probability $P(s)$ that any of the events described under *action 53* will result in the formation of a hazardous modified GMO

55 (Q). Is $P(r) \cdot P(s) > s$?

Tab. 12. (Continued)

---

56 (A). Determine the possibility $P(t)$ that the modified GMO will enter the food chain

57 (Q). Is $P(r) \cdot P(s) \cdot P(t) > t$?

58 (A). Determine the possibility $P(u)$ that the DNA originating from the GMO transforms the normal inhabitants of the s-system

59 (Q). Is $P(u) > u$?

60 (Q). Will the STO lyse with a probability $> v$ in the s-system?

61 (A). Determine the probability $P(w)$ that the STO will transfer its genetic information to the normal inhabitants of the s-system

62 (Q). Is $\{P(o) + P(u)\} \cdot P(w) > w$?

63 (R). The GMO can be released if the "right hand tree" starting with *question 42* leads to *result 72*

64 (A). Determine whether the transformed inhabitants (TTO) gain an advantage (or new property) over the normal inhabitants of the s-system: $E(i)$
Define $E(i)$ in either faster growth rates: $t'''''(g)$; better survival s''''' or higher production of certain metabolites ($P''''''(x)$, $x = 1, ...$)

65 (A). Determine the probability $P(y)$ that any of the events described under *action 64* will result in the formation of a hazardous TTO

66 (Q). Is $\{P(o) + P(u)\} \cdot P(w) \cdot P(y) > y$ or $\{P(o) + P(u)\} \cdot v \cdot P(a) \cdot P(y) > y$?

67 (A). Determine the probability $P(z)$ that the hazardous TTO will enter the food chain

68 (Q). Is $\{P(o) + P(u)\} \cdot P(w) \cdot P(y) \cdot P(z) > z$ or $\{P(o) + P(u)\} \cdot v \cdot P(a) \cdot P(y) \cdot P(z) > z$?

69 (Q). Will the tertiary transformed microorganism (TTO) change the ecosystem considerably?

70 (A). Determine the possibility $P(a)$ that the DNA originating from the GMO via STO transforms the normal inhabitants of the s-system

71 (Q). Is $\{P(o) + P(u)\} \cdot v \cdot P(a) > a$?

72 (R). Provided that the new genetic information of the (surviving) GMO still has its original configuration, the GMO can be released

---

strain to the primary recipient, or even whether it should include the probability that the primary recipient will pass its newly acquired genetic information to a secondary recipient. Other important but difficult questions to be answered are the acceptable probabilities of the transfer of this genetic material and how to evaluate these criteria. In the additional Tab. 13 very preliminary proposals for the various criteria are given. Many of the probabilities and numbers proposed in Tab. 13 are not based on experimental data. It is obvious that before GMOs in food products can be introduced, these proposed probabilities and numbers should be confirmed and, whenever the proposals are not correct, be replaced.

However, in an attempt to structure the research in this area and to avoid "political" interpretations of the research data obtained, this scheme and the corresponding probabilities and numbers are proposed.

4.6 Integrated Scheme to Evaluate Any rDNA Product

Based on the various decision schemes given above, a generalized decision scheme for all rDNA products has been developed (Fig. 15, Tab. 14). This scheme divides the various types of products into groups, and for each group one of the previously discussed schemes can be followed.

Although this scheme has not been authorized, it is useful during the development phase of an rDNA product as it points out questions that most probably will have to be answered to obtain the approval of official bodies. This will guide the research work in such a way that most of the questions can be answered straight away.

Tab. 13. Proposed Probabilities for the Release of GMOs in Food Products

Probability Function	Probability	Donor	Recipient	Events	
				Result	Others
$P(a)>a$	$a=10^{-16}$	GMO	Co-ferment cell	STO	
$P(e)>e$	$e=10^{-16}$	GMO	G/i tract cell	STO	
$P(f)>f$	$f=10^{-10}$	G/i tract cell	GMO	Modified GMO	
$\{P(e)+P(l)\}\cdot P(g)>g$ or $P(m)\cdot P(g)>g$	$g=10^{-23}$			STO	$\rightarrow$ Hazardous STO
$\{P(e)+P(l)\}\cdot P(i)>i$ or $P(m)\cdot P(i)>i$	$i=10^{-19}$	STO	G/i tract cell	TTO	
$P(e)\cdot P(i)\cdot P(j)>j$ or $P(m)\cdot P(i)\cdot P(j)>j$	$j=10^{-23}$			TTO	$\rightarrow$ Hazardous TTO
$P(f)\cdot P(k)>k$	$k=10^{-19}$			Mod. GMO	$\rightarrow$ Hazardous GMO
$P(l)>l$	$l=10^{-16}$	Lysed GMO	G/i tract cell	STO	
$P(m)>m$	$m=10^{-16}$	Inactivated GMO	G/i tract cell	STO	
$P(o)>o$	$o=10^{-16}$	GMO	Sewage cell	STO	
$\{P(o)+P(u)\}\cdot P(p)>p$	$p=10^{-19}$			STO	$\rightarrow$ Hazardous STO
$\{P(o)+P(u)\}\cdot P(q)>q$	$q=10^{-23}$			Haz. STO	$\rightarrow$ Food chain
$P(r)>r$	$r=10^{-10}$	Sewage cell	GMO	Mod. GMO	
$P(r)\cdot P(s)>s$	$s=10^{-19}$			Mod. GMO	$\rightarrow$ Hazardous GMO
$P(r)\cdot P(s)\cdot P(t)>t$	$t=10^{-23}$			Haz. mod. GMO	$\rightarrow$ Food chain
$P(u)>u$	$u=10^{-16}$	Lysed GMO	Sewage cell	STO	
$\{P(o)+P(u)\}\cdot P(w)>w$	$w=10^{-11}$	STO	Sewage cell	TTO	
$\{P(o)+P(u)\}\cdot P(w)\cdot P(y)>y$ or $\{P(o)+P(u)\}\cdot v\cdot P(a)\cdot P(y)>y$	$y=10^{-19}$			TTO	$\rightarrow$ Hazardous TTO
$\{P(o)+P(u)\}\cdot P(w)\cdot P(y)\cdot P(z)>z$ or $\{P(o)+P(u)\}\cdot v\cdot P(a)\cdot P(y)\cdot P(z)>z$	$z=10^{-23}$	Lysed STO	Sewage cell	Haz. TTO	$\rightarrow$ Food chain
$\{P(o)+P(u)\}\cdot v\cdot P(\alpha)>\alpha$	$\alpha=10^{-11}$			TTO	

$b=10^{-2}$ $c=10^{11}$

$d=10^{-1}$ $h=10^{-14}$

$n=10^{-1}$ $v=10^{-1}$

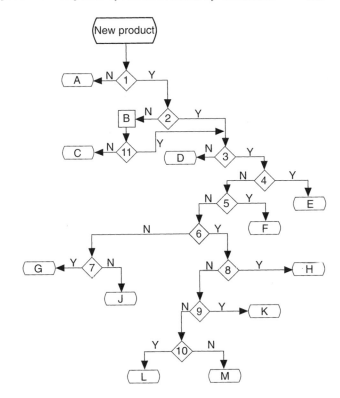

Fig. 15. Decision scheme to evaluate any rDNA product.

5 Some Ethical Aspects and Acceptance of rDNA Products by Consumers

One cannot discuss acceptance of rDNA products without taking moral or ethical aspects into account. In contrast to moral considerations which are often based on emotions, ethicists try to analyze in a scientific way the moral questions in detail and to formulate standards for societies to regulate their behavior. In a recent report for ICI STRAUGHAN (1992) divided ethical concern towards 1992 rDNA technology in three groups: Modern biotechnology is blasphemous, unnatural and disrespectful.

The blasphemous arguments are based on the view "God has created a perfect, natural order; for mankind to improve that order by manipulating DNA, the basic ingredient of all life, thereby crossing species boundaries instituted by God, is not merely presumptuous but sinful". Analyzing rDNA technology it is difficult to conclude that *rDNA technology* is blasphemous as it *is a careful and very limited imitation of what has been used by Nature* to develop the richness of biological diversity we know at present. Applying the laws of thermodynamics to the totality of living systems on earth, it can be forecasted that at least as long as the solar energy supply exceeds the energy necessary for maintenance of the totality of living systems, new (higher) ordered structures will be created (MOROWITZ, 1978). Moreover, the horizontal and vertical fluxes of genes have created and will continue to create novel forms of life. The most striking example of a horizontal gene flux in evolution is without any doubt the transfer of genetic systems of archaebacteria into spirochaete thereby forming primitive eukaryotic cells having organelles like mitochondria (Fig. 16).

Tab. 14. Integrated Scheme to Evaluate Any rDNA Product

Entry, Actions and Results
A. Not relevant to this scheme
B. Evaluate the safety of the non-modified product
C. Stop evaluation procedure because evidence is insufficient
D. Use approval systems for traditionally modified products
E. Use adapted form of the scheme for newly defined chemicals (Fig. 8)
F. Use normal approval systems for new pharmaceuticals
G. Follow scheme for newly defined chemicals and the normal procedures for the approval of personal care products
H. Use scheme for newly defined single food component (Fig. 8)
J. Not relevant to this scheme
K. Use scheme for products derived from animals (Fig. 9)
L. Use one of the schemes for products derived from plants (Fig. 10)
M. Use scheme for microbial products (Fig. 11)

Questions
1. Is the product new or modified or does it contain a new or modified compound?
2. Is the product derived from a source (strain, variety, cultivar) or bred from an approved source?
3. Does the source contain a new trait (=well defined genetic property/properties) obtained by rDNA techniques?
4. Has the product or the new or modified compound been produced by a process that involves products produced by rDNA techniques?
5. Is the product used as a pharmaceutical?
6. Is the product used as a food?
7. Is the product used for personal care?
8. Is the product a single well defined compound?
9. Is the product derived from animals?
10. Is the product derived from plants?
11. Has the non-modified product been approved?

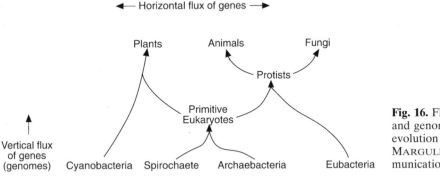

Fig. 16. Fluxes of genes and genomes during the evolution of Nature (L. MARGULIS, personal communication).

Also external factors have influenced the creation of the present living systems as do external factors at present. Without any rDNA technology in living systems, enzyme systems have been created to adapt these cells to the xenobiotics introduced by mankind (ALEXANDER, 1981). Finally the argument of boundaries between species is weak. Although the exact processes that resulted in the formation of eukaryotes and eubacteria most probably will remain obscure, circumstantial evidence is available that "natural boundaries" are rather a kinetic than a thermodynamic barrier. Since evolution has been going on for billions of years under quite varying conditions, even slow processes of transfer of genetic material from unrelated species will have occurred (see Figs. 2 and 16). Therefore, so-called natural boundaries can be considered more as an invention of biologists to divide the present biological world into often quite arbitrary groups.

The second concern in ethical discussions is the "unnaturalness" of rDNA technology. However, rDNA technology is just applying enzymes developed during evolution for the purpose of changing DNA. The immune system in mammals would not have been able to cope with the enormous number of foreign molecules, if it had not been based on cutting, joining and mutation of DNA sequences, in a superior but in principle similar way as molecular biologists do. It is very likely that in the near future it will be found that the immune system is not the only system using *in vivo* methods similar to rDNA technology. A candidate is the olfactory receptor system (R. AXEL, personal communication).

In discussions rDNA technology is often called disrespectful, because it is perceived as "not to abide by the law as set (or perhaps as seen to be used) by Nature itself". Another aspect of disrespect is that "even if no rules as set by Nature are violated, it cannot be denied that Nature has never been in the position to organize itself against the interests of mankind on purpose and consciously, rDNA technology enables man to do just that". As such the argument that "Nature has never been in the position to organize itself on purpose" is correct. However, it is not correct to introduce in the discussion on rDNA technology the argument that this technology will be used *against* the interests of mankind on purpose and consciously. What is in the interest of mankind is a different discussion.

The concern that rDNA technology as such is "disrespectful" cannot be counteracted by straight scientific arguments. STRAUGHAN (1992) quoted the German philosopher KANT, who argued that "respect required treating others as *ends*, never only as *means*". Accepting this view opens a rational way to discuss rDNA technology. Applying rDNA technology to correct genetic diseases falls certainly within the category "ends"; however-

er, using animals just as a means to produce rDNA products is in KANT's view "disrespectful". Until recently there has been no debate on "ends" and "means" for plants and microorganisms.

In fact, in ancient times the precursor form of our present civilization started with the cultivation of plants and the use of microorganisms to ferment many of these agricultural products to prolong the shelf life of these products. Since that time plant varieties have been selected and crossed to improve yield, to become resistant to certain harsh environmental conditions, etc. Similar procedures have been followed to select the most appropriate microbial strains for fermentation processes. Never in human history these procedures (changes of the genetic makeup of these organisms) have been a subject of ethical discussions. Also in more recent times the work of MENDEL and PASTEUR on these procedures have in spite of tremendous influence on human life not been regarded by the public as disrespectful. Of course they did not have any idea that in principle their work was based on manipulation of DNA.

According to the criteria of KANT, the results of classical breeding can be seen as "end" and in line with that view no debates on the ethical aspects of traditional breeding have been held and traditional breeding is widely accepted in all societies. It is very difficult to understand that traditional breeding techniques are accepted as "respectful" and the much more precise rDNA technology is considered as disrespectful. Having been involved in various aspects of rDNA technology for nearly 20 years, I have noticed that molecular biologists are deeply impressed by the elegance of the solutions Nature has found to solve problems, and the vast majority of molecular biologists are very respectful to living systems. This respect is the driving force of numerous molecular biologists to unravel the way Nature works and to use this knowledge to find very specific and effective medicines, to ensure that agriculture will provide safe, nutritious and affordable food supplies for the ever growing population of this world and to ensure that the environmental problems created by this increase in population and the desire for a higher quality of life for the population of the developing countries can be met. Respect also means that certain types of rDNA experiments with animals and certainly with humans will not be carried out.

Often discussions on the ethical aspects of rDNA technology are blurred by risk aspects. As explained in the previous sections, this should be discussed separately. Provided that the necessary control experiments are done carefully (which should be checked by independent bodies) for nearly all approved applications of rDNA technology, it can be proved that the hazard is so small that it can be considered as zero as can the risk. For the average consumer to accept modern biotechnology it is very important that the risk is very low. In a recent survey in the U.K. the influence of the risk on the public acceptance of rDNA technology has been measured (Tab. 15).

Tab. 15. Public Acceptability of Risks from the Release into the Environment of Genetically Engineered Organisms

Risk Level	Approve (%)	Do Not Approve (%)
Unknown	31	65
1:100	40	51
Unknown, but very remote	45	46
1:1000	55	37
1:10000	65	27
1:100000	71	21
1:1000000	74	18

The prosperity of the present societies of Western Europe, USA, Canada and Japan are mainly based on the exploitation of scientific discoveries and technological breakthroughs. However, besides these positive aspects the consumers in these societies have been confronted with a large number of unforeseen or neglected negative consequences of the application of these developments. Any discussion about the public acceptance of rDNA technology should take these positive and negative feelings of consumers into account. Therefore, it is essential that manufacturers estimate the benefits versus risks for

new technologies or even for new products. In the case of rDNA products, it is obvious that the benefit of this technology for the development of a new generation of very effective, precise medicines with minimal side effects is enormous. Nevertheless, the application of rDNA technology for the development of new medicines has not been accepted in all its aspects (e.g., gene therapy), mainly because the communication with the consumers has been insufficient and inadequate. The author is not aware of any systematic nationwide education program to teach students or consumers the benefits and risks of the various aspects of the application of rDNA technology. In an extensive analysis of the Dutch Institute for Consumer Research (HAMSTRA and FEENSTRA, 1989) only 57% of 1729 interviewed persons knew the word "biotechnology", and even of this 57% only 45% knew that insulin was made via biotechnology, whereas fewer than 10% of all interviewed persons knew that experiments are in progress on rDNA cells modified in such a way that they can inhibit certain forms of cancer.

In a recent analysis done by the University of Strathclyde (U.K.) the public was divided into seven specifically targeted subgroups. The attitude of the various subgroups towards biotechnology is extremely variable. As could be expected, certain applications of rDNA technology are not or hardly acceptable for the public: biological warfare (95%), changing the human physical appearance (84%), cloning prize cattle (72%), using viruses to attack crop pests (49%) and improving milk yield (47%). Medical applications are considered by most of the groups as acceptable (Tab. 16). Similar results were obtained in a Dutch investigation, although the acceptance of products made via rDNA technology that can contribute to an improvement of the environment scores higher (HEIJS and MIDDEN, 1993).

Labeling of products is only relevant if consumers are able to understand the information on the label. According to the view of the FDA, labeling should not be based on the way a certain product is obtained. This is or should be part of normal approval for agricultural practice or industrial processes, and if approved then labeling is unnecessary. This is

Tab. 16. Attitudes to Applications of Genetic Manipulation Adapted from (A) MARTIN and TAIT (1992) and (B) HEIJS and MIDDEN (1993)

		Comfortable	Neutral	Uncomfortable
Microbial production of bio-plastics	(B)	91	6	3
Cell fusion to improve crops	(B)	81	10	10
Curing diseases such as cancer	(A)	71	17	9.5
Extension shelf life tomatoes	(B)	71	11	19
Cleaning up oil slicks	(A)	65	20	13
Detoxifying industrial waste	(A)	65	20	13
Anti blood clotting enzymes produced by rats	(B)	65	14	22
Medical research	(A)	59	23	15
Making medicines	(A)	57	26	13
Making crops to grow in the Third World	(A)	54	25	19
Mastitis-resistant cows by genetic modification of cows	(B)	52	16	31
Producing disease-resistant crops	(A)	46	29	23
Chymosin production by yeast	(B)	43	30	27
Improving crop yields	(A)	39	31	29
Using viruses to attack crop pests	(A)	23	26	49
Improving milk yields	(A)	22	30	47
Cloning prize cattle	(A)	7.2	18	72
Changing human physical appearance	(A)	4.5	9.5	84
Producing hybrid animals	(A)	4.5	12	82
Biological warfare	(A)	1.9	2.7	95

common practice for nearly all food products, e.g., the label on a can of meat or fish never mentions that the product has received a heat treatment equivalent to F_0 of at least 5 (to ensure 12 decimal reduction of the risk of botulism), but the consumers trust the government to check the manufacturer of these cans. An argument against the view of the FDA is that certain consumers are in principle against a certain technique and these consumers have the right to know which technology has been applied to produce a product. However, this argument is not restricted to rDNA technology, but applies to all types of technologies. Without any doubt the label should contain information on the contents of the product in order to inform the consumer. Therefore, it seems logical that if a product still contains a foreign gene or gene product, it should be labeled. Accepting this view it is clear that also products derived from plants obtained via classical breeding or products containing mutated organisms should be labeled. However, in view of the general ignorance of the public about biotechnology it is dangerous to label products containing rDNA-derived products. Such labeling can repel consumers unneces-

sarily and can therefore slow down considerably the introduction of this technology even for applications that are considered by the public as desirable, such as reduction of the amount of chemicals used to protect crops.

6 Conclusions

It is very likely that rDNA technology will become the most important technology in the first part of the 21st century, because this technology will help us to unravel the secrets of Nature, and by doing so we can learn from Nature how to cope with diseases and large ecological problems, for Nature has frequently faced these problems and found magnificent solutions. Also for food products rDNA will become an important tool to reduce the use of chemicals in agriculture and energy and production of waste during processing. Moreover, rDNA will provide techniques to optimize the quality of food products and increase the naturalness of food products (VERRIPS, 1991).

The risk related to the introduction of new rDNA products should be assessed carefully. Always two aspects should be assessed separately. Firstly, the intrinsic toxicological or eco-toxicological hazard of the foreign gene(s) and the construction methods should be determined. If the foreign gene codes for a harmless protein and the host organism is safe (recognized as functional organism in fermented food; see, e.g., Tab. 4 of DFG Mitteilung XI, 1987) and all the aspects mentioned in Sects. 2.3, 2.4, 2.6–2.9 are carefully analyzed and no mistakes or unexpected events have been found, then the hazard of this transgenic organism can be considered as zero. In that case an assessment of the probability that the foreign gene present in the transgenic organism will be transferred into a wild organism is not necessary because whatever that probability is the hazard is zero, so the product of hazard·probability ($=$ risk) is zero. Nevertheless, a risk assessment for the eco-toxicological consequence will have to be made.

If the hazard analysis shows that the hazard is not zero, e.g., in case of a pharmaceutical product that encodes for an inhibitor of an enzymic reaction in malignant cells of a certain tissue but this inhibitor will also be able to block this reaction in healthy cells of other tissues or even other species, a full risk assessment should be carried out. The open question is then what will be an acceptable risk. No definitive answers to this question can be given. In the food industry the acceptable risk of microbial spoilage (in the order of one product unit in 10^{-4}) is much larger than the acceptable risk of microbial poisoning (ranging between about 10^{-6} for relative innocent poisonings to 10^{-12} for poisoning by botulinum toxin). Also the kind of hazard related to rDNA products and the acceptable risk will be related to each other. Moreover, as was explained earlier in this chapter, the risk assessment of GMOs will be more complicated than that of cell-free rDNA products, but the approaches are not fundamentally different.

For medicines made via rDNA technology, even a third factor will become important in this risk assessment, notably the benefit of the rDNA product to cure severely ill people.

The most important aspect of legislation of new scientific developments is clarity. Only if clear legislation is in place, will the manufacturers or seed companies know exactly under which conditions a new discovery can be commercialized and can they calculate whether their benefit versus cost ratio will be positive. Although improvements of the present legislation are certainly possible and harmonization of the legislation between the various countries is of great importance, in most countries there is legislation of sufficient clarity to enable introduction of this technology. However, nearly all these legislations are based on the assumption that rDNA technology is an intrinsically hazardous technology and, as described above, this cannot be justified by scientific data.

Governments should educate their citizens (especially pupils at schools) to ensure that they really understand this new technology and are able to make real choices. This is important for all new technologies but especially for rDNA technology, as this technology is closely related to the most fundamental processes of life. A better understanding of biology will certainly result in a better acceptance by the consumers.

Acknowledgement

The author is grateful for the contribution of ESTHER KOK (Rijks-Kwaliteitsinstituut voor Land- en Tuinbouwprodukten (RIKILT-DLO), Wageningen, The Netherlands) for preparation of Table 8 and W. HOEKSTRA (Rijksuniversiteit Utrecht, The Netherlands), E. KOK (RIKILT-DLO), O. KORVER, W. MUSTERS (Unilever Research, Vlaardingen), D. TOET (Gist-Brocades; at present Hercules, Rijswijk, The Netherlands), B. DE VET (Unilever, Rotterdam) and A. WEERKAMP (NIZO, Ede) for the valuable comments on the manuscript.

7 References

ACNFP (Advisory Committee on Novel Foods and Processes) (1991), in: *Guidelines and the Assessment of Novel Foods and Processes*, Department of Health, Report **38**. London: Department of Health.

ALEXANDER, M. (1981), *Science* **211**, 132–138.

ANDERSSON, S. G. E., KURLAND, C. G. (1990), *Microbiol. Rev.* **54**, 198–210.

ARBER, W. (1990), in: *Introduction of Genetically Modified Organisms into the Environment* (MOONEY, H. A., BERNARDI, G., Eds.), pp. 17–26, *Scope* **44**, New York: J. Wiley & Sons.

AVERY, O. T., MacLEOD, C. M., McCARTY, M. (1944), *J. Exp. Med.* **79**, 137–158.

BAIM, S. B., SHERMAN, F. (1988), *Mol. Cell. Biol.* **8**, 1591–1601.

BERG, P., BALTIMORE, D., BOYER, H. W., COHEN, S. N., DAVIS, R. W., HOGNESS, D. S., NATHANS, D., ROBLIN, R., WATSON, J. D., WEISMAN, C., ZINDER, N. D. (1974), *Science* **185**, 303.

BERGMANS, J. E. N., HOEKSTRA, W. P. M., JONKER, J., VAN VLOTEN-DOTING, L., WINKLER, K. C. (1992), in: *Erfelijke Veranderingen bij Bacterien, Planten en Dieren*, Chapter 3, Utrecht, The Netherlands: VCOGEM, Van Arkel.

BERRY, M. J., BANU, L., CHEN, Y., MANDEL, S. J., KIFFER, J. D., HARNEY, J. W., LARSEN, P. R. (1991), *Nature* **353**, 273–276.

BRIERLEY, I., DIGARD, P., INGLIS, S. C. (1989), *Cell* **57**, 537–547.

CALGENE Inc. (1991), *Patent* PCT WO 91/13972.

CAMPBELL, A. L. (1990), in: *Introduction of Genetically Modified Organisms into the Environment* (MOONEY, H. A., BERNARDI, G., Eds.), *Scope* **44**, New York: J. Wiley & Sons.

CRAIGEN, W. J., CASKEY, C. T. (1986), *Nature* **322**, 273–275.

CRAIGEN, W. J., COOK, R. H., TATE, W. P., CASKEY, C. T. (1985), *Proc. Natl. Acad. Sci. USA* **82**, 3616–3620.

DARNELL, J. E., DOOLITTLE, W. F. (1986), *Proc. Natl. Acad. Sci.* **83**, 1271–1275.

DAVIES, J. (1990), *Trends Biotechnol.* **8**, 198–203.

DFG (Deutsche Forschungsgemeinschaft) (1987), *Starterkulturen und Enzyme für die Lebensmitteltechnik*, Mitteilung XI der Senatskommission zur Prüfung von Lebensmittelzusatz- und -inhaltsstoffen, Weinheim: VCH.

ELLSTRAND, N. C. (1988), *Trends Biotechnol.* **6**, S30–S32.

EEC (European Economic Communities) (1990a), *Guidelines* **90**/219.

EEC (European Economic Communities) (1990b), *Guidelines* **90**/220.

FDA (Food and Drug Administration) (1992), *Federal Register* **57**, No. 104, May 29.

FRENKEN, L. J. G., BOS, W. J., VISSER, C., MÜLLER, W., TOMMASSEN, J., VERRIPS, C. T. (1993a), *Mol. Microbiol.* **9**, 579–589.

FRENKEN, L. J. G., DE GROOT, A., TOMMASSEN, J., VERRIPS, C. T. (1993b), *Mol. Microbiol.* **9**, 591–599.

GESTELAND, R. F., WEISS, R. B., ATKINS, J. F. (1992), *Science* **257**, 1640–1641.

GEZONDHEIDSRAAD (1992), *Produktveiligheid bij Nieuwe Biotechnologie*, Den Haag: Gezondheidsraad.

GIST-BROCADES (1982), *Patent* EP-B1-0.096.430.

GIST-BROCADES (1987), *Patent* EP-A-0.306.107.

GRAHAM, J. B., ISTOCK, C. A. (1978), *Mol. Gen. Genet.* **166**, 247–260.

GRIERSON, D., FRAY, R. G., HAMILTON, A. J., SMITH, C. J. S., WATSON, C. F. (1991), *Trends Biotechnol.* **9**, 122–123.

GUINEE, P. (1977), *Tweede Jaarverslag KNAW Commissie*, pp. 94–108, Amsterdam: KNAW.

HAMSTRA, A. M., FEENSTRA, M. H. (1989), *SWOKA Report* **85**, Den Haag: SWOKA.

HEIJS, W., MIDDEN, C. (1993), in: *Biotechnologie: Houdingen en Achtergronden*, Den Haag: Ministerie van Economische Zaken.

HEINEMANN, J. A., SPRAGUE, G. F. Jr. (1989), *Nature* **340**, 205–209.

HILL, K. E., LLOYD, R. S., YANG, J.-G., READ, R., BURK, R. F. (1991), *J. Biol. Chem.* **266**, 10050–10053.

HUANG, W. M., AO, S.-Z., CASJENS, S., ORLANDI, R., ZEIKUS, R., WEISS, R., WINGE, D., FANG, M. (1988), *Science* **239**, 1005–1012.

HWANG, C., SINSKEY, A. J., LODISH, H. F. (1992), *Science* **257**, 1496–1502.

IFBC (1990), *Regul. Toxicol. Pharmacol.* **12**, S1–196.

ISBERG, R. R., FALKOW, S. (1985), *Nature* **317**, 262–264.

ISRAEL, M. A., CHAN, H. W., ROWE, W. P., MARTIN, M. A. (1979), *Science* **203**, 883–887.

KOK, E. J., REYNAERTS, A., KUIPER, H. A. (1993), *Trends Food Sci. Technol.* **4**, 42–48.

KOZAK, M. (1983), *Microbiol. Rev.* **47**, 1–45.

LEVINE, M. M., KAPER, J. B., LOCKMAN, H. (1983), *Recomb. DNA Tech. Bull.* **6**, (3) 89–97.

LIANG, P., PARDEE, A. B. (1992), *Science* **257**, 967–974.

MARTIN, S., TAIT, J. (1992), in: *Biotechnology in Public: a Review of Recent Research* (DURANT, J., Ed.), London: Science Museum for the European Federation of Biotechnology.

MOROWITZ, H. J. (1978), in: *Foundations of Bioenergetics*, Chapter 12, New York: Academic Press.

MURRAY, A. W., SZOSTAK, J. W. (1983), *Cell* **34,** 961–970.

NTT (Nordic Working Group on Food Toxicology and Risk Assessment) (1991), in: *Food and New Biotechnology, Novelty, Safety and Control Aspects of Foods Made by New Biotechnology* (Nord 1991; **18**), Copenhagen: Nordic Council of Ministers.

OECD (Organization for Economic Cooperation and Development) (1991), Com/Env/DSTI/EC/BT **80**/(ADD).

OVERBEEKE N., TERMORSHUIZEN, G. H. M., GIU-SEPPIN, M. L. F., UNDERWOOD, D. R., VERRIPS, C. T. (1990), *Appl. Environ. Microbiol.* **56,** 1429–1434.

PARIZA, M. W., FOSTER, E. M. (1983), *J. Food Protect.* **46,** 453–468.

POSTMA, E. (1990), *Thesis*, Technical University Delft, The Netherlands.

SAYE, D. J., OGUNSEITAN, O. A., SAYLER, G. S., MILLER, R. V. (1987), *Appl. Environ. Microbiol.* **56,** 140–145.

SAYRE, P., MILLER, R. V. (1991), *Plasmid* **26,** 151–171.

SCHAAFF, I., HEINRICH, J., ZIMMERMANN, F. K. (1989), *Yeast* **5,** 285–290.

SCHELL, J. S. (1987), *Science* **237,** 1176–1182.

SMELT, J. P. P. M. (1980), *Thesis*, University of Utrecht, The Netherlands.

SMITH, G. R. (1988), *Cell* **58,** 807–809.

SMITH, C. J. S., WATSON, C. F., RAY, J., BIRD, C. R., MORRIS, P. C., SCHUCH, W., GRIERSON, D. (1988), *Nature* **334,** 724–726.

SPRAGUE, G. F. Jr. (1991), *Trends Genet.* **7,** 393–398.

STRAUGHAN, R. (1992), in: *Ethics, Morality and Crop Biotechnology*, Reading: ICI.

TEUBER, M. (1990), *Lebensm. Ind. Milchwirtsch.* **35,** 1118–1123.

TSUCHIHASHI, Z., KORNBERG, A. (1990), *Proc. Natl. Acad. Sci. USA* **87,** 2516–2520.

UNILEVER (1981), *Patent* EP-B0077109.

UNILEVER (1989), *Patent* WO-A-91/00920.

VERRIPS, C. T. (1991), *Food Biotechnol.* **5,** 347–364.

VOEDINGSRAAD (1993), *Advies inzake Biotechnologie*, Den Haag: Voedingsraad.

WELLS, J. A., ESTELL, D. A. (1988), *Trends Biol. Sci.* **13,** 291–297.

WHO/FAO (World Health Organization/Food and Agriculture Organization) (1991), in: *Strategies for Assessing the Safety of Foods Produced by Biotechnology*, New York – Geneva: WHO/FAO.

WOESE, C. R., KANDLER, O., WHEELIS, M. L. (1990), *Proc. Natl. Acad. Sci. USA* **87,** 4576–4579.

ZINONI, F., HEIDER, J., BÖCK, A. (1990), *Proc. Natl. Acad. Sci. USA* **87,** 4660–4664.

6 Strategic Regulations for Safe Development of Transgenic Plants

TERRY L. MEDLEY
SALLY L. McCAMMON

Washington, DC 20090-6464, USA

1 Introduction

In business, a strategic plan determines where an organization should be going. It is also used to assure that all organizational efforts are pointed in that same direction. There is an old saying that "if you don't know where you are going, any road will take you there" (BELOW, 1987). For agricultural biotechnology, strategic regulations are regulations which provide a framework or process for actions that lead to consistent and planned results. Therefore, they are regulations that are *developed* and *applied* in a strategic manner. Strategic regulations are developed and applied in a comprehensive manner to avoid being one-dimensional or limited to a single issue focus. Although consideration of risk versus safety is of paramount importance, the regulations should also consider their impact on other concerns such as product verification/utilization, safe technology transfer, economic competitiveness, international harmonization, and global needs/acceptance. Consequently, strategic regulations have a multi-dimensional focus. In our opinion, to ensure the best likelihood of success, strategic regulations should: (1) have identifiable science-based triggers that are consistent, easily understood, and transparent; (2) be effective and responsive as well as flexible and dynamic; and (3) meet domestic and international needs.

2 The Challenge

Biotechnology is an enabling technology with broad application to many different areas of industry and commerce. A recent report by the Committee on Life Sciences and Health of the Federal Coordinating Council for Science, Engineering, and Technology, *Biotechnology for the 21st Century: Realizing the Promise*, Washington, DC., June 30, 1992, broadly defines biotechnology to "include any technique that uses living organisms (or parts of organisms) to make or modify products to improve plants or animals or to develop microorganisms for specific use." The report further predicts that, "by the year 2000, the biotechnology industry is projected to have sales reaching $ 50 billion in the United States."

For American agriculture, biotechnology has the potential to increase productivity, enhance the environment, improve food safety and quality, and bolster US agricultural competitiveness (OTA, 1992). However, if biotechnology's vast potential, which is an outgrowth of substantial public and private investment, is to be of the greatest benefit, the development of appropriate and effective regulatory structures is a necessity.

The 1970s and 1980s in the United States have come to be associated with certain regulatory policy milestones for the development of modern biotechnology. The 1970s were associated with the concern over the safety of conducting experiments with recombinant DNA under conditions of adequate laboratory containment. This led to issuance of the National Institutes' of Health guidelines for research involving recombinant molecules (NIH, 1976). The 1980s have been associated with concern over the safety of conducting experiments using genetically modified organisms outside of a laboratory in the environment. This led to issuance of the Federal Coordinated Framework for the Regulation of Biotechnology (OSTP, 1986).

If the first three years are any kind of indicator, the 1990s promise to be no less significant in terms of their association with biotechnology regulatory policy milestones. At the United States Department of Agriculture (USDA), a petition process has been finalized for removing transgenic plants from further regulation (USDA/APHIS, 1993). The Food and Drug Administration (FDA) published its policy statement on food derived from transgenic plants (FDA, 1992). The Environmental Protection Agency (EPA) held an open meeting of the agency's Scientific Advisory Panel to assist in developing regulatory policy for plants genetically engineered to express pesticides (EPA, 1992). Many of these new plant varieties will be available in the 1990s. But their introductions will be under circumstances unlike any met by other new plant varieties because of the technology used to develop them. "Uncertainties over these new technologies raise questions of potential impacts on food safety and the environment, and possible economic and social costs. Nevertheless, there will be a push for ... biotechnology ... to be used commercially, adopted by industry, and accepted by the public. ... The *challenge*, however, will be whether government, industry, and the public can strike the proper balance of direction, oversight, and allow these technologies to flourish." (OTA, 1992, emphasis added).

In our opinion, strategic regulations have the greatest potential for creating a framework or process to meet this challenge. The following is a discussion of the Animal and Plant Health Inspection Service (APHIS), USDA, regulations for transgenic plants in such a strategic context.

3 Identifiable Science-Based Triggers

One of the most critical and controversial decisions associated with the development of regulations is identification of scope. In other words, what organisms will be covered or what are the "triggers" for inclusion. We must guard against different approaches to this issue being inaccurately characterized as a continuation of the process-versus-product debate. It is more appropriately characterized as a commitment by regulatory officials to the development of regulations that neither over-regulate nor under-regulate. In the United States we do not know of any Federal agency which seeks to regulate based upon process only, independent of risk or uncertainty (MEDLEY, 1990). For biotechnology in the United States, this issue is best demonstrated by the publication of a proposed scope document for regulation of biotechnology in 1990 and a very different final scope policy document in 1992. (OSTP, 1990, 1992). These two documents were very different and did little to resolve the scope debate or determine necessary triggers for inclusion. Regulations which are strategic in their development can avoid this initial barrier by using triggers for

inclusion that are consistent with adopted regulatory policies, based on sound science, and in themselves transparent.

3.1 Policy

Industry analysts as well as academic scientists have stressed the importance of US Federal programs and policies for the future of US biotechnology, particularly US Federal support for basic and applied research fundamental to the biotechnology industry (OLSON, 1986; ERNST & YOUNG, 1992). The US Federal investment in biotechnology research for Fiscal Year 1994 is approximately 4.3 billion US dollars. An adjunct to the US Federal investment in biotechnology is the Federal commitment to promote the safe development of products of biotechnology. Such a commitment requires that each Federal agency with authority to regulate such organisms and products implement risk-based regulatory requirements. These regulatory requirements should assure safety and facilitate technology development and utilization by removing regulatory uncertainty.

The Federal policy for the regulation of biotechnology was proposed on December 31, 1984 (OSTP, 1984), and published in final form on June 26, 1986 by the US Office of Science and Technology Policy (OSTP, 1986). The Federal policy is referred to as the "Coordinated Framework for Regulation of Biotechnology." The OSTP concluded that products of recombinant DNA technology will not differ fundamentally from unmodified organisms or from conventional products. Therefore, existing laws and programs are considered adequate for regulating the organisms and products developed by biotechnology.

The USDA policy on biotechnology was consistent with the policy established by the OSTP. USDA's policy document of December 1984 stated that existing statutes were adequate for regulating the products of agricultural biotechnology, and further, that USDA did not anticipate that such products would differ fundamentally from those produced by conventional techniques (OSTP, 1984, p. 50896). This position was reaffirmed in the final policy document of June 26, 1986 (OSTP, 1986, p. 23345).

The current Administration has framed its policy objectives in President CLINTON's and Vice President GORE's initiative, *Technology for America's Economic Growth, A New Direction to Build Economic Strength*, February 22, 1993, White House, Washington, DC. In this document, the Administration stated, in relevant part, the following objectives:
(1) We must ... turn ... to a regulatory policy that encourages innovation and achieves social objectives efficiently;
(2) We can promote technology as a catalyst for economic growth by ... directly supporting the development, commercialization, and deployment of new technology; [and]
(3) To improve the environment for private sector investment and create jobs, we will: ensure that Federal regulatory policy encourages investment in innovation and technology development that achieve the purposes of the regulation at the lowest possible cost.

These policy objectives are reinforced by Executive Order 12866 (*Fed. Reg.* **58**, 51735-51744, October 4, 1993) which states, in relevant part:
(4) In setting regulatory priorities, each agency shall consider to the extent reasonable, the degree and nature of the risks posed by various substances or activities within its jurisdiction.
(10) Each agency shall avoid regulations that are inconsistent, incompatible, or duplicative with its other regulations or those of other Federal agencies.
(11) Each agency shall tailor its regulations to impose the least burden on society, including individuals, businesses of differing size, and other entities (including small communities and governmental entities), consistent with obtaining the regulatory objectives, taking into account, among other things, and to the extent practicable, the cost of cumulative regulations.

However, one of the most prominent and essential aspects of the Clinton Administration's biotechnology policy is the requirement that there be more public access to the Federal decision-making process or as GREG SIMON, Domestic Policy Advisor to Vice President GORE, states "the importance of a regul-

atory system for biotechnology that operates in an open and fair way" (BROWNING, 1993).

3.2 Science

Strategic regulations for biotechnology should be developed and applied using sound science as a cornerstone. However, moving genes across natural barriers creates uncertainty for some scientists as well as for society in general. A major question is: what constitutes an "unreasonable" risk? Such risks normally fall into categories of risks to the food supply, human health, and the environment. Plant and animal pathogens are included as well as carcinogens. Thus, uncertainty is created in new cases or situations in which pathogens are used as part of the process of development and genes for pathogenesis are inserted into an organism.

The scientific community itself does not appear to be of a single mind as to the ramifications to the environment of moving genes from non-sexually compatible species and from pathogens into plants. In particular, the scientific establishment does not agree on the types of risk and what is low risk. For example, in 1989, the report by the National Research Council, *Field Testing Genetically Modified Organisms: Framework for Decisions*, emphasized familiarity with the "properties of the organism and the environment" in order to make a determination of risk (NRC, 1989). The Council stated that "[f]amiliar does not necessarily mean safe. Rather, to be familiar with the elements of an introduction means to have enough information to be able to judge the introduction's safety *or* risk." Enhanced weediness was cited as the major environmental issue with plants. With the addition of traits for herbicide tolerance or pest resistance, it was felt that crop plants would be unlikely to become weedy. However, the report also stated that "genetic modifications of only a few genes can produce a modified plant with significant, ecologically important alterations". They stated that, for initial tests, confinement was the "key to minimizing the environmental impact to nontarget species" (NRC, 1989).

In the same year, the Ecological Society of America published a paper entitled "The Planned Introduction of Genetically Engineered Organisms: Ecological Considerations and Recommendations" in which several factors were rated for their level of concern in a risk evaluation (TIEDJE et al., 1989). These included factors relating to the genetic alteration, the parental organism, and the phenotype of the altered organism.

Factors such as a new gene function being introduced to the parent organism, the source of the gene being from an unrelated organism, the vector being from an unrelated species or pathogen, and the parent being self-propagating all warranted scientific evaluation. Other scientists have raised concerns over the effects of introducing genes from plant pathogens into the plants themselves; especially, concerns about viral coat protein genes and viral nucleic acid recombination have been discussed (HULL, 1990; DE ZOETEN, 1991). It also appears to us that the view that one gene change cannot cause a change in pathogenesis is an oversimplification, given reports in the literature (PAYNE et al., 1987; WHEELER, 1975).

Therefore, strategic regulations for biotechnology biosafety reviews should focus on the scientific questions and the most efficient way of doing the review in the context of facilitating the safe application of the biotechnologies. Thus, a certification system for evaluating any safety issues while the variety is being developed, preferably at the initial field testing stage, is most appropriate (COOK, 1993).

Science must be the basis of the decisions that address the concerns associated with the application of biotechnology to agriculture. A necessary role for regulatory officials is to frame the questions and issues that science must answer. Science is the base by which regulatory officials can assure and build upon credibility, remain current, and assure a rational basis for decision-making. Science and process are inextricably linked for strategic regulations that evaluate biological programs and products.

In the area of biotechnology regulation, science without the process to frame the issues becomes overwhelming and misleading, while process without science is reduced to being bureaucratic, self-serving, and ineffective. Neither science nor regulation can afford to be in an all or none category. To adapt to new information and new needs, both types of systems need to work together to frame and address the concerns and requirements of the many components of our society.

3.3 Transparent Triggers

It is not enough that the triggers for strategic regulations be consistent with national biotechnology policies and are science-based, they must also be easy to understand and identify (transparent). Such transparency is necessary to enable those regulated to clearly understand the regulatory requirements. Transparent regulations should remove regulatory uncertainty, establish predictable standards and allow safe technology utilization and expanded development. For those that have concerns about the biotechnology activities being regulated, it should provide a clear or transparent window to observe and evaluate governmental oversight.

APHIS regulations, which were promulgated pursuant to authority granted by the Federal Plant Pest Act (FPPA), (7 U.S.C. 150aa–150jj) as amended, and the Plant Quarantine Act (PQA), (7 U.S.C. 151–164a, 166–167) as amended, regulate the introduction (importation, interstate movement, or release into the environment) of certain genetically engineered organisms and products. A genetically engineered organism is deemed a regulated article if either the donor organism, recipient organism, vector or vector agent used in engineering the organism belongs to one of the taxa listed in section 340.2 of the regulations and is also a plant pest; if it is unclassified; or, if APHIS has reason to believe that the genetically engineered organism presents a plant pest risk. (7 CFR part 340; see also McCAMMON and MEDLEY, 1990).

Prior to the introduction of a regulated article, a person is required under section 340.1 of the regulations to either (1) notify APHIS in accordance with section 340.3 or (2) obtain a permit in accordance with section 340.4. Under section 340.4, a permit is granted for a field trial when APHIS has determined that the conduct of the field trial, under the conditions specified by the applicant or stipulated by APHIS, does not pose a plant pest risk.

The FPPA gives USDA authority to regulate plant pests and other articles to prevent direct or indirect injury, disease, or damage to plants, plant products, and crops. The PQA provides an additional level of protection by enabling USDA to regulate the importation and movement of nursery stock and other plants which may harbor injurious pests or diseases, and requires that they be grown under certain conditions after importation. For certain genetically engineered organisms, field testing may be required to verify that they exhibit the expected biological properties, and to demonstrate that although derived using components from plant pests, they do not possess plant pest characteristics.

An organism is not subject to the regulatory requirements of 7 CFR part 340 when it is demonstrated not to present a plant pest risk (SCHECHTMAN, 1994).

A plant developed from a plant pest, using genetic engineering, is not called a plant pest, but rather a "regulated article". There has been misunderstanding as to the basic assumption of APHIS regulations. The regulations do not imply that part of a plant pest makes a plant pest or that plants or genes are pathogenic. The regulations instead have as a trigger that when plants are developed using biological vectors from pathogenic sources, use material from pathogenic sources, or pathogens are used as vector agents, that they should be evaluated to assure that there is not a plant pest risk. For example, APHIS *does* accept that Ti plasmids used to transform plants have been disarmed. APHIS does a review that allows a verification of the biology and procedures used; assesses the degree of uncertainty and familiarity, and allows the identification of any risks, should they be present and predictable.

Vectors are evaluated because plant pathogens are used to move genetic material into the plant (genetic materials from viruses and pathogens are used as promoters, terminators, polyadenylation signals, and enhancers; and genes from plant pathogens are inserted to obtain resistance to these pathogens). Evaluation includes the verification of the removal of plant pathogenic potential of biological vectors so that pathogenic properties do not become part of the heritable characteristics of new crop varieties. Strictly interpreted, concern is not over the use of viral promoters or disarmed Ti plasmids but over the elimination of genes for pathogenesis or unknown/non-characterized DNA. APHIS regulations do not assert that genetic traits would cause a plant to exhibit plant pest characteristics unless the plant itself was already a pest to other plants. However, they do allow verification that (1) a disarmed plasmid was used, (2) genes implicated in plant pathogenesis did not become part of the heritable characteristics of the plant, and (3) the introduced genetic material is inherited as a single Mendelian trait, or lack of genetic stability of the new plant variety is not expressed by the plants in the field.

APHIS has issued permits for field tests for plants containing various genes for insect tolerance, virus tolerance, fungal tolerance, herbicide tolerance, nutritional and value factors, heavy metal sequestration, pharmaceutical products, and selective markers. There are direct implications of genetic factors on plant health and the effects of toxicants on beneficial insect populations that are important to consider in evaluating impact on plants and the environment.

The regulations would not be based on sound science if assertions rather than evidence, or laboratory data alone, without field confirmation were used as the sole basis of decisions under the regulations. Molecular sequence information does not preclude the need for appropriate scientific evaluation and data collection in various environments in the field.

The use of the technology itself poses no unique hazards, i.e., genetic material inserted into a plant via this technology will behave no differently than genetic material inserted via any other technology. The risks in the environment will be the same; there is no evidence that unique hazards in the use of rDNA techniques are present. The risk is in the genes interacting in a new genetic background and with an organism in a new environment. In the case of the use of pathogenic material that has been reviewed under the FPPA and the PQA, pathogenic genes could not be introduced into the heritable characteristics of the plant without the use of the technique. *The plant pest source of the genetic material, not the technique, is the major basis for regulation.*

4 Effective and Responsive Regulations

The baseline for strategic regulations, similar to the baseline for any business, is whether the system is productive, i.e., does it work. When applied, regulations should not only address any safety concerns raised about the use of the technology but also meet the needs of those being regulated. Some members of the academic research community had the following to say to fellow researchers about the effectiveness of the APHIS biotechnology regulations: "our experiences with the permit process have been positive. We are all in agreement that the existent system is fast, friendly, and efficient, allowing us to concentrate our energy on experimentation rather than paperwork" (SHAW, 1992).

In addition to being effective and responsive, oversight in an area such as biotechnology must be dynamic, flexible, and constantly evolving. This flexible and dynamic focus is needed because the technologies and their applications are rapidly evolving. The structure of regulations should assure safety and facilitate technology development and utilization. Although regulations should be risk-based, the purpose of regulation itself can be multifold and include: (1) safety; (2) product verification; and (3) quality assurance. Regulation is at the intersection of scientific risk assessment and product evaluation and, in a new area like biotechnology, necessarily has to adjust and change rapidly with advances in both.

APHIS, like other US Federal agencies, regulates the products of biotechnology under its existing statutory authority. APHIS' broad historic authority to protect plant health is applicable to the regulation of certain plants as well as microorganisms developed through biotechnological processes. Equally as important is the statutory and legislative intent to empower APHIS (via the Secretary of Agriculture) to be the first line of defense in protecting US plant health. Hence, the statutes have built in flexibility to take those measures necessary to prevent direct or indirect risks to plants (7 U.S.C. 150aa–jj).

Almost six years have elapsed since APHIS first established regulations for certain transgenic plants. We are committed to assuring that our regulations are flexible and modified when appropriate to remove unnecessary restrictions or, if necessary, to add additional ones. Flexibility of the system to adapt based upon experience is essential for strategic regulations that enable the structure/operation of the regulations to be most effective and responsive.

Such regulatory systems should evolve and be well-designed. Thereby, strategic regulations stimulate rather than frustrate or inhibit innovation. Strategic regulations provide options for the safe introduction and commercialization of products of new technologies. They can stimulate private sector investment and innovation, while at the same time, reducing regulatory costs and complying with environmental/safety concerns and mandates.

APHIS, through strategic regulations, has attempted to develop a flexible regulatory structure that (1) evaluates on the basis of risk; (2) provides a means of specifying performance standards as far as criteria for exemption; (3) defines and documents risk or lack of risk; and (4) exempts regulated material from oversight efficiently once evidence of lack of risk is presented. All procedures followed depend upon the individual proposal itself which the applicant submits and APHIS reviews. Therefore, this regulatory structure is *applicant-driven* which assures that design standards (rigid protocols) are avoided and performance standards emphasized.

Our responsibility is to identify the risks and to make a decision as to their significance. We have attempted to design a system that evaluates the molecular biology and practices and knowledge from traditional agriculture in order to document conclusions of safety or acceptable uncertainty. Under the APHIS regulations, our analysis compares the new plant line with previous varieties and crops in evaluating hazards.

The initial stage of testing gives the most valuable information on the interaction of a gene in a new organism. The plant itself is the most sensitive test for the effects of a new gene in a new organism (e.g., assessing the interaction of the trait, the plant, and the environment).

From 1987 to 1992, the primary regulatory option for the planned introduction of a transgenic plant was the permit. The regulations stipulate that, once a complete permit request has been submitted, APHIS has 120 days in which to reach a decision whether to issue or deny a permit. The five major steps APHIS takes in this process are to: (1) evaluate relevant information (both that submitted by the permit applicant and that gathered by APHIS from other primary and secondary sources); (2) notify and consult with regulatory officials in States where the applicant proposes to field test; (3) provide, through a general announcement, an opportunity for public comment; (4) prepare a site-specific environmental assessment; and (5) reach a decision as to whether to issue or deny the permit (McCAMMON and MEDLEY, 1992).

APHIS considers a broad range of information for an environmental assessment, including details of the biology and characteristics of both the parent organism and the organism that provides the exogenous genetic material; factors associated with the modification itself, such as whether or not gene sequences from a plant pest are used, and if so, how they have been modified to ensure they pose no risk of plant disease; the expected characteristics and behavior of the modified organism; and data concerning the environment in which it would be field-tested. Such information includes, for example, the location and size of test plots, numbers of test organisms, nature of the nearby environment including the presence or absence of sexually compatible relatives, or the presence in the same county of threatened or endangered

species that could be impacted. On the basis of this analysis, APHIS decides whether or not to issue a Finding of No Significant Impact (FONSI). If a FONSI is reached, a permit is issued. If a FONSI is not reached, then the permit is denied, or issuance would be deferred pending results of a more extensive analysis, known as an Environmental Impact Statement (EIS). APHIS has not yet declined to issue a permit for a field test because of failure to reach a FONSI. However, we have had applicants to either withdraw a submission for a permit based upon initial review by APHIS or not submit a permit request based upon pre-submission discussions with the APHIS scientific reviewers.

After almost six years evaluating permits and considering the results of field trials under permit, experience demonstrated that categories could be defined for certain field tests that do not present plant pest risks, uncertainty or significant agricultural safety issues. Therefore, the type of regulatory analysis described above for the issuance of a permit was demonstrated to be unnecessary for such categories.

General procedures already existed for exemption of broad categories of organisms from regulation based upon scientific data, consistent with the internationally accepted definition for "case-by-case". That definition was adopted by the United States and other members of the Organization for Economic Cooperation and Development (OECD), and explicitly allows for exemptions based upon scientific data. "Case-by-case means an individual review of a proposal against assessment criteria which are relevant to the particular proposal; this is not intended to imply that every case will require review by a national or other authority since various classes of proposals may be excluded" (OECD, 1986). Many cases reviewed by APHIS are categorically excluded from coverage by the regulations. Some field tests are not reviewed or are reviewed only upon request.

The increase in field trials in the United States of transgenic plants or plant-associated microbes in recent years has been explosive. In anticipation of dramatic increases, on March 31, 1993, APHIS put in place two additional regulatory options, notification and petition, which became effective on April 30, 1993. APHIS now has four options to facilitate the safe introduction of transgenic plants during research, development, and commercialization: the organism is not a regulated article, a permit for movement or release, notification for movement or release, and petition for removal from further regulation (USDA/APHIS, 1993).

"Notification" and "petition" options provide for what amounts to streamlined review, or exemption from further regulatory review by APHIS, respectively. A field test conducted under notification allows the applicant, with APHIS acknowledgement, to commence a field trial within thirty days instead of the 120 days required under permit. Introduction under notification (section 340.3) requires that the introduction meets specified eligibility criteria and performance standards. The eligibility criteria impose limitations on the types of genetic modification that qualify for notification, and the performance standards impose limitations on how the introduction may be conducted. The notification option is presently restricted to field tests of new varieties of six crops: corn, potato, cotton, tomato, soybean, and tobacco. Although it is expected that other crops will become eligible for field testing under the notification option in the future, at the time this option was initially adopted these six crops captured approximately 85% of the field test permits being issued. The notification option is further limited by precluding certain types of modifications (e.g., those encoding genes for pharmaceutical compounds or from human or animal pathogens) and requiring field tests under notification to be conducted under speci-

fied performance standards that amount to genetic containment (MEDLEY, 1993). Field tests not falling within these constraints may still be conducted under permit, as before.

The petition option, for which the full term of art is "petition for determination of nonregulated status," allows an applicant to request that APHIS decides whether a given transgenic plant or microbe should no longer be considered a regulated article. The determination is, as with a permit, based on information provided by the applicant, which is considered with other information collected by APHIS. Once a petition is approved by APHIS there is no longer any need for APHIS review or approval for introduction of the article into American agriculture or commerce (MEDLEY, 1993). If food safety, pesticide or other regulatory questions remain, however, the mandatory requirements of relevant regulatory agencies, such as the FDA, EPA, and USDA's Food Safety and Inspection Service, must be met.

From 1987 through 1993, there have been 674 field tests of transgenic plants or plant-associated microbes approved by APHIS in the United States. Of these field tests 471 were conducted under permit, and 203 under notification, at 1739 sites. These sites have been located in 42 of the 50 States (most often in California, Illinois, and Iowa), and Puerto Rico.

These numbers reveal a substantial amount of research and development activity. The vast majority (85%) of activities are being carried out by commercial entities Academia ranks second, with 11%, and the Agricultural Research Service (ARS) of the USDA third with 4% (see Fig. 1).

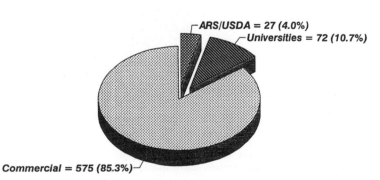

Fig. 1. Source of applicants for biotechnology release permits 1987–1993.

ARS/USDA = 27 (4.0%)

Universities = 72 (10.7%)

Commercial = 575 (85.3%)

Of the 34 crops or plant-associated microbes that have been field tested, by far the largest category has been new varieties of maize (corn), at 167. Soybeans and tomatoes are second and third, with about one hundred each, followed by the other crops presently eligible for field tests under notification, tobacco being lowest, at 35. The next two crops most often field-tested in the United States are cucurbits (melons or squash) and rapeseed, at 28 and 20, respectively (see Fig. 2).

The largest category is for crops with improved weed management qualities based on herbicide tolerance, comprising 32% of all field tests. This is followed by crops variously modified to resist losses to insect pests (most often by addition of the gene encoding the insecticidal protein derived from *Bacillus thuringiensis*), a category that embraces 21% of crops field tested. Modifications for resistance to viral diseases follow at 19%, and improvements to product quality at 16 percent (see Fig. 3).

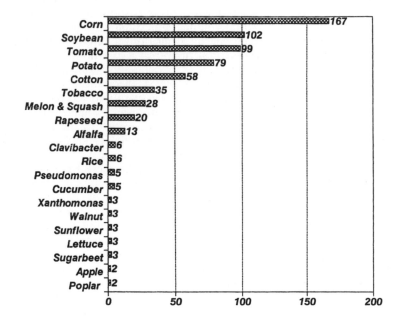

Fig. 2. Types of plants and microorganisms and number of permits issued and of notifications acknowledged 1987–1993.

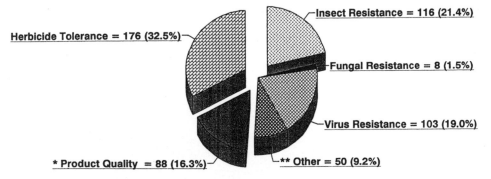

Fig. 3. Gene-phenotype summary of permits issued and notifications acknowledged 1987–1993.
*   Metabolic characteristics, modified ripening, oil modification, seed protein starch/solid increase, wheat germ agglutinate.
** Markers, male sterility, chiA, chicken lysozyme, DHDPAS, *Rhyzobium* R and C1 transcript activator (courtesy), *tn5* (marker), *tmr*, transposable elements, pharmaceutical products, plant stress tolerance.

5 Meeting Domestic and International Needs

Strategic regulations should provide a process by which domestic and international needs are met. A coordinated procedural and scientific review of field testing and development assures that plant products are developed, commercialized, and accepted. Regulations must not only assure safety, they also must provide a framework for evaluation and technology transfer. This framework includes the initial phases of testing, the developmental phases, commercialization, and international harmonization of evaluation. This will contribute to flourishing industries, global acceptance and use, and creation of jobs.

5.1 Domestic Needs

To meet the domestic challenge, regulations must take into account needs of the States, other Federal agencies such as the EPA and the FDA, as well as those of the public, including academic, consumer, environmental, and industry groups. The regulations must remove uncertainty and allow development and utilization through documenting informed decisions and evolution based upon experience. The notification and petition regulations evolved because of documented experience with the field testing of certain crops.

Domestic needs have been served through several avenues. USDA has developed a user-friendly system through publishing a User's Guide (USDA/APHIS, 1991), giving examples of applications for importation, movement, and release. Applications can now be submitted as well as processed, and permits issued, via computer technology. Decisions for release are based on information from workshops bringing together experts from various fields and affiliations (McCAMMON and DWYER, 1990; KEYSTONE, 1990). *Federal Register* notices alert the public to opportunities to comment on specific decisions. For example, prior to the decision to propose the notification and petition amendments to the APHIS regulations, the Agricultural Biotechnology Research Advisory Committee (ABRAC), in an open forum, reviewed the scientific basis for proceeding with a notification and petition process based upon documented experience with traditional and genetically engineered crops. Additionally, prior to the adoption of the rule amendments, comments were solicited from the general public (FDA, 1992).

Coordination across agencies is a necessity as agricultural, environmental, and food safety issues arise and as products develop. The decisions by one agency should support and build upon the decisions required by other agencies. The decisions required by the implementation of these biotechnologies cut across the traditional responsibilities of several regulatory jurisdictions. This requires an unprecedented integration of the entire decision-making process in order to assure that safety is assured, duplication is avoided, and the entire infrastructure is effective.

In a democracy, multipartisan support must be achieved in order to proceed wisely. Thus, industry, the scientific community, domestic regulatory groups, non-governmental organizations and other public interest groups, all need to be involved as systems and policies evolve (OTA, 1992). Regulations should allow decisions to be made in a transparent manner and these decisions must be informed. Data and input must come from many perspectives. The questions do not easily distill into "either/or", "right/wrong", "yes/no". Decisions of safety will not necessarily be the final answer, but these informed choices will be made on the basis of the best science available, and the process will be carried out in the open.

5.2 International Needs

The types of domestic concerns associated with the development of transgenic plants will most likely be the same types of concerns identified internationally. Due to the universal nature of such questions and concerns, products approved under strategic regulations should be generally accepted internationally. International harmonization will occur through compatible regulatory systems that allow reciprocity of evaluations of safety. International harmonization of approaches to commercialization means that "equivalent" approaches will be used in different countries. In this context, equivalent means "equal", not "the same". Facilitation of international trade, the prevention of trade barriers, and harmonization of international systems are all important results of a coordinated agreement as to approaches to review. Therefore, the strategic application of our USDA regulations has required that we focus on (1) integration of compatible national approaches; (2) coordination of national approaches; and (3) assurance that scientific principles are used for evaluation of organisms (MEDLEY, 1991).

A general process that can cover the needs of all countries for the use and regulation of biotechnologies based upon the technique or the final product alone is unlikely. The infrastructure, legal systems, types of products and testing, and other specific issues will be different for each country and region. The biotechnologies will implement already existing needs, conventional goals, and production requirements. The biotechnologies will affect agriculture in various ways – through plant products that enter or are tested in developing countries; through technologies that are used to develop products and commodities for local consumption or export; through export of products and commodities from the country; and through using the technologies to identify genetic resources and understand biological and ecological systems. The appropriate application of the biotechnologies will vary from country to country.

Thus, internationally it would be difficult to develop a general document addressing procedures for evaluating these products. Such a document would most likely result in the least common denominator being accepted and would be the most restrictive. Thus, international agreement on the general issues for products has been and will continue to be the most useful approach.

International agreement on consistent and compatible approaches for review should focus on the scientific questions and the most efficient way of doing the review. International agreement on the general issues for evaluation is essential for transfer of the technology to developing countries and for acceptance of commodities and products in international trade. In the Americas, the Inter-American Institute for Cooperation in Agriculture and the North American Plant Protection Organization have been forums for discussion of a wide range of issues relating to biotechnology. The United Nations Organizations have also hosted many discussions. Recently, the scientific approaches to the evaluation of plants, and especially transgenic plants, have been developed internationally in the Organization for Economic Cooperation and Development (OECD).

Internationally, several different venues have looked at and coordinated the approaches developed in each country based on scientifically identified issues. The United States frequently meets in bilateral discussions with its trading partners, Canada and Mexico, as well as with the European Community (now: European Union, EU). The decisions and experience developed by one country should support and build upon the decisions required by other countries. Recently, the USDA published its notification and petition procedures (USDA/APHIS, 1993). This was followed soon after by the EU Simplified Procedures (EEC, 1993) and the United Kingdom's "Fast Track Guidelines" (DOE/ACRE, 1994). Each of these documents uses information that builds upon both local and international experience. These procedures will facilitate international harmonization based upon experience.

Within the OECD, several documents relating to the scientific principles underlying the safe use of genetically modified organisms have been developed. These have included documents on environmental testing of plants and microorganisms, the industrial use of microorganisms, food safety, and monitoring of these organisms in the environment. The process of developing these documents requires the consensus of the member governments at each stage in the development of a document.

Evaluation of initial field tests of transgenic plants was treated by the OECD in "Good Developmental Principles (GDP): Guidance for the Design of Small-Scale Field Research with Genetically Modified Plants and Microorganisms" (OECD, 1992). GDP showed the rationale by which initial field tests could be evaluated and carried out safely. This determination is based on evaluating the plant, the site, and the agricultural practices.

The next document, treating the developmental testing of a potential new plant variety, is called "Scientific Considerations Pertaining to the Environmental Safety of the Scale-up of Crop Plants Developed by Biotechnology" (OECD, 1993a). Scale-up includes performance trials, advanced tests, demonstration trials, yield trials, and seed increase. The "scale-up" document shows how an evaluation is made to move from small-scale tests to developmental-scale tests for crop plants for traditional uses such as food, feed, fiber and flowers. The emphasis shifts from containment to identification of environmental issues. The environmental issues for crop plants can be framed, defined, and analyzed using the combined expertise and approaches of breeders, agricultural scientists, and regulatory scientists. These environmental issues are relevant at any scale. The issues will be the same no matter what the stage of the planting, whether initial tests, developmental tests, or commercial plantings. However, their probability of occurrence increases with increasing scale. Familiarity with the host plant, the environment, and the trait is the basis for this analysis. When transgenic plants were first being reviewed, they were being defined as exotic organisms in some countries. This document provides the basis for viewing transgenic plants as familiar organisms with a new trait.

Essentially, the basic premise of the "scale-up" document is that plant breeders have a tremendous amount of knowledge about how crop plants behave in the environments in which they are grown and also can identify issues of concern to agriculture, the agricultural environment, and the surrounding environments. However, plant breeders have not documented this knowledge previously in terms of safety. The "scale-up" document and the companion document "Traditional Crop Breeding Practices: A Historical Review to Serve as a Baseline for Assessing the Role of Modern Biotechnology" (OECD, 1993b), which describes the characteristics of and agricultural practices used with 17 agricultural crop plants, treats the issues of safety on the basis of our familiarity with these crop plants.

The commercialization phase of transgenic plants is now occurring. To assist in addressing commercialization issues, the USDA has funded a position for a consultant to work with the Environment and Agriculture Directorates in OECD. The consultant is working on a project which will compare the different policies of the OECD member countries towards development and commercialization of transgenic plants. The project will also explore ways that these new varieties will move into the traditional systems of seed certification and variety registration. The point at which food safety issues are evaluated will also be compared.

6 Conclusion

President CLINTON's technology initiative, articulated in the February 22, 1993, Report *Technology for America's Economic Growth, A New Direction to Build Economic Strength* stated that investing in technology was necessary for progress into the future and that this investment would require several components. The report emphasized that although the transitional Federal role in technology development has been limited to research initiatives, Government can play a key role in helping private firms. This can be accomplished while assuring a Government that is more productive and more responsive to the needs of its citizens. The report further stated that "regulatory policy can have a significant impact on the rate of technology development in energy, biotechnology ... regulatory agencies can affect the international competitiveness of the industries they oversee ... skillful support of new technologies can help businesses reduce costs while complying with ambitious environmental regulations".

After the seemingly endless debates of the 80s, the Clinton Administration hopes to use the 90s to develop a regime for commercializing biotechnology products, a regime that is reassuring to the public, predictable and expeditious for industry and free of unnecessary political hurdles (SIMON, 1993). Strategic regulations can assist in achieving these goals by (1) having identifiable science-based triggers that are consistent, easily understood, and transparent; (2) being effective and responsive as well as flexible and dynamic; and (3) meeting domestic and international needs.

Acknowledgements

To Dr. VAL GIDDINGS for analysis of data on field tests, and Mr. CLAYTON GIVENS for preparation of charts.

Disclaimer

TERRY L. MEDLEY, J.D., and SALLY L. MCCAMMON, PH.D., are Acting Associate Administrator and Science Advisor, respectively, APHIS, USDA. The views expressed in this chapter are those of the authors and do not necessarily represent those of the United States Government.

7 References

BELOW, P. J., MORRISEY, G. L., ACOMB, B. C. (1987), *The Executive Guide to Strategic Planning*, San Francisco: Jossey-Bass Publisher.

BROWNING, G., in person, GREG SIMON (1993), *The White House's Window on Biotechnology National Journal*, July 24, 1993.

COOK, R. J. (1993), *US Policy on Biotechnology: Implication for Public-Supported Research*, Department of Agricultural Communications, Texas A & M University, College Station, Texas, June 1993.

DE ZOETEN, G. A. (1991), Risk assessment: do we let history repeat itself? *Phytopathology* **81**, 585–586.

DOE/ACRE (1994) Guidance Note 2. *Fast Track Procedures for Certain GMO Releases*, Department of Environment in association with the Scottish Office, the Welsh Office, the Ministry of Agriculture, Fisheries and Food, and the Health and Safety Executive, UK, 6 pp.

EEC (European Economic Communities) (1993),

93/584/EEC, Commission Decision of 22 October 1993 establishing the criteria for simplified procedures concerning the deliberate release into the environment for genetically modified plants pursuant to Article 6(5) of *Council Directive 90/220/EEC, Off. J. Eur. Commun.* **L279**, 42–43.

EPA (Environmental Protection Agency) (1992), *FIFRA Scientific Advisory Panel – Subpanel on Plant Pesticides, Fed. Reg.* **57**, 55531.

ERNST & YOUNG, San Francisco (1992), *Biotech '93: Accelerating Commercialization*, September 1992; and *Biotech '92: Promise to Reality*, September 1991.

FDA (Food and Drug Administration) (1992), *Statement of Policy: Foods Derived from New Plant Varieties*; notice, *Fed. Reg.* **57**, 22984–23005.

HULL R. (1990), The use and misuse of viruses in cloning and expression in plants, *NATO ASI Ser.* **41**, 443.

Keystone, Colorado. (1990), *Workshop on Safeguards for Planned Introduction of Transgenic Corn and Wheat*, Conference Report.

MCCAMMON, S. L., DWYER, S. G. (Eds.) (1990), *Workshop on Safeguards for the Planned Introduction of Oilseed Crucifers*, Proceedings, Cornell University, Ithaca, NY, 33 pp.

MCCAMMON, S. L., MEDLEY T. L. (1990), Certification for the planned introduction of transgenic plants into the environment, in: *The Molecular and Cellular Biology of the Potato* (VAYDA, M. E., PARK W. D., Eds.), pp. 233–250, Wallingford, UK: CAB International.

MCCAMMON, S. L., MEDLEY, T. L. (1992), Commercialization review process for plant varieties developed through biotechnology, in: *The Biosafety Results of Field Tests of Genetically Modified Plants and Microorganisms* (CASPER, R., LANDSMANN J., Eds.), pp. 192–296, Braunschweig: Biologische Bundesanstalt für Land- und Forstwirtschaft.

MEDLEY, T. L. (1990), APHIS seeks to aid tech transfer while ensuring environmental safety, *Genet. Eng. News* **V10**, N8.

MEDLEY, T.L. (1991), Domestic and international regulatory responsibilities of USDA – global harmonization, Abstracts of Papers of the American Association for the Advancement of Science, *AAAS Publication* **91-02S;** Section 24-2, P56.

MEDLEY, T. L. (1993), USDA's oversight approach for plants produced through biotechnology, in: *Bio Europe '93 International Conference and Proceedings*, pp. 117–129, Senior Advisory Group on Biotechnology, Belgium.

NIH (National Institutes of Health) (1976), Department of Health, Education, and Welfare, *Recombinant DNA Research Guidelines; Fed. Reg.* **41,** 77.

NRC (National Research Council) (1987), *Agricultural Biotechnology: Strategies for National Competitiveness,* Washington, DC: National Academy Press, 205 pp.

NRC (National Research Council) (1989), *Field Testing Genetically Modified Organisms: Framework for Decisions,* Washington, DC: National Academy Press, 170 pp.

OECD (Organization for Economic Cooperation and Development) (1986), *Recombinant DNA Safety Considerations,* 69 pp., Paris: OECD Publications Service.

OECD (Organization for Economic Cooperation and Development) (1992), *Good Developmental Principles (GDP): Guidance for the Design of Small-Scale-Field Research with Genetically Modified Micro-organisms,* Paris: OECD Publications Service.

OECD (Organization for Economic Cooperation and Development) (1993a), *Safety Considerations for Biotechnology: Scale-up of Crop Plants,* 40 pp., Paris: OECD Publications Service.

OECD (Organization for Economic Cooperation and Development) (1993b), *Traditional Crop Breeding Practices: A Historical Review to Serve as a Baseline for Assessing the Role of Modern Biotechnology,* 235 pp., Paris: OECD Publications Service.

OLSON, S. (1986), *Biotechnology: An Industry Comes of Age,* National Academy of Sciences, Washington, DC: National Academy Press.

OSTP (Office of Science and Technology Policy) (1984), *Proposal for a Coordinated Framework for the Regulation of Biotechnology, Fed. Reg.* **49,** 50856–50907.

OSTP (Office of Science and Technology Policy) (1986), *Coordinated Framework for Regulation of Biotechnology: Announcement of Policy and Notice for Public Comment, Fed. Reg.* **51,** 23302–23393.

OSTP (Office of Science and Technology Policy) (1990), *Principles for Federal Oversight of Biotechnology: Planned Introduction into the Environment of Organisms with Modified Hereditary Traits, Fed. Reg.* **55,** 31118–31121.

OSTP (Office of Science and Technology Policy) (1992), *Exercise of Federal Oversight within Scope of Statutory Authority: Planned Introductions of Biotechnology Products into the Environment. Fed. Reg.* **57,** 6753–6762.

OTA (US Congress, Office of Technology Assessment) (1992), *A New Technological Era for American Agriculture, OTA-F-474,* Washington, DC: US Government Printing Office.

PAYNE, J. H., SCHOEDEL, C., KEEN, N. T., COLLMER, A. (1987), Multiplication and virulence in plant tissues of *Escherichia coli* clones producing pectate lyase isozymes PLb and PLe at high levels and of an *Erwinia chrysanthemi* mutant deficient in PLe, *J. Appl. Environ. Microbiol.* **53,** 2315–2320.

SCHECHTMAN, M. (Chief Preparer) (1994), *Response to Calgene Petition P93-196-01 for Determination of Non-regulated Status BXN Cotton,* USDA/APHIS, Hyattsville, MD, February 15, 1994.

SHAW, J. J., BEAUCHAMP, C., DANE, F., KRIEL, R. J. (1992), Securing a Permit from the United States Department of Agriculture for Field Work with Genetically Engineered Microbes: A Non-Prohibitory Process, *Microbial Releases,* **1,** 51–53.

SIMON, G. (1993), Technology, trade and economic Growth, in: *Bio Europe '93,* International Conference and Proceedings, pp. 17–20, Senior Advisory Group on Biotechnology, Belgium.

TIEDJE, J. M., COLWELL, R. K., GROSSMAN, Y. L., HODSON, R., LENSKI, R. E., MACK, N., REGAL, P. J. (1989), The planned introduction of genetically engineered organisms: ecological considerations and recommendations, *Ecology* **70,** 298–315.

WHEELER, H. (1975), *Plant Pathogenesis,* New York–Heidelberg–Berlin: Springer-Verlag, 106 pp.

USDA/APHIS (US Department of Agriculture/Animal and Plant Health Inspection Service (1987), *Introduction of Organisms and Products Altered or Produced through Genetic Engineering which are Plant Pests or which there is Reason to Believe are Plant Pests, Fed. Reg.* **52,** 22892–22915.

USDA/APHIS (US Department of Agriculture/Animal and Plant Health Inspection Service (1991), *User's Guide for Introducing Genetically Engineered Plants and Microorganisms, Tech. Bull.* **1783.**

USDA/APHIS (US Department of Agriculture/Animal and Plant Health Inspection Service) (1992), *Genetically Engineered Organisms and Products: Notification Procedures for the Introduction of Certain Regulated Articles; and Petition for Nonregulated Status: Proposed Rule, Fed. Reg.* **57,** 53036–53053.

USDA/APHIS (US Department of Agriculture/Animal and Plant Health Inspection Service) (1993), *Genetically Engineered Organisms and Products: Notification Procedures for the Introduction of Certain Regulated Articles; and Petition for Nonregulated Status: Final Rule, Fed. Reg.* **58,** 17044–17059.

7 Biomedicinical Product Development

JENS-PETER GREGERSEN

Marburg, Federal Republic of Germany

1 General and Commercial Aspects of Biomedicinal Product Development

1.1 Product Development Follows Different Rules than Research

Product development for biological pharmaceuticals might appear as a simple continuation of a research project after a sufficiently effective compound has been identified. From a scientific point of view there is no clear distinction between research and development. Until the clinical phase, product development utilizes the same basic skills and methods and – to a great extent – the experience of the same people who did the research. However, the rules change considerably as soon as innovative research turns into conservative testing of quality, safety, efficacy and into analytical and process development.

The decision to develop a pharmaceutical product transfers an active principle or compound from the resource-limited research laboratory into a new environment, the stage of development. In a multi-disciplinary approach the compound is subject to a variety of analytical, pharmacological, toxicological, and clinical tests and trials. From now on methods, results, documentation and the materials created are under strict scrutiny exerted by the developing organization itself and finally by several authorities. Research activities are hardly ever subject to such surveillance and control concerning, for example, the precision and consistency of methods and results. It usually takes time for a researcher to fully acknowlewdge and accept this fact.

Since there is no clear distinction between the research and development phase, projects frequently move into development while the people involved are unaware of the changed rules. This may result in neglecting essential aspects, in extra cost for the repetition of critical studies and in a considerable loss of time.

In larger organizations research and product development are usually separated in different units because of the fundamental differences between creative research and systematic, goal-oriented development and because development requires special knowledge and experience. In the field of biological pharmaceuticals, however, most companies and organizations are relatively small, since they traditionally operate on a national scale. They do not have several products under development which are subject to a similar development scheme. The methods are more complex, and the materials created in the research phase are directly passed on for development. Furthermore, biotechnological products are still in their infancy, and pharmaceuticals from biotechnological processes are not yet routine products. Thus, there are many reasons why researchers in these areas in industrial and non-industrial research organizations contribute to or perform product development. These scientists need to know more than just their science to understand their new task.

The general aspects of biomedicinal product development presented in this chapter are meant for researchers and scientists who are less experienced in product development. This chapter is also meant to prevent the often observed overly ambitious ideas about the commercialization of research results. Unrealistic expectations about the simplicity and duration of development may lead to inappropriate management decisions, which may even endanger the future of research-oriented enterprises. The success of biotechnology ventures does not only depend on good research but also on a realistic assessment of the potential products and on the ability to adopt the required skills to effectively develop research results into commercial products.

1.2 Commercial Chances and Risks of Pharmaceutical Development

Pharmaceutical development is expensive, and the risk that the project fails during this process is high. The following facts and figures may illustrate the costs and risks that may be expected. Although these figures do not differentiate between research and development and are not specific for biological pharmaceuticals (they represent mainly chemical entities for human drugs), the general picture is similar for all types of pharmaceuticals, and the implications for product development are essentially the same.

The average capitalized cost to develop a human pharmaceutical product in the USA amount to US $ 231 million (DiMASI et al., 1991). Comparable figures for 1982 are US $ 91 million (LANGLE and OCCELLI, 1983) and for the time between 1963 and 1975 US $ 54 million (HANSEN, 1979). On one hand, these considerable increases in cost, which tended to double within ten years, reflect the increasing difficulties to find new and commercially attractive drugs, since these figures include the cost of aborted projects. On the other hand, a substantial part of the rising cost is due to the increasing regulatory demands which must be met to achieve marketing approval.

The average time to develop a newly discovered compound increased continuously over the past decades from less than 4 years in 1960 to more than 10 years in 1989 (KARIA et al., 1992). As a consequence, the average effective patent protection for human and animal health products approved in the USA between 1984 and 1989 was only 10 years and 7 months, including the possibility of restoration of patent terms for medicinal products of up to 5 years (VANGELOS, 1991). Longer development times automatically reduce the effective patent protection period and thus significantly influence the revenue expectations.

Despite the sharp increase in development cost, the number of new products launched and their sales expectations remained fairly constant (LUMLEY and WALKER, 1992). It has been calculated that in the UK even the best selling 10% new drugs require longer than the effective life-span of their patents to recover the research and development cost; the majority of new pharmaceuticals cannot recoup these cost even after more than 20 years (PRENTIS et al., 1988).

Comparable figures with reasonable validity for animal health products are not available. Veterinary product development cost are definitely lower, but the revenues are also much lower. Due to rising registration demands, influenced by both the pharmaceutical and the agricultural sector, veterinary product development cost follow the same trend as human health products.

In general the animal health market has stagnated since 1980. The international competition has increased, and the more producers of biologicals and vaccines try to internationalize their products and sales in order to outbalance the price erosions by growth in sales volume. Whereas vaccine producers in the USA have been rather successful to penetrate European markets, European vaccines were prevented from access to the lucrative US market due to concerns about potential adventitious agents in these vaccines (e.g., FMD, BSE). At the same time European veterinary vaccines have to comply with increasing regulatory demands and the introduction of GMP (Good Manufacturing Practice) standards.

Relatively few really new and attractive products have been introduced in the animal health care market. The recently developed new and expensive human therapeutics do not play a major role in animal health care. Prophylactic vaccines against the major endemic diseases are already available and are difficult to improve for a competitive price. Several other interesting vaccine candidates are stuck in the research phase due to unsurpassable difficulties to achieve sufficient efficacy.

The continuing high interest and investment in pharmaceutical research cannot be attributed to guaranteed high revenues. Return on assets – a commonly used measure of industrial profitability – for the eight most successful health care companies in the USA in 1989 was approximately 16%, or 11% if research and development expenditures are included as gross assets for better comparability with other, less research-intensive industries (VANGELOS, 1991).

It seems that the general perception of high profits is strongly influenced by the commercial success of a few "blockbuster" products such as the H2-receptor antagonists, ivermectin or the recently launched erythropoeitin. These are the exceptions, whereas extremely hopeful candidates, rendering very moderate success, seem to be more the rule.

The pharmaceutical industry is much more research-oriented than other industries and has to cope with higher risks. Due to the high development cost and long development times, risks must be eliminated as early and consequently as possible. Risk elimination must already start in the research phase and before the expensive industrial development commences. High-risk research projects need much time, thus they must be kept small and have to focus only on the major risk factors. Research projects should be clearly separated – if not physically, at least in mind – from the more rigid and much more expensive development.

Once an attractive product has been identified, the time factor becomes extremely important; risk and cost are no longer the only relevant criteria. From now on speed and efficiency are of paramount importance. A minor planning error which results in a delay of a task on the critical path for only one month can delay the entire project by one month. Almost inevitably it also adds $\frac{1}{12}$ of the annual budget to the development cost. Thus, in a project with an average annual budget of $ 2.5 million, a delay like this can cost more than $ 200 000. Apart from these direct cost, later access to the market can cause much higher losses.

2 Market Research

2.1 Creative Market Research and Other Valuable Background Information

According to a study *Management of New Products* by Booz, Allen and Hamilton conducted in 1968, on average about 70–90% of the money spent in research and development of new products is used for products which are failures. No industrial organization would commit itself to an applied research or development project without investigating the general and market situation to reduce this failure rate. There is no reason why publicly funded research projects should not do the same. Starting a project without such an assessment means being more concerned about doing the project right than doing the right project.

Researchers (in academic institutes as well as in industry) tend to view market research in a very narrow sense. They do not perceive the potential of the approach, thus missing innovative ideas. Market research may, in fact, bring research ideas back to the earth and help avoid major mistakes. If used with a wider perspective and by a creative mind, it can help identify interesting options. Market research should add information and use it for a better assessment of chances, risks and options. It should not be used to reduce the project idea to a few simple figures. The usefulness of market research depends a good deal on intensive interaction between research and marketing and information exchange in both directions.

A basic assessment of important market and other environmental factors does not neccessarily require marketing experts. But if their advice is sought, it should be by an interactive process, not by presenting the idea for a verdict. During this process the marketing expert will probably identify the weak spots of the researcher's proposal. With reasonable background information the researcher (who else?) should also be able to recognize and understand the subjective assumptions (i.e., weaknesses) of the marketing expert's assessment and to challenge them where adequate.

The problem is that the additional information usually requires changes of the direction or the priorities of the project. Unless there are very convincing arguments, a researcher who is convinced of an idea is unlikely to consider such changes. The most convincing arguments are those which are developed from one's own considerations. Thus, researchers should do their own market assessment to develop an opinion about the chances, risks and best applications of their approach. (If they don't like the term market assessment, they may call it "SPA" for "strategic project assessment" if this sounds more reputable.) One needs no special skills to identify the major threats to and opportunities for a project. These are usually rather obvious, if some basic information is available and is considered critically.

At the beginning it must be kept in mind that it is very easy to define an "ideal product". But this ideal product is probably the most difficult one to develop or may be an unrealistic development target. Sooner or later it pays off to have identified all the options, including the lesser ones with a higher probability of success. Of course, the optimal product may be chosen as the aim of the project, but minor options should be included in the planning as a fall back position.

The more detailed the collected information is, the better it serves its purpose. Acquitting oneself of the task by only quoting a figure about the (ill-defined) "market potential" of a product (usually followed by a sum of some hundred millions or even some billions) obscures the matter rather than illuminating it. In many cases this figure is so meaningless that it would not result in any consequences for the project, if somebody simply changes the currency unit from British Pounds to Italian Lira.

Creative market research should initially collect a few basic figures on the occurrence of the disease or condition to be treated. Scientific review articles and epidemiological investigations as well as disease statistics may serve to provide these. As a next step information should be collected about current methods to diagnose and treat the disease or condition. This includes the products or compounds used, the frequency of their application, cost of the products used and the overall cost of the treatment. Most of this information can be obtained by a specific literature search in the medicinal field or from physicians and clinicians. Clinicians should also be asked for deficiencies in the current products or treatment schemes, such as side-reactions, lack of efficacy (in general or under certain conditions) and for other medicinal treatments (e.g., prevention or surgical techniques) which have an impact on pharmaceutical products in this area.

This background information should be used to identify the needs of the "market", i.e., of patients and physicians, of veterinarians and their clients or of diagnostic laboratories. It also serves the purpose of checking whether the project in mind has the required potential to improve the existing situation.

A simple assessment of the likely cost of the new or improved product, compared to the benefits that it provides, can be very illuminating, even if the figures are only rough estimates. Acceptable costs for the new product may be deduced from prices of comparable product classes or from specific products which are currently used to treat the disease in question. These prices are likely to give a reasonable estimate. Depending upon the estimated value of additional benefits of the new product, a premium can be added to the basic figure. This sort of simple calculation and estimate is particularly warranted for veterinary products and for products which will compete with already existing and established products. If the existing products or prophylactic treatment schemes cost only a few dollars, the active ingredient is only worth a few cents. To get a rough idea about the upper limit for the cost of the active component in these cases, the market price of the existing product may be split up by the following "rule of thumb calculation": about 40–50% for cost of sales and a profit margin (before tax), 20–30% for quality control, packaging and labeling and 10% for formulation. For lower priced products in a competitive environment the active ingredient may account for only about 10–30% of the market price.

If it is believed that the proposed new product can compete in terms of cost of manufacturing despite a more demanding process, or that the product offers significant advantages which justify a higher price, it will be necessary to present evidence for a high-yield system or to focus on a proof of the claimed advantages.

Live and vector vaccine approaches should take into account the epidemiological situation and health care schemes in the target countries. The efficacy of live vaccines and vectors may be affected by a pre-existing immunity in the population. Live vaccines can also interfere with screening tests, e.g., for salmonellosis in people who handle food or for bovine leukosis in eradication schemes.

Diagnostic tests may be rather useless, if the test result does not lead to any consequences, for example, when no effective treatment is available or because the test procedure takes too long to influence the choice of therapeutic options. In the latter case a much faster and simpler "bed-side test" (or cow-side test for veterinarians) may be the better solution.

The use of medicinal products and therapies in man is more driven by ethical and less by economic considerations. Direct commercial aspects such as cost of the product may be less critical. Instead, the assessment should concentrate more on safety and quality aspects and their scientific and economic ramifications. Which safety standards must be met by the product? Does it contain components with an unknown safety profile or a poor tolerance at the application site? What are the chances of replacing these and getting access to better components? Are adequate safety tests for attenuated or genetically modified microorganisms available or must/can these be established and will these tests give sufficient confidence in the safety of the product to justify later trials in human beings? Who would be interested and could contribute to address the safety issues, e.g., by characterization of the strain of microorganism used, by developing adequate models and tests? How early should these activities start or how long should the project proceed without addressing these very critical issues?

Not only live vaccines, but also therapeutic polypeptides and subunit vaccines have to meet very demanding safety requirements, before they can be tested in humans. As soon as a protein with potential as a vaccine component or a therapeutic drug has been identified, the questions arise of which systems and methods should be used to produce the active component, how to separate it from potentially harmful components and how to purify it to the required degree. Assuming that a purity of at least 95% and only a very limited number of known and defined impurities will be allowed for a human medicinal product, the expression and purification systems and yields warrant some careful considerations at

a very early stage. The wrong choice or taking what is readily available, instead of using the best option, can result in a loss of time, impractical patent applications and a loss of the competitive edge of one's research.

A list of aspects to be examined as part of a creative market research and far-sighted project planning is given in Tab. 1. Depending on the maturity of the project, it may not be necessary or possible to cover all aspects. However, the attitude should prevail that it is better to gather as much information as possible and then decide whether a specific piece of information is useful or not. Information never comes too early, but frequently too late!

Based upon this background information a thorough market assessment for a product to be developed identifies the potential strength and weaknesses of a new product as well as those of competitive programs or therapeutic schemes. This assessment should include the patent situation and the competitiveness of the organization in terms of the skills and resources to develop, manufacture and sell the product when assessing the chances and risks of the intended development product. From this information one or a series of potential products can be identified, defined by their key characteristics (their product profile) and can be compared.

2.2 The Product Profile

A product profile should be established for each development product or candidate. The product profile describes essential product qualities such as the indication, form of presentation, intended application and use pattern, efficacy, safety, and other critical aspects, e.g., patent restrictions and cost limitations for the compound or for the formulated product. The expected date of product launch and sales expectations as well as the expected development cost are important elements of the product profile. Tab. 2 summarizes and explains the aspects which should be covered by a product profile.

Based on the product profile a cost–benefit analysis can be calculated. Such an analysis helps in deciding whether, under the given circumstances, a potential product is suffi-

Tab. 1. Points to Consider in Strategic and Market-Oriented Planning of Research and Development for Biological Medicinal Products

---

General Aspects
Occurrence of the disease or condition: geographic distribution and frequency
Current diagnostic and therapeutic measures
Deficiencies of current treatment: efficacy, side-effects, cost, practicability
Potential improvements of existing methods
Acceptance of potential improvements: safety, cost, practicability
Interaction with health care and eradication schemes

Efficacy Aspects
Potential indications, scientific feasibility and risks
Degree of efficacy required
Systems and models to demonstrate efficacy

Safety Aspects
Potentially harmful effects of the proposed product
Safety requirements: similar or better than existing products
Systems and models to assess safety

Commercial Aspects
Product prices and cost of current treatment
Frequency of application of existing products
Likely cost of the proposed new product and treatment
Cost/benefit relationship

Patent Aspects
Existing patents and applications in the field
Gaps and chances for an own patent application
Availability and costs of patent support
Cost/benefit of patenting

Project-Related Aspects
Other research and development projects in this area: scientific competitiveness of the own project
Specific skills and methods required
Collaborators for specific problems ot to extend the scope of the project
Who would be interested in the project/product and provide practical or financial support?
Funding organizations to be approached: interest and user groups, government and special research funds, supranational funding organizations, industry

---

ciently attractive to develop and which indication for the new compound is more or less attractive. It is mainly a means to prioritize and to compare the particular development product with other products under development or under consideration. This may be necessary since several development candidates may compete for the same resources such as personnel, development budget, capital investment, and production capacity.

It should be noted that cost–benefit calculations for the same development product may vary considerably between different or-

ganizations (HODDER and RIGGS, 1991). First, the calculation methods may be different. Second, the assumptions that go into these calculations, e.g., the expected sales of a certain product, may vary considerably. These figures depend on company-specific factors such as available sales force and presence on relevant markets, as well as on certain personal opinions and other subjective judgements. Apart from this, it is obvious that already existing manufacturing facilities without the need for a major investment can significantly influence the result of a cost–bene-

Tab. 2. Contents of a Development Product Profile[a]

I. Essential Product Characteristics

Indication	Condition or disease to be treated. Target species for veterinary products
Presentation	Type of product and formulation, e.g., live or inactivated vaccine, vector vaccine, type of vector, liquid or lyophilized product, adjuvant, special formulations
Application/use pattern	Mode and frequency of application, e.g., route of application, single or repeated use, duration of use, booster intervals
Efficacy	Specified or in comparison to other products or treatment schemes
Safety/tolerability	Specific demands, e.g., standards set by other products or treatments

II. Commercial Aspects

Cost of development	Per year until product launch, updated during development
Cost of product	Active ingredient, formulation and packaging, cost limitations
Investment costs	Manufacturing plant and equipment
Sales expectations	Differentiated by medicinal indications, by major countries
Expected date of launch	Month and year for major countries
Patent protection	Duration in major countries, quality of patent protection
Major product advantages	Major selling arguments, strategic importance
Major development risks	Reasons for failure, e.g., efficacy, safety, regulatory risks, competing products or developments
Cost–benefit analysis	With annotations on critical limitations and including the important assumptions upon which the analysis is based

[a] The product profile should address all essential aspects of the product and the development. It describes the agreed development task and assists in management decisions.

fit analysis. Thus, a cost–benefit analysis is a scenario and the assumptions that formed the basis of the described situation should be specified and carefully taken into account, before general conclusions are drawn.

Besides the aspect of providing information to assist in decision making, a precise product profile has several other benefits. It creates a common vision about the direction of the development by defining the aim unmistakably to all people involved, thus assisting in tight project planning without the need for endless discussions about the goals. (About the importance of a common vision for projects see also HARDAKER and WARD, 1991.) Critical aspects of the project are exposed at the beginning, and measures to solve these problems can be included in the development plan at the best, most convenient and cost-effective time. Defining the product profile helps to recognize changes in critical parameters (e.g., in efficacy) and deviation from the decided development route earlier, so that counter-measures can be initiated before these become too costly. The later changes are initiated, the more expensive they are.

Product profiles frequently tend to describe an ideal or almost ideal product because this is much easier and less controversial to define. In order to become a meaningful base for planning and decision making, the essential product characteristics in a product profile should be described as minimum criteria which define what must be achieved and where the product must not fall short. Only minimum criteria are useful to define potential milestones for the development plan which in turn are used as an objective guide for decisions to continue a project or to abandon it, if the milestones cannot be achieved. Desirable goals for further improvements can be added where appropriate.

Of course, a product profile is not a one-sided instruction. It requires input from research, preclinical and clinical development, marketing and sales departments and is developed by an iterative, stepwise procedure.

3 Registration Requirements

Registration regulations for medicinal products describe the quality, safety, efficacy and formal requirements for market approval. During the last few decades pharmaceutical regulations in the western world have become so complex and specific that the task of observing them cannot simply be delegated to registration specialists. Registration specialists should be consulted already during the planning phase of a development project and should be able to give advice which laws, directives and guidelines apply in a specific case and to what degree guidelines must be adhered to. Everybody involved in product development who develops analytical methods and processes or contributes data on the quality, safety and efficacy of the product candidate needs to know the specific directives and guidelines concerning her or his field of expertise. For effective development, one has to consider the details of registration requirements already during the planning of all major tasks. Otherwise months and years can be lost because trials and experiments have to be added or repeated, after the development has already been finished and the registration dossiers were handed in.

The registration of medicinal products is governed by relative values which cannot be clearly defined or limited, for example, by ethical considerations and the endeavor to ensure the best possible quality and safety of pharmaceuticals. As a consequence, one cannot expect that registration guidelines give definite specifications for all major product characteristics and define exactly what must be done or – which is probably more important – needs not to be done. Registration requirements are flexible and change with time.

What is possible and achievable soon becomes a standard by which subsequent products will be measured. This leads to continuous and desirable improvements and requires continuous attention of those who develop and register products. Since there is no limit, very costly and undesirable extremes may be another consequence. According to today's standards, the most successful and beneficial pharmaceutical products, such as the vaccinia virus vaccine, live poliovirus vaccine and other live vaccines, would probably not be registrable for a wide-spread use in humans.

The following section gives only a basic understanding of regulatory requirements by pointing out the common and essential elements which are applicable to most products. For a more profound but reasonably short overview with special emphasis on the technical rather than the formal requirements and for information about the specific guidelines covering biomedicines, the interested reader is referred to GREGERSEN (1994).

3.1 Three Basic Elements

Market approval or registration of a pharmaceutical product depends on the fulfillment of three main requirements:

- Quality
- Safety
- Efficacy

This sounds fairly simple, but these three basic terms must be interpreted in a very wide sense; every single aspect within these categories must be substantiated and proven under practical conditions. The scope and definitions change with time and with scientific, pharmaceutical, and methodological progress. The definitions for quality, safety and efficacy vary between product groups, countries, authorities and between people within authorities. In fact, there are no exact definitions at all.

Quality – as far as it can be judged independently of safety and efficacy – is in principle assessed by a comparison with the state of the art, i.e., with similar products. Products with inferior quality than other products may need improvements before they can be registered; products which surpass current standards will raise the general standards. Safety and efficacy, too, are no absolute values which can be easily defined. Market approval depends on a reasonable relationship between safety and efficacy, between risks and benefits.

Quality, safety and efficacy are no independent criteria. The quality can significantly affect the safety and efficacy. A highly efficacious and beneficial product may be acceptable, even if it has serious side-effects. Keeping these interdependencies in mind, product development must address and emphasize the three basic registration requirements in the same order as listed above, starting with the product quality, although project risks suggest a different strategy which concentrates more on efficacy. Without a defined quality, all safety and efficacy tests may be irrelevant, and without an adequate safety, efficacy trials under natural conditions cannot be carried out.

3.1.1 Quality

Quality does not only comprise the final product itself and its attributes that can be tested. The high variability, in particular of biological products, demands that all variables of the manufacturing process are evaluated. Product quality commences with the starting materials and their attributes, runs through the entire manufacturing process and extends to the manufacturing facility and its infrastructure. Everything that can affect the product quality must be assessed, tested and documented. For example, if one intends to harvest a product continuously from a culture, it must be specified what "continuously" means. If it is intended to harvest a hundred times over a hundred days, one has to prove that the product is still the same after the hundredth harvest – and probably even beyond that limit – to add some safety margin.

Of course, it is necessary to validate all in-process control and product quality control tests, i.e., to substantiate that the tests really measure what they are supposed to do and within which limits.

All claims on product safety and efficacy depend on the specified quality; if the quality is not consistent, safety and efficacy may not be the same. Safety and efficacy data for the product are only valid if they were established with material of the specified quality. The practical implications are that test material must meet the relevant specifications of the final product before safety and efficacy data can be generated. For biological products, this means that the test substances have to be made by a process which must not deviate from the final manufacturing process in essential elements. If the manufacturing process is changed during the development, all previous data with earlier material may be worthless, unless bioequivalence can be proven.

3.1.2 Safety

Safety of a product comprises any aspect of its production, use and disposal. This includes the manufacturing premises and their environments, starting materials (especially cells and microorganisms), the active component of the product as well as metabolites, impurities and the excipients. Excipients are components which are added to the product, for example, preservatives, vaccine adjuvants, stabilizers, emulsifiers and release controlling substances.

Safety of the manufacturing process and the facilities is covered by a variety of regulations concerning the layout of facilities, work organization, procedures and related controls and inspections. General national regulations on work safety and biohazard control measures, specific regulations for manufacturing pharmaceutical products and quasi-legal GMP requirements have to be applied. These are an inherent part of the registration dossier and market approval to ensure consistent quality and safety of the product and its manufacturing.

Safety of the product itself for the target organism, the user (who applies it) or the environment is addressed by a range of preclinical and clinical assessments which depend on the product and its use pattern. The range of safety features to be assessed includes local and systemic tolerance, acute and chronic toxicity, immunotoxicity and, for veterinary medicinal products, also the ecotoxicity. Tests for mutagenicity, tumorigenicity and reproductive toxicity may also be required for certain products or indications.

For safety assessment of a pharmaceutical product in the target organism it is necessary to know the action and fate of the product and its components within the body, after it has been administered. This knowledge derives from pharmacodynamic and pharmacokinetic studies, the results of which will influence the choice and setup of safety tests to be conducted. Pharmacodynamic studies investigate the influence of pharmaceutical substances on the human or animal organism, such as their mode of action and mechanism of side effects along with dose-effect relationships. Pharmacokinetic studies address the influence of the organism on absorption, distribution, metabolism, and elimination (*ADME*) of pharmaceutical products in adequate models and in the target species. Pharmacokinetic studies for biomedicines (especially metabolism and elimination studies) are usually less intensive than for chemical drugs.

For biological products the (chemical) excipients are often overlooked. They, too, may have effects or side-effects and can leave residues in animals that end up in human food. Their choice for a development product must be considered carefully. Either one uses ingredients with a well established quality and safety profile or a variety of analytical and safety data have to be established for these excipients. Safety studies for new excipients may cost more than the safety assessment of the biological product itself.

It is essential to demonstrate that pharmaceuticals for food-producing animals do not have harmful effects on humans, particularly if residues of the active component or of excipients and of their metabolites may occur in meat, milk or eggs. This means that extensive studies regarding the safety for humans and residue depletion studies must be performed with these products.

As a general rule, products that are used only once or a few times have to undergo fewer safety tests than products which will administered regularly or over a longer time. Less – if any – side-reactions are tolerated for products intended to be used in healthy individuals (e.g., vaccines) than for products which are used in diseased individuals. In any case the benefits must clearly outreach any potential side-effects.

3.1.3 Efficacy

A pharmaceutical product must fulfill its purpose under all recommended conditions and it must do so reproducibly. Experience shows that efficacy in controlled experiments does not guarantee that the product does the same and to the same degree under practical conditions. Pharmaceuticals for humans must be tested in human beings for their pharmacodynamic and pharmacokinetic characteristics to prove their effectiveness and safety. These clinical trials are normally carried out in three stages, each phase involving an increased number of patients (see Tab. 4 below). Clinical trials in humans are usually the most critical, most expensive, and most time-consuming developmental step and can account for more than one half of the development costs.

Concerning the proof of efficacy for animal health products, there is often the misconception that controlled experiments in the target species (pen trials or a challenge experiment under controlled conditions) which show that the product is efficacious, are sufficient to register a product. Animal health products also have to undergo clinical or field trials to prove their efficacy and safety under practical conditions. Quite often, these trials reveal several hitherto unrecognized weaknesses of the product. However, clinical trials for veterinary medicinal products are much less critical, less expensive, and usually less time-consuming than human clinical trials. Due to the preclinical pharmacological, efficacy and safety tests performed in the target species, the clinical trials are mainly designed to confirm preclinical efficacy data and to provide a broader basis for the safety evaluation.

Clinical trials are an important base of efficacy claims for the product. As a logical consequence, all indications for which the product is recommended, all proposed routes of application, treatment schemes, doses, relevant age groups and, in the case of veterinary products, all recommended species must be tested. For an effective development it is often better to initially concentrate on only one or a few of the potential indications to reduce the risk and to get the product on the market earlier, than to follow up all variants at the same time.

4 Technical Aspects of Product Development

Medicinal products must have a consistent quality in order to be reliably effective and safe. Achieving consistency in all aspects is a cumbersome technical and analytical task which takes much time and which is usually not the main virtue of research work. Scientists without experience in product development tend to neglect the technical aspects which lead to a consistent product quality.

Increasing registration requirements for the product and the production process as well as environmental issues also add a significant technical and analytical dimension to the development. This particularly applies to products derived from biotechnology. Compared to conventional biological products, the final product is usually much better defined and its manufacturing process can be better controlled. But it requires the application of a variety of quite sophisticated techniques to do so.

For most new biotechnological products, processes and analytical methods have to be newly established. This represents a major and costly part of product development. In newly founded biotechnology firms pertinent technical experience and practical knowledge about the application of regulatory requirements is usually missing and must be developed. Early information and consideration of these aspects may considerably shorten this learning process.

4.1 Process Development and Manufacturing

The active ingredient of a potential product, as it was identified in research, is initially produced in minute amounts and by a process which in most cases is not acceptable for a pharmaceutical product or for large-scale production. Process development represents a link between research and manufacturing and adapts the research methods to the needs of production or develops new methods where necessary. Process developers must take multiple aspects into consideration (Tab. 3). Their task would be much easier, if researchers were aware of these aspects.

In the worst case process development can imply that an entirely new procedure to obtain the active ingredient must be found and established. Frequently master seed stocks (cells, viruses, bacteria) must be newly generated and tested, since those from research are not suitable because of their quality (contaminants, inhomogeneity) or unclear history and documentation.

Unacceptable starting materials used in research must be replaced by those which meet the regulatory quality and safety criteria. Reference to standard volumes of pharmaceutical ingredients and excipients should be made to decide whether a certain ingredient can be used. Standards are described in the pharmacopoeias of relevant countries, in *The Extra Pharmacopoeia* (MARTINDALE, 1993) and in the *Handbook of Pharmaceutical Excipients* (AMERICAN PHARMACEUTICAL ASSOCIATION/THE PHARMACEUTICAL SOCIETY OF GREAT BRITAIN, 1986). The German *Lexikon der Hilfsstoffe* (FIEDLER, 1989) also provides many useful informations about acceptable excipients and about their applications and safety. Components used and approved for human food may also be considered as potential ingredients for certain applications. Hazardous components cannot be used at all or have to be removed during the process. Special conditions apply to the use of starting material of human or animal origin to avoid the presence of adventitious agents in the final product.

Tab. 3. Process Development for Biological Pharmaceuticals: Major Points to Consider in Process Development

---

I. Starting Materials and Excipients

Availability: appropriate quantities, regular supply
Quality: specifications and consistency (compliance with pharmacopoeia specifications if these exist)
Safety acceptable for a pharmaceutical product and freedom of adventitious agents, e.g., human serum components, bovine serum!
Masters seed cultures: quality, documentation, suitability

II. Safety and Environmental Aspects

Work safety, e.g., organic solvents, aerosols containing microorganisms
Contamination from the environment
Contamination of the environment
Cross-contamination and separation of activities
Waste material decontamination and disposal

III. Legal and Regulatory Aspects

Registration requirements for product and process
Licencing of facilities
Good Manufacturing Practice (GMP)
Infringement of existing patents?

IV. Economical Aspects

Cost of goods
Yields
Recovery rates
Investment into equipment/facility

Process developers should check existing patents and patent applications in order to avoid patent infringement with any use or process patent. On the other hand, there may be possibilities to patent the particular process newly developed or critical elements of it. Economic aspects such as yields from cell culture or fermentation, product recovery after purification and cost of materials and equipment need constant attention. Production cost forecasts should be calculated and updated regularly.

A variety of product safety, work safety, environmental safety aspects and other legal aspects must be taken into consideration when a process is designed. Good Manufacturing Practice standards for facilities, equipment, and working procedures must be met. GMP approval is subject to inspection and requires constant attention and updating. Extensive documentation must be provided to comply with these requirements.

Manufacturing facilities are approved for their specific purpose, e.g., for the manufacturing of one particular product. If it is intended to manufacture another product in the same facilities, this must also be approved and requires very strict measures to avoid any cross-contamination and mix-ups. Either the facilities have to be separated, or production runs for different products have to be performed at different times with intensive cleaning and decontamination in between.

The preferred option is of course to develop a manufacturing process for an already existing manufacturing plant. If adequate and approved facilities are not available, the possibility should be considered to develop the product in cooperation with somebody who has free capacities in a suitable plant. (Many conventional vaccine producers have more capacities than they need.) If this alternative is unacceptable, considerable investment and time will be required to establish manufacturing facilities. They have to be planned, built, equipped and approved first, and it will be necessary to gain sufficient experience with the new plant and the manufacturing process in these facilities.

The use of existing manufacturing facilities also overcomes another common limiting factor in product development: the availability of product for test purposes and clinical trials. All critical tests must be done with a product which is essentially the same as the final product, if the data from these tests are to be used for the registration application. Research material of dubious quality is definitely not appropriate.

It may take months or years to develop a controlled process which reproducibly gives a product with adequate quality. As a consequence, process development must start very early in the development phase. The product specifications should not simply evolve from the process; the main parameters should be fixed beforehand. This does not only provide clear objectives for process development, it also makes sure that no time is wasted by testing products of inadequate quality which may render the data obtained useless for registration purposes.

In some cases it may be unavoidable to change relevant product specifications. If major pharmacological and safety tests have already been done, it may not be necessary to repeat all of these. Comparative tests (bioequivalence studies) with the earlier and the new product should be performed to check whether both are equivalent in all critical aspects. In the case of higher standards, (e.g., for purity) this may be relatively simple, lowering these standards, however, will be difficult to justify and requires thorough studies about the nature and the effect of the additional impurities.

4.2 Analytical Development and Quality Assurance

Quality assurance schemes for medicinal products have evolved mainly from experience. Whereas initial quality controls were carried out only with the final product – with elimination of batches which did not pass the test – nowadays quality assurance covers the entire manufacturing process and continuously develops towards a more holistic system. Presently quality standards cover everything that goes into the product (starting materials, excipients, the active ingredient and its by-products), everything that could come into contact with the product (e.g., air, water to clean vessels, packaging materials) or has the potential to influence product quality without being noticed by the usual checks (e.g., personal qualification, management responsibilities, the validity of methods, documentation). The given examples may illustrate the general scope of modern quality assurance for pharmaceutical products.

Analytical procedures have to be developed for the active ingredient as well as for starting materials and the excipients used in the final formulation. These tests should be able to specify and confirm the identity, purity, potency, stability and consistency of these materials. If significant impurities, degradation products or critical metabolites occur, analytical methods for these will also be required.

In-process control methods must be devised and developed to observe all relevant steps of the manufacturing process in order to have adequate control over the inherent variants and to detect potentially harmful contaminants at a stage where they may be easier to trace.

The spectrum of tests to be employed comprises methods from microbiology, immunology, protein and peptide biochemistry, genetic engineering and physico-chemical methods as well as animal experimentation. If these exist, for example, in pharmacopoeias, standard method descriptions have to be followed.

Critical steps of the manufacturing process and analytical methods have to be validated carefully. Process validation will be required, for example, for the inactivation or elimination of potentially pathogenic microorganisms (e.g., viruses in transformed cell cultures) from the product. This can be done by running spiked samples through the process or through a smaller laboratory version of the process which exactly mimics all relevant parameters. Volumes, tests, and test sensitivity must be considered carefully and statistically for such experiments.

Analytical methods are validated by investigating specificity, sensitivity, detection limits, quantification limits, accuracy (of the individual result), precision (variation between different tests), applicability and practicability under laboratory conditions and the robustness (susceptibility to interference) of the method. This enables the organization itself to employ these tests much more consciously and with better results, but it also serves the purpose of enabling the registration authorities to use these tests for the regular batch control tests.

Official guidelines by the regulatory authorities for the validation of analytical methods and processes are available for consultation.

Good Manufacturing Practice (GMP) standards are a further step towards the concept of "total quality". Implementing and maintaining a GMP status requires commitment of the entire organization and special and constant attention by those who are in charge of quality assurance. Because of the complexity and wide scope of the subject and the amount of paperwork (some translate GMP as "Give Me Paper") extra personnel or external consultants will most likely be required.

GMP includes for example:

- The organization, management structure, personal qualifications and training
- Standard Operating Procedures (SOPs) with appropriate documentation and implementation systems ensuring their effective application
- Production, packaging, labeling, handling, testing and approval of starting materials, intermediate and final product
- Construction, infrastructure and maintenance of buildings and equipment
- Regular inspections and self-inspections and many more details.

Whereas GMP as well as GLP (Good Laboratory Practice) and GCP (Good Clinical Practice) are mainly acting on the operational or technical level, other quality assurance standards have been established by the International Organization of Standardization (ISO) to assure quality at the organizational level. Inadequate organizations with unclear tasks and responsibilities can severely influence the quality of products and services.

In the near future further quality assurance elements as recommended by the quality guidelines ISO 9000-9004 will certainly also be applied to pharmaceutical developers, manufacturers and to sales organizations. A formal accreditation of compliance with these standards requires the implementation of quality assurance schemes covering, for example, the following aspects of a business:

- Management responsibility and commitment
- Organizational structures with explicitly defined and delegated responsibilities
- Quality assurance systems covering all functions in procurement, production, control of production, handling, storage, product identification, packaging, delivery, marketing, after-sales servicing, product supervision and "design" (design = research and development)
- Identification of non-conformity and corrective actions
- Means to ensure product traceability, facilitating recall and planned investigation of products or services suspected of having unsafe features
- Use of statistical methods to enhance and maintain quality at all stages and of all activities in the product cycle.

It is to be expected that by the year 2000 or earlier compliance with ISO quality management standards will be obligatory for staying in business. The personnel cost of implementing and maintaining these quality assurance systems should not be underestimated.

5 Planning and Managing Product Development

The development of a modern pharmaceutical product is a complex task which is further complicated for biological or biotechnological products by the potential variabilities of the biological production system and the resulting consequences. Therefore, apart from considerations of adequate skills and resources, careful and very detailed planning and controlling is imperative for the success of development.

Those who are unexperienced with the tools of project planning and managing may refer to one of the numerous books about project managing. Although these books are usually aiming at more technical applications, the basic techniques do not differ and are of course also applicable to other areas. Alternatively one can buy a project management software package and learn the application of these tools directly by using it in the planning process.

Knowledge of the essential registration requirements is mandatory for the planning of a pharmaceutical development project. If these are neglected, the development process will be like navigation in unknown waters without a compass or a map. The risk of a shipwreck is high, in any case the journey will take much longer and will cost more.

Provided that a reasonable and useful product profile (see Tab. 2) has been established, the general aim of the project and the most relevant objectives have already been decided. The objectives which serve as orientation points in the project will most likely address the exact indication, efficacy, the type of product, its formulation and application scheme. Another important general objective in the product profile may be the expected price limit for the production of the active ingredient or the formulated product. Specific tasks and measurable success criteria for process development and manufacturing, e.g., in terms of yields and recovery after purification, can be deduced from this price limit. Other very specific objectives and the majority of project tasks relate to specific registration requirements.

It may take several months to establish a workable development plan, and most of this time will be required to collect the neccessary information. At the beginning the project plan will probably be a rather simple outline which then grows continuously while the project progresses. A project master plan which covers the major sections, milestones and decision points can be established far ahead and – if prepared thoroughly – will remain essentially unchanged from then on. More specific and detailed planning and updating is done in subordinated plans for the individual sections and must be performed continuously.

Development plans must be much more detailed than research plans. The simple reason for this is that the commercial environment in which development takes place requires a much stricter planning, use and control of time, budgets, and manpower.

Significant time savings can, for example, be achieved by preparing tasks and trials early and by careful and extensive evaluation of trials. Often time is lost, since in negligent project plans resources for the planning and evaluation phase of larger studies as well as certain aspects, which can be tested simultaneously and do not require separate experiments, are ignored. This applies especially to clinical trials which need considerable time to get started and where it is possible to collect data on efficacy, pharmacology, safety and tolerability aspects at the same time.

Many tasks have to be performed according to very detailed legal and quasi-legal standards which may differ from country to country. Relevant countries' guidelines must be checked and considered in planning the individual tasks. Studies performed according to GLP standards, animal trials and clinical studies can only be performed, after internal or external examinations of protocols or approval procedures have been passed. In this situation it is essential that all people involved start their activities with precise and detailed instruction and are given sufficient time for a careful preparation of their tasks. Good project plans include the preparation of relevant studies and trials as specific tasks in order to avoid delays.

Effective project controlling also needs detailed planning. Under normal circumstances the plans must be exact enough to allow progress to be monitored at intervals of 2–4 weeks. (Some industries and companies have much tighter schemes!) In critical situations or for activities on the critical path (if delayed, activities on the critical path delay the entire project), more frequent checks and corrections may be necessary. Where possible, progress should be monitored before a task is finished. Requesting information about the progress halfway through a job can also create a better awareness for time limits and target dates. If problems are likely, means to solve them can be planned and implemented before these problems cause major concern.

5.1 Risk-Oriented Planning

The most difficult and controversial part in planning and managing of a development process is to keep the right balance between the three cornerstones cost, risk, and time. Using more time reduces the risk but increases the cost and delays the product launch. Controller and marketing manager will raise their protest. Taking more risks increases the possibility of not meeting the milestones, and everybody is concerned. There is no way to do it right for everyone!

The willingness to take risks can be a key element for a rapid and cost-effective development program. Making assumptions about the likely outcome of a piece of work and commencing the next step which depends on a particular result before it is available, does increase the risk, but it also permits to shorten the development time considerably.

The most relevant risk to be considered during the development of a biological pharmaceutical is the efficacy and safety in the target organism. Despite extensive preclinical tests, many product candidates are abandoned after first tests in humans. The problems encountered at this stage are often due to pharmacological or immunological differences between the model used in the preclinical tests and humans. Thus, it is necessary to plan Phase I clinical trials as early as possible. Similarly most veterinary development projects fail due to a lack of efficacy and unexpected adverse effects which remain unnoticed until the product is tested in the target species. Veterinary products should be tested in the target species as soon as sufficient product is available. Tests under practical conditions should follow as soon as possible. Quite often major progress, but also unexpected negative results, come from experiments in humans (or primates) or in target animal species.

Besides the already emphasized importance of tests in the target organism, the importance of animal and *in vitro* models for pharmaceutical product development have to be pointed out. In most cases the available range of models and *in vitro* tests needs to be extended for the development phase, because the models used in research are not adequate nor sufficient. Considering the fact that all potential facets and variations of a new active component (different conformations, concentrations, formulations the mode of action, immunology, pharmacokinetic and safety and efficacy under various conditions) should be studied, the availability of models can be an important asset for a fast and cost-effective product development.

Models allow certain aspects to be investigated in more depth, but they reflect only a part of the entire situation. Therefore, they need to be validated by comparing their results with those in the target organism. Models should be considered as a valuable addition, not as a replacement for tests in the target organism.

The economic feasibility of a product can be another major risk which needs continuous attention during the development. This is especially true for animal health projects and for all the products which will compete with simple and effective alternative treatments, for example, with existing vaccines. Factors directly affecting the economic assessment are those listed in the product profile under "commercial aspects" (cf. Tab. 2). A project manager will have to control the development cost and pay special attention to details like yields and recovery rates during the process development. Critical limits for the manufacturing should be addressed by setting specific objectives.

One should not forget that the economic aspects of a project depend on the underlying assumptions about the product's characteristics. Less efficacy than assumed or a slightly compromised tolerability may seriously affect a product's performance and viability on the market.

5.2 Product Development Phases

As a starting point for the planning of a development project, Tabs. 4 and 5 propose and summarize a phased scheme of a risk-oriented development plan. Several important milestones and decision points have already been included. Each phase must fulfill certain criteria in order to be accomplished before the next phase may start. Management decides whether a project is advanced enough

to proceed to the next phase, since this decision is usually associated with the involvement of more people in different departments or of external participants and has significant financial implications. The data and information needed for an assessment of progress and criteria for the decision should be known beforehand. Most of these criteria, which are included in Tab. 5 as proposed milestones (which need to be specified in detail for a given project), address the critical aspects of the project and the question what should be done to reduce these risks and to allow the project to enter the next phase.

The first management decision will be to forward a research proposal and component to development. This decision should not be made if the active principle (substance) is not

Tab. 4. Product Development Phases, Main Risks and Tasks

Project Phase	Main Risks	Main Project Tasks
Research	Scientific feasibility, efficacy	Proof of a reproducible active principle Achievement of sufficient efficacy
Pre-development	Efficacy, safety, economic feasibility	Confirmation of efficacy in best available models or in target species Limited, orientating safety studies Establishment of a small-scale process, scaleable and with acceptable quality Calculation of economic parameters
Preclinical development	Efficaccy, safety	Evaluation of efficacy for all indications Full pharmacological, immunological and safety evaluation Process development and validation Analytical development and validation
Clinical development	Efficacy, safety and tolerability	Phase I: Dose-finding, pharmacologic action, metabolism, side-effects: Intensive studies in patients or healthy individuals, usually 20–80 subjects[a] Phase II: Controlled studies on effectiveness and on side-effects: usually no more than several hundred subjects[a] Phase III: Expanded, controlled and uncontrolled trials on efficacy, safety, risk–benefit relationship, practicability: usually several hundred to several thousand subjects[a]
Registration	Formalities, safety, quality	Updating/improvement of dossier Additional safety studies Additional quality assurance tests of validations
Postmarketing development	Safety, acceptance and practicability	Drug monitoring/pharmacovigilance Further safety assessment Postmarketing trials

[a] Numbers quoted from 21 CFR 312.21 (USA)

Tab. 5. Product Development Phases, Milestones and Decisions

Milestones	Major Decisions
Research Active component characterized Efficacy proven	
	≫ Decision to enter pre-development Decision on indications
Pre-development Efficacy confirmed in reliable model(s) or in the target species for veterinary products No safety risk identified Scaleable process available; yields, cost and quality acceptable Favorable economic feasibility study	
	≫ Decision to enter preclinical development Decision on target countries Decision on (preliminary) product specifications Decision on manufacturing site and method
Preclinical development Proven efficacy for all indications Acceptable safety/tolerability for all indications/conditions Process details according to plan: scale, yields, cost, product specifications	
	≫ Decision on product specifications Decision on final production process Decision on indications to be pursued Decision to enter clinical development
Prelicencing serials passed quality controls; trial material available	
	≫ Decision to start clinical trials
Clinical development Successful completion of Phase I Successful completion of Phase II Successful completion of Phase III	
	≫ Decision to file registration
Registration Obtain market approval	
	≫ Decision to market product

Decisions after completion of a phase should not exclude that *preparations* for later activities can be made before that decision. The actual activity in question (e.g., clinical trials, product sales), may only start with full approval.

yet available or insufficiently defined. A recombinant antigen which is incompletely characterized most likely needs much more research before it can be considered a development candidate. It should stay in the less costly research laboratory until sufficient efficacy with a clearly characterized antigen is achieved. There would be no point in testing safety features or to develop a process for a component if its essential features are still unknown. The ability to complete the product profile to a reasonable degree could serve as a prerequisite for a decision to enter development.

An important element, which is frequently used in drug development, is the introduction of a *pre-development phase* which serves the purpose of reducing major project risks before the full development commences and before extensive resources are allocated to the project. It can also serve as a time buffer to complete missing points of the product profile. In order not to delay the project too much, this project phase must have a limited duration of usually one year, but not more than two years. As far as possible, the crucial risk factors efficacy, safety, and economical risks should be investigated, e.g., by performing efficacy trials under the best available conditions. These could be target species experiments for veterinary indications and a range of animal model studies or perhaps even a small study in primates for human indications. Short-term safety tests can be performed to indicate where potential problems may be and which range of toxicity tests will address them. Most likely the methods used in research to generate the active component are unsuitable for later production, and the pre-development phase may be used to identify, investigate, and calculate better options.

The introduction of defined project phases and management decisions before each new phase mainly intends to create a clear basis and an overall agreement before major new activities are initiated. These management tools should be handled pragmatically in order not to cause unnecessary complications and time delays. For example, it could be necessary to approve preparations for clinical trials months before the preclinical development has been finished and before the formal decision to enter clinical trials is envisaged. The planning of such trials and the recruitment of trial cooperators and trial subjects can take a long time, which would be lost if decision procedures were not flexible enough.

5.3 Decision Making

Decisions are only the visible result of a process which is preceded by the generation of data, writing of reports and statements, exchange of information, meetings, and discussions. For larger development projects, in which many people are involved and decisions imply significant expenses and investment, this process can take a considerable time. If decisions are well planned and prepared, they represent an important element of stability and support. If applied in the wrong way, they may severely hamper the progress of a project.

A well-founded decision to enter the development phase of a product usually takes several months. Often delays occur because the decision is not anticipated and the necessary facts, figures, and reports are not collected in advance. But also if the decision-making individuals are not prepared and do not indicate early enough what kind of information and prerequisites they need to make the decision, months could be lost. For example, the existence of an acceptable product profile (cf. Tab. 2) including the required information and evaluations may be used as a basis for a development decision.

Major decisions can and must be anticipated and prepared by both the decision makers and by those who's further work depends on the decision. The major decisions for a development project, as proposed in Tab. 5, should be integrated in the project plan along with the necessary tasks for their preparation. Management should provide guidance on the information needed for such decisions. Most larger companies have established routine schemes for the major project decisions and for the necessary documentation. If applied pragmatically, such schemes can shorten decision procedures considerably.

Minor decisions should not be made by the top management. Considering the number of people who are directly or indirectly involved, the time spent and the cost of this time at work, a decision can be more expensive than the expenditure for the object of the decision (WITTE, 1969). If the suspicion arises that this could be the case, the project organization is probably top-heavy, and delegation is done without giving appropriate authority.

Delegating responsibility to a project manager and to project team members also includes the delegation of decisions to them. Extent and limits of the delegated responsibility and authority should be defined in the job descriptions. Only if there is trust and confidence in their capabilities, responsibility and decision making can be delegated.

Unless there are important and very convincing arguments which have not been taken into account, the subordinate's decision should be respected. Accepting a subordinate's decision can sometimes be very difficult for the superior manager, particularly if (s)he would have decided differently and the decision is criticized and must be defended. Overruling decisions of subordinates or facing them with already made decisions which would have fallen into their direct responsibility can seriously affect the working relationship.

Approval of a project requires the consent of many people, but a single "no" can cause the project to collapse. This mechanism has the potential of being misused by individuals to undermine projects they personally do not like. Risky projects (projects are by definition risky) are frequently surrounded by people who question the entire project by emphasizing one or the other inherent risk. If their concern has been recognized before and was considered during previous decisions, such criticism seems inappropriate and the critics should be requested to provide more constructive proposals.

Decisions to stop a project are the most difficult decisions. Although perseverance and full support by everybody are essential for the success of any project, there is a point where "the plug must be pulled". A feeling of personal failure if the project fails, selective perception and reporting of information, major setbacks being considered as temporary problems, and the hope to recoup at least part of the investment are strong forces which hinder a reasonable decision at the right time. STAW and ROSS (1991) have discussed the personal interests of people involved in such situations in more detail.

Simple recipes do not exist to prevent the project dragging on in a hopeless situation. But things can be done to make decisions to stop a project more rationally and less painfully for those affected by it.

First, the feeling of personal failure must be eliminated. Many organizations still consider the fear of failure and direct or indirect punishment (status, salary, no further promotion) as an important driving force – according to the "carrot and stick" method to keep a donkey moving. This method seems hardly adequate for the needs of project work with responsible and highly skilled individuals.

Project groups and their managers need recognition of their work and praise if they did a good job. This also applies to a project group which could not solve the scientific or economic problems of the project despite all reasonable efforts. The only reason to blame somebody personally would be if she or he refused to do what could have been done. In this case the person needs to be retrained to solve the problem.

Second, honest reporting also of negative results must be encouraged. The best encouragement is a modest, rational response and the offer of active help if unexpected problems arise.

Finally, one must not accept the argument that too much money and effort already went into the project and that it needs only some further investment to rescue it. Decisions must be based on the future perspective of the project. The past, including all money spent, does not count. The project should be evaluated in exactly the same way as a new project proposal and using the same kind of information. If in the new scenario, chances and risks, cost and time to develop the product do not justify the continuation of the project, it should be stopped based upon objective and rational arguments, but not because someone failed.

Project decisions tend to take a long time, and even projects which eventually finish up successfully are sometimes subject to several critical decisions during their life-span. In these times the individuals in the project group may frequently ask themselves, whether their efforts are worth it. The rule should be that as long as there is no official decision to discontinue a project, everyone should continue unharmed by the ongoing discussion.

5.4 The Project Manager

The success of any larger and complex project depends to a great extent upon an effective project management. Development projects for medicinal products need project managers with a variety of skills and personal qualities to cope with the very demanding job (see Tab. 6). A good working knowledge in different scientific fields (e.g., medicine, microbiology, immunology, protein biochemistry, pharmacology, toxicology) as well as in technical subjects (e.g., biotechnology, process development, analytical methods, formulation, manufacturing) are valuable assets for the planning and management of a project. Project managers must be prepared to continuously learn about these aspects. It is not sufficient to rely entirely on the specialists, since good decisions and judgement need overview. Furthermore, it could be helpful to know the subject in detail, if resources and time performances have to be negotiated.

Management skills, certain personal characteristics and psychological knowledge are required to coordinate and control the project and to maintain progress despite the unevitable setbacks. Project managers must always think ahead, sense potential problems before they actually appear and devise measure to avoid them. Nevertheless, they must be prepared to spend much of their time by solving unexpected problems and conflicts and defending their project against criticism which is raised from various sides as soon as a project does not proceed smoothly. Project managers are always in the most prominent position when problems are tackled. They are the scapegoat and need a high level of resistance against frustration. Success will be claimed by everyone.

Project managers need clear and adequate organizational structures with delegation of defined duties and authority. Delegating the responsibility for a project to the project manager means that the "ability to respond" (in French: responsabilité) must be given. Resources and authority should be at the project manager's disposition to react according to the needs of a situation. For the benefit of the project, this must include the full authority to use the planned and agreed resources plus a certain degree of flexibility in special cases – without the need for extensive discussions, negotiation, and approval procedures.

The assistance and support of all project group members is essential for a project manager. Considering the fact that project managers have to impose very unpopular time, cost and quality control measures on their work, some individuals in the group may have difficulties with accepting this.

Project managers can fulfill their task only by delegating work to other members of the project group. (There will always be enough left for them to do.) The presence of unqualified or less agile project team members reveals itself by the repeated inability of certain individuals to achieve agreed objectives and milestones or much earlier by the fact that

Tab. 6. Responsibilities of a Project Manager

Planning
Development of a project concept
Definition of objectives, milestones, and decision points
Definition of project tasks
Scheduling of project tasks
Planning of resources
Replanning and updating of plans

Coordination
Identification of shortages
Reallocation of resources
Negotiation of use of resources versus time performance
Preparation and initiation of major decisions
Organization and chairing of project meetings
Special tasks as required ("putting out fires")
Development and provision of missing skills

Controlling
Cost monitoring
Compliance with time frames (milestones)
Compliance with quality criteria (objectives)
Effective use of resources
Adequacy of documentation

Information
Effective communication among project group members
Initiation of scientific and technical reports and documentation
Progress reports
Project presentation and justification to the management

others (usually the project managers) have to do essential parts of their job. Another indication for weak spots in the project group are people who's planning and reporting needs constant assistance and who do not actively and in time collect the information which they need for their work.

If permanent pressure must be applied to ensure that objectives are met completely and in time, this could either indicate that necessary skills are missing or that the project plan was too ambitious and unachievable. Since the individuals should know best what they can or cannot achieve in a certain time, this can be avoided by a delegation of the planning of details (and provision of assistance where necessary) to the functional groups and to the individual project members. If the problems remain the same, the project manager faces the very difficult task of changing the general attitude and motivation of the project group.

But who motivates the project manager?

For further reading about this subject the classic article *The Project Manager* by PAUL GADDIS (1991) as well as the *Harvard Business Review*, volume *Motivation* are recommended (see References).

5.5 Organizational Structures and Motivation

An organization which focuses entirely on only one project should not have major internal conflicts about resources. Such a "task force" represents a largely independent group within an organization with far-reaching competences, which is created temporarily and in exceptional situations to solve a critical problem or for very large projects.

Under normal circumstances different departments or groups with different skills contribute to one or several projects besides their own routine work, e.g., in research. Somebody within these departments is nominated as project leader and continues to report to the head of his department. Quite often the different leaders of these functional departments actually lead the project, each of them with a particular personal opinion and prefer-

ence. An overall responsibility and a coordinating person with adequate authority is missing.

A common solution to accomodate the needs of development projects is the creation of a matrix management structure. In a matrix organization a project manager is appointed, who reports directly to the higher management and, as far as it concerns the project, is usually on an equal level with the functional managers who control the resources. The authority of the project manager reaches across the functional departments.

A matrix organization clearly puts more emphasis on project work. Since in a commercially oriented organization research projects with a higher priority than development projects rarely exist, it also strengthens development activities. In other words, if the functional departments conduct research and development, conflicts between these two activities are usually resolved by giving the higher priority to the development project.

The matrix organization does not abolish resource and priority conflicts between different projects and their managers. If necessary, these conflicts have to be resolved on a higher level, either by the upper management or by a project steering committee. This steering committee acts as project mandator, approves project plans, decides in case of conflicts and has the overall responsibility for product development. The project managers should be members of the steering committee.

Matrix organizations provide for an efficient use of resources and a better coordination across the functional departments (BARTLETT and GHOSHAL, 1990). However, much time must be spent on proper planning, coordination, negotiation and finding agreements about details. Project team members report at least to two bosses and have to organize, negotiate and prioritize their work between their functional manager and one or several project managers.

The individual in a matrix organization can be caught in a difficult situation and bears the unpleasant consequences if the structure does not function properly. This, as well as the inherent instability of the system, may result in a situation where the individuals disregard projects and set their own priorities.

Tab. 7. Main Motivation and Dissatisfaction Factors on the Job[a]

Motivation results from:
1. Achievement
2. Recognition
3. Work itself
4. Responsibility
5. Advancement
> Motivation is achieved on the individual level!

Dissatisfaction results from:
1. Company policy and administration
2. Supervision
3. Relationship with supervisor
4. Work conditions
5. Salary
> Dissatisfaction must be avoided on the organizational level!

[a] The five most important factors on the job which motivate or lead to dissatisfaction, ranked according to their importance.

Organizational structures and administration have a significant potential for dissatisfaction among the staff and thus may severely influence the overall performance of the organization. Tab. 7 lists the five major factors which can result in positive motivation or dissatisfaction at work. It was summarized from 12 different investigations and covers all hierarchical levels and jobs in various organizations (HERZBERG, 1986). The message given by this table is rather clear: Dissatisfaction seems to be caused mainly by internal structures, supervisors, and general work conditions. Discontent about the salary ranks relatively low on this negative side of the list (and does not appear among the motivating factors!). Changing the dissatisfaction factors which are criticized only by an individual person seems difficult because this would in most cases affect the entire organization. If, however, the majority of staff complains about such aspects, changes must be considered seriously.

Motivation can quite simply and effectively be created by the delegation of recognizable sections of work or independent parts of a project along with adequate authority to make own decisions concerning these parts. This allows the individual to assume responsibility, creates more interest in a given job and

makes sure that the individual's achievements become visible and are recognized.

Motivation appears to resemble biotechnological products: they both have a tremendous potential and are inexpensive and highly effective. In practice, however, their availablility is rather limited because their application is more difficult than expected.

Acknowledgements

I owe special thanks to GARY COBON and JIM HUNGERFORD who critically reviewed draft versions of this chapter and made many important suggestions for its improvement.

My colleagues at Behringwerke AG, Biotech Australia Pty. Ltd., and Hoechst AG contributed greatly to this work by providing good practical examples and competent advice at many occasions. At this place I wish to thank all those who shared their experience with me and who's cooperation I enjoyed during the past decade as an important source of learning and motivation.

The responsibility for the content of this chapter lies entirely with the author. The views and opinions expressed are those of the author and are not necessarily those of the persons and companies mentioned above.

6 References

AMERICAN PHARMACEUTICAL ASSOCIATION/THE PHARMACEUTICAL SOCIETY OF GREAT BRITAIN (1986), *Handbook of Pharmaceutical Excipients*, Washington, DC–London.

BARTLETT, C.A., GHOSHAL, S. (1990), *Matrix Management: Not a Structure, a Frame in Mind*, HBR 90401, Boston: Harvard Business School Press.

DIMASI, J.A., HANSEN, R.W., GRABOWSKI, H.G., LASAGNA, L. (1991), The cost of innovation in pharmaceutical industry: new drug R&D cost estimates, *J. Health Econ.* **10**, 107–142.

FIEDLER, H.P. (1989), *Lexikon der Hilfsstoffe für Pharmazie, Kosmetik und angrenzende Gebiete.* (Lexicon of Excipients for Pharmaceuticals, Cosmetics and Related Areas), Aulendorf/Germany: Editio Cantor.

GADDIS, P.O. (1991), The project manager, in: *Project Management*, HBR 90053, pp. 29–37. Boston: Harvard Business School Press.

GREGERSEN, J.P. (1994), *Research and Development of Vaccines and Pharmaceuticals from Biotechnology: A Guide for Effective Project Management, Patenting and Registration*, Weinheim – New York – Basel – Cambridge – Tokyo: VCH.

HANSEN, R.W. (1979), The pharmaceutical development process: Estimates of development costs and times and the effect of proposed regulatory changes, in: *Issues in Pharmaceutical Economics* (CHIEN, R. I., Ed.), pp. 151–186, Cambridge, MA: Lexington Books.

HARDAKER, M., WARD, B.K. (1991), How to make a team work, in: *Project Management*, HBR 90053, pp. 39–44, Boston: Harvard Business School Press.

Harvard Business Review, (1991 or latest editions), Article Collection No. 90010: *Motivation*, Boston: Harvard Business School Press.

HERZBERG, F. (1986), *One More Time: How Do You Motivate Employees?* HBR 68108, Jan.-Febr. 1986. Boston: Harvard Business School Press.

HODDER, J.E., RIGGS, H.E. (1991), Pitfalls in evaluating risky projects, in: *Project Management*, HBR 90053, pp. 9–28, Boston: Harvard Business School Press.

KARIA, R., LIS, Y., WALKER, S.R. (1992), The erosion of effective patent life – an international comparison, in: *Medicines, Regulation and Risk* (GRIFFIN, J. P. Ed.), 2nd Ed., pp. 287–302, Belfast: The Queen's University of Belfast.

LANGLE, L., OCCELLI, R. (1983), The cost of a new drug, *J. d'Economie Medicale* **1**, 77-106.

LUMLEY, C.E., WALKER, S.R. (1992), The cost and risk of pharmaceutical research, in: *Medicines, Regulation and Risk* (GRIFFIN, J. P., Ed.), 2nd Ed., pp. 303-318, Belfast: The Queen's University of Belfast.

MARTINDALE (1993), *The Extra Pharmacopoeia*, 30th Ed., London: The Pharmaceutical Press.

MEHRMANN, E., WIRTZ, T. (1992), *Effizientes Projektmanagement*. Düsseldorf: ECON Verlag.

PRENTIS, R.A., WALKER, S.R., HEARD, D.D., TUCKER, A.M. (1988), R&D investment and pharmaceutical innovation in the UK, *Managerial and Decision Economics* **9**, 197–203.

STAW, B.M., ROSS, J. (1991), Knowing when to pull the plug, in: *Project Management*, HBR 90053, pp. 57–63, Boston: Harvard Business School Press.

VANGELOS, P.R. (1991), Are prescription drug prices high? *Science* **252**, 1080-1084.

WITTE, E. (1969), *Mikroskopie einer unternehmerischen Entscheidung*, IBM 19 (193), 490 (quoted after MEHRMANN and WIRTZ).

8 Regulations for Recombinant DNA Research, Product Development and Production in the US, Japan and Europe

– Analogies, Disparities, Competitiveness –

DIETER BRAUER

Frankfurt am Main, Federal Republic of Germany

HORST DIETER SCHLUMBERGER

Wuppertal, Federal Republic of Germany

1 Introduction

This chapter compares regulations of recombinant DNA (rDNA) operations in research, product development and production of the United States of America and of Japan with those of the European Union. The European regulations are enacted in two rDNA-specific EC Council Directives, 90/219/EEC and 90/220/EEC (EEC, 1990a, b). The description of the US and Japanese regulations are based on the respective guidelines, interviews and material provided by various institutions such as universities, government agencies and industry. The objective of this report is to point to the differences in regulations for recombinant DNA (rDNA) operations with respect to research, large-scale research, development, production and to the deliberate release of recombinant organisms. Procedures for the approval of products and licensing (e.g., for drugs, diagnostics, foods and seeds) are not considered.

The *United States* did not adopt any specific legislation to regulate genetically modified organisms (GMO) or their products. The handling of GMOs is controlled by guidelines issued by the National Institutes of Health (NIH) which have been continuously adjusted to scientific experience and progress. The government policy to supervise rDNA activities and their products applies existing legislation through three government agencies: the Food and Drug Administration (FDA), the Animal and Plant Health Inspection Service (APHIS) of the US Department of Agriculture (USDA) and the Environmental Protection Agency (EPA). In recent measures to further simplify the application of biotechnology, the US government has decided to adhere to a risk-based approach to guide review. This risk-based approach of regulatory control is scientifically sound and properly protects public health and the environment against potential hazards, and moreover, it avoids obstructing safe innovations.

Japan has extensive experience in handling microorganisms in biotechnological procedures as well as in monitoring biotechnological production processes. In this context, the long-standing practice has proven the value of employing flexible guidelines instead of rigid laws which – as a rule – are difficult to adapt to scientific progress. Japan has developed a review system for recombinant DNA technology which can be easily handled and which is highly flexible. It is based on guidelines and has been simplified during the past years. Japan very closely follows the OECD recommendations. The OECD emphasized as early as 1986, on the basis of worldwide experience, that novel risks are not involved in experiments and operations with genetically modified organisms. Depending on the nature of the recombinant DNA experimentation and the properties of the product, a variety of guidelines issued by different ministries are to be considered. In the core areas – i.e., with regard to classification and evaluation of recombinant DNA operations, as well as to the resulting containment levels – the different guidelines are identical.

Despite the worldwide safety record of recombinant DNA technology, i.e., no rDNA-specific accidents have been reported although many thousands of experiments have been carried out in thousands of laboratories and production plants, the *European Union* (EU) has chosen a technology-specific approach for the regulation of recombinant DNA technology. In contrast to the United States and Japan, it is not the product and its application, that is primarily regarded as a risk, but the methods by which products are derived. This legal approach is technically described as "horizontal" as opposed to a "vertical", product-oriented regulatory system. This leads to an overlap of several approval procedures for any biotechnological product; it is time-consuming and expensive and hampers European competitiveness. The European regulations are also extremely unusual in that they are designed to regulate imagined rather than real risks. At the time of generation of the respective EC Council Directives in 1987, the responsibility for the regulation of biotechnology shifted from the experienced Directorate General XII (Research) and Directorate General III (Economy and Drug Approval) to the Directorate General XI (Environment). In Europe, the Directives disregarded the international experience which had led 1986 to the statement of the OECD that "it is expected that any risk associated with the application of rDNA organisms may be assessed in generally the same way as those with non-rDNA modified organisms." In Europe, universities, research institutions as well as industry repeatedly argued in vain that there is no scientific evidence that recombinant DNA technology would pose unique risks to human health or to the environment.

2 USA

Biotechnology consists of a set of different processes and methods of which the most powerful tool is gene technology. The production of organisms with new molecular techniques may or may not pose risks, depending on the characteristics of the modified organism and the type of application. The National Research Council (NRC, 1989; OSTP, 1992a) has extensively reviewed the potential risks of introducing genetically modified organisms into the environment. The Council reached the general conclusion that organisms, that have been genetically modified, are not *per se* of inherently greater risk than unmodified organisms.

In August 1990, President BUSH approved 'Four Principles of Regulatory Review for Biotechnology' (OSTP, 1992b) which can be applied as general rule of regulatory review, as follows:

1. Federal government regulatory oversight should focus on the characteristics and risks of the biotechnology product – not the process by which it is created.
2. For biotechnology products that require review, regulatory review should be designed to minimize regulatory burden while assuring protection of public health and welfare.
3. Regulatory programs should be designed to accommodate the rapid advances in biotechnology. Performance-based standards are, therefore, generally preferred over design standards.

A "performance-based standard" sets the ends or goals to be achieved, rather than specifying the means to achieve it, e.g., through a design standard. For example, a performance-based standard for containment would permit alternative biological approaches for assuring containment in place of a design-based standard requiring specific physical barriers.

4. In order to create opportunities for the application of innovative new biotechnology products, all regulations in environmental and health areas – whether or not they address biotechnology – should use performance standards rather than specifying rigid controls or specific designs for compliance.

"Design-based" requirement may preclude use of biotechnology products even when such approaches may be less costly or more effective. For example, a requirement to employ specific pollution control equipment would prevent the use of innovative biotechnology pollution remediation or control techniques.

The new administration is continuing this policy. According to a policy statement by President CLINTON and Vice-President GORE (1993), "regulatory policy can have a significant impact on the rate of technology development in energy, biotechnology, pharmaceuticals, telecommunications, and many other areas." Therefore, the US regulatory policy will be reviewed in order to "ensure, that unnecessary obstacles to technical innovation are removed and that priorities are attached to programs introducing technology to help reduce the cost of regulatory compliance."

2.1 The NIH Guidelines

After the moratorium of Asilomar in 1975, the "Guidelines for Research Involving Recombinant DNA Molecules" were promulgated by the National Institutes of Health (NIH) in June 1976. These guidelines were revised several times whereby some relaxation of primarily restrictive provisions were achieved according to the state of science and technology. The latest complete version of the NIH Guidelines was published on May 7, 1986 (NIH, 1986a). On July 18, 1991 the appendix K of the NIH Guidelines was introduced addressing "Physical Containment for Large-Scale Uses of Organisms Containing Recombinant DNA Molecules" (NIH, 1991).

Based on the Biological Control Act (1902), the original Food and Drugs Act (1906) and the Federal Food, Drug and Cosmetic Act (FDCA) (1938), the Food and Drug Administration (FDA) is the regulatory agency which approves rDNA products, except recombinant pesticides, plants and animals.

The NIH Guidelines apply to all rDNA research conducted at or sponsored by any institution receiving financial support from the Federal government. Although the NIH Guidelines are not binding for private companies, they have created a mechanism of compliance for the private industrial sector. Furthermore, industry is required to comply with general safety standards issued by the Occupational Safety and Health Administration (OSHA) of the US Department of Labor. Appendix K of the NIH Guidelines which concerns the large-scale use of rDNA organisms, is particularly relevant to the private sector (NIH, 1991). Some cities such as Cambridge and Worcester, Massachusetts, and Berkeley, California, have introduced local ordinances that require universities and industry to observe the NIH Guidelines. There is no evidence that private companies in the United States did not follow and comply with the NIH Guidelines.

2.2 Committees and Offices Involved in Review Processes

According to the NIH Guidelines, rDNA experiments are divided into four classes:

A. Experiments which require specific "Recombinant DNA Advisory Committee (RAC)" review and approval by the "Institutional Biosafety Committee (IBC)" and the Director, NIH, before initiation of the experiment

B. Experiments which require IBC approval before initiation of the experiment
C. Experiments which require IBC notification at the time of initiation of the experiment
D. Experiments which are exempt from the procedure of the NIH Guidelines.

To determine the different types of rDNA experiments with respect to safety classification and safety measures, different committees, offices and services are assigned for review.

2.2.1 Director of the NIH, Recombinant DNA Advisory Committee (RAC) and NIH Office of Recombinant DNA Activities (ORDA)

Under the NIH Guidelines, the Director, NIH, is the final decision maker. He can overrule the advice of the RAC, as was the case with the first somatic gene therapy experiments. Every action taken by the Director must meet the goal of achieving "no significant risk to health or the environment" by any rDNA experiment.

The Recombinant DNA Advisory Committee (RAC) advises the Director, NIH, concerning rDNA research and meets three to four times a year. The committee consists of 25 members appointed by the Secretary, Health and Human Services, including the chair. At least 14 members are selected from authorities knowledgeable in the fields of molecular biology or rDNA or other scientific fields. At least six members shall be knowledgeable in applicable law, standards of professional conduct, the environment, public and occupational health or related fields. Representatives of Federal agencies shall serve as non-voting members. The chairman can install subcommittees with additional experts in order to investigate issues further. RAC meetings, including the agenda, are announced in the *Federal Register* and are open to public comment. The meetings are open to the public. The specific functions and operation procedures of the RAC are defined in the NIH Guidelines.

The Office of Recombinant DNA Activities (ORDA) serves as a focal point for information on rDNA activities and provides advice to all within or outside NIH including scientific and industrial institutions, Biosafety Officers, Principal Investigators, Federal agencies, State and local governments and the private sector. In the case of experiments that require both IBC approval and RAC review, the IBC has to submit the necessary information to the ORDA which decides, whether RAC review is to be initiated or not. ORDA is responsible for reviewing and approving IBC membership.

The Federal Interagency Advisory Committee on Recombinant DNA Research (Interagency Committee) is composed of representatives from approximately 20 agencies, provides additional oversight and coordinates all federal rDNA activities. Its members are non-voting members of the RAC.

2.2.2 Institutional Biosafety Committee (IBC)

The IBC supervises all rDNA work of an individual institution that is involved in rDNA research, development or production and often also assumes the function as a general safety committee including general occupational safety and health, and environmental protection. The IBC shall comprise no fewer than five members, so selected that they collectively have experience and expertise to assess the safety of rDNA experiments and any potential risk to public health or the environment. Large organizations usually establish an IBC with more than five members. The biological safety officer (BSO) is mandatory when research is conducted at the Biosafety level 3 (BL3), Biosafety level 4 (BL4), and at all large-scale operations. The BSO serves as a member of the IBC. For institutions that have to comply with the NIH Guidelines, two members shall represent interests of the community with respect to health and protection of the environment and shall not be affiliated with the institution. At federally funded insti-

tutions, these non-affiliated members are usually appointed by the local Board of Health. Commonly, private companies have established an IBC and usually appoint two additional members from the community or experts from other public non-affiliated organizations (e.g., universities). In the few sites with specific city ordinances (i.e., Cambridge, Worcester, Berkeley) and/or State laws that regulate rDNA operation, private companies are obliged to establish an IBC as requested by the NIH Guidelines.

Except for experiments which require review by the Federal Recombinant DNA Advisory Committee, only the IBC reviews and approves experiments or must be notified of rDNA projects. The IBC has the competence to deal with *all* safety issues under the NIH Guidelines. The IBC meets regularly and has to submit an annual report to ORDA containing a list of the IBC members and background information of each member on the request of ORDA. The registration documents submitted to the IBC for review are, however, not communicated to ORDA and, thus, remain confidential. Upon request, only those documents on research projects must be released under the Freedom of Information Act which are submitted to or received from Federal funding agencies. On request of the applicant, data involving important intellectual property rights or production know-how are kept confidential.

2.3 Biosafety Levels, Classification and Review of Experiments

The NIH Guidelines introduced four biosafety levels (BL1–BL4) which consist of combinations of laboratory techniques and practices and safety equipment appropriate to the potential hazards derived from the organisms used. Based on the "Classification of Etiological Agents on the Basis of Hazard" published by the Center for Disease Control (CDC) in 1974, a list of classified microorganisms is included in the NIH Guidelines. These organisms are divided into five classes whereby non-pathogenic organisms are in Class 1 and pathogenic agents are in Classes 2, 3, and 4. Class 5 agents comprise disease agents which are forbidden entry into the US by law, by US Department of Agriculture policy or which may not be studied in the US except at specified facilities (NIH Guidelines Appendix B-III – Classification of Microorganisms on the Basis of Hazard), e.g., the WHO Collaborating Center for Smallpox Research in Atlanta.

It must be emphasized that the NIH Guidelines are based on existing experience and established approaches to the containment of pathogenic organisms. In addition, two levels of biological containment were introduced which limit the survival of a host–vector system and its dissemination in the environment. Recombinant DNA experiments are classified into four categories initially assessed, judged and based on the experience obtained with similar non-modified organisms. Meanwhile, a substantial amount of experience with organisms containing rDNA molecules has accumulated. In addition to the NIH Guidelines, the National Cancer Institutes recommend three safety levels for research with oncogenic viruses which correspond to the categories BL2, BL3 and BL4.

A   Experiments that require review of the Recombinant DNA Advisory Committee, approval by the IBC and the Director, NIH, before initiation are:

A-1 Deliberate formation of recombinant DNAs containing genes for the biosynthesis of toxic molecules.

Lethality for vertebrates at an LD_{50} of less than 100 nanograms per kilogram body weight, e.g., microbial toxins such as the botulinum toxin, tetanus toxin, *Shigella dysenteriae* neurotoxin. Specific approval has been given for the cloning in *E. coli* K12 DNAs containing genes coding for the biosynthesis of toxic molecules which are lethal to vertebrates at 100 nanograms to 100 micrograms per kilogram body weight. Containment levels for these experiments are also specified in Appendix F of the NIH Guidelines (NIH, 1986b).

A-2 Deliberate release into the environment of any organism containing recombinant DNA, except certain plants, as specified in Appendix L of the NIH Guidelines (NIH, 1986b).

A-3 Deliberate transfer of a drug-resistance trait to microorganisms that are not known to acquire it naturally, if such acquisition could compromise the use of the drug to control disease agents in human or veterinary medicine or agriculture.

A-4 Deliberate transfer of recombinant DNA or DNA or RNA derived from recombinant DNA into human subjects. The requirement for RAC review should not be considered to preempt any other review of experiments with human subjects. Institutional Review Board (IRB) review of the proposal should be completed before submission to NIH.

This approval procedure of the RAC thus applies only to new types and certain specified experiments, e.g., with highly toxic molecules or the deliberate transfer of recombinant molecules into human subjects (e.g., somatic gene therapy). Such experiments may only be initiated after review by the RAC and approval of both the Director, NIH, and the IBC.

B   Experiments that require IBC approval before initiation
For experiments that require IBC approval before initiation, a short registration document must be submitted to the IBC which contains a description of: (i) the source(s) of DNA; (ii) the nature of the inserted DNA; (iii) the hosts and vectors to be used; (iv) whether a deliberate attempt will be made to obtain expression of a gene, and what protein will be produced; and (v) the containment conditions specified in the NIH Guidelines. The IBC has to review and approve the proposal before the experiment is initiated. Experiments which fall under this procedure can be carried out at BL2, BL3 and BL4 containment. For work with Class 5 agents an additional permit must be obtained from the US Department of Agriculture (USDA).

C   Experiments that require IBC notification simultaneously with the initiation of experiments

A registration document containing the information listed under B must be submitted to the IBC, which in turn shall review the proposal. However, review prior to initiation is not needed. The experiments which fall into this category can be carried out at containment level BL1 and comprise those in which all components derive from non-pathogenic prokaryotes and non-pathogenic lower eukaryotes.

D   Exempt experiments
Certain rDNA molecules are exempt from registration and approval by the IBC. Examples are listed in the NIH Guidelines, e.g., DNA molecules which consist entirely of DNA segments from different species that exchange DNA by known physiological processes or DNA from a prokaryotic host when propagated only in the same host organism (self-cloning).

It is estimated that approximately 80% of experiments reviewed by the RAC are devoted to the application of rDNA technology to humans, e.g., by gene transfer protocols and somatic gene therapy. More than 99% of the experiments are classified into categories B, C and D. While experiments falling into category C can be initiated at the time of filing the notification, experiments classified into category B usually can be initiated within a few days after filing the registration form to the IBC. IBCs from universities and private companies are not required to send copies of the registration documents to ORDA, thus rDNA experiments classified into the categories B, C and D are not published, as is often assumed and incorrectly stated by European regulatory authorities and politicians.

Experiments that require BL4 containment can be reviewed and approved by the IBC. Such experiments are – unique in the United States – prohibited by ordinance in the city of Cambridge, MA.

It is probably noteworthy that the NIH Guidelines reflect the "state of the art" and can be rapidly adapted to scientific progress and are considered "never (to) be complete or final, since all conceivable experiments involving recombinant DNA cannot be fore-

seen. Therefore, it is the responsibility of the institution and those associated with it to adhere to the intent of the NIH Guidelines as well as to their specifics".

2.4 Large-Scale Uses of Organisms Containing Recombinant DNA Molecules (Research and Production)

Appendix K of the NIH Guidelines (NIH, 1991) specifies physical containment guidance for large-scale research or production involving viable organisms containing DNA molecules (>10 liters of culture fluid). This applies to both large-scale research and production activities, and addresses potential hazards that may be associated with rDNA organisms. Other potential hazards accompanying large-scale cultivation of genetically modified organisms such as toxic properties of the products, physical, mechanical or chemical aspects of downstream processing must be considered separately. For large-scale operations, the NIH Guidelines apply with the following minor additions:

• The institution shall appoint a Biological Safety Officer (BSO) with duties specified in the NIH Guidelines (section IV-B-4).
• The institution shall establish and maintain a health surveillance program for personnel engaged in activities that require BL2 and BL3 containment.

For large-scale operations four physical containment levels are established which are referred to as Good Large Scale Practice (GLSP), BL1-LS, BL2-LS and BL3-LS. No provisions are made for large-scale research or production requiring BL4 containment. If necessary, the requirements will be established by the NIH on an individual basis.

Depending on the containment level

• discharges containing viable recombinant organisms may be handled according to governmental environmental regulations (GLSP) or need to be inactivated by a validatcd inactivation procedure (BL1-LS to BL3-LS),

- containment equipment has to reduce the potential for escape of viable organisms (BL1-LS) or to prevent the escape of viable organisms (BL2-LS and BL3-LS).

Organizational and technical safety specifications are provided in the NIH Guidelines and are summarized in Tab. 1.

To qualify for GLSP, an organism must meet certain criteria such as non-pathogenicity and built-in environmental limitations, and must have an extended history of safe large-scale use (OECD, 1986). With respect to

Tab. 1. Comparison of Technical and Organizational Measures of Different Biosafety Levels at Large-Scale Practices (adapted from NIH Guidelines[a] according to MEAGHER, 1992)

Criterion	GLSP	BL1-LS	BL2-LS	BL3-LS
Provide written instruction and training of personnel to reduce exposure to biological, chemical, or physical agents	■	■	■	■
Provide changing and handwashing facilities and protective clothing	■	■	■	■
Prohibit eating, drinking, smoking, mouth pipetting and applying cosmetics in the facility	■	■	■	■
Develop emergency plans for handling large spills	■	■	■	■
Inactivate waste solutions and material in accordance with their biohazard potential	■	■	■	■
Use a validated protocol to sterilize closed vessels		■	■	■
Operate in closed systems		■	■	■
Control aerosols by engineering or procedural controls to prevent or minimize release of organisms during sampling, addition of materials, transfer of cultured cells, and removal of materials, products, and effluents from system[b]			■	■
Treat exhaust gases from a closed system[c]			■	■
Ensure medical surveillance			■	■
Rotate seals and other penetrations into a closed system designed to prevent or minimize leakage			■	■
Incorporate monitoring or sensing devices to monitor integrity of containment			■	■
Validate integrity testing of closed systems			■	■
Ensure permanent identification of closed systems			■	■
Post a universal biohazard sign on each closed system			■	■
Require controlled access to facility (double doors, air lock)				■
Maintain closed system at as low a pressure as possible to ensure integrity of containment features				■
Design walls, ceiling, and floors to permit ready cleaning and decontamination				■
Protect utilities and services against contamination				■
Design controlled area to preclude release of culture fluid into the environment in the event of an accident				■

[a] Appendix K–I – V, NIH Guidelines, *Fed. Reg.* **56,** 33.178 (1991)
[b] In GLSP, minimize by procedure; in BL1-LS, minimize by engineering
[c] In BL1-LS, minimize

practical applications of large-scale operations, this means that the construction of facilities is conducted according to the permit system outlined in Sect. 2.6 and that the IBC assesses the potential risks and determines the measures necessary to ensure occupational and public safety. It is the IBC that determines the procedures to be followed in the case of an accident, e.g., containment and measures to handle a large spill.

At the few city sites (Cambridge and Worcester, Massachusetts) where local ordinances are in place, additional permits for rDNA large-scale operations must be obtained. The permit requires compliance with the NIH Guidelines and is issued commonly on an annual basis and, usually, is renewed automatically. In an amendment of the Municipal Code of the City of Cambridge such permits can since February 22, 1993 be obtained without a public hearing.

2.5 Construction of Buildings for Recombinant DNA Operations

There are no regulations specific for this rDNA technology at Federal and State level with regard to construction of buildings, i.e., laboratories, pilot (up-scaling) and production plants, to conduct research, product development and production. In general, any building owner has to acquire a variety of permits from the competent State or city authorities according to respective Federal and State laws and city ordinances. These permits include safety issues with regard to construction, fire prevention, electricity, waste and waste water management. At this level, there are no specific requirements for technical construction of laboratories or other facilities, e.g., specific containment requirements, equipment or even specific demands for the intended use of the building. It is, however, required that the characteristics and the expected amount of waste water are outlined in order to obtain the necessary permit for the connection to public waste water processing plants.

A certified architect, the construction company and representatives of the owner submit a construction plan of the building, and plans for technical installations, i.e., electricity, plumbing etc., to the competent city department in order to obtain the permits. The building permit is usually granted within 1 to 3 weeks so that construction work can start immediately. For obtaining other permits at State or Federal level, e.g., FDA approval for experimental animal facilities, more time may be needed. However, these permits are usually obtained at the time when the construction work is concluded. Partial permits can be obtained for construction work, electrical installations, plumbing, etc. This speeds up construction work and helps to reduce costs.

During the different stages of construction the progress of construction is supervised by inspectors from the respective State or city authorities. After completion of the building, the occupation and the use of the building is granted by issue of the "Certificate of Occupancy".

There are no differences in the administrative approval procedures for setting up facilities for chemical, bacteriological, immunological or recombinant DNA operations, in almost all locations in the US. Cities like Cambridge and Worcester (Massachusetts), and Berkeley (California), require an additional permit for working with recombinant DNA methods. The permits have to be obtained from the respective Boards of Health according to local rDNA-specific ordinances. The ordinances obligate the applicant to meet the provisions of the NIH Guidelines on recombinant DNA technology. This permit is closely linked with the operation and the obligations of the Institutional Biosafety Committee and is usually obtained within 30 days. It is valid for a whole building and is not temporally restricted or restricted to certain specified rooms or floors. It may even be effective for a large organization such as a university.

2.6 Product Approval and Production Facility

In the United States, approval of products by the responsible agency (e.g., FDA, APHIS) is necessary for the marketing of products and is based on safety investigations of the product. There is no permit required with regard to specific demands for the application of rDNA methods. Only the general building permits for the production facility are required.

It needs to be emphasized that the FDA is not involved at the stage of the construction of experimental or large-scale production facilities. FDA inspects production facilities when products are to be approved for a Product License Application (PLA) to place a product onto the market, e.g., drugs, diagnostics, food additives and certain foods. At this stage, performance-based regulations that are not specific for the production method, such as validation of downstream processing, good manufacturing practice for production of clinical trial supplies, virus inactivation, and other recommendations that may be laid down in a variety of "Points to consider...", are employed. These investigations and measures are based on a decision of the Congress to provide FDA under the Food, Drug and Cosmetic Act (FDCA) with the authority necessary to regulate foods, drugs, diagnostic kits, devices and cosmetics. In addition, under the provisions of the Public Health Service Act, FDA also regulates and approves biological products.

Biological products for use in humans are, therefore, subject to the provisions of the FDCA and, thus, under the responsibility of FDA's Center for Biologics Evaluation and Research (CBER). Biological drugs for animals are regulated under the Virus, Serum and Toxin Act (VSTA). These products are administered by the US Department of Agriculture and the FDA Center for Food Safety and require reviews for food additives at the pre-marketing level. Genetically modified crops, especially the expression products of introduced genes are defined as an "additive" and, therefore, need regulation.

When a production facility is larger than 10 000 sq. ft (ca. 1000 m^2), the general licensing and approval regulations require that FDA considers the general environmental impact of applications before taking any final action. For this, FDA demands the manufacturer to submit an Environmental Assessment Certificate (EAC) under the National Environmental Policy Act (NEPA) of 1969 which must include a description of the containment procedures used to protect the workers, the environment and the product, and the system used for solid, liquid and gaseous waste disposal. This is a general requirement for all production facilities and is not specific for productions using GMOs. Similarly, a petition must be filed and approved by USDA/APHIS prior to the marketing of recombinant crops which demonstrates that the respective crop possesses properties that do not jeopardize the environment and can safely be released and marketed.

A review of the procedures for product approval is not within the scope of this chapter and will not be further discussed.

2.7 Shipment and Transport of Organisms Containing rDNA Materials

The NIH Guidelines currently require that "rDNA molecules" shall not be transferred to other investigators or institutions unless their facilities and techniques have been assured to be adequate by their local IBC. To comply with this requirement, the following procedures must be followed:

- Before receiving rDNA materials or other potentially hazardous agents, the investigator must have approval from the IBC that allows the use of the host–vector–donor recombinant system in question unless such systems are exempt.
- Because the labeling requirements for cargo transport change frequently, the investigator should contact the IBC regularly before shipping potentially hazardous and, therefore, regulated agents including rDNA material.
- When transporting rDNA material locally, it must be packaged under the provisions of the NIH Guidelines. Interstate transport of recombinant plants, soil bacteria, plant and animal pathogens requires a permit issued by

USDA/APHIS. Such a permit is usually obtained within 10 days and can also be used for multiple shipments. When plants or plant materials are shipped, they must be shipped in such a way that the viable plant material is unlikely to be disseminated while in transit and must be maintained at the destination facility in such a way that there is no release into the environment.

2.8 Field Testing of Recombinant Organisms

Field testing of recombinant organisms requires the permission of the Animal and Plant Health Service (APHIS) of the US Department of Agriculture (USDA) or the Environmental Protection Agency (EPA) when recombinant microorganisms are to be released. The application for the field test must contain the data about the applicant, in the case of a foreign applicant also the US-based cooperator. APHIS works out a document in which risks are assessed and which serves as the basis for the permit. Permits for field trials are regularly published in the *Federal Register*. The published lists of permits contain the registration number, the receiving date and the date of issue, the applicant, the recipient organism, the coding gene and the state of the release.

APHIS also informs the state of the intended field trial and coordinates possible modifications of the trial. In some states, e.g., North Carolina, that have further regulations on rDNA, a State permit is required. The State regulatory system must parallel the Federal regulations. In other states, a specific notification is necessary (e.g., Wisconsin, Hawaii) and in Minnesota, West Virginia and Oklahoma the Federal permit is necessary to comply with State regulations.

Plants that contain recombinant genes from the same species, in the sense of "self-cloning", are not under regulation. However, a researcher informs APHIS through a letter that the intended release does not fall under the regulations, and APHIS returns a "letter of agreement" usually within 10 days. This is a voluntary procedure that has become a common custom.

New options for field trials with recombinant crop plants were established by USDA/APHIS (1993). These new provisions amend the existing regulations executed by APHIS through a notification procedure for the introduction of certain plants and furthermore provides a petition process allowing the determination that plants are no longer regulated articles. The US Department of Agriculture has, based upon experience, adjusted and supplemented its system for regulating the field testing of new plants, and thus encourages further progress and innovation in this area of new biotechnology (PAYNE, 1992).

The new rules are based on the fact that the number of documented field trials with genetically engineered organisms approved by regulatory agencies of the US greatly exceed approvals of any other country and, particularly, that ample experience has accumulated. The experience with field releases verifies that these trials have proved to be safe and did not give rise to plant pests or environmental risks. In particular, the data show that there are the same kinds of ecological concerns with genetically engineered crops as are associated with other non-modified plants (i.e., weediness, competitiveness, toxicity). Up to September 1993, 465 permits for field trials on 1208 different sites had been issued. During the time period between March and September 1993, when the notification procedure was enacted, 105 notifications at 276 sites were received (MEDLEY, 1993; Figs. 1 and 2). Comparable to the categories A–D defined by the NIH Guidelines, there are four different procedures for the use of recombinant plants outside greenhouses regulated by USDA/APHIS:

A. Field tests which need a permit

To obtain a permit, the applicant has to comply with the user guides and in particular has to file the necessary forms and its appendices (usually approximately 15 to 20 pages). APHIS assesses the environmental impact from the data obtained. The issue of a permit

No of permits & notifications

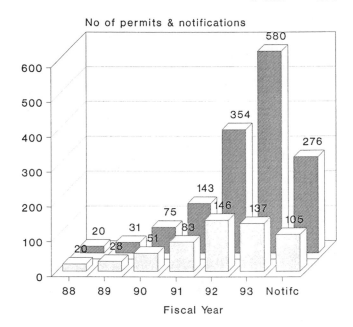

Fig. 1. Release permits and release sites in the US 1988–1993 (Source: MEDLEY, 1993).

Fiscal Year

☐ Permits ▨ Sites

Organisms released

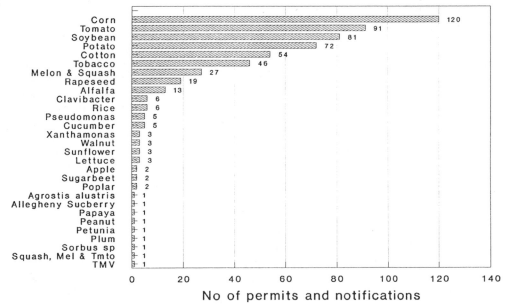

No of permits and notifications

Fig. 2. Number and kind of crops released in the US 1988–1993 (Source: MEDLEY, 1993).

is published in the *Federal Register* and information on the permit must be released upon request under the Freedom of Information Act. If the necessary criteria are met by the applicant, the permit has to be issued within 120 days; the average time of approval is usually 90 days. Only North Carolina requires a separate permit by State law whereby no additional data are requested, but a public hearing may be carried out. Only if scientific questions are raised by the public, will a hearing take place. To our knowledge, this has happened only two or three times. This law may, however, expire in September 1995. With the permit issued by APHIS, field trials can be conducted in all states of the United States.

For field tests on more than 10 acres with transgenic plants expressing a pesticide phenotype (e.g., BT cotton), a voluntary experimental use permit issued by the Environmental Protection Agency is recommended in addition to the APHIS permit. EPA is responsible for regulating the application of pesticides and for issuing permits for field trials with recombinant microorganisms (not vaccines). Regulatory procedures of EPA are not reviewed in this chapter.

After the termination of a field test a "termination report" has to be filed with APHIS.

B. Field tests which need only notification

This applies to formerly regulated articles for introduction under the notification procedure without a permit. A regulated article is an organism that has been genetically engineered – via recombinant DNA techniques – from a donor organism, recipient organism, vector or vector agent, any of which is a plant pest, or contains plant pest components. Other genetically engineered organisms may be regulated articles, if they have been genetically engineered using an unclassified organism or if the Director, Biotechnology, Biologics, and Environmental Protection (BBEP), determines that the genetically engineered organism meets the definition of a regulated article because of plant pest potential.

The notification 30 days prior to the field test must include the name, address, trade names of the plant, etc. and has to be filed with the Director, BBEP of APHIS. The regulated articles eligible for introduction into the environment under the notification procedure must meet the following requirements and performance standards:

1. The regulated article is one of the following plant species: corn, cotton, potato, soybean, tobacco, tomato and certain additional plant species to be determined by the Director, BBEP.
2. The introduced genetic material is stably integrated in the plant genome.
3. The function of the introduced genetic material is known and its expression in the regulated article does not result in a plant disease.
4. The introduced material does not (i) cause the production of an infectious entity, or (ii) encode substances that are known or likely to be toxic to non-target organisms known or likely to feed or live on the plant species, or (iii) encode products intended for pharmaceutical use.
5. To ensure that introduced genetic sequences do not pose a significant risk of the creation of any new plant virus, they must be (i) non-coding regulatory sequences of known function, (ii) sense or antisense genetic constructs derived from viral coat protein genes from plant viruses that are prevalent and endemic in the area where the introduction will occur and that infect plants of the same host species, or (iii) antisense genetic constructs from non-capsid viral genes from plant viruses that are prevalent and endemic in the area where the introduction will occur and that infect plants of the same host species.
6. The plant has not been modified to contain the following genetic material from animal or human pathogens: (i) any nucleic acid sequence derived from animal or human virus, or (ii) coding sequences whose products are known or likely causal agents of disease in animals or humans.

C. Field tests which are exempt from regulations

This applies to "self-cloning" of plant genes, but it is recommended, and is a practised custom, to consult APHIS on a voluntary basis. In such cases APHIS recognizes the "non-regulated" status in written form in a "letter of agreement" usually within ten days.

D. Marketing of recombinant plants

For this purpose, a petition for determination of the non-regulated status for an article has to be filed with the Director, BBEP. This petition can be filed already at early stages when the breeding program is started in accordance with procedures and formats specified. The required information comprises biological data of the narrowest taxonomic groupings applicable, experimental data, details of the inserted genetic material and a description, why the article in question is unlikely to pose a greater plant pest risk than the unmodified organism from which it was derived.

After the filling of a completed petition, APHIS published a notice in the *Federal Register* allowing public comment within a 60 day period and consults with other Federal agencies. Within 180 days after receiving the petition, the Director, BBEP, furnishes either an approval or a denial in written form. The first petition granted concerned the "Flavr Savr™" tomato developed by Calgene Inc., California.

APHIS assures that State authorities are kept informed about pending applications which may involve the field testing and/or introduction of genetically engineered plants within the State by using the existing working relationship with the Department of Agriculture of the State. The provision for notification by APHIS of the State officials is written into USDA's regulation.

Companies applying for field testing of genetically engineered organisms usually inform the relevant city and/or State officials before receiving the permit by APHIS. Only for first field trials, the public is extensively informed by the applicants.

Except for the first field testing of genetically engineered microorganisms (i.e., "ice-minus" *Pseudomonas syringae* in California) and unlike the responses from opponents of releases in the Netherlands and in Germany, no vandalism occurred at any of the sites where releases of GMOs were conducted.

3 Japan

3.1 Introduction

Japan regulates recombinant DNA technology (rDNA) by a variety of guidelines which places the properties of genetically modified organisms (GMO) to the fore (Fig. 3) and follows very closely the OECD recommendations (OECD, 1986) and the NIH Guidelines. The OECD recommendations state that "there is no scientific basis for specific regulation of the use of recombinant DNA organisms". Apart from the Ministry of the Environment, all the ministries involved have enacted guidelines to supervise recombinant DNA operations:

1. Science and Technology Agency (STA): Guideline for rDNA Experiments (first version 1979)
2. Ministry of Education (MoEd): Guideline for rDNA Experiments in Universities and other related Research Institutes (first version 1979)
3. Ministry of International Trade and Industry (MITI): Guideline for Industrial Application of rDNA Technology (first version 1986)
4. Ministry of Health and Welfare (MHW): Guideline for Manufacturing Drugs etc. by Application of rDNA Technology (first version 1986)
5. Ministry of Health and Welfare: Guideline for manufacturing Foods and Food Additives by application of rDNA

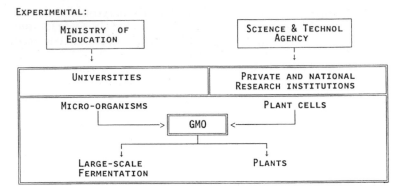

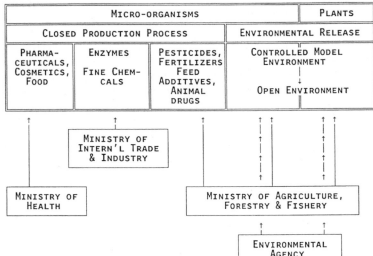

Fig. 3. Framework of guidelines for regulation of recombinant DNA technology in Japan.

Techniques (first version 1991)
6. Ministry of Health and Welfare: Guideline for Safety Assessment of Foods and Food Additives produced by rDNA Techniques (first version 1991)
7. Ministry of Health and Welfare and Ministry of Education: Japanese Guidelines for Gene Therapy Clinical Research (first version February 1994)
8. Ministry of Agriculture, Forestry and Fisheries (MAFF): Guideline for Application of rDNA organisms in Agriculture, Forestry, Fisheries and other related Industries (first version 1989).

Product authorizations are effected in accordance with the relevant guidelines of the responsible ministries. The different guidelines contain identical criteria for classifying genetically modified organisms and for containment conditions, and have been revised and continuously simplified since their introduction. In 1990, the Ministry of the Environment (MoE) proposed legislation for regulating field trials with GMO. This suggestion was, however, not pursued as a result of the negative responses of other ministries, universities (Tab. 2), the Japan Health Foundation and the Federation of Economic Organizations (Keidandren). At present, the Ministry of the Environment considers to issue a guideline for the use of recombinant organisms in 'bioremediation', 'leaching' or 'biomining'. The policy of the Japanese govern-

Tab. 2. Appeal of 45 Professors Regarding Safety of rDNA Technology (September 1990, Science Council of JBA)

1. rDNA technology was developed in 1974 and is presently an essential technology for health, food and environment.
2. During the past 16 years, a lot of scientific knowledge has been accumulated, where no risk was observed.
3. Scientists have come to common agreement that "there is no particular risk in rDNA technology".
4. The Guideline of NIH as well as other countries has been relaxed greatly as a result of confirmation of the safety of the technology.
5. In Japan, scientists followed the guidelines of the Japanese government and more than 30000 experiments were carried out. Not even the smallest risk was observed.
6. The initial guidelines of the Japanese government for rDNA experiments and industrial application were relaxed. Moreover, we request further relaxation of the guidelines and the issuance of guidelines for large-scale experiments.
7. In Japan there are moves to submit a "genetic engineering bill" to the diet. We are very concerned that the movement might be based not only on the scientific knowledge accumulated for the past 16 years but also on the product basis and might adversely affect the development of rDNA technology.

ment towards biotechnology and recombinant DNA technology has been outlined in an "Environment White Paper" (MINISTRY OF THE ENVIRONMENT, 1991). This document clearly states that there are no particular environmental risks associated with the development and exploitation of biotechnology and recombinant DNA technology.

As an interesting fact and unlike customs in other countries, it is necessary to obtain consensus with all parties concerned before legislation may be enacted; decisions based on a majority are evidently not sufficient in Japan. The achievement of a consensus is usually more time-consuming than attaining decisions in the US or in Europe. Decisions that are supported by all participants are usually funded on higher investment levels and are difficult to be overruled by changing

majorities. This may also explain that the Japan Bioindustry Association (JBA), supported by MITI and about 200 enterprises, started a rather costly nation-wide 10-year program to achieve understanding and acceptance of the use of biotechnology and recombinant DNA technology on a broad scale in 1991.

At the end of 1992, the Japanese market for products produced by recombinant DNA technology amounted to over 5 billion DM. By the end of 1990, 76 products from 31 companies had been approved under the jurisdiction of the Ministry of Health and Welfare. Of these, 72 obtained "Good Industrial Large Scale Practice (GILSP)" status, and the remainder were classified as containment level 1. At present, more than 60 recombinant pharmaceuticals are in clinical trials, and in 1993 about 40 different products were at the stage of approval (Tab. 3). As in the USA, research and development in Japan is carried out in numerous areas of business including many small enterprises.

At present, there is a lack of experience in Japan with regard to field trials of recombinant organisms. To date, only one field trial has been carried out with a TMV-resistant tomato variety. However, four further permits for field trials for rice, melon and petunia were granted in February 1993, and about 20 further projects have by now reached the stage of experiments in "open" greenhouses. Japan strictly follows the OECD recommendations and strives to reach the same high level in field trials as the United States.

3.2 Recombinant DNA Technology in Research, Product Development and Production

In contrast to the USA and European states, different State authorities are responsible for research in Japan: the Science and Technology Agency (STA) for application-oriented research, and the Ministry of Education for research at universities. About 67% of rDNA research is carried out by the industry, approximately 18% at universities and 13% in independent research institutions.

Tab. 3. Actual Cases of Industrialization of rDNA Technology and Experimental Applications

1. Industrial Application		(as of September 1991)
Ministry	Product	Cases
Ministry of International Trade and Industry (Total 190)	Chemical agents	110
	Enzymes	62
	Amino acids	15
	Others	3
Ministry of Health and Welfare (Total 80)	Diagnosis, observation	25
	Medical treatment	47
	Vaccine	8
Ministry of Agriculture, Forestry and Fishery (Total 5)	Amino acids	3
	Medical treatment	1
	Application in the simulated model environment	1
Total		275

2. Experimental Application		(as of March 1990)
Ministry and Agency		Cases
Science and Technology Agency	Outside the standard[a]	1442
	Within the standard[b]	10526
Ministry of Education	Outside the standard[a]	about 2000
	Within the standard[b]	21061

[a] Cases for which individual examination is necessary
[b] Cases for which the standard is stipulated in the guideline

The Science and Technology Agency (Tab. 4), which acts under the responsibility of the Prime Minister, plays a special role in coordinating and supporting science and technology.

The share of the Science and Technology Agency of the budget of "Science and Technology by Ministries and Agencies" which amounted to about 33 billion DM in 1992, is correspondingly high (see Tab. 5).

The tasks and the importance of the Science and Technology Agency must principally be assessed in association with the complementing tasks of MITI, and with the coordination of activities of these two governmental institutions.

The guidelines of the Science and Technology Agency and the Ministry of Education are exclusively used for work with recombinant microorganisms, plants and animals in laboratories or closed greenhouses in research and development. The guidelines of MITI, MAFF and MHW are used for commercial work and for field trials.

3.2.1 Containment Categories, Classification and Assessment of Experiments

The Science and Technology Agency's "Guidelines for Recombinant DNA Experiments" were enacted by the Prime Minister on August 27, 1979 and have been simplified nine times since then. The core of the guidelines is the differentiation of recombinant DNA work into "standard experiments" and "non-standard experiments". A standard experiment includes experiments with volumes up to 20 liters in the containment levels P1 to P4. Large-scale experiments in the containment levels LS-C, LS-1 and LS-2 are non-standard experiments (Tab. 6). It is intended to give up the differentiation of rDNA work by volume at the text guideline update. In addition, biological safety measures B1 and B2 have been defined (Tab. 7). For categorizing rDNA operations, the organisms are listed into four risk categories, and the resulting

Tab. 4. The Role of the Science and Technology Agency (STA)

Examples of Main Activities

1. Exploratory Research for Advanced Technology
   - Precursory research for embryonic science and technology system
   - Frontier research program
   - Development of new technology transfer
2. Promotion of Science and Technology Aiming at More Affluent Life Styles
   - Promotion of human genome analysis
   - Solution of problems closely related to living
3. Playing Active Roles in International Society through Science and Technology
   - The human frontier science program
   - International thermonuclear experimental reactor project
   - Space station program
4. Promotion of Science and Technology Administration
   - Planning and formulation of science and technology policy
   - Overall coordination functions in science and technology administration
   - Promoting science and technology policy research
5. Promoting Research and Development in Advanced Fields of Science and Technology
   - Nuclear energy, nuclear safety
   - Ocean development, aeronautical technology
   - Earth science and technology
   - Disaster prevention
   - Materials science and technology
   - Life sciences

Functions of the STA

Operation/Coordinating/Monitoring of:
Science and Technology Policy Bureau
Science and Technology Promotion Bureau
Research and Development Bureau
Atomic Bureau, Nuclear Safety Bureau
Advisory bodies
Institutes
Public corporations
Research and development activities in major countries
Administrative structure and technology in Japan

---

containment levels are described in detail in the guidelines. The guidelines describe the criteria which define the classification of a "non-standard" experiment:

- Work with non-characterized microorganisms, if their non-pathogenicity has *not* yet been proven,
- work with genes which code for proteins that are toxic for vertebrates,
- work with host organisms or vectors and certain work with DNA sequences which correspond to Risk groups 3 and 4 or DNA sequences from these organisms (described in appendices 1-(1), 2-(4) and 3-(4) of the guidelines),
- field trials,
- large-scale experiments which correspond to biosafety levels P3 or P4 and which do not satisfy certain criteria defined by the guidelines.

The preparation of characterized proteins having "useful functions", coded for by host–vector systems defined in the STA guidelines corresponding to a containment level of P1 or P2, is categorized as a "standard experiment" and, thus, does not require any further approval by the Science and Technology Agency.

All other operations are classified as "standard experiments". Comparable with the mode of operation of the "Institutional Biosafety Committees" in the USA, "standard experiments" in Japan either are communicated to the Biosafety Committee of the institution by the scientist at the beginning of the project or may be started only *after* approval by this Biosafety Committee of the respective institution. "Non-standard experiments" can be started after the approval of both, the research institution *and* the Science and Technology Agency (Fig. 4). Prior to approval, the Science and Technology Agency arranges a review by the "Recombinant DNA Advisory Committee" in the "Council for Science and Technology". This committee consists of 14 scientists and meets at intervals of six weeks.

The containment criteria summarized in Tabs. 8 and 9 permit the categorization of the above-described experiments (notification or approval for P1, P2, P3 level and non-standard experiments) with microorganisms or mammalian cells for laboratory and large-scale work. The containment conditions for

Tab. 5. Share of the Budget of Science and Technology by Ministers and Agencies

Authorities	Budget Share (%)
Ministry of Education	46.5
Science and Technology Agency	25.8
Ministry of International Trade and Industry	12.1
Defense Agency	5.9
Ministry of Agriculture, Forestry and Fisheries	3.6
Ministry of Health and Welfare	2.9
Ministry of Post and Telecommunication	1.5
Others	1.7
	100.0

Tab. 6. Levels of Physical Containment in Large-Scale Fermentation

	Structure	Fermenters	Mechanisms of Exhaust from Fermenters	Accommodation of Installation for Treatment of Recombinant Organisms (Centrifuge etc.)
LS-C	Facility having well-maintained large-scale fermenters or equivalent facilities	Well-maintained	Designed to minimize release of recombinant organisms	None
LS-1	Well-equipped experimental facilities using various tightly containing apparatuses such as large-scale fermenters	Designed to prevent release of recombinant organisms and to facilitate interior sterilization without opening the vessel	Sterilizing filter or other sterilizing means	Safety cabinet or equivalent means
LS-2	The same as above	The same as above, and especially parts subjected to direct contact with fermenter, such as revolving seal or tubings, must be designed with adequate precaution to prevent leaks of recombinant organisms	Microbe-removing performance better or equivalent to HEPA filter, or other sterilizing means	Class-II safety cabinet or equivalent means
			Means to monitor air-tightness of the apparatus during large-scale culture experiments	

recombinant DNA work with viruses are recorded in Tab. 10. Similar tables for projects with animals, plants and parasites are also part of the guidelines. The operator who is responsible for installations and work subject of the Science and Technology Agency guidelines reports to the Science and Technology Agency annually on all R&D work. These reports are kept confidential by the Science and Technology Agency and are not available for inspection by third parties.

In practice, the Japanese guideline system ensures that the research and development projects can either be started immediately or within a few days or, if the Science and Technology Agency has to review a project, within a few weeks after notification.

Tab. 7. B1 and B2 Level Host–Vector Systems

B1 Level Host–Vector System

a. EK1. The host–vector system composed of *Escherichia coli* K12 strain, which is genetically and physiologically well understood, nontoxic, and poorly viable under natural conditions, and its induced strain as host, and plasmid or bacteriophage as vector that are inconjugatable and not transferable to other bacteria. The host must not have conjugatable plasmids or conventionally introduced bacteriophages

b. SC1. The host–vector system composed of a laboratory-maintained strain of yeast, *Saccharomyces cerevisiae*, as host, and plasmid or minichromosome as vector

c. BS1. The host–vector system composed of an induced strain of *Bacillus subtilis* Marburg 168 which has mutation of multinutritional requirement for amino acids or nucleic acid bases as host, and plasmid or bacteriophage that is inconjugatable and not transferable to other bacteria

d. The host–vector system composed of animal or plant cells (excluding those for differentiation) as host (excluding those cases to produce infectious viruses)

B2 Level Host–Vector System

a. EK2. Those systems belonging to EK1, and composed of a genetically defective host, which has very low viability except under special incubating conditions and a highly host-dependent vector, which has very low transferability to other cells. The system must have a survival ratio for recombinant cells of 10 parts per billion after 24 h of incubation except for special conditions.

3.2.2 Construction of Facilities for Research (STA and MHW Guidelines)

As in the USA, building permits for facilities to carry out general research including recombinant DNA experiments are issued by local governments. These permits are based on the "Construction Standard Law". Within the scope of the authorization process, the general and local regulations are taken into consideration, and competent authorities are involved, e.g., environmental, health and water authorities. The owner is obliged to post a sign at the building site outlining and describing the construction plans and, if appropriate, the construction authority can order a local hearing for the installation which to date has rarely occurred. Close cooperation between the central and prefectural authorities is traditional in Japan. Documents relating to the authorization procedure are forwarded to the responsible ministry (Science and Technology Agency, MITI or MHW) on a voluntary basis for inspection and examination, e.g., of the containment level and safety measures. The central authority notifies the local authority promptly of its position, so that construction permits for rDNA research facilities can be, as a rule, obtained within 3–4 weeks.

Since this procedure has proven to function well, 45 of the 47 prefectures have decided not to introduce any provisions for authorizing facilities to carry out recombinant DNA technology. Exceptions are the prefectures of Aichi (since June 1, 1990) and Kanagawa (since September 1, 1993) in which rDNA operation permits are required for rDNA production facilities. However, these are not required if the internal Safety Committee or the Science and Technology Agency approval procedures are applied.

3.2.3 Construction of Facilities for Production (MHW and MITI Guidelines)

The administrative procedure for the installation of production facilities is similar to that of installing research and experimental facilities by local building permits. The competent ministries are responsible for licensing products, using the respective guidelines (Fig. 3): MHW for recombinant pharmaceuticals and foodstuff additives by consulting the "Pharmaceutical Affairs Bureau" or the "Food and Sanitation Investigation Council". Other industrial products are subject to MITI guidelines. Since there are not yet any marketable products of recombinant organisms, i.e., foods such as the "Flavr Savr™" tomato in the USA, the implementation of respective guidelines has not been considered to be necessary.

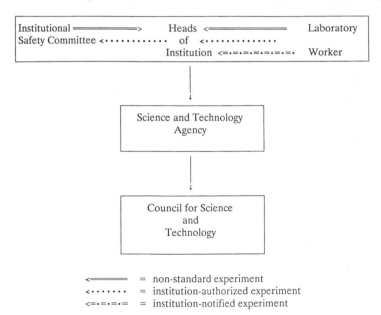

Fig. 4. Procedures for rDNA experiments in Japan.

In contrast to the Food and Drug Administration (FDA) of the USA, special guidelines have been issued in Japan for the licensing of corresponding recombinant DNA products. These guidelines are to assess the properties of recombinant organisms but do not discriminate rDNA methods, thereby encouraging the development of recombinant DNA products. With regard to their tenor, the different guidelines are practically identical as far as technical specifications and classification of rDNA organisms are concerned. Moreover, they implement the OECD recommendations.

The core points of the guidelines are to define the tasks of the "Manufacturing Safety Committee" for which the operator has to assume responsibility, and the classification of the production processes into one of several categories: GILSP (Good Industrial Large Scale Practice), C1, C2 and C3 (Tabs. 11, 12 and 13). It must be emphasized that in the guidelines no particular technical and organizational safety measures are stipulated for recombinant organisms apart from the application of good microbiological practice with regard to GILSP, i.e., application of harmless, non-pathogenic organisms. Under GILSP, recombinant organisms are treated exactly like "natural" microorganisms for production of antibiotics, in the fermentation industry, or for processing milk products. Since the Japanese guidelines are oriented towards the manufacture of safe products, and not to the registration of recombinant DNA operations, as is the case in the EU system, the administrative procedures and practices are greatly simplified.

For assessing the various products, "rDNA Product Approval Committees" have been established by MITI and by MHW on the basis of experience, thereby striving for rapid procedural administration. In Japan, it is common that administrators assume responsibility in industry on rather high levels after their retirement at the age of 55. Detailed knowledge of administrative practice is, therefore, available in the industry and leads to the rapid handling of administrative issues.

3.2.4 Field Trials (MAFF Guidelines)

The "Guidelines for Application of rDNA Organisms in Agriculture, Forestry, Fisheries,

Tab. 8. Containment Levels in Experiments Using Microorganisms or Cultured Mammalian Cells (Below 20 L Volume)

Host-Vector System	DNA Donor	Animal (P2)	Plant (P1)	Attached[a] Table 1-(1) (P3)	Attached[a] (Table 1-(2) (P2)	Attached[a] Table 1-(3) (P1)
EK2 series		P1	Institution notified experiment (P1)	P2	P1	Institution notified experiment (P1)
EK1 series, SC1 series, BS1 series		P2	Institution notified experiment (P1)	P3	P2	Institution notified experiment (P1)
Host-vector system listed in the left column of Attached[a] Table 3		P2	P1	P3	P2	P1
Cultured cell (host) (excepting those for differentiation purposes)	Attached[a] Table 1-(1) [vector]	Non-standard experiment	Non-standard experiment	Non-standard experiment	Non-standard experiment	Non-standard experiment
	Attached[a] Table 1-(2) [vector]	P2	P2	P3	P2	P2
	Attached[a] Table 1-(3) [vector]	P2	P1	P3	P2	P1
Those using Attached[a] Table 1-(1) as either host or vector		Non-standard experiment	Non-standard experiment	Non-standard experiment	Non-standard experiment	Non-standard experiment
Those using Attached[a] Table 1-(2) as either host or vector (excepting those using Attached[a] Table 1-(1))		P2	P2	P3	P2	P2
Host-vector system composed solely of Attached[a] Table 1-(3)		P2	P1	P3	P2	P1

Experiment using microbes which are not identified to species and not confirmed of non-pathogenicity is classified as non-standard experiment.
Those with containment level description only are institution-authorized experiments.
When all DNAs derived from viruses or other organisms have promoter or terminator function only and do not generate infectious virus particles, Attached[a] Table 1 is substituted by Attached[a] Table 2 or Attached[a] Table 3.
[a] So-called Attached Tables mentioned refer to the original source.

the Food Industry and other related Industries" issued in 1989, include all work with recombinant organisms (Fig. 3). This also includes assessment of environmental impact. The guidelines were changed in April 1992 in order to allow the commercial use of small recombinant experimental animals. At the same time, the definition of 'Standard Isolated Fields' was introduced which has led to a facilitation of field trials with transgenic plants.

Tab. 9. Containment Levels in Experiments Using Microorganisms

Host-Vector System	DNA Donor	Animal (P2)	Plant (P1)	Attached[a] Table 1-(1) (P3)	Attached[a] (Table 1-(2)) (P2)	Attached[a] Table 1-(3) (P1)
EK2 series		LS-1	LS-1	LS-2	LS-1	LS-1
EK1 series, SC1 series, BS1 series		LS-2	LS-1	Non-standard experiment	LS-2	LS-1
Host-vector system listed in the left column of Attached[a] Table 3		Non-standard experiment	Non-standard experiment	Non-standard experiment	Non-standard experiment	LS-1
Cultured Cell (host) (excepting those for different-iation purposes)	Attached[a] Table 1-(1) [vector]	Non-standard experiment	Non-standard experiment	Non-standard experiment	Non-standard experiment	Non-standard experiment
	Attached[a] Table 1-(2) [vector]	LS-2	LS-2	Non-standard experiment	LS-2	LS-2
	Attached[a] Table 1-(3) [vector]	LS-2	LS-1	Non-standard experiment	LS-2	LS-1
Those using Attached[a] Table 1-(1) as either host or vector		Non-standard experiment	Non-standard experiment	Non-standard experiment	Non-standard experiment	Non-standard experiment
Those using Attached[a] Table 1-(2) as either host or vector (excepting those using Attached[a] Table 1-(1))		Non-standard experiment	Non-standard experiment	Non-standard experiment	Non-standard experiment	Non-standard experiment
Host-vector system composed solely of Attached[a] Table 1-(3)		Non-standard experiment	Non-standard experiment	Non-standard experiment	Non-standard experiment	Non-standard experiment

DNAs derived from DNA donor must be those having specific useful functions as described in the text. Experiment using microbes which are not identified to species and not confirmed of non-pathogenicity is classified as non-standard experiment.

Those with containment level description only are institution-authorized experiments.

When all DNAs derived from viruses or other organisms have promoter or terminator function only and do not generate infectious virus particles, Attached[a] Table 1 is substituted by Attached[a] Table 2 or Attached[a] Table 3.

[a] So-called Attached Tables refer to the original source.

Guidelines for the commercial use of transgenic fish, cattle and sheep, etc. are in preparation.

In conjunction with the import of transgenic animals, the importer of such animals must carry out examination of the animals and of the conditions under which they are to be kept, as prescribed by the guidelines. Field trials may only be carried out in Japan with the permission of MAFF, and the respective prefecture is to be informed about such trials. Authorization is based on the assumption that the results of experiments carried out before do not indicate any risks. For the sequence of experimental stages in field trials see Tab. 14.

Tab. 10. Containment Levels in Experiments Using Viruses

DNA Donor / Host-Vector System	Attached[a] Table 2-(1) or Attached[a] Table 3-(1) (P3-B1) (P2-B2)	Attached[a] Table 2-(2) or Attached[a] Table 3-(2) (P2-B1) (P1-B2)	Attached[a] Table 2-(3) or Attached[a] Table 3-(3) (P1-B1) (P1-B2)	Attached[a] Table 2-(4) or Attached[a] Table 3-(4) (Non-standard)	(Excluding cases in which these are the sole DNA donor)				
					Animal	Plant	Attached[a] Table 1-(1)	Attached[a] Table 1-(2)	Attached[a] Table 1-(3)
B1 Series Host–Vector System	P3	P2	P1	Non-standard experiment	P2	P1	P3	P2	P1
B2 Series Host–Vector System	P2	P1	P1	Non-standard experiment	P1	P1	P2	P1	P1
Experiment Using Host–Vector System Other than Authorized Host–Vector System	Non-standard experiment	Non-standard experiment	Non-standard experiment	Non-standard experiment	Non-standard experiment	Non-standard experiment	Non-standard experiment	Non-standard experiment	Non-standard experiment

Experiments using microbes which are not identified to species and not confirmed of non-pathogenicity are non-standard experiments.
Experiments for which only containment level is described are institution-authorized experiments.
[a] So-called Attached Tables refer to the original source.

Tab. 11. Criteria of Properties of GMOs Required by GILSP

Host	1. Non-pathogenic
	2. No adventitious agents (virus etc.)
	3. Extended history of safe industrial use or built-in environmental limitations permitting optimal growth in industrial setting but limited survival without adverse consequences in environment
Vector/ Inserted Genes	1. Well characterized and free from known harmful sequences
	2. Limited in size as much as possible to the DNA required to perform the intended function
	Should not increase the stability of the construct in the environment (unless that is requirement of the intended function)
	3. Should be poorly mobilizable
	4. Should not transfer any resistance markers to microorganisms not known to acquire them naturally (if such acquisition could compromise use of drug to control disease agents)
Recombinants	1. Non-pathogenic
	2. As safe in industrial setting as host organism, with limited survival without adverse consequences in environment

Tab. 12. Criteria of Properties of GMOs Required for Manufacturing in Containment Categories 1, 2 and 3

	Recombinant
Category 1	Non-pathogenic, excluding recombinants corresponding to GILSP
Category 2	Rarely develops a disease through infections in humans and has preventive measures and effective therapy
Category 3	Is pathogenic in human and requires very careful handling. Infections and diseases, even if caused, are comparatively less dangerous, and preventive measures and effective therapy

Note: Criteria of properties of recombinants, falling into Categories 2 and 3, are given as a reference.

The guidelines differentiate between field trials with transgenic plants in simulated model environments carried out on 'Standard Isolated Fields', from extensive and commercial field applications without any containment measures. The prefectures which were interested in identifying areas for field trials asked the MAFF to introduce a definition for Standard Isolated Fields: a Standard Isolated Field is defined by a fence of approximately 150 cm in height, facility for incineration within this area and equipment for cleaning the applied machinery. In addition, the guidelines define the technical details for prevention of pollen transfer, transport of transgenic plants and waste treatment. They contain a questionnaire for assessing the properties of the organism to be released.

Prior to granting an authorization, the MAFF obtains a review of the "rDNA Application Special Committee" which consists of 17 scientists of different science areas and meets four times a year (Tab. 15).

Authorizations for small-scale field trials are granted within 3–4 months (Tab. 16). MAFF expects to be able to approve future applications for commercial purposes within a comparable time. By April 1993, 13 permits had been granted within the framework of the guidelines. On the basis of preliminary enquiries from universities, industry and the different prefectures, MAFF expects that the number and type of applications will increase drastically in the next few years.

4 European Union

The Council of Ministers decided in 1986, in view of the harmonization of the European Communities, to realize the European Union (EU) by the year 1993. Among many fields of

Tab. 13. Control Level, Buildings and Facilities

	Containment Categories		
	1	2	3
1. Viable organisms should be handled in a system which physically separates the process from the environment (closed system)	Yes	Yes	Yes
2. Exhaust gases from the closed system should be treated so as to:	Minimize release	Prevent release	Prevent release
3. Sample collection, addition of materials to a closed system and transfer of viable organisms to another closed system, should be performed so as to:	Minimize release	Prevent release	Prevent release
4. Bulk culture fluids should not be removed from the closed system unless the viable organisms have been:	Inactivated by validated means	Inactivated by validated chemical or physical means	Inactivated by validated chemical or physical means
5. Seals should be designed so as to:	Minimize release	Prevent release	Prevent release
6. Closed systems should be located within a controlled area	Optional	Optional	Yes, and purpose built
a) Biohazard signs should be posted	Optional	Yes	Yes
b) Access should be restricted to nominated personnel only	Optional	Yes	Yes, via an airlock
c) Personnel should wear protective clothing	Yes, work clothing	Yes	A complete change
d) Decontamination and washing facilities should be provided for personnel	Yes	Yes	Yes
e) Personnel should shower before leaving the controlled area	No	Optional	Yes
f) Effluent from sinks and showers should be collected and inactivated before release	No	Optional	Yes
g) The controlled area should be adequately ventilated to minimize air contamination	Optional	Optional	Yes
h) The controlled area should be maintained at an air pressure negative to atmosphere	No	Optional	Yes
i) Input air and extract air to the controlled area should be HEPA filtered	No	Optional	Yes
j) The controlled area should be designed to contain spillage of the entire contents of the closed systems	No	Optional	Yes
k) The controlled area should be sealable to permit fumigation	No	Optional	Yes
7. Effluent treatment before final discharge	Inactivated by validated means	Inactivated by validated chemical or physical means	Inactivated by validated chemical or physical means

Categories 2 and 3 are given as reference for the standards of facilities, equipment and control level for more advanced containment.

Tab. 14. Sequence of Experimental Stages and Responsibilities in Field Trials

Type of Experiment	Responsibility
I. Experiments in a "closed system"	STA
II. Experiments in a "semi-closed system" (greenhouses without "containment" equipment)	STA
III. "Small-scale" field trials	MAFF
IV. Extensive, commercial applications	MAFF

Tab. 15. Members of the rDNA Application Special Committee

Speciality	No. of Experts
Molecular Biology	5
Medicine and Veterinary Medicine	3
Plant Biotechnology	3
Animal Breeding	2
Nutrition	2
Ecology	2

Tab. 16. Release Permits Approved until 1993 under the MAFF Guidelines

Year	Number	Plants	Micro-organ.	Animals
1989	3	0	3	0
1990	2	1	1	0
1991	2	1	1	0
1992	18	4	2	12
	25	6	7	12

application that were proposed for harmonization, applications of rDNA technologies were granted priority. The Directive 90/219/ EEC concerning the "contained use of genetically modified microorganisms" (EEC, 1990a) and the Directive 90/220/EEC on the "deliberate release of genetically modified organisms into the environment" (EEC, 1990b) were approved by the Council on April 23,

1990. A further Directive (90/679/EEC) on the "protection of workers from risks related to exposure to biological agents at work" was enacted on November 26, 1990 (EEC, 1990c). All three Directives relate to living microorganisms or organisms, the 'contained use' and the 'deliberate release' Directives deal with genetically modified organisms only.

The implementation of the 'contained use' and 'deliberate release' Directives into national legislation should have been completed within 18 months (i.e., October 23, 1991) whereby different types of national implementation are to be achieved. The Directive concerning the 'contained use of genetically modified micro-organisms' was decided under Article 130s of the treaty establishing the European Economic Community (EEC), and thus, represent minimum demands that can be exceeded by national legislation. The 'deliberate release' Directive was enacted under Article 100a of the afore-mentioned treaty and, therefore, must fulfill the provisions of the Directive and cannot be superseeded by national legislation.

4.1 Directive 90/219/EEC on the "Contained Use of Genetically Modified Microorganisms"

4.1.1 Scope of the Directive

Subject to this Directive are microorganisms whose genetic material has been altered in a way that does not occur naturally by mating and/or natural recombination. The methods and processes leading to such genetic modifications are listed in Annex I A of the Directive 90/219/EEC as standard:

(1) recombinant DNA techniques using vector systems as defined in the Recommendation 82/472/EEC
(2) techniques involving the direct introduction into a microorganism of heritable material prepared outsite the microorganism including micro-injection, macro-injection and micro-encapsulation
(3) cell fusion or hybridizing techniques where living cells with new combinations

of heritable genetic material are formed by the fusion of two or more cells by means of methods that do not occur naturally.

Techniques that are *not* considered to result in genetic modifications assuming that they do not involve the use of recombinant DNA molecules or of genetically modified organisms:

(1) *in vitro* fertilization
(2) conjugation, transduction, transformation or any other natural process
(3) polyploidy induction.

Under the assumption that they do not involve the use of genetically modified microorganisms as recipient or parental organism, techniques of genetic modifications to be excluded from the Directive are:

(1) mutagenesis
(2) the construction and use of somatic animal hybridoma cells (e.g., for the production of monoclonal antibodies)
(3) cell fusion including protoplast fusion of cells from plants which can be produced from traditional breeding methods
(4) self-cloning of non-pathogenic naturally occurring microorganisms which fulfill the criteria of Group I for recipient microorganisms.

4.1.2 Classification of Microorganisms and Operations

The Directive distinguishes between two groups of microorganisms (Art. 4) and two types of operations (Art. 2) which consequently leads to different procedures of notification or permission (Art. 9 and 10).

Microorganisms are classified into:

Group I microorganisms which are harmless and present no risk for human health or the environment (corresponding to WHO Risk group I)
Group II microorganisms that do *not* belong into Group I;

these microorganisms correspond to WHO Risk groups II–IV representing low, moderate or high risk to human health and the environment, respectively.

Operations:

Type A Type A operations mean any operation used for teaching research, development, or non-industrial or non-commercial purposes and which is of small scale; as an example, Directive 90/219/EEC sets 10 liters culture volume or less
Type B Type B operations mean any operation other than a Type A operation.

The user is required to assess the risk to human health and the environment prior to the submission of notifications or permits for DNA operations of any type (Art. 6). In making such an assessment the user shall take due account of the parameters set out in Annex III of the Directive, as far as they are relevant, for any genetically modified microorganism he is proposing to use. Annex III demands detailed information about the donor and recipient organisms, nature and source of the vectors employed, about stability of the microorganisms in terms of genetic traits, toxic or allergenic effects, possible interactions with the environment and methods of detection and decontamination.

4.1.3 Procedures and Duration of Obtaining a Permit

The competent authority responsible for issuing permits for rDNA operations in closed systems is installed by the Member States. The competent authority receives a submission for initiating up rDNA work in writing and confirms its receipt. It has to examine the conformity of the application with the requirements of the Directive, the completeness of the necessary information and the accuracy of the risk assessment. The competent authority reviews measures of waste treatment, safety and emergencies. If necessary, the competent authority has the right to make specific injunctions, to ask for further information and to change conditions under which the application in a closed system is to be carried out. It can also limit the time range of an experiment.

Recombinant DNA experiments in contained systems must be communicated to the national authorities according to Art. 8, 9 and 10 of the Directive. Whether a notification or a permission procedure has to be applied depends on the type of organisms to be used and on the type of operation involved. The national authority has to act upon the notification of an rDNA experiment within 90 days after submission. For notifications, passing of the time limit of 90 days can be taken as an agreement, whereas the proceedings for a permit require a written statement by the authority before work can be assumed.

The *first notification* of rDNA experiments or the first use of a facility with organisms of the *Group I for Type A operations* (research) must be submitted to the competent national authority, including all information listed in the Annex V A of the Directive (Art. 8), no fewer than three months before an experiment is intended to start. This Annex requests information about the persons responsible for the operations, their qualifications and training. Furthermore, a description of the facility and of the proposed work as well as characterization of the organisms to be used, including an evaluation of potential risks involved for human health and the environment has to be submitted. For *further* rDNA research projects involving organisms of Group I, only recording the rDNA work carried out is required. These records must be made available to the competent authority upon request.

The first notification of rDNA experiments involving organisms of *Group I for Type B operations* (industrial or commercial) must be submitted three months before commencing the operation (Art. 9). For this purpose, information in writing in addition to that described in Annex V A must be submitted: employed recipient and donor organisms, origin of the genetic material, purpose of the application, expected results and fermentation volume (Annex V B). For *further* Type B operations with organisms of Group I by the same user in the same facility, a notification with a term of two months is assigned. Work may be started after the expiration of this time period or earlier, if permission is granted by the national authority.

The first realization of experiments with organisms of *Group II* needs a permit. The national authority must be notified three months prior to the commencement of work (Art. 10). All information required for *Type A operations* (Annex V B) and further information with regard to technical details of the facility is to be submitted. Furthermore, details about the handling of the microorganisms, the climatic conditions at the facility and potential hazards arising from the location, the intended containment category, waste treatment and safety measures to be implemented as well as planned protection measures and measures for supervision and control must be supplied (Annex V C). For *Type B operations*, extensive data in addition to the information requested in Annex V B and V C about responsible persons and the maximum number of persons working in the facility, the location and setup of the facility, the used microorganisms, plans for waste treatment, prevention of accidents and emergency measures must be included. *Further experiments* of Type B operations with Group II organisms require written notification two months in advance, and written consent of the competent authority is necessary for commencement.

If the competent authority requires further information from the user, the time to submit this information is not subtracted from the legal time limit for granting of authorization. This also holds in the case of a public hearing. Tab. 17 provides a summary of the different legal procedures and their terms for processing. Public hearings during the process of the acquisition of a permit rest within the authority of the Member States.

Tab. 17. Summary of the Procedures and Terms for Notifications and Permits for the Contained Use of GMOs

Microorganisms	Research Projects Type A Operations		Commercial/Industrial Projects Type B Operations	
	First Operation	Further Operation	First Operation	Further Operations
Group I	Notification 3 months	Record keeping —	Permit 3 months	Notification 2 months
Group II	Permit 3 months	Notification 2 months	Permit 3 months	Permit 3 months

4.2 Directive 90/220/EEC on the "Deliberate Release into the Environment of Genetically Modified Organisms"

Subject of the Directive 90/220/EEC are GMOs that are intended to be deliberately released into the environment as well as products containing or consisting of GMOs which are intended to be placed on the market. This Directive, similarly to the 'contained use' Directive, also deals with organisms whose genetic material has been altered in a way that does not occur naturally by mating and/or natural recombination. Whereas the 'contained use' Directive relates to modified microorganisms only, the 'deliberate release' Directive covers *all* genetically modified organisms.

All methods of genetic modification falling into the scope of the 'deliberate release' Directive are the same as described in the 'contained use' Directive. Provided that no genetically modified organisms are involved as donor or recipient, the following procedures are exempt from the Directive:

- mutagenesis
- cell fusion (including protoplast fusion) of plant cells where the resulting organisms can also be produced by traditional breeding methods.

4.2.1 Procedures and Duration of Obtaining a Permit for the 'Deliberate Release' of GMOs

A submission for an experimental field trial must contain the following extensive information described in Annex II of the Directive. The following list is largely summarized:

(1) Name of person(s) responsible for planning and carrying out the release including the names of those responsible for supervision, monitoring and safety, in particular, name and qualifications of the responsible scientists and the personnel involved in carrying out the release
(2) detailed characteristics of the donors, the recipient or parental organisms as well as the characteristics of the vectors used
(3) detailed characteristics of the modified organism, description of inserts and/or vector construction including their sequence and purity
(4) detailed description of the final GMO including health considerations
(5) information on the release with description of the site, size, quantity of GMOs and frequency and duration of the release including worker's protection and techniques foreseen for elimination or inactivation of GMOs at the end of the experiment
(6) information on the environment of the release site and the wider environment including geographical, geological and climatic characteristics and description of

target and non-target ecosystems likely to be affected

(7) information relating to interactions between the GMOs and environment including survival, multiplication, dissemination and genetic transfer capability of GMOs and their potential environmental impact

(8) information on monitoring, control, waste treatment and emergency response plans.

According to Art. 5, the release of any GMOs on the territory of a Member State is to be submitted to the competent authority of the respective Member State.

After acknowledgement of the submission, the competent authority is to examine the petition for compliance with the Directive and assess potential environmental risks within 90 days. If additional information is needed or a public hearing is necessary, the time period during which the competent authorities are awaiting further information or carry out a public hearing is not taken into account. The decision whether or not to carry out a public hearing is determined by the respective Member State.

Within 30 days after a submission, the national authorities shall send to the Commission a summary of the notification received. The Commission immediately forwards this summary to other Member States, which may, within 30 days, ask for further information or present observations through the Commission or directly. The competent authorities may take the observations of the Member States into consideration and inform the petitioner within the 90 day term of its decision in written form.

4.2.2 Procedure and Duration of the 'Placing on the Market' of Products Containing or Consisting of GMOs

Before a product, consisting of or containing GMOs, is placed on the market, the manufacturer or the importer shall submit a peti- tion to the competent authority of the Member State where the product is intended to be placed on the market for the first time. The notification must contain all information called for in Annex II of the Directive. The information has also to take into account the geographic and climatic diversity of intended release sites. Additional information according to Annex III is required and must be included in the notification:

(1) name of the product and names of GMO therein

(2) name of the manufacturer or distributor and his address in the Community

(3) specificity of the product, exact conditions of use including, when appropriate, the type of environment and/or geographical area(s) of the Community for which the product is suited

(4) the type of expected use: industry, agriculture and skilled trades, consumer use by public at large.

Further information shall be provided, when relevant, in addition to the above-mentioned instruction in accordance with Art. 11:

(5) measures to be taken in case of unintended release or misuse

(6) specific instructions or recommendations for storage and handling

(7) estimated production in and/or imports to the Community

(8) proposed packaging which must be appropriate so as the avoid unintended release of GMOs during storage, or at a later stage

(9) proposed labeling. This must include, at least in summarized form, the information referred to in points 1, 2, 3, 5 and 6.

Within 90 days after the notification date, the competent national authority must either pass the dossier, together with a favorable recommendation, on to the Commission or inform the petitioner that a permit is withheld because the proposed release does not fulfill the provisions of the Directive. The dossier forwarded to the Commission will include a summary of the notification as well as a description of the conditions under which the competent authority would license the proposed product for the placing on the market. The Commission forwards this dossier *immediately* to the competent authorities of the Member States. If Member States do not object within 60 days, the competent authority may grant the permit. Should other Member States raise substantiated objections against the placing on the market of a product and a consent cannot be reached within a time period of 60 days, the Commission decides on the petition.

Public hearings are not scheduled for the placing on the market of products containing or consisting of genetically modified organisms. Once a product is licensed by written consent, it may be used without further notification throughout the European Union provided the specific conditions of use with regard to environmental facts are strictly observed. If there are, however, reasons to assume that an approved product presents a risk to human health or the environment, it may be transitorily restricted or prohibited from use or marketing.

In addition to the specified time limits, a prolonged period of time must be expected for the final passing of a resolution of the Commission or the national authorities. The total time for this procedure is expected to exceed 150 days.

4.3 Directive 90/679/EEC on the "Protection of Workers from Risks Related to Exposure to Biological Agents at Work"

A further Council Directive for the "protection of workers from risks related to exposure to biological agents at work" was enacted on November 26, 1990 and was to be implemented into national legislation within three years by the Member States of the European Union. This Directive, based on Article 118a of the EEC Treaty provides also minimum demands and deals with natural, non-modified and genetically modified *pathogenic* microorganisms. The Directive applies to the deliberate and the unintentional use of or exposure to pathogenic microorganisms, including their toxicity or their allergenic potential. Each activity of workers exposed or potentially exposed to biological agents must be assessed for potential risk by the employer. A list of activities for the unintended use of biological agents is provided in Annex I of the Directive (Tab. 18).

Tab. 18. Indicative List of Activities According to Annex I and Article 4 (2) of Directive 90/679/EEC

1. Work in food production plants
2. Work in agriculture
3. Work activities where there is contact with animals and/or products of animal origin
4. Work in health care, including isolation and *post mortem* units
5. Work in clinical, veterinary and diagnostic laboratories, excluding diagnostic microbiological laboratories
6. Work in refuse disposal plants
7. Work in sewage purification installations

This Directive must be considered as "horizontal" and is a non-discriminating European legislation for occupational health and safety that makes relating provisions in the technology-based Directive 90/219 unnecessary.

Furthermore, "vertical" legislation for areas of application of products or product groups manufactured by methods of recombinant DNA technology are specified in the Directive 91/414/EEC (EEC, 1991a) that relates to the placing on the market of pesticides, regulations on the authorization of drugs and on 'novel food' (EEC, 1990d) which is still in the draft stage, at present.

5 Concluding Remarks

In the *United States* specific legislations for operations with recombinant DNA techniques were not enacted and the regulatory structure of the US is based on existing legislation. Recombinant DNA operations are carried out under the NIH Guidelines. Recombinant DNA projects are reviewed and approved by the local Institutional Biosafety Committee (IBC) in self-responsibility for all biosafety levels. Such IBC are not foreseen in the European regulations. Only very few rDNA operations have to be reviewed by the Recombinant DNA Advisory Committee and approved by the Director, NIH. These include, for example, the production of toxins and the application of rDNA techniques in humans. Construction of facilities is approved by the competent State and local governments independent of the research or production technology applied and of product approval by Federal agencies. The NIH Guidelines for recombinant DNA technology in closed systems and the USDA rules for field trials represent the "state of the art" with regard to safe use of genetically modified organisms. The US regulatory system is risk-based and focuses on the characteristics and potential risks of *the product* and *not* on *the process* by which it is created.

The accumulated experience with genetically modified organisms by all US regulatory agencies did not provide any evidence that a unique hazard potential exists, neither through the use of rDNA techniques nor through the transfer of genes between unrelated organisms. This scientific view and the experience is shared in Europe by the scientific community and the industry. Besides the political and the financial support of the government, this science-based approach of regulation of rDNA operations has greatly contributed to the rapid progress of modern biotechnology and the leadership of the United States in this field. It has been clearly demonstrated that this system of self-responsibility and compliance is able to provide safety and at the same time allows fast scientific progress. The US regulatory system – with regard to flexibility, short processing and approval times of research projects and the rapid adaptation of regulations to scientific progress – is pragmatic, scientifically appropriate, and thus, highly competitive.

In *Japan*, no specific laws for the construction of research and production facilities and recombinant DNA operations in research, development and production have been enacted. Instead, a rather complex system of

guidelines of different ministries ensures reliable and rapid examination of each project proposal. All recombinant DNA work is registered under the guidelines. These guidelines are product-based and do not lead to discrimination between recombinant DNA work and classical biotechnological processes with non-modified microorganisms. The Japanese regulatory system provides great flexibility and permits rapid adaptation to scientific progress. For the user, it provides fast approval of projects, minimizes bureaucratic burdens and permits wide applications within the industry. The experience gathered in Japan – corresponding to the USA and the OECD – has not presented any evidence that a unique potential for risk arises from the use of recombinant DNA methods or from the exchange of genes between unrelated organisms.

Similar to the US system, the Japanese system of guidelines also permits a rapid and flexible exploitation of modern biotechnology. The application of recombinant DNA technology is not only politically encouraged by the Japanese government, but is also supported by numerous programs coordinated by different ministries and by continuous adjustment of the guidelines.

In the *European Union*, three Directives were enacted by the Commission in 1990 that deal with the 'contained use' of GMO, the 'deliberate release' of GMO and the 'protection of workers from risks of biological agents at work'. All three Directives relate to living organisms or microorganisms, the 'contained use' and the 'deliberate release' Directives deal only with genetically modified organisms.

In Europe recombinant DNA technology is primarily considered as potentially hazardous, in contrast to the United States and Japan. Europe has adopted a catastrophe scenario of new biotechnology. The safety of biotechnological methods and processes has to be proven case by case in rather complex and time-consuming regulatory procedures. Paradoxically, this has also to be carried out for work with organisms that are classified by definition as harmless. The "philosophy" to control methods and processes must inevitably lead to an outsized bureaucratic apparatus and in consequence to a slow and costly procedure to control a whole technology field instead of primarily controlling the products where real risks may emerge. This different approach to control the potential risks of research methods or production processes certainly impairs European competitiveness with regard to the United States and Japan.

On the other hand, from a regulatory point of view all non-living *products of biotechnology* are treated in the same way as products of any other production process. This doubling of regulatory controls by controlling research methods and the production processes *and* the product certainly does not contribute to Europe's competitiveness.

Astonishingly, biotechnology in the pharmaceutical research and development domain is supervised by the Environment Directorate – a development that is dangerous to European competitiveness and scientific progress because non-experts are involved. It is difficult to understand that, for example, medical experimentation should be guided by a body that was set up to protect the environment. Moreover, it is hard to believe that such an institution would be able to support and promote scientific progress of a key technology that it regards primarily as catastrophe-prone.

A good example of how the basic philosophy of these Directives led to over-regulation and to difficulties that hamper an efficient utilization of modern biotechnology is the German Gene Technology Act that has recently been amended because of the remarkable constraints it has imposed on science and industry.

In Germany, about 80% of all research projects at universities and industry are carried out at biosafety level S1 which, by definition, does not impose any risks to humans and to the environment. Further 15% of projects are classified into biosafety level S2 which, according to the state of science and technology, provides only moderate risks for human health or the environment. These two biosafety levels of no or moderate risks were under heavy administrative control and produced bureaucratic burdens that significantly inhibited scientific projects. When the reasons for these burdens were analyzed, one came to the conclusion that the administrative execution of the law certainly contained some "home-made" elements – due to the federal structure of Germany – but the main problems are caused by the provisions of the 'contained use' Directive.

It is difficult to understand why, in Europe, the handling of genetically modified organisms is considered to be more dangerous than the handling of natural microorganisms. Risks occur only when pathogenic microorganisms are involved in the experimentation. Genetic modification of pathogens does not automatically render them more dangerous. Safety assessment is made precautionary on a case-by-case basis and involves analysis of all elements of a recombinant DNA experiment, i.e., the recipient host organism, the vector and the coding DNA sequence that is transferred. The experience of more than 100 years in handling pathogenic microorganisms ensures a high degree of safety as long as a variety of well proven precautionary safety measures are observed. Moreover, in thousands and thousands of experiments and productions worldwide, no accidents specific for rDNA technology have been reported. Here again, European over-regulation contrasts sharply to the American and Japanese approach to biotechnology: recombinant DNA technology does not present risks that are any different from those presented by the unmodified, "natural" host organism. Non-scientific considerations of regulating biotechnology provide thus a serious barrier for research and scientific education.

The described legal and regulatory environments of the United States and Japan give a clear competitive advantage over companies based in Europe where even basic research projects are weighed down with bureaucracy that considerably inhibits scientific progress. In the United States and Japan, universities and industry have the opportunity to conduct their research under their own responsibility at least with microorganisms of no and only moderate risk for humans and the environment.

On the contrary, the EU system reflects the view of the EU Commission that rDNA techniques require – without exemption – regulatory review and control specific for all rDNA operations. It has to be expected that the existing EC system is neither competitive nor will it be adaptable to scientific progress in a reasonable period of time. Therefore, it needs to be replaced by a regulatory system that is based on the characteristics and the properties of the recombinant organisms and their products similar to the US system, and not on the methods by which these products are manufactured. The different regulatory systems of the United States, Japan and the European Union are compared and summarized in Tab. 19.

During the recent process of amending the German Gene Technology Act, the Bundestag has urged the Federal Government to negotiate with the EU Commission for change of the respective Directives to restore competitiveness for the German and European biotechnology industry (DEUTSCHER BUNDESTAG, 1992).

Similarly, the House of Lords (Select Committee 'Science and Technology') has recently extensively reviewed the regulations for recombinant DNA technology of the United Kingdom with regard to the competitiveness of the British industry. These regulations are, of course, based on the EC Directives 'contained use' and 'deliberate release'.

The Select Committee undertook this enquiry following allegations by industry that recent changes in the regulations governing biotechnology both in contained use and following deliberate release into the environment were likely to place United Kingdom industry at a competitive disadvantage, particularly in comparison with non-EU competitor countries like the United States and Japan. The Committee also found that modern biotechnology is an exciting and continually evolving set of applications of molecular and cell biology. Application of modern biotechnology has already led to medicinal and other products, extensive agricultural applications are imminent as well as somatic gene therapy for hitherto incurable hereditary conditions. Early fears about GMOs in 'contained use' turned out to be unfounded. As a principle, existing laboratory (Good Microbiological Practice, GMP) and industrial (Good Industrial Large-Scale Practice, GILSP) practices provide sufficient safeguards under the pur-

view of institutional biosafety committees except where pathogens are involved.

In framing the EU Directives, the European Commission took an excessively precautionary line which, in terms of scientific knowledge, was already out of date when they were being prepared in the late 1980s. Advice to that effect appears to have been ignored. It is vital for the future development of biotechnology regulation that Commission policy be coherent.

The Committee on Science and Technology found that the current European regulatory regime is unscientific in that both sets of regulations fail to discriminate between activities involving real risk and those which do not. Moreover, they scrutinize any activity of genetic modification which forms part of the process of making a product rather than the better targeted and more economical method of regulating the product. The regulations are, therefore, bureaucratic, costly and time-consuming and are an unnecessary burden to academic research and industry alike.

Other factors such as investment and intellectual property rights are equally if not more important in determining the competitiveness in modern biotechnology. Any regulations that cannot be justified on scientific or public interest grounds must be viewed critically and the respective EC Directives must be revised.

The benefits of biotechnology are well proven. Biotechnology and biotechnological products are likely to yield enormous future benefits to mankind. Biotechnology is a growing area, and in all areas where biotechnology has applications people should be able to exploit its economic benefits. Subject of regulations may be necessary to meet identifiable disbenefits and preserve safety. With a few exceptions involving bacterial or viral vectors, deliberate release of GMOs is not inherently dangerous.

The Committee recommends especially that the EC Directive 'contained use' should be amended to substitute the system of *risk assessment* in place of the present classifications. Pending restoration of a risk-based system, under the current contained use regulations the use of safe organisms (Group I) should be subject only to notification whatev-

Tab. 19. Comparison of Regulations on rDNA Technology: United States (USA), Japan (JA), European Union (EU)

Criteria	USA	JA	EU	Comment
Risk Assessment:				

US/JA	EU (90/219)	(90/679)
GLP/GILSP	I	1
Risk Group 1	II	2
Risk Group 2	II	3
Risk Group 3	II	4

Criteria	USA	JA	EU	Comment
GLP/GILSP	exempt	exempt	notification	
Risk Group 1	IBC	IBC	CA	
Risk Group 2	IBC	IBC	CA	
Risk Group 3	IBC	no exp.	CA	
Risk-based regulations	yes	yes	no	
Technology-based regulations	no	no	yes	
Product-based, non-discriminating regulations	yes	yes	no	
Standard experiments	yes*	yes*	no	* Approval by IBC
Non-standard experiments	yes*	yes*	no	* Approval by RAC
"Small-volume" provision	(no)	no	yes	
Delay (60–90 days) by legal provisions	no	no	yes	
Duplication of approvals	no	no	yes	90/219/EEC, 90/220/EEC, product approval
Overlapping regulations with inconsistencies	no	no	yes	90/219/EEC, 90/679/EEC
Regulations specific for rDNA products	no	no	yes	
Regulations easy to modify	yes	yes	no*	* EU: European Parliament and Council of Ministers
Authorization for GMO field trials	yes	yes	yes	
Simplified procedure for field trials (30 days notice)	yes	no	no	

GLP, Good Laboratory Practice; GILSP, Good Industrial Large Scale Practice; CA, Competent Authority; IBC, Institutional Biosafety Committee; RAC, Recombinant DNA Advisory Committee er the scale of operations. The "Byzantine structure" of the 'deliberate release' Directive should also be amended to permit a vastly accelerated and simplified procedure of notification. Applications under the existing regulations should be processed in no more than 30 days.

A GMO-derived product should be subject only to a sectoral regulatory regime under existing product legislation. Evolution from process-based to product-based regulation should be accelerated rapidly. No case can be made for the universal generic labeling of GMO-derived foods or food constituents. Calls for such labeling should be rejected.

Process-based regulation should be retained for research and development in those limited areas where regulation is required, i.e., work involving pathogenic organisms, and for deliberate release of GMOs outside the low risk category. Work on further process-based EC Draft Directives should cease immediately. The implementation of the EU Directives on which the regulations are based is so uneven as to create inequalities even within the Community. A "Fourth Hurdle" of socio-economic need must not be introduced as an additional criterion in product regulation of biotechnology.

The Committee further points out that promotion of *public understanding* is important and that biotechnology products will ultimately gain public acceptance because they are desirable and reliable. Education in schools is one of the most important methods of introducing familiarity with the concepts of biotechnology. In short term, scientists and industry with the support of government have the chief responsibility for promoting wider public understanding of biotechnology by appropriate means.

In a recently published White Paper (EEC, 1993) on "Growth, Competitiveness, Employment", the Commission states that "regulations concerning the safety of applications of the new biotechnology is necessary to ensure harmonization, safety, and public acceptance. However, the current horizontal approach is unfavourably perceived by scientists and industry as introducing constrains on basic and applied research and its diffusion and hence having unfavourable effects on EC competi-

tiveness." Although, the Commission intends "to make full use of the possibilities which exist in the present regulatory framework on flexibility and simplification of procedures as well as technical adaptation. To sustain a high level of "environmental protection and to underpin public acceptance, it is important to reinforce and pool the scientific support for regulations", the correction of some procedural and technical details is not enough to retain European competitiveness. It is the basic "philosophy" of the legislation on modern biotechnology which must be changed from the present method- and process-oriented regulatory framework to a science-based, product-oriented regulatory structure. If the EU does not wish to be left in the 20th century while USA and Japan move steadily forward into the 21st, the Commission must urgently reconsider its environmental approach to biotechnology and call for science-based regulations on products.

6 References

CLINTON, W., GORE, A. (1993), *Technology for America's Economic Growth, A New Direction to Build Economic Strength*, Policy Statement, February 22, 1993.

DEUTSCHER BUNDESTAG (German Federal Parliament) (1992), 12. Wahlperiode: Beschlußempfehlung und Bericht des Ausschusses für Forschung, Technologie und Technikfolgenabschätzung (20. Ausschuß), *Drucksache* **12**/3658 (December 6, 1992).

EEC (1990a), Council Directive of 23 April 1990 on the contained use of genetically modified microorganisms (90/219/EEC), *Off. J. Eur. Commun.* No. L 117/1, May 8, 1990.

EEC (1990b), Council Directive of 23 April 1990 on the deliberate release into the environment of genetically modified organisms (90/220/EEC), *Off. J. Eur. Commun.* No. L 117/15, May 8, 1990.

EEC (1990c), Council Directive of 26 November 1990 on the protection of workers from risks related to exposure to biological agents at work (seventh individual Directive within the meaning of Article 16 (1) of Directive 89/391/EEC (90/679/EEC), *Off. J. Eur. Commun.* No. L 374/1, December 31, 1990.

EEC (1990d), Novel Food – 92/C190/4 Proposal of the Council, *Off. J. Eur. Commun.* No. C 190/3, July 29, 1990.

EEC (1991), Council Directive of 15 July 1991 concerning the placing of plant protection products on the market (91/414/EEC), *Off. J. Eur. Commun.* No. L 230/1, August 19, 1991.

EEC (1993), Commission of the European Communities: Growth, Competitiveness, Employment. The Challenge and Ways Forward into the 21st Century, *Bull. Eur. Commun., Suppl.* **6**/93.

MEAGHER, M. M. (1992), Design of a university-based biosafety level 2, large-scale fermentation facility, *BioPharm* **30** (October).

MEDLEY, T. L. (1993), *Regulation of Genetically Engineered Plants – Status Report*, Bioindustry Organization (BIO), Food and Agriculture Section (September 23, 1993).

MINISTRY OF THE ENVIRONMENT (1991), *The 1991 Quality of the Environment in Japan.*

NIH (National Institutes of Health) (1986a), Department of Health and Human Services, Guidelines for Research Involving Recombinant DNA Molecules; Notice, *Fed. Reg.* **51**, 16.358–16.985, No. 88 (May 7, 1986).

NIH (National Institutes of Health) (1986b), Appendix L, NIH Guidelines: Release into the Environment of Certain Plants, *Fed. Reg.* **51**, 16.984, No. 88 (May 7, 1986).

NIH (National Institutes of Health) (1991), Amendment of Appendix K of the NIH Guidelines: Physical Containment for Large-Scale Uses of Organisms Containing Recombinant

DNA Molecules, *Fed. Reg.* **56**, 33.178, No. 138 (July 18, 1991).

NRC (National Research Council) (1989), *Field Testing Genetically Modified Organisms.*

OECD (Organisation for Economic Cooperation and Development) (1986), *Recombinant DNA Safety Considerations*, Safety considerations for industrial, agricultural and environmental applications of organisms derived by recombinant DNA techniques, Paris: OECD.

OECD (1992), *Safety Considerations for Biotechnology 1992*, Paris: OECD.

OSTP (Office of Science and Technology (Policy) (1992a), Exercise of Federal Oversight Within the Scope of Statutory Authority: Planned Introductions of Biotechnology Products into the Environment, *Fed. Reg.* **57**, 6.755, No. 39 (February 27, 1992).

OSTP (Office of Science and Technology Policy) (1992b), Exercise of Federal Oversight Within the Scope of Statutory Authority: Planned Introductions of Biotechnology Products into the Environment, *Fed. Reg.* **57**, 6.780.

PAYNE, J. (1992), *USDA's Biotechnology Regulatory Relief Initiative*, USDA Backgrounder. News Division, Office of Public Affairs, US Department of Agriculture (October 30, 1992).

USDA/APHIS (US Department of Agriculture, Animal and Plant Health Service) (1993), Genetically Engineered Organisms and Products; Notification Procedures for the Introduction of Certain Regulated Articles; and Petition for Nonregulated Status, *Fed. Reg.* **58**, 17.044, No. 60 (March 31, 1993).

III. Intellectual Property and Bioinformatics

9 Biotechnology and Intellectual Property

JOSEPH STRAUS

München, Federal Republic of Germany,
Ithaca, NY 14853-4901, USA

1 Notion of Intellectual Property as Legal and Economic Instrument

In order to promote technical, economic and social development, society has for centuries made use of a special, inexpensive, but effective legal instrument, that of protection of inventions and other creations of human intellect. Despite the changes that have taken place and the differences in the economic and social parameters, the main objectives of protection have basically remained unchanged since its beginnings in the system of inventors' privileges instituted at the close of the Middle Ages. That is to say, creations of the mind are given recognition by the grant of exclusive rights limited in time, the inventor is rewarded for his/her achievements that are useful to society, industry is encouraged to invent, invest and innovate and, finally, disclosure and dissemination of technical knowledge are promoted (BEIER and STRAUS, 1977; BEIER, 1980).

In this area of law in which economic and legal motives of the indicated kind are determinant, the implications of biotechnology in general and of genetic engineering more specifically are to be viewed under aspects that in many respects differ from those of the debate on safety and avoidance of risks where criteria of ethics and morals dominate the stage. The prime aim here is not to regulate but to promote the new technology by opening and adapting the existing and proven instruments of industrial property protection, such as patents and plant breeders' rights, to the new technology. Although the patent system and the plant breeders' rights system are not in their entirety limited to the promotion of technological progress and the granting of protection, there is but little place in these systems for ethical and moral judgements (BEIER and STRAUS, 1986).

2 Intellectual Property Rights as Exclusive Rights

A patent is an exclusive right, limited in time, to exploit an invention, that is to say an instruction for solving a (technical) problem by planned action (German Federal Supreme Court, 1969), provided the solution found is novel, involves an inventive step (is non-obvious) and is susceptible of industrial application (useful), i.e., meets the patentability criteria as provided under practically all patent laws. Third parties may not use the invention without the permission of the patent owner. Where the invention also manifests itself physically, the exclusive rights granted by the patent also extend to the physical form, such as a living organism, be it a microorganism, plant, or animal. If the owner of the proprietary right of the physical form is not the same person as the owner of the right of the immaterial good which is the invention and of which the physical form is a manifestation, the holder of each must respect the rights of the other.

The granting of a patent does not confer on the patent owner the right to make use of the patented invention. He or she may do so if it does not bring him or her into conflict with the law, for example, with health or animal welfare legislation or with rights of third parties (STRAUS, 1992). By excluding from the effects of the patent acts done for experimental purposes relating to the subject matter of the patented invention, legislators attempt to secure that patents by and large cannot inhibit the scientific and technological progress: experiments aimed also at perfecting, improving and further developing protected inventions under most patent laws do not constitute acts of patent infringement (HANTMAN, 1985; CHROCZIEL, 1986; STRAUS, 1993a).

Historically a patent has always represented a bargain between the inventor and the society: in exchange for the disclosure of his/her invention, the inventor is granted an exclusive right! Thus, the *quid pro quo* for granting the patent is the disclosure of the invention, of the solution for which protection is claimed. This disclosure must be sufficiently clear for the invention to be reproduced by anyone skilled in the art. Taking into account the specific problems posed to inventors in their attempts to repeatably describe inventions in the field of living matter, patent laws have adjusted the disclosure requirements: if the written description is not sufficient to reproduce the invention, the inventor may and must furnish further information by depositing biological material in depository institutions which are publicly accessible, long-term and impartial. International legal instruments, such as the Budapest Treaty on the International Recognition of the Deposit of Microorganisms for the Purposes of Patent Procedure of 1977, have been adopted in order to enable the inventor to comply with the disclosure requirement in a number of countries by making one deposit in an internationally recognized depository (BEIER et al., 1985; STRAUS and MOUFANG, 1990). Special rules such as the Rule 28 of the Implementing Regulation of European Patent Convention (EPC) relating to deposit and release conditions have been implemented in international conventions or national laws. However, in most countries the courts have provided for the necessary adjustment of the enabling disclosure requirement (STRAUS and MOUFANG, 1990). As a consequence, within 18 months from the first filing date at the latest, e.g., patent laws of all European countries and the EPC guarantee transparency. From that point onward the contents of a patent application, and also any deposited material are available to the public, and experiments by third parties aimed at improving and perfecting the invention are both possible and permitted in patent law terms. Under the US patent law system, the point in time is the patent grant which, however, on average occurs also some 20 months from the filing date (STRAUS and MOUFANG, 1990). Thus, while it is certainly true that expectation of profit discourages open discussions of technical details during the critical phase before patent filing (KEVLES and HOOD, 1992), it is also true that the situation is worse if scientists producing technically and commercially applicable research results have no prospects for patent protection. Then secrecy is the only alternative and in most cases the alternative which is actually used (BRENNER, 1990).

3 Patenting of Living Organisms – Brief Historical Retrospect

Contrary to some widely disseminated views, patenting of living organisms has a long-standing tradition. The first known patent for a living organism, a yeast ("ferment de poche") was granted in Finland in 1843 (LOMMI, 1989), and the United States Patent and Trademark Office granted to LOUIS PASTEUR as early as 1873 a patent with claims for, among others, a "yeast free from organic germs of disease as an article of manufacture" (FEDERICO, 1937). Patents were granted in the US for an antitoxic serum in 1877, for a bacterial vaccine in 1904 and for a viral vaccine in 1916 (COOPER, 1982). In Germany, e.g., patents were granted from the very beginning, i.e., from 1877 onward, for culturing of yeasts, the preparation of bread and beer, and the production of vinegar. Later on, patents were granted for the fermentative production of butyl alcohol and acetone with the aid of bacteria. After the discovery of the antibiotic penicillin, patents were also granted for aureomycin, streptomycin, tetracycline, etc., produced with the aid of known microorganisms or those to be discovered through fermentative methods. In the thirties of this century, the German Patent Office started to allow claims for agricultural cultivating processes, as well as for new varieties of plants. Later on, similar practices were applied by the patent offices of Belgium, France, Hungary, Italy, Japan and Sweden. However, in many countries plants were not admitted to patent protection primarily because of the difficulties of the breeders to meet the enabling disclosure requirement in the written specification (STRAUS, 1985). Already in 1930 the United States of America extended patent protection to asexually reproduced plants, by adopting a special Plant Patent Act which has eventually become a part of the Patent Act – 35 USC (ROSSMAN, 1931; COOPER, 1982).

It was well before the event of genetic engineering that the German Federal Supreme

Court in a at that time undisputed decision (1 IIC 136 [1970] "Red Dove") in 1969 explicitly declared animal breeding methods and their resulting products patentable subject matter. When referring to the concept of invention to be applied for in the field of biology, the court stated *inter alia*:

"We are concerned with the question whether biological phenomena and forces can be treated the same way as those of a technical nature. In this connection it is again of little importance what the legislator in 1877 considered to be 'technology', but rather how the biological phenomena and forces are to be understood and classified in the present state of sciences... A teaching to methodically utilize controllable natural forces to achieve a causal, perceivable result that could be considered patentable provided that teaching meets the general prerequisites of industrial application" (1 IIC 138 [1970]).

Whereas the "Red Dove" decision of the German Federal Supreme Court has remained virtually unnoted outside professional circles, in 1980 a decision handed down by the US Supreme Court has attracted wide attention and had far reaching legal and economic implications. The Supreme Court, after having reviewed the historical development of statutory and case law, ruled in its widely known decision "Diamond vs. Chakrabarty" concerning a human-made genetically engineered bacterium capable of breaking down multiple components of crude oil, that CHA-KRABARTY's claim was "not to a hitherto unknown natural phenomenon, but to non-naturally occurring manufacture or composition of matter – a product of human ingenuity having distinctive name, character, and use". The claim was thus to a new bacterium, which has potential for significant utility, with markedly different characteristics from any found in nature, thereby qualifying the bacterium as patentable subject matter under Section 101 of the United States Patent Law. Moreover, the Supreme Court stated that "Congress recognized that relevant distinction was not between living and inanimate things, but between products of nature, whether living or not, and human made inventions." (206 USPQ 193 [1980]).

4 Plant Breeders' Rights

Although patents had been granted for plants and plant varieties in a number of countries since the 1930s, the fact that this was not the case in many others let efforts to grant special protection for new varieties of plants emerging in national laws continue. These efforts finally led, on an international level, to the conclusion of the "International Convention for the Protection of New Varieties of Plants" in Paris on 2nd December 1961 (the so-called UPOV Convention).

Bearing in mind the difficulties which plant breeders had encountered in attempting to patent plant varieties, the UPOV Convention adopted new requirements for protection, tailored to the characteristics of the special subject matter: under the scheme of this convention the new plant variety, in order to be eligible for protection, must be clearly distinguishable by one or more important characteristics from any other variety whose existence is a matter of common knowledge at the time when protection is applied for; it must be sufficiently homogeneous, having regard to particular features of its sexual reproduction or vegetative propagation, and it must be stable in its essential characteristics (Articles 6a–d). The novelty concept as adopted under this convention is a very liberal one, providing *inter alia* a remarkable grace period for the breeder. The convention, as revised in 1972 and 1978 and still in force for all member countries, on the other hand, restricts the rights of the breeder to commercial production and marketing of the propagating material, as such of the protected variety. Thus, it does not cover the commercializing of the end products, e.g., neither fruits, leaves, roots, etc., nor seeds sold for sowing and subsequent growing into a crop of wheat to be sold for milling into flour (BEIER et al., 1985). Moreover, the system of plant breeders' rights under the 1978 UPOV Act is governed by the principle of independence known as "breeders' privilege", since no authorization by the breeder for the utilization of the protected variety as an initial source of variation for the purpose of creating new varieties as well as for the marketing of such varieties is required

(Article 5 (3)). In other words, according to this scheme a biologist, who has introduced a specific gene coding for an important trait in his or her protected variety, is either entitled to prohibit the use of this variety for breeding new varieties or to act against commercializing of such varieties, even though their specific commercial advantages were due to that very gene (STRAUS, 1987a). Notwithstanding the intent that the "breeders' privilege" should primarily secure a broad basis for plant breeding activities and thus provide for more genetic diversity, it seemingly often leads only to a remarkable increase of small variations of existing varieties of rather minor if any value in terms of genetic diversity (BERLAND and LEWONTIN, 1986; STRAUS, 1987a).

Four further aspects of plant breeders' rights as established under the UPOV Acts of 1972 and 1978 are worthwhile to be emphasized: First, the member states, although obliged to granting breeders tailored protection based on uniform and clearly defined principles, need only in the long term extend this protection to a limited number of genera or species of plants – after 8 years to at least 24 (Article 4 (3) (b) (iii)). Due to the fact that most of the member states took advantage of this provision, this has led to a patchwork-like protection, often resulting in no protection whatsoever for breeders of numerous species in a number of member states (STRAUS, 1987b).

Second, as a result of the limited scope of protection, farmers enjoy the right to resow seeds saved from their own harvest, free of any authorization and free of charge – so-called farmers' privilege.

Third, Article 2 (1) of the Convention leaves the choice as to the avenue of protection by way of "either a special title of protection or of a patent" to the contracting states, but it declares explicitly that they may not provide both forms of protection for one and the same botanical species, so-called prohibition of double protection. In order to facilitate accession of the United States to the convention, however, Article 37 of the 1978 Act introduced exceptional rules for protection under two forms (STRAUS, 1984; ADLER, 1986).

Fourth, a plant breeders' right based upon a plant variety certificate under the UPOV scheme relates always to a specific variety only and thus cannot offer adequate protection for generic, i.e., generally applicable technological advances in biotechnology (CRESPI, 1992).

At the end of 1993, i.e., after 32 years of existence, 24 countries were members of the UPOV Convention: Australia (1989), Belgium (1976), Canada (1991), Czech Republic (1993), Denmark (1968), Finland (1993), France (1971), Germany (1968), Hungary (1983), Ireland (1981), Israel (1979), Italy (1977), Japan (1982), The Netherlands (1968), New Zealand (1981), Norway (1993), Poland (1989), Slovak Republic (1993), South Africa (1977), Spain (1980), Sweden (1971), Switzerland (1977), United Kingdom (1968), and the United States of America (1981).

5 International Efforts to Meet the Needs Resulting from New Scientific Developments

5.1 OECD Initiative

The first international governmental organization which took note of and reacted to the upcoming new biotechnology was the Organization for Economic Cooperation and Development (OECD). Based mainly on official replies to a questionnaire from 19 of 24 member countries, a review prepared by a group of experts published in 1985 has above all discerned considerable differences in pertinent national laws and stressed the general need for international harmonization. It specifically recommended *inter alia*: to allow claims directed to microorganisms; to accept the deposit of microorganisms in a recognized culture collection as an alternative or complementary means of securing the disclosure and the repeatability of the invention; to improve the position of the applicant depositor with respect to the release of microorganisms either by postponing the release until the patent grant or by introducing a system of limited access for the public to the deposited material between the publication of application and patent grant, i.e., the "expert solution", as known under the Rule 28 EPC Implementing Regulation; and to seek solutions to give more effective protection to new plants arising from genetic engineering methods, preferably by allowing the innovator the choice of the type of protection most appropriate to secure a proper return in his investment, i.e., the choice between patents and special breeders' rights along the lines of the UPOV Convention. It constituted a matter of major concern to the group that such a choice was not allowed under the UPOV rules and that the situation was further aggravated by the patent law provisions of the European countries and the EPC. These have expressly excluded plant or animal varieties and essentially biological processes for the production of plants or animals from patent protection, as a result of legal developments which had taken place long before the rise of the modern biotechnology (BEIER et al., 1985).

Subsequently international developments were marked by parallel efforts of the World Intellectual Property Organization (WIPO), the International Union for the Protection of New Varieties of Plants (UPOV), and the Commission of the European Communities, aimed at taking up the suggestions of the OECD group, each in its own field of competence.

5.2 WIPO Efforts

Between 1984 and 1987 a WIPO Committee of Experts on Biotechnological Inventions and Industrial Property dealt with the issue. WIPO commissioned a number of studies (e.g., STRAUS, 1985) and sent out questionnaires to governments and intergovernmental organizations as well as to non-governmental organizations. The outcome of these efforts was a report comprised of three parts, the first part dealing with the existing situation with respect to the availability and scope of protection for biotechnological inventions

and the system of deposit of microorganisms, the second part summarizing the suggestions for improvements by non-governmental organizations and the third part containing nineteen Suggested Solutions, prepared by the International Bureau (WIPO, 1986, 1988). The overall tendency of WIPO's Suggested Solutions was to ensure as narrow as possible an application of the existing exclusionary patent law provisions and interpretation of patent laws which would adequately take into account the particularities and special needs of biotechnological inventions and their creators. Although the work of WIPO was well received, in view of the existing differences in national patent laws, however, the Suggested Solutions have never been formally adopted either by the Committee of Experts nor by any other body of the Paris Union. The essential merit of the WIPO efforts is to be seen in the improved understanding of the patent law problems accompanying the rise of modern biotechnology by all persons concerned. The most concrete result of the Suggested Solutions, however, was the fact that they constituted the basis and starting point for the still pending legislative initiative of the EC Commission on the adoption of a Directive on the Legal Protection of Biotechnological Inventions.

5.3 UPOV Achievements

From the outset of the international discussions UPOV also has closely watched the impact of biotechnology on the basic philosophy underlying the UPOV Convention. The first result was summarized in a study entitled "The Protection of Plant Varieties and the Debate on Biotechnological Inventions" (UPOV, 1985) and eventually led to the UPOV Convention revision work which was accomplished in March 1991 by the adoption of a New Revised Act (UPOV, 1992), which, however, is not yet in force. The adopted revised Act must be viewed as a compromise between, on one hand, the desire to safeguard the predominance or even exclusivity of the UPOV type of protection in the field of plants and, on the other hand, the need to modernize and improve the system so as to meet not only the changed situation resulting from modern plant biotechnology, but also to overcome old and familiar inadequacies and shortcomings.

The most important changes in the New Act can be summarized as follows: The ban on double protection, i.e., the prohibition to protect plant varieties by plant breeders' rights and patents has been removed completely. Contracting parties, under the New Act, after a transitional period of five years after accession to this Act, must extend protection to all plant genera and species (Article 3), which will terminate the patchwork-like protection allowed under the present Act. The principle of national treatment may no longer be limited by that of reciprocity. The scope of protection has been considerably broadened and now, under specific conditions, can also cover harvested material, including entire plants and parts of plants and even certain products made directly from harvested material of the protected variety. By broadening the scope of protection to varieties essentially derived from the protected variety, the principle of independence has experienced limitations. Similar observations can be made in respect to the "farmers' privilege". Although this exception is maintained, it is explicitly called "optional exemption", which contracting parties may provide for only within reasonable limits and subject to the self-guarding of the legitimate interests of the breeder (BYRNE, 1992). The New Act has introduced a new and comprehensive definition of "plant variety" which would provide for a fruitful coexistence with patent protection (GREENGRASS, 1991; TESCHEMACHER, 1992). The Act continues to provide for the breeder the possibility to get protected not only varieties resulting from his or her breeding activities, but also those found (discovered) in nature. However, the 1991 UPOV Act does not alter the basic principle of the Convention that a title granted under its scheme relates to a specific plant variety, and to it only, and thus does not provide any protection for generic, i.e., generally applicable inventions in the field of biotechnology (CRESPI, 1992; TESCHEMACHER, 1992; STRAUS, 1993c). In the latter context it should be noted that under the UPOV Convention neither breeding processes nor any other processes are eligible for protection.

5.4 EC Commission's Draft Legislation

5.4.1 Proposal for a Council Directive on the Legal Protection of Biotechnological Inventions

The main purpose of the proposal for a Council Directive on the Legal Protection of Biotechnological Inventions, which the EC Commission presented in October 1988 (EC COMMISSION, 1988) was: "To establish, harmonize, clarify and improve standards for protecting biotechnological inventions in order to foster the overall innovatory potential and competitiveness of Community, Science and Industry in this important field of technology." The proposal addresses such essential issues as: patentability of living matter, the effects of the exclusion from patentability of plant and animal varieties and essentially biological processes for the production of plants or animals as provided under Article 53 (b) EPC and respective national laws of the member states, the scope of patent protection for inventions in living matter, and the sufficiency of disclosure including all aspects of the deposit and release of biological material. Since the original proposal in respect to issues of ethics and morals has entirely relied on national patent law provisions of the member states corresponding to Article 53 (a) EPC, which excludes from patentability inventions, the publication or exploitation of which would be contrary to public order or morality, it was faced with strong opposition from the part of the European Parliament and had to be revised and amended. Moreover, the European Parliament strongly urged the Commission to introduce into the proposal the so-called farmers' privilege.

After years of extensive discussions and controversies, the last draft was accomplished in December, 1993 in the EC Council (EC COMMISSION, 1993). The content of this draft may be presented as follows: Under the unchanged general principle of the draft that subject matter of an invention shall not be considered unpatentable merely on the grounds that it is composed of, uses or is applied to biological material, the latter being any material containing genetic information and capable of self-reproducing or capable of being reproduced in a biological system (Article 2 (1) (2)). This rule applies to plants and animals as well as to their parts, which are explicitly declared patentable, as long as a plant or animal variety is not at stake (Article 3). Article 2 (3) now takes into account the public concerns on ethics and morality by adding to the already known rule of Article 53 (a) EPC that on that basis *inter alia* shall be unpatentable: (a) the human body or parts of the human body as such; (b) processes for modifying the genetic identity of the human body contrary to the dignity of man; and (c) processes for modifying the genetic identity of animals which are likely to cause them suffering or physical handicaps without any substantial benefit to man or animal, and animals resulting from such processes. In order to meet the concerns of the biotechnology community that the provision of Article 2 (3) (a) (b) (c) might lead to the exclusion of inventions from patentability, which until now have been patented under the EPC, new legally binding Recitals have been introduced into the draft proposal. Thus, the new Recital 10 first emphasizes the general principle that the ownership of human beings is prohibited and therefore, the human body or parts of the human body as such, for example, a gene, protein or cell in the natural state in the human body, which includes germ cells and products resulting directly from conception, must be excluded from patentability, but then the Recital clarifies that "isolated parts of the human body should not be unpatentable merely because of their human origin, it being understood that the parts of the human body from which such isolated parts are derived are excluded from patentability." If adopted by the Council, this would secure the acceptance of the patent granting practice of the EPO throughout the EC member states. In reflecting the public debate and the concerns of most scientists involved about ethical issues related to applications which claim protection

for DNA stretches with not yet known functions, as in the case of the application of the National Institutes of Health (NIH) of June 1991, which has been published on January 1, 1993 as PCT Publication Number WO 933-00353, and which discloses 3421 sequences that contain a total of 724 837 nucleotides (ROBERTS, 1991, 1992; ADLER, 1992; KILEY, 1992; MAEBIUS, 1992; EVANS, 1993), the new Recital 10[bis] now states: "Whereas, however, isolated human nucleic acids having no described application other than the expected properties attributable to any such nucleic acid, for example, it's ability to be used as a probe or a primer for synthesis of further copies of nucleic acids, should be unpatentable."

In respect to the patentability of inventions related to the so-called somatic and germ line therapy, it should be noted that the above mentioned provisions of Article 2 (3) (b) (c) stay firmly on the ground that methods of treatment of human or animal body by surgery or therapy or of diagnosis practiced on human or animal bodies are not patentable. If, however, not practiced on human or animal body and if the human gene therapy is not contrary to the dignity of man, somatic as well as germ gene therapy methods should be eligible for patent protection. This results from the Recitals (11a) and (11b), which, however, make clear that such patents would "in no way imply recognition of the legitimacy of what is known as germ gene therapy" and which continue to read "and whereas, if a patent were to be granted in this field, the national or Community authorization procedures applying to this type of therapy would have to be observed before any use of this therapy." While this latter phrase is indeed self-evident in view of the general principles of patent law, it leaves somewhat open the important question whether, e.g., under the German Patent Act such methods, if not contrary to the dignity of man, had to be patented in spite of the fact that the German Human Embryo Protection Act of December 1990 expressly prohibits any manipulation of human germ cells as criminal offense (STRAUS, 1992).

As regards the scope of protection, the pertinent provisions of the proposal are aimed at securing the legitimate interests of the process patent owner not only in respect to the first generation products of his or her patented process but also in respect to subsequent generations, which are treated as products directly obtained by the patented process. Such protection covers also plant or animal varieties originally produced by the patented process (Article 10 (2)). The patentee of a product patent related to a product consisting of or containing particular genetic information is offered protection which extends to any material in which the said product has been incorporated and in which said genetic information is contained or expressed (Article 10a). In the context of the scope of protection, two further provisions should be mentioned: on the one hand, Article 11, which expressly clarifies that the scope of protection of a patent does not extend to biological material, which has been produced by generative or vegetative propagation, if acts of that propagation result from the use for which the biological material was brought on the market and provided that material so produced is not used for subsequent propagation. On the other hand, Article 13 introduces the so-called farmers' privilege which is linked expressly to the rules which will govern that privilege under the future Community Plant Variety Rights System. In other words, Article 13 introduces this privilege only in respect to the effects of patents on plants, but not on animals.

5.4.2 Proposal for a Community Regulation on Community Plant Variety Rights

Whereas in the field of patents the Commission Efforts are aimed at harmonizing national patent laws, in the field of plant variety rights, the Commission has envisioned the establishment of a Community Plant Variety Right, corresponding to the Community Patent as designed under the Community Patent Convention of 1975 and the instruments adopted in 1989 thereto, but not yet in force. The proposal, which the Commission presented in August 1990 (EC COMMISSION, 1990),

therefore contains provisions regulating all aspects of such a system, in particular the substantive law, the effects of the new right, the granting procedure as well as the organization of the system. Article 1 claims that the Community Plant Variety Rights system as established under Regulation is "the sole and exclusive form of Community industrial property rights for plant varieties as a result of breeding or discovery". Moreover, Article 89 prohibits cumulative protection, i.e., varieties that are the subject matter of Community Plant Variety Rights shall not be patented nor be the subject matter of National Plant Variety Rights. Any rights granted contrary to this provision shall be ineffective. Thus, the EC Legislator contrary to the situation under the US law and in spite of the fact that alternative and even cumulative protection of new varieties of plants would perfectly fit into the 1990 UPOV scheme is about to maintain the "double protection" ban.

6 GATT 1994 Agreement

Eventually the issue of the protection of biotechnological inventions has also been addressed in the Agreement on Trade Related Aspects of Intellectual Property Rights, Including Trade in Counterfeit Goods adopted on December 15, 1993 within the General Agreement on Tariffs and Trade (GATT, 1994) scheduled to enter into force in 1995.

As regards the subject matter eligible for patent protection, Article 27 (1) declares that patents shall be available for any inventions, whether products or processes in all fields of technology, provided that they are new, involve an inventive step (are "non-obvious") and are capable of industrial application ("useful"). Very much along the lines of the EPC and the last version of the proposed EC Directive on the Protection of Biotechnological Inventions, however, paragraph 2 of this provision allows members to exclude from patentability inventions, "the prevention within their territory of the commercial exploitation is necessary to protect *order public* or morality, including to protect human, animal or plant life or health or to avoid serious prejudice to the environment, provided that such exclusion is not made merely because the exploitation is prohibited by domestic law." Moreover, members may also exclude from patentability diagnostic, therapeutic and surgical methods for the treatment of humans or animals (Article 27 (3) (a)), as well as plants and animals other than microorganisms, and essentially biological processes for the production of plants or animals other than non-biological and microbiological processes (Article 27 (3) (b) first sentence). As regards the protection of plant varieties, however, members are obliged to provide protection either by patents or by an effective *sui generis* system or by any combination thereof (Article 27 (3) (b) second sentence). The same provision stipulates that it shall be reviewed four years after the entry into force of the Agreement Establishing the World Trade Organization (WTO).

Thus, whereas it is to welcome that in the future intellectual property rights in microbiological processes and their resulting products seem secured and this seems to be true also for non-biological and microbiological processes for the production of plants or animals, product patents might remain unavailable for innovations in animals and plants in many GATT member states. For innovations in plant varieties most probably plant breeders' rights along the lines of the UPOV Convention will be introduced. But animal breeders and biotechnologists working in animals will continue to lack effective protection in many countries. This must be viewed as a quite deplorable situation, since a remarkable opportunity has been missed to get rid of some of outdated provisions primarily to be found in patent laws of European countries and the EPC. Instead, the legal situation in Europe will remain deadlocked and might very well attract followers outside Europe. But leaving higher life forms, like transgenic animals, outside of patent or any other protection will probably have effects exactly adverse to those expected. It will barely result in the expected free availability of propagating material but rather make such material to a so-called hot commodity with strict limitations of access (STRAUS, 1992).

7 The Practice of Patent Offices

In contrast to the cumbersome progress of developing the statutory law, patent offices which have been challenged by numerous applications in all fields of biotechnology had to act promptly. Here only the practices of the US PTO and the EPO will be presented. However, it should be noted that also many other patent offices do act very much along the same lines.

7.1 US PTO Practice

In applying the basic rules, which the US Supreme Court has developed in the 1980 "Chakrabarty case" (see Sect. 3), the US Patent and Trademark Office (PTO) issued numerous patents not only for all kinds of biotechnological processes, such as for the process of recombinant DNA technique, microinjection or hybridoma technique (COOPER, 1982; STRAUS, 1985), but also for full length gene sequences, viruses, plasmids, cell lines and monoclonal antibodies and starting with the "Hibberd case" in 1985 (227 USPQ 443) for plants too. Eventually, in 1987 the PTO in the decision "ex parte Allen" (2 USPQ 2d 1425) announced that "non-naturally occurring human multicellular living organisms, including animals" were considered patentable subject matter within the scope of 35 USC 101, but refused the application due to its lack of inventiveness. The first patent for a transgenic animal, the "Harvard onco mouse" was issued by PTO on April 12, 1988 (US Patent number 4,736,966) (LESSER 1989; MARKEY, 1989; MOUFANG, 1989). Since the first announcement by the PTO on animal patenting, a number of bills aimed at imposing a moratorium on animal patenting or at limiting the effects of animal patents in respect to farmers have been introduced into the Congress. Neither bill, however, was adopted (DELEVIE, 1992). By mid of 1992 some 145 patent applications involving new animal life forms were pending before the PTO, which in December 1992, issued 3 further mouse patents

(US PTO, 1993). As regards the reported application of the National Institutes of Health (see Sect. 5.4.1) claiming protection for DNA stretches with not yet known functions, it has been finally rejected by the PTO examiner mostly due to the lack of utility and/or novelty, and the NIH announced that it will not appeal the rejection. Moreover, the NIH will withdraw another patent application claiming 4448 additional sequences (*NIH News Release*, February 11, 1994). In respect to the US situation it should be noted that whereas the patent granting practice of the PTO relating to biotechnological processes and biological materials, such as genes, hybridoma, cell lines, etc., has already been tested and upheld by courts (e.g., CAFC, 231 USPQ 81ss. – "Hybertech Inc. vs. Monoclonal Antibodies", CAFC, 15 USPQ 2d 139 – "Hormone Research Foundation, Inc. vs. Genentech, Inc."; CAFC, 18 USPQ 2d 1016 – "Amgen Inc. v. Chugai Pharmaceutical Co.") this, however, is not yet the case with patents granted on higher life forms.

7.2 European Patent Office

The European Patent Office (EPO) also followed suit and granted along the same lines as the US PTO numerous patents for all kinds of biological material, such as human lymphoblastoid cell lines (EP No. 0113,769 B1), a DNA molecule capable of inducing the expression in unicellular hosts of a polypeptide displaying immunological or biological activity of human beta-interferon (EP No. 0041,313 B1), a human hepatocyte culture process (EP No. 0143,809 B1), the molecular cloning and characterization of a gene sequence coding for human relaxine (EP No. 0101,301 B1), various interferons, e.g., alpha type (EP No. 032,134 B1), granulocyte colony stimulating factor (EP No. 0237,545 B1), or erythropoietin (EP No. 0148,605 B1). Claims issued on genes covered both the genomic and the complimentary DNA (cDNA). Since in all known cases of patents granted for DNA sequences the complete DNA sequence of the respective gene has been disclosed and the protein for which the gene is coding as well as its useful properties indicated, the main problems to be overcome by applicants were those of the novelty and the inventive step requirement (SZABO, 1990, JAENICHEN, 1993). The fact that discoveries as such are not regarded inventions under Article 52 (2) EPC, however, did not pose substantial problems; in the case of cDNA sequences this problem does not exist at all, since they do not occur in nature. But also as far as genomic DNA sequences are concerned, the prevailing view and the patent office practice is that human technical intervention is required to recognize them and to produce them in a reproducible manner. Subject matter of such application, therefore, is an invention and not a mere discovery (EPO Examination Guidelines Part C-IV, 2.1).

Different categories of living organisms with respect to their patentability are offered different treatment under the EPC. Whereas microbiological processes or the products thereof are explicitly declared patentable subject matter (Article 53 (b) second-half sentence EPC), plant or animal varieties and essentially biological processes for the production of plants or animals, although recognized in principle as inventions, are excluded from patent protection (Article 53 (b) first-half sentence EPC). Since with respect to microorganisms the only severe problem, namely the acceptance of deposits of strains in order to meet the enabling disclosure requirement for the product protection, too, has been resolved at an early stage of development of the European patent system (EPO Examination Guidelines, Part C-IV, 3.5 as amended in 1982), there is no need for any special discussion on patenting of microorganisms here.

As far as higher life forms are concerned, the EPO was first faced with patent applications claiming protection for inventions in plants. Claims were directed to plants, plant cells, or plant tissue, modified by the use of either rDNA or somatic cells hybridization techniques. For long these applications have been predominantly questioned under the Article 53 (b) EPC provision. Therefore, the question the EPO had to answer was, whether the respective claims were directed to plant varieties and whether protection to be granted would extend to plant varieties, irrespective of the concrete wording of the claims.

The principles still valid with respect to the granting of patents for genetic inventions in plants or animals have been established by a decision of the Technical Board of Appeal in 1983 (1984 OJ EPO 112- "Propagating Material CIBA-GEIGY"). The Board has first clarified that no general exclusion of inventions in this field of animate nature can be inferred from the EPC. For the interpretation of the term "plant variety", the Board referred to the definition originally used in the UPOV Convention 1961 Act. It stated that Article 53 (b) EPC prohibits only the patenting of plants or their propagating material in the genetically fixed form of the plant variety, i.e., multiplicity of plants which are largely the same in their characteristics and remain the same within specific tolerance after every propagation or every propagation cycle, and that the very wording of that provision precludes the equation of plants and plant varieties which would also be at variance with general sense of provision. In an unpublished decision of June 5, 1992 (Patent Application No. 84,302,533-"LUBRIZOL Genetics Inc.") claims were directed not only to DNA shuttle vectors, but also to a plant cell, and to a plant or to plant tissue grown from that plant cell. Therein, the Opposition Division has confirmed the patent granting practice developed on the basis of the "CIBA-GEIGY" decision also for the time to come after the 1991 UPOV Act, in which a new definition of plant variety has been adopted and from which the present ban on double protection has been removed, has entered into force. Since the claims at stake referred to plant cells, plants, and plant tissue in general and not to a specifically defined narrow group of plants, they were not directed to plant varieties as defined in the new UPOV Act. As to the interpretation of claims, it was stressed that they must be interpreted in the light of description and if the description clearly teaches a broad applicability of the invention to a large group of plants, it is not permissible to interpret such a claim narrowly as referring to a plant variety. On the other hand, it was pointed out, if the description taught a narrow applicability of the invention to a specific plant variety, then a corresponding claim of a generic form would be inadmissable as being directed to a plant variety. Thus, the practice of the EPO secures the highly needed and most important generic protection of inventions in plants, which does not exist under the special plant variety protection scheme along the lines of the UPOV Convention.

In a decision of March 30, 1993 the German Federal Supreme Court explicitly stated that as regards patentability of biological inventions in general and those of plants in particular, the principles to be applied are those which had been developed in respect to chemical substances. Thus, product claims to plants are allowed, if the subject matter claimed is unequivocally identified by characteristics which are apt to external or internal perception. If this is impossible or totally impractical, then the plant can be described by unequivocally distinguishable and reliably identifiable parameters of its characteristics. Should this also prove impossible, the plant can be characterized by its production process, and thus can be granted the so-called *product by process* product protection, provided that method can be reproduced by the person skilled in the art on the basis of the written description (1993 GRUR 651- "Tetraploid camomile").

As regards the exclusion of essentially biological processes for the production of plants or animals, the Board of Appeal in the "Hybrid plants-LUBRIZOL" decision of November 10, 1988 (21 IIC 361 [1991]) took the view that whether or not a (non-microbiological) process is to be considered as "essentially biological" within the meaning of Article 53 (b) EPC has to be judged on the basis of the essence of the invention taking into account the totality of human intervention and its impacts on the result achieved. When discussing the particularities for the process for breeding hybrid plants at stake the Board went on saying: "The required fundamental alteration of the character of a known process for the production of plants may lie either in the features of the process, i.e., in its constituent parts, as in the special sequence of the process steps if a multi step process is claimed. In some cases the effect of this can be seen in the result. In the present case, which presents a multi step process, each single step as such may be characterized as biological in a scientific sense.

However, instead of the traditional approach of creating a single new crossing first and trying to propagate the individual result afterwards, the specific arrangement of the steps as presented above provides a process with a reversed sequence: it multiplies the parent plants by cloning and then crosses the cloned and thus derived "parent lines" on a large scale repeatably to provide the desired resulting hybrid population. This arrangement of steps is decisive for the invention and presents the desired control of the special result in spite of the fact that at least one of the parents is heterozygous. The facts of the present case clearly indicate that the claimed processes for the preparation of hybrid plants represent an essential modification of known biological and classical breeding processes, and the efficiency and high yield associated with the product show important technological character." The process therefore was not an "essentially biological" one, and thus patentable.

The Technical Board of Appeal of EPO was first confronted with the issue of patentability of inventions in animals and in processes for their production in the "Harvard Onco Mouse" case, in which protection was claimed for a method for producing a transgenic non-human mammalian animal having an increased probability of developing neoplasms and for transgenic animals produced by the said method. In its decision of October 3, 1990 (22 IIC 74) the Board in respect to Article 53 (b) EPC first ruled that the principles, which had been developed for inventions in plants were valid also for inventions in animals. Thus, the exception to patentability under Article 53 (b) applies to certain categories of animals only, but not to animals as such. For the time being, however, it is still an open question, how those categories of animals, namely "animal varieties", "races animales", and "Tierarten", respectively, are to be defined. This decision made it also completely clear that the processes employed namely that of rDNA and subsequent microinjection were not essentially biological ones and that product by process claims directed to animals were admissible. Moreover, it explicitly stated that patents were grantable for animals produced by a microbiological process.

The Board also examined whether the invention at stake in the "Harvard Onco Mouse" case was excluded from patenting under Article 53 (a) EPC, which excludes from patent protection inventions the publication or exploitation of which would be contrary to public order or morality. Its respective considerations eventually resulted in an instruction that a test has to be performed on case by case basis, whereby a careful weighing-up of the suffering of animals and possible risks to the environment on one hand, and the invention's usefulness to mankind on the other has to take place. The Examining Division has performed the necessary tests and found in its decision of April 3, 1992 (24 IIC 103 [1993] – "Onco Mouse–HARVARD III") that the benefits to the medical research of laboratory mice with accelerated tumor development prevail over the suffering inflicted, and potential risks for the environment virtually were not existent. Consequently, the patent was granted but has been opposed by a great number of opponents.

Although especially in the context of the "Onco Mouse" case quite convincing, the reasoning of the Technical Board of Appeal on the application and interpretation of Article 53 (a) EPC gives reason for considerable concerns when viewed from a more general patent law perspective. If the "weighing-up test" had to be applied to inventions in all fields of technology, this would no doubt most seriously affect the work of patent offices and the patent system in general. How could patent offices ever grant patents for guns, bullets, bombs, gases such as tear gas, which are specifically designed to inflict harm on humanity or animals (BYRNE, 1993)? Considerations of that kind seemingly led the Opposition Division of the EPO in a decision of February 25, 1993 (24 IIC 624 – "Patents for Plant Life Forms-GREENPEACE") to refuse to apply the "weighing-up test" in a case in which a patent for a herbicide-resistant plant was at stake. In deciding on a opposition of Greenpeace, the Opposition Division referred to the EPO Examination Guidelines and stated that the "Onco Mouse" decision did not supersede the general approach that Article 53 (a) EPC is to be invoked only in rare cases. The "Onco Mouse" case was such an exceptional case relating only to the patenting of animals used as test models in an assay where suffering is unavoidable. According to the reasons for the decision, opposition proceedings before the EPO are not a proper forum for discussing the pros and cons of the genetic engineering of plants in general or the plants in the particular case. In view of the fact that Greenpeace claimed the necessity to perform a risk–benefit assessment of the invention in question, the Division also recalled that the assessment of risks and the consequent regulation of the exploitation of the invention were a matter for other bodies to consider and were outside the scope of the EPO. Moreover, the Opposition Division pointed out that even if potential risks had to be taken into account in this context, the burden of proof in the opposition proceedings was with the opponent. But Greenpeace has not been able to prove, or at least render highly plausible his risks allegations. It appears that this approach presents a highly needed correlative to the "Onco Mouse" weighing-up test, retaining Article 53 (a) EPC its exceptional character within the context of the EPC and also within our legal order.

8 Intellectual Property and Biodiversity

During the last twenty years or so patents and plant breeders' rights have been facing vigorous criticism from developing countries under two main aspects. On one hand, in the seventies, under the influence of the in the meantime already abandoned philosophy of the United Nations Conference on Trade and Development (UNCTAD, 1975), patents were viewed as at best non-detrimental but by no means beneficial to developing countries (STRAUS, 1989). On the other hand, the complaint was repeatedly made that patents and plant breeders' rights eventually lead to reduction of biological diversity. Linked to the latter issue is also the long-lasting controversy between the North and the South on access, exploitation and preservation of world ge-

netic resources (MOONEY, 1983; PLUCKNETT et. al., 1987; STRAUS, 1988; CORREA, 1992). Since this sort of criticism of intellectual property rights continues (MACER, Chapter 4 this volume), it seems appropriate to touch briefly upon it.

As far as the past and current legal situation is concerned, it is essential to note that in most if not in all countries with rich genetic resources protection of intellectual property in biotechnology virtually does not exist (EVENSON and EVENSON, 1987; CORREA, 1990). Interestingly, most of those countries are not even members of the Paris Convention for the Protection of Industrial Property of 1983: thus far, the only remaining white spots on the map of this convention are located in areas of rich genetic diversity, e.g., India, Pakistan, Thailand, Bolivia, Ecuador, Peru, and Venezuela. As far as objections of this kind had been raised in the past, e.g., as regards India, they have no justification whatsoever, simply in view of the fact that there was no such protection in those countries. In view of the principle of territoriality, which governs the entire area of intellectual property, nobody in protection-free countries was prevented from reproducing, using, selling, etc. of biological materials patented in the developed world. The same is true for plant varieties protected under the UPOV plant variety protection scheme.

As regards the possible future impacts of industrial property rights on biological diversity, i.e., if and when the respective countries have introduced such rights in biotechnology, one has always to be aware of the fact that these rights "are only the expression of a fundamental choice based essentially on a certain number of socio-economic considerations and reasonings, and that statutes conferring these exclusive rights are but technical instruments, the configuration of which is essentially dictated by the specificity of their object (in casu living matter) and by the logic of market economy. Any decision about the legal instrument requires a clear statement on the goal, i.e., whether and to what extent the effects on biological diversity are (not) admissible taking into account the other effects of the protection of biological innovation through exclusive industrial property rights" (BERG-

MANS, 1990). Thus, the impact of intellectual property rights on biological diversity cannot be judged adequately in an isolated way. Rather it must be seen in the context of the entire complexity of a given national economy. It is the responsibility of the respective legislature to prevent excessive use of single successful varieties whether protected or not, e.g., by prescribing either minimum percentage of landraces to be used in addition to the new varieties, for which, e.g., subsidies and other financial incentives could be provided, or by providing measures for in *situ* or *ex situ* (gene banks) conservation. Having regard to past experience, every country should be able to avoid vulnerable monocultures in agriculture (STRAUS, 1993b).

Finally, the economically and politically important issue on access, exploitation and preservation of world genetic resources, which are primarily concentrated in the tropics (RAVEN and WILSON, 1992) is to be reviewed. First of all it has to be realized that new technologies have always led to sometimes painful changes in structure of national economies. Those who dispose of new technologies, capable of replacing old ones, have always used their advantages, eventually to their and their national economies' benefit. In some countries, even in most developed ones, whole branches of industry have disappeared, as, for instance, motorcycles in the United Kingdom or entertainment electronics in the USA. It would be irresponsible to expect something different in the field of biotechnology. It therefore does not help to deplore the fact that the use of enzymatic conversion of corn starch into high-fructose corn syrup causes serious damage to the economies of sugar exporting nations in the hope that the use of the new technology be prevented (SASSON, 1988; MACER, Chapter 4 this volume). Sugar exporting nations would be ill-advised to concentrate on objecting the use of the new technology, since this technology, once developed, will be used, regardless of any objections if the market will favor it. The only relief-promising approach is a realistic evaluation of the changed situation and of the new potential this new situation offers.

When discussing the issue on access and exploitation of world genetic resources, one has first to realize that plants and animals as well as microorganisms as genotypes, i.e., information embodied in the genetic constitution of plant or animal species, not only conform to the definition of public good but rather can possess exclusivity, too (SEDJO, 1992). The latter holds true if the access to them is limited by either tangible property ownership or by intellectual property rights, such as patents or plant breeders' rights. Since, however, tangible property in self-replicable material by the very nature of the latter tends to face very narrow limits, developments, which led to patenting plants and animals as well as microorganisms and other biological material under the patent law provisions of the developed world, deserve particular attention. These developments have made genetic resources potentially direct or more probably indirect subject matter of intellectual property (STRAUS, 1993b; KADIDAL, 1993). This changed situation is well reflected by new forms of cooperation between companies and universities interested in prospecting for pharmaceutically useful chemicals in plants, animals and insects of the tropics (EISNER, 1989/90) and the public or private institutions in countries disposing of such genetic resources. The first agreement of the kind, which has become known, was an agreement between Merck and Company, Inc. and a Costa Rican biodiversity research organization called Instituto Nacional de Biodiversidad (INBio) concluded in 1991. INBio is in the process of cataloging all plants and animals in Costa Rica. In exchange for the right to screen this catalog of plants and animals for biological activity, Merck is paying US dollars one million as down payment. In case Merck does discover (invent) a marketable product based on the information obtained, it will retain all patent rights, but will pay INBio a royalty which may be somewhere between one percent and three percent. Ten percent of the down payment and fifty percent of the INBio share in royalties will be invested in conservation in Costa Rica (STONE, 1992; STRAUS, 1993b). Meanwhile international commitment to chemical prospecting has increased. In December 1993 in Washington, the International Cooperative Biodiversity Group (ICBG) program was established, aimed at funding research partnership between private and public institutions in various countries, and pharmaceutical companies. The financial means are provided by the National Institutes of Health (NIH), the National Science Foundation (NSF), and the US Agency for International Development (USAID). At present, Argentina, Cameroon, Chile, Costa Rica, Mexico, Nigeria, and Peru participate in the program. The companies involved are American Cyanamid Corporation, Bristol-Myers Squibb, Monsanto, Shaman Pharmaceuticals. Moreover, a number of US and foreign universities, botanical gardens, museums, and conservation organizations are also involved in these efforts (EISNER, 1994). In this new context a "Biotic Exploration Fund" is proposed. According to the suggestion made, the financial means for the fund should be supplied from the revenues industry and public institutions involved will recoup from the sales of products developed on the basis of the results of chemical prospecting. The funds should be used for further chemical prospecting work, as well as for conservation purposes (EISNER and BEIRING, 1994).

Notwithstanding, all deficiencies which characterize the Convention on Biological Diversity signed on June 5, 1992 in Rio de Janeiro in particular in the context of its provisions relating to intellectual property (STRAUS, 1993b), the fact alone that under this convention for the first time the principle of sovereignty of states over their "genetic resources" has replaced the formerly advocated doctrine that genetic resources are mankind's heritage and consequently should be available without restrictions (LOMMI, 1989), may be viewed as a landmark. Since once it is fully realized that genetic resources as "genotypes" represent information embodied in the genetic constitution of plant and animal species and other biological material, which is not only transferred to the progeny but is also "automatically" multiplied by natural reproduction processes, one should also realize that in practice tangible property in genetic resources can provide adequate return only via patents and other intellectual property rights. Thus, the essential issue can only be how to provide effective intellectual property

rights for genetic resources *per se* and their derivatives (KADIDAL, 1993). Limitations of intellectual property rights in biotechnology are not needed, but rather their establishment and improvement in order to enable those who develop new plant or animal varieties, pharmaceuticals or chemicals, based on genetic resources, to share their profits with the owner of those resources on a mutually agreed basis. It can hardly be questioned that an effective protection of intellectual property rights also in countries which dispose of rich genetic resources would facilitate the conclusion of respective agreements. It would also enable the owner of genetic resources to negotiate more favorable terms for the access to those resources, not to mention that it would also offer incentives for indigenous research and production activities. It is essential to note that especially in agricultural research results of biotechnology have to be adapted to local soil and climate conditions. Thus, adequate protection as an incentive for local investments in development and research from what source so ever is needed in the country itself (STRAUS, 1993b).

9 References

ADLER, R. G. (1986), Can patents coexist with breeders' rights? Development in U.S. and international biotechnology, *Int. Rev. Ind. Prop. Copyright Law (IIC)* **17**, 195–227.

ADLER, R. G. (1992), Genome research: Fulfilling the public's expectations for knowledge and commercialization, *Science* **257**, 908–914.

BEIER, F. K. (1980), The significance of the patent system for technical, economic and social progress, *Int. Rev. Ind. Prop. Copyright Law (IIC)* **11**, 563–584.

BEIER, F K., STRAUS, J. (1977), The patent system and its informational function – yesterday and today, *Int. Rev. Ind. Prop. Copyright Law (IIC)* **8**, 387–406.

BEIER, F. K., STRAUS, J. (1986), Genetic engineering and industrial property, *Ind. Prop.*, 447–459.

BEIER, F. K., CRESPI, R. S., STRAUS J. (1985), *Biotechnology and Patent Protection – An International Review,* Paris: OECD.

BERGMANS, B. (1990), Industrial property and biological diversity of plant and animal species, *J. Pat. Trademark Off. Soc. (JPTOS)* **72,** 600–609.

BERLAND, J. P., LEWONTIN, R. (1986), Breeders' rights and patenting life forms, *Nature* **322,** 785–788.

BRENNER, S. (1990), *Human Genetic Information: Science, Law and Ethics,* Statement at Ciba Foundation Symposium 149, pp. 145–146, New York – Brisbane – Toronto – Singapore: John Wiley & Sons.

BYRNE, N. (1992), *Commentary on the Substantive Law of the 1991 UPOV Convention,* London: University of London, Centre for Commercial Law Studies.

BYRNE, N. (1993), Patents for biological inventions in the European Community, *World Patent Information,* 77–80.

CHROCZIEL, P. (1986), Die Benutzung patentierter Erfindungen zu Versuchs- und Forschungszwecken, Köln – Berlin – Bonn – München: Carl Heymanns Verlag KG.

COLLARD, C. (1993), Limited patent protection for proteins, *Can. Intellect. Prop. Rev.* **10,** 25–33.

COOPER, I. P. (1982), *Biotechnology and the Law,* New York: Boardman.

CORREA, C. M. (1990), Patentes y biotecnologia: Opciones para America Latina, *Rev. Derecho Ind.* **34,** 5–53.

CORREA, C. M. (1992), Biological resources and intellectual property rights, *Eur. Intellect. Prop. Rev. (EIPR),* 154–157.

CRESPI, R. S. (1992), Patents and plant variety rights: Is there an interface problem?, *Int. Rev. Ind. Prop. Copyright Law (IIC)* **23,** 168–184.

DELEVIE, H. A. (1992), Animal patenting; probing the limits of U.S. patent laws, *J. Pat. Trademark Off. Soc.* **74,** 492–509.

EC COMMISSION (1988), *Proposal for a Council Directive on the Legal Protection of Biotechnological Inventions,* Doc. COM (88) 496 Final – SYN 159 = 20 IIC 55ss. (1989) = OJ EC No. EC-Commission (1993), C10 1989, 3.

EC COMMISSION (1990), *Proposal for a Council Regulation (EEC) on Community Plant Variety Rights,* Doc. COM (90) 347 Final of August 30, 1990.

EC COMMISSION (1993), *Directive 94/EC of the European Parliament and of the Council on the Legal Protection of Biotechnological Inventions,* Doc. COM 4065/94.

EISNER, T. (1989/90), Prospecting for Nature's chemical riches, *Issues Sci. Technol.,* 31–34.

EISNER, T., (1994), *Chemical Prospecting: A Global Imperative* (forthcoming).

EISNER, T., BEIRING, E. A. (1994), Biotic Exploration Fund – Protecting biodiversity through chemical prospecting, *Bioscience* **44** (2), 95–98.

EVANS, L. (1993), Intellectual property issues: Getting to gripps with the patent problem, *Genome Digest – HUGO Eur. News* **1** (2), 4–7.

EVENSON, R. E., EVENSON, D. D., (1987), Private sector agricultural invention in developing countries, in: *Policy for Agricultural Research,* (RUTTAN, V. W., PRAY, C. E., Eds.), pp. 469–511, Boulder, CO: Westview Press.

FEDERICO, P. J. (1937), Louis Pasteur's patents, *Science* **86,** 327–328.

GATT (General Agreement on Tariffs and Trade) (1994), Agreement on trade-related aspects of intellectual property rights, including trade counterfeit goods (TRIPS), *GRUR International Part (Gewerblicher Rechtsschutz und Urheberrecht),* 128–140.

GREENGRASS, B. (1991), The 1991 Act of the UPOV Convention, *Eur. Intellect. Prop. Rev. (EIPR),* 466–472.

HANTMAN, R. D. (1985), Experimental use as an exception to patent infringement, *J. Pat. Trademark Off. Soc. (JPTOS)* **67,** 617–644.

JAENICHEN, H. R. (1993), *The European Patent Office's Case Law on the Patentability of Biotechnology Inventions,* Köln – Berlin – Bonn – München: Carl Heymanns Verlag KG.

KADIDAL, S. (1993), Plants, poverty, and pharmaceutical patents, *Yale Law J.* **103,** 223–258.

KEVLES, D. J., HOOD, L. (1992), Reflections, in: *The Code of Codes* (KEVLES, D. J., HOOD, L., Eds.), pp. 300–328, Cambridge, MA – London: Harvard University Press.

KILEY, T. D. (1992), Patents on random complimentary DNA fragments?, *Science* **257,** 915–918.

LESSER, W. H. (1989), *Animal Patents: The Legal, Economic and Social Issues,* Basingstoke, UK: Macmillan Publishers Ltd.

LOMMI, H. (1989), Problems related to legal protection of plants, *NIR (Nordic Intellectual Property Law Review, Sweden),* 1–13.

MAEBIUS, S. B. (1992), Novel DNA sequences and the utility requirement: The human genome initiative, *J. Pat. Trademark Off. Soc. (JPTOS)* **74,** 651–658.

MARKEY, H. T. (1989), Patentability of animals in the United States, *Int. Rev. Ind. Prop. Copyright Law (IIC)* **20,** 372–389.

MOONEY, P. (1983), The law of the seed. Another development and plant genetic resources, *Dev. Dialogue,* 1–2.

MOUFANG, R. (1989), Patentability of genetic inventions in animals, *Int. Rev. Ind. Prop. Copyright Law (IIC)* **20,** 823–846.

PLUCKNETT, D. L., SMITH, N. J. H., WILLIAMS, J. T., ANISHETTY, N. M. (1987), *Gene Banks and the World's Food,* Princeton, NJ: Princeton University Press.

RAVEN, P. H., WILSON, E. D. (1992), A fifty year plan for biodiversity surveys, *Science* **285**, 1099–1100.

ROBERTS, L. (1991), Genome patent fight erupts, *Science* **254**, 184–186.

ROBERTS, L. (1992), NIH genes patent round two, *Science* **255**, 912–913.

ROSSMAN, J (1931), Plant patents, *J. Pat. Trademark Off. Soc. (JPOS)* **13**, 7–21.

SASSON, A. (1988), *Biotechnologies and Development,* Paris: UNESCO.

SEDJO, R. A. (1992), Property rights, genetic resources, and biotechnological change, *J. Law Econ.* **35** (1), 199–213.

STONE, R. (1992), The biodiversity treaty: Pandora's box or fair deal?, *Science* **256**, 1624.

STRAUS, J. (1984), Patent protection for new varieties of plants produced by genetic engineering – Should "double protection" be prohibited?, *Int. Rev. Ind. Prop. Copyright Law (IIC)* **15**, 426–442.

STRAUS, J. (1985), *Industrial Property Protection of Biotechnological Inventions – Analysis of Certain Basic Issues,* Geneva: WIPO Doc. BIG/281.

STRAUS, J. (1987a), The principle of "dependence" under patents and plant breeders' rights, *Ind. Prop.,* 433–443.

STRAUS, J. (1987b), The relationship between plant variety protection and patent protection for biotechnological inventions from an international viewpoint, *Int. Rev. Ind. Prop. Copyright Law (IIC)* **16**, 723–737.

STRAUS, J. (1988), Plant biotechnology, industrial property and plant genetic resources, *Intellectual Property in Asia and the Pacific,* **21**, March-June, 41–49.

STRAUS, J. (1989), Patent protection in developing countries – overview, in: *Equitable Patent Protection for Developing World* (LESSER, W. H., et. al., Eds.), Cornell Agricultural Economics Staff Paper, 89–36 (18 pp.).

STRAUS, J. (1992), Biotechnologische Erfindungen – Ihr Schutz und seine Grenzen, *GRUR (Gewerblicher Rechtsschutz und Urheberrecht),* 252–266.

STRAUS, J. (1993a), Zur Zulässigkeit klinischer Untersuchungen am Gegenstand abhängiger Verbesserungserfindungen, *GRUR,* 308–319.

STRAUS, J. (1993b), The Rio Biodiversity Convention and intellectual property, *Int. Rev. Ind. Prop. Copyright Law (IIC)* **24**, 602–615.

STRAUS, J. (1993c), Pflanzenpatente und Sortenschutz – Friedliche Koexistenz, *GRUR,* 794–801.

STRAUS, J., MOUFANG, R. (1990), *Deposit and Release of Biological Material for the Purposes of Patent Procedure – Industrial and Tangible Property Issues,* Baden-Baden: Nomos Verlagsgesellschaft.

SZABO, G. S. A. (1990), Patent protection of biotechnological inventions – European perspectives, *Int. Rev. Ind. Prop. Copyright Law (IIC)* **21**, 468–479.

TESCHEMACHER, R. (1992), Die Schnittstelle zwischen Patent- und Sortenschutzrecht nach der Revision des UPOV-Übereinkommens von 1991, in: *Festschrift Rudolf Nirk zum 70. Geburtstag* (BRUCHHAUSEN, K. et al., Eds.), pp. 1005–1015, München: Ch. Beck'sche Verlagsbuchhandlung.

UNCTAD (United Nations Conference on Trade and Development) (1975), *The Role of the Patent System in the Transfer of Technology to Developing Countries,* New York: United Nations.

UPOV (Union for the Protection of New Varieties of Plants) (1985), *The Protection of Plant Varieties and the Debate on Biotechnological Inventions,* UPOV Doc. INF/11 Geneva.

UPOV (Union for the Protection of New Varieties of Plants) (1992), *Records of the Diplomatic Conference for the Revision of the International Convention for the Protection of New Varieties of Plants,* Geneva 1991.

US PTO (US Patent and Trademark Office) (1993), Report on the issue of three patents for genetically engineered mice, *Pat. Trademark Copyright J. (PTCJ)* **45**, 159.

WIPO (World Intellectual Property Organization) (1986), Report on industrial property protection of biotechnological inventions, *Ind. Prop.,* 252.

WIPO (World Intellectual Property Organization) (1988), Report on the third session of the Committee of Experts on Biotechnological Inventions and Industrial Property, *Ind. Prop.,* 104.

10 Patent Applications for Biomedicinal Products

JENS-PETER GREGERSEN

Marburg, Federal Republic of Germany

1 Introduction

A patent is certainly the most adequate means to protect an invention, provided that a product derived from this invention can be developed and sold on the market within the foreseeable future and if one expects major financial revenues from this product. These expected revenues should be high enough to recoup the cost of product development and to justify the expense of obtaining and maintaining a patent. However, the majority of patented pharmaceuticals cannot to recoup the cost of product development during the lifespan of their patents (PRENTIS et al., 1988). Thus, patents do not guarantee high profits, but – depending on their quality – patents can significantly improve the chances of commercial success of innovative products. The quality of patents does not only depend on the quality of the research which led to the patented invention, but also on the quality of the patent application and a skillful patent strategy.

This chapter describes the main characteristics of a patentable invention and of a patent application with special emphasis on patents on medicinal products of biological or recombinant origin. It is intended to help the inexperienced applicant in deciding if and at which stage of research a patent application will be adequate or most successful. This chapter will also assist in drafting a patent application with an optimal patenting strategy. For researchers in non-profit organizations, it might be of interest to read about the opportunities for converting their intellectual property into money or into funds for further research.

As far as the subject allows, only the general principles will be outlined which apply to patents in the field of biomedical products, rather than dealing with specific cases (as the interpretation of their patentability changes with time). For practical reasons emphasis will be laid on the patent policies of the European Patent Organization (EPO) and of the USA, since these are the most advanced in this field and strongly influence other countries' policies. Patents are usually applied first in the USA and in EPO countries, as they

represent major markets. Important deviations from the EPO's or USA's policies in other countries will be mentioned and summarized where necessary. Alternatives to patents will be briefly discussed at the end of this introduction.

1.1 The Purpose of a Patent

The basic purpose of the patent system is to provide incentives for innovation, industrial development and industrial investment. A patent entitles its owner to exclude others from the commercial exploitation of the patented invention for a limited time. In exchange for this temporary monopoly the public receives an accurate and detailed description of the invention, which makes the knowledge available to the public at an early point of time and, after the patent has expired, enables others to exploit the invention. A patent does not only contain patent claims in order to exclude others from using the invention commercially, the patent specification also contains a technical teaching to solve a problem and thus contributes to the advancement of science and technology. A patent cannot be granted for something that is not innovative or is already in the public domain. Novelty, (industrial) utility and inventiveness are the three prerequisites for a patentable invention which must be fulfilled to ensure that a patent serves its purpose.

Owning a valid patent alone does usually not generate profit. Selling the patent rights and ultimately manufacturing, selling or using a potential invention may eventually yield a profit. A patent owner will only enjoy the full benefit of a patented product if the patent is not easily circumvented and is strong enough to be enforced against infringers. However, the main purpose of a patent is not to prevent others from doing something, it is also necessary that a patent is used. Patent rights can be lost if the patent is not used in a constructive manner.

The quality of a patent, i.e., its scope, the chances to get it granted and its enforceability, is to a great extent dependent on the quality and depth of the underlying research. The most cunning patent description and wording of the patent claims cannot cover weaknesses in the research upon which a patent is based.

Due to the commercial interest in patents, their basic purpose as a means to promote science and technology tends to be neglected. Patent examiners might appear pedantic or narrow-minded when they are reviewing a patent application, but it should be kept in mind that they do so because they are protecting the interests of the public in an attempt to grant patent rights which are balanced against the contribution to innovation.

1.2 Alternatives to Patents

Most national laws provide specific legal protection of intellectual property not only for patents but also for registered designs and copyrights. Registered designs refer to the visual appearance of an industrial product. They are not suitable to protect scientific inventions aiming to achieve a certain effect or to fulfill a certain function. Copyright was conventionally intended for products of the arts, but nowadays it is also extended to instruction manuals, architectural drawings, computer software and (under special legislation) to integrated circuit layouts.

The idea of extending copyright to the information embodied by DNA or RNA molecules seems to suggest itself. However, copyright is limited to the information itself or to a certain form in which the information is presented. Copyright is not intended as a tool to control the *use* of the information. If genomic sequences would be subject to copyright provisions, the owner of a copyright could theoretically control all products derived from the use of these sequences by granting or withholding licences on the reproduction of the information. Copyright provisions would not exclude that the sequences may even be a product of the author's imagination or of a computer program which generates random sequences. Without major adaptations, copyright provisions thus seem to be inadequate or useless for genomic information.

An important alternative to patents for certain minor inventions in the field of biology and biotechnology is to keep the information secret and to control its release by confidentiality agreements. It can be very useful for a company to keep a manufacturing process as a trade secret and enjoy the benefits of the developed technology as long as possible, rather than releasing patent information which might stimulate others to work around it.

Keeping research results secret for a long time is hardly a realistic option for academic scientists. However, if it is intended to patent research work, it is essential to make this decision early and to keep an invention secret until its full potential has been explored. Publications of new results in short intervals and a later ad-hoc decision to file a patent application on the same subject will, at best, result in weak patents. In the worst case, premature publications by the inventor can render patent claims or the entire patent invalid.

When patents for registered medicinal products expire, others may apply for a registration of an essentially similar product by referring to the pharmacological, toxicological and clinical data of the first applicant and without providing own data on these parts. For certain products second registrations of imitations are excluded for a limited period of time.

In the EEC "high-technology" products, including biotechnological products, which are registered according to the centralized European registration procedure, are protected from second registrations for a period of ten years (EEC COUNCIL DIRECTIVES 87/21/EEC and 87/22/EEC). Similarly in the USA "orphan drugs" for rare diseases or conditions are protected from second registrations for a period of seven years. For commercial enterprises these provisions may be considered as an addition to patent protection but not as an alternative.

2 Basic Requirements for a Patentable Invention

2.1 Novelty

A patentable invention must be new. Novelty excludes the "state of the art" (also "prior art") which is what was "made available to the public by means of a written or oral description, by use, or in any other way, before the date of filing of the ... patent application" (European Patent Convention, EPC, Article 54). Under the EPC the content of an earlier European patent application (which is not yet published) is also considered as state of the art.

All scientific publications and other written articles, oral presentations, interviews, potentially even casual non-confidential talks among researchers prior to the filing date of a patent application can be detrimental to patentability. This includes publications by the inventor anywhere in the world, therefore, it is essential to check all publications before they are released to determine whether they contain anything that could be used in a patent application. Since most researchers have a confidentiality clause in their employment contract, discussions among colleagues within the same institution usually do not make the discussed matter "available to the public".

The most diligent study of all available information may not rule out the case that a patent application lacks novelty, because the matter to be patented was already subject to an earlier patent application which was not known to the inventor at that time. Patent applications in most countries are published 18 months after filing.

According to the United States patent law, "A person shall be entitled to a patent unless (a) the invention was known or used by others in this country, or patented or described in a printed publication in this or a foreign country, before the *invention* thereof by the applicant for patent or (b) the invention was patented or described in a printed publication in this or a foreign country more than one year prior to the *date of the application* for patent in the United States". (United States Code, 35 USC, § 102; emphasis added). The differentiation between the point of time when an invention was made and when the patent was filed (according to the EPC only the filing date of a patent application is relevant) gives several advantages to inventors in the USA: An earlier state of art will be applied when their invention is judged in terms of novelty, and they may publish the contents of their patent application up to one year before filing. Other more important advantages of this paragraph of the US patent law will be discussed below in context with the priority date.

2.2 Non-obviousness

It is not necessary for a patentable invention to be the result of an ingenious idea. Article 56 EPC requires an "inventive step" which is "not obvious to a person skilled in the art", regarding the state of the art at the time the application was filed. Similarly in the USA "... a patent may not be obtained ... if the differences between the subject matter sought to be patented and the prior art are such that the subject matter as a whole would have been obvious at the time the invention was made to a person having ordinary skill in the art to which said subject matter pertains" (35 USC, § 103).

Despite the conformity of both definitions (not obvious regarding the prior art to a person skilled in the art), there is considerable room to speculate how these terms are defined in each individual case. Almost inevitably these terms are interpreted differently by the applicant and the patent examiner during the prosecution of a patent application. In the absence of official general explanations for "non-obviousness to a person skilled in the art" the following attributes may help to decide whether an alleged invention meets this requirement (see also VOSSIUS, 1982).

For the average expert in the field an invention, for example, may

- be surprising or unexpected,
- be unpredictable,
- be unconventional or against a general prejudice of experts,

- solve a major problem or a problem that existed for a long time,
- teach a method that has been considered not feasible,
- demonstrate an advantage over prior techniques,
- be the result of a serendipity, e.g., the selection of a microorganism, cell strain or monoclonal antibody with new, favorable attributes (selection invention),
- present a new use for a known product.

The last point of the list seems to be in contradiction with the requirement for novelty. However, novelty must apply to the patented matter, which in this case would be the new use. The inventiveness lies in the fact that the new use for a known product was not obvious in light of what was known in the field.

The first really unbiased "persons having ordinary skill in the art" to judge a patent application will be the patent examiners. If they come to the conclusion that an invention is obvious (for example because it combines two pieces of well known prior art), this may be due to the application failing to clearly emphasize the inventive step and might be overcome by providing further evidence and appropriate arguments. However, it would be better if such doubts were not raised at all. To avoid unnecessary arguments about the inventiveness of an invention, it is advisable to strongly emphasize the inventive step in applications for a patent. Almost invariably patent applications contain a separate paragraph which explains the non-obvious and surprising step of the invention, for example by quoting previous, unsuccessful attempts to solve the problem.

2.3 Utility or Industrial Applicability

Patentable inventions must be useful, "reduced to practice" (35 USC, § 102) and, by legal definition in most countries, also amenable to an industrial application. The German concept of a patentable invention describes it as "Lehre zum technischen Handeln" (teaching of a technical operation). The Polish law even requires a strictly technical character of

the invention and currently excludes from patentability any biological product as well as pharmaceuticals and chemicals, even if these derive from a technical process. The technical process which is used to make such products, however, is patentable in almost all countries, irrespective of whether these refuse to grant patents on certain or all products of nature. According to the general purpose of a patent, its utility must be described in such detail that the average expert is able to reproduce the invention. Thus the descriptive part of a patent application closely resembles the materials and methods section of a scientific paper. This requirement of utility prevents patents on mere theories.

The famous example of SAMUEL MORSE's patent application for the telegraph is particularly well suited to illustrate the limitation to patentability set by the utility requirement. As an extension to his invention, MORSE applied for a claim as follows: "I do not propose to limit myself to the specific machinery ... described in the foregoing specification and claims; the essence of my invention being the use of the motive power of the electric or galvanic current, which I call electro magnetism, however developed, for making or printing intelligible characters, signs or letters ..." (quoted after BENT et al., 1987). This extensive claim which literally includes still non-existing appliances ("however developed") was not accepted. Theoretically such a claim would have given MORSE patent rights on electric typewriters, photocopiers and even computers, had they been developed within the patent life.

Every patent applicant will, of course, try to extend the scope of his patent as far as possible by claiming all obvious and imaginable applications besides those which are described. Whereas the obvious extensions are likely to be accepted, speculative applications are probably claims on theories or on a principle of nature, which are excluded from patentability. Claims on theories without utility in the present form would be against the spirit of patents. They would be an attempt to control future inventions in this field and thus are not acceptable.

The first isolation and characterization of a viral protein from infected cell cultures may be sufficient to apply successfully for a patent on a serological diagnostic test based upon this antigen. If an ELISA test is described in detail, including a few results which prove that this test really fulfills the claimed purpose, it seems legitimate to extend the patent claims to variations of the described ELISA method. Most likely, extensions to other well known serological tests can also be included in the patent, since it can be assumed that the average expert in the field will be able to apply the patented protein in these methods without undue experimentation. Further claims, e.g., for a vaccine, would need to be substantiated and reduced to practice, at least by the ability to induce an immune response and some evidence that this immune response is protective in an animal model or a meaningful *in vitro* test.

It is currently unclear whether, based on the current state of the art, a patent following the above-mentioned example could be extended to recombinant derivatives of the native protein. One might argue that, once the native protein is known and accessible, it needs no inventiveness to sequence the amino acids for parts of this protein, synthesize the corresponding DNAs, use these as probes to identify and isolate the entire coding sequence of the protein, which is then inserted into a suitable expression system to produce the protein in any desired form and quantity. Experience, however, teaches that it still requires some non-obvious steps and usually more than a limited degree of experimentation (often even a stroke of luck) to get there and to achieve the desired utility with recombinant polypeptides. For a vaccine it may be necessary to find and express the important epitopes in an appropriate (still unknown) way and to develop adequate purification and further processing protocols (with unpredictable technical problems) to achieve the desired and claimed protective effect in the target species. The fact that some methods are obvious to try does not mean that the desired result is also obvious!

The degree to which utility limits a patent depends not only on the enabling descriptions or on the current technical standards. Another aspect to be considered is the characterization of the claimed substance matter. If a native protein can be defined by its exact amino acid sequence, the chances are much better to cover any recombinant form of the same protein or of one with minor, irrelevant variations by a patent on the native molecule. In this case and with today's technical standards, it may be argued that the knowledge of the amino acid sequence avoids many of the uncertainties and unpredictable problems mentioned above. The problem of limitations and possible extensions of patent claims for biological molecules is discussed in more detail in Sect. 3.3 on patent claims.

2.4 Inventions, Discoveries and Products of Nature

Inventions are patentable, but in almost all countries mere discoveries and products of nature as such are not. The distinction between invention and discovery, however, is difficult to define and new technological developments, e.g., in the field of biotechnology, may require new interpretations of the official definitions.

According to the European point of view, discoveries only recognize something which exists, but remained unnoticed. Thus *products of nature as such* can be discovered but not invented. A product of nature may be very useful, but its discovery lacks technical teaching and an inventive step, thus it cannot be patented. This does not mean that products *occurring in nature* cannot be patented. If, by a technical operation, a product of nature is isolated, enriched, purified or somehow changed into a hitherto unknown form and can be utilized, it can be patented. For example, the discovery of a natural antibiotic effect alone is not sufficient to be patented, whereas an antibiotic which is isolated from a microorganism or which is utilized by feeding whole microorganisms from a fermentation process to animals may represent a patentable invention.

The first identification of a certain DNA sequence coding for a defined polypeptide which can be used for a diagnostic or therapeutic purpose is in principle patentable. The construction of a link between the DNA sequence and the functional polypeptide is the result of a mental act which might be considered as non-obvious and inventive. Utility can be provided by a technical description of how the polypeptide was expressed, purified and by evidence that it can be used as a diagnostic or therapeutic agent.

For a detailed discussion and interpretation of discoveries and products of nature the interested reader is referred to UTERMANN (1978), BENT et al. (1987), BENT (1982) and PLANT et al. (1982). Recent events, however, have stimulated new discussions on this subject and it remains to be seen whether the interpretation of the above definitions will be changed.

The Human Genome Project and the decision of the US National Institutes of Health (NIH) in 1991 to file patent applications on randomly selected partial cDNA sequences of unknown function [potentially useful, for example, "as genetic markers for forensic identification or for tissue typing" (utilities quoted after EISENBERG, 1992)] challenge hitherto valid interpretations of the terms discoveries, products of nature and utility. One might argue that the Human Genome Project "reads" and describes products of nature using an established methodology and without an inventive element. Even the term "discovery" may be considered as too sophisticated for the result of a repeated routine process which produces new – but hardly surprising – results, since the function of the DNA remains unknown.

Besides doubt about the inventiveness of these patent applications, there appears to be a lack of utility (a new technical teaching) of the claimed subject matter. The mentioned utility for such a vast number of cDNA sequences will most likely be based upon theoretical examples without a constructive reduction to practice. There is hardly any doubt that the intention is to use the given examples of utility to fulfill a requirement only formally but to claim patent rights on all potential utilities.

With the European Patent Convention in mind, these patent applications appear hopeless. But rather pragmatic or missing definitions provided by the US Code do not exclude discoveries from being patented. The

USC states that "patentability shall not be negotiated by the manner in which the invention was made" (USC, Title 35, § 103). Based on this pragmatic definition of inventiveness and generally low requirements to prove utility, it appears less hopeless to get the above-mentioned patent application granted. This is especially true if there is a political moment involved: The applicant of these patents are "The United States of America", and in France and the UK similar patent applications were filed as a response to the US policy. This controversial issue was discussed in three articles by ADLER, EISENBERG, and by KILEY published in *Science* in 1992.

Decisions on these or similar patent applications will hopefully be made with the basic purpose or spirit of patents in mind. But even if these patent applications would be granted, it is questionable whether commercial applications can be developed within the life-span of these patents.

2.5 Patentable Inventions and Exclusions

As a general rule, the US policy on patentable subject matters represents an extreme with the most liberal definitions: "Whoever invents or discovers any new and useful process, machine, manufacture or composition of matter, or any new and useful improvement thereof may obtain a patent therefor (35 USC, § 101).

At the other end of the spectrum, a number of countries still exclude several inventions – mainly on food and medicines – from patentability. (For a discussion on the underlying ethical and commercial considerations see, for example, TOBIAS, 1992.) In the last few years many of those countries have abolished these exclusions, others also consider changing their policies.

The standard set by the EPC lies between these extremes. Many countries throughout the world have similar policies or have adopted the EPC statutes on patentable subject matter either literally or with minor modifications. It is likely that, wherever future changes of patent statutes are expected (e.g., in Eastern European and South American countries), these will also adopt at least major parts of the EPC standards.

The EPC defines patentable inventions mainly by exclusions. Parts of Articles 52 and 53 EPC are quoted here since they also reflect the view and the exclusions valid for most other countries:

"(1) European patents shall be granted for any inventions which are susceptible of industrial application, which are new and which involve an inventive step.

(2) The following in particular shall not be regarded as inventions within the meaning of paragraph 1:

   a) discoveries, scientific theories and mathematical methods;
   b) aesthetic creations;
   c) schemes, rules and methods for performing mental acts, playing games or doing business and programmes for computers;
   d) presentations of information."

"(4) Methods for treatment of the human and animal body by surgery or therapy or diagnostic methods practised on the human or animal body shall not be regarded as inventions which are susceptible of industrial application within the meaning of paragraph 1. This provision shall not apply to products, in particular substances or compositions, for use in any of these methods." (Article 52, § 1, 2, and 4)

Article 53 EPC also excludes "plant or animal varieties or essential biological processes for the production of plants or animals; this provision does not apply to microbiological processes or the products thereof".

The European Patent Convention explicitly demands an industrial application and excludes therapeutical and diagnostic methods practised on the human or animal body (but not therapeutics and diagnostics or related instruments) and plant and animal varieties (but not microorganisms). None of these exclusions exist in the USA where patents are permitted on basically anything that is man-made, including agricultural or breeding methods as well as on clinical therapeutic and diagnostic methods.

Definitions like those quoted above for patentable inventions in the EPC, which are mainly based on exclusions, always appear somewhat unsatisfactory. A list of examples for principally patentable subject matters in the field of biological or biotechnological medicines may provide a more positive orientation (see Tab. 1). It should be noted that until recently numerous countries still excluded many of these inventions.

Several countries do not accept patents on pharmaceuticals and medicines or provide only limited patent protection. Based on latest available information, Tab. 2 summarizes

Tab. 1. Patentable Biological Subject Matter[a]

Microorganisms Including Viruses
   Specific strains of microorganisms
   Manipulated/engineered strains of microorganisms
   Mutants and variants of microorganisms
   Particular formulations of specific strains or cultures

Macromolecules
   Peptides and proteins
   Enzymes and hormones
   Specific antibodies, monoclonal antibodies, antibody conjugates
   Adjuvants

Cells
   Cell cultures
   Hybrid cells

Recombinant DNA
   Isolates sequences, isolated genes
   Promoters, plasmids, expression vectors
   Transformants and genetically engineered cell cultures

Process Inventions
   Processes preparing the subjects listed above
   Processes using the subjects listed above

Products derived from processes using objects
   listed above

[a] Under the general provisions of novelty, non-obviousness and utility/technical applicability the above examples of biological subject matter with biomedicinal applications may be considered as patentable. For products which also occur in nature, this applies to a form which is isolated from nature or changed from how they exist in nature. See also Tab. 2 for country-specific exclusions.

Tab. 2. Statutory Exclusions of Inventions from Patentability

---

Pharmaceuticals/Medicines

South and Central America:

    Argentina[a], Costa Rica, Cuba[b], Guatemala, Nicaragua, Uruguay. Andean Group (Bolivia, Columbia, Ecuador, Peru, Venezuela): inventions relating to pharmaceutical products appearing on the list of essential medicines of the WHO.

    Brazil[a] and Honduras: chemical-pharmaceutical substances *and* processes.

    Bermudas: patents which are mischievous or "inconvenient" to the state may not be granted.

Asia: India[a], Iran, Iraq, North Korea, Syria, Thailand, Turkey, Vietnam. People's Republic of China[a]: chemical pharmaceuticals only.

Africa: Egypt, Ghana, Libya, Somalia, Tanger, Tunesia, Zambia. Zimbabwe and Malawi: mixtures of known ingredients. (*No* exclusions in OAPI member countries and in Algeria, Kenya, South Africa, Uganda.)

Europe: Bulgaria[b], Hungary, Iceland[a], Romania[b], Poland.

Chemicals (patentability excluded or limited)

    Brazil[a], Bulgaria[b], P.R. China[a], Cuba[b], Hungary, India[a], North Korea, Poland, Romania, Vietnam.

Therapeutic and Diagnostic Methods Practised on Humans or Animals

    Not patentable in almost all countries except in the USA and the Philippines.

Microorganisms and Biological or Biotechnological Products

    Bulgaria[b], Cuba[b], Romania[b].

    Poland: strict *technical* character of the patent required.

---

[a] Changes are under discussion
[b] Certificate of inventorship issued
Processes for making the excluded substance matter are patentable in most countries.
Collected from JACOBS (1993) and updated by own information sources

---

restrictions which are relevant to pharmaceutical products.

Special provisions for a variety of inventions were established in communist states. In these cases, "Certificates of Inventorship" were issued. The right of exploiting the invention commercially was usually exercised by the state. Following major political changes in recent years, most of these certificates were abolished and replaced by full patent protection.

Almost all countries and regional patent conventions have provisions that do not allow patents on the grounds that these are contrary to, e.g., national laws, principles of humanity, public morality, public health, public safety and for other ethical reasons.

Tabs. 1 and 2 are intended to give an indication about possibilities and restrictions for patents on biological subject matter. Of course these tables should not be used as legal advice. For more specific information it will be necessary to check the latest statutory provisions, patent office policies and probably even judicial decisions in the individual countries with a specific invention in mind. This should be left to patent specialists in the specific field.

2.6 Product, Process and Use Patents

The separation of patent claims into those on a product, a process or on use is not only a theoretical division; in practice these three categories are of great importance for the enforcement of a patent. They differ considerably in their effectiveness to exclude competitors from using the invention.

The *product patent* category comprises

- substances, e.g., adjuvants, isolated proteins or microorganisms, a cDNA sequence coding for a certain protein, (genetically engineered) cell lines
- composition of matter, e.g., composite vaccine stabilizers, peptide conjugates, cell culture media
- apparatuses and devices, e.g., a vaccination gun, a pulse release system.

Product patents are the key patents and generally offer the most effective protection against potential competitors. This is especially true for end products which are sold to the consumer or user, since infringement is easily detected. Less protective and enforceable are product patents on intermediate or starting products which are used during a manufacturing process. In many cases the commercial end product does not reveal the fact that a patented starting or intermediate product was used. The situation becomes obscure, when a patented product (e.g., PCR tools) was used only during the development of a commercial product. However, in most of the latter cases, patent claims which are broad enough to cover the end product of such a development stage will rarely be granted.

Process patents cover, for example, methods to prepare substances. Possible examples in the context of biomedicines are, e.g., isolation or purification methods, cell culture techniques, attenuation schemes and other more directed genetic manipulations of microorganisms, cloning techniques, and expression techniques for recombinant polypeptides. Since process patents are usually admitted in countries where certain products cannot be patented (see Tab. 2), process claims are a possibility to protect inventions in these countries.

Process patents can be difficult to enforce due to the fact that an infringement is less overt. Furthermore, process patents are easier to circumvent. Once a new and attractive method is disclosed, it stimulates others to invent around the disclosed process, which is legal and even in line with the purpose of the patent system. As for all patent categories, this can be prevented by a thorough exploration of possible variations of the invention along with an adequate description, which allows broader patent claims. But even the best possible process patent does not preclude others from exercising the invention in a country where patent protection is not available, no patent application was made, the particular invention was not patentable, or an available patent cannot be properly enforced. Since a mere process patent does not always cover the product derived from this process, the product may be imported and sold in countries which are covered by the process patent. Certain South and Central American states as well as some Eastern European countries with restricted patentability of pharmaceuticals and chemicals are the base of companies which exploit this situation, particularly with regard to products covered only by process patents.

Whenever products which are made by an invented process are new, useful and non-obvious (e.g., have advantages over existing products), the products derived from the inventive process should be included in the patent claims. Combined process and product patents are quite common, in fact many countries including the USA and the EPO extend patent protection to the products directly obtained from a patented process.

If a conventional vaccine for a certain disease exists, a new process/product patent would, for example, claim a process for the manufacture of a recombinant vaccine and the resulting new vaccine itself. Patent claims for the vaccine could be deduced from a higher purity, an improved efficacy or a reduced risk of adventitious agents. Of course, these advantages should be substantiated by experimental data or by conclusive evidence from the literature.

Process patents on improved and critical manufacturing steps, filed and granted after the initial product patent, can be used as an effective means to extend the time of patent protection for a product indirectly and at a lower level.

Use patents comprise the application of a product or process for a particular purpose. Examples for pure use patents which are independent of product or process patents are:

- the use of a known compound in a fermentation process, e.g., to enhance yields,
- the use of a known excipient or pharmaceutical as an adjuvant,
- the application of a known expression system to produce a protein.

Independent use patents offer the weakest patent protection. They almost invite others to seek possibilities to circumvent them, and they do not cover the products that result from the invented new use. Wherever the necessary requirements are fulfilled, product patents should be applied for. In the first example given above, it seems hardly justified to construct product claims for (all) products made by the modified fermentation process. The excipient with adjuvant effects could probably be patented as a product patent, claiming "vaccines containing substance X as a novel adjuvant" with the non-obvious advantage of a better tolerability or efficacy over existing adjuvants. As for the third example, any advantageous characteristic of the resulting product may be used to justify a product patent.

In the USA mere use patents are not accepted. If possible, these inventions should be specified as process patents, e.g., as a "method of use".

2.7 Dependent Patents

A new patent may depend on other, earlier patents by using the entire earlier invention or elements thereof to achieve the desired result. Classic examples are improvements on earlier inventions. Pioneer inventions may stimulate a whole series of dependent patent applications. When granted, dependent patents cannot be worked without infringing the pioneering patent. Thus the patentee of the dependent patent needs approval by the owner of the prior patent. This may be achieved by buying a licence or by a mutual cross-licence. In some countries there are legal provisions to prevent the owner of superior patents from blocking other, inferior patents. Licences must be given but are, of course, subject to remuneration by the licencee. Alternatively the owner of the prior patent may in return receive a licence on the dependent patent (reciprocal licences). If both parties cannot reach an agreement, the conditions and licence fees will be set by a court.

3 The Patent Application

3.1 The Patent Description

A patent application must fulfill different purposes. First, it must provide the necessary information to prove the novelty, non-obviousness and utility of the invention in order to successfully pass examination by the patent office and, if required, to defend the invention against opposition and infringement by third parties. Second, it must describe the invention in sufficient detail to enable others to reproduce the invention. Third, it must specify what exactly the applicant intends to claim.

As for scientific papers, a certain scheme has been shown to be most suitable to meet the necessary requirements, and applicants or scientists who draft a patent application for their invention are well advised to adhere to this scheme. Studying earlier patents or patent applications will enable an applicant to become acquainted with the usual form of patents. Alternatively applicants can have patent attorneys prepare the application for them.

The title of a patent describes the invented subject matter and usually also the patent category (e.g.: Therapeutic agent against disease X and process for preparing the same) and is followed by a short abstract identifying the field of the invention.

The background of the invention or prior art is then described and discussed and is often quoted from earlier patents or patent applications which may be newer than scientific publications. This requires a diligently performed patent and literature search to set the invention apart from the prior art and to determine the potential novelty of the invention, as well as indicating the allowable breadth for patent claims. It should be kept in mind that granted patents or patent claims can be denied at any time during the life of a patent, if obtained by false or incorrect statements.

A short paragraph usually outlines the unsolved problem, which the invention addresses, and emphasizes the inventive, nonobvious element of the invention. This is followed by a detailed description of the invention. It is a statutory requirement that the invention must be disclosed completely so "as to enable any person skilled in the art ... to make and to use the same, ... and shall set forth the best mode contemplated by the inventor of carrying out his invention" (35 USC, § 112; similar wording in EPC, Article 83).

All materials used, including microorganisms and cells, should be identified according to a generally accepted nomenclature. The origin or method of isolation must be revealed, and all necessary characteristics defining any biological material should be given. If necessary, drawings or formulas may be added.

Due to the diversity of microorganisms or cells it may be impossible for a person skilled in the art to repeat the invented solution without undue experimentation (a certain degree of experimentation to gain experience is acceptable) or without some inventiveness of his own. It is, therefore, mandatory in such cases to deposit a sample of the microorganisms or cells at an approved depository institution in order to provide the necessary enabling disclosure and to meet the requirements of a patent application (see also Sect. 3.2 on deposition of microorganisms).

If certain parameters can be varied, the possible range of these variations should be experimentally explored and specified in the patent description, followed by an indication of the preferred range as demanded by 35 USC, § 112 as quoted above. A description of the general methodology is not sufficient to describe and enable the invention. Specific details must be given which is generally done by describing examples and results. The broader the scope of the patent, the more specific examples will be required to teach the invention and to show its utility. It is self-evident that the methodology must reproducibly give the desired results.

A patent application ends by specifying what is to be protected in one or more claims. The exact wording of these claims is of utmost importance for the enforceability of the patent. A separate paragraph below is dedicated to the patent claims.

Patent applications are usually written in the official language of the country where the application is filed, with translations being provided where necessary. European patent (EP) applications must be written in one of the three official languages, English, French or German, unless nationals of an EPC member country choose to file an application in the official language of their country along with a translation.

3.2 Deposition of Microorganisms

Inventions involving microorganisms or cell cultures are often difficult to characterize sufficiently in writing to enable the average expert to reproduce the isolation, construction, attenuation or other processes performed with these materials. However, this does not release the applicant for a patent from the essential obligation to disclose the invention clearly and completely.

A practical way to solve this dilemma is to deposit the microorganism or cell culture at a depository which will provide samples upon request. According to the Budapest Treaty on the "International Recognition of the Deposit of Microorganisms for the Purposes of Patent Procedure", such deposits can be made at internationally recognized depository authorities.

The applicant is not obliged to deposit specimen(s) in different countries individually; a single deposit can be sufficient. Having confirmed that a certain depository is recognized by the patent office of the country where the patent application is due to be filed and that the depository is able to handle the specimens in question (for safety reasons certain samples may not be acceptable in some institutions, and some countries may not accept certain biological samples for quarantine reasons) the deposit must be made by the date of filing of the application. The institution where the deposit was made, the accession number which identifies it and the date of the deposit must be mentioned in the patent description. The depository must be authorized to dispense samples of the deposited specimen to the authorities involved in the examination of the patent application and to third parties upon request. Access to the samples by third parties (e.g., interested researchers, but also opponents) may be restricted until the publication date of the patent application (EPC members and other states) or until the patent is granted (USA). As long as the patent is valid, the patent owner may make arrangements with those requesting a sample that the sample is used for experimental purposes only and may not be passed on to another party.

Viable samples must be maintained at least during the life of the patent, usually for 30 years or 5 years after the last sample has been requested – whatever the latter is. If the depository becomes unable to maintain the deposited specimen or to supply samples, a new, identical sample must be deposited within 3 months. This indirectly requires that the patent owners maintain samples themselves or deposit them at different institutions.

Selected international depository authorities under the Budapest Treaty accepting bacteria, yeasts, fungi, viruses, and strains containing recombinant DNA molecules or isolated DNA preparations are listed below.

ATCC, American Type Culture Collection, 12301 Parklawn Drive, Rockville, MD 20852, USA

AGAL, Australian Government Analytical Laboratories, 1 Suakin Street, Pymble, NSW 2073, Australia

CBS, Centraalbureau voor Schimmelcultures, Osterstraat 1, Postbus 273, NL-3740 AG Baarn, The Netherlands

CNCM, Collection Nationale de Cultures de Micro-Organismes, Institut Pasteur, Rue de Docteur Roux 25-28, F-75724 Paris Cedex 15, France

DSM, Deutsche Sammlung von Mikroorganismen und Zellkulturen, Mascheroder Weg 1b, D-38124 Braunschweig, Germany

NIBH, National Institute of Bioscience and Human Technology, Ministry of International Trade and Industry, 1-3, Higashi, 1-Chome, Yatabe-machi, Tsukuba-gun, Ibaraki-ken 305, Japan

A complete list of approved, international depositories is given in the PCT APPLICANT'S GUIDE (1993) (see References); detailed information about conditions and restrictions is summarized in the *World Directory of Collection of Cultures of Microorganisms* (STAINES et al., 1986).

3.3 Patent Claims

A patent must end with one or more patent claims which describe the patented subject matter as unambiguously as possible. Patent claims may be considered as the definition of the patent scope for legal purposes. The precedent patent description has to justify all aspects of the patent claims and will be used for this purpose during the examination process. But only in cases of doubt will details of the patent description be used later on to interpret the patent claims. Thus, patent claims must be formulated with the utmost care. The three basic requirements novelty, non-obviousness and utility have to apply to anything that is covered by the claims and should be used to check the drafted claims. On the other hand, patent claims should read as broadly as possible to cover all patentable aspects of the invention. In practice, this conflict between the inventor's intention to keep patent claims as broad as possible and the public's interest to grant patents only on what has actually been invented, disclosed and reduced to practice, has led to the situation that patent applications contain rather broad claims which are subsequently narrowed down during the examination process. Since patent examiners on their own initiative cannot recommend to extend the scope of the proposed patent claims, it is probably best to reach for a maximum in the patent application and revise the claims according to the objections from a patent office. However, it is useless to go too far, e.g., by extending patent claims to mere principles, products of nature or undisclosed analogous subjects as well as to known and obvious things.

Japan and the USA formally limit the protection of a patent to what is literally defined by the claims. In the USA "equivalents" are also covered by a patent, i.e., infringers using equivalents may be sued. Equivalence is defined by the US Supreme Court as doing "substantially the same thing in substantially the same way to get substantially the same result" (Graver vs. Linde, 339 U.S. 605, 608, 1950) as a patented invention without literally infringing it.

The "Doctrine of Equivalents" applies to the interpretation of claims to establish infringement of existing patent rights in those cases where there is no literal infringement. Anything that comes under the definition of equivalents would have been patentable at the priority date (and should be included in the claims). Thus the Doctrine cannot be used as an argument to negotiate broader claims for a patent at the time of application or prosecution (KUSHAN, 1992).

The requirement to exactly define and limit the scope of a patent has implications for patents on biological molecules because it leads to patents which can be easily circumvented. This is illustrated by the following example: A polypeptide X is claimed and defined by its complementary DNA (cDNA) sequence. An equivalent according to the above-mentioned US definition and covered by this claim, would be the *same* polypeptide which is expressed by a modified cDNA, containing different base sequences but coding for *identical* amino acids. Polypeptide X with an exchange of amino acids would currently not be considered as equivalent, since innumerable analogs would be possible and the patent description would be unable to specify the scope of the patent. For example, over 3600 different analogs of the erythropoeitin molecule can be made by substituting at only a single amino acid position, and over a million different analogs can be made by substituting three amino acids (Amgen vs. Chugai, as discussed in KUSHAN, 1992). Furthermore, it is possible that an exchange of amino acids might alter the entire molecule in its three-dimensional structure and function. A molecule with such an exchange would not be substantially the same.

This opens up many possibilities for introducing minor variations into patented polypeptides with the aim of circumventing existing patents. One way to avoid this or at least to make it more difficult, would be to test and patent variants having the same effect or to identify the functional region or the most relevant epitopes and claim these specifically. This can be an endless task and does not seem practical for most molecules.

Another possibility would be to define the molecule to be patented by its function and claim forms of this polypeptide having the same function. Highly innovative patents will have better chances with this approach than less innovative ones. In the USA this may be considered as an attempt to claim patent rights on the end result rather than on the particular means by which the result can be achieved. Except for pioneering inventions, functional claims without sufficient qualifying structural elements are usually rejected, since these do not define the patented novelty with a reasonable degree of particularity and distinctness (see *Manual of Patent Examinating Procedure*, MPEP, 1988).

Due to the different policies in the European countries, the chances of achieving a broader patent protection, for example, based on functional claims, are somewhat better. The European Patent Convention partly reflects the German patent policy with stronger emphasis on the protection of the basic idea of the invention ("Erfindungsgedanke"). According to the EPC patent claims should not be applied by the strict literal meaning of the wording used in the claims – nor should claims serve only as guidelines. Of course this does not mean that a proof of technical applicability or an exact description of the patented subject matter is not required.

In practice, broader claims in European patent applications on biological substance matter have a better chance of being accepted, although they extend, for example, beyond the disclosed DNA or amino acid sequence (e.g., to related molecules from other species which can reasonably be expected to fulfill the same purpose). Acceptability will be improved if the extensions are supported by at least some experimental evidence. For claims extending to not exactly defined DNA or amino acid sequences, such evidence may be presented by means of cross-hybridization results or by showing cross-reactivity of antigens.

Giving numerous examples for the exact wording of successful patent claims would require discussing these at length along with the disclosed description and the state of the art at a particular time. An idea of patent claim formulation for a specific invention can be obtained by studying recently granted patents which are related to one's invention and by seeking advice from specialized patent attorneys.

If patent claims on highly variable biological molecules are restricted to the literal wording of the patent claims and if functional language is rejected, these patents may become useless. Many conventional biological pharmaceuticals were considered as generics because, under the existing patent policies, they suffered from insufficient patent enforceability. Technical advantages over competitors, kept as trade secrets, were more effective than patents.

Modern products from biotechnological processes will probably be even more affected by insufficient patent protection. They usually require higher investment into research and development and thus need better protection against product piracy than mere trade secrets. It is likely that imitators of successful products will emerge as soon as the technical risks are eliminated by pioneers and analogous products can be developed without the initial high rate of failure experienced by the first developer. This may take ten years or more, but the encouragement for imitating products is dependent upon the current policies and the present approach to patent claims and their scope for biological inventions. It appears that the EPC guarantees a slightly more adequate patent protection for this eventuality than does the US system.

3.4 Filing a Patent Application

A draft of the patent application by the inventor usually requires several amendments and must be transferred into the correct form before it can be filed at a patent office. The scientific part of a patent relies largely on the inventor(s), but patent specialists should be consulted to help drafting the claims, to decide on the countries where an application should be made and to attend to the formal aspects. These formal aspects (abstract length, paper format, filing procedures, authorization documents, priority date declaration, payment of fees, time limits, etc.) are sometimes extremely detailed and differ from country to country.

After submission to a patent office, the file is examined for some basic formal requirements and, if adequate, is given a filing date. A first indication about the chances of a patent application may be obtained from a search report which is issued, for example, by the European Patent Office and also for international patent applications under the Patent Cooperation Treaty (see below). In most cases, the search report will list a number of earlier patents or publications and indicate whether these interfere with the proposed claims of the examined file. As a consequence of the search report the applicant can amend the proposed claims before the examination is initiated.

A substantive examination for patentability (novelty, utility, inventiveness) will be performed by a specialized patent examiner or a group of examiners, and an official report will be issued. Quite frequently the official report indicates a complete rejection of all proposed claims. The applicant has the opportunity to review the references cited in the official report and comment on the validity of the conclusions drawn. Most patents and publications quoted should already be known to the inventor and should have been considered in the "background of the invention" section. If these appear in the report, it is a good indication that the invention was not convincingly set apart from the prior art. Further evidence (i.e., more detailed arguments or further supportive literature) can be provided, illustrating that the invention is novel over the prior art. (Objections on the grounds of a lack of utility are rare.) The patent office will consider the response and, if matters remain outstanding, will respond with further official action. Some countries have general time limits for this process, some other countries limit the number of rounds of official action and response by the applicant. If the initially proposed broad claims cannot be defended successfully during this process, it will be necessary to amend them according to the examiner's objections, or to abandon individual claims or even the entire patent application.

The prosecution procedure of official action and responses may be repeated in all of the individual countries where a patent application was filed. Fortunately, international patent cooperation treaties exist to shorten and simplify this awkward process for both sides.

The *International Patent Cooperation Treaty (PCT)* created the opportunity to file a patent application almost world-wide (most countries accept PCT applications) at international patent registration offices which are usually identical with the national patent offices. A search report summarizing relevant prior art will be issued and, if requested, a preliminary examination considering the prior art is performed. The preliminary international examination provides a non-binding opinion whether the claimed invention appears to be novel, inventive and industrially applicable. It does not investigate the patentability according to any national law.

The *International Patent Bureau* in Geneva initiates the submission of the application in all designated member states, for which the application is intended, thus saving a considerable amount of time, effort and cost. However, due to the differences of international patent laws, the definitive substantial examination and granting of a patent applied for under the PCT is still in the hands of the individual countries.

A "PCT Applicant's Guide" is issued by the International Patent Organization in Geneva.

The *European Patent Organization (EPO)* provides a system for the application and granting of a patent for all designated member states at only one authority. Members of the EPO are Austria, Belgium, Denmark, France, Germany, Greece, Ireland, Italy, Liechtenstein, Luxembourg, Monaco, The Netherlands, Portugal, Spain, Sweden, Switzerland and the United Kingdom. Norway has not yet ratified the European Patent Convention.

Patents granted by the *European Patent Office* in Munich and in The Hague confer on its proprietors the same rights as a national patent. Since the contracting countries have adapted their national laws, only one examination process takes place, and only one common certificate is issued. Applications filed under the PCT system may also designate EPO countries.

Similar to the European Patent Convention are two African organizations, joining English speaking countries or former French colonies by a common patent application and examination system.

The *African Regional Industrial Property Organisation (ARIPO)* has a central registry in Harare, Zimbabwe, which examines patent applications and notifies member states on favorable examinations. Member states can object to granting a patent in their territory, a provision which is necessary because member countries have no common policy on certain statutory exclusions of patentable subject matter. Member states of the ARIPO are: Botswana, Ghana, Kenya, Lesoto, Malawi, Sudan, Swaziland, Uganda, Zambia, Zimbabwe. Sierra Leone, Somalia and Tanzania have not yet signed the Harare Protocol.

The *Organisation Africaine de la Propriété Intellectuelle (OAPI)*, also referred to in English as African Intellectual Property Organisation (AIPO), with its central office in Yaounde, Cameroon, issues only one single patent which is valid in all member countries. The general patent policy in the OAPI is comparable to the EPC. Member states of the OAPI are: Benin, Burkina Faso, Cameroon, Central African Republic, Chad, Congo, Ivory Coast, Gabon, Guinea, Mali, Mauretania, Niger, Senegal, and Togo.

International (PCT), European (EP) and most other national patent applications are published 18 months after the priority date. PCT and EP applications are published along with the search report. (If a search report is not available at that time, it will be published separately.) This publication offers third parties the possibility of inspecting the file upon request and to inform the patent offices about their related observations. These must be considered by the patent office and will be transmitted to the applicant. Considering the fact that the time between the patent application and the granting is normally around five years, the publication of an application also provides a means to inform the public as early as possible about the technical progress of the invention which, after all, is one of the main purposes of the patent system. After being granted, the final patent description is published again. Within a limited time an opposition can be filed by any interested person. The window of time to oppose is between 9 months for EP patents and only 3 months for Australian patents. Patent applications in the USA are not published before they are finally granted. As a consequence, attacking these patents is possible at any time thereafter by requesting re-examination by the US Patent Office or before a court.

Although it is theoretically possible for an individual to file a patent application, it is in most cases necessary and strongly recommended to entrust a patent attorney with this task. Only a specialist can oversee all the formal and legal requirements of the patenting process and make sure that fees are paid in time and time limits to respond to the patent office will be kept so that patent rights are not lost due to formal mistakes and errors. Due to the different situation in other countries and the requirement to provide a local contact person, the patent attorney on his part will liaise with colleagues overseas to handle foreign patent applications. This delegation of work has, of course, its price but the applicant will soon appreciate the advice and support provided by a specialized patent attorney. Nevertheless, the inventor must be prepared to spend some time during the next few years defending a patent application by scientific arguments.

3.5 Priority of Patents and Continuation-in-Part

A patent application is assigned a priority date for the art disclosed in the application. Any aspect of the patent application which was known to the public before the priority date cannot be claimed as an invention. From the priority date onwards the knowledge contained in a patent application must be considered as prior art and excludes others from any attempt to patent the same subject matter. In general the first filing of a patent application establishes the priority. Referring back to the initial priority date, further applications in other countries can be made within a period of usually twelve months. Applications first filed in the European Patent Office or in individual PCT member states can also be used to claim a priority date for a PCT application.

Although the same basic rules on the priority apply to US patent applications, two special provisions of the US patent laws should be mentioned which have some influence on the effective priority date of an invention. Until a US patent is issued a "Continuation-in-Part"(CIP) may be filed by the applicant. The same or a similar invention is refiled with new information which may enable the inventor to change or extend the patent claims and to better define the subject matter and its use. If the CIP is based on the same invention and not on later art than the original application, the priority date of the first application may be maintained. To retain the priority is, in fact, the intention of most CIP applications. New patent claims which are based upon information added in the CIP ("new matter") will get a separate, later priority.

Similarly, in Australia a provisional patent application can be lodged to establish priority; the complete application must be filed within twelve months.

A second exception in the USA which influences the effective priority date of an invention (but not the official priority of the patent!) is the so-called swearing-back according to the Code of Federal Regulations, Title 37 (37 CFR, 1.131). If there is an interference with other patents or patent applications, the US patent authority will ask the applicant of USA-derived inventions to provide detailed information on the actual time of conception of the invention, reduction to practice and on all further steps which finally led to the current application. Proof of those activities can be provided by laboratory notebooks and other relevant documents, e.g., those which prove the involvement of a patent attorney during the process of drafting and filing the application in question. This information will be considered in order to assess the effective time when the invention was made and when it was reduced to practice in order to fulfill the requirements for patentability. This "Rule 131" only applies to inventions made and reduced to practice in the USA. Applicants from abroad have a clear disadvantage because they can only claim the official priority date in cases of conflicts in the USA. This, along with the long time lag until a US patent is finally published, urges applicants from abroad to file patent applications in the USA as early as possible, probably exploiting the possibilities of a Continuation-in-Part later on.

3.6 Duration of Patent Protection

Patents may be valid for 10 up to 20 years, depending on the individual country. With a few exceptions the patent duration starts at the time of filing of the application. EPO and OAPI countries as well as most other countries, including Canada, the Commonwealth of Independent States, the Czech and Slovak Republics, South Korea and Mexico, offer patent protection for 20 years from the date of filing. Patents in the USA are valid for 17 years after issue. In Japan patent duration is limited to 15 years after issue with a maximum of 20 years between the filing of the application and the end of the patent term. Patent protection in Australia and New Zealand

is 16 years from the filing date. Countries which offer a shorter patent protection (10–15 years) are mainly South and Central American countries. These countries also exclude patents on medicines and pharmaceuticals (see Tab. 2). In the last few years many countries with short patent durations and other restrictions have changed their patent laws and in most cases have adopted many of the EPC standards.

In almost all countries (except the USA) there are legal provisions to prevent the abuse of patent rights by either not using ("working") the invention or by not granting licences to others to work the invention. If the holder of a patent fails to prove that the patent is worked in the country within a reasonable time (usually after 3 to 5 years) or does not provide sufficient evidence that the patent could not have been worked reasonably (e.g., due to regulatory conditions for pharmaceuticals), this may result in a compulsory licence being granted to a third party prepared to work the patent, at terms set by the authorities. Import of a patented product is not accepted by some countries as being sufficient to comply with the requirement to work a patent.

3.7 Extension of Patent Terms for Pharmaceuticals

Patents which are related to pharmaceuticals can provide a significantly shorter effective protection for the resulting products, since a considerable time of the patent life may pass before all conditions for the market approval of pharmaceutical products are fulfilled. For such cases, some countries extend the patent duration of pharmaceutical patents to cover or offset the time which is lost during the registration of the related product. Possible extensions are equivalent to the time required to get market approval, but are not meant to compensate for time losses associated with the technical hurdles of product development.

Extensions of patent duration for pharmaceuticals are available in the following countries:

Australia: maximum 4 years, only for human pharmaceuticals.
EPO countries: maximum 5 years.
Japan: maximum 5 years, if working delayed for more than 2 years.
New Zealand: no specific time (and product) limitations.
USA: no specific time limitation.

It should be noted that applications for a patent term extension must be handed in as early as within 60 days (USA) or 6 months (EPO) from product approval.

3.8 Opposition against Patents

In most countries the formal process of patenting includes several regulations and steps which enable third parties to prevent the grant of patents which are not justified. These steps are: early publication of a patent application, publication of the search report, access to the files upon request, provision of an objection period after publication of a patent grant and finally the possibility to submit objections against granted patents.

During the early stages of patent examination anyone may submit information relating to a patent application that has been published. This information will be considered by the patent office and will also be forwarded to the applicant. Since most patent claims are modified during the examination procedure and many patent applications are withdrawn, it seems more relevant to file an opposition against a patent as it is published after acceptance – as far as this is still necessary.

A patent opposition may be based upon evidence which shows that an invention lacks the basic requirements for a patentable invention and may aim for the invention as a whole or for individual claims or the scope of such claims. Lack of novelty may be indicated by all kinds of earlier publications or other disclosures of the invention to the public. As already mentioned, prior publication by the inventor or applicant also interferes with novelty and can be used as an argument against their own patent application. This does not fully apply to US inventions, the content of which can be published by the inventor within twelve months before filing a patent. Novelty of an invention may also be contested by the proof that the invention was already in use, for example, as a manufacturing process which was kept secret.

Oppositions on the grounds of obviousness also depend on earlier publications, patents, and other evidence which demonstrates that the invention could have been deduced by a person skilled in the art without an inventive step. Objections against the utility of a patent are rare, because in most cases it will be very difficult to call into question that the subject of the invention will (also in future) be of no use.

Evidence that the patent description does not sufficiently enable others to reproduce the invention to the extent given by the claims can be used successfully to render an entire patent or certain patent claims invalid. In practice this would be a very expensive and difficult exercise, however, this situation can emerge, if two competitors simultaneously work on the same invention.

The inability (as well as the unwillingness) of the owner of a granted patent to use or "work" the patent may result in a loss of privileges provided by the patent. Whereas some countries require a proof of working in regular intervals, other countries take actions against the owner of such a patent only upon request of a third party. This includes cases in which a third party is in a position to practice and exploit an invention while the owner of the patent is not (yet) able to use it. In such a situation a compulsory license will be given to the objecting third party.

3.9 Patent Costs

The costs of filing and maintaining a patent vary, of course, from country to country and depend on the complexity of any particular case. Thus, only a few examples and estimates will be given below.

Applications for a national patent or at the International Patent Office may cost up to an equivalent of US $ 3000 per application for a search report and the examination. A similar amount can be assumed for the involvement of a patent attorney as long as only formalities are concerned. Creative work by a patent attorney office (e.g., formulation of the claim or of defending statements) as well as translations result in the extra cost of about US $ 100–300 equivalent per hour, depending on the qualification of the person who deals with it. Thus, the effective cost of the patent application may rise up to US $ 8000-10000 per country. Annuities for a national patent are raised as a constant rate per year or increase during the life span of a patent. The figure of US § 100–300 per year and country (or per patent certificate from international patent organizations) may serve as a clue for what must be expected.

Patent office fees may be reduced to 50 or 60% of the normal fee, if the applicant agrees to a "license of right" notation, in which case everyone may obtain a license to exploit the patent. If no agreement can be found with the patentee on the terms of such a license, the terms will be set by the court.

While the above-mentioned estimates assume that the patenting process is rather straightforward, considerable expenses must be anticipated if an opposition has been filed, or if, in the USA, an interference is to be expected. Due to the complicated procedure, especially of a US patent interference, several ten or hundred thousand dollars are spent readily and the entire cost of an interference procedure may well exceed one or even several millions, especially if the case is fought in court. Direct negotiations between the parties involved may settle such cases easier, provided that compromises are acceptable.

3.10 Patent Information

Any serious research effort directed towards achieving a commercial reward and patent protection needs assistance from adequate information services. Besides the scientific literature, related patents and patent applications must be available and can be traced through patent data bases and patent libraries. Computerized patent data bases (Tab. 3) usually contain all front page information of a patent or patent application, such as patent number, applicant, inventor, filing and priority date, title of the invention and the abstract, in some cases also the patent claims. Full copies of selected patents can be ordered from national patent libraries or from the patent office.

If a scientific library does not provide an online patent and literature service, the patent office library may be approached directly. Patent attorneys may also offer a patent search service. Furthermore, contacts with companies which are interested in the field may be established in order to use their patent information network.

Compared with the cost of fruitless re-inventions, the expenses of patent searches are minute and comparable to scientific literature

Tab. 3. Major Patent Data Bases

File	Host	Scope	From
EPIDOS (INPADOC)	Orbit, EPO Vienna	international	1968
CA-FILE	STN	international	1967
WPI/WPIL	Orbit, Dialog, Telesystems Questel	international	1963
CLAIMS/CITATION	Dialog	USA	1947
CLAIMS/US Patent Abstr.	Dialog, Orbit, STN	USA	1950
JAPIO	Orbit	Japan	1976
PATOS	Bertelsmann	Germany	1968
PATDATA	BRS	USA	1975
FPAT	Telesystems Questel	France	1969
PATDPA	STN	Germany	1981
EPAT	Telesystems Questel	Europe	1978
BIOTECH ABSTRACTS	Orbit	international	1982

Check List for Patent Applicants	yes	no
1. Has a complete and up-to-date patent and literature search been conducted?	○	○
2. Does the invention meet the requirements of novelty and non-obviousness?	○	○
3. Does the invention have a commercial or industrial applicability?	○	○
4. Are the intended patent claims sufficiently broad to avoid an easy circumvention of the patent?	○	○
5. Can a commercial product based upon the invention be developed and marketed before the life-span of the patent expires?	○	○
6. Is (are) the resulting product(s) likely to render sufficient profit to recoup the patenting and development cost?	○	○
7. Is the invention patentable in those countries which represent the major markets for the resulting products?	○	○
8. Is the applicant of the patent prepared to pay the expenses of patenting and to actively sell the invention or are there potential licensing partners who are prepared to promote further development and marketing of the resulting product(s)?	○	○

searches. The information obtained may be extremely valuable considering the fact that patent descriptions usually are very detailed and may also refer to alternative methods and possible variations of important parameters.

3.11 Check List for Prospective Patent Applicants

Considering all aspects discussed above, the applicant for a patent should respond positively to the questions in the check list above. These may help to decide whether an invention should be patented at the current stage.

Negative or questionable answers indicate weaknesses, which may result in unsuccessful patent applications or in weak or unattractive patents. Activities for improvement should be considered, e.g., by discussing the proposal with patent and marketing specialists or by additional, targeted research.

4 Selling an Invention, Licenses and Royalties

Most inventors are not independent private persons but employees of a company, institute, university, or the state. Inventions emerge from work for which they are paid or were the objective of their work. Thus, the resulting patent rights belong to the employer. (However, the inventor(s) must be named on a patent.) Many employment contracts for researchers contain specific provisions which transfer the rights to commercially exploit inventions to the employer.

The question arises whether there is any incentive for an employed researcher to patent an invention, apart from the contractual obligation to do so as an employee. Filing and defending a patent often results in a lot of additional work for the inventor which in the case of a patent defense may be related to work done months or years before, while the inventor's research proceeded in the meantime. Publications on the subject and on later improvements may be held back in order to avoid supporting potential competitors. Thus, the interest of a researcher to patent an invention interferes with the interest to publish and to turn the attention to new research fields. Specific measures by the employer, for example, a bonus for a successful patent application or other awards, but at least an adequate recognition of the achievement should be considered as incentive for inventors. In this context it is noteworthy that a rather unique law in Germany entitles inventors to compensation payments by the employer.

If the employer is not interested in making use of the invention, the inventor is usually free to file a patent application himself and at his own expense. Otherwise, the employer will act as applicant, in which case it is most likely that a legal/patent/license department and a patent attorney will provide the necessary support. Many universities and government funded institutes have established independent groups, departments or even companies in order to commercially exploit the inventive potential – especially in biotechnology – directly. These organizations may also provide service and advice in patenting.

In the absence of adequate support to file and defend patents, individuals or institutes may decide to approach companies with the aim of licensing the invention. This is often done already before the draft stage of a patent application. The invention as a whole may be offered to the company, in which case the company may act as applicant or licensee of the patent and will request exclusive exploitation rights in exchange for the payment of patenting costs and for later royalties. Continuous payment of the patenting expenses by the company will ensure that the industrial partner of such an agreement does not lose interest in the project but nevertheless keeps the patent. As an alternative joint ownership of the patent can be negotiated. Provisions may be included in an agreement to the effect that the license of the patent shall be terminated if the company does not wish to exercise these rights or does not exert serious efforts to develop or market the invention.

The main advantages of such agreements are that the inventor avoids the patent costs in a stage where the chances of getting a patent are still unclear and that the industrial partner has early access to the invention and may gain a significant advantage over competitors. These agreements also often link the two parties together in a common research and development project in which both have a strong interest in a fast and efficient commercialization.

Joint development agreements in which both partners share effort and cost can be of significant benefit for both sides because of the necessity to cooperate early and closely and to exchange information in two directions. If researchers agree to participate in further research and especially in the development-related activities which are funded by the industrial partner, it must be anticipated that tight time schedules must be met and that funds are dependent on the accomplishment of certain tasks or measurable milestones.

In the case of major inventions the inventor may prefer to apply for a patent without foreign participation and try to sell non-exclusive patent rights to different industrial companies or to negotiate country-specific or use- and indication-specific exclusive agreements. At first sight this looks much more attractive for the inventor and may be possible for important inventions which can be split into reasonably sized, separate market segments. Any attempt to do so with minor inventions will most likely result in much less interest of the industrial partners and lower royalty rates for the inventor. Similar development cost will be calculated against smaller market shares due to restricted market access and immediate competitors, and the overall benefit for all will be smaller.

Possible variations of license contracts are lump sum payments (the license is sold for a single, larger payment), royalties on sales, and combinations of up-front payments or option payments in certain intervals with royalties.

Of course, the strength of the patent strongly influences the position of the inventor during licensing negotiations. However, in order to improve one's position it is not recommended that negotiations are delayed until a patent is granted. The long time lag between application and issue of a patent would delay the development of the related product and could lead to the loss of a major competitive advantage. The effective patent protection would be shortened and less valuable. A favorable search report, however, may be useful to considerably strengthen the position of the patent applicant.

Reasonable agreements which are satisfactory to both contract partners for the entire life-span of the patent are the result of negotiations in good faith. The industrial partner needs to know all technical details and the scientific background of the invention, its gaps and technical risks, and the inventor's party should be informed about the potential market expectations and especially on the product requirements on which these are based. If definite products can be envisaged, estimates of the development and manufacturing cost for the product and information on trade relations of the company, transfer prices, sales cost, and customary trade margins, may be helpful to establish optimal agreements for both sides.

For those who believe that after deduction of all cost the profit margins are still "huge", it may be necessary to mention that for real key patents (which cover a whole new area and are rare in the field of biological or biotechnological inventions) the usual license agreement aims at a 50% profit share. This may probably result in royalty rates of the order of 10% of the sales. Standard royalties for "dominant" patents that cover a whole product class, are more likely to be around 5%. Less significant patents may render 2–3% royalties and improvements of existing methods or products may even be valued below this. The above figures are mentioned to give potential inventors an idea about the dimension of royalties in general. They are not intended to serve as a guideline for license negotiations, which should be based on detailed information and sound calculations.

Acknowledgements

I gratefully acknowledge the efforts of GARY COBON, GERARD HENRY and ROSALIND KALDOR of Biotech Australia Pty. Ltd. and WERNER SCHMID of Hoechst AG who critically reviewed the manuscript and made many important suggestions for its improvement.

5 References

ADLER, R. G. (1992), Genome research: fulfilling the public's expectations for knowledge and commercialisation, *Science* **257**, 908–914.

BENT, S. A. (1982), Patent Protection of DNA Molecules, *J. Pat. Off. Soc. (JPOS)* **64**, 60–86.

BENT, S. A., SCHWAB, R. L., COULIN, D. G., JEFFERY, D. D. (1987), *Intellectual Property Rights in Biotechnology Worldwide*, New York: MacMillan Publishers, UK: Stockton Press (country-specific information not always up-to-date).

CFR (Code of Federal Regulations) Title 37 – *Patents, Trademarks, and Copyrights*, Parts 1 – 199, Patent and Trademark Office, US Department of Commerce, Washington, DC: Government Printing Office.

EEC COUNCIL DIRECTIVE 87/21/EEC of 22 December 1986 amending Directive 65/65/EEC on the approximation of provisions laid down by law, regulation or administrative action relating to proprietary medicinal products;

EEC COUNCIL DIRECTIVE 87/22/EEC of 22 December 1986 on the approximation of national measures relating to the placing on the market of high technology medicinal products, particularly those derived from biotechnology.

Both available in: *The Rules Governing Medicinal Products in the European Community*, Vol. I (1989), Luxembourg: Office des Publications Officielles des Communautés Européennes.

EISENBERG, R. S. (1992), Genes, patents and product development, *Science* **257**, 903–908.

EPC (European Patent Convention) Appendix B3 in: JACOBS (1993).

JACOBS A. J., updated by MOROWITZ N. H. (Ed.) (1993), *Patents Throughout the World*, 4th Ed., New York: Clark Boardman Callaghan (looseleaf edition for updating provides latest country-specific details).

KILEY, T. D. (1992), Patents on random complementary DNA fragments? *Science* **257**, 915–918.

KUSHAN, J. P. (1992), Protein patents and the doctrine of equivalents: Limits on the expansion of patent rights (Comment), *High Technol. Law J.* **6**, 109–148.

MPEP (Manual of Patent Examining Procedure) (1988), 5th. Ed., Rev.9, 706.03 c, d. US Department of Commerce, Patent and Trademark Office, Washington, DC: Government Printing Office.

PCT APPLICANT'S GUIDE (1993), Geneva: The International Bureau of WIPO.

PLANT, D. W., REIMERS, N. J., ZINDER, N. D., (Eds.) (1982), *Patenting of Life Forms; Banbury*

Rep. **10,** Cold Spring Harbor, NY: Cold Spring Harbor Laboratories. (Specific information on regulations in several countries are not up-to-date.)

PRENTIS, R. A., WALKER, S. R., HEARD, D. D., TUCKER, A. M. (1988), R&D investment and pharmaceutical innovation in the UK, *Managerial and Decision Economics* **9,** 197-203.

SAMUELS, J. M. (1991), *Patent, Trademark and Copyright Laws* (in the USA), Washington, DC: BNA Books.

STAINES, J. E., MCGOWAN, V. F., SKERMAN, V. B. D. (Eds.) (1986), *World Directory of Collection of Cultures of Microorganisms*, 3rd Ed., Brisbane: World Data Center, University of Queensland.

TOBIAS, G. (1992), Patent protection – the Argentinian devide, *Scrip Magazine*, Nov., 26–28.

USC (United States Code) Title 35 – *Patents*, Washington, DC: Government Printing Office.

UTERMANN, J. (1978), Reflections on patent protection of products of nature. Part One: *Int. Rev. Int. Property Copyright Law* **9** (5), 409–421; Part Two: ibd. **9** (6), 523–541.

VOSSIUS, V. (1982), The patenting of life forms under the European Patent Convention and German Patent Law: Patentable inventions in the field of genetic manipulations, in: *Patenting of Life Forms; Banbury Rep.* **10** (PLANT, D. W., REIMERS, N. J., ZINDER, N. D., Eds.), Cold Spring Harbor, NY: Cold Spring Harbor Laboratories.

11 Databases in Biotechnology

ELEONORE POETZSCH

Berlin, Federal Republic of Germany

1 Introduction

The availability of high-quality, up-to-date and comprehensive information is an important requirement in biotechnology and its applications in the ever widening fields of medicine, pharmacy, agriculture, food industry and the environmental sciences. Advances in biotechnology, especially in genome research, depend to an increasing extent on which required information is made available and how it is used. The growing amount of data, especially the genome projects, deliver an enormous number of sequence data, meaning that collecting, processing and disseminating these data is only possible with the help of modern information technology and international cooperation.

Modern biotechnology is highly information-dependent and uses a wide variety of information sources and information technologies. The consequences of rapid developments in biotechnology with its tremendous volume of data and in informatics with its potential for processing and using data are:

1. In relation to research –
   * the creation of a new scientific field: Bioinformatics
2. In relation to infrastructures –
   * very large databases and smaller, more highly specialized databases
   * highly sophisticated software for processing and using these databases
   * efficient communication networks for access to databases and information exchange
   * establishment of information centers for collection, processing and distribution of information
   * comprehensive information services

Many databases are available in various types of media: Online via a number of networks and hosts, or on CD-ROM, diskette, magnetic tape, or as a printed version. The overall growth in the online database industry during the past years can be traced through the statistics: 300 online databases were registered in the 1979 edition of *Directory of Online Databases*. In the 1993 edition, the *Gale Directory of Databases* registered profiles of more than 5200 online databases, among them about 250 with relevance to biotechnology, and more than 3200 database products offered in portable form, among them about 150 with relevance to biotechnology (MARCACCIO, 1993). The number of records stored in these databases increased from 52 million (1975) to about 5 billion (1993).

2 Overview of Databases Relevant to Biotechnology

The compilations of databases in the fields of biotechnology (CRAFTS-LIGHTY, 1986; POETZSCH, 1986, 1988; ALSTON and COOMBS, 1992) and the investigations of the very different information needs of users reflecting the multidisciplinary character of biotechnology bear witness to the diversity of the necessary information and of the wide variety of offered databases. Derived from the results of these investigations, we can state that the following types of information are necessary for biotechnology:

Factual information. Especially sequence information (nucleic acid sequences, protein sequences) is of greatest importance to biotechnology, but also map information, structure information, and property information. Commercial and financial information, especially for the biotechnology industry, play an increasing role.

Bibliographic information. In addition to literature information, patent information is important to biotechnology.

Referral information (Directory information), including information on research institutions, Research & Development projects, companies, products, culture collections.

Full-text information with the complete text from journals, newsletters, biotechnology and related regulations, encyclopedias, market research reports.

Databases relevant to biotechnology can be classified

- according to subject areas:
  The subject area is usually the determining point for user selection of a database. In this connection, an important factor influencing biotechnology information is the interdisciplinarity of biotechnology with various sciences, application areas and related fields.
- according to the type of stored information:
  Factual databases, bibliographic databases, referral databases, full-text databases.

2.1 Factual Databases

The close link between biotechnology and information technology is particularly evident in relation to nucleic acid and protein acid sequences stored in factual databases. In the literature, authors report that the DNA sequence database will become as important as the Periodic Table of Elements. Factual databases are most important to biotechnology because these databases serve not only as an information tool but as a direct research tool. Factual databases provide a research instrument which exists at the interface between subject area and information technology, whereby the scientist increasingly assumes the role of producer and user of information, and bear witness to the increasing influence of information technology on the research process.

Depending on the type of stored information, the most important factual databases in biotechnology are the following:

Nucleotide sequences:
GenBank (USA)
EMBL Data Library (EC, UK)
DNA Data Bank of Japan (Japan)
GENESEQ-Patents Sequence Database (UK)
REGISTRY file (USA)
MEDLINE Molecular Sequence Data (USA)
RNA Data Bank (Germany)
VectorBank (USA)
dbEST (USA)

Protein sequences:
PIR Protein Sequence Database (USA, in collaboration with Germany and Japan)
SWISS-PROT Protein Sequence Database (Switzerland)
REGISTRY file (USA)
GENESEQ-Patents Sequence Database (UK)
GenPept (USA)
PseqIP (France)

Species-specific mapping data:
Genome Data Base (GDB, USA)
Online Mendelian Inheritance in Man (OMIM, USA)
Genomic Database of the Mouse (GBASE, USA)
The Encyclopedia of the Mouse Genome (USA)
Escherichia coli Genetic Stock Center (CGSC, USA)
FlyBase (USA)
EcoMap (USA)
ACEDB (a *Caenorhabditis elegans* database, UK)
AAtDB (*Arabidopsis thaliana* database, USA)
Plant Genome Database (PGB, USA)

Structures:
Protein Data Bank (USA)
Cambridge Structural Database (UK)
CarbBank (Carbohydrate Structure Database, USA)
BEILSTEIN (Germany)
REGISTRY file (USA)
CASREACT (USA)
ChemInform RX (Germany)

Restriction enzymes:
REBASE (Restriction Enzyme Databank, USA)

Enzymes:
BRENDA (BRaunschweiger ENzyme DAtabase, Germany)
DBEMP (DataBase on Enzymes and Metabolic Pathways, Russia)

Microorganisms/culture collections:
DSM Catalogue (Germany)
MSDN Central Directory (UK)
MINE (EC, Germany)
ATCC Catalogues (USA)
MiCIS (UK)

Cell cultures/hybridomas:
Immunoclone Database (France in collaboration with Germany and UK)
INTERLAB Network: Cell Line Data Base, B Line Data Base, Molecular Probe Data Base, ImmunoClone Data Base, Genotoxicity and Carcinogenicity Database (all Italy)

In the following, the most important sequence databases (GenBank, EMBL Data Library, PIR International) are described in more detail.

GenBank (Genetic Sequences Databank)
Producer: National Center for Biotechnology Information, USA
Contents: GenBank contains DNA and RNA sequences, bibliographic citations, related information such as sequence descriptions, source organisms, sequence lengths etc.; and software packages for using GenBank.
Access: Online, e.g., via STN International, GENIUSnet, BIONET Online Service and FTP access; CD-ROM; Magnetic Tape
Cooperation: EMBL European Molecular Biology Laboratory, DDBJ DNA Database of Japan
Information Sources: Scientific Journals, Direct Data Submission
Additional Service: E-mail Servers, NCBI Data Repository, NCBI Newsletter

EMBL (EBI) Data Library
Producers: European Molecular Biology Laboratory, Heidelberg, Germany (until August, 1994); European Bioinformatics Institute, Hinxton Park, Cambridge, UK

Products and services available from the EMBL (EBI) Data Library:

- *Databases on CD-ROM:* EMBL NUCLEOTIDE SEQUENCE DATABASE, SWISS-PROT PROTEIN SEQUENCE DATABASE and 31 related databases: PROSITE (Protein pattern database), ENZYME (Database of EC nomenclature), ECP (*E. coli* map database), EPD (Eukaryotic promotor database), REBASE (Restriction enzyme database), FlyBase (*Drosophila* genetic map database), TFD (Transcription factor database), TRNA (tRNA sequences), RRNA (Small subunit rRNA sequences), BERLIN (5S rRNA sequences), KABAT (Proteins of immunological interest) etc., and software for search and retrieval of data (EMBL-Search, CD-SEQ)
- *Network File Servers* using Electronic Mail, FTP, Gopher server
- *European Molecular Biology Network (EMBnet):* The main activity of EMBnet is the daily distribution of all new sequence data via the computer network to 16 nationally mandated nodes.
- *Sequence Searching Services:* BLITZ, Mail-Quicksearch and Mail-FastA are services that allow external users to search the DNA and protein sequence databases via electronic mail.
- *Sequence Data Submission:* AUTHORIN SOFTWARE PACKAGE, Submission Form (Computer-readable copies or printed copies), Data Submission by electronic mail or by post.

PIR International Protein Sequence Database
Producers: Protein Information Resource (PIR) at the National Biomedical Research Foundation (NBRF), USA; Martinsried Institute for Protein Sequences (MIPS) at the Max Planck Institute for Biochemistry, Germany; Japan International Protein Information Database (JIPID) at the Science University of Tokyo, Japan
Contents: The PIR contains descriptions of partial and whole protein sequences including function of protein, taxonomy, sequence features of biological interest, how sequence was experimentally determined, unambiguously determined residues within the sequence, and citations to relevant literature.
Data Input: PIR (USA) provides approx. 50% of data; MIPS (Germany) provides approx. 35% of data; JIPID (Japan) provides approx. 15% of data.
Access: CD-ROM ("Atlas of Protein and Genomic Sequences"), Magnetic Tape, Online via EMBnet, BIONET and through other networks on a number of file servers

PIR International includes the following databases:

PIR1 (annotated and classified entries), PIR2 (preliminary entries), PIR3 (unverified entries), PATCHX (yet unprocessed by PIR), MIPSH (Yeast Protein Sequences), ECON (*E. coli* data set), Alignment Database, NRL-3D Sequence-Structure Database, and software for processing sequence data
Sources: Scientific Journals, Direct Data Submission, EMBL Data Library

2.2 Bibliographic Databases

Bibliographic databases contain citations, mostly with abstracts, to the published literature, i.e., journal articles, patents, reports, dissertations, conference proceedings, books, etc. There are about 100 bibliographic databases relevant to biotechnology. Especially the access to patent information is of great importance to everyone working in the field of biotechnology. Patents have been called the lifeblood of the biotechnology industry.

Major bibliographic databases in biotechnology and related fields are:

BIOSIS Previews, CAS ONLINE, DERWENT Biotechnology Abstracts, Chemical Engineering and Biotechnology Abstracts, CSA Life Sciences Collection, BioBusiness, BioCommerce Abstracts and Directory, PASCAL: Biotechnologies, Biotechnology Citation Index, BioExpress, Current Awareness in Biological Sciences and others, together with

- **application-related databases** in
  Medicine and Pharmacy: MEDLINE, EMBASE, CANCERLIT, AIDS Database, AIDSLINE, International Pharmaceutical Abstracts, DERWENT Drug File, Pharmline, Bioethicsline
  Agriculture and Nutrition: CAB ABSTRACTS, AGRIS, AGRICOLA, AgBiotech News and Information, DERWENT Crop Protection File, Food Science and Technology Abstracts, Foods adlibra

Environment: POLLUAB, ULIDAT, ENVIROLINE, TOXLINE, AQUASCI
Chemistry: Chemical Abstracts, Chemical Business NewsBase, Chemical Industry Notes, Analytical Abstracts
Engineering: COMPENDEX, INSPEC, VtB Verfahrenstechnische Berichte
- **multidisciplinary databases relevant to biotechnology:**
  Science Citation Index, Current Contents, and others
- **patents databases relevant to biotechnology:**
  DERWENT Biotechnology Abstracts, DERWENT World Patents Index, PATDPA, PATOSEP, PATOSWO, INPADOC, Current Patents, PATFULL, JAPIO, CLAIMS, Drug Patents International

2.3 Referral Databases

Referral databases (Directory databases) contain information on research activities, Research & Development projects, institution and company profiles, products, and services. Especially in connection with the necessity of business information in biotechnology, those databases containing company information are of great importance. There are about 50 referral databases with relevance to biotechnology and a number of printed directories which are also available as databases.

- **Biotechnology:** WHO-WHAT-WHERE in Biosciences and Biotechnology (WWW/BIKE, Germany; printed versions: *Bio-Technologie Das Jahr- und Adreßbuch* (POETZSCH, 1993) and *Biotechnology Directory Eastern Europe* (LÜCKE and POETZSCH, 1993), BIOREP (BIOtechnology Research Projects, EC), BioCommerce Abstracts and Directory (UK; printed version: *The U.K. Biotechnology Handbook*), BEST Biotech (USA), BIOTEC (Cuba), Leading Biotechnology Companies (USA), Federal Bio-Technology Transfer Directory (USA)
- **Multidisciplinary** (with information on biotechnology): CORDIS (EC), Corporate Technology Database (USA), ICC Directory (UK), Japanese Corporate Directory (Japan), Directory of French Companies

(France), Who Supplies What? (Germany), American Business Directory (USA), Research Centers and Services Directory (USA), Who's Who in Technology? (USA)

- **Pharmacy:** Pharmaprojects (UK), Pharmacontacts (UK), AIDSDRUGS (USA)
- **Agriculture:** AGREP (EC), CRIS/USDA (USA), TEKTRAN (USA), European Directory of Agrochemical Products (UK)
- **Environment:** UFORDAT (Germany), DETEQ (Germany), DEQUIP (Germany)
- **Information sources:** Directory for Biotechnology Information Resources (USA), Listing of Molecular Biology Databases (USA), Information Sources in Biotechnology (Germany), I'M Guide (EC), CUADRA/GALE's Database Directory (USA)

2.4 Full-Text Databases

Full-text databases contain the complete text of original publications, e.g., journal articles, newsletters, newspapers, regulatory documents, encyclopedias, market research reports.

- **Databases containing texts of journals and newsletters,** e.g., European Biotechnology Information Service, Biotech Knowledge Sources, Biotechnology Investment Opportunities, Biotech Business, Biotechnology Newswatch, BioWorld Online (USA), Japan Report: Biotechnology, Genetic Technology News, Applied Genetics News, ADIS DrugNews, Drug Information Fulltext, MEDTEXT, Pharmaceutical and Healthcare Industry News, AIDS Newsletter
- **Databases containing (complete) texts of statutes and legislations,** e.g., DIOGENES, Federal Register, Biological Monitoring Database
- **Databases containing (complete) texts of encyclopedias,** e.g., IMSWorld New Product Launch Letter, KIRK-OTHMER Online, The MERCK INDEX Online
- **Databases containing (complete) texts of market research reports,** e.g., MARKETFULL, MarkIntel, KOBRA

3 Hosts Offering Biotechnology Databases

Databases may be searched online through appropriate networks and gateways by direct connection or through a host (online service, vendor). Access to a wide range of biotechnology databases is available to anyone equipped with a computer fitted with a modem linked through a telecommunications network to hosts offering biotechnology databases. It is necessary to have a license agreement with hosts which then authorize the user a password. With this password, the user has access to the host computer and can request access to the database of interest. If the user has access to Internet, he can directly access many databases in genetics and molecular biology.

There are many international hosts offering databases for biotechnology. The most important hosts are:

- **DIALOG Information Services, Inc., USA** offering more than 450 online databases and CD-ROMs (DIALOG OnDisc).
  In the fields of biosciences and biotechnology: BIOSIS, LIFE SCIENCES Collection, DERWENT Biotechnology Abstracts, CEABA, PASCAL, BioBusiness, BioCommerce Abstracts and Directory, MEDLINE, EMBASE, Pharmline, CA, FSTA, CAB Abstracts, Enviroline, World Patents Index, Science Citation Index, Predicasts, CLAIMS, etc.
- **Data-Star, Switzerland** offering more than 300 online databases.
  In the fields of biosciences and biotechnology: BIOSIS, CEABA, BioBusiness, BIKE, BioCommerce Abstracts and Directory, DERWENT Biotechnology Abstracts, Immunoclone Database, MEDLINE, EMBASE, Pharmline, CA, FSTA, CAB Abstracts, Enviroline, Science Citation Index, Predicasts, etc.
- **STN International (Scientific and Technical Information Network with Service Centers in Europe, USA and Japan)** offering more than 160 online databases.
  In the fields of biosciences and biotechnology: BIOSIS, CEABA, DERWENT Bio-

technology Abstracts, World Patents Index, BIOBUSINESS, LIFESCI, MEDLINE, CA, EMBASE, GenBank, FSTA, AQUASCI, CABA, etc.

- **ESA-IRS Information Retrieval Service, Italy**
  offering more than 200 online databases.
  In the fields of biosciences and biotechnology: BIOSIS, CAB Abstracts, CA, PASCAL, FSTA, INSPEC, AGRIS, ASFA, ENVIROLINE, etc.

- **DIMDI (German Institute for Medical Information and Documentation, Köln, Germany)**
  offering more than 80 online databases.
  In the fields of biosciences and biotechnology: BIOSIS, MEDLINE, EMBASE, BIKE, IMMUNOCLONE DATABASE, CANCERLIT, AIDSLINE, CAB Abstracts, AGRICOLA, AGRIS, FSTA, BIOETHICSLINE, SCISEARCH, CURRENT CONTENTS, etc.

Special hosts for biosciences and biotechnology:

- **Microbial Strain Data Network (MSDN)** is an international network of microbial, cell line, and biotechnology information resources and offers among others the following databases and services: MSDN Directory, MiCIS, WDC Database, Hybridoma Databank, Animal Virus Information System, Brasilian Tropical Database; Databases with Catalogue Information (ATCC Catalogues, DSM Catalogue, NCYC/NCFB Catalogue, ECACC Catalogue, CBS/NCC Catalogue, Czech Catalogues, Russian Catalogues, etc.); Databases with Newsletter Information (Biotech Knowledge Sources, European Biotechnology Information Service, Biotechnology Courses); Databases with Nomenclature Information (DSM Bacterial Nomenclature Database, Bacterial Nomenclature); Bulletin Board (Biodiversity Information Network, Information Resource on Release of Organisms); Electronic Mail Service.

- **GENIUSnet at the German Cancer Research Center, Heidelberg, Germany** offers more than 30 online databases in the fields of biosciences and biotechnology: EMBL Data Library, GenBank, PIR International, SWISS-PROT, Protein Data Bank, Genome Data Base, OMIM Online Mendelian Inheritance, VecBase, HIV-Base, HIV-Prot, IF-Prot, 5s RNA DataBank, AluBase, EPD, ReBase, etc.

- **ICECC Information Center for European Culture Collections, Braunschweig, Germany** offers more than 10 online databases in the fields of biosciences and biotechnology: MiCIS, DSM, MINE, etc.

- **Human Genome Information Resource (HGIR, USA)** develops and offers databases containing information on nucleotide probes, restriction fragments, physical and genetic maps, etc.

- **Life Science Network from BIOSIS (USA)** provides access to more than 80 online databases in the fields of biosciences and biotechnology: BIOSIS Previews, BioExpress, BioBusiness, ATCC Catalogues Database, Biology Digest Online, CA Search, Life Sciences Collection, MEDLINE, World Patents Index, etc.

- **Biosafety Information Network & Advisory Service** (BINAS, United Nations Industrial Development Organization) as an information network which will help meet an increasing need to standardize regulatory procedures for release of genetically modified organisms into the environment. The service will develop databases on existing guidelines, regulations and standards for the use and release of genetically engineered organisms.

- **BioLine Publications, UK** provides online access through the Internet network to journals, reports, conference proceedings and newsletters in biosciences and biotechnology.

Which special services are offered by these hosts?

Most hosts also offer an extensive range of support services to assist in searching:

- Advanced front-end software packages that help searching online
- Cross-searching of complementary databases
- Help Desk to assist in online searching

- Comprehensive range of user aids and manuals
- Document Delivery Service for articles and patents that are referenced
- Seminars, training programs and workshops
- Electronic Mail Systems that allow sending and receipt of text messages online.

Each host has agents in most countries who provide technical support and training. They provide information to all those who are concerned with online access to databases in the field of biotechnology.

4 Information Centers with Information Services for Biotechnology

In addition to hosts (see Sect. 3) and database producers, there are a number of information centers offering information services in the fields of biotechnology. Information services for biotechnology include: Production and marketing of databases, publications, information brokerage, workshops and seminars, consultancy services, document delivery, etc.

Some examples for information centers offering information services for biotechnology are:

USA: National Center for Biotechnology Information at the National Library of Medicine, Plant Genome Data and Information Center at the National Agricultural Library, BIOSciences Information Service, Institute for Biotechnology Information at the North Carolina Biotechnology Center, Human Genome Information Resource

UK: The British Library Biotechnology Information Service, European Bioinformatics Institute, BioCommerce Data Ltd, Royal Society of Chemistry Information Services

Germany: FachInformationsZentrum CHEMIE (FIZ CHEMIE, Information Center for Chemistry), Berlin

Italy: International Center for Genetic Engineering and Biotechnology

Switzerland: Biocomputing Biocenter, University of Basel

India: Biotechnology Information System Network (BTNET) with distributed information centers

5 International Organizations Involved in Biotechnology Information

- **Committee on Data for Science and Technology (CODATA)** with CODATA Task Groups and Commissions related to biotechnology: Biological Macromolecules, Standardized Terminology for Access to Biological Data
- **International Centre for Genetic Engineering and Biotechnology Trieste and New Delhi (ICGEB)** as an international organization of the United Nations and established to facilitate the dissemination and regulation of biotechnology worldwide with particular regard to the needs of the developing world. ICGEB operates the ICGEBnet as an information network that provides free services to users in 47 countries worldwide. ICGEBnet's services include online access to a large number of molecular biology databases and a wide variety of analysis software. In addition, ICGEBnet provides E-Mail services and training facilities.
- **European Biotechnology Information Strategic Forum (BTSF)** with eleven European organizations involved in biotechnology information: CAB International (UK), Excerpta Medica (NL), INIST: PASCAL (France), Pergamon Press (UK), Springer-Verlag (Germany), The Royal Society of Chemistry (UK), Wolters-Kluwer Academic Publishers (NL), EMBL Data Library (EC, UK), INSERM/CERDIC (France), J. Wiley & Sons Ltd. (UK), and DERWENT (UK). The forum has the following tasks:
  - identify and characterize user needs through interactions with industry plat-

forms, advisory groups, market sectors and individual purchasers/users;

- measure these needs against commercial, technical and political limitations;
- examine potential solutions to such market-led questions;
- develop systems for the support of non-commercial products;
- strengthen the European information industry's response to international market and political demands in the field of international bioinformatics;
- establish the BTSF as the formal, permanent focal point for European biotechnology information.

6 The Relationship Between Biotechnology Databases and Bioinformatics

Bioinformatics represents an interdisciplinary compiling of biosciences and biotechnology with mathematics and informatics. Databases, especially factual databases as for example nucleotide and protein sequence databases, are one of the most important tools in bioinformatics. Bioinformatics includes different subject areas, e.g., development of databases and software. The use of sequence and structure information stored in databases opens together with new concepts, such as neural networks, new approaches in molecular modeling for structure prediction and description, theories and methods of 3D-modeling and of optimizing macromolecules, knowledge-based sequence analysis and prediction of protein folding, development of knowledge-based systems and artificial intelligence methods for applications in genome research, protein design, drug design, etc.

European activities in bioinformatics:

With the aim to improve the development of bioinformatics in Europe, in November 1990

the European Chemical Industry Foundation (CEFIC) recommended to the Commission of the European Communities the establishment of a European Nucleotide Sequence Center. After long discussion, in 1993 the decision has been made to locate the new European Bioinformatics Institute (EBI) at the Genome Campus near Cambridge in the UK. The EBI will work to:

- ensure that the Data Library continues, develops in response to advancing biological research, and is capable of exploiting advances in informatics;
- reinforce areas which have been neglected, particularly training and user support;
- make the voice of European bioinformatics heard in the global arena;
- increase the effectiveness of dispersed, high-quality European research and service by entering into collaborations with centers of expertise throughout Europe.

The EBI will provide the EMBL Nucleotide Sequence Database and the SWISS-PROT Protein Sequence Database; support and distribute other databases in collaboration with European scientists; help to coordinate the EMBnet nodes and molecular biology network services; carry out research and development in the application of information technology, actively tracking advances to explore their utility to biological problems; provide quality user support; and ensure that Europeans are globally competitive in the profession of bioinformatics.

U.S. activities in bioinformatics:

In connection with the U.S. Human Genome Program and the U.S. Plant Genome Research Program many large-scale bioinformatics projects were initialized, especially related to the development and improvement of databases for sequence and map data, and software for sequencing and mapping as well as for the creation and use of databases (cf. Sect. 2).

7 Problems of Biotechnology Information

1. Underestimation of the value of information which finds its expression in
   - an underdeveloped information awareness and a dangerous ignorance concerning the use of information;
   - an insufficient support for the production and maintenance of databases;
   - an insufficient appreciation of "Information" as a "Knowledge" resource with the result that there is a flood of data, but a shortage of knowledge;
   - the interfaces between research and information are insufficiently developed;
   - an underdeveloped information management.
2. Lack of reputation of biotechnology information as a science or as a part of a science
   "Young fields of research often experience the lack of recognition. Bioinformatics needs solid foundation and funds to be accepted as a field of research on its own" (MEWES, 1990). At present, research and development for future generations of databases are seriously neglected.
3. Financing
   USA: Substantial public funding in long-term programs with the result that the USA plays the leading role in the production of databases, and there is a European dependency on American database producers. US Government and industry provide huge sums for the establishment of biotechnology databases.
   Europe: Insufficient support of particular databases without a long-term strategy. The willingness to cooperate internationally does not meet the requirements.
   Germany: The necessary pre-conditions for the setting-up of biotechnology databases are insufficiently provided.
   As a result, there are great differences between the funding of biotechnology databases and bioinformatics in the USA and Europe, and international cooperation does not meet requirements, with some exceptions, e.g., PIR International.

4. Growth of data – Growth of databases
   The main problem of molecular biology databases is the exponential growth of biological data, especially the growth in the volume of nucleotide and protein sequence data. Large-scale genome sequencing projects have resulted in tremendous increases in the volume of sequence data which can only be managed by automated storage, processing and interpretation techniques. With respect to the growth of databases, this means that at present the size of the GenBank doubles every fifteen months.
   Dimensions: The sequence of a simple viral genome may be written on a single page; a bacterial genome would require a book with 1000 pages and the sequence of the human genome would require a library of 1000 books, each with 1000 pages.
5. Relation (disproportion) between
   the quantitative aspect: volume of data, e.g., the tremendous increase in genetic data, and
   the qualitative aspect: reliability of data with respect to their current relevance and accuracy; interpretability of data with respect to their complexity, structure–function relationships, cause–effect relationships.
   The fact that the creation and up-dating of databases is a time-consuming and expensive process on the one hand, and a necessity to ensure the currency and coverage of the databases on the other hand, often leads to decreasing quality of databases (the "quick and dirty" principle).

8 Conclusions for Biotechnology Information

8.1 Establishing Pre-conditions

1. **Creation of a new information consciousness and information behavior**
   Supporting the development of a new attitude and relationship to information:
   - Information is an absolutely necessary component of research

- Information has its price.

2. **Interdisciplinary cooperation**

Recognizing the interdisciplinary character of biotechnology, increased attention is being devoted to establish

- a close interdisciplinary cooperation among scientists from different fields;
- a broad linkage of biosciences and biotechnology with efficient informatics methods;
- a close interdisciplinary cooperation of all participants in research, industry, government, and in public;
- productive interaction between producers and users of biotechnology information.

3. **International cooperation**

Proceeding from the fact that significant databases in the field of biotechnology such as nucleotide sequence, protein sequence, hybridoma and culture collections databases can only be created through international cooperation, it is imperative that all countries work together. The enhancement of international cooperation with the purpose of

- increasing and coordinating information exchange,
- establishing international information centers for collection, processing and exchange of data,
- exploiting the opportunities of international networks being developed

demands a number of activities, e.g., standardization of the structure of databases (common syntactic and semantic formats for the exchange of sequences and other related data) and of the vocabulary and nomenclature. Another problem which must be solved is the organization of data transfer and the very important consideration of legal aspects of information transfer and database copyrights.

4. **Establishment of a European infrastructure of biotechnology information**

- Creation of the European Bioinformatics Institute (EBI) and expansion of the EBI or creation of a "European Centre for Biotechnology Information" that meet information needs which are not sequence-linked;
- European participation in the construction of international databases and improvement in the cooperation between Western and Eastern Europe.

5. **Necessity of support and investment**

at national, European and international levels and from the government, industry and international organizations

- to ensure the position of European database producers, providers and hosts, but also
- to participate in international information exchange with the aim to achieve a greater integration of national, European and international activities in the field of biotechnology information.

6. **Creation of educational opportunities**

To promote bioinformatics and biotechnology information, special emphasis must be placed on the creation, extension and improvement of educational opportunities. It is necessary to recognize bioinformatics as its own field in education (at undergraduate and postgraduate level) and to organize training courses and workshops in order to teach the research community in the use of the growing body of information.

7. **Improvements in information management and marketing for biotechnology information products**

New product development teams, to:

- define the new product's characteristics,
- define the future needs of users and monitor the progress of efforts undertaken to meet them,
- improve the synergy between the public and private sector in the information market and the scientific areas involved in biotechnology information,
- examine the impact, costs and acceptability of new technologies,
- develop position statements on such matters as standards and legal aspects of information, including copyright.

8.2 Further Development of Biological Databases

The further development of biological databases via the application of informatics will be connected with

- the increased integration of biosciences (biotechnology) and information (technologies),
- the development of advanced strategies for the handling of large volumes of data,
- improved methods for the use and evaluation of databases,
- integrated data and knowledge bases for various types of information,
- the extension of access methods, and
- the focus on the quality of databases:
  (1) The *currency* of the database determined by the time it takes to make new information available in a database and by the possibility of rapid access to the database. The quickest possible way is the direct data submission from the author to one of the (sequence) databases via E-mail or diskette.
  (2) The *reliability and accuracy* of data and databases connected with the necessity to integrate evaluation algorithms, to expand the syntactic and semantic representation possibilities, etc. In the future, various forms of representation patterns will have to be combined in order to be able to adequately and completely represent objects' properties, relations and interrelations, and to transfer information into knowledge.
  (3) The *complexity* of data and databases connected with the necessity to cross-link different (types of) information which are part of different (types of) databases, to combine different forms of information representation, to extend the numeric data by means of supplementary, descriptive information, to use standardized or easily translatable formats which must be interconnected in order to integrate individual databases in a global concept. The integration of databases from various producers and structures in systems which have a single, unified administration and allow a homogeneous access to the various heterogeneous data present. For bibliographic databases, the Commission of the European Communities recommends the creation of a "Common Core Database" by the bunching of central biotechnology databases for the prevention of duplicates, overlapping, etc.
  (4) A *highly sophisticated software* is required to make the knowledge contained implicitly in large databases explicitly available. New approaches as neural networks, pattern recognition methods, knowledge retrieval, combination of imaging and artificial intelligence techniques, etc. will help to improve the processing of different kinds of information.
  (5) *Improvement of the access to biotechnology databases:* It is necessary to have several tools which allow users to ask a wide variety of questions about sequences, ranging from sequence analysis to sequence comparison, and to access a wide variety of databases and software products in a user-friendly manner.

8.3 Improvements in Information and Communication Infrastructure

It is necessary to establish databases in specialized areas which are not yet covered, together with referral database systems containing information of the type "Who-What-Where" building one of the bases of information brokerage; and serving as worldwide guides which stewards the user through the information landscape to the desired information. Highly developed information networks and services are required for the creation and use of databases as well as for communication purposes.

9 Use of Databases in Biotechnology

With the large databases we have a huge information potential at our disposal, but until now we are not able to utilize it to a high degree. It is necessary to improve the use of databases which depends on the tasks:

- **In research:**
  Databases serve as a retrieval tool,
  as a research tool (e.g., protein design, drug design),
  as an evaluation tool (e.g., Science Citation Index).
- **In industry:**
  Databases serve as a retrieval tool,
  as an aid in decision-making and awareness of competitors,
  as tools for product and technology information and evaluation,
  as tools for sales force productivity.
- **In the public eye:**
  Databases serve to help increase the public perception of biotechnology.

Examples of use of databases in research:

1. Protein design
   Knowledge of the protein structure and of specific structure–function relationships offers the possibilities of their targeted modification or their targeted construction, thus leading to molecules with modified activities or other new properties for, e.g., medical applications.
   Required databases:
   Protein sequence databases (e.g., PIR), structure databases (e.g., Protein Data Bank),
   and software developed for the use of these databases and especially for modeling.
2. Drug design
   The development of pharmacologically active substances with the aid of computers by quantitative structure–activity analyses, multivariate data analyses, and molecular modeling: The modeling begins with the creation of databases containing experimental data concerning the properties of chemical and biological systems. They deliver information concerning the structure of molecules or their sites of action in the organism.
   Required databases:
   Structure databases (e.g., Cambridge Structural Database),
   drug databases,
   and software developed for the use of these databases and especially for modeling.

Increased use of databases for

- basic research (e.g., in genome research),
- applied research (e.g., for molecular modeling),
- technology transfer and improvement of innovations (e.g., to transfer research results to practical commercial applications and to help to introduce innovative technologies to the marketplace),
- coordination of national and international research programs (e.g., for the avoidance of duplicate work),
- competition analysis (e.g., for information concerning the marketability of products),
- improving the public perception of biotechnology (e.g., for the objective assessment of experiments in genetic engineering).

The use of databases and other information sources as an aid to increase the *public perception* of biotechnology: The lack of public acceptance of biotechnology is also a result of the lack of information, since there is a direct link between information and attitude. Researchers in the public perception of biotechnology agree that attempts to improve access to scientific information are highly desirable (GRINDLEY and BENNETT, 1993). Specific measures should be taken to enhance public perception mostly through the availability of objective information, especially in connection with biotechnology's impact on human health through the development of new pharmaceutical, diagnostic, and other medical products.

Conclusions for biotechnology information

The development and the use of databases become increasingly important to biotechnology in research and development, in industry, and in the public eye. The particular importance of biotechnology information is especially obvious in connection with the necessity of molecular biology databases. This can be seen from the high value of information in the frame of genome projects, as the U.S. Human Genome Program and the U.S. Plant Genome Research Program, which include large-scale projects combining mapping and sequencing with data collection and distribution.

The "information highway" exists *per se* in form of a highly developed information infrastructure, and the political changes brought the advantage that we have no unbridgeable frontiers. The task of biotechnology is to draw a map, not only in genome research, but also in the field of biotechnology information and bioinformatics with the aim of targeting the information highway. The combination of biotechnology and information technology is a challenge for both fields and should be a strategy for the future.

10 References

ALSTON, Y., COOMBS, J. (1992), *Biosciences, Information Sources and Services*, New York: Stockton Press.

CRAFTS-LIGHTY, A. (1986), *Information Sources in Biotechnology*, Weinheim: VCH Verlagsgesellschaft.

GRINDLEY, J. N., BENNETT, D. J. (1993), Public perception and the socio-economic integration of biotechnology, in: *Biotechnologia* **20**, 89–102.

LÜCKE, E.-M., POETZSCH, E. (1993), *Biotechnology Directory Eastern Europe*, Berlin–New York: de Gruyter.

MARCACCIO, K. Y. (1993), *Gale Directory of Databases*, Vol. 1: *Online Databases*, Detroit: Gale Research Inc.

MEWES, H.-W. (1990), *Workshop Computer Applications in Biosciences*, Book of Abstracts, p. 11, Martinsried.

POETZSCH, E. (1986), *Faktographische Informationsfonds zur Biotechnologie*, Berlin: WIZ.

POETZSCH, E. (1988), *Verzeichnis faktographischer Datenbasen zur Biotechnologie*, Berlin: WIZ.

POETZSCH, E. (1993), *BioTechnologie Das Jahr- und Adreßbuch 93/94*, Berlin: polycom Verlagsgesellschaft.

IV. Biotechnology in a Developing World

12 Commercial Biotechnology: Developing World Prospects

GEORGE T. TZOTZOS

Vienna, Austria

MARION LEOPOLD

Montreal, Canada

Abstract

The evolution of biotechnology from the centuries old brewing and breeding practices to the engineering based fermentation technologies of the 1950s and finally to the recombinant technologies of today, represents a quantum shift in qualitative terms. Biotechnology being sector-specific until recently became, in a remarkably short time, an all-pervasive technology. It evolved from being empirical in nature, to technology/manufacturing-based, to, finally, science-intensive. The range of its potential applications expands continuously with scientific knowledge and encompasses all sectors of economic activity.

The changing technological scene influences, inevitably, the public image of biotechnology (KENNEDY, 1991; BUD, 1989; FLEISING, 1989). Early day fermentation biotechnology was perceived to offer unimaginable possibilities for sustainable techno-economic development. For the developing world this meant the utilization of excess biomass to provide food and energy for ever increasing populations. Single cell protein and continuous fermentation technologies were promoted in international fora as magic bullets for socioeconomic development. Biotechnology, thus, received an idealistic image. This image was not to last long. It died with the failure of single cell protein to become a viable commercial product.

The advent of molecular biology in the mid 1970s and its translation into what came to be known as "new" biotechnology (henceforth referred to simply as biotechnology) demanded venture capital that could only be raised if its commercial potential were sufficiently attractive. The prevailing image of biotechnology is now that of an all-pervasive, profit-generating technology playing a strategic role in maintaining and enhancing national competitiveness in an environment of global economic interaction.

This chapter considers the extent to which the disjunction of priorities between the industrial and developing world (i.e., corporate profit generation vs. social benefit) influences the potential of biotechnology to solve health and food supply related problems in the developing world. At the same time, it examines whether aspirations for technological "leapfrogging" through biotechnology stand on firm ground.

The focus is on the major factors that influence the transition of science into technology and commerce and on assessing the impact of transferring biotechnology to the developing world. The scientific background of the sector-specific applications of advanced biotechnology has been consciously underplayed, as there already exists a vast volume of relevant literature. Technological innovations that have revolutionized practice at the scientific bench also receive little attention, as the emphasis is on what turns bioscience into an economic activity rather than on the scientific ingredients of the technology. Access to the tools of the technology does not necessarily mean ability to master the "trade".

A major problem in addressing the subject matter is the huge diversity at the country and regional level in terms of human and material resources, as well as in terms of socioeconomic systems. Such variation implies different sets of technological priorities. To overcome this problem the common denominator approach has been followed. That is the need of almost all developing countries to alleviate disease and secure food supply. The focus being on the technology rather than the science and the fact that the first commercial results of advanced biotechnology come from the health and agricultural areas have facilitated the task of writing this chapter.

Health and agricultural biotechnology are, therefore, treated as cases in point to demonstrate the technological entry barriers, technology transfer and potential socioeconomic impacts. While recognizing the fact that other biotechnological applications (e.g., environmental, industrial) may be appropriate for the developing world, the thematic focus of the volume at hand does not justify anything other than paying lip service to them.

1 Introduction

In the late 1970s and early 1980s, biotechnology was hailed as offering the potential to deal with some of the major problems of developing countries. Improved health care, agricultural and energy supplies were among the critical goals that biotechnology could help meet. At the same time, biotechnology appeared particularly appropriate for technological leapfrogging by the developing countries, for enhancing national income and international competitiveness.

To date, little has been accomplished by way of fulfilling these social and economic expectations. Only a few developing countries – mostly top-tier – have an incipient biotechnology sector or the research and development (R & D) component thereof; and even in many of these instances, the future holds uncertainties. As for the "ground level" social benefits of biotechnology, they have yet to make themselves felt.

The difficulties in developing a biotechnology capability and in capturing its social and private returns are not restricted to less-developed countries (LDCs). Indeed, even in the industrial economies, competitive entry into biotechnology is a lengthy, costly, and risky undertaking. And although the "industry" is gaining momentum, failure rates among dedicated biotechnology companies remain high, and financing uncertain.

In the following, we consider some of the major requirements for and obstacles to successful entry into biotechnology (SERCOVICH and LEOPOLD, 1991; LEOPOLD 1993; SERCOVICH, 1993). As far as these apply to developed and developing countries alike, we touch upon both groups. The interest in looking at the experience of the industrialized economies lies in the fact that they offer an overview of the issues involved in the biotechnology innovation process. Because biotechnology is a high-stakes, internationally competitive game, the leading countries have become increasingly aggressive in their attempts at strengthening capabilities and overcoming obstacles. Finally, the country- and company-level strategies and activities of these countries cannot but impact entry conditions in the developing world, all the more so because of growing international competitive rivalry. Developing-country-specific considerations are, of course, dealt with as such.

With respect to North-South international technology transfer in biotechnology, irrespective of its scope, the recipient country must have the technological, social and institutional capabilities to make competitive use of the technology. From this point of view, too, understanding the ingredients of successful biotechnology innovation is critical. Lest this new set of generic technologies should be just one more area where developing-countries' attempts at industrial capability-building ultimately fail, the agents of technological change – public and private, national and international – should definitely integrate these considerations into their strategic thinking.

After addressing the above issues (Sect. 2), we focus specifically on two application sectors that are of primary importance for developing countries, and that also represent the most advanced areas of biotechnology: health care (Sect. 3) and agriculture (Sect. 4). An attempt is made to assess biotechnology's potential impact – or absence thereof – as a source of economic growth and social benefit.

It should be borne in mind that while a "common denominator" approach has been adopted in examining conditions for and barriers to commercial entry into biotechnology, the latter are subject to considerable contextual variability. Requirements tend to combine differently from one country or region to another (see, e.g., BURRILL and ROBERTS, 1992). Similarly, the weight of entry deterrents varies according to many factors, including country, application sector, company size, learning- and experience curves, and the macro-economic environment. User-industry specificity of technical and institutional requirements and entry barriers are increasing, as the tools of biotechnology are themselves progressively adapted to various application sectors (SERCOVICH and LEOPOLD, 1991; U.S. CONGRESS, 1988). Cross-country differences are based on a wide range of factors, including level of development, resource endowment, and previous experience in technology capability-building.

2 Competitive Entry into Biotechnology

2.1 Science and Technology

A driving force of advanced biotechnology is the exceptional dynamism of the science–technology interface; the technology is both highly dependent upon its scientific underpinnings and constantly pushing the knowledge base forward, thus creating the grounds for further technological breakthroughs. At the same time, the state-of-the-art is tied to a complex network of linkages and feedbacks between scientific and engineering disciplines. Whether it be the contribution of massively parallel-computing technology to the optimization of drug design, advances born of the upstream and downstream collaboration between biologists and chemical engineers, the creation of spill-over benefits between biomedical research and animal science research, or, more fundamentally, the biotechnology-generated exponential growth and channeling of knowledge in molecular biology, interactions, synergies and cross-fertilization have a major impact on the shaping of biotechnology's trajectory and on innovation lead times.

The speed at which new knowledge is put to practical use has increased considerably, as has the precision of breakthrough techniques, with an attendant increase in the number and efficacy of products in the pipeline. In the case of biopharmaceuticals, technology-driven improvements in product specificity and efficacy-to-safety ratios could shorten commercial time lags (CASDIN, 1992), although their effect may be offset by other factors, such as relative product-cost effectiveness.

As bioscience and biotechnology gain in complexity and become more synergistic, human resource requirements also become more demanding. Molecular biologists face new challenges as their field evolves. Furthermore, there is a growing need for scientists and engineers capable of working in cross-disciplinary research teams and/or having themselves multi-disciplinary expertise. These skill requirements are, in turn, leading to publicly and privately funded innovative education and training.

The above trends in science and technology bring into sharp focus certain issues related to developing-country attempts at upstream entry. In particular, given the speed at which the scientific and technological frontier is receding, the point of potential entry into research tends, in all but the most advanced LDCs, to become increasingly removed from that frontier. Moreover, countries seeking to be close-followers will probably have to keep their sights narrow with respect to process- and product-targeting, even if they have to trade-off potential downstream scope economies. Finally, because of the wide and growing knowledge gap between developed and developing countries, the latter clearly have an interest in gearing an important part of their biotechnology research capability-building towards the effective exploitation of existing knowledge rather than the generation of new knowledge. (It has been convincingly argued that in knowledge-intensive sectors, the effective utilization of scientific knowledge provides greater returns to countries than the production of that knowledge. See, e.g. MOWERY and ROSENBERG, 1989; WORLD BANK, 1993.)

This, of course, in no way diminishes the need for developing countries to build domestic skills and capabilities in bioscience and biotechnology. On the contrary, it is in those countries that have the strongest science and technology base that – other innovation-related requirements notwithstanding – the chance of effectively exploiting existing knowledge is the greatest. And conversely, using, assimilating and adapting new knowledge is part of the cumulative learning process that will increase a country's potential for upgrading its R & D capability.

However, whatever the means, entering into biotechnology R & D alone cannot lead to commercial market entry. For successful innovation to occur, developing countries, like the industrialized countries, must make effective use of the output of R & D. This involves the ability to overcome various innovation-related threshold barriers and to bring into play numerous institutional and organizational capabilities.

2.2 Threshold Factors

2.2.1 Research and Development

R & D-related costs weigh heavily upon biotechnology companies, particularly upon dedicated firms and in the pharmaceutical application sector. From July 1991 through June 1992, R & D spending by public US biopharmaceutical companies represented 68% of their product sales and more than 40% of total cost (ERNST and YOUNG, 1992). Although absolute R & D costs are well below those of the conventional pharmaceutical industry, where productivity rates are lower and clinical development times longer, the fact that most biopharmaceutical firms do not yet generate significant product sales revenues creates a financial burden relatively greater than that of the highly profitable pharmaceutical majors. Given their inability to recoup sunk R & D investments through timely product sales, most dedicated firms must seek out alternative forms of financing, including various types of collaboration with the pharmaceutical industry.

In the case of developing countries, accessing and efficiently exploiting available knowledge should partially offset high R & D threshold barriers. But by no means does the imitator's advantage automatically turn R & D into a costless, risk-free and timely undertaking; the availability of complementary technological, physical, human and institutional assets will determine the degree of potential savings to the follower. Furthermore, even lower R & D costs will not be of much help if access to capital is not also assured.

2.2.2 Production

Because of the science-driven nature of biotechnology and because production volumes have tended to be small and returns high, scale-up and downstream processing efficiencies have not been given much attention until recently. However, with production volumes increasing, the future of market prices uncertain, and inter-firm and international competition on the rise, production-cost effectiveness has moved up rapidly on the list of company- and country-level priorities. (As in other areas of innovation, Japan was early to recognize the importance of downstream capabilities. The Asian newly industrialized countries [NICs] appear to have followed suit.)

With respect to purification and recovery – by far the most costly steps in large-scale recombinant protein production – attempts to increase efficiency have yielded several promising results (on the latter, see FULTON, 1992). Similarly, breakthroughs in upstream processing include protein-producing transgenic animals, which offer potentially competitive alternatives to conventional fermenters and to bioreactors. Efforts at generating scope-economies in the production of biologics have led to the proposed design of multi-use manufacturing installations.

These and other attempts at improving production-cost effectiveness have involved overcoming important process design- and production-engineering obstacles, which in turn has depended upon the availability of high-level skills and capabilities, particularly in bioprocess engineering. Downstream technological mastery is also necessary in order to assure optimal decision-making with respect to techniques.

Agricultural biotechnology has its own production-related thresholds; these include agronomical as well as technological problems. Although, as in manufacturing, scale-up capabilities are required early on (for small-scale trials), the difficulties encountered in the various phases of the growth cycle relate to issues such as reproducing growth chamber results under field conditions, coping with the seasonality factor, etc. As in pharmaceuticals, maximizing yields, and reducing waste are important considerations, but most such problems tend to be confronted upstream, with molecular biologists designing appropriate traits into host cells.

The scarcity of appropriate skills and capabilities represents a severe competitive disadvantage for most developing countries in most application sectors. Furthermore, increasing user-industry specificity of engineering skills will not facilitate things; the danger of spreading thin already scarce qualified resources makes it all the more important that developing countries set priorities in biotechnology. Finally, it should be noted that accessing technological know-how in manufacturing has become more difficult under conditions of intense global competition.

2.2.3 Marketing and Distribution

In all but a few applications (e.g., diagnostic kits, supplier sectors) health care biotechnology is still a supply-led industry, and marketing thresholds remain considerable. Contributing to the latter are high relative and absolute product prices, narrow markets, and growing but still comparatively low production volumes. Marketing as well as distribution barriers are additional reasons why most biopharmaceutical firms resort to partnering or licensing arrangements with established corporations; sufficiently broad market access is critical to R & D investment paybacks.

In agricultural biotechnology, the picture is somewhat less clear; some products are supply-led (by science or industry), while others are market-driven. Furthermore, certain commodity sectors are potentially subject to subsidizing or to competition from subsidized products. All of these factors affect the significance of marketing as a threshold barrier.

In most developing countries, marketing and distribution thresholds, like many other entry-related issues, are linked to the larger question of market inefficiencies and inadequate policy responses. In developing countries with well-working markets and a comparatively stronger demand-pull bias (the Asian NICs), marketing and distribution thresholds are presumably relatively low (at least for domestic firms).

2.3 Social and Institutional Innovations and Entry Barriers

At the institutional level, factors such as rigidities, traditions, learning- and experience curves have delayed the commercialization of products and processes of biotechnology. In particular, lengthy market lead times have been repeatedly attributed to the retarding effects of the regulatory process. One would be hard put to deny the time lags generated by these externalities; after all, by mid 1992, only 14 biopharmaceuticals and no transgenic plants had been commercially marketed (BIENZ-TADMOR, 1993; OECD, 1992).

This being said, the considerable attention being given to these barriers may be creating some distortionary effects when it comes to assessing the overall role of institutional, organizational and social factors in the innovation process. In the industrialized countries, technical innovation in biotechnology has been accompanied by a series of dynamic adjustments at the institutional and social levels. These include innovative approaches to technology transfer and to financing; structural and strategic reorganization at the industry level; skill and capability upgrading through novel training mechanisms; and important shifts in science and industrial policy. On balance, these adjustments, which tend to complement one another, have created a positive institutional climate for biotechnology innovation, although cross-country differences exist, and certain systemic rigidities and other retarding factors remain to be overcome in all countries.

With respect to developing countries seeking to enter into commercial biotechnology, it is extremely important that they take stock of the institutional and social adjustment processes underway in the industrialized world, since without such adjustments – and assuming that the appropriate institutions do indeed exist and function properly – it is difficult to imagine the acquisition of a commercial capability in biotechnology.

In the following, we briefly examine some of the areas in which the institutional and social dynamics of biotechnology innovation play themselves out.

2.3.1 Horizontal Technology Transfer

R & D collaboration among university, industry and government has been critical in supporting the development of bioscience and biotechnology. The modalities and mechanisms of collaboration have varied across countries, in accordance with prevailing public and private sector traditions; in most countries, the trend has been towards greater inter-institutional cooperation.

In Japan, which lacks an entrepreneurial culture, government has become increasingly active in encouraging university–industry linkages; the Ministry of Education, Science and Culture sponsors individual collaborative projects, while MITI (Ministry of International Trade and Industry) has set up major technology centers such as the Protein Engineering Research Institute, where industry and university scientists and engineers join forces (KINOSHITA, 1993). In the U.S., a particularly successful approach to accelerating innovation in advanced technology has emerged in the form of nationally decentralized centers of knowledge generation and transfer; dedicated firms and established corporations gather around major research universities and centers of excellence, and with the financial and logistic backing of state-government and venture capital, create a situation conductive to the rapid diffusion and commercial application of university-based ideas. The success of the approach is such that foreign companies are also buying into these U.S. zones (and, by the same occasion, transforming them into centers of international technology transfer) (see KELLY et al., 1992).

In contrast to these dynamic situations, the vast majority of developing countries lack most of the key ingredients involved in bringing knowledge out of the laboratory. Save exception, academia, industry and government have neither the tradition, the means and mechanisms, the skills and capabilities, nor the determination required to create and take advantage of working institutional interfaces. (Brazil and Costa Rica are among the few countries where university-industry collaboration is cultivated on a notable scale.) The absence of ties between public research institutes and industry is well illustrated by the fact that the former do not have specialized technology transfer offices, while the latter lack technology acquisition departments (ZILINSKAS, 1993; conclusions are based on a limited number of country-level case studies, but appear to be widely applicable). In the absence of such linkages, much scientific output (be it indigenous or imported) is simply wasted.

2.3.2 Industry Structure and Company Strategy

Although the industrial trajectory of biotechnology is both user-specific and country-specific, certain overall structural patterns can be identified. In particular, large established corporations either already dominate the industry (Japan, European Economic Community), or are gradually increasing their control of the latter (USA). In the U.S., traditional pharmaceutical and agrichemical corporations are looking to biotechnology as a solution to problems such as decreases in productivity, expirations of major patents, and social and private pressures on cost–benefit ratios. Dedicated biotechnology firms will continue to provide R & D services, and a limited number of them may develop into fully integrated companies. In several instances this may be the effect of accelerated commercialization strategies, whereby firms engage in a variety of activities (e.g., acquisition of generic business, manufacturing facilities and marketing networks) to generate revenues and gain experience while their flagship products are still in the pipeline (ERNST and YOUNG, 1992). However, because of shortcomings in the areas of manufacturing, commercialization and financing, most of them will remain dependent upon collaborative undertakings with large companies or be acquired by the latter. (Although alliances *among* dedicated firms are also on the rise, these usually concern R & D, and do not change the overall trends with respect to industry structure.) As science-push forces become relatively less important, the role of small firms in shaping the direction of change will also decrease.

With regard to developing-country user-industries and companies, much hope – probably too much – has been placed on cooperation with the dedicated firms of the industrialized countries. For one thing, as just mentioned, the future trajectory of the latter is, in most instances, subject to considerable uncertainty. Secondly, trade-offs for developing countries are not always clear: does the technology being transferred correspond to national and company-level priorities and capabilities? Has it gone beyond the experimental phase? Finally, it would appear that the biotechnology firms themselves have lost interest in developing countries partnerships, at least in the area of R & D. To that effect, it has been reported that between 1986 and 1991, R & D arrangements between U.S. dedicated biotechnology firms and developing country partners fell from 20 percent to 3 percent of total cross-country R & D arrangements (WAGNER, 1992; survey universe: 268 firms in 1986; 263 in 1991. According to ERNST and YOUNG data, there were 1231 U.S. dedicated biotech firms in 1991).

If this trend is to be reversed, which is far from certain, it will be necessary for the public and private agents of technological change in developing countries to surmount the kinds of technological, institutional and market obstacles that are causing the shift. At the same time, because of the aforementioned caveats, among others, it is obviously necessary to keep options open with regard to sources of science and technology acquisition. The actual size and shape of the biotechnology industry in the developing world will depend both on country-level needs, capabilities and policies and on developments in the industrialized economies.

2.3.3 Capital Base

Because biotechnology is highly capital-intensive, innovation is critically linked to financial options and strategies at both the country- and company levels. In the U.S., financial considerations are particularly weighty because of the large number of startup firms. Since heavy up-front R & D costs are not offset by timely product sales, these firms strongly depend upon public equity markets as well as on various other capital-raising options, such as venture capital, domestic and cross-border strategic alliances and R & D partnerships. Notwithstanding the variety of potential financial sources, the unpredictability of equity markets may enhance the rate of industry restructuring, at least in the short and medium term. This tendency may in turn be tempered by a reorganizing of the financial markets themselves as well as by a more coherent biopolicy context; in both areas signs of change are beginning to emerge. (Public equity investors interested in developing "patient capital" markets are starting to make themselves felt, and the U.S. government seems to be moving towards integrating industrial policy in areas of advanced technology.)

In countries where large established corporations already dominate the biotechnology industry, capital supply is less of an issue, although externalities (e.g., market failure, social conditions) and risk factors may, in some instances, still keep the costs of innovation above what even some of wealthiest corporations are prepared to invest.

In all but a limited number of NICs, developing country firms face difficulties in securing a capital base for biotechnology innovation. Financial institutions such as risk capital markets are nonexistent or do not work well; governments do not utilize or under-utilize financial instruments to promote domestic capability-building; the scope for interfirm collaboration is limited; and international competition for financing is on the rise. This leaves the donor agencies and international organizations, but technical assistance moneys cannot be expected to replace other sources of capital.

2.3.4 Human Resource Base

As biotechnology evolves, so do the types and the spectrum of skills and capabilities required not only to supply R & D, but to assure its conversion into competitive products and processes. As mentioned above, the number of skills to be mastered, the difficulty in mastering them, and their application-sector specificity, increase as the scientific frontier recedes. In addition to first-class scientists and engineers capable of working in interdisciplinary teams, successful biotechnology innovation requires high-level managerial skills and qualified government regulators.

In the industrialized countries, education and training have become critical to overcoming human resource thresholds and bottlenecks. Major efforts at upgrading and broadening skills are presently underway in both publicly and privately funded university laboratories dedicated to biotechnology research, and in government-sponsored research centers. Similarly the training of managers is geared towards dual career paths; i.e., professional managers with the technical proficiency necessary to integrate the multiple functions of biotechnology innovation, while assuring productive multi-disciplinary teamwork. Finally innovative approaches have been taken by government and industry to increase the numbers and qualifications of government agents working in the fields of biosafety regulation and patenting.

In the case of developing countries, the problem is not always one of finding first-rate molecular biologists. Several of the more advanced LDCs can boast such expertise, and, indeed, may have reputable centers of excellence. In many of these countries the problem is rather that: (1) there is not a *critical mass* of qualified scientists; (2) scientific endeavors are strongly biased towards fundamental research; (3) there appears to be little interest in developing the skills and capabilities required to move from the laboratory to the marketplace. Although these trends are beginning to change in several developing countries, to date only the Asian NICs have led a systematic effort at mastering downstream skills and capabilities, and at creating the institutional and organizational framework conducive to such mastery. The vast majority of developing countries suffer from a critical shortage of the key human and institutional capabilities needed to engage in competitive biotechnology innovation.

2.3.5 National Policy[a]

The ways in which and the extent to which government policy is used to support or pro-

mote biotechnology vary considerably across countries, and much debate has arisen as to the relative effectiveness of these differing approaches and practices. How, for instance, does technology-targeting measure up to user-industry targeting or a more laissez-faire stance? What are appropriate levels of infant industry protection? Whatever the usefulness of such debate – which is often flawed by its disregard for general country-level differences –, it is clear that the questioning itself is part of an attempt by the industrialized countries to optimize the use of policy instruments in view of strengthening their competitive position in biotechnology. This attempt involves science and technology, industrial, trade and even economic policy, as well as government interaction with the private sector. As international competitive rivalry increases, government appears willing to intervene closer to the market. In the U.S., for instance, it has been suggested that, because of their potential for generating non-appropriabilities, scale-up processes be eligible for government financing. (The PRESIDENT'S COUNCIL ON COMPETITIVENESS, 1991).

In the majority of developing countries the question of national policy towards biotechnology is singularly important. Given widespread market failure, an unfavorable macroeconomic environment, as well as the host of technical, social and institutional weaknesses discussed in this chapter, there is much that governments can and must do if entry into biotechnology is to be considered seriously. (This is not to say that government policy can *replace* a working market.) In many instances policy initiatives are already being undertaken, with varying degrees of success.

The Asian NICs have been particularly skillful in deploying industrial policy to promote commercial biotechnology. They use grants, direct capital investment, fiscal stimulants, and support skill-building and downstream development (SERCOVICH, 1993; YUAN, 1988). Korea, Singapore and Taiwan have made systematic use of foreign technology licensing and the repatriation of scientists trained abroad to strengthen their relatively weak research bases (U.S. CONGRESS, 1991).

In Latin America, most governments have tended to confine their activities to R & D

---

[a] Given their importance as instruments of biopolicy, regulation and intellectual property protection are treated separately in Sects. 2.3.6 and 2.3.7.

support, treating science and technology policy as if it were industrial policy. Although some of the larger economies are beginning to develop a more comprehensive approach to biopolicy, it will be some time before this shift begins to make itself felt in the marketplace.

A critical policy challenge facing developing countries concerns the ability to be selective in the promotion of biotechnology, so as to optimize the social and economic return on scarce resources. Such decisions must, of course, be compatible with overall economic policies and development strategies.

2.3.6 Regulation

Biosafety regulation is widely recognized as a significant barrier to innovation in biotechnology. In the pharmaceutical and agriculture application-sectors, it is often cited as the most important obstacle to timely market entry. This situation is rooted in a complex interplay of factors, including: science-based bottlenecks related to risk assessment; disagreements among policy-makers as to the optimal risk–benefit equation; jurisdictional turf wars; a lack of qualified regulators to handle an increasingly overcharged product pipeline; insufficient financing and infrastructure; pressures from public interest groups. Furthermore, the lack of internal and international regulatory harmonization is considered to hamper foreign market penetration. (For a more detailed discussion of these and related issues, see LEOPOLD, 1993.)

The development, implementation and enforcement of rules governing biosafety have considerably retarded biotechnology innovation in critical application sectors, and the overall effect of this retarding process has been largely negative (at least from the point of view of private costs and benefits). Regulation-related development costs, risks and uncertainties, and time lags have affected the direction of innovation both within and across application sectors, and have forced many companies to fold or to enter into partnerships; indeed, according to some doomsday scenarios, these constraints have cast a shadow over the very competitiveness of biotechnology.

Although biosafety regulation has indeed caused major delays in the biotech innovation process, it should be noted that technological breakthroughs and learning-curve factors are beginning to shorten lead times in several instances. Furthermore, to continue approaching the regulatory externality in a purely "finger-pointing" manner misses another and ultimately more significant point. Namely, regulatory institutions and mechanisms have been and are going through an important process of adjustment to meet the needs of biotechnology; i.e., the situation is by no means a static one. Thus, for instance, the U.S. government recently passed legislation authorizing the Food and Drug Administration to apply a system of user fees to the pharmaceutical and biopharmaceutical industries; by increasing agency revenues and regulatory staff this measure should allow for greater regulatory productivity and shortened product approval times. Because it will receive a disproportionately high number of new staff members, the biotechnology sector may reap greater benefits from the new measure than the traditional pharmaceutical industry (FOX, 1992). In the U.S. there has also been a tendency to loosen up rules governing the testing and approval of the products of agricultural biotechnology. In this instance, the possibility of a "public opinion" backlash should not be excluded; this could offset gains in lead times, at least in the short run.

The speed at which regulatory hurdles are overcome is largely country-specific; implementing change appears to be more cumbersome in Japan and in the EEC (particularly in certain key countries). But given the highly competitive international environment in which biotechnology is evolving, it is probably a question of time until these countries also put in place institutional and public policy mechanisms allowing for a more timely regulatory process. None of this should obscure the weight of new scientific knowledge in determining risk evaluation time frames.

Such adjustments should also favor greater international harmonization of risk assessment methodologies and regulations. Presently, for example, products which have been granted approval in one country may be subject to regulatory control by countries to which they are exported. This may be appropriate in cases where risk assessment considerations call for it (e.g., transgenic plants introduced in different ecosystems), or, again, where the exporting country's regulatory system or practices do not meet international standards. In other instances, redundant control may merely serve as a non-tariff trade barrier.

Developing countries present regulatory problems of their own. In most of these countries, neither biosafety rules nor monitoring procedures are in place. With the exception of very few countries, regulatory infrastructure is non-existent or, at most, rudimentary. The establishment and implementation of a regulatory system demands expertise and imposes monetary requirements beyond the means of most developing countries. This situation compromises the ability of the latter to set up joint ventures with partners from the industrialized world and further weakens the competitive potential of developing-country biotechnology. Clearly, strong government measures are needed to counter the effects of market failure in the area of biosafety; but in those countries where there is a chronic lack of appropriate institutional and social capabilities, these measures will not be forthcoming.

2.3.7 Intellectual Property Protection (IPP)

IPP, and patenting in particular, have created high barriers to entry into biotechnology. As in the case of biosafety regulation, these barriers are associated with long lead times and heavy costs, and a lack of internal and international harmonization. Also as in the case of regulation, time lags stem, *inter alia*, from experience-curve bottlenecks created by a rapidly advancing science base; from unresolved questions about scope (of biosafety rules, of property protection); and from skill shortages (with important competition for human resources coming from the private sector). Finally, in the two areas,

cross-country procedural and substantive differences affect choices as to which markets to enter (LEOPOLD, 1993).

In the industrialized countries, institutional, organizational and procedural adjustments aimed at overcoming IPP-related obstacles have been made. In the U.S., for instance, the Patent and Trademark Office has initiated a program aimed at reducing patent backlog bottlenecks. Measures include the creation of an examining group that deals uniquely with biotechnology; the granting of special engineering salary rates for new examiners, enhancing update training (U.S. GENERAL ACCOUNTING OFFICE, 1989). This set of initiatives has been complemented by private sector activities, with the Industrial Biotechnology Association setting up its own institute for training biotechnology examiners. With regard to the international harmonization of patent law, there has been debate in the U.S. concerning the possibility of shifting from the first-to-invent system in effect in the U.S. and Canada, to the first-to-file practice that prevails in most of the world.

For many developing countries IPRs have been an important battleground in their attempt to hold their own against foreign interests. Concerns that strong protection could lead to quasi-monopolistic market control by foreign companies were among the considerations that led several LDCs to weaken their IPP systems, particularly during the 1970s. In the area of pharmaceuticals, a number of countries put an effective ban on patents, or threatened to demand compulsory licenses on foreign patents after their registration.

Since the mid-80s, when the U.S. and the E.C. began linking intellectual property rights to trade (threatening retaliatory measures against countries whose IPP regimes were deemed unacceptable), and "trade-related intellectual property" (TRIP) issues became a key theme at the Uruguay Round of the GATT, many developing countries have been reconsidering their positions on IPRs.

As to the economic and social gains that developing countries might derive from strengthened intellectual property regimes, the question is still largely debated (see PRIMO BRAGA, 1990). For instance, although advances in technology have increased the potential economic value of knowledge-appropriation, making international IP considerations all the more important to industrialized economies (PRIMO BRAGA, 1990, p. 72), accessing newly-disclosed proprietary knowledge may be of little economic value to those developing countries that lack the skills and capabilities required to move scientific output into the economic arena (see above; SERCOVICH and LEOPOLD, 1991; SERCOVICH, 1993). Similarly, with regard to social issues, although harmonizing international patent law is generally held to be beneficial to the consumer, it is less than clear as to how this benefit plays itself out among the different segments of developing country populations.

Notwithstanding these and other caveats, an increasing number of developing countries are considering stronger patent protection in the area of biotechnology. These countries, which include Argentina, Brazil, Costa Rica, Chile, Mexico, Venezuela (ERNST and YOUNG, 1991), are presumably seeking to create more favorable conditions for the transfer of biotechnologies, and, possibly, additional incentives for domestic biotechnology R & D and investment capital; the effects of strong international pressure as a reason for these initiatives should, however, not be dismissed. In the majority of developing countries, any concessions on IPP are likely to be related to more general economic considerations (e.g., increasing inward capital flows, encouraging more technology transfer, avoiding trade penalties). An unresolved question concerns the ability of these countries to set up an efficient and effective IPP system; enforcement requirements, in particular, may be difficult to meet.

2.4 Summary

As the above overview makes quite clear, commercial entry into biotechnology is anything but the leap-froggers dream it was initially though to be; there is, as one observer put it, "no shortcut to biotechnology nirvana" (KINOSHITA, 1993). Reaching the market competitively involves mastering a wide range of scientific, engineering, industrial, commercial and institutional capabilities.

In the developing world, top-tier Asian NICs, Brazil and Mexico are among the few countries to have a working understanding of the complex and multiple requirements that must be met in order to develop an economically viable biotechnology industry. In the Asian NICs, moreover, efforts to promote the technology have been fostered by government's extensive use of industrial policy instruments.

Most developing countries find themselves in the difficult situation of lacking many if not all of the critical ingredients that go into building a biotechnology capability. This does not exclude the possibility of individual public or private sector initiatives in individual countries, including some of the least-developed countries. But on the whole such initiatives tend to be isolated and partial. In fact, for the great majority of developing countries, biotechnology, like the other advanced generic technologies, will contribute to widening the growth gap between themselves and the industrialized world. It will also accentuate the difference between them and those developing countries that are effectively exploiting knowledge-intensive technologies.

Various forms of international technical cooperation, including assistance in technology transfer and adaptation (see below) and in information network-building, will undoubtedly help certain countries to acquire and exploit specific capabilities in biotechnology. An extremely timely service that international organizations can render is to assist governments in setting realistic priorities in biotechnology (which application sectors, which techniques, etc.). These choices should be based on country-level needs and markets, and on existing scientific, technological, industrial and institutional capabilities and experience. International organizations can also help develop appropriate industrial policy guidelines.

In the following two sections, an attempt is made to assess biotechnology's potential economic and social impact on developing countries in the areas of health care and agriculture. What does the technology offer in these application sectors? How does this square with developing countries' needs and demand? What are the prospects for developing country entry into either or both of these application sectors? What is the present role of international assistance with respect to capability-building and/or to product procurement?

3 Health Care

3.1 Technology Assessment

Biotechnology has completely reshaped industrial approaches towards the development of new health care products. Better understanding of the correlation between the structure and function of biomolecules coupled with an ever-increasing ability to target biomolecules to specific organs have allowed enormously greater flexibility in developing new diagnostics, therapeutics and disease prevention agents. The rationalization of the innovation process in health care product development created a niche for the dedicated biotechnology companies which capitalized on their innovation capability. Revenues of the health sector biotechnology start-ups jumped from US $ 1.8 billion in 1989 to over US $ 5 billion in 1991. This still represents a very small percentage of the total turnover of the pharmaceutical industry which, in 1989, was estimated to be in the region of US $ 150–180 billion. Biopharmaceuticals (i.e., proteins produced by recombinant DNA or hybridoma technology) make up almost all of the products of "advanced" biotechnology developed to date. In 1989, biopharmaceuticals made up one out of every three clinically tested new drugs (BIENZ-TADMOR et al., 1992).

The cost efficiency of the health care dedicated biotechnology firms is 2.3 times higher than that of the traditional pharmaceutical industry. A recent Tufts University study estimates the development costs of a new drug by traditional methods at an average of US $ 231 million. The corresponding figure drops to US $ 100 million for biotechnologically derived products (ERNST and YOUNG, 1992, p. 7).

Biopharmaceuticals have been shown to have much higher clinical success rates, the direct result of which may be cost efficiency, due to shorter development times and higher approval rates (BIENZ-TADMOR et al., 1992). As the period of regulatory review (6–8 years) of drugs which successfully pass clinical trials corresponds on average to half of their patent life, the shorter the development time, the longer the effective time of proprietary rights. Cost efficiency of drug development through biotechnology and the new possibilities of therapeutic intervention based on biopharmaceuticals have driven all major pharmaceutical companies to develop a biotechnology capability. This has been achieved through a variety of means (see Sect. 2.3.3).

Likewise, in disease prevention, the new technologies promise fundamental changes in R & D approaches. Conventional vaccine development involves isolation and purification of the pathogenic agent and its subsequent introduction into a susceptible host to observe disease progression. The final step is the "packaging" of the pathogen, as a mutated or whole-killed mimic that does not produce disease, into an effective adjuvant (GRIFFITHS, 1991). This is an empirical trial and error process that can delay product development significantly, as has been the case with the diphtheria and whole-cell pertussis vaccines. The process pays little attention to the *in vivo* mechanisms of pathogenesis, focusing mainly on the *in vitro* expression of pathogenic antigens.

Advanced biotechnology promises not only to streamline vaccine development, but to expand its scope beyond the use of killed strains or analogs to pathogens (e.g., substitution of small pox by its cow pox analogs). Disadvantages, such as contamination or incomplete killing in the manufacturing process can be eliminated, and it is also feasible, in theory, to develop multivalent, synthetic, as well as heterologous (multiple) vaccines.

Despite these prospects, the number of commercial vaccine producers worldwide has declined dramatically in the last two decades (LIEW, 1988, p. 81). In practice, recombinant vaccine technology is still in its infancy. Only one recombinant vaccine (hepatitis B) is in widespread use (EDGINGTON, 1992). The situation is compounded by economic, regulatory and to a lesser extent technical considerations. These are thought to be:

1. Low profitability margins. Prices are under pressure as the principal consumers of vaccines are public programs.
2. The premanufacturing costs of vaccine development. These have been estimated by the US Institute of Medicine to be in the range of US $ 30–50 million for each new vaccine. (Research to discover a candidate vaccine covers only a fraction of these costs. Development costs involve production of adequate amounts of vaccine for quality control and clinical trials, the running of the trials themselves, licensing and assessment of the scale-up requirements.)
3. The time-consuming regulatory process for the approval of new vaccines, as in the case of biopharmaceuticals. The regulatory process is notoriously slow, although for serious diseases, such as AIDS, candidate vaccines can be considered through an accelerated approval process. (Concern has been expressed that such "fast track" approaches could potentially pollute population samples during field trials. This in turn, may lead to relocation of field trials to developing countries. Political sensitivities ("vaccine maker's guinea pigs") may be triggered as a result of this.)
4. Liability issues. Manufacturer's liability is an additional disincentive, although courts in some cases (e.g., USA) recognize limits to such liability (GRIFFITHS, 1991).

To evaluate the potential impact of the new generation of health care products on developing countries, it is instructive to look into: the causes of avoidable mortality and morbidity in the Third World; development trends in biopharmaceuticals and vaccines; technology transfer issues and the role of the international development agencies.

3.2 Health Care Problems and Priorities

The global distribution of illness and death from preventable causes is given in Tab. 1. More than 90% of the world's preventable mortality occurs in developing countries. Children are the principal victims of infectious diseases. In Africa, Asia and Latin America, the combination of acute respiratory infections, diarrheal diseases and other diseases such as malaria, tetanus, meningitis and typhoid kills, annually, an estimated 14 million children under the age of five. If infectious diseases are not eradicated, health problems can only become accentuated. In recent years, there has been a noticeable trend in epidemiological transition from infectious diseases (pre-transitional) to non-infectious/degenerative diseases (post-transitional) due to improving standards of living and increasing life expectancy. It is predicted that as the epidemiological transition continues, the developing world will be challenged to cope with both pre- and post-transitional diseases (COMMISSION OF HEALTH RESEARCH FOR DEVELOPMENT, 1990, p. 37).

Health care policies are predominantly dictated by demographic and economic pressures. The magnitude of the health problems of the developing world calls for the implementation of low-cost, high-impact programs with emphasis on prevention rather than therapeutic intervention.

Early diagnosis and wide vaccination programs offer advantages over other public health care practices. In general, they are less costly, easier to implement and reach much larger populations. In the developing world, the implementation of a preventive health program excluding wide vaccination is not a viable approach. Other measures, such as the improvement of sanitation, of the quality of food and water supplies, and the control of infectivity vectors, can and must be applied concomitantly. By themselves, however, they do not constitute comprehensive health care strategies. The complete eradication of infectivity vectors, for example, is not always feasible, and the improvement of sanitation and of the food and water supplies requires high

Tab. 1. Estimated Annual Number of Deaths by Cause and in Industrialized and Developing Countries, 1990 (thousands of persons)

Health Problem	Industrialized Countries	Developing Countries
Pre-transition		
Infectious and Parasitic Diseases	(478)	(17000)
Acute respiratory infections	367	6500
Diarrheal diseases		4180
Tuberculosis	38	3300
Malaria		1000
Other infections and parasitic diseases	73	1700
Measles		220
Pertussis		100
Post-transition		
Circulatory, Degenerative	(12174)	(19952)
Malignant neoplasma	2700	2421
Chronic lung disease	388	2500
Maternal causes		500
Perinatal conditions	86	3030
Injury and poisoning	570	2400
Other and unknown causes	1930	3670
Diseases of the circulatory system	6500	5431

Modified from WHO (1992)

capital investments. The application of therapeutic measures to the treatment of infectious diseases involves high costs resulting from the administration of multiple drug doses and from infrastructural and logistic requirements. Simple life-saving measures such as oral rehydration, in the case of diarrheal diseases, are easy to implement in principle, but often fail in practice, because of the lack of public education programs.

To assess the potential impact of health care biotechnology on developing countries, it is necessary to review the availability of technologies with respect to problems and priorities.

3.3 Technology Trends

Tabs. 2 and 3 present the therapeutic and preventive indications of the major biotechnology-derived drugs and vaccines. Tab. 3 also gives the priorities for vaccine development in the Third World. It is noteworthy that almost all biotechnology-derived drugs and vaccines target the treatment and preven-

tion of the major health problems of the industrialized world, namely cardiovascular disease, neurological disorders and cancer.

While conventional medical technologies are quite effective in the prevention and treatment of pre-transition infectious diseases, there are still cases, however, where some diseases (e.g., parasitic) have eluded effective treatment. Nevertheless and despite the urgency to engage in new drug and vaccine development, little activity is underway.

3.4 Technology Transfer

The reason for this lack of commercial interest is that pre-transition infectious diseases are the "diseases of the poor". The comparison of the data presented in Tabs. 2 and 3 demonstrates clearly that commercial product devolopment is primarily driven by the market. Because of market failure in developing countries only 5% of the global investment in health research, which amounted to US $ 30 billion in 1986, is devoted to health problems affecting them (Fig. 1). The development

Tab. 2. Therapeutic and Preventive Indications of Biopharmaceutics

Biopharmceutical	Indication
Human insulin (1982)	Diabetes
Human growth hormone (1985)	Human growth hormone deficiency in children
Interferon-alpha 2b (1986)	Hairy cell leukemia,
	genital warts
	AIDS-related Kaposi's sarcoma
Muromonab CD3 (1986)	Kidney transplant rejection
Interferon-alpha 2a (1986)	Hairy cell leukemia,
	AIDS-related Kaposi's sarcoma
Tissue plasminogen activator (1987)	Acute myocardial infarction,
	pulmonary embolism
Erythropoietin-alpha (1989)	Dialysis
CMV Immune globulin (1990)	CMV disease associated with kidney transplantation
BCG live (1990)	CIS of the urinary bladder

Modified from ERNST and YOUNG (1991)

Tab. 3. Therapeutic and Preventive Indications of Vaccines

Vaccines Developed	Indication	Third World Vaccine Priorities
Hepatitis B vaccine	Hepatitis B prevention	*Streptococcus pneumoniae*
Haemophilus B vaccine	Haemophilus influenza type B prevention	*Plasmodium* spp.
Hepatitis B vaccine	Hepatitis B prevention	Rotavirus
Interferon-gamma	Chronic granulomatous disease	*Salmonella typhimurium*
Erythropoietin	Anemia associated with AZT/AIDS	*Shigella* spp.
Factor IX	Hemophilia B	
BCG vaccine	CIS of the urinary bladder	
Interferon-alpha	Hepatitis C	
G-CSF	Adjunct to chemotherapy	
GM-CSF	Bone marrow transplant	
Glucocerebrosidase	Type I Gaucher's disease	

Modified from ERNST and YOUNG (1991) and ROBINS and FREEMAN (1988)

costs of new drugs render public sector involvement completely prohibitive. Little wonder then that the pharmaceutical market is dominated by 50 multinational companies, while three quarters of all medicines are consumed in the industrialized world. Although, approximately 20% of the world production of pharmaceuticals takes place in developing countries, two thirds of this is produced by subsidiary franchises of multinationals which operate mainly as market outlets (BALLANCE et al., 1992).

The industry has a fairly significant presence in some of the larger developing countries, but its indigenous component is, in general, small in size, and has a limited range of strategic options and little innovation capability. Traditionally, the industry has focused primarily on the production of patent-expired products. Its stagnation, over the last decade, is attributable to many factors, not the least of which are:

- the lack of interface between government and industry
- adverse pricing policies
- shrinking drug lists for public reimbursement

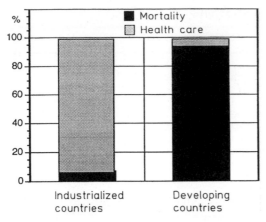

Fig. 1. World health: contrast in premature mortality burden and allocation of health research funds. Modified from COMMISSION OF HEALTH RESEARCH FOR DEVELOPMENT (1990).

- shift in consumer preferences to imported drugs (BALLANCE et al., 1992).

As a result, countries with large domestic markets and sizeable pharmaceutical industries (Brazil, India, Mexico, etc.) have tended to protect their markets through the adoption of a policy of weak patents (e.g., patenting of process instead of product) and the control of prices and distribution systems.

As discussed in Sect. 2, pharmaceutical companies seeking to expand their technology base by building a capability in biotechnology face a number of entry barriers, such as,

- lack of scientific, technical and managerial skills
- limited access to capital markets
- appropriate fiscal and regulatory environment
- lack of risk-capital culture
- market inefficiencies.

As far as the private sector in developing countries is concerned, the prospects of profit making are remote as long as effective demand is limited. As for national health authorities, their sparce resources dictate the application of traditional treatment protocols, which still offer scope for large-scale health improvement programs. The fragmentation of the public health care system is an additional disincentive for introducing less conventional health care approaches.

Therefore, the number of options open to the indigenous industry for technology acquisition is limited. Product licensing, for example, does not appear to be very attractive, as royalties may range from 1–5% for research reagents, 5–10% for therapeutic products, monoclonals and vaccines, and may go as high as 15% for bulk product sales (CORREA, 1991).

Alternatively, it has been suggested that a more appropriate short-term strategy would be to license technological know-how from the smaller biotechnology start-ups which rely, to a large extent, on such licenses to finance their R & D (CORREA, 1991). While this approach may offer solutions upstream, it does not account for downstream capability for product development and marketing. In any case, even this possibility seems to be exhausted considering the present trends of strategic alliances among the owners of the technology (see Sect. 2.3.2).

Transfer of technology through domestic private sector involvement seems, thus, unlikely. Possible exceptions are the area of diagnostics and some specialized market niches.

In the case of diagnostics a great deal of the technology is available in the public domain and development costs are relatively low. A considerable amount of technological adaptation is, however, necessary in order to develop diagnostics that are not dependent on laboratory infrastructure (i.e., analytical instrumentation, communications). This is because the mass dissemination of diagnostic technologies in most developing countries requires that these be dispensed by field medical practitioners and paramedical services rather than by centralized laboratory services.

Technological innovation/adaptation may also be driven by existing and emerging market niches (Asian Pacific Rim, larger developing countries). The use of tissue and cell culture technologies to derive herbal drugs is often quoted as a case in point. The intermediate level of sophistication of these technologies renders their acquisition relatively easy and, therefore, within the reach of a good number of developing countries, particularly those with entrepreneurial capability and market potential.

Another possible technological entry point is the "prospecting" of chemicals and pharmaceuticals from indigenous genetic resources. While the technological and scientific demands of, e.g., separation and structural analysis of biologically active components extend beyond the capability of most developing countries, it is conceivable that some may acquire expertise in intermediate technologies (e.g., screening of indigenous germ plasm for biological and chemical activity). "Prospecting" is known to be taking place in some countries (Costa Rica, Malaysia, Philippines) (KOMEN, 1991) through commercial contracts between national institutions and universities and companies in industrial countries. Although this may result in substantial returns for these countries, there is little evidence that technology transfer can take place in a major way.

No matter what the technology transfer barriers are, new generation products will eventually find their way into the developing world. The indigenous industry may be involved in manufacturing and distribution, but it is most unlikely it will be involved in product development (with the exceptions mentioned above). Market penetration will, however, depend on the purchasing power of the middle classes. On the basis of present trends, dissemination of health care biotechnology products is likely to occur earlier in the Asian Pacific Rim and the larger Latin American countries. Africa may remain excluded for some time to come.

Public institutions (national, international development agencies, etc.) have to play a role in technology transfer, particularly where this is considered to provide competitive advantages over the existing technologies. The following section presents the challenges facing agencies involved in health care work.

3.5 Public Institutions: Possible Options

For reasons mentioned above and the dearth of public funding, international agencies focus their strategies almost entirely on immunization programs. The primary target of the latter are children, as they are the main victims of infectious diseases.

Prime consideration in all of these vaccination programs is given to measures that would mitigate infrastructural and logistic difficulties, such as the, often, inadequate delivery systems and insufficient level of public education. The main concern is, however, the gap that exists between the priorities of commercial vaccine developers and the needs of the developing world (see Tab. 3).

Despite difficulties, the global immunization programs supported by the United Nations over the last 25 years have had astonishing results. Smallpox has been completely eradicated and almost 50% of the world's children are immunized against measles, diphtheria, pertussis, tetanus, polio and tuberculosis, as a result of the Expanded Programme on Immunization (EPI) of the World Health Organization.

EPI managed to acquire vaccines at extraordinarily low prices through a system of international bidding. The price for a single dose of each of the six vaccines is US $ 0.05, whereas for all doses of all six vaccines it does not exceed US $ 0.50. The logistic support of the vaccination program costs 10 times more per child. This includes delivery, vaccine protection through a cold chain and administration of vaccines. It has been estimated that one billion doses per year are needed to deliver all six vaccines to all the children of the world. The basic strategy employed in the bidding system is to wait for prices to drop to adequately low levels and then buy in bulk. This approach can be effective for those vaccines that have already been developed and have a large enough market in industrialized countries or, alternatively, a market for travelers to the developing world. The essence of the approach is to purchase surplus goods.

The need for new vaccines for diseases that are endemic in the Third World necessitates new approaches of vaccine development. To overcome the compound effects resulting from the lack of funds, logistic and infrastructural support, the new-generation vaccines will be required to be multivalent, heat-stable and suitable for oral administration in one or two rather than multiple numbers of doses. These are precisely the set objectives of Children's Vaccine Initiative (CVI) (WHO, 1991) launched in 1991 by WHO, UNICEF, the World Bank and the Rockefeller Foundation to accelerate production of new vaccines.

While advanced biotechnology is likely to provide most of the technological solutions, the economics of vaccine development mentioned above impose additional limitations. In view of these, international agencies engaged in global immunization programs may, thus, opt for:

1. Securing purchases of an agreed volume of a given vaccine over a number of years at a fixed price close to market value. However, no international development agency has the mandate to make such long-term financial commitments.
2. Establishing a public institution to develop and manufacture its own vaccines. The problem with this option is that international donors are adverse to creating new institutions. The preferred policy is networking of existing ones.
3. Setting-up development and production units in countries/regions with large population for a small range of products. This is a long-term strategy, and the management of such a scheme may prove difficult in practice. Funding agencies and the World Bank, for examples are increasingly committed to meeting bankers' standards (return on investment).
4. Securing the development costs for the manufacturer and then guaranteeing purchase at prices little over the production costs. This "front-end strategy" would imply spending money in the industrialized world, since the know-how is to be found there. This kind of approach may find opposition from some of the policy makers of the developing world (ROBINS and FREEMAN, 1988).

Long-term commitment to R & D funding is an absolute prerequisite in all cases. This is not likely to happen without the involvement of commercial manufacturers. The major challenge is to assure the latter of acceptable returns on investment. These considerations would tend to favor the fourth of the options mentioned above. Striking a balance between private and social benefits is the major challenge for policy makers in government, development agencies and industry.

In conclusion, the impacts of biotechnology on improving health care in the developing world can only be beneficial. The magnitude of the benefits and the time horizons of integration of the new therapeutic and preventive protocols will depend on socioeconomics. Health care, however, is not to be the likely window of entry into commercial biotechnology in most instances.

4 Agricultural Biotechnology

4.1 Technology Assessment

The application of molecular technologies in plant agriculture aims at developing new higher-yielding crop varieties through improvements in:

- biotic and abiotic stress tolerance
- nitrogen-fixing ability and adaptability under low chemical input conditions
- nutritional content and product processing quality.

Early problems related to gene transfer, expression and stability are being rapidly resolved. It is anticipated that gene transfer will be routinely performed in all the major crops in the next 2–3 years. It has also been demonstrated that introduced genes are inheritable, stable and are expressed in a predictable fashion. Furthermore, the identification and isolation of agronomically useful genes has been facilitated enormously by the refinement and automation of restriction fragment length polymorphism (RFLP)-based techniques.

Population expansion at the projected levels and increasing land erosion pose a real threat to food security in most of the developing world. The annual growth of world agricultural output would need to be commensurate to that achieved over the last 40 years (2.43% from 1950 to 1985) to meet the demand of a population which some expect to double by the year 2030. The routine introduction of molecular and cellular techniques in plant and animal agriculture was widely perceived to be the dawn of a new era for agriculture. In fact, it was hailed by many people as the "biorevolution" by analogy to the "green revolution". In retrospect, such expectations, including forecasts for productivity increases of up to 20–40% in the six-year span between 1986 and 1992 (RATAFIA and PURINTON, 1988) have proven to be wildly overoptimistic. The rapidly expanding knowledge base and the refinement of techniques have not been enough to accelerate, in a profound way, the commercial development of transgenic plants, animals and products thereof. It is now held that rather than revolutionizing farming practices, biotechnology will serve as a catalyst for the next evolutionary phase of the agro-based industry, generating considerable – but no massive – productivity increases and qualitative improvements in the agricultural sector (FRALEY, 1992; BUTTEL, 1991). This phase will be characterized by the intensive utilization of the tools of molecular biology and the integration of advanced information and engineering technologies in the production systems.

Molecular technologies will, thus, be increasingly integrated into conventional agronomic programs, complementing rather than superseding existing technologies.

The first commercial transgenic crops have been developed by modification of important agronomic traits, such as insect resistance, disease resistance and other traits that are controlled by a single gene. The next technological generation will produce transgenic plants with modified agronomic traits controlled polygenically, such as those determining photosynthetic ability and abiotic stress tolerance. In the short and medium term, therefore, productivity increases are expected from crop protection and improved nutrient uptake, rather than from the enhancement of the photosynthetic ability of crops, which, if possible, would indeed constitute a revolutionary scientific and technological development.

In the livestock sector, single-gene products such as bovine and porcine somatotropin have been known to increase productivity considerably. Consumer acceptance of such products is a key factor for their dissemination. The development of new animal diagnostics and vaccines has received a new impetus by human health care technological "spin-offs". A number of diagnostic systems based on monoclonal antibodies are already available. In the short term, new vaccines for cattle and poultry are expected to have a major impact in increasing productivity, though lesser but still important improvements in animal performance are likely to be derived from the development of more flexible feed formulations (GRAHAM and INBORR, 1991). In livestock production, the range of biotechnologies, as well as some of the major problems and potential biotechnology related solutions have been reviewed in depth elsewhere (PERSLEY, 1991).

As time goes by, technological maturation is attested by the increasing number of applications for field trials of agro-biotechnology products. The number of permits for environmental releases issued by the U.S. Department of Agriculture (USDA) increased at an annual rate of about 50% between 1989 and 1991 (ERNST and YOUNG, 1991, p. 19). More than 700 field trials of such products have been recorded in the OECD database (BIO-TRACK).

The application of biotechnology in agriculture is almost completely dominated by the private sector. This is in sharp contrast with the "green revolution", which was driven by public investment. The consolidation of the agro-chemical industry and its vertical integration with the seed, food processing, fine chemical and pharmaceutical industries have made the acquisition of generic technologies (biotechnology, informatics) major strategic objectives as these can be applied across and within sectors, thus increasing competitive advantage and market share.

Technology development has been driven by the need of the agro-industry to meet the changing demands of their primary markets, namely:

- the qualitative rather than quantitative shift in food consumption patterns
- food safety
- ecological/environmental sustainability of agricultural production systems.

Changing consumer habits have increased the pressure on the agrofood industry to diversify to crops with improved processing characteristics (e.g., high fiber content, improved amino acid composition). In livestock production, a good part of R & D is aimed at improving the protein to fat ratio. The high individual value of cattle has turned them into the primary target for the application of the new technologies. Likewise in crops, market pull has directed R & D effort towards high-value horticultural and industrial plantation crops, cereals and to a lesser extent grain legumes. Similarly, it is the commercial interest of the agro-chemical industry in increasing the economic life-time of herbicides that has driven the development of herbicide-resistant crops. Herbicides constitute over 50% of the total pesticide usage in the industrialized countries (GRESSELL, 1992).

It is, therefore, not surprising that crops of major importance for the developing world (roots, tubers and plantains) as well as hardwood, have received minimal attention (HODGSON, 1992). Even for crops of global importance (e.g., maize), R & D focuses mainly on the temperate varieties (BRENNER, 1991). Crop varieties grown in tropical or subtropical environments, where different biotic and abiotic stresses apply, have remained outside the mainstream of research. The availability of new technologies for a number of important food and cash crops for developing countries is illustrated in Tab. 4.

Scientific capability to conduct agricultural research and development at the frontier is lacking in most developing countries. A recent case study in maize demonstrates that such capacity is absent even in the top-tier developing countries (BRENNER, 1991). It has been assessed, for example, that in the case of Latin America and the Carribean only two laboratories have state-of-the-art research capability and, perhaps, up to another 25 moderate ability (JAFFÉ and TRIGO, 1992). There is some expertise in intermediate technologies (e.g., fermentation, micropropagation, embryo transfer, etc.). The situation appears to be similar in Asia. The overriding problem for developing countries is, thus, the insufficient integration of biotechnology in research institutions and the agrofood industry. Research fund allocations are extremely low

Tab. 4. Availability of New Technologies for Selected Crops

Crop	New Diagnostics	Rapid Propagation System	Transformation	Regeneration	Time Frame (years)
Banana/plantain	+	+	−	+	5–10
Cassava	+	+	−	−	5–10
Cocoa	+	−	−	−	>10
Coconut	+	−	−	−	>10
Coffee	+	+	−	+	5–10
Oilpalm	+	+	−	−	>10
Potato	+	+	+	+	< 5
Rapeseed	+	+	+	+	< 5
Rice	+	+	+	+	< 5
Wheat	+	+	−	−	>10

Source: PERSLEY (1990)

both in Asia and Latin America for reasons that will become evident below (BRUMBY, 1991; JAFFÉ and TRIGO, 1992).

4.2 Possible Impacts on Developing Countries

Food sufficiency is a stated policy objective in many countries, and biotechnology is called upon to help achieve more stable levels of production in harsh and often fragile environments. Capitalizing on the technological opportunities, however, depends on the ability of individual countries to introduce drastic changes in economic, market and distribution systems. For example, it has been long recognized that the problem of food distribution is as important as that of production. Distribution is a bottleneck in many countries that are, or could potentially be, in a situation of food sufficiency.

On the technology front itself, the tools of biotechnology provide agriculture with unique opportunities to meet the food demands of a rapidly growing population. Moreover, substantial productivity increases and improvement of crop quality could have a profound impact on those national economies which rely predominantly on agriculture for GNP growth and international trade.

For many developing countries, agro-biotechnology may still offer the only window of opportunity for technological advancement. This is because of the range of available technologies of varying levels of sophistication which can be utilized according to the needs and priorities of the highly diversified agricultural systems in the developing world. Unlike health care biotechnology, market forces are not necessarily prohibitive to investments in agricultural biotechnology R & D. With respect to demand, there are emerging markets with the requisite purchasing power for new products (Asia, Latin America). There are also specialized market niches where consumer demand is not likely to be met by the major multinational industries. In addition, the wealth of genetic resources in large parts of the developing world may offer distinctive commercial advantages for the adoption of the new technologies.

Entry barriers to agricultural biotechnology will rise with the strengthening of the intellectual property protection, the domination of the major marketing and distribution networks of the multinational companies and the limited ability of developing countries to integrate biotechnology with information and engineering systems of ever increasing complexity. Moreover, in most countries fragmentation of the agricultural production systems, in which the role of small land holding and marginal land cultivation is dominant, does not allow for economies of scale and, therefore, represents an additional barrier. Finally, the still considerable scope to achieve productivity increases through existing technologies and agricultural management may often discourage the introduction of new technologies.

Any attempt, therefore, to make generalizations about the potential socioeconomic impact of agricultural biotechnology is practically impossible. The heterogeneity of developing countries in terms of agro-climatic conditions, crops of importance, access to germplasm, technological capability, size of markets, socioeconomic advancement and so forth does not allow a systematic approach. The likely impacts will be compounded by country level ability to make an entry into agricultural biotechnology and by international market forces.

With respect to technological potential, it is held that probably no more than ten countries have the capacity to integrate micropropagation and recombinant DNA technologies into their agriculture-based industries (UNCTAD, 1991). These countries have the potential to adapt knowledge-intensive systems to local needs and conditions. The likely points of entry are the adaptation of existing technologies for "orphan crops" (see below), the development and application of molecular diagnostic tests for disease control, the utilization of RFLP-based technologies in plant breeding programs and the deployment of tissue culture technologies. The development of transgenic "orphan crops", even if it is through adaptation, would require public capital funding and a fair amount of synergism among public and private institutions. Most of the technologies mentioned above are not knowledge-intensive and require low investment costs. This makes them particularly attractive to local industries and brings them within the reach of a large number of countries.

As regards international market trends, global economic restructuring and structural adjustment programs have generated an intensification in the production of export cash crops. The latter, being the raw materials for sweeteners, medicinals, coloring and flavoring agents, etc. are vulnerable to substitution by industrial products derived with the help of state-of-the-art biotechnology. New technologies are already introducing shifts in comparative quality and price advantages for a large number of products (e.g., starch, proteins, oils), inevitably leading to increased competition among agricultural commodities. It is, therefore, necessary for developing countries to restore the balance through diversification of agricultural production towards food crops and high-added-value crops. This would have the advantage of meeting the food demand of the expanding population, while keeping rural populations from migrating to urban centers.

A recent study attempts to estimate the impact of micropropagation and recombinant DNA technologies on the trade of major export crops (UNCTAD, 1991). The estimate takes into account the projected time frame in which these technologies are expected to produce commercial results for the crops under consideration. Tissue culture technologies are expected to affect the value of export crops by approximately US $ 42 billion in the next five to ten year period. The corresponding figure for genetic engineering is in the region of US $ 24 billion. There are several affected individual commodities which make up over 5% of the total trade for a considerable number of countries. These are shown in Tab. 5.

The uncertainties involved in quantifying some of the major parameters make the estimates of the study very approximate. Furthermore, there is no attempt to qualify whether trade in these crops will be adversely or positively affected. Other studies predict substitution of agricultural commodities by the *in vitro* propagation of plants (OECD, 1989). It is predicted that this is likely to have adverse socioeconomic effects on a number of coun-

Tab. 5. Time Frame of Application of Two Main Categories of Biotechnology to Selected Commodities and Corresponding Values of Affected Exports of Developing Countries and Territories

Time Frame for Routine Use	Value of Exports (US $ billions)	Commodities Affected (no. of developing-country/territory exporters)
A. Tissue and Cell Culture Techniques		
Up to 1995	20.9	Coffee (28), bananas/plantains (16), rice (6), rubber (5), tobacco (2), vanilla (2), cassava (1), potatoes (1)
1995–2000	21.2	Sugar cane/sugar beet (16), cocoa (15), tea (4), soybeans (3), oilpalm (3), wheat (3), maize (1), sunflower (1), pineapple (0), sorghum (0), barley (0), sweet potatoes (0), yams (0)
After 2000	3.4	Cotton (15), coconut (10), rapeseed (0), millet (0)
B. Transgenic Plants		
Up to 1995	6.4	Rubber (5), tobacco (2), maize (1), potatoes (1), tomatoes (0)
1995–2000	17.5	Sugar beet (16), cotton (15), bananas/plantains (16), rice (6), soybeans (3), cassava (1), sunflower (1), barley (0), rapeseed (0), sweet potatoes (0), yams (0)
After 2000	21.7	Coffee (27), sugar cane (16), cocoa (15), coconut (10), tea (4), oilpalm (3), wheat and flour (3), pineapple (0), sorghum (0), millet (0)

Modified from UNCTAD (1991)
Figures in parentheses refer to countries for which share of commodity exports in total exports is above 5%

tries dependent upon exports of natural products. In the case of pyrethrum, for example, which affects the income of approximately 100 000 farmers in Kenya, commercial activity in *in vitro* propagation of the plant may soon verify such predictions (SIHANYA, 1992).

The shift to more technology-oriented, less labor-intensive agricultural production systems will likely have negative impacts, in the short term, on the labor market. Agriculture is the main employer of the economically active population and has a major role as low-cost labor absorbing technology. Labor displacement is likely to be accentuated in the poorer countries in which small-scale farming is the rule. Sustenance farmers have limited ability to adopt integrated technology packages and are, therefore, quite vulnerable to the new production methods. It has been argued, for example, that increased use of herbicides may not only have adverse economic and environmental impacts but will also inevitably lead to labor displacement as weeding is a major source of women employment in least developed countries.

These concerns have to be balanced against the need to introduce new technologies in order to meet rising food demand. Food productivity is likely to be enhanced through plant protection rather than by increasing the photosynthetic ability of crops at least in the medium term. Weeds are perhaps the greatest cause of agricultural productivity loss. It has been estimated that hundreds of millions of people in Africa and Asia lose over half of their agricultural yields because of weed infestation. The latter is thought to be a major factor for land abandonment, ecosystem invasion and urbanization. Considering that 100% agricultural sustainability is an unattainable target, there is need to make the right technology choices. Herbicide resistant crops offer unique prospects for productivity increases to meet the rising food demand. In addition increased use of herbicides may not be an unmanageable environmental problem. Herbicides are significantly less toxic than insecticides, which are much more widely used in the developing world (GRESSEL, 1992).

4.3 Technology Transfer

Hybrid seeds are expected to be one of the main sources of technology dissemination. This is because hybrid seeds embody the technological innovation into the final agricultural product and can be used in ways that are familiar to the end users. The new genetic engineering tools will speed up breeding time and considerably increase profitability margins. Return on R & D investment is secured as hybrid seed does away with problems associated with intellectual property protection. Farmers cannot re-sow harvested seeds as the latter lose their vigor after one to two cropping cycles.

Notwithstanding the above advantages the superiority in productivity of hybrids over open pollinated varieties will have to be clearly ascertained. In the case of maize, for example, their comparative advantage has not been clearly demonstrated (BRENNER, 1991). The major disadvantage of hybrid seeds is associated with their costs which may render them inaccessible to all but the large farm owners.

The role of the agro-industry as the principal driving force of technological innovation and the tendency to strengthen intellectual property rights will undoubtedly have important repercussions on the diffusion of the new technologies in the developing world. The high degree of industrial concentration limits the range of options for technology acquisition.

The seed industry is under increasing pressure of acquisition by the agro-chemical industry. Over 400 such acquisitions took place in the period 1970 to 1990. Despite this, it still remains relatively dispersed and small in size. Seeds are still undervalued, although they are thought to account for 50% of the increase of agricultural productivity in recent years (UNCTAD, 1991).

In the developing world the presence and capability of the indigenous seed industry is variable. FAO (Food and Agriculture Organization) has been surveying the industry for a number of years. Latin America and to a lesser extent Asia have some industrial infrastructure, while in Africa it is rudimentary or non-existent. There is an increasing number of companies acquiring subsidiaries in developing countries to capitalize on the growing domestic markets and to take advantage of the inherent botanical characteristics of seed which are often dependent on local agroclimatic conditions.

The strengthening of patent law and its extension to cover plant varieties and genetic material (individual genes, gene constructs, cell lines) are likely to slow down the rate and increase the cost of technology transfer.

Proprietary considerations may not, however, be important in the case of the so-called "orphan commodities", i.e., crops of little or no commercial interest to the industrial world. This lack of interest should make possible the acquisition of technology spinoffs. It should, for example, be possible to acquire genetic constructs (cassettes) developed for a crop of commercial interest and adapt them for a particular "orphan crop".

In all but few instances where the indigenous plant breeding industry has the capability, technology adaptation can only be met by public research. The role of the National and International Agricultural Research Centers (IARCs) in this respect is crucial. It is, therefore, of utmost importance that they overcome a certain number of problems. The IARCs are active in commodity oriented R & D, genetic resource conservation, policy studies and technology transfer and are considered to have been the driving force of the "green revolution". A number of them are involved in biotechnology research albeit to varying degrees of sophistication. The commodity-oriented centers are focusing on "orphan commodities" and crops of little interest to the industrialized world. The main focus of their work is presented in Tab. 6.

At present IARCs not only do need in-house expertise in biotechnology, but also expertise and appropriate institutional mandates to negotiate technology transfer from the private sector (see Sect. 4.4). For example, the current practice of the IARCs to distribute freely genetic material is considered to be incompatible with patent protection. Public institutions in industrial countries, which have traditionally been the primary sources of

Tab. 6. Consultative Group on International Agricultural Research (CGIAR) Centers: Focus of Activities

Center	Focus
CIAT (Centro Internacional de Agricultura Tropical, Colombia)	Cassava, common beans
CIMMYT (Centro Internacional de Mejoramiento de Maiz y Trigo, Mexico)	Wheat, maize
CIP (Centro Internacional de la Papa, Peru)	Potato, sweet potato
ICARDA (International Centre for Agricultural Research in the Dry Areas, Syria)	Lentil, faba bean
ICRISAT (International Crops Research Institute for the Semi-Arid Tropics, India)	Chickpea, groundnut/pearl millet
IITA (International Institute of Tropical Agriculture, Nigeria)	Cassava, sweet potato/yam, cooking bananas/plantains, cowpea
ILCA (International Livestock Center for Africa, Ethiopia)	Forage crops
ILRAD/ILCA (International Laboratory for Research on Animal Diseases, Kenya)	Livestock, vaccine development
IRRI (International Rice Research Institute, Philippines)	Rice
WARDA/IRRI (West African Rice Development Association, Côte d'Ivoire)	Rice

Source: *Biotechnology and Development Monitor* (1989) **1**, 15–16

technology for the IARCs, are reluctant to provide know-how in view of their own increased dependence on funding from private sources and the resulting obligation to respect intellectual property (COHEN and CHAMBERS, 1992).

The intellectual property issue has equally bearings on culture collections. Germ plasm banks were originally established to preserve genetic material from endangered areas. Advanced biotechnology has turned them from "biological museums" into resources for the utilization of genetic material. The International Board for Plant Genetic Resources (IBPGR) was established to coordinate international efforts connected with the conservation of plant genetic resources. IBPGR, like the IARCs, operates under the auspices of the Consultative Group on International Agricultural Research (CGIAR). CGIAR is cosponsored by the World Bank, FAO and the United Nations Development Programme (UNDP). It is a consortium of government, private and public institutions. IBPGR's programs added nearly 200000 samples of germplasm (accessions) to collections placed in gene banks in approximately 100 countries. Advanced biotechnology and the patentability of genes have made them strategic assets to their holders who have the overall jurisdiction over them. It has been estimated by the Rural Advancement Fund International (RAFI) that approximately 50% of IBPGR's collections are housed in national gene banks in industrial countries, compared to 29% and 21% in developing countries and the International Agricultural Research Centers (IARCs), respectively. The changing international attitudes toward intellectual property rights are already impinging on the IARCs whose collections were originally distributed freely to plant breeders. Partnering with third party collaborators has obliged IARCs to share rights in future patents (KOMEN, 1990). This development goes against the norm that wants institutions under the UN umbrella shying away from proprietary rights. This is not necessarily a negative development. It allows IARCs to return to developing countries, directly or indirectly, the benefits accruing from the commercial exploitation of genes obtained from them. The current practice of free access to IARCs' germplasm may be open to abuse. It could allow private companies to patent material obtained from the IARCs without the latter receiving any share of the profits.

Finally, another form of technology diffusion through extension services (disembodied technology) aimed at supporting local R & D and breeding programs in resolving technological and other bottlenecks does not appear to be adequately explored as a technology transfer mechanism by the public and private sectors in developing countries (BRENNER, 1991).

4.4 The Role of Public Institutions in Overcoming the Barriers

As mentioned earlier, the ability to identify, select and adapt technology to local socioeconomic environments is limited to a few countries. Even in these countries there appears little or no opportunity to enter biotechnology at the frontier. The increasingly proprietary nature of agro-biotechnologies and the dominance of the private sector as the main supplier of technology and capital means that transfer of know-how will depend highly on the ability of the technology buyers to forge partnerships with the technology sellers.

The role of International Institutions in catalyzing such partnerships is deemed to be a key role. They could serve as the main conduits of technology transfer from technology owners to the national agricultural systems. Moreover, they have relative strengths that can entice the private sector from entering into such partnerships. These are thought to be able to test and verify the viability of new genetic constructs in field test conditions, plant breeding programs and different agroclimatic environments. Finally, they are often the repositories of germ plasm that would otherwise be inaccessible to the private sector.

Partnerships between the International Public Institutions and the private sector require reciprocity which, as mentioned earlier, may be incompatible with the mandate of the former. It is, therefore, considered that unless these centers are able to guarantee the intellectual property rights of prospective partners, they would be excluded from any benefits that may result from the application of the new technologies.

In addition the testing of new transgenic plant varieties requires compliance with safety regulations/guidelines, and research centers will also have to acquire capability to conduct field trials according to set standards.

Finally the need for human resource development is a manifest policy priority in view of its direct relationship with the ability to make technology choices. The IARCs are an important resource in strengthening the educational capability of the national agricultural systems.

It is obvious from the above that public resources will have to be allocated to strengthen the IARCs in order to meet the new technological and institutional challenges. Donor agencies will be called upon to meet the additional investments in these centers. They are thought to have a vested interest to do so. Otherwise the North–South technology gap will cause further imbalance in the global agricultural production systems (COHEN and CHAMBERS, 1992).

5 References

BALLANCE, R., POGANY, J., FORSTNER, H. (1992), *The World's Pharmaceutical Industries*, pp. 22–28, Great Britain: Edward Elgar Publ.

BIENZ-TADMOR, B. (1993), Biopharmaceuticals go to market: Patterns of worldwide development, *Bio/Technology* **11** (2), 168.

BIENZ-TADMOR, B., et al. (1992), Biopharmaceuticals and conventional drugs: Clinical success rates, *Bio/Technology* **10** (5), 521–525.

BRENNER, C. (1991), *Biotechnology and Developing Country Agriculture: The Case of Maize*, Paris: OECD.

BRUMBY, P. (1991), An international perspective on agricultural biotechnology in Asia, in: *Biotechnology for Asian Agriculture* (GETUBIG, Jr., I. P., Ed.), p. 308, Kuala Lumpur: Asian and Pacific Development Centre.

BUD, R. (1989), Janus-faced biotechnology: a historical perspective, *TIBTECH* **7**, 230–233.

BURRILL, G. S., ROBERTS, W. J. (1992), Biotechnology and economic development: The winning formula, *Bio/Technology* **10** (6), 647–653.

BUTTEL, F. H. (1991), The socioeconomic impact of biotechnologies on developing countries, in: *Plant Biotechnologies for Developing Countries* (SASSON, A., COSTARINI, V., Eds.), pp. 101–107, Washington, DC: CTA/FAO.

CASDIN, J. (1992), Taking stock, *Bio/Technology* **10** (10), 1084.

COHEN, J. I., CHAMBERS, J. A. (1992), Industry and public sector cooperation in biotechnology for crop improvement, in: *Biotechnology and Crop Improvement in Asia* (MOSS, J. P., Ed.), pp. 17–29, India: International Crops Research Institute for the Semi-Arid Tropics.

COMMISSION OF HEALTH RESEARCH FOR DEVELOPMENT (1990), *Health Research: Essential Link to Equity in Development*, Oxford: Oxford University Press.

CORREA, C. M. (1991), Developing private biopharmaceutical capacity in developing countries, *Biotechnol. Dev. Monitor* **9** (December), 7–8.

EDGINGTON, S. M. (1992), Biotech vaccines' problematic promise, *Bio/Technology* **10** (7), 763–766.

ERNST and YOUNG (1991), *Biotech '92: Promise to Reality: An Industry Annual Report*, San Francisco: Ernst and Young.

ERNST and YOUNG (1992), *Biotech 93: Accelerating Commercialization*, San Francisco: Ernst and Young.

FLEISING, U. (1989), Risk and culture in biotechnology, *TIBTECH* **7**, 52–57.

FOX, J. L. (1992), Congress okays FDA user fees, *Bio/Technology* **10** (11), 1409.

FRALEY, R. (1992), Sustaining the food supply, *Bio/Technology* **10** (1), 40–43.

FULTON, S. P. (1992), Large-scale processing and high-throughput perfusion chromatography, *Bio/Technology* **10** (6), 635.

GRAHAM, H., INBORR, J. (1991), Enzymes in monogastric feeding, *Agro Food Industry Hi-Tech* **2** (1), 45–48.

GRESSELL, J. (1992), Genetically-engineered herbicide-resistant crops – a moral imperative for world food production, *Agro Food Industry Hi-Tech* **3** (6), 2–7.

GRIFFITHS, E. (1991), Environmental regulation of bacterial virulence – implications for vaccine design and production, *TIBTECH* **9**, 309–314.

HODGSON, J. (1992), The hard reality, *Bio/Technology* **10** (3), 250.

JAFFÉ, W. R., TRIGO, E. J. (1992), Agribiotechnology in the developing world: Issues, trends and policy, in: *Biotechnology R & D Trends: Science Policy for Development*, Conference Proceedings, in preparation.

KELLY, K., et al. (1992), *Business Week*, October 19, 80–84.

KENNEDY, M. J. (1991), The evolution of the world 'Biotechnology', *TIBTECH* **9**, 218–220.

KINOSHITA, J. (1993), Is Japan a boon or a burden to US industry's leadership?, *Science* **259**, 597–598.

KOMEN, J. (1990), IBPGR and genetic diversity, *Biotechnol. Dev. Monitor* **4** (September), 11–12.

KOMEN, J. (1991), Screening plants for new drugs, *Biotechnol. Dev. Monitor* **9** (December), 4–6.

LEOPOLD, M. (1993), The Commercialization of Biotechnology: The Shifting Frontier, *Issues on Commercial Biotechnology*, Vienna: UNIDO.

LIEW, F. Y. (1988), Biotechnology of vaccine development, *Biotechnol. Genet. Eng. Rev.* **8**, 53–95.

MOWERY, D. C., ROSENBERG, N. (1989), *Technology and the Pursuit of Economic Growth*, Cambridge: Cambridge University Press.

OECD (1989), *Biotechnology Economic and Wider Impacts*, pp. 89–90, Paris: OECD.

OECD (1992), *Biotechnology, Agriculture and Food*, Paris: OECD.

PERSLEY, G. J. (1991), *Beyond Mendel's Garden: Biotechnology in the Service of World Agriculture*, CAB International, U.K.

PRESIDENT'S COUNCIL ON COMPETITIVENESS (1991), *Report on National Biotechnology Policy*, Washington, DC.

PRIMO BRAGA, C. A. (1990), The developing country case for and against intellectual property protection, in: *Strengthening Protection of Intellectual Property in Developing Countries: A Survey of the Literature* (SIEBECK, W. E., Ed.), World Bank Discussion Papers, no. 112, Washington, DC.

RATAFIA, M., PURINTON, T. (1988), World agricultural markets, *Bio/Technology* **6** (3), 280–281.

ROBINS, A., FREEMAN, P. (1988), Obstacles to developing vaccines for the Third World, *Sci. Am.* November, 90–95.

SERCOVICH, F. C. (1993), Industrial Biotechnology Policy: Guidelines for Semi-Industrial Countries, *Issues on Commercial Biotechnology*, Vienna: UNIDO.

SERCOVICH, F. C., LEOPOLD, M. (1991), *Developing Countries and the New Biotechnology: Market Entry and Industrial Policy*, Ottawa: International Development Research Centre.

SIHANYA, B. M. (1992), Kenya's pyrethrum export under pressure, *Biotechnol. Dev. Monitor* **13** (December), 11.

UNCTAD (United Nations Conference on Trade and Development) (1991), *Trade and Development Aspects and Implications of New and Emerging Technologies: The Case of Biotechnology*.

U.S. CONGRESS (Office of Technology Assessment) (1988), *New Developments in Biotechnology: U.S. Investment in Biotechnology*, OTA-BA-360, Springfield.

U.S. CONGRESS (Office of Technology Assessment) (1991), *Biotechnology in a Global Economy*, OTA-BA-494, Washington, DC.

U.S. GENERAL ACCOUNTING OFFICE (1989), *Biotechnology: Backlog of Patent Applications*, GAO/RCED-89-120BR.

WAGNER, C. K. (1992), International R & D is the rule, *Bio/Technology* **10** (5), 529–531.

WHO (World Health Organization) (1991), *Press Release/63/17*, December.

WHO (World Health Organization) (1992), *Global Health Situation and Projections: Estimates*, HST/92.2, Geneva.

WORLD BANK (1993), *Marshaling Knowledge for Development: Main Messages and Principal Arguments for the Proposed WDR 1993, Mimeo*.

YUAN, R. T. (1988), *Biotechnology in Singapore, South Korea, and Taiwan*, Great Britain, U.S.A.: MacMillan Publishers.

ZILINSKAS, R. A. (1993), Bridging the gap between research and applications in the developing countries, *World J. Microbiol. Biotechnol.* **9** (2), forthcoming.

13 Biotechnology in the Asian-Pacific Region

ROLF D. SCHMID

Stuttgart, Germany

BONG-HYUN CHUNG

Taejon, South Korea

ALAN J. JONES

Bonn, Germany

SUSONO SAONO

Bogor, Indonesia

JEANNIE SCRIVEN

Braunschweig, Germany

JANE H. J. TSAI

Göttingen, Germany

1 Introduction

The nations which are surveyed in this chapter (Fig. 1) constitute about 1/4 of the world population (three among them – China, Indonesia and Japan – are among the six most populous nations of the world), and they generate 1/5 of the world's gross national products (GNP), with the probability of raising their combined wealth to 1/4 of the world's within a generation. The cultural, religious and historical background of these nations is highly diverse, ranging from the Buddhist/ Confucian traditions in Northern Asia, with their strong focus on education, to the highly heterogeneous traditions of the ASEAN member states where Buddhist, Hindu, Muslim and Christian influences blend. The cultures of Australia and New Zealand, finally,

Fig. 1. Map of the nations analyzed in this survey.

are largely based on British traditions, a notion which explicitly includes their system of education and science.

Only three of the nations described in this survey (Japan, Australia and New Zealand) are members of the Organisation of Economic Co-operation and Development (OECD), and all of them have developed their foundations for science and technology within the past 150 years, or less. Thus, their traditions in "Western"-style science and technology are young, compared to Europe or North America. However, Japan, Australia and New Zealand for long have emphasized the importance of science and technology for their economic growth, and China and smaller regional powers (e.g., Korea, Singapore) are clearly following this lead (Tab. 1).

Biotechnology in all these countries has a long tradition through the development of fermented foods and drinks. Sake and soy sauce in China and Japan, kimchi in Korea and tempé in Indonesia are just a few examples of traditional products obtained through

Saccharomyces, Aspergillus or *Rhizopus* fermentation. As modern biotechnology has provided even more powerful tools such as genetic engineering, cell biology and protein engineering, with huge economic prospects for medicine and agriculture, the economic giant of the hemisphere, Japan, has long turned her attention to the development of "bioindustries". Industry and government alike are concerned with the manufacture of value-added biotechnological products such as drugs, diagnostic kits and specialties ranging from artificial sweeteners to seed varieties improved by tissue culture or genetic engineering. Japanese bioproducts compete on the global market place, and her R&D capacities are second to none. Other countries in the hemisphere are following this lead. Notably, Taiwan, South Korea and Singapore, often dubbed the "little dragons", "little tigers" or, more recently, "little orchids", have established both the industrial and the R&D infrastructure to challenge Japan and other industrialized countries. A third group, notably

Tab. 1. Population, GDP and Status of Science and Technology of Selected Asian-Pacific Nations[a]

	Population (Million, 1991)	Area (1000 km²)	GDP (Bill. US-$, 1991)	GDP per capita (US-$, 1991)	Real Growth Rate (%) 1980–1990	Real Growth Rate (%) 1988–1990	R&D Expenditure (% of GDP, 1990)	Number of Scientists and Engineers, 1990 (×1000)	Number of Scientists and Engineers, 1990 (per 1000 Inhabitants)
PR China	1156.0	9597	372.9	323	7.9	2.4	0.93	397.2	0.3
Indonesia	187.8	1905	116.4	620	4.1	5.6	0.2	125	0.4
Japan	123.9	378	3346.4	27005	3.5	3.8	3.08	484	3.9
Korea (ROK)	43.1	99	283.9	6561	8.9	7.7	1.92	66.2	1.6
Philippines	63.9	300	45.2	720	−1.5	2.4	0.12		
Thailand	56.5	513	79.4	1402	6.9	10.0	0.25	14.0	0.3
Taiwan	20.6	36	179.8	8746	8.1	6.2	1.70	64.0	3.1
Malaysia	17.3	330	41.5	2340	5.7	7.6	1		
Australia	17.3	7713	295.8	17059	1.7	0.7	1.34	42.4	2.4
Hong Kong	5.7	1.1	82.5	14348	5.5	1.5			
New Zealand	3.4	271	43.2	12680	0.6	2.1	2.02	4.9	1.4
Singapore	3	0.6	39.9	14487	5.7	8.0	1.0	4.3	1.4
USA	251.7	9373	5677.5	22468	2.2	0.9	2.74	949.3	3.8
European Union	345.7	2371	4856.5[b]	14938[b]	2.2	3.3[b]	1.99	609.5	1.8
Germany	79.8	357	1574.3	24553	2.2	3.7	2.75	176.4	2.8

[a] As various data sources have been used to compile this table, some of the data may be inconsistent. Any interpretation should thus be limited to qualitative trends.
[b] 1989

Tab. 2. Publications (a) and Patents (b) of Asian-Pacific Nations in Selected Areas of Biotechnology and Environmental Research

Tab. 2a. Publications

Country	Population (Million, 1991)	Scientists and Engineers (1000, 1990)	Publications on Fermentation			Publications on Molecular Genetics			Publications on Food			Publications on Air Pollution		
			1983	1988	1993	1983	1988	1993	1983	1988	1993	1983	1988	1993
PR China	1156.0	397.2	115	265	282	40	199	427	176	376	505	153	253	361
Indonesia	125		–	1	4	–	–	2	6	5	3	–	–	1
Japan	123.9	484	1042	1664	1473	670	1707	3268	1789	2331	2704	1369	1508	1962
Korea	43.1	66.2	36	55	78	8	58	158	120	101	147	9	13	26
Philippines	63.9		4	1	4	–	–	4	31	4	10	1	2	3
Thailand	56.5	14.0	9	3	8	–	4	12	14	8	13	–	2	1
Taiwan	20.6	64.0	16	20	45	3	32	97	63	74	108	13	23	54
Malaysia	18.3		–	1	6	–	–	3	23	10	26	1	5	2
Australia	17.3	42.4	42	11	34	143	293	513	139	112	143	51	59	65
Hong Kong	5.7		1	–	2	–	8	11	1	1	8	2	8	8
New Zealand	3.4	4.9	16	18	10	9	28	84	57	40	72	7	10	18
Singapore	3	4.3	–	4	9	–	2	63	6	4	9	2	10	17
USA	251.7	949.3	570	804	772	3828	8004	12143	2084	1879	2050	2646	2828	2655
Germany	80.0	176.4	308	515	290	391	1020	1581	390	1020	1510	883	1473	1294
World	5389		4133	5648	4656	7313	15980	26984	9619	10031	11248	8415	9952	10184

Search in CAS on-line, sections "Fermentation and Bioindustrial Chemistry", "Biochemical Genetics", "Food and Feed Chemistry" and "Air Pollution and Industrial Hygiene"

Tab. 2b. Patents

Country	Population (Million, 1991)	Scientists and Engineers (1000, 1990)	Patents on Fermentation			Patents on Molecular Genetics			Patents on Food			Patents on Air Pollution Control		
			1983	1988	1993	1983	1988	1993	1983	1988	1993	1983	1988	1993
PR China	1156.0	397.2	–	35	47	–	–	13	–	52	75	–	15	47
Indonesia	125	125	–	–	–	–	–	–	–	–	–	–	–	–
Japan	123.9	484	630	1098	939	70	329	437	569	1210	1371	681	983	1159
Korea	43.1	66.2	1	3	4	–	4	3	–	2	4	–	–	1
Philippines	63.9		–	–	–	–	–	–	–	–	–	–	–	–
Thailand	56.5	14.0	–	–	–	–	–	–	–	–	–	–	–	–
Taiwan	20.6	64.0	–	–	2	–	–	–	–	1	1	–	–	1
Malaysia	18.3		–	–	–	–	–	–	–	–	–	–	–	–
Australia	17.3	42.4	2	11	1	1	22	24	3	8	9	2	–	1
Hong Kong	5.7		–	–	–	–	1	–	–	–	–	–	–	1
New Zealand	3.4	4.9	–	–	–	–	1	1	1	1	–	–	–	1
Singapore	3	4.3	–	–	–	–	–	–	–	–	–	–	–	–
USA	251.7	949.3	159	211	246	87	360	671	226	260	218	135	171	216
Germany	80.0	176.4	80	169	85	11	80	62	11	80	60	139	361	198
World	5389		1142	183	1534	201	957	1519	1160	2084	1997	1132	1807	1858

China and the larger member states of ASEAN (Indonesia, Thailand, Malaysia and the Philippines), has given priority to the potentials of biotechnology for improving public health and food supplies. Finally, the only two OECD members in the Southern hemisphere, Australia and New Zealand, blessed by large stretches of fertile land and a soundly based science and technology infrastructure, have made biotechnology one of their priority areas for the development of indigenous industries. While all nations included in this survey have thus taken care to support the growth of biotechnology industries, the status achieved so far is quite different. This is evident from the number of pertinent scientific publications and patents, as summarized in Tab. 2. The table clearly emphasizes the regional leadership of Japan.

For developing and industrial nations alike, public health and agricultural yields are important targets for the application of the techniques and products of modern biotech-

nology. The position of the Asian-Pacific nations with respect to these areas is indicated in Tab. 3.

These figures show that many countries in the Asian Pacific hemisphere are in a state of rapid transformation from agricultural to industry- and service-based economies. They also indicate the advanced position of the OECD members Japan, Australia and New Zealand, and the economic basis for Korea's aspiration to OECD membership, with regard to the development of their indigenous resources. Clearly, the status of science and technology (S&T) achieved by each nation varies greatly, as will be outlined in the more detailed discussions to follow. In the whole region, however, biotechnology has achieved a high priority status as a key technology for the development of future industries, and the underlying methods for S&T in this area are strongly developed. Genetic engineering plays a key role in this endeavor, as can be seen from the growing number of publica-

Tab. 3. Medical and Agricultural Position of Major Countries in the Asian-Pacific Region

Nation	Population (Million)	Life Expectancy at Birth (Years)		Agriculture's Share in GDP[a] (%)		Agricultural Population (1000, 1990)	Agricultural Area[b] (1000 ha, 1990)	Agricultural Area[b] per Capita (ha)
		1970	1990	1970	1990			
PR China	1156.0	59	70	35	27	768396	96563	0.13
Indonesia	187.8	47	62	45	22	81845	22000	0.27
Japan	123.9	72	79	6	3	7572	4596	0.61
Philippines	63.9	57	64	30	22	29024	7970	0.24
Thailand	56.5	58	66	26	16	33675	22140	0.66
Korea (ROK)	43.1	60	70	26	9	9573	2109	0.22
Taiwan	20.6	60	75	15	4	4262	890	1.00
Malaysia	18.3	62	70	29	~20	5417	4880	0.90
Australia	17.3	71	77	6	4	853	48919	57.35
Hong Kong	5.7	70	78	2	0	68	7	0.10
New Zealand	3.4	72	75	n.i.	10	310	412	1.33
Singapore	3	68	74	2	0	27	1	0.04
USA	251.7	71	76	3	2	6583	189915	28.85
European Union	345.7	71	76	7	6	18530	78180	4.22
Germany	79.8	71	75	3	2	1935[c]	7492[c]	3.87[c]

[a] GDP: gross domestic product
[b] arable and permanent crop land only
[c] before unification with German Democratic Republic

tions and patents in all the nations under survey, and from the establishment of pertinent regulatory systems. However, the regulatory status of genetic engineering varies strongly, while there is standard OECD practice in Japan, Australia and New Zealand, there is nil, or nearly nil, in China and the ASEAN nations.

In summary, the Asian-Pacific region is becoming a strong contender to the development of biotechnology in North America and Europe, and it is desirable and timely to strengthen the links to this region through cooperative projects in industry and the public domain.

2 Biotechnology in Japan

Rolf D. Schmid

While Japan had a late entry in the industrial revolution, the impact of science and technology (S&T) on her social and economic development was very great. Initiated by the government after the Meiji revolution (1866), advanced S&T was introduced from Europe and America by sending students abroad, inviting Western scientists and engineers to Japan and by establishing schools and universities throughout the country. In the post-war period, the educational infrastructure was reshaped, mostly according to US models.

As technology was considered an indispensable factor to compete in the global marketplace, the percentage of GNP (gross national product) invested in S&T rose steadily, from 1.8% in 1970 to 3.05% in 1991. The nature of this activity, however, differs significantly from leading Western nations, as the private sector of Japan spends 81.7% of total S&T expenditure (USA: 56.7%; Germany: 60.4%). Accordingly, basic research in Japan adds up to only 12.9% of total, compared to 15.8% in the USA and 19.3% in Germany (all data for 1991), and has decreased further during the recent economic recession.

Indeed, in post-war Japan, a strong position in applied research and development was considered vital for catching up with real or presumed leads of the Western industrialized nations. On the contrary, results of fundamental research could be easily introduced from foreign countries through the networks of the international scientific community. Eventually, this choice of priority left Japan with a "free rider" image and recently has led her leaders to advocate strong efforts towards a more fundamental and internationalized research system.

Biotechnology can serve to illustrate these post-war trends. A few companies with fermentation know-how, based in the food or pharmaceutical business (e.g., Ajinomoto, Kyowa Hakko, Takeda, Tanabe Seiyaku, Mitsubishi Kasei, Wakunaga Pharmaceuticals, Suntory) had developed a comfortable market niche for fermentation products which, in 1975, comprised over 1% of the GNP (including alcoholic beverages). Major products, besides beer and sake, were antibiotics, glutamate, lysine and chiral organic acids, and innovation was focused on enhancing productivity, e.g., by "metabolic design" (glutamic acid, Kyowa Hakko) or bioreactor technology based on immobilized enzymes (amino acids, Tanabe Seiyaku). Those new techniques, however, which laid the foundation to the "modern biotechnologies", in particular genetic engineering and cell biology, were invented outside Japan. As these developments were fueled by US venture companies, who filed international patents in this area, Japanese industry around 1980 suddenly found itself under a foreign "patent attack" in a highly innovative field (in 1980, 45% of all Japanese patents in genetic engineering were of foreign origin). This led to a concerted action by government, industry and academia, which announced biotechnology as one of the three pillars of future industries. Within a few years, the key methods of the "modern biotechnologies" were then broadly established and led to a "bio-boom" in Japan, which climaxed around 1990, and still is strong: while the recent economic depression caused many companies to focus on quicker return on investments, and R&D budgets dropped by 26.9% in 1992 and 32.1% by 1993, R&D budgets for biotechnology continued to increase by 5% in 1991, by 1% in 1992 and by 11.9% in 1993.

The industry

Today, about 250–300 Japanese companies are actively involved in biotechnology R&D. Apart from pharmaceutical, food and chemical companies, about 40 companies in the heavy industries alone, e.g., steel, metals, cement, petrochemicals, automobile and shipbuilding manufacturers, have set up bio-businesses or bio-related R&D centers, as the Japanese style of company management, especially the seniority system and lifetime employment, is a strong stimulant to diversify: private companies must grow continuously, in order to satisfy the requirements of their young employees for higher salaries and positions.

The largest companies in the field of pharmaceuticals, chemicals and food & beverages, and their research budgets are indicated in Tab. 4.

The market for biotechnological products in 1993 was 606.5 billion yen (retail price), equivalent to 6 billion US-$. Major bio-products are summarized in Tab. 5.

According to a survey of the Ministry of International Trade and Industry (MITI) in June 1993, the bioindustry market is expected to expand to 3 trillion yen in 2000 and 10 trillion yen in 2010 (corresponding to about 30 and 100 billion US-$) – a growth rate far beyond GNP growth.

In 1993, 1962 out of Japan's 9033 researchers involved in genetic engineering worked in companies. According to a survey of the Japan Bioindustry Association (JBA) in December 1992, the roof organization of Japanese bioindustry with a membership of over 230 companies, the industrialization of recombinant DNA procedures comprised 313 cases: 53 drugs, 31 diagnostic agents, 126 chemical agents, 22 recombinant products were registered for industrial production, and 83 applications for 41 different products were filed. At present, pharmaceuticals and diagnostics dominate in the market for bio-products. Recently, however, the development of food and agricultural products based on genetic engineering has gained momentum: as the self-sufficiency of Japan in food consumption is only 47% (calorie-base), the food and agricultural companies depend strongly on

Tab. 4. Top 5 Japanese Pharmaceutical, Chemical and Food Companies, and Their R&D Expenditures in Biotechnology

Company	Total Sales (FY 1993) (Billion Yen)	R&D Expenditure (FY 1993) (Billion Yen)
Pharmaceuticals		
Takeda	562	57.5
Sankyo	396	36.9
Yamanouchi	260	31.6
Shionogi	230	26.8
Fujisawa	235	30.4
Chemicals, Paints, Textiles, Paper		
Asahi Chemical	937	48.0
Mitsubishi Kasei[a]	697→840	40.8→50.8
Sumitomo Chemical	545	38.6
Kao	620	35.2
Toray	529	31.7
Foods and Beverages		
Japan Tobacco	2702	22[b]
Kirin Brewery	1346	16.2
Suntory	800	169[b]
Asahi Breweries	756	n. a.
Ajinomoto	574	20.6

[a] to be merged, by the end of 1994, with Mitsubishi Petrochemicals ("Mitsubishi Chemical Corporation")
[b] Fy (1991)

high-technology innovation in order to compete with foreign suppliers. Whereas the total number of field trials with recombinant plants is still small compared to the United States (more than 300 field trials), there are up to now ten experiments in contained fields in Japan:

Ministry of Agriculture and Fisheries	virus-resistant rice, melon and tomato
Japan Tobacco	virus-resistant tobacco
Rice Breeding Research Institute	low-protein rice
Plantech Research Institute	virus-resistant rice
Mitsui Toatsu	allergen-free rice
Hokkaido Green Bio-Institute	virus-resistant potato
Suntory	virus-resistant petunia
Nihon Monsanto	herbicide-resistant soybean

Tab. 5. Major Japanese Bio-Products

Product	1992 (100 Mill. Yen)	1993 (100 Mill. Yen)
Recombinant Products	**3076**	**4215**
Interferon-alpha	370	835
Erythropoietin	550	630
Growth hormone	500	570
Granulocyte-stimulating factor	400	470
Human insulin	184	200
Detergents with recombinant enzymes	600	800
Recombinant lipase for paper treatment	200	400
Other recombinant products	282	310
Cell Fusion Products	**404**	**453**
Monoclonal antibodies	380	430
Cell Culture Products	**803**	**1397**
Interferon-alpha	390	810
Interferon-beta	204	380
Plant seeds from tissue culture	123	120
Other cell culture products	86	87
Total	**4283**	**6066**

The first commercial food and agricultural products based on recombinant technology are expected to be launched on the Japanese market by 1996/97. Tissue culture techniques and embryo transfer, on the contrary, are already much further advanced: in 1991, 60% of all strawberries and 80% of all carnations were produced using virus-free seedlings generated by tissue culture, and over 7000 calves were born using embryo transfer.

Many companies have built beautiful and well-equipped industrial research centers dedicated to life-science research – a sharp contrast to the usually simple architecture and equipment of research facilities in the public domain. Whereas industrial research centers are often located within the central research facility of the company, science parks and "science cities" have recently attracted a considerable number of Japanese and foreign companies. Notable examples are the Tsukuba Science Parks near Tokyo, where companies such as Eisai, Fujisawa, Kyowa Hakko, Yamanouchi, Takeda, Upjohn and Sandoz have set up bio-related R&D facilities, the Senri Science Park near Osaka, and the Kansai Science City, with major research centers of Bayer, Kanegauchi and others. More R&D centers are and will be built in the 26 "science cities" presently under completion or construction throughout Japan.

There are virtually no venture companies in Japan. As in the USA venture companies play a crucial role in seeding new technology through strategic alliances with larger companies, there have been attempts in Japan to enhance the foundation of "research companies". Since social life in Japan is group-oriented, and the implementation of new methods is traditionally carried out by "research associations" (see below), where up to 40 companies temporarily collaborate, such "research companies" are set up by industrial consortia rather than by individual companies. At present, more than 30 biotechnology research companies are in operation, based on seed money from the government. Their targets cover areas as diverse as biosensors, artificial skin or turf grass. It remains to be seen whether this very Japanese approach will be able to match the high-flying, individualist and aggressive performance of US venture companies.

The government

Japan has been a centrally governed nation through most of its history, and accordingly the central government exerts an important role in shaping the economy. As science and technology in Japan (and throughout Asia) is predominantly seen as an infrastructural measure to enhance the economy, the influence of the government is strong in these fields. However, this is less a "top-down" process than it seems at first glance, since the action taken by the government is strongly influenced in turn by numerous "councils", where leaders from industry, universities, the media and the government consult. The "recommendations" of these councils are in fact the informal basis for legislation and consequent government activities. Similar to all other decision-making processes in Japan, the S&T-related "councils", with their hundreds of sub-organizations, build the country's consensus on where and how to move in the field of science and technology.

The structure of S&T administration by the central government of Japan is indicated in Fig. 2. As can be seen from this chart, the formal coordination of the nation's S&T rests with the Science and Technology Agency (STA), an organization formally headed by the Prime Minister (through the Council for Science and Technology). While STA controls the largest S&T budget among the governmental agencies, several other ministries compete with STA for leadership. Notably the Ministry of International Trade and Industry (MITI), the Ministry of Health and Welfare (MHW), the Ministry of Agriculture, Forestry and Fisheries (MAFF) and the Ministry of Education and Science (Monbusho) (which all have older traditions than the STA), have for long established strong activities in science and technology, and they all operate their own research centers. The biotechnology-related budget of these ministries over the past few years is indicated in Tab. 6.

Apart from establishing suitable regulatory systems, a major task of the ministries, in the past, was "guidance" of "their" industries (e.g., pharmaceutical industry by MHW, heavy industries by MITI, food industries by MAFF). This was, and is, often carried out by forming industrial consortia ("research associations", more recently research companies) (see below). The national research institutes of the ministries, shown in Tab. 6 in the case of biotechnology, form important cornerstones in disseminating new projects and technology. They participate in the start-up phases and provide training facilities for personnel from medium and small enterprises. About 40% of all government scientists working in the national research centers are presently located in Tsukuba Scientific Town, a science city 60 km from Tokyo where STA, MITI and MAFF have moved most of their research institutes; but there is a tendency to spread governmental S&T activities more evenly to "technopolis" cities throughout Japan, in order to counterbalance the ever-increasing conurbation of Tokyo, Osaka and Nagoya.

With increasing internationalization of industry, and with a tendency to global networks in S&T, the Council of Science and Technology in 1992 issued a recommendation to step up the internationalization of Japan in this area. As a result, Japan's governmental research institutes became more open to foreign researchers, and foreign companies started to participate in national Japanese research programs. A major thrust of Japan in this direction is the "Human Frontier Science Program", introduced in 1987. The program is focused on fundamental research in two areas, namely brain research and biological functions, and has a headquarter in Strassburg, France. The majority of the projects and workshops, which are funded in the order of 4 billion yen p.a., is in the favor of research groups outside Japan.

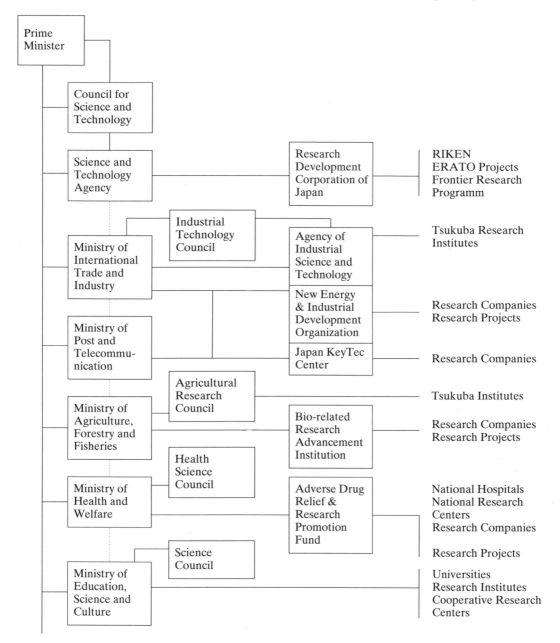

Fig. 2. Structure of S&T in Japan, with emphasis on life sciences.

Tab. 6. Life Science-Related Budgets of Major Japanes Ministries, and Major Programs Relating to Biotechnology

Ministry or Agency	Budget, FY 1993 (Billion Yen)	Major Institutes Working on Biotechnology	Major Programs in Biotechnology
Science and Technology Agency (STA)	28.5	RIKEN (main institute and Tsukuba branch)	Human genome analysis ERATO projects Frontier research programs
Ministry of International Trade and Industry (MITI)	8.4	National Institute of Bioscience and Human-Technology National Institute for Advanced Interdisciplinary Science	Biochips, micromachine technology, glycotechnology, marine biotechnology, genome analysis, biological CO_2 fixation Bioremediation
Ministry of Health and Welfare (MHW)	50.5	National Institute of Health National Institute of Hygienic Sciences National Cancer Center National Hospitals	Genome projects relating to gerontology Anticancer project Gene therapy projects AIDS prevention Food safety
Ministry of Agriculture, Forestry and Fisheries (MAFF)	10.1	Food Research Institute National Institute of Agrobiological Resources National Institute of Agro-Environmental Sciences	Rice genome project Genome studies in agricultural animals Tissue culture for plant and tree propagation Embryo transfer project Biosensors for the food industry
Ministry of Science and Education (Monbusho)	20.7	National Universities, Cooperative Research Centers	Genome projects Cancer-related projects Joint research with industry
Environmental Agency (EA)	2.0	Environmental Research Institute	Environmental impact of biotechnology Release of recombinant organisms Bioremediation
Other ministries	0.5		
Total budget	120.8		

Universities

The Ministry of Science and Culture (Monbusho) is in charge of 97 national universities, with 78 attached research institutes. It also controls the curricula of the 411 private and public universities of Japan. Monbusho without doubt is a strong player in Japanese S&T, but its simultaneous responsibility for the Japanese school system, which in the opinion of

experts waits for urgent reforms, seems to reduce its manoeuvrability in science affairs. Thus, other ministries such as STA and MITI have taken on much of the initiative to improve the conditions for basic research. Examples are: STA's "Frontier Research Program" and ERATO projects; STA's and MITI's "Human Frontier Science Program"; the foundation of a basic research-oriented MITI institute "National Institute for Advanced Interdisciplinary Science" in its research complex in Tsukuba; and the definition of "Centers of Excellence" among national research centers.

The number of scientists leaving their university with a Ph.D. degree is small, in the order of 20%. A science master course is completed, as a rule, at the age of 22, and most graduates at this age enter a company where they are further trained. In special cases, they later return to a university for a doctoral course, financed by their company, in order to learn special techniques and foster the company's relations to a university.

The major national universities are those of Tokyo, Osaka, Kyoto and Fukuoka (Kyushu), which contribute about 50% of all scientific publications originating from Japan. However, biotechnology is a popular research target in many universities, and some have introduced curricula specializing in biotechnology. Recently, 23 "Cooperative Research Centers" have been established throughout Japan, where cooperative projects with local industry are being carried out. However, the politically most important mission of academia in Japan is to provide advice to the government, through the countless committees under the "Science Council" (see Fig. 3). Another key function for information exchange resides in the learned societies, of which there are over 500 in the science field. The largest and most influential society for the biotechnologies is the "Japan Society for Bioscience, Biotechnology and Agrochemistry" (formerly "Society for Agricultural Chemistry"), with a membership of over 11 000 (655 member companies), but numerous others, e.g., the "Japanese Biochemical Society" (over 10 000 members), the "Society of Fermentation Technology" (2500 members), the "Protein Engineering Society" (500 members), contribute to form consensus about the positioning of Japan in S&T.

Associations, research associations and research companies

A unique feature of S&T in Japan is the importance of industrial research consortia. In fact, opinion polls among industry leaders repeatedly showed that the organization of collective development of enabling technologies was considered one of the pre-eminent responsibilites of the government – more important than even tax deductions. This attitude contrasts sharply to the Western business community where individual approaches to competitiveness are clearly favored.

Associations. In the field of biotechnology, MITI, later also the MHW and MAFF, started to organize associations as early as 1980. Their function is similar to the US lobbies. They become involved in initiating and harmonizing regulations and standards, providing training, disseminating information and acting on needs as diverse as organizing exhibitions at home and abroad, or enhancing public perception.

The major biotechnology associations are indicated in Tab. 7.

Research associations. The function of a research association is to introduce or jointly develop new industrial methods through a temporary industrial consortium. As in 1993 more than 100 research associations had been formed in Japan, it can thus be claimed that RAs have played an important role in government-organized technology changes. Established by MITI in 1968, and emulated by other ministries later, these research consortia have covered fields as diverse as improved jet engine construction, deep-sea manganese nodule mining, 5th generation computers or biosensors for the food industry. A selection of research associations important for the field of biotechnology and their industrial participants is summarized in Tab. 8.

While these and other research associations generated a wealth of information, and certainly helped to rapidly disseminate new methods throughout the industrial research community, there were drawbacks such as the question of intellectual property rights (usually the rights belong to the government – thus JITA, the pertinent body of MITI, owns more than 20 000 patents). Recently, a new format of industrial R&D consortia was organized.

Research companies. Organized again by the "guiding" ministry, but heavily co-financed by a consortium of at least 3, sometimes over 100 firms, these companies are typically founded for a lifespan of 7 to 10

Tab. 7. Major Biotechnology Associations in Japan

Name of Association	Abbreviation	Guiding Ministry	Type of Industry	Number of Participating Companies
Japan Bioindustry Association (formerly Bioechnology Development Center)	JBA (formerly BIDEC)	MITI	Heavy industry, chemicals, fermentation, energy, pharmaceuticals	about 230
Japan Health Science Foundation	JHSF	MHW	Pharmaceuticals, food	about 180
Society for Techno-Innovation of Agriculture, Forestry and Fisheries	STAFF	MAFF	Food, agriculture, fisheries	about 160

Tab. 8. Some Japanese Research Associations Relating to Biotechnology

Name of Research Association (RA) Financial System	Main Target	Organizing Ministry	Number of Participating Companies	Duration
RA for biotechnology	Bioreactors, large-scale cell technology, genetic engineering Functional protein complexes	MITI	14	1980–90 1988–98
Aqua Renaissance '90	Bioreactor and biosensor technology for compact sewage plants	MITI	21	1984–90
"Bioremediation Network Laboratory"	Ex-situ microbial treatment of polluted soil from industrial sites	EA	13, plus 7 foreign companies	1990–99
RA for food bioreactor	Value-added carbohydrates, proteins and lipids through immobilized biocatalysts	MAFF	55	1982–88

EA = Environmental Agency

years. They are usually located in a building specially constructed for this purpose. Financial support is not limited to one ministry, but includes various national and local bodies, e.g., the Japan Development Bank or the Hokkaido Development Board. Intellectual property is shared among member companies. Some examples are indicated in Tab. 9.

The project management and capitalization of research companies and research associations in the field of biotechnology is carried out by the following organizations:

STA: a) Research Development Corporation of Japan (RDCJ)
       b) Frontier Research Program
MITI: a) New Energy and Industrial Technology Development Organization (NEDO)
       b) Japan Key-Tec Center (jointly with Ministry of Post and Telecommunication)

MHW: Adverse Drug Sufferings Relief and Research Promotion Fund (ADSRRPF)
MAFF: Bio-oriented Technology Research Advancement Institution (BRAIN)

A typical financial scheme, for BRAIN, is shown in Fig. 3. Nearly the same systems apply for the project management bodies of the other ministries.

While most of the research associations and research companies are focused on industrial targets, there is an increasing number of cooperative projects in the basic sciences. The STA has taken an early lead in this area, through its ERATO and Frontier Science Projects.

ERATO projects. Started as a program in 1981, the format of these projects relies on a project coordinator of high scientific reputation, who obtains laboratories and generous financial support for a limited period (usually

Tab. 9. Japanese Research Companies Active in the Field of Biotechnology

Name of Company	Participating Industry	Main Target	Duration (years)	Guiding Ministry
Protein Engineering Research Institute, Osaka	13 companies	Establishment of protein engineering techniques	10	MITI, through Japan Key-Tec Center
Research Institute for Environmental Technology for the Earth (RITE), Kyoto	60 companies	Global CO_2 reduction, bioreactor technology, biodegradable polymers etc.	10	MITI, thorugh NEDO
Marine Biotechnology Institute, 3 locations	24 companies	Biochemicals of marine origin	10	MITI, through NEDO
Drug Delivery Systems Institute	7 companies	Targeted drug delivery systems	7	MHW, through ADSRRPF
Institute for Advanced Skin Research	3 companies	Artificial skin systems	7	MHW, through ADSRRPF
Biosensor Laboratories Co., Ltd.	4 companies	Optoelectronic sensors for *in-vivo* monitoring	7	MHW, through ADSRRPF
Seatechs Co.	3 companies	Aquaculture or large fish	5	MAFF, through BRAIN
Turf Grass Co.	4 companies	Recombinant herbicide-resistant turf grass	7	MAFF, through BRAIN

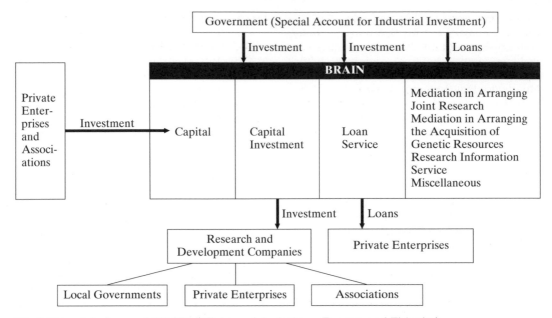

Fig. 3. Financial scheme of BRAIN (Ministry of Agriculture, Forestry and Fisheries).

5 years) to investigate basic questions. His group is composed of usually about 30 scientists, with delegates from universities, research centers and companies. At present, there are 18 ERATO projects, and 15 more have been completed. The annual budget is around 3 billion yen. Projects relating to biotechnology are:

Frontier Research Program. Established in 1986 at the RIKEN research institute of the STA in new laboratories built for this purpose, this program is internationally open and includes foreign project leaders. The annual budget is about 2 billion yen. The program presently comprises

Fusetani Biofouling Project	bioadhesives
Okayama Cell Switching Project	mechanism of cell growth and division
Nagayama Protein Array Project	2D protein crystals
Torii Nutrient-Stasis Project	impact of nutrition on behavior and brain
Ikeda GenoSphere Project	introns in the human genome
Mizutani Plant Ecochemicals Project	ecologically active plant metabolites
Hotani Molecular Assembly Project	bacterial flagella aggregation mechanisms
Inaba Biophoton Project (completed)	ultraweak light emissions in biology
Horikoshi Superbugs Project (completed)	extremophilic microorganisms

Plant homeostasis research	photoreception, signal transduction, plant hormones
Glycobiology research	glyco cell biology, glyco molecular biology, glyco technology
Frontier materials research	nano-electronics materials, nano-photonics materials, exotic nanomaterials
Brain mechanisms of mind and behavior	neural information processing, neural networks, neural systems
Photodynamics research (Sendai)	sub-millimeter waves, photophysics, organometallic photodynamics, photobiology

Contextual measures. Regulations

Japan is a nation with a patriarchical history. Thus, Japan has evolved into a strongly regulated country, but at the same time citizens' and consumers' protection are vital issues. Consensus building among fractions is, on the other hand, a time-consuming and often difficult routine of Japanese life. In the field of genetic engineering, the US lead, as issued 1983 in the form of the NIH guidelines, was closely followed, but the rivalries among ministries, who exerted "guidance" for their industries or universities proper, has led to a number of diverse ministerial guidelines concerning genetic engineering, which only in 1991 were unified to a national guideline (issued by the STA). This guideline is very similar to the pertinent US guidelines. Attempts of the Environmental Agency, in 1991, to issue genetic engineering legislation were rejected.

Patents

Japan is one of the constituting parties of the "Budapest Convention" which regulates the deposit of microorganisms and other forms of life in the framework of patent applications.

Depositories

The main depository for microorganisms (over 8000 strains) is the Japan Collection of Microorganisms (JCM), located at STA's RIKEN research institute, which also serves as the world data center for the World Federation for Culture Collections (WFCC). There are separate depositories for patent strains at the MITI National Institute of Bioscience and Human Technology in Tsukuba, for cell lines at the Monbusho National Institute of Genetics in Mishima, and for germ plasm at MAFF's gene bank at the National Institute of Agrobiological Resources in Tsukuba. In addition, there are several privately organized depositories, the largest (over 8000 strains) being the Institute of Fermentation in Osaka (IFO), originally founded by the Takeda Company.

Information networks

For general S&T information, the most important system is JICST (Japan Information Center for Science and Technology), an organization under the STA which, through a national network connected to overseas hosts in Europe and the USA, provides about 8 million documents on-line. JICST is also preparing to become the national host for genome sequence data from national research centers and companies.

Present tendencies. Genome sequencing

Compared to the USA, Japan had a late entry into genome sequencing. A major project was carried out at STA's Tsukuba Life Science Center of RIKEN, where several companies such as Hitachi, Seiko and Mitsui Knowledge Industry collaborated in the development of high-speed robots for gene sequencing. Since 1993, governmental activities for genome sequencing have been greatly enhanced. Major projects are summarized in Tab. 10.

Global environment projects. As Japan is located in the windfall of emerging Asian industrialized nations such as Korea and China, acid rain and other forms of air pollution have become a major concern to the nation. During recent years, Japan has taken a range of steps to counterbalance these effects by political and technological measures.

Tab. 10. Major Japanese Projects in Genome Sequencing

Ministry	Project
STA	Human Genome Program, JICST Genome Database
Monbusho	Research projects at university level
MITI	Center of DNA Analysis at Yokohama: Genome of industrially relevant microorganisms
MAFF	Rice genome program, *Arabidopsis* genome, various vegetables and fruits
MHW	Human genomes with relevance to an ageing population (e.g., Alzheimer)

Documentation. The National Institute of Science Policy, an organization under the STA, continuously collects data on SO_x, NO_x and CO_2 in Asia. The National Aeronautics and Space Development Agency (NASDA), also under STA, will launch its first ADEOS satellite in 1995, with the goal to carry out continuous surveillance of environmental parameters in Asia from space.

"New Earth '21" Program. Based on a report of the Environmental Agency in May 1992, the program focuses on a range of measures to promote environmentally friendly technologies and to absorb environmental CO_2. The central facility for this program is the Research Institute for Environmental Technology for the Earth (RITE), located in Kansai Science City near Kyoto, where a consortium composed of a large number of industries, under MITI guidance, carries out research on biodegradable polymers, improved bioreactor technology, and enhanced chemical or biological CO_2 assimilation.

International Center for Environmental Technology Transfer (ICETT). Founded in 1990 through an initiative of Yokkaichi city and MITI, and strongly supported by ODA funds, this institute has the goal to train 10000 engineers and technicians from developing nations in environmental technology. So far, workshops and training courses abroad have reached about 1000 engineers and technicians, mainly from and in the developing Asian nations.

International Tropical Resource Institute. The main target of this project is the conservation of biological diversity through the establishment of reginal centers for tropical resources, with a strong financial commitment from Japan.

3 Biotechnology in the Republic of Korea

Bong-Hyun Chung and Rolf D. Schmid

Located between two powerful neighbors, China and Japan, Korea has managed over twelve hundred years to keep her independence, except for brief interruptions during the height of the first Mongol empire in the 13th century and the annexation by Japan in the 20th century. While Korea is part of a broader East Asian culture centered on China, she has managed to keep her national identity and has made significant contributions to the culture of mankind, e.g., through the invention of printing technology with movable metal types in the 12th century. After half a century of extreme disarray, following the annexation through Japan in 1910 and the Korean war of 1950–53 which split the country in two nations, the Republic of Korea (ROK, "Korea") entered a period of rapid economic growth. Today, the Korean economy is larger than that of all but eight OECD countries, and is the thirteenth largest export nation in the world. Some pertinent statistical data are summarized in Tab. 11.

During its post-war growth, the Korean economy has become a manufacturing and service-oriented economy in 1990, only 18.3% of the GNP was achieved by agriculture, 27.3% by the manufacturing industries and 54.4% by the service sector. Whereas the four leading export items in 1961 were iron ore, tungsten, raw silk and anthracite (40% of total export), these were, in 1991, electronic products, textiles/garments, steel products and ships (60% of total exports). Whereas until 1990 the USA was the major export market, followed by Japan and the EEC, the exports to other Asian nations have sharply risen since 1990, and in 1992 contributed about 1/3 to the overall value.

Korea's period of rapid economic growth started with a population that was well-educated by the standards of most developing countries. As early as 1960, primary school attendance was almost universal (95%). By 1990, 35% of the population had received a high school education. The number of students in four-year colleges and universities nearly doubled during the 1980s. By 1990, the share of college graduates in the adult population had doubled from 6 to 12% compared to 1980. Standardized international tests suggest that the quality of education in Korea is high, even in comparison to OECD countries. The Korean government is further emphasiz-

Tab. 11. Demographic and Economic Data of Some Northern Asian Countries

	South Korea (ROK)	North Korea (DPRK)	Japan	China
Population (1000, 1990)	42.9	21.8	123.6	1156.0
Area (1000 km^2)	99	120.5	378	9597
Inhabitants per km^2	432	181	327	120
GNP (billion US-$, 1990)	242	23.2	2940 4	373
GNP per capita (US-$, 1990)	5659	1064	23800 1	323
Average annual volume growth over previous 5 years	9.7	−0.5%	9	6.8
Exports as % GDP (1990)	26.6	3.5	9.8	21.6
Rural population (% of population, 1990)	27.9	40.2	3.0	39.8
Total fertility rate (births per woman)	1.8	2.3	1.7	2.5

ing the role of higher education, e.g., through offering free middle and high school education, and through changing school curricula in favor of science and mathematics. At the university level, subsidies will be provided to the private sector colleges (which account for 75 percent of total enrolment) as a means to raise the proportion of science and engineering graduates to 55% of the total (1992: 40%). Another form of government intervention in the labor market is the requirement that private enterprises with more than 150 employees either train their own employees or pay a levy (0.67% of their payroll on average) to the government. When the system was introduced in 1977, 33% of the firms opted for the levy – in 1992, this figure had increased to 80%. Out of 2700 companies, 500 train their own employees, while 2200 firms, mainly smaller ones, have opted for the training levy. The Korean government spends an additional W 170 billion (about 0.1% of the GDP) for this purpose.

The industry

In the period after the Korean war, the government started industrialization based on proven technologies at their mature stage, taking advantage of a well-trained but low-waged labor market. At this point, large firms, *chaebols,* were created as an instrument to allow an economy of scale and in turn to develop "strategic industries" as leaders for export and the economy in general. The government helped strongly in the capitalization as well as in subsequent diversification of these trusts, but, in contrast to other countries who also promote big businesses, exercised strict discipline over these chaebols by penalizing poor performers and rewarding only good ones. Thus, the government repeatedly refused to bail out large-scale, badly managed, bankrupt firms in otherwise healthy industries, and instead selected better managed chaebols to take them over. After maturation of the Korean economy, however, the government now follows a policy of "economic democratization", promoting small and medium-sized industries, in particular technology-based firms. By this move, it is intended to remedy the imbalance between large and small enterprises. However, the combined sales of the five largest chaebols (Lucky-Goldstar, Hyundai, Daewoo, Samsung and Sunkyung) increased from 12.8% of GNP (1975) to 52.4% (1984).

A number of these trusts became involved in the fermentation business, which started by producing fermented foods and drinks such as soy sauce, beer and rice wine, and later advanced to the production of amino acids, enzymes and antibiotics. The products of tradi-

Tab. 12. Biotechnological Products of Korea, 1980

Product	Value 1980 (1000 US-$)	Major Producer
Alcoholic beverages	1 286 000	Oriental Brewery Co., Jin Ro Ltd., Chosun Brewery Co.
Fermented soy-bean products	32 800	Samyang Rood Co., Whayoung Co.
Kimchi	34 300	Doosan Farmland Co.
Dairy products	168 500	Hankuk Yakult Milk Products, Bing-grae Co., Seoul Dairy Cooperative
Amino acids	77 100	Miwon Co., Cheil Sugar Inc.
Antibiotics	167 700	Cong Kun Dang Corp., Dongmyong Industries, Yuhan Chemicals
Mushrooms	57 100	
Starch sugars	51 400	Sun Hill Glucose Co.
Enzymes	14 300	Pacific Chemical Industrial Co., Dong-A Pharmaceutical Co.
Biological products	18 600	
Vitamins and organic acids	700	Shinhand Milling Co.
Total	1 906 500	

tional biotechnology in 1980, before the advent of genetic engineering in Korea, are summarized in Tab. 12.

In 1982, after the successful operation of genetic engineering-based venture companies in the USA, 19 companies joined to form KOGERA, the Korea Genetic Engineering Research Association. The member companies and their interests in the area of biotechnology are summarized in Tab. 13.

The professional manpower of KOGERA increased from 64 (7 PhDs) in 1982 to 415 (48 PhDs) in 1990, and investments rose from about 5 billion to 54 billion Won (about 80 million US-$) in the same period.

Simultaneously, Korean companies invested in US ventures or produced and/or sold biotechnological products under license from US companies. Tab. 14 gives some examples.

The value of imported recombinant pharmaceuticals was 16.6 billion Won (about 23 million US-$) in 1991, as compared to sales of 34.7 billion Won (about 47 million US-$) on domestic fermentation products, mostly of non-recombinant origin.

The government

The Korean government is interventionist by nature and only recently has liberalized to some extent. The *Presidential Council for Science and Technology* is the supreme advisory body to the president. It has 11 members, including representatives of universities, government research institutes, and private industry. The present chairman is SHANGHI RHEE, former minister of science and technology. The council meets every other week and its chairman reports to the president on a monthly basis. The General Committee of Science and Technology, which coordinates activities of the various ministries, is chaired by the prime minister. The vice chairman is the minister of the Economic Planning Board. A total of 14 ministries are represented in the committee. The structure of S&T, pertinent to biotechnology, in the Korean government is summarized in Fig. 4.

The *Science and Technology Policy Institute (STEPI)*, affiliated to KIST under the Ministry of Science and Technology, has recently been strengthened and in the Korean language is now called the *Science and Technology Policy and Management Institute*. STEPI is responsible for managing the *Highly Advanced National Project (HAN)*, an inter-

Tab. 13. Member Companies of the Korea Genetic Engineering Association (KOGERA)

Company	Sales (Won Bill. 1992)	No. of Employees	Major Product Lines	Focus in Biotechnology R&D
Cheil Foods & Chemicals (Samsung group)	1264.8	5 5306	Sugar, condiment, pharmaceuticals	Interferon, EPO, hepatitis B vaccine, oligosaccharides
Chong Kung Dang	101.2	1459	Pharmaceuticals	Antibiotics, EPO, cyclodextrin
Cungs's Foods			Soy milk, barley flakes, juice	Functional foods, bioconversion
Dong-A Pharmaceutical	188.3	2049	Chemotherapeutic agents, enzymes, diagnostics, antibiotics	Hepatitis B vaccines, diagnostic kits, CSF, glutathione
Dong Shin Pharmaceutical	38.5	576	Plasma fractions, antibiotics, vaccines,	Vaccines, TNF
Hangkuk Yagurt			fermented milk, instant food	Lactic acid fermentation, renning, anticancer agents
Il Dong Pharmaceutical	34.9	987	Pharmaceuticals, vitamins, intermediates	Humanized, monoclonal antibodies, anticancer agents, *B. thuringiensis* toxin, lactic acid
KOHAP	660.4	1971	Nylon yarn, polyester yarn	Biopolymers, EPO, fermenters
Kolon Industries	775.6	5476	Nylon textures, polyester films	Cancer diagnostics, steroids, biopolymers
Korea Explosives			Explosives, fine chemicals	Antibiotics, *B. thuringiensis* toxin
Korea Green Cross	127.9	1196	Microbial products, plasma fractions, diagnostics	Hepatitis B vaccine, diagnostic kits, EPO
Lucky	2096.0	11 718	Plastics, petrochemicals, specialty chemicals, pharmaceuticals	Interferon, CSD, human growth hormone
Miwon	415.2	3647	Condiments, amino acids, polyester resins	Amino acids, recombinant phenylalanine, sweeteners, pigments
Oriental Brewery (Doosan group)	460.2	1969	Beer, wine, food processing	Yeast improvement, embryo transfer, virus-free hop
Pacific Chemical	451.0	5190	Cosmetics, food, enzymes	Enzymes, hyaluronic acid, collagen, cyclodextrin
Yuhan	127.7	1539	Pharmaceuticals, veterinary drugs, cosmetics	Collagen, human gonadotropin, proteases
Yukong	4710.9	5838	Petroleum, petrochemical products	Biopolymers, amino acids, wastewater treatment
Yung Jin Pharmaceutical Industries	98.1	1238	Antibiotics, cosmetics, drinks	Microbial products

Tab. 14. Korean–USA Ties in the Field of Biotechnology (Selection)

Korean Company	US Partner or Licenser	Venture/Product	Year of Start/Sale
Lucky Ltd.	Chiron Corp.	Lucky Biotech Corp.	1984
Cheil Food & Chemcials	Eugenetech International		1984
Korean Green Cross	Alpha Omega Labs.		1987
Miwon, Inc.	Princeton Biomeditech. Corp.		1988
Daewong Lilly Pharm.	Eli Lilly	Insulin, human growth hormone	1987, 1989
Korea Essex	Schering Plough	α-Interferon	1987
Korea Green Cross	Serono	β-Interferon	1991
Cilag Korea	Cilag International	Immune repressor, EPO	1988, 1990
Dong-A Pharm	Merck	Hepatitis B vaccine	1990
Korea Abbott	Abbott	Diagnostic kits for hepatitis B, various cancers	1988–90
Dong Shin Pharmaceuticals	Abbott	Pregnancy test kit	1988
Kolon Pharmaceuticals	Centocor	Diagnostics for colon cancer	1988
Lucky Ltd.	Quidel	Ovulation/pregnancy test	1989
Kang Nam Industries	Ecostar International	Biodegradable plastics	1992

ministry project (see below), and other national projects, and has a direct reporting channel to the Ministry of Science and Technology.

The funding of biotechnology has grown strongly in the past 10 years, as summarized in Tab. 15.

The most comprehensive program for S&T, including biotechnology, is with the *Ministry of Science and Technology (MOST)*. It was established in 1967 with the objective to drive scientific research programs in government research institutes, support university research, and assist commercial R&D in industries. It is maintained, however, that MOST's long-term science and technology policies were largely ignored by the ministries shaping industrial policies, and were not integrated into national development plans. Due to government priorities and an industry structure focused on the cost-efficient production of mature products invented abroad, in

turn-key plants engineered and imported from the industrialized nations, there was little demand for Korean efforts in R&D (in 1970, there was just one corporate R&D center in Korea). Aggressive recruitment of overseas-trained Korean scientists and engineers, or tax incentives and preferential financing for corporate R&D activities faltered due to the easy means of acquiring and assimilating foreign technologies then available from many sources. In spite of these drawbacks, MOST managed to set up 21 research institutes; its most important research facility in biotechnology is the *Genetic Engineering Research Institute (GERI)* (see below). When the economic environment changed in the 1980s, through protectionism in Western export markets, rapid wage rises and competition from lower wage neighbors, the time became ripe for a major shift in technology policy. In 1987, MOST adopted a "Long-term Science and Technology Development Plan

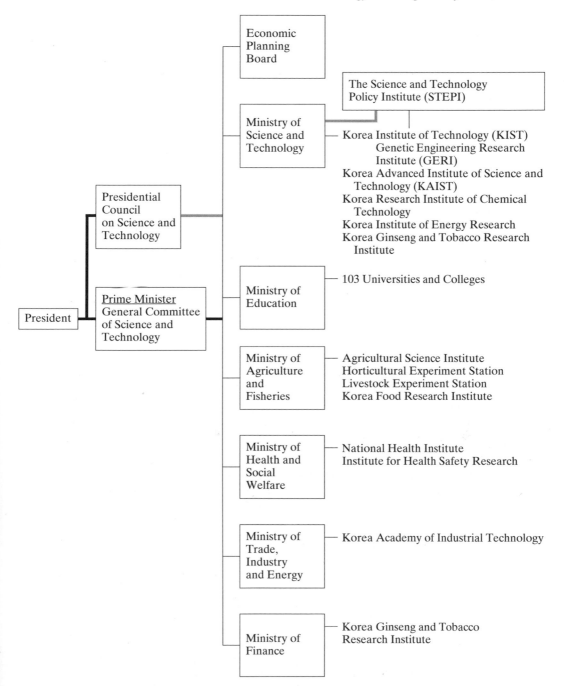

Fig. 4. Structure of S&T in Korea, with emphasis on life sciences.

Tab. 15. R&D Investments in Biotechnology by the Korean Government and Industry, 1982–1990 (Million Won)

	1982	1984	1986	1988	1990
Ministry of Science and Technology (MOST)	70	860	1660	2420	7370
Ministry of Agriculture and Fisheries (MAF)	0	480	520	890	1170
Ministry of Education (ME)	0	50	600	930	1290
Ministry of Health and Social Welfare (MHSW)	0	20	110	120	120
Total Government	70	1410	2890	4360	9950
Korean Traders Association (KTA)	0	0	20	30	0
Private Industry	3150	12130	16400	25630	35690
Grand Total	3220	13540	19310	30020	45640

Toward the Year 2000", defining biotechnology as one of the priority areas. Launched in 1992, the *Highly Advanced National Project (HAN)* aims at the development of competitive R&D capabilities in eleven critical technologies, one of which is for new functional biomaterials (biotechnology). The program has been dubbed "G-7 project", because its goal is to catch up with the advanced technology of Japan and leading Western nations by 2001. The Korean government and industry plan to spend more than 6000 million US-$ on the project over the next ten years. This plan also sets the goal for Korea to share 2% of the global biotechnology markets by 2001. Within the biotechnology sector of the HAN, MOST adopted the following R&D programs:

- Screening technologies for biomaterials (cell regulating factors, bioactive lead structures)
- Renovating technologies for biomaterials (protein engineering, transgenic animals and plants, new enzyme technology)
- Process technology for biomaterials (biodegradable polymers, recombinant proteins, new sweeteners and lipid food, cell culture and bioreactor technology)

The supervising institutions are:

Cell Regulating Factors	Genetic Engineering Research Institute (GERI)
Bioactive Lead Compounds	GERI
Protein Engineering	GERI
Transgenic Plants and Animals	GERI
New Enzyme Technology	GERI
Biodegradable Polymers	Korean Institute of Science and Technology (KIST)
Recombinant Proteins	GERI
New Sweeteners and Lipid Food	Korea Food Research Institute (KFRI)
Cell Culture and Bioreactor	The Korean Genetic Engineering Research Association Technology (KOGERA)

The budget proposed for these targets is provided jointly by the government and by industry, as shown in Tab. 16.

MOST is an important funding agency for the universities as well, through its Korea Science and Engineering Foundation (KOSEF). In 1991, KOSEF supported 7 biotechnology-related Science and 3 Engineering Research Centers in the universities (total funding: 5.95 million US-$), and 55 biotechnology-related research projects (0.46 million US-$).

The *Ministry of Agriculture and Fishery (MAF)* is focused on the support of agriculture and the food industry. The Office of Ru-

Tab. 16. R&D Investment for Biotechnology within the HAN Project (Million Won)

	1992	1993	1994–1997	1998–2002	2003–2007
Government (MOST)	3840	5001	111 000	260 000	790 000
Industry	2326	3074	366 000	790 000	1 650 000
Total	6166	8075	477 000	1 050 000	2 440 000

1 US-$ = 800 Won

ral Development in the MAF maintains 10 research institutes, with a biotechnology budget of about Won 50 000 million (about US-$ 62.5 million) for biotechnology-related R&D in 1993.

The *Ministry of Education (ME)* is in charge of the 103 Korean universities, and responsible for basic research and manpower training. It is also a major funding source for the universities, through the Korca Research Foundation and the Korea Higher Education Foundation.

The *Ministry of Health and Social Welfare (MHSW)* has two research institutes under its jurisdiction and provides grants for basic research in the universities.

Whereas the *Ministry of Trade, Industry and Energy (MTIE)* was not yet directly involved in governmental S&T policy, due to the functions of MOST, it successfully established, in 1989, its *Korea Academy of Industrial Technology* and, concomitantly with industry's growing emphasis on S&T, is beginning to take a more active role in future technology policies.

National Research Institutes. Under the administration of the Ministry of Science and Technology (MOST), the leading national research institution is the *Korea Institute of Science and Technology (KIST)*. In 1985, the Genetic Engineering Promotion Law provided the basis to establish at KIST the *Genetic Engineering Research Institute (GERI),* located since 1991 at Daeduk Science Town, Taejon. GERI's staff is over 300 (about 90 Ph.D.s), and the budget for 1993 was US-$ 20.4 million. In its 9 divisions, all fields of molecular genetics, protein chemistry, protein engineering, microbial genetics and ecology, enzyme engineering and fermentation are being addressed. It also functions as the official culture collection laboratory of Korea.

Further institutes of MOST involved in some areas of biotechnology R&D are:

Korea Advanced Institute of Science and Technology (KAIST). Founded in 1971 by the Ministry of Science and Technology (MOST) as a graduate university, this is one of the most prestigious graduate schools in Korea and considered to be Korea's "Center of Excellence", especially in the area of science and engineering. It is the only university in Korea that is not under the control of the Ministry of Education. Its total staff are 318 professors, 2031 Ph.D. students and 1223 M.S. students, who are well funded: research contracts in 1992 totaled Won 22 billion (about US-$ 30 million, or on average just under US-$ 100 000 for each faculty member). In its three departments related to biotechnology (Dept. of Biotechnology, Dept. of Life Science and Dept. of Chemical Engineering), there are 24 professors who are focused on biotechnology. KAIST runs a Bio-Process Engineering Research Center (BPERC), supported by KOSEF.

Korea Research Institute of Chemical Technology. It had an applied biology division focused on the development of new bioproducts and materials. Recently, it became a part of GERI.

Korea Atomic Energy Research Institute (KAERI) at Daeduk Science Town. This was founded in 1973 as the principal research institute of atomic energy, and runs a Radiation Biology Department Engineering Laboratory with 7 staff members.

The *Korea Institute of Energy Research* at Daeduk Science Town has a biomass research division with 6 staff members.

The *Korea Ginseng and Tobacco Research Institute* at Daeduk Science Town carries out several projects in genetic engineering. It is managed by the Ministry of Finance.

The Ministry of Agriculture and Fisheries (MAF) operates two institutes in Suwon which carry out major projects in biotechnology: the *Agricultural Science Institute* has a staff of 200 (50 Ph.D.s), and the *Horticultural Experiment Station* totals 139 coworkers, including 20 Ph.D.s.

Nine other research organizations under the MAF are also involved in some form or other of biotechnology-related R&D, e.g., the *Livestock Experiment Station* in Suwon (total 335 employees, 31 Ph.D.s).

Under the Ministry of Health and Social Welfare (MHSW), the major research institute in the life sciences is the *National Health Institute* in Seoul. In its wide range of programs, about 20 are related to biotechnology, ranging from AIDS to the development of monoclonal antibodies. The *Institute for Health Safety Research* in Seoul specializes in testing new medical and pharmaceutical products, including drugs developed by recombinant technology.

Universities

The Republic of Korea has 103 universities and colleges, most of which have some departments related to biotechnology, e.g., biology, biochemistry, food science or agricultural chemistry. In addition, 24 universities have recently added new biotechnology-related departments such as Departments of Genetic Engineering etc. Seoul National University (SNU) is considered the best university in Korea. An extremely small number of students gain admission to SNU every year after fierce competition. In the three science and engineering-related colleges of SNU (engineering, natural science and agricultural and life science), 49 out of the 449 professors, and 135 out of the 1098 Ph.D. students are focused on biotechnology. Recently, SNU renamed its Department of Zoology into Department of Molecular Biology. Currently, 18 universities maintain genetic engineering research centers, which are usually staffed by members from several departments. They received total government grants of 1.37 million US-$ in 1990. Lately, some universities have created inter-university research centers in order to promote collaborative research in this area on the regional and national level. The number of scientists and engineers in the life sciences, employed at Korean universities, is summarized in Tab. 17.

Associations

KOGERA, the Korean Genetic Engineering Research Association founded in 1982, was mentioned already. In 1991, under the guidance of the Ministry of Science and Technology (MOST), 50 companies from the pharmaceutical, chemical, food and environmental business plus 5 government research centers, including the Genetic Engineering Research Institute (GERI), formed the *Bioindustry Association of Korea (BAK)*. The mission of BAK is to promote the development and industrialization of biotechnology, to participate in the foundation of industry, to collect and disseminate information, and to improve international cooperation. The companies associated within BAK are indicated in Tab. 18.

Plant Biotechnology. Korea has established several projects in the field of plant biotech-

Tab. 17. Number of Scientists and Engineers in the Life Science, Genetic Engineering and Biotechnology at Korean Universities

Field	1987	1988	1989	1990	1991	1992
Science	295	398	420	411	425	438
Engineering	172	198	230	238	223	280
Medical Science	227	235	266	268	246	280
Agriculture and Fishery	125	130	144	160	150	147
Total	819	961	1050	1077	1034	1145

Tab. 18. Member Companies and Institutes of the Bioindustry Association of Korea (BAK) (1992)

B. Braun Korea	Korea Green Cross Corp.
Boryung Pharmaceutical Co.	Korea Green Pharmaceutical
Cheil Sugar Inc.	Korea Institute of Industry and Technology Information
Cheil Synthetic Textiles Co.	mation
Cho Seon Pharmaceutical and Trading Co.	Korea Research Institute of Chemical Technology
Choung Kun Dang Corp.	Korea Zinc Co.
Coong Ang Chemical Co.	Kwang Dong Pharmaceutical
Dae Woo Engineering Co.	Lucky Ltd.
Dae Woong Pharmaceutical	Miwon Co., Ltd.
Dong Bank Corp.	Miwon Foods Co.
Dong Shin Pharmaceutical Co.	Nhong Shim Co., Ltd.
Dong Suh Foods Corp.	Oriental Chemical Industry
Dong-A Pharmaceutical Co.	Ottugi Foods Co.
Doo San Technical Center	Pacific Chemical Co.
Dr. Chung's Food Co.	Pohang Iron and Steel Co.
Genetic Engineering Research Institute	Pulmoun Foods Co.
Green Cross Medical Equipment Corp.	Sam Yang Co.
Han Kuk Yakult Milk Products	Sampyo Foods Co.
Han Mi Pharmaceutical Ind.	Shinhan Scientific Co.
Honam Oil Refinery Co.	Soo Do Chemical Co.
Hoong A Engineering Co.	Sun Hill Glucose Co.
Il Dong Pharmaceutica Co.	Tainan Sugar Industiral Co.
Jindo Industries, Inc.	Tong Yang Nylon Co.
Jindo Ltd.	Woo Sung Co.
Julia Ltd.	Yu Kong Ltd.
Kohap Ltd.	Yhuan Chemical Co.
Kolon Industries, Inc.	Yuhan Corp.
Korea Explosive Co.	Yungjin Pharmaceutical Ind.
Korea Fermentor Co.	

nology. Clonal propagation methods were introduced as early as the mid-1960s, and recombinant DNA technology in plants was studied since the mid-1980s. Ornamental crops constitute a major field of industrial application, and at present 1–2 million flowers (mainly *Gerbera*, carnations and *Gypsophila*) are commercially produced per year via micropropagation. Vegetables such as potatoes and green onions constitute another field of industrial application, and it is estimated that over 15 million potato microtubers are presently produced. Transgenic rice, bermuda grass, hot pepper, potatoes and many other vegetables, e.g., watermelon, Chinese cabbage and Ginseng, have been studied. Leading research centers in this area are (a) the *Plant Molecular Biology and Biotechnology Research Center* at Gyeongsang National University, (b) the *Agricultural Biotechnology Institute* at the Rural Development Administra-

tion of MAF, and (c) *the Bioresources Research Group* at GERI.

Contextual measures. Patents

Korea introduced a patent system in 1946, and joined the Budapest convention in 1988. The national depository for microorganisms, plasmids etc. is the Korean Collection for Type Cultures (KCTC) at GERI, Taeduk Science Town. In addition, there are two major depositories, namely the Korean Culture Center of Microorganisms (KCCM) and the Korean Cell Line Research Foundation (KCLRF), both in Seoul. The number of biotechnology-related patents (ICP classification C12) issued in Korea increased from 111 (1984–1986) to 267 in 1992; about half of this number was filed by foreign companies or their subsidiaries. About half of all Korean patents on biotechnology were filed by just 5 companies and institutes: Lucky Ltd. (33 applications in 1992), Cheil Sugar Inc. and GERI (11 each), Miwon Co. and Pacific Chemical Co.

Genetic engineering regulations

So far, there are no regulations such as recombinant DNA guidelines for the biotechnology industry. However, the Ministry of Health and Social Welfare is preparing a draft of such regulation which will soon be issued.

Research information networks

At Daeduk Science Town and at Seoul, two major scientific information centers have been installed and linked to other national and international services through KREO-Net. One is the Korea Institute of Industry

Technology Information (KINITI) in Seoul, which is managed by the Ministry of Trade, Industry and Energy and services mainly the information on trade and industry. The other is the Korea Research and Development Information Center (KORDIC) at Daeduk Science Town supported by the MOST, which services the information on the scientific R&D.

4 Biotechnology in the People's Republic of China

Jane H. J. Tsai

Mainland China (Fig. 5) has an area of 9.6 million km^2 and 1.2 billion inhabitants. Due to the large number of people, the Chinese government has always been under pressure

Fig. 5. Map of the People's Republic of China.

Tab. 19. History of Biotechnology R&D in China

Stage	Period	Major Targets and Products
1	before 1950	Traditional fermentation products such as soy sauce, sake etc.
2	1950–1978	Antibiotics, vitamins, amino acids, enzymes, plant tissue culture
3	1978–1990	• establish new biotechnological techniques through joint ventures with foreign institutes • train scientists abroad • establish CNCBD (1983)
4	1990–	Focus on • genetic engineering • cell biology • modern fermentation procedures

to organize food supplies and medical care. Since the "open door"- and the "market economy" policies were proclaimed in 1987 and 1993, respectively, China's economical growth rate has increased several fold and ranked top of the global list last year. At present, the Chinese government is increasingly combining its efforts to commercialize results of biotechnology R&D with its "open door" policy towards foreign investors.

Historically, biotechnology in China can be grouped into four major stages, as illustrated in Tab. 19.

Based on a report of the State Science and Technology Commission (SSTC), the market for biotechnological products in 1990 is summarized in Tab. 20.

China is the world's largest producer of antibiotics with an annual output of over 10000 tons. There are over 40 enzyme manufacturing factories with an annual output of over 2000 tons.

In 1986, R&D on biotechnology gained official priority status in China's government planning. Scientists, local and international entrepreneurs looking for business in the biotechnology field are presently backed up by a new funding system and by extensive support for technology transfer. China's huge domestic market attracts worldwide attention. Despite difficulties imminent in its socialist political system, China's genetic engineers have recently managed to achieve the following results:

• Genetically engineered wheat was planted on 65000 ha of land.
• From 1986 to 1992, the cultivated area of hybrid rice species occupied about 127 million ha, resulting in a harvest of 200 billion kilogram rice, which represented an increase of 5–10% in yield.
• A genetically engineered strain of *Alfalfa* grew on 10000 ha of land, with a harvest increase of 1620 tons.
• TMV-resistant tobacco, cultivated in an area of 13000 ha, provided the government with RMB 400 million in income tax, equivalent to 9–10% of the total national income tax.

Tab. 20. Market of Biotechnological Products in China, 1990

Product	Annual Production (Tons)	Sales Value (Million RMB)
Traditional biotechnological products	–	6600
Antibiotics	10000	300
Amino acids	75000	–
Enzymes	2000	–

RMB = Ren-Ming-Bi, equivalent to Chinese Yuan

- Transgenic calves and fish were successfully bred.
- Recombinant vaccines for hepatitis-B and hemorrhage fever, and α-interferon and interleukin-2 were commercialized.

The government

Starting from 1980, biotechnology was identified to be one of the top high-tech priorities in China, and any business associated with biotechnology was entitled to a number of government incentives. Major government programs include

- incentives on reduced income tax (reduction to 15% in certain economic zones and coastal cities, in comparison to 24% elsewhere), exemption of income taxes for the first three years of production,
- the establishment of a patent system ("intellectual property protection law"), and
- the foundation of "science-based economic zones" (SEZ).

Moreover, the government has permitted foreign companies to hold between 25% and 100% of equity in Chinese companies. China has devoted considerable capital investments for the development of biotechnology. Government funds are available in three forms: grants, loans and equity, for financing both domestic and foreign investments. Funding for R&D in this particular field has increased by a factor of 25 during the past ten years.

The attempts to liberalize the national market led to an enhanced dependency of China's new biotechnology start-ups on the industrialized countries. Most orginal research still remains in agriculture, but biomedical research and protein engineering have been defined as new targets in the government research policy.

In March 1985, the government announced several programs to improve R&D. These were:

- "7.5 Plan" (1985–1990)
- "8.5 Plan" (1990–1995)

- "863 Plan" (since 1986)
- "Spark Plan" (since 1985)
- "Torch Plan" (1988–1990).

Among these plans, the "863 Plan", also known as the "High Technology, Research and Development Program", has medium- and long-term goals, aiming at the frontiers of global high-tech development and adapting the principle of "limited target but strong focus" in selected subjects. About 3000 Chinese scientists and technicians, and more than 200 institutions with about 200 projects, were involved in the programs and plans mentioned above. In addition, the *National Science Fund* provided RMB 100 million for scientific research, of which 30% is used for innovations in biotechnology. The funding of the "Torch Plan" is either derived from banks, or is provided through grants from CNCBD, resulting in a budget of RMB 300–500 million annually.

In the framework of this program, the following institutions were established:

China National Center for Biotechnology (CNCBD) in Beijing, established in 1983. The center is not only China's top maker of biotechnology policy, but also the authority in reviewing research grants.

China–EC Biotechnology Center (CEBC) in Beijing, founded in 1991; it is organized in a similar way, and acts as the link between China and EU countries, being jointly managed by CNCBD and the DG-XII of the European Commission. It is responsible for the co-operation and implementation of all biotechnology-related international scientific cooperation activities.

China New-Investment Enterprise Company (CIEC), established by the government in 1987, and affiliated to the State Science and Technology Commission (SSTC) and the People's Bank of China. CIEC is one of the very few companies which will grant loans in foreign currency and provide equity for new businesses.

Shanghai Research Center of Biotechnology, Chinese Academy of Science (CAS), was founded in 1991 by the government. The government investment for the center and its pilot plant, affiliated to the Institute of Microbiology of the CAS in Beijing, was about US-

$ 15 million. The center gives priorities to genetic engineering of microorganisms and cell biology, as well as enzyme engineering. The center has established a gene bank.

China United Biotechnology Corporation (CUBC) was approved in 1992 and is administered by the *State Science and Technology Commission (SSTC)*. It is an independent enterprise for biological science and technology development and is registered at the China Industrial Commercial Administration Bureau. Its role is to link scientific research and industrial applications. CUBC collaborates extensively with bioindustries at home and abroad. Its major business is to a) handle biotechnology trade, b) establish biotechnological and related industries by joint ventures, c) organize exchange and cooperation in the field of biotechnology, d) engage in indenture and consignment of domestic and overseas bioproducts, e) act as an agent of overseas manufacturers, and f) provide consultancy and services relating to bio-products and biotechnology trading.

The economic development of China is centrally controlled by the political system. If one wishes to enter into an academic or commercial relationship in China, it is thus highly important to find the appropriate Chinese partner in the pertinent administration unit. This is not always simple due to the decentralized organization of the State Council. The relationship between government authorities and research institutions relevant to biotechnology is schematically illustrated in Fig. 6.

On the national level, top researchers are presently encouraged to serve as employees or consultants to enterprises. Top researchers now often offer their expertise to foreign companies, which results in an increased influence of top-class scientists.

National State Key Laboratories and Science Economic Zones. As a measure to improve on biotechnological research, the so-called *State Key Laboratories* were recently established (Tab. 21).

In order to provide an appropriate infrastructure for both national and foreign investments, central and local governments have created several so-called *Science Economic Zones (SEZ)* or *High-Tech Science Parks*

(HTSP), which have a semi-autonomous political and economical status. The best-known SEZ or HTSP are listed in Tab. 22.

Industrial activities: joint ventures and newly established biotechnological companies

Joint ventures with foreign partners are encouraged by the government. Representative joint projects are shown in Tab. 23.

In order to promote commercialization of biotechnology in China, the CNCBD, several research institutes and universities have jointly set up several high-tech enterprises which are listed in Tab. 24.

Regulations

The Chinese Patent Office was established in 1984 and came into function in 1985. Safety protocols for genetic engineering and biotechnology R&D were prepared by the SSTC and came into effect in December 1993. These regulations establish guidelines for safe pilot plant and laboratory research work, and for the safe manufacture and field release of GMOs. They also cover the transportation, destruction and disposal of GMOs and all related materials. In order to implement these biosafety regulations, the China National Biosafety Commission (CNBC) was established, and functions under the direct leadership of the Chinese State Council. The secretariat of the CNBC, which deals with the daily administration of the Commission, is located in the National Center for Biotechnology Development (CNCBD).

The rules stipulate that any living GMO waste must be treated or destroyed prior to its entering the environment, and that all experimental records must be readily available for a period of ten years. Regulations cover all genetically engineered, or DNA manipulated plants, animals and microorganisms. Traditional methods of hybrid breeding, chemical fertilization and tissue culture are not included in these guidelines.

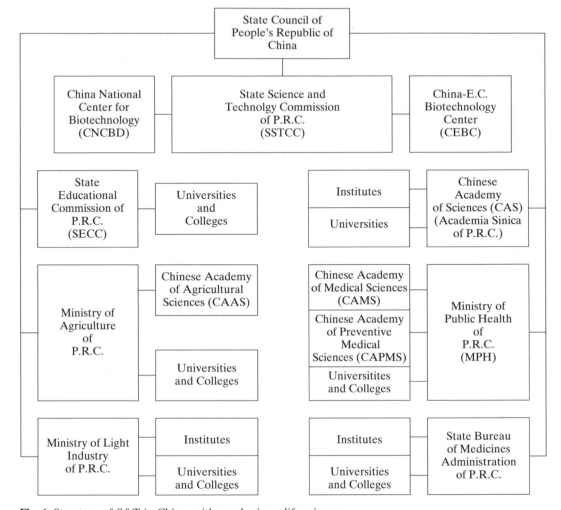

Fig. 6. Structure of S&T in China, with emphasis on life sciences.

Environmental Protection Law

China is planning to spend 0.7% of its annual GNP in order to improve the environment. In fact, the Chinese people have paid very little attention to this problem so far.

Conclusion

Although the "open-door" policy is an important step to stimulate biotechnology R&D in China, there are also serious drawbacks for the industrialization of results, e.g., there is a danger of unnecessary and overlapping investments in new areas of biotechnology R&D despite the limited national market. Moreover, the strong links between the Chinese political system and economical development could hamper the commercialization of biotechnology. Problems such as the lack of financial resources, the slow transfer of technology at the national level, and the lack of well-trained researchers must be taken into serious consideration.

Tab. 21. "National State Key Laboratories" Related to Biotechnology

National State Key Laboratory	Affiliated Institutions	Location	Government Authority
Plant Molecular Biology and Genetics Laboratory	Shanghai Plant Physiology Institute	Shanghai	CAS
Division of Fish Genetics and Breeding	Institute of Hydrobiology	Hubei	CAS
Plant Cell and Chromosome Laboratory	Institute of Genetics	Beijing	CAS
Biochemical Engineering Laboratory	Institute of Chemical Metallurgy	Beijing	CAS
National Laboratory of Protein and Plant Genetic Engineering	Peking University	Beijing	SEC
Laboratory of Genetic Engineering	Fudan University	Shanghai	SEC
Laboratory of Enzymology	Jilin University	Changchun	SEC
Biomembrane and Membrane Biotechnology Laboratory	Institute of Zoology; Peking University; Qinghua University	Beijing	CAS/SEC
Laboratory for Biological Safety	Zhongshan University	Guangong	SEC
Medical and Pharmaceutical Biotech Laboratory	Nanjing University	Nanjing	SEC
Laboratory for Agricultural Ecology	Lanzhou University	Lanzhou	SEC
Laboratory for Biosensor Research	East China University of Chemical Technology	Shanghai	SEC
Laboratory of Fermentation Technology	Shandong University	Jinan	SEC
Laboratory of Viral Genetic Engineering	Institute of Virology, Chinese Academy of Preventive Medicine	Beijing	MPH
Laboratory of Natural Pharmacopoeia	Beijing Medical University	Beijing	MPH
Laboratory for Cancer Genetics	Shanghai Cancer Research Institute	Shanghai	MPH
Molecular Oncology Research Laboratory	Institute of Oncology, Chinese Academy of Medical Sciences	Beijing	MPH
Laboratory for Experimental Hematology	Institute of Hematology, Chinese Academy of Medical Science	Tianjin	MPH
National Major Laboratory of Medical Genetics	Hunan Medical University	Changsha	MPH
Laboratory of Molecular Biology	Institute for Basic Medical Research, Chinese Academy of Science	Beijing	MPH
Laboratory of Medical Neurology	Shanghai Medical University	Shanghai	MPH
Laboratory for Nuclear Medicine	Jiangsu Provincial Institute for Nuclear Medicine	Wuxi	MPH
Laboratory of Agricultural Biotechnology	Beijing University of Agriculture	Beijing	MA
Laboratory for Crop Genetics	Huazhong University of Agriculture	Wuhan	MA
Laboratory for Plant Insects Pathology	Institute of the Chinese Academy of Science	Beijing	MA
State Key Biotech Laboratory for Tropical Crops	Institute of Huanan Tropic Crops	Hainan	
National Veterinary Biotechnology Laboratory	Harbin Veterinary Research Institute, Chinese Academy of Agricultural Science	Harbin	MA
The National Laboratory of Fish Sperm Resources and Biotechnology	Chinese Academy of Fishery Science; Changjian Fishery Research Institute	Shashi	MA

CAS   Chinese Academy of Science
SEC   State Educational Committee
MPH   Ministry of Public Health
MA   Ministry of Agriculture

Tab. 22. Science Economic Zones/High-Tech Science Parks (Selection)

Beijing High-Tech	Beijing	Nanin	Guangxi
Wuxi High-Tech	Jiangsu	Zhenghou	Henen
Nanking High-Tech	Nanjing	Shijiazhuang	Hebei
Shenyang	Liaonin	Lanzhou	Gansu
Dalain	Liaonin	Pudong	Shanghai
Tienjin	Hebei	Caoheijing	Shanghai
Chengdu	Sichuan	Hainan	Guandong
Chonggin	Sichuan	Qanming	Yuannan
Weihai Torch Park	Shandong	Taiyuan	Shaanxi
Qindu	Shandong	Ulumugi	Xinjiang
Jinan	Shandong	Baotou	Inner Mongolia
Zhongshan Torch	Guandong		
Guanshou	Guandong		
Shenzhen	Guandong		
Zhuhai	Guandong		
Changchuan	Jilin		
Harbin	Heilongjiang		
Dachin	Heilongjiang		
Changsha	Hunan		
Fuzhou	Fujan		
Xiamen Torch	Fujan		
Hefei	Anhui		
Hangzhou	Zhejiang		

Tab. 23. Joint Venture Projects and Their Partners

Project	Foreign Partner	Subject/Chinese Partner
Hepatitis-B vaccine	Pharmacia (Sweden)	Large-scale preparation of HB vaccine
Rice genome project	Rockefeller Foundation (USA)	16 projects within "863 plan". US-$ 1.5 million per year support, plus training of young Chinese scientists abroad
ELISA test kit	Biochemical Systems (Austria)	Atomic Energy Institute
Hepatitis-B vaccine	Merck, Sharp & Dohme (USA)	Recombinant HB vaccine, with Beijing Institute of Biological Products and Kang-Xin Biotechnology Corporation, Shenzhen
Blood substitute (hemoglobin)	International Hemoglobin Technology Division (Canada)	Shanghai Research Center of Biotechnology, Shanghai
Various Chinese drugs (Hismanal, Vermos, Motillium etc.)	Johnson Pharmaceutical Comp. (Belgium)	Proxince Shanxi Pharmaceutical Co., Xian
Pharmaceuticals	Abbott Co. (USA)	Ninbo Abbott Biotechnology

Tab. 24. China's New Biotechnology Companies

Company	Location
China United Biotechnology Corp.	Shenzhen, Guandong
Kexing Biological Products Corp.	Shenzhen, Guandong
Sino-Tech Products Corp.	Shenzhen, Guandong
New-Life Pharmaceutical Corp.	Shenzhen, Guandong
Kangtai Biological Products	Shenzhen, Guandong
Natural Instant Health Products Co.	Shenzhen, Guandong
B.S. Medical and Biochemical Co.	Shenzhen, Guandong
Weko Bio-tech Ltd.	Shenzhen, Guandong
Membrane Separation & Biotechnology Co.	Shenzhen, Guandong
Tian Wang Intern. Pharmaceutical Co.	Hainan, Hainan
North China Pharmaceutical Corp.	Hebei
Changchung Institute of Biological Products	Changchun, Jilin
Wuxi Enzyme Factory	Wuxi, Jiangsu
Sino-American Biotechnology Co.	Loyang, Henan
East China Biotechnology Engineering Co.	Shanghai
Yuan Dong Biotechnology Co.	Shanghai
Huxain Biotechnology Co.	Shanghai
Beijing Biotechnology Co.	Beijing
Hubei Institute of Biotechnology	Wuhan, Hubei

5 Biotechnology in the Republic of China (Taiwan)

Jane H. J. Tsai

Introduction

The Republic of China in Taiwan has an area of 36000 km² and twenty million inhabitants. With an average growth rate of 9% since 1952, the island has been very prosperous and has accumulated foreign reserves of US-$ 980 billion. Most companies in Taiwan are small to medium size in comparison to the industrialized nations. Expansion of research and development has therefore become one of the main responsibilities of the government: undertaking joint ventures with foreign companies and research institutions were the main aims of the government at an early stage. However, recent rise in labor costs, shortage of available industrial land, substantial appreciation of currency and strong pressure for environmental protection are rendering the traditional manufacturing business no longer competitive. Development of high technology has therefore raised the hope for the nation to maintain the rapid economic growth and to become one of the industrialized nations by the year 2000. High-tech oriented biotechnology has been identified by the government and the business sector since 1982 as one of eight strategic areas of S&T.

Biotechnology business opportunities are entitled to a number of incentive programs created by the government: they include tax exemptions for up to five years, government loans to subsidize 50% of the cost of the development for industrial products, duty exemption on imported machinery and materials for the purpose of manufacturing, and market information service. Consequently, the government has created several institutions to handle R&D and to work with the industrial sector in order to promote bioindustry. Among these institutions are the *Biomedical Institute* and the *Molecular Biology Institute* at Nankang which were established for basic research and for training; both are fully supported by government funds. The development of industrial processes is carried out mainly at the *Development Center of Bio-*

technology *(DCB)* in Taipei, and by the *Union Chemical Laboratories (UCL)* of the *Industrial Technology Research Institute (ITRI)* at Hsinchu. Both institutions are non-profit organizations and are operated semi-privately. Their funding sources can be either obtained from government agencies or from the private sector. In addition, the *Food Industry Research and Development Institute (FIRDI)* at Hsinchu has been developed into an excellent R&D center and the major depository for microorganisms and plasmids.

There are at least twenty academic laboratories, medical schools or agricultural research centers which are involved in biotechnological research. All national universities and their research centers are fully supported by the government. Among them, *Academia Sinica* leads the field. In 1994, Nobel Laureate Professor Y. C. LEE returned from the USA and became head of the institute. Other universities such as *National Taiwan University* at Taipei, the *National Tsing-Hua University* at Hsingchu, the *National Chung-Hsing University* at Taichung and the *National Cheng-Kung University* at Tainan also play an active part in biotechnogical research.

In 1980, the *Science-based Industrial Park (SIP)* was established in Hsinchu by the National Science Council (NSC). By 1992, nearly one hundred high-tech companies had been set up within SIP benefiting from various incentive packages for S&T-related investments.

In addition to these efforts of the government, the private sector has established thirteen biotechnology-related venture capital firms since 1986. The cumulative initial equity of these funds meanwhile exceeds US-$ 250 million. In 1990, the government announced that private companies may apply for public R&D funding on a competitive basis. Such policy has resulted in many new research laboratories and investments in private companies. Since 1992, the output of the "bioindustry" has been US-$ 27 billion p.a., 17% of total industries. This production figure comprises traditional industries of agriculture, food, pharmaceutical and specialty products.

Industrial biotechnology R&D efforts in government-supported institutions have been organized into six major programs, namely

- pharmaceutical products
- agricultural products
- specialty chemicals
- biological agents used in pollution control
- biosensors, and
- strain collection and improvement.

Combined R&D expenditures of industry and the government in the agricultural and medical sciences were in excess of US-$ 300 million in the FY 1989-90. 33% of this budget went to the "upstream" research bodies, mainly the universities and Academica Sinica; 50% went to government and quasi-government laboratories ("midstream"); and 17% went to "downstream" institutions, predominantly private and public enterprises. The government funded 86%, and industry 14% of R&D.

The industry

Taiwan is a leading manufacturer of fermentation products such as glutamate and antibiotics. The combined annual sales of fermentation products and biological products was estimated to be US-$ 2500 million and 100 million, respectively (1992). Monosodium glutamate alone accounts for US-$ 300 million per year in sales, generated by 4 production plants of 3 companies. The brewing and sugar industry, a major driving force for new biotech investments, amounts to over 80% of the total sales. Sales figures are summarized in Tab. 25.

The most important companies related to biotechnology are those based on traditional fermentation technologies, as outlined in Tab. 26.

The total market value of biotechnological products was estimated by the *National Science Council* to be in the order of US-$ 46 million in 1987, with an increase to US-$ 186 million in 1990 and US-$ 807 million in 1996.

Recently, several companies focused on the "new biotechnologies" were set up. They are listed in Tab. 27.

Tab. 25. Sales Figures of Fermented Products and Biological Products, 1988–90

Type of Product	Sales (Billion NT-$)			Number of Manufacturers
	1988	1989	1990	
Agricultural products	313.5	329	314	70084
Food and related goods	336	352	371	3101
Speciality chemicals	31	33.5	36.3	454

Tab. 26. Taiwanese Companies Selling Biological or Fermentation Products

Company	Sales (US-$ Million) 1989 and Products	
China New Pharmaceutical Co.	20	Gluconic acid, clinical enzymes
Kaohsiung Chemical Co.	3	Animal vaccines
Pan Lab, Inc.	13	Strain improvement service
Polyamine Corp.	10	Urokinase, bromelain, selenium yeast
President Enterprise	613	Soy sauce, yoghurt
San-Fu Chemical Co.	33	Citric acid
Ta-Tung Enterprise	1151	Cyclodextrin, enzymes
Tai-Tung Enterprise	6	Urokinase
Taiwan Cyanamid	20	Tetracyclines
Taiwan Green Cross	3	Immunoglobins, factor VIII
Taiwan Serum Co.	4	Animal vaccines
Taiwan Sugar Co.	685	Lysine, *Torula* yeast, ethanol
Taiwan Tobacco and Wine Monopoly Bureau	2915	Beer, rice wine, distilled spirits
Ve-Dan Corp.	176	Glutamate, *Chlorella,* blue-green algae
Ve Wong	173	Glutamate, soy sauce, cephalosporin
Wei-Chuan Food	288	Milk products, glutamate, soy sauce, yoghurt

Tab. 27. Taiwanese New Biotechnology Companies

Company	Year Founded	Number of Researchers	Value of Assets (US-$ Million, 1989)	R&D Budget (NT-$ Million)
Tai-Fu Pharmaceutical Co.	1982	6	6.0	4–5
Tai-Da Pharmaceutical Co.	1982	4	1.5	4
General Biologicals Corp.	1984	10	12.0	10
BGH Biochemical Co.	1984	3	0.5	3
Life Guard Pharmaceutical Inc.	1984	5	28.0	–
Ever New Biotech Co.	1984	10	8.0	5
Search Biol. Technology Group	1987		1.5	–
King Car Biotech Co.	1988	3	1.5	–
Ming-Sheng Pharm. Technology Inc.	1988	7	1.5	3–5
Taiwan Biotech Inc.	1989	5	3.0	10
Hwa-Yang Pharm. Technology Inc.	1989	3	1.5	10–20
Grand Biotech. Corp.	1989	2	5.0	10
Total			70.0	

Tab. 28. Member Companies of BIDEA

USI Far East Corp.	Yung Shin Pharmaceutical Ind. Co., Ltd.
Eternal Chemical Co., Ltd.	Development Center for Biotechnology
Y. F. Chemical Corp.	Ve Wong Corp.
Taiwan Pharmaceutical Industry Association	Wei Chuan Foods Corp.
King Car Biotechnology Industrial Co., Ltd.	Chui Shui Che Foods Mfg. Co., Ltd.
Taitung Enterprise Corp.	China Chemical Synthesis Industrial Co., Ltd.
Tung Hai Fermentation Industrial Corp.	Food Industry R&D Institute
Industrial Technology Research Institute	Cyanamid Taiwan Corp.
Shinung Corp.	Crossbond Corp.
President Enterprise Corp.	Hanaqua International Corp.
Lifeguard Pharmaceutical Inc.	Polyamine (Taiwan) Corp.
Taiwan Monosodium Glutamate Manufacturers	Standard Chem. & Pharm. Co., Ltd.

However, exports of high-tech biotechnological products at present play only a minor role in Taiwan's trade.

In 1989, the *Taiwan Bioindustry Development Association (BIDEA)* was formed for the promotion of Taiwanese bioindustry. The 26 member companies of BIDEA are summarized in Tab. 28.

BIDEA includes several public research institutions and some 200 individual members. The main task of BIDEA is to carry out research and investigations into biotechnology and bioindustry, to gather information on a national and international scale, to participate in educational measures, hold symposia, to plan and promote large-scale projects, through interaction with policy-makers, and to enhance international cooperation. BIDEA has close contacts to JBA in Japan.

The government

The government's role in promoting the development of biotechnology is illustrated in Tab. 29.

The government has strongly stimulated R&D in the field of biotechnology. The pertinent budgets for the period of 1988–1992 are indicated in Tab. 30.

The largest part of the budget has been spent through the *Development Center of Biotechnology (DCB)*, the major governmental research institute for R&D in biotechnology. DCB is located on the campus of the National Taiwan University in Taipei, Taiwan's foremost university. It occupies a total space of about 10000 m^2, including a pilot plant of 700 m^2 which is among the most advanced pilot plants in the world. DCB is operated by 300 employees, of whom 15% hold Ph.D. degrees, 70% M.S. degrees, and 12% with B.S. degrees. The research areas of DCB are summarized in Tab. 31.

In addition to funding R&D at universities and research centers, Taiwan's government has initiated a range of measures for the promotion of technology-based industries.

- In 1983, the *Industrial Development Bureau (IDB)*, a body under the Ministry of Economic Affairs, began a subsidiary program for new products and processes. Government funds were made available to cover the cost of development as well as the interests on capital borrowed to finance a project; such funds are provided by the Executive Yuan (Cabinet), private industry and royalties from successful projects, ranging from 1% to 4% of the profits. In addition, a technology-oriented company may claim for tax exemption on production equipment, which usually accounts for up to 10–20% of the total investment. Once production of a high-technology product has begun, the company can claim exemption from any tax payment on this product for five consecutive years. As a result of these measures, the maximum rate of taxation for a technology-based company can be as low as 22%, as compared to 33% for other types of business.

Tab. 29. The Framework for the Promotion of Biotechnology in Taiwan

		Basic Research	Applied Research	Technology Development	Commer-cialization
Promotion	Government	Academia Sinica, Science Council (NSC)			
			National Council of Agriculture Department of Health		
Execution	Universities, research institutes	Academia Sinica			
			Agriculture Experimental Research Institute		
		University institutes			
	Non-profit organizations Industry Hsinchu Science Park	DCB ITRI, FIRDI			
				Public and private industries	

NSC	National Science Council
MOEA	Ministry of Economic Affairs
MND	Ministry of National Defense
MOC	Ministry of Communication
DCB	Development Center for Biotechnology
ITRI	Industry Technology Research Institute
FIRDI	Food Industry Research and Development Institute

Tab. 30. Budget for Governmental R&D in Biotechnology (NT-$ Million)

Organization	1988	1989	1990	1991	1992
Academia Sinica	20	22	24	26	92
National Science Council	42.5	42.5	55	60	200
Council of Agriculture	40	42	44	46	172
Ministry of Health	20	20	25	35	100
Development Center for Biotechnology (DCB)	270	280	290	290	1136
Total (NT-$)	392.5	506.5	538.0	557.0	2000
Total (US-$)	12.7	16.3	17.4	18.0	64.5

1 US-$ = 30.77 NT-$

- The *Bank of Communication,* Taiwan's official development bank, provides up to 70% on loans in order to finance high-tech investments in the medium- and long term.
- The *China Development Corporation* is a privately run organization and also offers low-interest loans.
- The government further assists new high-tech companies by providing up to 49% of equity and granting up to 25% ownership to its founders for the use of new technology.
- The statute of investments by foreign nationals permits the repatriation of profits from approved investments. Capital can be repatriated at a rate of 20% per year after the start-up.
- Venture capital organizations have recently been admitted to Taiwan. They

Tab. 31. Research Areas of the Development Center of Biotechnology (DCB), Taipei

Division	Area of R&D	Current Projects
Molecular biology	Gene cloning, gene expression on large scale, structure of nucleic acids	HBV recombinant vaccines, animal vaccines, NPC diagnostic kits, hepatitis C diagnostic kits, DNA probes
Microbiology	Industrial microbiology, continuous bioreactor design, environmental microbiology, bioremediation	Strain improvement, microbial insecticides, industrial enzymes, solid-state bioremediation
Cell biology and immunology	Antigen-specific T-cell clones, human T-T hybriclones and lymphokines, humanized antibodies	T-cell culture and cloning, mouse-mouse-B-cell hybridomas, bispecific humanized antibodies
Biochemistry	Protein engineering, enzyme immobilization, enzyme reactor design, biosensor technology, amino acid derivatives and peptide products	Aspartam production by enzymatic synthesis, amino acids via enzymic conversion, design and synthesis of growth hormones, piezoelectric biosensors
Agricultural biochemistry	Plant tissue culture technology, biological control of plant diseases and pests, new pesticide technology, diagnosis of agricultural pollutants	Evaluation and development of microbial pesticides, transgenic plants, diagnostic systems for plant diseases
Process development	Large-scale cell culture technology, process design and economic evaluation, pilot and production plant engineering	Recombinant hepatitis B vaccine, bioinsecticide and biopesticide production process, mammalian cell and hybridoma production on large scale, pilot plant services
Information services	Updated information about biotechnology R&D for the Taiwanese bioindustry	Market analysis, strategic analysis for product development, databases, technology transfer
Pharmaceutical R&D	New chemical lead structures for pharmaceuticals, bulk and generic drugs R&D	Antibiotics, anticancer, anti-viral and anti-hypertensive agents, drug safety and metabolism, formulation

can invest abroad, under the condition that they must transfer the technology back to Taiwan.

Regulations

In 1987, the government introduced a major revision of the *patent law*. Under this regulation, foreign firms are able to file suits against patent infringements and request penalties for the violation of patent laws. *Genetic engineering regulations* date back to September, 1989. They are very similar to the pertinent regulations of the USA. In order to reduce environmental pollution, the cabinet in 1987 established the *Environment Protection Administration (EPA)* directly under the cabinet. The EPA is responsible for introducing

legislative proposals and implementing national laws designed to protect the environment and natural resources. R&D of the EPA has been conducted through grants, cooperative agreements and research contracts with universities and other private institutes. The total budget (including government protection plans) was NT-\$ 61.5 billion (approx. US-\$ 2.46 billion) in 1990. The estimated investments for the environment from FY 1991–1997 will be U.S. \$ 37 billion, according to the six-year National Development Plan.

6 Biotechnology in the Member States of ASEAN

6.1 Indonesia

Susono Saono

General

Indonesia is a tropical country with a total area of 9819317 km², which comprises 1919317 km² of land consisting of 667 islands and 7900000 km² of sea (including the Exclusive Economic Zone). It has a wet season with an average temperature of 22 to 28°C, and a mean annual rainfall of 700 mm. Indonesia has a rich flora and fauna – about 10% of all living species in the world are represented in Indonesia.

With a total population of 183 million, it is the fourth largest country in the world in terms of population. Its population growth rate presently is 1.8%, with a life expectancy of 55 years for males and 58 years for females. Eighty percent of the Indonesians are muslim, but christianity, hinduism and buddhism are also present.

Economy

Indonesia's economy has grown dramatically during the last 25 years. The nation's gross domestic product (GDP) reached a level of Rp 273 trillion in 1993, and growth in GDP averaged 7% in real terms over a 25-year period. Inflation rate is between 6.0 and 7.0%. The agricultural sector tends to decrease, e.g., from 19.5% of GDP in 1991 to 18.5% in 1992. Over the last five years, productivity in manufacturing increased in real terms by 18%. Exports in 1993 were US-$ 38.4 billion, and imports US-$ 34.9 billion, leading to a trade surplus of US-$ 3.5 billion. The leading items are:

Exports: oil & gas, textiles/garments, and plywood (to Japan, the EU and the USA)

Imports: raw materials, capital goods, and consumer goods (from Japan, the USA, Singapore and Germany)

Indonesia welcomes foreign investment, particularly in export-oriented industries. Foreign companies normally establish joint ventures, but may take 100% equity in projects worth at least US$ 50 million or in those located in remote areas. Investments are approved by the *Investment Coordinating Board*, which provides specialist help to foreign companies in taking advantage of incentives such as duty-free allowances. The direct investment data reveal a rapidly expanding Indonesian economy, if still concentrating on low-technology industries. Total foreign investment in 1992 was US-$ 10.3 billion, an increase from US-$ 8.8 billion in 1991. Leading sources of foreign investment are Taiwan, Japan, Britain, and Hong Kong, while leading targets of foreign investment are chemicals, mining, and hotels. During the last few years, tourism has become another important industry. In 1992, 3 million tourists generated receipts of US-$ 3.2 billion.

Research and Development

The basic policy of the government on R&D is to use them as a means to improve planning and to accelerate national development. There are two categories of R&D programs: "short-term" and "long-term" programs. In the "short-term" programs, priority is given to R&D activities in the fields of agriculture, industry, energy, and mineral resources, with emphasis on labor absorption, the use of local materials, and the balance of payments. In the "long-term" programs, emphasis is on human and natural resources and the environment. The implementation of this policy is carried out by research institutes and universities, whose research activities are coordinated by the State Minister for Research and Technology. As most of R&D activities are conducted by government institutions and universities, the government is still the biggest source of funding (Tab. 32).

Indonesia's spending on R&D in 1991 was Rp 500 billion, equivalent to 0.2% of the

Tab. 32. Sources of Funding and Performers of R&D

	Govern-ment	Indu-stry	Others
Source (%)	80	19	1
Performer (%)	62	33	5

country's GDP. The breakdown of R&D expenditure is summarized in Tab. 33.

Within the non-ministerial research institutes (LPND), R&D accounted for about 25% of the total budget. The most important of the LPNDs, the *Agency for National Atomic Energy (BATAN),* had the largest R&D budget, with 35% of the LPND total. The high level of R&D support for BATAN reflects the government's policy aimed at developing an indigenous nuclear energy capability.

In 1991, about one half of the nonsalary R&D funds of public universities was directed toward engineering or agricultural sciences (80%) and to the medical and physical sciences (20%).

The R&D workforce of Indonesia is summarized in Tab. 34.

The industry

The participation of the private sector in R&D, and in the business of biotechnology, is still insignificant. A limited number of large companies, mostly joint venture companies, are active in the production of flavoring substances such as soy sauce and monosodium glutamate (MSG), beer, and lactic acid bever-

Tab. 33. Use of R&D Expenditure

Institution Performing R&D	Budget 1993/94 (Billion Rp)		
	Routine[a]	Development[b]	Total
No-Department/Ministerial Institutions (LPND)			
National Aerospace Agency (LAPAN)	0	6.828	6.828
Indonesian Institute of Sciences (LIPI)	16.967	26.663	43.630
National Atomic Energy Agency (BATAN)	29.753	22.211	51.964
Agency for Technology Assessment and Application (BPPT)	3.871	21.468	25.339
National Survey and Mapping Coordinative Agency (BAKOSURTANAL)	0	16.948	16.948
Central Bureau of Statistics (BPS)	0	15.420	15.420
National Archives	0	0.351	0.351
National Library	0	2.307	2.307
Environmental Control Agency (BAPEDAL)	2.667	4.454	7.121
Others	0.382	21.376	21.758
Sub-total	53.640	138.026	191.666
Departmental/Ministerial Institutions			
Ministry of Agriculture	27.054	55.644	82.698
Ministry of Industry	10.580	12.584	23.164
Ministry of Mining and Energy	34.442	6.817	41.259
Ministry of Education and Culture	0	116.301	116.301
Ministry of Health	0	36.216	36.216
Ministry of Forestry	3.941	7.341	11.282
Others	14.471	128.238	142.709
Sub-total	90.488	363.141	453.629
Grand total	144.128	501.167	645.295

[a] excluding budget for training and services
[b] excluding capital (building and equipment)

Tab. 34. R&D Workforce in Indonesia, 1993

Training Level	Government	Industry	University	Total
Diploma (D3)	99 055	15 375	75	114 505
Bachelor (S1)	80 945	17 620	15 250	113 815
Masters (S2)	4360	425	2400	7185
Ph.D. (S3)	3215	230	775	4220
Total	187 575	33 650	18 500	239 725

Tab. 35. Companies Active in Biotechnology

Company	Location	Ownership	Products
Perum Bio Farma	Bandung	State enterprise	Vaccines, sera, diagnostics
P. T. Kalbe Farma	Jakarta	Indonesian	Pharmaceuticals, diagnostics
P. T. Meiji Indonesia Pharmaceutical Industries	Jakarta	Japanese	Antibiotics
P. T. Rhône-Poulenc Indonesia Pharma	Jakarta	French	Pharmaceuticals, vaccines
P. T. Sandoz Biochemie Farma Indonesia	Jakarta	Swiss	Antibiotics
Pusat Veterinaria Farma	Surabaya	State enterprise	Vaccines, antigens
P. T. Sasa Inti	Probolinggo	Indonesian	Glutamic acid
P. T. Ajinomoto	Mojokerto	Japanese	Glutamic acid
P. T. Miwon Indonesia	Gresik	Korean	Glutamic acid
P. T. Indo Acidatama	Surakarta	Indonesian	Ethanol
Perusahaan Daerah Aneka Kimia	Surabaya	State enterprise	Ethanol
Rhizogin Indonesia	Jakarta/Bogor	Indonesian	*Rhizobium* starter cultures

ages. Others have started with the production of human and animal vaccines, sera, and diagnostic kits for a number of diseases. Most of the R&D is carried out by their overseas partners. Many medium and small companies are active in the production of traditional fermented foods and beverages such as tempé. None of them, however, is conducting R&D to improve their products. Major companies active in biotechnology are listed in Tab. 35.

The government

The main body responsible for S&T is the Ministry of State for Science and Technology (MSST).

Biotechnology has received a high priority in the national S&T development program. To that end, a *National Committee on the De-velopment of Biotechnology (NCDB)* was established by the MSST. The Committee has the following functions:

- to prepare and formulate a national biotechnology policy and development program to assist national development,
- to give guidance and encouragement in the development of bioindustry and its supporting R&D and human resources,
- to give directions for the establishment of national, regional, as well as an international network of cooperation on biotechnology, and
- to monitor the implementation of the national policy.

To implement this policy, a program has been formulated in 1990, with the following priorities

(1) production of fine chemicals and pharmaceuticals such as antibiotics, amino acids, vitamins,
(2) mass production through micropropagation of industrial, horticultural, and forestry plant species,
(3) improvement of food crops quality, in particular rice and soybean,
(4) improvement of beef and dairy cattle quality through embryo transfer, and
(5) production of various diagnostics and vaccines for human and animal diseases.

The program is implemented by several "Centers of Excellence":

● Centers of Excellence on Agricultural Biotechnology I and II, coordinated by the *Central Research Institute for Food Crops (CRIFC)* and the *R&D Center for Biotechnology (LIPI)*, respectively, both in Bogor
● Center of Excellence on Health Biotechnology, coordinated by the *Medical Faculty of the University of Indonesia* in Jakarta,
● Center of Excellence on Industrial Biotechnology, coordinated by the *Agency for Technology Assessment and Application (BPPT)* in Jakarta.

Each of these centers has the task to set up a network of institutions active in its particular field.

Besides the NCDB, there is also a *National Commission for the Preservation of Plant and Animal Genetic Resources,* which was set up by the Minister of Agriculture. Although there is no formal working arrangement with the NCDB, a close cooperation exists at the operational level because a number of R&D institutes are represented in both.

Research institutes

There is a considerable number of research institutes concerned with studies relating to biotechnology. They are summarized in Tab. 36.

Universites

The main task of the universities is in education and training. Research is only conducted in laboratories and institutes attached to different faculties of the universities. Some universities, however, have specific research institutes which operate independently of the faculties. Administratively, state universities are under the jurisdiction of the Ministry of Education and Culture (see Fig. 7), but in their activities, including research, they enjoy a considerably high degree of autonomy. There are over 15000 Science and Engineering faculties in Indonesia's public universities: about 20% of the 1.5 million students at public and private universities are enrolled in the natural sciences and engineering. The public universities play an important role in S&E education, enrolling nearly 1/2 of all students in this field. Public universities on Java continue to enroll more S&E students than do those on the other islands, but the enrollment rates on the other islands are increasing faster than on Java. The number of S&E degrees awarded to the 22-year-old population in Indonesia was about 4 per 1000. Major university faculties active in biotechnology R&D are summarized in Tab. 37.

Contextural measures

Indonesia has recognized the value of microbial and cell culture collections. At least 6 culture collections provide reliable services:

1. the Veterinary Research Institute's (Balitvet) culture collection for animal pathogenic microorganisms,
2. the Bandung Institute of Technology's (ITB) culture collection for industrial microorganisms,
3. the Medical Faculty of the University of Indonesia's culture collection for human pathogenic microorganisms,
4. the Agricultural Faculty of Gajah Mada University's culture collection for soil microorganisms,
5. the Faculty of Science of the University of Indonesia's culture collection for industrial molds, and

Tab. 36. Research Institutes Concerned with R&D in Biotechnology

Institute	Location	Supervision	Targets
Indonesian Sugar Research Institute (ISRI)	Pasuruan	MA	Dextranase, xanthan gum, sugarcane breeding, wastewater treatment, genetic engineering techniques
Central Research Institute for Food Crops (CRIFC), Laboratory of Plant Biotechnology	Bogor	MA	Molecular genetics of rice diseases, cell and tissue culture, nitrogen fixation, bio-fertilizers, bioconversion
Marihat Research Center	Pematang Siantar	MA	Tissue culture on cocoa, tobacco, rattan, vanilla, oil palm etc.
Research Institute for Animal Production	Ciawi-Bogor	MA	Feed improvement using fermentation, mannanase, embryo transfer, phytase, cassava-protein
Research Institute for Veterinary Science	Bogor	MA	Cloning of veterinary toxins, veterinary immunology, monoclonal antibodies
Institute for R&D of Agro-based Industry (IRDABI)	Bogor	MI	Industrial biotechnology, fermentation of soybean curd whey, food quality control
Center for the Assessment and Application of Technology (BPPT)	Jakarta	BPPT	Antibiotics production; plant, fish and livestock production; vitamin, enzyme and amino acid production
R&D Center for Applied Chemistry (RDCAC), Indonesian Institute of Sciences (LIPI)	Bandung	LIPI	Bioconversion of solasodine, waste water treatment, fermentation, tempé
R&D Center for Biotechnology (RDCB), Indonesian Institute of Sciences (LIPI)	Cibinong-Bogor	LIPI	Fermentation and enzyme technology, SCP, plant biotechnology for tropical fruits, embryo transfer technology
Eijkman Institute for Molecular Biology	Jakarta	MSRT	Mitochondrial DNA mutations in human diseases, ageing process, energy-transducing systems, thalassemia, diagnostic kit for Dengue hemorrhagic fever

6. the R&D Center for Biotechnology's culture collection for economically important microorganisms.

In 1989, a patent law was passed which provides patent protection for 14 years. With respect to biotechnological invention, however, the law is not too supportive because no patent will be granted to (a) any process for the production of foods and drinks for human and animal consumption, (b) foods and drinks for human and animal consumption, (c) new plant varieties and animals, (d) any process for the production of new plants and animals or their products.

Although no special regulation on the release of genetically modified organisms (GMOs) has been drafted, there are several laws which could be used to safeguard the en- vironment and society against possible hazards due to the release of GMOs. Recently, in August 1993, a *Guideline on Genetic Engineering Research* has been released by the State Ministry on Research and Technology. While the emphasis of this guideline is on the requirements for and control of research on genetic engineering, it provides to some extent additional protection against possible hazards of GMOs release.

6.2 Philippines

Jeannie Scriven

The Philippines, made up of more than 7000 islands, has a population of 67 million (1994) which is expected to grow to 75 million

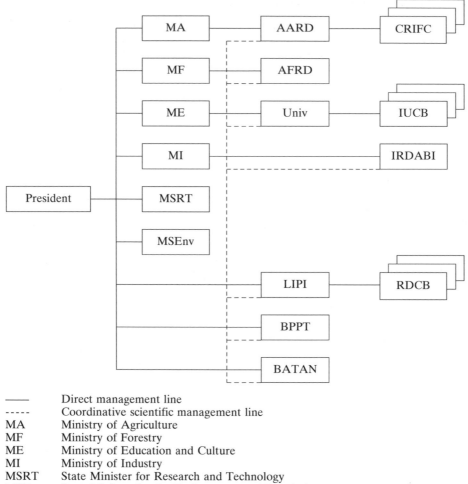

———   Direct management line
- - - - -   Coordinative scientific management line
MA   Ministry of Agriculture
MF   Ministry of Forestry
ME   Ministry of Education and Culture
MI   Ministry of Industry
MSRT   State Minister for Research and Technology
MSEnv   State Minister for the Environment and Population
LIPI   Indonesian Institute of Sciences
BPPT   Agency for the Development and Application of Technology
BATAN   National Atomic Energy Agency
AARD   Agency for Agricultural Research and Development
AFRD   Agency for Forestry Research and Development
Univ   University
CRIFC   Central Research Institute for Food Crops
IUCB   Inter University Center on Biotechnology
IRDABI   Institute for R&D on Agrobased Industry
RDCB   R&D Center for Biotechnology

Fig. 7. Structure of S&T in Indonesia, with emphasis on life sciences.

Tab. 37. University Faculties with Major Activities in Biotechnology

University	Location	Field of Study
Faculty of Pharmacy, University of Airlangga	Surabaya	Plant cell cultures, biotransformation with plant cells, rat hepatocyte cultures
Food and Nutrition Development and Research Center (FANDARC), Gadjah Mada University (UGM)	Yogyakarta	Biopreservation, lactic acid bacteria, cell fusion among *Aspergillus* strains, monoclonal antibodies for aflatoxin
Inter University Center for Biotechnology, Gajah Mada University	Yogyakarta	Genetic analysis of Waardenburg syndrome, Thalassemia, Dengue viral antigens, diagnostic tools based on PCR, erythromycin and BT toxin production
Inter University Center on Biotechnology, Institut Teknologi Bandung	Bandung	Microbial fermentation, enzyme technology, genetic engineering, biological waste-water treatment
School of Medicine, University of Airlangga	Surabaya	Reproductive health, infectious diseases, cancer and degenerative diseases, forensic serology
Department of Microbiology, University of Indonesia	Jakarta	Dengue virus diagnostics, *Salmonella* diagnosis, hepatitis C research
Faculty of Agriculture, Gajah Mada University	Yogyakarta	Baculo virus detection, CVPD-free citrus seedlings, PCR technology, SMZ coat protein genetics for virus-free soybean stocks, food biotechnology
Inter University Center on Biotechnology, Institut Pertanian Bogor	Bogor	Improvement of plant productivity by tissue culture, embryo transfer, microbial biotechnology, waste treatment, culture collection

by the year 2000. In the fifties, the Philippines was the second richest country in Asia, behind Japan. During the eighties, the economy grew by an average of 1.5% but the population grew by 2.5% per annum. Currently, more than 40% of the population live below the poverty line. Foreign debts – incurred under MARCOS are in the region of US-$ 34 thousand million. 40% of the state budget is used to finance these debt repayments. Gross domestic product (GDP) quarterly growth in mid 1994 was 3.8% compared to stagnation in GDP growth between 1991/92 and 2.3% in 1993. This is due to two factors – political stability under General RAMOS and better electrical power supplies. During 1993 in Manila, there were "brown-outs", which meant that there was no electricity for 10–12 hours daily. However, a bright spot on the economic front is electronics, instrumentation and control industries, which generated more than a third of total exports in 1994. Expenditure for

R&D is 0.12% of GDP, below the UNESCO recommendation of 1% for developing countries.

The Philippines has an agricultural-based economy. However, it is blessed with a diversity of organisms available, which may be possible sources for natural products in the future. The country produces some 30 million metric tons of organic waste per year, including corn stalks, rice straw, coconut husks and shells. However, transport costs may limit commercial exploitation of these raw materials. Annual production of bagasses is estimated to be about 6.4 million metric tons while that of coconut shells and husks is 7.5 million tons. There is urgent need for the production of planting materials by means of tissue culture techniques to help speed up the country's coconut replanting program.

There is no home Philippines pharmaceutical industry; it is entirely dependent on imports. There is a limited volume of anti-diphtheria, and anti-tetanus vaccines produced by the Department of Health. As far as biotechnology activities in the private sector are concerned, there is some production of animal vaccines at commercial level by Biologics Corporation, Riverdale Tryco and the Bureau of Animal Industry. Haemorrhagic septicemia vaccines are produced and it is planned to commercialize swine plague and hog cholera vaccines. Current research includes DNA probes for hepatitis B virus, pregnancy testing kits, a monoclonal antibody against Schistosomia japonicum, experimental production of penicillin and antifungals and rifamycin. Biotechnological products include Zymax, a probiotic and poultry additive using yeast and bacteria, a *Bacillus thuringiensis* powder to control vegetable pests and malaria, a mycorrhizal inoculant for pine seedlings, and nata de coco.

As far as the agrochemical and food sectors are concerned, there is little R&D in biotechnology, which is characterized by such firms as the San Miguel Corporation, a multinational in the food and beverage trade. The UFC company recently built a ten thousand liter fermentor for its vinegar production.

Science and technology (S&T) activities are monitored by the *Department of Science and Technology (DOST)*. The DOST in turn guides five councils: Agriculture and Forestry, Health, Advanced S&T (PCASTRD), Industry and Energy, and Aquatic Management. DOST also directs the following research institutes: Industrial Technology Development Institute, Food and Nutrition Research Institute, Forest Product Research and Development Institute, Philippine Nuclear Research Institute. The bulk of S&T manpower is to be found in academic institutions:

S&T Manpower	% of Total
State colleges and universities	47
Department of Agriculture	27
Department of Environment and Natural Resources	5.7
Department of Science and Technology	2
Other commodity research institutes	13
Miscellaneous	5

In 1990, the country had some 58 doctoral degree holders and 150 personnel with master degrees engaged in biotechnology research and development. In 1990, the government drew up a Master Science plan through DOST; it identified biotechnology as a priority area. The *Philippine Council for Advanced Science and Technology Research and Development (PCASTRD)* prepared a strategy, which was to develop agriculture, aquaculture, health, industry and the environment. In agriculture, fertilizer substitutes, tissue culture, biological control agents, animal production and improvement, and tissue culture for secondary metabolites were to be given attention. Diagnostics and vaccines, coconut tissue culture, tailored fats from coconut oil and the application of biotechnology in reforestation were to support productivity in agriculture. Six projects are to be implemented between 1991 and 1996 including penicillin production.

A major center of R&D activity is at the University of Los Banos, at the *National Institutes of Biotechnology and Applied Microbiology (BIOTECH)*. The siting of the International Rice Research Institute (IRRI) nearby provides scientific impetus and technological resources in the Los Banos region. In late 1994, IRRI produced a "superrice" hybrid, which could improve yields by 25%. The following are areas of expertise at BIOTECH: molecular biology (isozyme markers for gene mapping in mungbean, high insecticidal activity against rice pests in *Bacillus thuringiensis*), micropropagation (tissue culture studies of crops), microbial control of pests and diseases, microbial fertilizer/biofertilizer, and legume-specific inoculants, and food/feed fermentation bioconversion. BIOTECH is also developing alternatives to conventional products and processes, such as mycorrhizal tablets for reforestation, cell and protoplast culture of orchids, lysine production, inoculants, pasteurella vaccine, and mushrooms. Research on medicinal plants under the "herbal medicine program" is being carried out. BIOTECH maintains a culture collection for safekeeping and research uses. The Philippines Coconut Authority has a gene bank for coconuts and IRRI maintains rice plant cultures. The University of the Philippines at Diliman concentrates on medical and industrial research, such as economically important plant and fish species.

Regarding financing biotechnology development, there are soft loans available to the private sector to commercialize products through the Development Bank of the Philippines, with the endorsement of DOST. Research funding comes mostly from international sources, such as the the Japan Society for the Promotion of Science, which provides exchange scientist programs. The Philippines is currently debating a change in its patent laws. Patents may be granted by the Bureau of Patents. However, the critical evaluation of patent applications is limited since the expertise to assess is lacking. There are regulations governing the traffic of microorganisms and biotechnology experimentation formulated by DOST, based on US and Australian guidelines.

6.3 Malaysia

Jeannie Scriven

Average economic growth over the period 1970–1990 was a steady 6.7% and currently the Malaysian economy is growing by over 8% per annum. Gross domestic product (GDP) per capita lies in the region of US-\$ 2570 with a population of approximately 20 millions. However, factory owners are complaining of labor shortages, and other Asian countries, (including Vietnam, Indonesia and India), can supply cheaper workers. There is a danger that Malaysia may soon become too rich to qualify for tariff exemption under GATT. The problem facing Malaysia now is to keep its cost-base low enough to compete with other countries. There is an inflation rate of 3.7% (1993). The economic indicators during July 1994 were an increase of 8.4% in GDP, and growth of 10.6% in industrial production when compared with the previous year. GDP is projected to grow at 7–8% per annum into the next century. Malaysia's investment in R&D, at present below 1% of GDP, is expected to reach 1.6% by 1995.

The economy is changing from being predominantly agriculturally based – rubber, palm oil, cocoa and forestry represented 18.7% of GDP in 1990 – to being manufacturing and service based (for example, semiconductors). It is hoped that manufactured goods will make up 80% of exports by the year 2000. One example of successful industrialization is that Malaysia has launched its own motor industry with penetration of 60% of total automobile sales. Malaysia has declared that it wishes to achieve the status of an industrial nation by the year 2020 (Vision 2020 strategic plan).

In 1987, a National Biotechnology Committee was established and a Biotechnology Program coordinated under the auspices of the *Malaysian National Council for Scientific Research and Development (MPKSN)*. The MPKSN reported in 1993 that, when compared with European universities, about 1–2 researchers comprise the biotechnology research base in a Malaysian university, as against 8–12 at a European university. A peer review system was also absent. As a result of the findings, the MPKSN decided to establish a National Biotechnology Center, and to set up a *Bioprocessing-Prototype Manufacturing Facility (BPM)*. This is intended to provide large-scale process facilities. Five priority investment areas for the center were identified: diagnostics, plant tissue culture scale-up, biotechnology of oleochemicals, natural products screening and development, and environmental biotechnology.

The MPKSN commissioned a study to evaluate current biotechnology activities and their potential for commercialization. It came to the conclusion that Malaysian industry and the private sector do not generally use biotechnology, but that certain sectors could certainly benefit, including pharmaceuticals, vaccines, diagnostics, food, novel plants etc. As for industrial biotechnology, there was a very low level of activity. Few industries existed for pharmaceuticals, diagnostics and advanced food production; industry was not involved in publicly funded grant schemes and biotechnology needed to be promoted. The study also found that there was lack of awareness in industry as to opportunities available and there was inadequate organization/structure. It also found that research was "isolated", there was little contact between universities and industry, there was an inadequate base of science students and lack of knowledge about patents and rights. It was felt that R&D funding was not operating efficiently.

All research programs undertaken by the 24 research institutes and eight universities are overseen by the MPKSN. For the current Five Year Plan (1991–95) RM 600 million has been provided. Two important government departments with regard to biotechnology are the *Malaysian Technology Development Corporation* and the *Malaysian Industrial Development Authority.* A Technology park is being set up adjacent to the main North-South Highway linking 5 universities and 8 institutes, including the Palm Oil and Standard and Industrial Research Institutes.

There is some success in transferring institutional research into private sector manufacturing practice. A food-based Newcastle Disease vaccine, developed by Universiti Pertanian Malaysia, and the production of diagnostic kits for Dengue fever and Japanese encephalitis by Universiti Sains Malaysia, have found partners in the private sector. Also the technology to deal with mill effluents in the palm-oil agroindustry using a bioreactor will be commercialized. Large plantation companies, such as Guthrie Chemara, have small R&D sections, made up in this instance by one senior scientist and assistant. Their role is micropropagation or cloning of palm-oil seedlings. Such private sector activity is often to meet in-house needs. However, the palm-oil industry has been set back in the field of cloning oil-palm seedlings, because of the abnormalities in the fruits of tissue culture-derived oil palms

Areas of relevance to biotechnology in the private sector are food processing, agriculture, aquaculture, pharmaceuticals and waste utilization. Fermentation technologies are used in tapioca and sago palm starch and vegetable oil processing, while enzymatic processes are used in the production of fruit juices, lipases and fats. International companies, relying heavily on foreign technology, characterize the food sector. Research for the rubber industry is carried out at the Rubber Research Institute of Malaysia (RRIM) and has focused on genetic diversity. Because of

the waste generated by rubber, coconut, cocoa and palm-oil estates, the government has introduced discharge regulations. There is interest in producing biofertilizers and in using microorganisms to deal with the waste slurry. The medical and veterinary pharmaceutical market is growing by 10% annually; products include diagnostic kits, vaccines, antibiotics and hormones.

There are some government fiscal industrial incentives including building allowances, plant and machinery capital allowances and tax relief for R&D companies. It is intended to implement biosafety guidelines in 1994 which deal with the release of genetically modified organisms (GMOs) into the environment.

6.4 Singapore

Jeannie Scriven

With a population of almost 3 million, Singapore's economic base consists of electronic manufacturing (mainly discdrives), communications, financial and banking services, transportation and petro-chemical refineries. Between 1988 and 1992, the economy grew on average at 8.2%. Gross domestic product (GDP) per capita for 1992 was S-$ 26604 (=US-$ 16627 equivalent). Gross expenditure on R&D was 1.27% of GDP in 1992.

The government is highly interventionist in its economic policy; a strategic business unit for the National Biotechnology Program was established in 1988, with the aim of founding a technology base for biotechnology. It was also to develop manpower. Singapore Bio-Innovations was founded by the *Economic Development Board (EDB)* in 1990 to encourage industrial biotechnology activities in Singapore. Four categories of projects are targeted: early startup, existing biotechnology companies, overseas promising biotechnology projects and new foreign biotechnology projects which are to be established in Singapore.

During 1992, Research and Development (R&D) spending was as follows:

	S$ millions	% of total
Private sector biotechnology firms	577.6	61
Higher education sector	156	16
Public research institutes	110	12
Government sector	105	11

There is assistance from the *National Science and Technology Board (NSTB)* to help promote R&D in the private sector, with financial grants and project costs. Private companies can also take advantage of other business development incentives including pioneer status (exemption of corporate tax). In the *Singapore Biotechnology Directory* published in 1994, the private sector is represented by 26 firms with more than 50 employees. However, the directory did not list any major Japanese firms, subsidiaries or trading supplies or even Beechams (pharmaceuticals). The NSTB is responsible for R&D promotion, but is not involved in medical research. Its R&D research budget for 1993 was S$ 350 million. It funds 5 research institutes and 4 research centers, whose recurrent annual budgets are given in parentheses:

GIMT GINTIC Institute of Manufacturing Technology (S$ 6 million)
IMCB Institute of Molecular and Cell Biology (S$ 25 million)
IME   Institute of Microelectronics (S$ 17 million)
ISS   Institute of System Science (S$ 18 million)
ITI   Information Technology Institute (S$ 16 million)
CRISP Center for Remote Imaging and Sensing Processsing
CWC   Center for Wireless Communication
MTC   Magnetics Technology Center
MSRC National Supercomputing Research Center

The NSTB formulated a National Technology Plan, which has two targets to be reached by 1995: national expenditure on R&D should reach 2% of GDP and the ratio of research scientists and engineers should be 40 for every ten thousand workers. A S$ 2 billion fund was provided to support R&D, to fund research, to develop manpower and to build a suitable infrastructure. Biotechnology is one of the areas targeted.

Much of the R&D in the higher education sector takes place at the *National University of Singapore (NUS)*. The chemical engineering department at the NUS is the only department graduating chemical engineers and researching in large-scale production and purification techniques. In order to meet manpower training needs as well as development in the field of fermentation technology, the *Bioprocessing Technology Unit (BTU)* of the NUS was established in 1990. It was equipped to the tune of S$ 26 million, with much of the fermentation apparatus coming from B. Braun Biotech (Germany). Its aim is to "facilitate the application of research in biotechnology and to meet the process development needs of the local biotech industries". The unit is funded ultimately by the EDB. It seeks to promote multidisciplinary collaborative research between university and industry.

As far as manpower training is concerned, a Massachusetts Institute of Technology–NUS international workshop on "Fermentation and Bioprocess Technology" has been held. BTU was also joint organizer of the 3rd Asia-Pacific Biochemical Engineering Conference during 1994. The technical sessions included topics such as Bioproducts from Prokaryotes, and from Eukaryotes, Biocatalysis/Protein Engineering, Bioprocess Monitoring, Modelling and Control, and Bioremediation and Decontamination. Current activities include protein/peptide and DNA/oligonucleotide sequencing and synthesis. The unit is prepared to take on contract jobs for industry in fermentation, cell culture and product purification. The BTU will provide custom-made peptides in immunological assays, DNA and peptide sequencing, and synthetic oligonucleotide primers and probes.

The *Institute of Molecular and Cell Biology (IMCB)* was inaugurated in 1987, with the aims of developing a modern research culture for biological and biomedical sciences and to train scientists. The research topics are as follows: Cell Regulation, Molecular Genetics/

Protein Engineering, Plant Molecular Biology, Molecular Neurobiology, Tumor Immunology/Virology, and Microsequencing and Peptide Synthesis service. The institute has 160 scientists, half of whom have Ph.D. degrees. The institute has research ventures with the following biotechnology companies: Glaxo UK (brain research), Amylin Corp. USA (diabetes research) Glead USA (antisense technology) Genzyme USA (TNF-β), Millipore (protein microsequencing/peptide synthesis), Bayer AG Germany (mosquitocidal toxin) and Singapore Bio-Innovations.

During 1994, the IMCB officially opened a Center for Natural Product Research, in order to discover novel lead compounds for the development of new drugs. The funding for the center is staged over 10 years; Glaxo will contribute S$ 30 million, the EDB S$ 10 million and the IMCB (through the NSTB) will contribute S$ 20 million in infrastructure and research support. The staffing will rise from 15 scientists to 30 by 1995.

There are two government departments which are important for biotechnology development – the *Primary Production Department (PPD)* and the *Singapore Institute of Standards and Industrial Research (SISIR)*. The former (PPD) is affiliated to the Ministry of National Development and its terms of reference are to supply safe, wholesome and quality foods, the safeguarding of the health of animals, fish and plants, and as a center of excellence for tropical agrotechnology services and to support trade in primary production. To this end, it is involved in projects such as hydroponic farming technology, soilborne diseases of trees, poultry disease and integrated pest management strategies. It has an R&D staff of 50, five of whom have doctorates. Its total budget is S$ 35.3 million.

SISIR is affiliated to the Ministry of Trade and Industry, and is involved in standardization, certification, quality systems, quality promotion, consultancy and training activities. It is also responsible for measurement standards, technology transfer, materials technology, patent and technical information. It has a staff of 570, and a total budget of S$ 46 million. Singapore adheres to the International Standards Organization 9000 regulations. There is no official culture collection in Singapore.

As part of the Information Technology 2000 masterplan, a science and technology network called Technet was founded to provide computing support. The NSTB helps with the expenses incurred in applying for patents and encourages the commercialization of technology developed overseas. Financial assistance is available; up to 50% of costs to a limit of S$ 30000. There is also funding available for patent applications in the US.

6.5 Thailand

Jeannie Scriven

Thailand has a population of 58.6 million (1993) with modest population growth prospects. Its annual gross domestic product (GDP) per capita is US-$ 1402 (1991) with a real annual growth rate from 1987 to 1991 of 11.2%. Compared to its neighbors, Thailand is ahead of Indonesia and the Philippines but behind Malaysia and Singapore. GDP quarterly growth in mid 1994 was up 7.5% on 1993, and industrial production increased by 10.7%, while consumer prices increased by 5.4% Agro-industry is the economic base of the country (tapioca, rubber, fruits, rice, sugar-cane, fish and poultry) while manufacturing exports (food and beverages, tobacco, textiles, computing components, construction materials and transport equipment) increased by 12%. Tourism provides a steady inflow into the balance of payments. The inflation rate for 1994 was 4%.

Thailand made a commitment to biotechnology in 1983 with the establishment of the National Center for Genetic Engineering and Biotechnology (NCGEB) formed under the Ministry of Science, Technology and the Environment (MOSTE). Between 1984 and 1987, the budget increased from 7 million Baht (approx. US-$ 300 thousand) to 29 million Baht (approx. US-$ 1.2 million). In 1985, supported by US–Thai cooperation, the Science and Technology Development Board was founded, with the task of underpinning development in priority areas, including biotechnology. This Board was merged with the NCGEB and two other national centers to form the National Science and Technology

Development Agency (NSTDA) in 1992. The current budget (1994) for NSTDA is 490 million Baht (approx. US-$ 19 million) with 132 million Baht (approx. US-$ 5 million) earmarked for NCGEB.

The National Research Council of Thailand is responsible for the formulation of R&D policy and supports research funding in science and technology. Most research is carried out in the universities. Current R&D expenditure by the Thai Government is about 0.25% of its GNP, equivalent to US-$ 100 million. About 20% of this amount is spent on biosciences and biotechnology.

The NCGEB has initiated national projects including the Biotechnology Information Network and a pilot plant for biochemical engineering, situated at Mahidol University. The King Mongkut's Institute of Technology at Thonburi also has a pilot plant up to the 1000 liter-scale at its disposal. Major research directions are presently:

- the improvements of crops, in particular rice,
- the development of biological fertilizers and pesticides,
- products related to human and animal health (vaccines, therapeutics),
- energy-related targets (e.g., recycling of biomass, methane generation),
- environmental R&D.

In the private sector, Thailand has Asia's largest beef and dairy cattle ranches and is its largest producer of poultry. Under the Five Year Development Plan (1992–1996), the government hopes to increase agriculture technology investments in the private sector.

Thailand's pharmaceutical industry has a yearly turnover of more than US-$ 1.2 billion, mostly in packaging and the formulation of imported products. For the future, there are business chances in the development of speciality chemicals, diagnostics, vaccines, antibiotics, amino acids and other food-related biotechnical products. Unfortunately, the private sector has not yet taken advantage of new biotechnological techniques such as genetic engineering or hybridoma. Also, the relatively weak infrastructure (difficulties in transport and communications) and occasion-

al water shortages do not encourage developments in the private sector.

Most R&D work in biotechnology in the public sector is carried out in some 20 institutions with almost 500 scientists. Four universities produce between 50–100 graduates in biotechnology per annum. There is some biotechnology activity at Chiang Mai University in the north of the country and at the Prince of Songkla University at Hat Yai, close to the Malaysian border.

As far as contextual measures are concerned, the government introduced tax exemptions for R&D expenditure in order to encourage biotechnology. In 1979, a patent system and standards were established. Currently there are no guidelines as to the production of biochemicals. The Thai Industrial Standards Institute issues specifications for industrial products, including biotechnology. The Government's Department of Medical Science monitors standards and quality control for biological products, including vaccines. The NCGEB has commissioned a study to consider what steps to take with regard to laboratory safety, field trials and environmental release. Patent law is particularly weak, since the Government fears that foreign firms will be the first to corner the market in local genetic resources if rigorous laws are introduced. A culture collection is available at the Thailand Institute of Scientific and Technological Research (TISTR) with its Microbiological Resources Center.

As an example of a biotechnology-transfer project for the benefit of the private sector, the Carl Duisberg Gesellschaft financed a soy sauce quality control center for Thai producers: this is augmented by an executive club. It is hoped that the manufacturers will carry the costs of the quality control center. However, the difficulties of financing such a venture illustrate how fragile biotechnology development is when the manufacturing base is limited.

In the field of international cooperation, NSTDA cooperates with many countries, for example, the USA, Japan, Australia, Britain, Holland and the ASEAN countries. In 1990, the EC began joint EC/ASEAN projects; in this framework, a group at Mahidol University is collaborating in the protein engineering of *Pseudomonas* lipase with Novo in Denmark, the University of Milan and the GBF, Braunschweig in Germany.

7 Biotechnology in Australia

Alan J. Jones and Rolf D. Schmid

Australia, one of the few OECD members in the Asian-pacific trading zone, is hampered by a sparce population (17 million) on a vast territory, similar in size to the USA or China. As a consequence, it suffers from small, highly competitive national markets, from long continental and intercontinental distances, and from severe competition by its Pacific neighbors. Its advantages reside in a wealth of resources, mainly in agriculture and mining, and in a well-trained population whose education is mostly based on British traditions.

After its ties to the United Kingdom were reduced as a consequence of the formation of the European Community, Australia has continuously strengthened the trade with her Asian-Pacific neighbors. In 1990/91, only 12% of its trade was with the EEC, but 28% was with Japan, 30% other Asian nations, and 11% with the USA.

The Australian economy is largely based on services (>60% of its GDP) and primary industries (agriculture and mining, 20% of GDP), whereas its manufacturing industries are of only modest but growing magnitude (15% of GDP). As a result, industrial R&D is limited and accounts for only 40% of national R&D, which in turn is among the lowest of the OECD countries (1.2% of GDP). However, the country has developed excellent capacities for S&T in the public domain, as indicated by, e.g., overproportional international citation rates in areas such as chemistry, biology or the earth sciences.

In its brief post-colonial history, Australia's governments have invested considerable efforts to create a national manufacturing industry, by taking advantage of the public S&T

sector. Biotechnology, with its various applications in the medical, agricultural and mining markets, is considered a key technology to support such goals. However, while some impressive results have been achieved, the "critical mass" of the internal markets may still be too small to support significant industrial developments on an international scale.

The industry

The Australian and New Zealand Biotechnology Directory, in its latest version of 1993, lists 178 Australian companies active in biotechnology, of which 102 are manufacturers (Tab. 38).

Tab. 38. Australian Biotechnology Companies: Areas of Biotechnology Business

Major Focus on Manufacturing of	Number of Companies
Biologicals	21
Pharmaceuticals	19
Diagnostics	17
Vaccines	5
Food and beverages	10
Plant agriculture	9
Aquaculture	3
Environment	7
Medical devices	2
Animal biotechnology	9

Concomitant to Australia's population distribution, 3 out of 4 companies are located in New South Wales or Victoria, i.e., in the metropolitan areas of Sydney and Melbourne. Most of the companies are medium or small enterprises, and only 10 employ more than 60 people. Others are subsidiaries of large international firms. In many instances, biotechnology is just one area of interest for such companies, and the number of "dedicated biotechnology companies", by the US definition, is 27. The total industrial workforce of professionally qualified staff in this field is estimated to be around 1400. Tab. 39 gives a list of those companies with >40 professionals in biotechnology R&D.

Tab. 39. Major Companies in Australian Biotechnology

Company	Main Business	Australian (A)/ Foreign (F)	Turnover (Million $A, 1992)	Professionals in R&D
Burns Philp & Co.	Food, yeast and retailer company	A	2500	180
Commonwealth Serum Laboratories	Vaccines, antibiotics, pharmaceuticals	A	152	175
Biotech Australia	Vaccines, veterinary products, diagnostics	F (Hoechst AG)		85
Institute of Drug Technology Australia	Pharmaceuticals, diagnostics	A		60
AGEN Biomedical	Diagnostics	A		47
Arthur Webster	Veterinary vaccines	A	some 10	45
South Australian Brewing	Alcoholic beverages	A		40

In view of this limited capacity, and of the high standard of public research, the Government has established several incentives for the creation of R&D-driven companies. Apart from a 150% tax concession for R&D, used by about 1600 companies since its initiation in 1985, there are Grant Schemes, including a focus on strategic technologies as in the former Biotechnology Generic Grant Scheme, now operated under AusIndustry. In 1987, the "Factor (F) Scheme" was inaugurated as a tool to develop a pharmaceutical industry in Australia. In a country with strong traditions in social security and healthcare, but also with a time-consuming drug approval system and harsh price regulations, the program refers to the sixth (factor f) out of eight factors to be considered by the Pharmaceutical Benefits Pricing Authority when negotiating prices paid to manufacturers of pharmaceuticals, predominantly of ethical drugs. It allows for higher notional prices to be paid to a company in return for approved programs of development and a significant commitment to production, research and development inside Australia. Apart from these subsidies, more direct attempts have been made to capitalize on Australian public R&D. Some examples are:

Biotech Australia Pty Ltd. was founded in 1979 and acquired, first in 1981 by the mining company CRA as a strategic investment, sec-ond in 1989 Hoechst AG took a 50% interest, and in 1993 Hoechst acquired full ownership. Hoechst prides itself with the philosophy that Biotech Australia will maintain its own identity. Product developments have occurred with assistance from Government grant schemes including numerous collaborations with Australia's premier research institutions. Biotech Australia has specialized in animal health products and food diagnostics, but has an increasing commitment in the human health area. The recent establishment of a new plasminogen PAI-2 pilot plant and a substantial increase in R&D expenditure attests to this. The company also holds key patents in inhibin biochemistry.

BRESATEC Ltd. started in 1982 from the Biochemistry Department of the University of Adelaide and became a world leader on porcine growth hormone. Recently, the company focused on transgenic animals and biochemical reagents. The company became public in 1987, and American Cyanamid and others took over the majority of shares. Turnover at present is around $A 2 million.

BIOTA Scientific Management (BSM) Pty Ltd. was founded 1985 by CSIRO with a focus on a molecular designed inhibitor to viral neuraminidase (influenza cure). Glaxo recently invested $A 4 million in the company.

AMRAD Co. Ltd. was established in 1986 by a $A 50 million investment of the Victorian government as the commercial arm of Australia's leading biomedical research institutes. Due to licensing agreements and international alliances based on the products of its institutes, pharmaceutical sales topped $A 30 million in 1992.

Gene Shears Pty Ltd. was founded in 1989 jointly by CSIRO and Groupe Limagrain, a French-based seed company, in order to commercially exploit a CSIRO patent on ribozymes. In 1991, Johnson & Johnson joined the group.

The government

The S&T policy of Australia is shaped by many bodies, and the funding is appropriated mainly by the central government and industry, as schematically shown in Fig. 8.

The country's history was shaped by a colonial government which, though liberal, still largely displays interventionist behavior. This is in particular true for economic affairs. As S&T in industrialized nations is a major factor for building the economy, the develop-

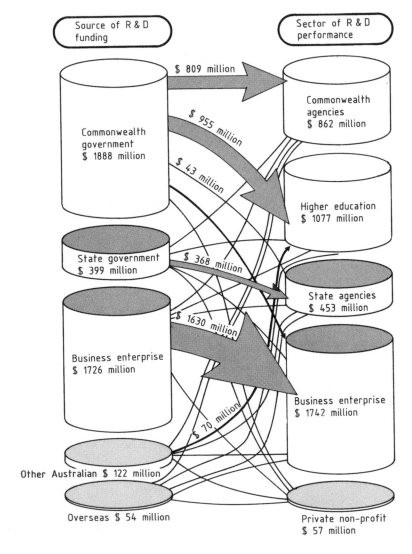

Fig. 8. Flow of funding for Australian R&D (1988–1989).

ment of the S&T policy of Australia is the responsibility of the Minister for Industry, Science and Technology, whose budget is located in the Industry, Science and Technology portfolio. Government support for industrial R&D, the budgets for the major government research agencies, e.g., of the CSIRO, AIMS, ANSTO and the Co-operative Research Centres (CRC, see below), are also located in this portfolio. In 1989, a Prime Minister's Science and Engineering Council, and a Coordination Committee on Science and Technology were established as discussion and evaluation panels. The latter is headed by the *Chief Scientist,* an institution borrowed from the British S&T system. Since the Chief Scientist is also the CEO of the Council, his function is pivotal in science policy coordination and development.

ASTEC, the *Australian Science and Technology Council,* is interlinked with the Science and Engineering Council, but forms an independent source of advice, since its members are leading academics and business executives.

The *Australian Research Council (ARC)* under the Ministry for Employment, Education and Training, and the National Health and Medical Research Council (NH & MRC) are the major bodies for university research grants.

National research centers/CSIRO

Among the national research centers, the Commonwealth Scientific and Industrial Research Organization (CSIRO) is by far the largest (followed by the Defence Science and Technology Organization, DSTO). It is structured into 6 institutes and 35 divisions, which cover all major economic activities in the country. This includes several divisions related to biotechnology, as outlined in Tab. 40.

Total CSIRO staff in 1992 was about 7300, and total expenditure was about $A 650 million, with about 25% coming from industry and competitive government funding. In its 5 years' strategic plans, the development of manufacturing industries, animal and plant production, and R&D programs on health and the environment take a lead position. Accordingly, CSIRO spends about 17% of Australia's R&D budget on plant production, and 26% on animal production. It shares large parts (>30%) of the national R&D budget in rural-based manufacturing (food, beverages, fibers and textiles, wood products) and in environmental research.

CSIRO has established a wide range of cooperative R&D programs with industry and with universities, notably through the Cooperative Research Centres (CRC's, see below). In Sydney, the *Biomolecular Research Institute* was jointly founded in 1990 by the CSI-

Tab. 40. Biotechnology-related Divisions of CSIRO and Their Major Targets (Examples)

CSIRO Institute	Division	Location	Major Areas
Institute of Industrial Technologies	Biomolecular Engineering	North Ryde Laboratory, Parkville Laboratory (180 staff)	Protein X-ray crystallography, protein engineering, gene therapy, virus biology, receptors and cytokines, vascular cell technology, biomaterials, recombinant vaccines
Institute of Animal Production and Processing	Animal Health	Geelong (200 staff)	Diagnostic center for veterinary diseases, vaccine development
Institute of Plant Production and Processing	Plant Industry	Canberra (400 staff)	Breeding programs including transgenic plants, with emphasis on wheat, barley, rice, cotton, oilseed, potato and tomato, pasture, sugar cane, eucalyptus, and soil improvement

RO, next to its Parkville Laboratories of Bio-molecular Engineering, and by the Strategic Research Foundation of the Victorian Government, as a specialist facility for molecular imaging and rational drug design. The staff is around 60, with sophisticated equipment in X-ray crystallography, NMR and biocomputing. The aim of the institute is to discover, synthesize and develop novel therapeutic compounds of commercial value in the treatment of important diseases such as HIV, hepatitis B and C, influenza and cancer. Some of the institutes are linked to the local universities.

CSIRO also provides a range of educational programs relating to biotechnology and other science areas, e.g.

- the CSIRO Double Helix Science Club (for elementary schools)
- the CSIRO Science Education Centers (for school children)
- the CSIRO Women in Science Project (for school girls)
- the CSIRO Student Research Scheme (for secondary students)
- the BHP Science Awards (for school teachers), and
- the CSIRO Scholarships.

Several research centers outside the CSIRO, with significant acitivities in various areas of biotechnology, should also be mentioned.

The *Australian Institute of Marine Sciences* in Townsville is located close to the Great Barrier Reef. Its 106 professional staff is focused on marine biotechnology, environmental studies, mariculture, genetic studies of wild and cultured stocks and bioactive metabolites. The *Australian Wine Research Institute* in Adelaide is financed by the wine-making industry of Australia, through a subsidy of $A 6 per ton of pressed grapes. Major activities include grape flavor, selection and improvement of oenological significance by application of molecular biology, and optimization of malolactic fermentation in wine. At the *Sugar Research Institute* in Brisbane, the capital of the sugar belt of Australia, various aspects of sugar biotechnology and biological treatment of wastes from sugar refineries are

studied. A range of outstanding biomedical research institutes is located along the "Parkville Strip", near Melbourne. Examples are the *Baker Medical Research Institute,* and the *Ludwig Institute for Cancer Research,* which has a strong focus on IL-6, EGF and GM-CSF-receptor research. The *Peter MacCallum Cancer Institute* is targeting R&D at haemopoiesis, stem cell and tumor biology, whereas the *Macfarlane Burnet Centre for Medical Research* has a strong emphasis on virology research including HIV, HTLV-1, hepatitis viruses, rubella, herpes viruses, respiratory viruses, and papilloma viruses. The *St. Vincent's Institute of Medical Research* is specialized in bone cell biology, protein chemistry, including protein crystallography, hypertension research, and peptide hormone design. The *Howard Florey Institute of Experimental Physiology and Medicine* has its main activities in the molecular biology of peptide and protein hormones, whereas the *Murdoch Institute for Research into Birth Defects* emphasizes genetic services and human genetics research; and the *Walter and Eliza Hall Institute of Medical Research,* with its professional staff of about 80, is a major research center for genetic, molecular and cellular aspects of immunity and cancer, including diabetes and other auto-immune diseases. Sydney also offers strong support to the biomedical area, through, e.g., the *Garvan Institute of Medical Research,* with its emphasis on cancer, metabolic diseases, bone and neurological disorders, and the *Heart Research Institute,* strongly focused on molecular biology (mammalian cell cultures, cytokines, antioxidants), whereas the leading institution in Brisbane is the *Queensland Institute of Medical Research,* with its *Sir Albert Sakzewski Virus Research Laboratory.*

Universities

In 1992, Australia counted 32 national universities. About 5% of an age group entered university, and the total number of students was 560 000. About 5% of these were foreign students who paid education fees, half of them from Indonesia, Malaysia and Singapore. About 2/3 of all students went to the 16 universities located in populous New South

Wales, Victoria and Canberra. While there is no formal ranking of the universities, >50% of all competitive grants were won by just 5 universities, namely

- the University of Melbourne
- the University of New South Wales (Sydney)
- the University of Sydney
- the University of Queensland (Brisbane), and
- the University of Adelaide

Monash University in Melbourne is considered a further top place in science and engineering education.

Science and engineering are preferred fields of studies: at present they receive about 80% of all competititve grants from the Australian Research Council (ARC), the major funding agency for academic research, and the biological sciences take the lead. Biotechnology-related training and research plays an important role in many faculties, and representative examples are shown in Tab. 41.

In an attempt to commercialize results of academic R&D, the government has initiated several plans:

- Since 1989, several hundred "research centres" have been formed at the 32 universities, to enhance collaborative links within the university and to commercial partners.
- Many universities participate in the Co-operative Research Centres' program, initiated in 1990 (see below).
- Most universities have also established commercial arms. These are all members of ATICCA, the Australian Tertiary Institutions Commercial Companies Association. As an example, ANUTECH Pty Ltd., the commercial transfer point of the Australian National University (ANU) in Canberra, markets about 70 inventions of ANU personnel, and has grown to a turnover of about $A 40 million per annum.
- In 1992, the formation of the Australian Technology Group (ATG) was announced, built on the model of the British Technology Group in the UK, a body

for the commercialization of public sector R&D. The initial investment by the government was A$ 300000.

Cooperative research centers

In 1990, the Australian government introduced a new scheme to develop core capacity in areas of joint strategic research. A limited number of participant institutions (5–10) from industry, research centers and universities coordinate to form a "Co-operative Research Centre" (CRC), managed by a project leader, but not centralized as a single working unit. Since indigenous industrial R&D in Australia is relatively weak, transnational corporations are also accepted as participants in the scheme, but the percentage of public R&D in the CRCs is high. At present, 52 CRCs have been established; the central government contributes up to 50% of the total cost, the remaining funds come from the participating organizations. The 52 CRCs will receive over $A 870 million from the Government over the initial contract period of 7 years, or about 3% of Australia's national S&T budget. 10 new CRCs will be announced in the next and final round. CRCs have been formed in six broad fields of research, namely manufacturing technology, information and communications technology, mining and energy, agriculture and rural-based manufacturing, environment, and medical science and technology. The following CRCs include aspects of biotechnology:

- CRC for Molecular Engineering and Technology
- CRC for Industrial Plant Biopolymers
- CRC for Plant Science
- CRC for Waste Management and Pollution Control
- CRC for Tissue Growth and Repair
- CRC for Cellular Growth Factors
- CRC for Biopharmaceutical Research
- CRC for Eye Research and Technology
- CRC for Cardiac Technology
- CRC for Premium Quality Wool
- CRC for the Cattle and Beef Industry (Meat Quality)
- CRC for Aquaculture

Tab. 41. Biotechnology-related R&D Activities at Australian Universities (Samples)

University	Institute/Center	Research Targets
University of Queensland, Brisbane	Centre for Molecular Biology and Biotechnology	Gene regulation and expression in pathogenesis and embryogenesis
	Centre for Drug Design and Development	Computer-assisted design of enzyme inhibitors
	Dept. of Chemical Engineering	Fermentation, down-stream processing, genetic engineering
Queensland Institute of Technology, Brisbane	Dept. of Medical Laboratory Science	DNA-based diagnostics
University of New South Wales	Dept. of Biotechnology	Microbial and animal cell fermentation, downstream processing, enzyme reactors, genetic engineering
Australian National University, Canberra	Research School of Biological Sciences	Plant molecular biology, crop biotechnology, host–pathogen interactions
La Trobe University, Melbourne	Centre for Protein and Enzyme Technology	Enzyme technology, downstream processing, protein design
Monash University, Melbourne	Centre for Bioprocess Technology	Design of downstream processing steps
	Dept. of Microbiology	Metabolic design for bioremediation
University of Melbourne	Plant Cell Biology Research Centre	Molecular biology of plants, transgenic plants
	Dept. of Biochemistry and Microbiology	Microbial transformations, vaccines
Royal Melbourne Institute of Technology, Melbourne	Dept. of Applied Biology and Chemical Engineering	Biosensors
University of Adelaide	Waite Agricultural Research Institute	Molecular plant biotechnology
Flinders University of South Australia, Adelaide	School of Biological Sciences	Biotechnology for coal liquefaction, genome sequencing
Murdoch University, Perth	Dept. of Biochemistry	Mass production of algae

- CRC for Sustainable Cotton Production
- CRC for International Floriculture
- CRC for Food Industry Innovation
- CRC for Vaccine Technology
- CRC for Viticulture

As an example, the *CRC for Molecular Engineering and Technology* in Sydney aims to develop new sensing and diagnostic techniques for healthcare, therapeutics, process control and environmental monitoring markets. Research programs are focusing on making sensors which are more sensitive, reliable and selective than those currently available. There are three research programs covering the development of molecular sensing devices: molecular recognition, molecular switching and studies of cellular biochemistry. Center partners are the School of Chemistry and Department of Biochemistry at Sydney

University, the Department of Clinical Biochemistry at the Royal Prince Alfred Hospital; CSIRO's Divisions of Biomolecular Engineering, Applied Physics and Food Processing; and the AMBRI consortium, comprising AWA Ltd., Bioclone Australia Pty Ltd., CSIRO, Memtec Ltd. and Nucleus Ltd. The center plans to provide specialist graduate and postgraduate training. It has federal funding of A$ 2.2 million a year.

Patenting and genetic engineering
regulations

Australia is a member of the Budapest Treaty on the deposit of patented forms of life. The national depository is the National Bureau of Standards Laboratory (NSBL) in Sydney. In 1987, the validity of patents for pharmaceuticals was extended from 16 to 20 years, in order to compensate for a slow drug registration system.

As evident from the publication profile in Asia (Tab. 2), Australia's activities in genetic engineering are high. A survey on genetic experiments and releases of genetically modified plants and microorganisms is given in Tab. 42.

Three major regulations concerning genetic engineering were issued by the Genetic Manipulation Advisory Committee (GMAC), a part-time body established in 1988 and predominantly consisting of scientific experts.

Guideline 1 covers small-scale genetic manipulation work (<10 L, contained animals; 1981–90: 1755 cases). There is an exemption for genetic work on debilitated *E. coli* K12, and for self-cloning experiments.

Guideline 2 covers large-scale work with recombinant DNA (>10 L; 1981–90: 15

cases). It contains a GILSP category for work within the guidelines but of negligible risk. Such work requires only GILSP standard procedures.

Guideline 3 covers the planned release of recombinant DNA organisms.

A fourth guideline concerning transgenic animals was recently issued by a joint working party. Deliberate release of transgenic animals and plants exceeds 16 and includes tomatoes, cotton and alfalfa.

Over 82 institutional biosafety committees watch over the correct applications and adherence to these guidelines, which form a voluntary code. There can be sanctions, which are more punitive for publicly funded institutions than for privately funded research bodies. Continuing breaches of substantive requirements could result in an enquiry under public health and occupational health and safety legislation.

Acknowledgements

Generous financial support for collecting and compiling the information used in this survey has been provided by the Bundesministerium für Forschung und Technologie (BMFT) in Bonn, through its Project Group BEO in Jülich, Germany.

R. D. SCHMID wishes to express his sincere gratitude to Prof em. SABURO FUKUI, Kyoto University, to Prof. ISAO KARUBE, RCAST, The University of Tokyo, and to Mr. and Mrs. M. KAWASHIMA, Kurihashi, for invaluable support and help in assessing the situation of biotechnology in Japan.

B. H. CHUNG wants to express his appreciation to Mr. HO MIN CHANG in GERI for advice and support.

Tab. 42. Genetic Experiments and Release of Genetically Modified Organisms in Australia

	Proposals	Approvals	Current (Aug. 5, 1994)
Small-scale recombinant experiments		3394	1974
Large-scale (>10 liters) experiments	29	24	24
Releases of genetically modified organisms	39	27	27

JEANNIE SCRIVEN wishes to thank leading ASEAN scientists for their help – Dr. YONG-YUTH (Thailand), Dr. PHAN (Philippines) – and Alumni of the GBF's International Training Programme in ASEAN countries for their invaluable support.

JANE H. J. TSAI wants to express her sincere thanks to Dr. DING YONG (NCBD), Prof. AO SHI-ZHOU and Prof. LIANG-WAN YONG (AS) and to Prof. HSIN TSAI (DCB) for continuing support.

Personal Remark: The Authors

ROLF D. SCHMID holds a Dr. degree in Chemistry and is Professor of Technical Biochemistry at the Center for Bioprocess Engineering, University of Stuttgart, Germany.

JANE H. J. TSAI holds a Ph.D. degree in Biochemistry and is President of the Chinese–German Professional Association at Göttingen, Germany.

JEANNIE SCRIVEN holds a B. A. degree in political science and is the Chief Organizer of the International Training Programme in Biotechnology of the Gesellschaft für Biotechnologische Forschung (GBF) in Braunschweig, Germany.

SUSONO SAONO holds a Ph.D. degree in Microbiology and is Senior Research Staff of the R&D Center for Biotechnology at the Indonesian Institute of Sciences (LIPI) at Bogor, Indonesia.

BONG HYUN CHUNG holds a Ph.D. degree in Biochemical Engineering and is a Senior Research Scientist in the bioprocess engineering group at the Genetic Engineering Research Institute of KIST at Taeduk Science Town in Taejon, South Korea.

ALAN J. JONES holds a Ph.D. degree in Chemistry and a B. A. degree in Economics, and is currently Counsellor of Industry, Science and Technology at the Australian Embassy in Bonn, Germany.

8 References

Bibliography to Section 1

CHEMICAL ABSTRACTS SERVICE (1994), CA SEARCH, Columbus, OH: American Chemical Society.

FAO (Food, Agriculture Organization of the United Nations) (1991), *FAO Yearbook Production 1991*, Vol. 45, Statistics Series No. 104, Rome.

KEIZAI KOHO CENTER (1994), *Japan 1994, An International Comparison*, Tokyo: Taiheisha Ltd. ISBN 4-87605-025-2.

OECD (Organization for Economic Co-operation, Development) (1994), *Main Science and Technology Indicators 1994/1*, Paris: OECD Publications Service. ISBN 0-8213-1977-9.

SCIENCE AND TECHNOLOGY AGENCY (1994), *Indicators of Science and Technology, 1994*, Tokyo: Science and Technology Statistics Office. ISBN 4-17-152269-2.

THE WORLD BANK (1991), *The World Bank Atlas 1991*, Washington DC. ISBN 1-08213-1977-9.

Bibliography to Section 2 (Japan)

THE GOVERNMENT OF JAPAN (1990), *Action Program to Arrest Global Warming*.

KEIZAI KOHO CENTER (1994), *Japan 1994, An International Comparison*, Tokyo: Taiheisha Ltd. ISBN 4-87605-025-2.

MIYATA, M. (1994), Toward a new era in biotechnology, *Sci. Technol. Jpn.*, 10–11.

NIKKEI BIOTECH (1994), *Nikkei Baionenkan 94* (Bio Yearbook), Tokyo. ISBN 4-8222-0822-2.

SCHMID, R. D. (1988), Tsukuba, Kansai Science City and 26 mal Technopolis: Wissenschaftsstädte in Japan, *Chem. Uns. Zeit*, 149.

SCHMID, R. D. (1991), *Biotechnology in Japan*, Heidelberg: Springer Verlag. ISBN 3-540-53554-3.

SCHMID, R. D. (1991), Verbundforschung in Japan, *Nachr. Chem. Tech. Lab.*, 39.

SCHMID, R. D. (1993), Globale Umweltprojekte in Japan, *Nachr. Chem. Lab.* **41**, 428–434.

SCIENCE AND TECHNOLOGY AGENCY (1991), *White Paper on Science and Technology 1991*. ISBN 4-88890-213-5.

TOYO KEIZAI INC. (1994), *Japan Company Handbook*, First Section, Autumn 1994, Tokyo. ISSN 0288-9307.

Bibliography to Section 3 (Korea)

CHO, M. J. (1990), Present status of genetic engineering and biotechnology in South Korea, *Crit. Rev. Biotechnol.* **10**, 47–67.

GWYNNE, P. Directing technology in Asia's 'Dragons', *Perspectives,* 12–15.

HAHM, KYONG-SOO (1992), *Present Status of Biotechnology R&D in Korea.* Genetic Engineering Research Institute.

KIM, L., DAHLMAN, C. J. (1992), *Technology Policy for Industrialization: An Integrative Framework and Korea's Experience,* pp. 437–451, Amsterdam: Elsevier Science Publishers B.V.

LIU, J. R. (1994), *Current Status of Plant Biotechnology in Korea,* Biotech 2000 Symposium and Biofair, Seoul, Korea.

OECD (Organization for Economic Co-operation and Development) (1994), *OECD Economic Surveys: Korea,* Paris: OECD Publications Service. ISBN 92-64-14129-4.

PARK, Y. H. (1994), *The Second Korea-US Science and Technology Cooperation Forum,* May, 25, 1994, Washington, D.C.

SWINBANKS, D. (1993), What road ahead for Korean science and technology? *Nature* **346,** 377–384.

TOYO KEIZAI INC. (1993), *Asian Company Handbook* 1993, Tokyo.

Bibliography to Section 4 (China)

CHINA NATIONAL CENTER FOR BIOTECHNOLOGY (1992), *Biotechnology in Progress – Biotech "863"* (in Chinese), Beijing.

HAN YING-SHAN (1991), Developing biotechnology opportunities in China, *Bio-Technology* **9,** 711–712.

ITIS BIOTECHNOLOGY INFORMATION SERVICE REPORT (1992), *Biotechnology in Mainland China* (in Chinese), Vol. 4, Taipei: DCB.

MANG KE-QIANG (1991), China's biotechnology in progress, *Bio-Technology* **9,** 705–709.

PEI, S. Y. (1992), Development and achievement of biotechnology in China (in Chinese), *Bioindustry* **2**(4), 273–280.

STATE SCIENCE, TECHNOLOGY COMMISSION (1992), *High-Technology Research and Development Programme of China.*

Bibliography to Section 5 (Taiwan)

DCB INFORMATION SERVICE (1994), *Bulletin on Biotechnology and Drug Industries* (in Chinese), Vol. **2**(6), pp. 54–60.

MAU, D. G. (1992), Development of biotechnology industry in developing countries: The Taiwan experience, *J. Am. Chem. Soc.* **1054,** 559–562.

SOONG, TAI-SEN (1991), Current industrial biotechnology development in Taiwan, *Agro-Industry Hi-Tech* **2**(2), 11–17.

SU, YUAN-CHI (1992), Development of environment technology in Taiwan, *Bioindustry* **3**(1,2), 13–20.

Bibliography to Section 6.1 (Indonesia)

ANONYMOUS (1993), *Science and technology Indicators of Indonesia.* BPPT, Ristek, LIPI. Jakarta.

ASIAN BUSINESS (1993), *Indonesia Statistics.*

CBS (Central Bureau of Statistics) (1992), *Statistical Yearbook of Indonesia,* 664 pp., Jakarta.

KOMEN, J., PERSLEY, G. (1993), *Agricultural Biotechnology in Developing Countries. A Cross-Country Review* (ISNAR Res. Rept. 2), 45 pp.

MENTERI NEGARA RISET DAN TEKNOLOGI (MNRT) (1988), Direktori Bioteknologi Indonesia (*Indonesian Biotechnology Directory*), 296 pp. Jakarta.

MENTERI NEGARA RISET DAN TEKNOLOGI (MNRT) (1990), Kebijaksanaan Pengembangan Bioteknologi di Indonesia (*Policy for the Development of Biotechnology in Indonesia*), 37 pp. Jakarta.

MENTERI NEGARA RISET DAN TEKNOLOGI (MNRT) (1993), Pedoman Penelitian Rekayasa Genetika di Indonesia (*Guidelines for Genetic Engineering Research in Indonesia*), 33 pp., Jakarta.

PAPITEK-LIPI (1994), *Routine and Development Budget.*

PERSLEY, G. J. (1990), Beyond Mendel's garden: Biotechnology in the service of world agriculture, *Biotechnology in Agriculture* No. 1, (CAB International).

SAONO, S. et al. (Eds.) (1986), *Traditional Food Fermentation as Industrial Resources in ASCA Countries,* (The Indonesian Institute of Sciences (LIPI)), 259 pp. Jakarta.

SAONO, S. et al. (Eds.) (1986), *A Concise Handbook of Indigenous Fermented Foods in the ASCA Countries* (The Indonesian Institute of Sciences (LIPI)), 237 pp., Jakarta.

SAONO, S. (1991), Pusat-pusat Informasi dan Koleksi Jasad Renik di Indonesia (*Information Centers and Collection of Cultures in Indonesia*), (Komisi Pelestarian Plasma Nutfah Nasional (National Commission for the Conservation of Genetic Resources), 52 pp. Bogor.

SAONO, S. (1991), Pengembangan Bioteknologi Guna Menunjang Pembangunan yang Lestari Dalam Kurun Waktu 25 Mendatang (*Development of Biotechnology in Support of Sustainable Development for the Next 25 Years*) (J. Mikrobiol. Ind. 1(2)), pp. 4–9.

SAONO, S. (1992), *Applications of Biological Nitrogen Fixation Technology in Southeast Asia,* in: SILVEIRA, M. P. W. (Ed.): ATAS (Advanced Technology Assessment System) Expanding the Capacity to Produce Food, UN, New York pp. 204–211.

SAONO, S. (1994), *Non-medical Application and*

Control of Microorganisms in Indonesia, in KO-MAGATA, K., et al. (Eds.): International Workshop on Application and Control of Microorganisms in Asia, Proceedings, S.T.A., Riken J.I.S.T.E.C., Tokyo.

SASTRAPRADJA, D. S., et al. (Eds.) (1989), Keanekaragaman Hayati Untuk Kelangsungan Hidup Bangsa (Biodiversity for National Survival), Pusat Penelitian dan Pengembangan Bioteknologi-LIPI, 98 pp., Bogor.

SOEDIBYO (Ed.) (1994), Date Anggaran Dan Kegiatan Iptek Departemen Dan Non-Departemen (Data on Budget and Activities of Departments and Non-Departments), (PAPIPI-LIPI), Jakarta, 152 pp.

STATE OF THE NATION ADDRESS OF THE PRESIDENT OF THE RI (1993) State Secretariat, Jakarta.

UNDANG-UNDANG (1989), Tentang Patent (RI Laws No. 6, 1989 on Patent), Jakarta.

WORLD BANK-ISNAR-AIDAB-ACIAR (1989), *Biotechnology Study Project Papers.* Summary of Country Reports, pp. 43.

Bibliography to Section 6.2 (Philippines)

CAPITAN, S. S., PHAN, V. Q., DALMACIO, I. F. (1993), The status of animal biotechnology in the Philippines, *Philippine J. Biotechnol.* **4,** 117–128.

CRUZ, D., NAVARRO, R. E., M. J. (18–29 Jan. 1994), Country Report: Agricultural Productivity Through Biotechnology in the Philippines, paper presented at the study meeting on biotechnology applications in agriculture, Asian Productivity Organisation, Tokyo.

Hannoversche Allgemeine Zeitung (1994), Der kranke Mann Asiens, July, 23, 1994.

UNIDO (1993), *Genetic Engineering and Biotechnology Monitor,* Issue No. 43, March.

Bibliography to Section 6.3 (Malaysia)

ANONYMOUS (1994), Malaysia's problems with prosperity, *The Economist* **332,** No. 7872, 59–60.

SASSON, A. (1993), *Biotechnologies in Developing Countries – Present and Future* (UNESCO, Ed.), Paris.

NIK ISMAIL, N. D., ZAKRI, A. H. (1992), Status of commercialization of biotechnology in Malaysia, in: *Commercialization of Biotechnology in a Developing Economy.* (ZAKRI, A. H., Ed.), Bangi Malaysia: Penerbit UKM Publishers.

NIK ISMAIL, N. D., EMBI, M. N. (1994), *Strengths and Weaknesses of Agriculture, Biotechnology* – Papers from the study meeting on Biotech Applications in Agriculture, Tokyo.

Bibliography to Section 6.4 (Singapore)

ANONYMOUS (1993/94), *Country Profile – Singapore,* Economist Intelligence Unit.

NATIONAL SCIENCE AND TECHNOLOGY BOARD (1992), *National Survey of R&D in Singapore.*

PUBLICITY DIVISION, MINISTRY OF INFORMATION AND THE ARTS (1993), *Singapore 1993 – A Review of 1992.*

TAN, T. W., KHOO, H. E., BYRNE, S. P., *Singapore Biotechnology Directory.*

Bibliography to Section 6.5 (Thailand)

DEPARTMENT OF ECONOMIC RESEARCH, BANK OF THAILAND (1994), *Thai Economic Performances in 1993 and Prospects for 1994.*

PIETRZYK, J. Z. (1992), Biotechnology in South East Asia and opportunities for foreign investment, *Biotech Forum Europe* **9** (10).

YUTHAVONG, BHUWAPATHANAPUN, Y. S. (1992), Biotechnology in Thailand, *ASEAN J. Sci. Technol. Dev.* **9** (1) 1–10.

Bibliography to Section 7 (Australia)

ANONYMOUS (1992), *CRC Compendium,* Canberra: Australian Government Publishing Service.

AUSTRALIAN BIOTECHNOLOGY ASSOCIATION LTD. (1993), *Australian & New Zealand Biotechnology Directory 1993,* Melbourne: E. & M. Bradley Pty Ltd.

AUSTRALIAN SCIENCE AND TECHNOLOGY COUNCIL (1989), *Profile of Australian Science,* Canberra: Australian Government Publishing Service.

CENTRE FOR RESEARCH POLICY (1992), *Report No. 7: Concentration and Collaboration: Research Centres in the Australian University System.*

DITAC (1992), *Australian Science and Innovation Resources Brief.*

JONES, A. J. (1987), *Selecting Technologies for the Future: A Discussion Paper,* Canberra: Canberra Publishing and Printing Co.

THE PARLIAMENT OF THE COMMONWEALTH OF AUSTRALIA (1992), *Genetic Manipulation: The Treath of the Glory?* Canberra: Australian Government Publishing Service.

14 Biotechnology and Biological Diversity

Masao Kawai

Tokyo 105, Japan

Introduction

The diversity of organisms which inhabit tropical regions is much greater than that of other geographical regions, and this diversity is particularly remarkable in tropical forests. However, tropical forests are decreasing rapidly due to various kinds of economic activity such as commercial logging and development, so there is now concern that valuable biological and genetic resources are being lost and that the global environment is deteriorating.

The conservation and restoration of tropical forests is being demanded by the world community. Yet tropical forests are valuable resources for the countries which possess them, so it is not possible to only consider a conservation–protection option. It is necessary therefore to achieve some sort of compatibility or balance between the conservation and utilization of tropical forests.

This chapter examines, from a global point of view, various aspects of tropical forests, such as biological diversity, the present state of the forests, movements afoot to protect the forests, and the activities of research organizations. Furthermore, after investigating measures to achieve compatibility between the conservation and utilization of tropical forests, as well as the potential for biotechnology to contribute to these measures, long-term policy proposals are made within a framework of international cooperation.

This chapter is a summary of a report prepared by a committee established in 1991 by the Japan Bioindustry Association (JBA) by researchers and policy planners from related fields. The authors of the respective sections are as follows.

List of Contributors

(Numbers in the parentheses mean section number(s) to which the contributor made his contribution.)

FUKUI, KATSUYOSHI (4, 6)
  National Museum of Ethnology
HARADA, HIROSHI (7)
  Institute of Biological Sciences, University of Tsukuba

INOUE, TAMIJI (1, 6)
  Center for Ecological Research, Kyoto University
ISHIKAWA, FUJIO (7, 8)
  Japan Bioindustry Association
IZAWA, KOUSEI (5)
  Faculty of Education, Miyagi University of Education
KAKEYA, MAKOTO (4, 5)
  Center for African Area Studies, Kyoto University
KANOH, TAKAYOSHI (5)
  Primate Research Institute, Kyoto University
KAWAI, MASAO (5, 8)
  Japan Monkey Center
KOMAGATA, KAZUO (4, 6)
  Faculty of Agriculture, Tokyo University of Agriculture
KOSHIMIZU, KOICHI (4)
  Faculty of Agriculture, Kyoto University
MAEDA, HIDEKATSU (4, 7)
  National Institute of Bioscience and Human-Technology (MITI)
MATSUMOTO, TADAO (3)
  College of Arts and Sciences, University of Tokyo
MIYACHI, SHIGETOH (4, 5, 7)
  Marine Biotechnology Institute Co. Ltd.
OGINO, KAZUHIKO (1, 5, 8)
  College of Agriculture, Ehime University
OHIGASHI, HAJIME (4)
  Faculty of Agriculture, Kyoto University
OHSAWA, HIDEYUKI (5, 6)
  Primate Research Institute, Kyoto University
TABATA, MAMORU (4, 7)
  Faculty of Pharmaceutical Sciences, Kyoto University
TAKAYA, YOSHIKAZU (2)
  Center for Southeast Asian Studies, Kyoto University
TAKENAKA, OSAMU (7)
  Primate Research Institute, Kyoto University
YAMADA, ISAMU (3, 5)
  Center for Southeast Asian Studies, Kyoto University
YAMAMOTO, NORIO (4, 5, 6)
  National Museum of Ethnology

Other Contributors

HOSOKAWA, MIKIO
  Director, MITI
JINTANA, VIPAK
  Graduate School, Ehime University
MAEDA, NARIFUMI
  Center for Southeast Asian Studies, Kyoto
  University
MASUDA, AKIHIRO
  Director, MITI
MASUDA, MASARU
  Director, MITI
MATSUZAWA, TETSURO
  Primate Research Institute, Kyoto University
MIYAWAKI, AKIRA
  Institute of Environmental Science and
  Technology, Yokohama National University
NAKAMURA, TAKEHISA
  Faculty of Agriculture, Tokyo University of
  Agriculture
NISHIZAWA, YOSHIHIKO
  Adviser, Sumitomo Chemical Co., Ltd.

MITI: The Ministry of International Trade
and Industry

1 Biological Diversity in the Tropics

Biological diversity was attained through the process of the evolution of organisms. Its understanding therefore must be based upon the basic theory of evolution. Historically, it has been thought that species are independent units and that different species would not merge with each other (genealogical tree). Recently, however, it has been found at the DNA level that there is an exchange of genes even among species which are genealogically separate, and that this is the mechanism by which new species are formed (genealogical network). According to Darwin's Theory of Natural Selection, the driving force for evolution is intra- and inter-species competition; however, recently the importance of coexistence among different organisms has become clear. It has been found that tree growth depends upon the symbiotic relationships between the tree and ectotrophic mycorrhiza which provide nutrient salts and insects that carry pollen. In this way it can be seen that a species is not an independent unit, and that symbiosis is important for the continued existence of the species. Therefore, biological diversity must be conserved based upon the premise that the entire ecosystem is a system in which such symbiotic relationships are involved.

The high degree of biological diversity of tropical forests is also evident from the fact that while tropical forests only account for about 3% of the Earth's surface area, they contain more than 50% of the species. In order to comprehend the biological diversity of the tropics, it is necessary to consider the natural history of the Earth. The last Ice Age, which reached its peak 20000 years ago, brought about a dry period in the tropics which resulted in the shrinkage and disruption of the tropical forests. Many organisms most likely became extinct during this period of upheaval. However, it must also be realized that the differentiation of species was accelerated by the dislocations which occurred. This drying out was particularly severe in the tropical regions of Africa and Central and South America. In Southeast Asia, the connection of land masses due to the drop in sea level at this time is the reason why the biological diversity of the area is not advanced even though the region is presently made up of islands.

Several theories have been proposed to explain the biological diversity of tropical forests. When discussing the conservation of biological diversity, it is important to consider the hypothesis of disturbance. Tropical forests undergo spatial disturbances of different magnitudes (forest fires, falling trees), as well as temporal disturbances which may occur over periods of anywhere from 4 years to hundreds of years. It has been observed that forests which experience a suitable degree of disturbance exhibit a high level of biological diversity. Humans have been living in tropical forests for thousands of years without causing a decrease in the biological diversity of the forests. This is because the scale and frequency of the disturbances caused by humans before the advent of modern civilization were similar to or even less than those of natural disturbances.

The destruction of tropical forests in the last several decades due to modern civilization has resulted in more disruption than which occurred during the last Ice Age. What is urgently needed is the establishment of protected areas of tropical forests where human impact can be kept to a minimum. Modern ecology is not yet at a level where it can show standards to establish protected areas. However, knowledge concerning groups of islands does offer some standards related to the extent of protected areas, the diversity of the habitats, and the species diversity which the protected areas can maintain. Using this information, a system must be set up with which to create protected areas and monitor biological diversity. It is possible to use methods such as dynamic planning techniques and adaptive management techniques for this purpose.

2 Humans and Nature in Tropical Regions

Tropical regions are diverse. Therefore, in order to accurately discuss tropical regions, they must be classified according to the special features of each particular region. It is not enough to base this classification of the regions only upon ecology, but rather, one should include social and cultural aspects as well because they are also related to the conservation and use of biological resources.

The islands of Southeast Asia can be assumed to consist of a single tropical zone in terms of the social, cultural and ecological classification alluded to above. The ecological characteristics of insular Southeast Asia then is one of an archipelago of tropical rain forests.

The people of this insular tropical rain forest region traditionally lived by harvesting forest products and practicing slash and burn agriculture. The harvesting and transporta-

tion of the forest products were carried out through cooperation between sailors from around the world and the Malays. Further inland relatively isolated farmers lived by practicing slash and burn rice farming. They respected the gods of the forest, and lived with the cosmic conception to live in harmony with the other living creatures of the forest.

With the arrival of the age of colonization, the capitalist economy thrust itself upon the tropical islands of Southeast Asia, and introduced plantations, commercial logging and rice cultivation to these traditional societies. The rubber plantations of peninsular Malaysia are a typical example of plantations that were established in Southeast Asia. Many Indian people from India were relocated to Malaysia by the British colonial government to work as laborers on the rubber plantations which produced rubber to meet the demands of the American automobile industry. Today, the problem of shifting farmers moving into lands which have been logged for timber production has become serious. In order to find a solution to this problem, it would appear that countermeasures which resemble a "Basic Law for Disaster Prevention" is required. Indonesia, which gained its independence after the Second World War, is putting effort into expanding rice cultivation, and subsequently the number of farmers is increasing. However, these farmers have a strong propensity towards risk-taking.

Two important points emerge from the above. The first concerns the definition of what exactly a "region" is. A region is created from competition between two worlds, the first an inner world that has its foundation in its native ecology, and the second which is an outside civilization that is demanding the inner world's transformation. The inner world of insular Southeast Asia has maintained a balanced state based upon a traditional way of life. The disappearance of tropical forests, which is a global problem, has been brought about by the foreign civilization called capitalism. This point must be thoroughly understood. The second point to note is that the inner world of the islands of Southeast Asia possesses valuable civilization and historical assets which are considered to be of great importance in constructing the global order of the 21st century. That is because the indigenous people in the inner world have respect for the gods of the forest, and live in harmony with all the living creatures. The risk-taking vitality is also a valuable asset.

3 Current Situation Concerning the Conservation and Utilization of Tropical Forests

Tropical rain forests are not simple forests, but rather are very diverse. In Southeast Asia there are mangrove forests situated along the seashore in lowland marshy areas, peat swamp forests, and moist fresh water forests. In lowland areas, mixed Dipterocarpaceae forests occupy a large portion of the entire forest area up to about 1500 meters above sea level, while above this are montane forests and subalpine forest zones. The geographical conditions of each are different and the types of trees which grow are diverse, so a high degree of knowledge about tropical forests is needed when discussing them. All of these forests have been disturbed to some degree by humans. In order to conserve genetic resources under these sorts of circumstances, the natural characteristics of tropical regions, as well as the human involvement and biological features must be understood from a long-term point of view. There are a multitude of different insects in tropical rain forest ecosystems which are fundamentally dependent on plants for their survival. However, at the present time, a mere 10% of the insects have been taxonomically classified. Very little is known about their ecology and behavior. Among the diverse array of insects which inhabit tropical forests, those which consume wood-fiber (xylophagy) and detritus play a significant role in the recycling of materials in the ecosystem. These insects function as the degraders of dead plant material, and they have intimate

symbiotic relationships with microorganisms such as bacteria, fungi, and protozoa. The role played by termites is especially important, since they contribute greatly to the recycling of carbon and nitrogen. Better policies for the conservation and utilization of tropical forests can be brought forth by deepening our understanding of the interactions among the plants, animals and microorganisms in this kind of ecosystem.

4 Useful Products Found in Tropical Forests

The range of secondary chemical metabolites of tropical forest plants which diversify in response to the species, ecological and genetic diversity of tropical forests, is also extremely diverse. Furthermore, it has recently been shown that these chemical substances function as chemical signals such as allelopathy(s), pheromones, phytoalexins and other compounds between plants and other plants, plants and animals, and plants and microorganisms. These metabolites influence the natural environment and play an important role in the structure of tropical ecosystems. In other words, tropical forests exist as forests because of their biological diversity, and they are treasure troves containing a majority of the Earth's resources of genes and physiologically active substances.

The biological resources of tropical forests have been effectively used in a variety of ways, including as primary and secondary products such as materials for everyday use, timber, industrial products, foods, drugs and agricultural chemicals. It should be noted in particular that there have been many discoveries of active drug ingredients that are vital to modern medicine being isolated from tropical plants. Several examples of the success of recent investigative research can be given, such as the isolation of the alkaloid anti-cancer agent vinblastine from the rosy periwinkle, the antiphlogistic compound cortisone, and diosgenin, a substance isolated from Mexican mountain potatoes that is a raw material used in the production of contraceptive pills. Artemisin, a compound which exhibits high efficacy against chloroquine-resistant malaria, has been isolated from a traditional folk medicine, and saponin, a substance which kills the snail that is the intermediary host for African schistosomes, has been isolated from an African leguminous plant. Thus, these compounds are helping to improve the health of the local inhabitants. Catechol, obtained from the catechu plant of tropical Asia, has been marketed in approximately 40 countries in Europe and elsewhere for treatment of liver disease.

It must be borne in mind that the biological resources now being used are the direct or indirect cultural heritage of the inhabitants of tropical forests in Southeast Asia, Africa and Central and South America. The slash and burn farming Hanunoo tribe in the Philippines, the hunter-gatherer Efe pygmy tribe in Northeastern Zaire and the slash and burn farming Lese tribe are examples of indigenous people who have been living in harmony with nature, in particular plants, and making use of the abundance of the forests. These "forest people" possess a remarkable amount of knowledge concerning the useful plants and animals of the tropical forest. However, changes in their way of life are occurring along with the wave of development that has been happening, and they are increasingly coming into contact with surrounding cultures. As a result, there is the danger that the valuable traditional knowledge which has been passed down from generation to generation is disappearing along with the tropical forests. In this respect, while the conservation of tropical forests is of course desirable, it is also hoped that biotechnology, which has been developing at a rapid pace in recent years, can be applied to the utilization of tropical forests. In other words, biotechnology should be used for the utilization of tropical forests, starting with basic research using small quantities of living material, to high-level applied research. This work should be carried out by obtaining and conserving the tropical forest of a particular area that is to be used as an experimental research station for biological diversity studies through cooperation with the local inhabitants, and by exploring the forest for the specific living phenomena that are utilized as cultural traditions. Through such endeavors, in the end not only can valuable results be expected related to the development of value-added pharmaceuticals, highly selective agricultural chemicals, and biochemicals, but it is also possible to conserve genetic resources as well as breed and select useful species of organisms. It is not unrealistic to say that these achievements will contribute to improving the lifestyles and economies of the local inhabitants and the conservation of tropical forests.

5 Research Facilities and Organizations in Tropical Areas

The majority of countries located in tropical regions are developing countries, which makes research activities into the "conservation and utilization of biological diversity" economically and politically difficult. These countries recognize the need for conducting this type of research, however, they lack both the funding and human resources required to effectively deal with the situation. While this situation is common to these countries, it should be emphasized that the various countries in Africa, Latin America and Southeast Asia each possess regional and national characteristics which reflect differences with respect to the number of research facilities, the state of their facilities, level of activity, and so on.

Among Southeast Asian countries there is a variety of needs and demands concerning tropical forest resources, ranging from that of Malaysia which is attempting to develop economically by exploiting its forests for timber, to Thailand which has adopted a policy of banning commercial logging in order to protect the environment. There are also differences with respect to the type of research activities being carried out. For example, Malaysia is pressing ahead with research that focuses mainly on forestry, while Thailand is putting its effort into the non-destructive utilization of forests and afforestation research, as well as the introduction of biotechnology.

The East African nation of Kenya is the location for the headquarters of the United Nations Environment Program (UNEP) and three agricultural-related international research organizations. The facilities and budgets are inferior to those of other international organizations, and research using biotechnology is showing little progress. National research organizations are conducting research on realistic topics using assistance provided by industrialized countries. In Zaire which still has a large area of intact tropical rain forest, research activities have stagnated and researchers' salaries have been delayed because of the inflation and political instability which has been increasing in recent years.

Several South American countries are troubled by economic crises and anti-government guerilla insurgency campaigns. In 1991, Columbia introduced a new constitution which declared the need to conserve the country's ecosystems and the obligations of the Columbian people to protect the ecosystems. National universities and research facilities in Columbia have carried out several tropical research projects, mainly with funding and personnel assistance from overseas countries. Peru, however, continues to be tormented by economic problems and poverty, and is plagued by terrorist attacks and poor social order. The Peruvian government and researchers in Peru have been unable to really start any work on the conservation and utilization of genetic resources, although activities are being carried out at the Nature Conservation Foundation and the CIP (International Potato Center) which were set up with assistance from overseas. On the other hand, Venezuela's SADA (Autonomous Service for the Environment Development of Amazon Federal Territories) organization is conducting some unique studies in the Amazon on the non-destructive development of tropical forests, and has received some attention for its achievements to date.

It can be seen that the situation surrounding research facilities varies depending upon the region and country, and is deeply dependent on financial and personnel assistance from industrialized countries. It is also necessary to realize that international research organizations have an important role to play. Japan has much to learn from these individual situations in order that it can come up with policies which are accepted by the international community, contribute to supporting the completion of self-supporting research systems in these developing countries, and build sustainable cooperative relationships.

Japanese researchers are deeply committed to the local community, and have carried out activities such as surveys, cooperative research and nature conservation. A major portion of them have been accumulated through "hands-on" research activities. The Korup Project in Cameroon, which originated due to an appeal by the primatologist Dr. GARTLAN, is an example of an internationally well known comprehensive nature conservation plan. One way for Japan to make a unique international contribution is to continue activities using the accumulated experience and achievements of Japanese researchers, while at the same time learning from examples such as the Korup Project.

6 International Trends Related to the Conservation and Utilization of Biological Resources

There is growing recognition in recent years of the importance of biological resources, and it is widely known that internationally there is a strong movement towards the conservation and utilization of these resources. It appears that there are three spheres of function, i.e., professionals, politics and the public, which are intricately involved in this issue. It is probably reasonable to conclude that the confusion which occurred at the Washington Treaty meeting held in Japan recently was due to their participation. A problem which should essentially be addressed from an ecological standpoint was swayed by those trying to protect their own individual interests and by North-South issues which have become manifest recently.

In actual fact, humans have been aware of the importance of biological genetic resources ever since they started to domesticate animals and cultivate plants. Humans have been enjoying nature's abundance by controlling plant and animal reproduction, and by conserving and fostering the diverse changes which resulted. This way of living in direct contact with the natural world was changed, due to the industrial revolution, in a direction in which modern science developed nature and industry started to utilize the diverse array of biological resources. The advances made in microbial engineering symbolize biotechnology's progress, while pharmaceuticals such as antibiotics which have contributed greatly to mankind can be described as gifts of nature that were born in the laboratory. Because of this, when one talks about the utilization of biological and genetic resources in industrialized nations, generally it is the field of biotechnology which comes to mind. There is an ever increasing international movement towards the utilization of biological and genetic resources led by industrialized countries through their advanced science.

Many storage facilities for microorganisms have been established, and a network has been formed among them.

In the agricultural field from 1971 and onwards, research institutes were set up in each region of the world, centering on developing countries, by the Consultative Group on International Agricultural Research (CGIAR). As of 1991 there was a total of 17 such institutions, including the International Plant Genetic Resources Institute (IPGRI), the International Livestock Center for Africa (ILCA), and the International Potato Center (CIP). At these research institutions researchers from around the world and local staff cooperate to improve the food situation in developing nations according to the founding ideology. Today, these institutions are expanding their field of activity to include protection of the global environment and studies on the diversity of genetic resources through the networks of each institution.

The United States has been cooperating with Latin American countries to systematically research tropical regions since the beginning of this century, and has accumulated a vast amount of knowhow concerning tropical research. Two organizations which are playing a central role are the Smithsonian Institution and the Organization for Tropical Studies (OTS).

In 1986 an international symposium on the conservation and utilization of genetic resources was held in Ethiopia. The meeting was assisted by the German foundation Deutsche Gesellschaft für Technische Zusammenarbeit (GTZ) and the International Development Research Center (IDRC) of Canada. The International Ethnobiology Conference held in Mexico in 1992 provided new direction for the conservation and utilization of biological diversity.

Some countries are fortunate enough to be endowed with rich biological diversity while others are not. Even if there are differences between developing and industrialized nations with respect to awareness and approaches to the issues of conservation and utilization of biological resources, they are matters which are important for all humans that have fallen upon the shoulders of current civilization. An attitude to support such steady research activities as mentioned above among nations and international institutions for a long period from now on can be seen in every aspect of new international moves.

7 Present State and Potential of Biotechnology for the Conservation and Utilization of Tropical Forests

Tropical forests are treasure troves of species and biological resources. The conservation of tropical forests is a subject that is extremely important for the entire world so that global ecosystems can be supported and the sustained utilization of the diversity of the organisms contained therein achieved. Tropical forests play an important role in the fixation and recycling of the greenhouse gas carbon dioxide, so from this point of view their conservation is also important.

The biological and genetic resources of the tropical forests have been used by humans for thousands of years until now, and advances in science and technology will likely increase the opportunities and scope of utilization. This therefore makes the conservation of tropical forests even more essential. In particular, the use of biological resources contained in tropical areas is indispensable for the people who inhabit the regions, and the compatibility of conservation and utilization is the key to successful conservation and utilization.

What kind of role will biotechnology play in the establishment of this compatibility, and indeed can it play a role? The basic concepts are as follows. (1) Research into the ecosystems of tropical regions from a global viewpoint, including animals, plants and microorganisms; (2) carry out basic research on the search, collection, identification, classification and storage of biological species and genetic resources, and accumulation of the results. These two will serve as the basics with which to proceed with studies on biological diversity. Based on these, (3) the conservation of tropical forests and the utilization of biological diversity will be advanced in a manner in which they are compatible with one another. Finally, (4) these activities will be carried out sustainably in harmony with local communities and through education of the people and training of personnel.

The present state of biotechnology in several countries situated in tropical regions has been summarized from a rather broad view. Brazilian universities and other institutions are conducting biotechnology research and application in a wide range of fields, such as industrial fermentation, enzymes, and the fixation of nitrogen by algae and microorganisms. In Thailand, Indonesia and Malaysia, plant cell and tissue culture technology is being applied to the breeding and propagation of plant species. Biotechnology in the fields of foods and agriculture has already attained a considerable level. The forestry policies of these countries are diverse, ranging from a complete ban on commercial logging in Thailand to Indonesia's policy of exploiting forests in remote rural areas before establishing plantations. In Kenya and Tanzania, it can be seen that research into the diseases of domestic animals and primate research are special features of their biotechnology programs. While biotechnology is viewed as being important, its introduction seems to be a topic for the future and not the immediate present.

Animal, plant and microbial biotechnology, within the framework mentioned at the beginning of this section, are expected to contribute significantly to achieving the balance between the conservation and utilization of tropical forests through the continuation of fundamental research and the accumulation of these results. It is important to pay attention to the fact that these various species of plants, animals and microorganisms, through their symbiotic relationships, are mutually intertwined and related to constitute an ecosystem.

8 Summary and Proposals

8.1 The Need to Conserve Species as a Global Environmental Problem

In recent years, the natural environment is being destroyed by industrial and social activities due to the rapid advances made in science and technology, as well as the population explosion. In tropical regions, the disappearance of tropical forests is particularly advanced, and many species are vanishing. This loss of a huge amount of unused biological resources is not only an economic loss, but is also a reckless act which amounts to the human-mediated destruction of the irretrievable process of evolution. As tropical rain forests are the habitat of primates, the destruction of these forests is akin to the destruction of our own place of origin, and can be described as accelerating the loss of humanity.

The diversity of life on the planet has contributed to the lifestyle of humans for thousands of years in the form of daily commodities, drugs and so on. In recent years particularly, the usefulness of biological species has been recognized through the development of new biotechnology, and the utilization of these species is being expected in the fields of foods, agriculture, medicine, chemicals and environmental conservation. The preservation of species is an urgent issue which needs to be addressed so that the future potential of these species can be realized.

The issue of conservation of biological diversity has now come to be recognized as a global environmental problem. At the United Nations Conference on Environment and Development (UNCED) which was held in Rio de Janeiro, Brazil, in June 1992, biodiversity and its conservation was a major topic, and more than 100 countries signed the Biodiversity Convention. The conservation of infinitely valuable biological diversity, together with its sustainable use, is an urgent issue for all of mankind. At UNCED, biological diversity and biotechnology were taken up in the action program called Agenda 21 and the relationship between biological diversity and biotechnology as well as means for the transfer of technology was revealed. Biotechnology is one powerful method to conserve and utilize tropical forests.

8.2 Sustainable Utilization of Genetic Resources

In tropical areas the genetic resources are located in moist zones, arid zones, highland zones, and aquatic environments like oceans and rivers. In the moist tropics, tropical rain forests are the most important. Tropical forests account for a mere 3.3% of the total surface area of the Earth, yet they contain 50% of all species. Approximately 1.4 million species have thus far been identified on Earth. However, it is assumed that between 10 and 30 million species exist in this region. This shows that tropical rain forests are indeed treasure troves of unknown genetic resources. Agriculture and the raising of livestock are carried out in arid tropical zones, and these

regions are an important place for genetic resources of cultivated plants and domesticated livestock, as well as being the habitat for an abundance of wildlife. Highland regions, such as typically in Ethiopia and Latin America, occupy an important position as the source of many types of plants cultivated by humans. For example, there are several thousand different varieties of potato species known, and they are an important gene pool for plants that can be used for cultivation.

Tropical regions possess a massive amount of useful resources, most of which are presently unknown. Therefore conservation policies, especially for tropical rain forests, should be a top priority. Protected regions must comprise large areas because biological diversity is dependent upon the complex mutual interaction among plants, animals and microorganisms. It is not possible to conserve an ecosystem if the protected area is too small. In arid tropical regions, national parks serve the function of conserving the ecosystem. For example, Kenya's Zabo National Park, whose ecosystem conservation is fairly well established, occupies an area of about 18000 km^2. Policies which improve areas that are presently protected in each country would be beneficial, however, measures similar to those for arid regions have not been adopted for the protection of tropical forests. First, protected areas must be set up, and then measures for the non-destructive use of the area should be introduced. Cultural and natural resources should both be regarded as important, useful resources. The local inhabitants of tropical forests possess detailed traditional knowledge concerning the use of the forests. For example, about 3000 different species of plants are used in Africa, illustrating that ethnosciences are essential in the utilization of resources.

8.3 The Promotion of Cooperation between Developing and Industrialized Countries Concerning Genetic Resources

Areas to be conserved are largely located in developing countries, and industrialized nations must not be egocentric with regard to the establishment and utilization of protected areas. Thorough consideration must be given to recycling and/or returning profits to the local community. It is necessary to come up with anthropological and educational policies in order to eliminate cultural friction with the local population, let alone political and economical issues. We must show respect for the fact that the local people utilize the forest in a non-destructive way that creates harmony among the natural environment, humans and culture. It is important to learn from their traditional way of life, and to adopt policies which sustain their ways. Industrialized nations in Europe and North America have implemented policies based upon the above points, and the Korup Project in Cameroon is one such example. SADA in Venezuela and the forest conservation methods employed by Thailand have received praise from other countries for the results achieved, and these cooperative accomplishments can be even more successful with assistance and cooperation from industrialized nations.

Developing nations, such as those situated in tropical regions, which possess an abundant array of species, are now starting to re-evaluate their species as "genetic resources" which have economic value. In the future it may become more difficult for foreign countries to gain access to these genetic resources, as the developing nations become more aware of their intrinsic value. Therefore, it is necessary to join the supplying countries to develop a framework in which an arrangement for the conservation and sustainable utilization of the genetic resources can be worked out.

Strengthening the cooperative relationship between industrialized countries that are the "possessors of technology" due to their high level of biotechnology expertise, and the developing nations which are the "possessors of resources" in the form of abundant species, especially in tropical regions, is important for developing countries which wish to promote regional economic development through the utilization of their country's biological resources.

In order to solve such a problem, it is necessary to construct a new framework so that the biological resources of tropical regions can be effectively utilized, through the transfer of technology for the isolation and accumulation of species, so that developing nations can gain access to biotechnology.

Countries in Europe and North America have already taken a national approach to this important subject. The Smithsonian Institution and the Research Institute for Biological Diversity in Costa Rica are two typical examples. Under such circumstances, an urgent issue for Japan also, at the national level, is the conservation and development of sustainable utilization of species in the tropical and other regions of developing nations.

8.4 Necessary Framework

In view of what has been discussed above, it is necessary to establish an "International Tropical Bioresources Institute" (tentative name) in order to attempt to conserve the biological diversity of tropical forests, as well as to develop technology for the effective utilization of the biological resources and the transfer of such technology. A global-scale research network will be constructed by establishing regional centers in each of the three main tropical regions and of conservation centers in each individual country in order to internationally link the research being carried out in the three tropical regions.

The International Tropical Bioresources Institute will function as a center for the international research network (cf. Fig. 1). It will conduct a variety of activities concerning tropical forests, including ecological research on biodiversity, the development of techniques for the conservation and propagation of biological resources, improvement of the museum-like methods for isolation and accumulation of biological resources, anthropological studies for clarification of ethnoscience, and biosciences for the efficient utilization of biological resources. The research institute will strive to publicize and exhibit the achieved results, as well as transfer newly developed techniques.

The research institute must have a system for international research cooperation and will consult with an international advisory

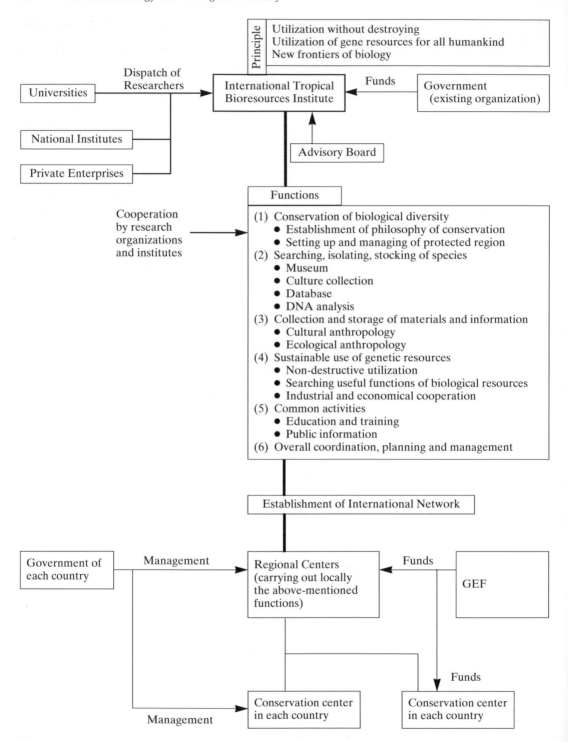

Fig. 1. Suggested concept for the International Tropical Bioresources Institute.

panel for basic operating policy issues. The director of the institute and his or her administrative staff will be responsible for the day-to-day operation of the institute. The structure of the institute will consist of the following parts: a research department (ecology, anthropology, biosciences) mainly engaged in field and laboratory work, e.g., search for resources; a department mainly dealing with environmental conservation and management, e.g., bioresource reserves; a museum department for the classification and reposition of specimens; a database department for the collection and storage of research results; and an education and training department which will publicize results and be involved in the transfer of technology.

The regional centers will plan, draft and carry out research projects according to the particular resource situation of each region.

The proposals made in this chapter concerning the sustainable development of tropical bioresources through conservation and non-destructive utilization of biodiversity are increasingly gaining international acceptance. However, further efforts are necessary for industrialized countries to be closely involved in the resource policies of countries which possess the resources and to build a truly international network.

V. Public Attitudes and Political Responses

15 "Oui" or "Non" to Biotechnology: The Other French Referendum

ANNETTE MILLET

Paris, France

(Translation by DECLAN BUTLER)

How do the French perceive biotechnology and what do they expect it to bring? The collective French psyche does not appear to be systematically opposed to biotechnology, nor bowled over in blind faith, but rather, hungry for understanding and decidedly for controls. This at least is the take-home message of polls on how the French people perceive biotechnology. The French government began cultivating biotechnology in the early 80s with its commitment to the "mobilisateur" program. Biotechnology has since blossomed, but growing alongside, in the minds of politicians, industrialists and scientists, has been the nagging question of whether the public would embrace the fruits of the new technology.

Fear and suspicion are frequently the knee-jerk responses to all new technologies: biotechnology is no different; the science itself poses almost as many questions as it answers. But finding out what the French public thinks about biotechnology has turned out to be a question in itself. This public shows no consensus of opinion. Perhaps because an entity "French public" does not exist as such, but rather a plurality of educational backgrounds, cultures and opinions.

Biotechnology is hardly an easy topic. Even the experts can't agree on what it is. Moreover, while the layman hears much talk of the pros and cons of biotechnology, he cannot yet relate this to tangible products and benefits; biotechnology has yet to produce a compact disc or a fax. It is hardly surprising that the man on the street is confused. The few studies of public opinion available confirm this. Whatever our opinion about such polls, they do highlight the main trends: a mixture of hopes and fears; a massive demand for control; and a thirst for information. At the time of writing, French politicians are airing these issues in a parliamentary debate on the genetic engineering bill.

How Informed are the French?

Let's start with a disclaimer. Deciphering and comparing the existing opinion polls on biotechnology is a dangerous and dubious business. Firstly, most poll questions are phrased too generally. The second problem is the word itself (which civil servant mistakenly figured one snappy term "biotechnology" could adequately be used to describe the multiplicity of techniques and numerous applications of – well – "biotechnology"?). Inconsistent terminology is the next weakness of such questionnaires. This sometimes leads to contradictory conclusions, and also makes it difficult to extract a global overview – a poll of polls. In short, quantitative poll data are often unreliable and must be interpreted in the light of qualitative studies which explore the images and emotions evoked by biotechnology.

Not Everybody's Cup of Tea

Science and technology do not figure highly in the French cultural stakes – the Ministry of Research and Technology discovered this in a 1989 study. But the French are curious about science. Just how curious, was the aim of a poll carried out in 1991 by the City of Science and Industry (CSI) (itself one of Paris' biggest attractions) on its 5th birthday (1) (see under "Notes" at the end of this chapter). Science and technology interested over half of those asked (54%). Less than music (74%), sport (60%) and literature (55%), but more than the creative arts (39%) and politics (34%). This was the upside; the downside – 70% of the respondents admitted they did not know very much about it (the French – modest?).

So much for science, what about biotechnology in particular? In 1991, the European Commission, under the guidance of MARK CANTLEY and CUBE (Concertation Unit on Biotechnology in Europe), carried out a massive opinion poll of public opinion of biotechnology within the Community: *Eurobarome-*

ter (2). They sought in particular to discover European's degree of knowledge and understanding of these diffusive technologies. The result: the French rank just above the European average (3), but below the Germans, Dutch, British and Danes.

Brave Unknown World

In reality, the French are often ignorant of biotechnology. Almost 40% of those questioned are "don't knows" or "wrong answers" (4). But the situation is getting better with time, reflecting the development of the sector and its diffusion by the press. In 1980 over 64% of our countrymen had never heard of biotechnology, according to a survey by SOFRES and JOSETTE ALIA (*Nouvel Observateur*) carried out as part of a project on biological risks called for by M. GIRAUD, the then Minister of Industry.

The French "man in the street" remains a biotechnology philistine; not that astonishing really, when we consider that biotechnology is not yet part of day-to-day life. And remember also that among industrialized countries France stands out as the one where the written press is least read; almost a third of the French never read a newspaper. Most people rely mainly on television for information.

Over 80% of people interviewed by the City of Science and Industry regularly or occasionally watched science programs. Unfortunately, the frequency of prime-time science programs has been reduced in the last few years. Those which remain are not surprisingly mostly health care-oriented, the appliance-of-science closest to the hearts of the public.

More Information Leads to Greater Convictions

The more people know about biotechnology, the more they like it, concluded *Eurobarometer*. In other words they are more likely to appreciate its potential benefits. But it also appeared that opponents of biotech know as much about it as its supporters!: a phenomenon also evident in the Sofres study. Asked

about genetic manipulation at the beginning of the decade, the French fell more or less into three equal cohorts: 36% opposed; 33% in favor; and 31% don't knows. The positive attitude went hand in hand with a high level of education (5). But behind the supporters' interest and curiosity lurked an ambivalence and an acute awareness of the dangers, since 68% of this category referred to possible risks compared with 45% for the entire group. In short, the most informed were the most positive towards biotechnology, but among them – and in particular among the researchers themselves – were to be found biotechnology's most hostile opponents. So much for the much touted thesis that better biotech information was the panacea of public acceptance (6).

Informing the public is a preoccupation of practically all the biotech protagonists. But to inform them of what? The accent has perhaps too often been on the socioeconomic benefits expected, while failing to appreciate that such imagined benefits are intangible to most people. Does this perhaps partly explain the loss of confidence which the scientific, technological and industrial preachers suffer today? Otherwise it is clear that scientific facts cannot be transmitted in its entirety and complexity. Is it not then as much the philosophy of science which needs to be explained as much as the products of research? Don't we need to imagine new ways to disseminate the culture of science and technology? The process is underway. The creation of the City of Science and Industry is a bold example of this. On a more ephemeral level, the nationwide *Science en Fête* initiative, launched in June 1992 by the Ministry of Research and Space, opened to the public the doors of laboratories throughout France enabling lay-people to meet researchers directly.

Leaving the state of knowledge aside, another question is how research, its objectives and the process of development are perceived. This aspect is fundamental, because the orientation of research policy and industrial strategy may ultimately be influenced by public opinion.

Do the French Look on the Bright Side?

Do research objectives, as conceived by those involved, correspond to their image in the public eye? *Eurobarometer* has already clearly shown that the words of industrialists and politicians are now regarded with suspicion. A mistrust, which extends also, although to a lesser extent, to the researchers themselves. But does this mistrust extend to the researchers' work itself? Do Europeans, and the French in particular, think that science will improve their way of life (Fig. 1)? Does it represent progress? While the overall answer is yes (the optimism of the French ranks third highest in the European Community, behind the Spanish and Dutch, and about the same as the Italians), genetic engineering scores much lower: only 47% think it is a good thing (7).

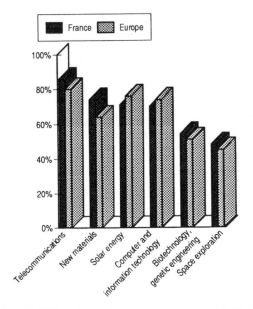

Fig. 1. Will science improve your way of life? (Source: *Eurobarometer* 35.1)
Question: Science and technology change the way we live. Here is a list of areas in which new technologies are currently developing. For each of these areas, do you think it will improve our way of life in the next 20 years?

The Magic of Words

All assessments of technology perception have to take into consideration the emotive effects of the vocabulary used. Genetic manipulation, genetic engineering, and biotechnology are the three expressions which have been most frequently used over the past ten years (both in the press and various studies) to denote the life science-based technologies. As early as 1980, the term genetic manipulation conjured up images of hopes, mainly in the therapeutic sector – and fears. Positive and negative connotations jostled together with (in order of frequency): risk, progress, hope, fear, curing, sorcerer's apprentice, destruction, madness. And also: monster, superman, horror, distress ... Little has changed since then. A 1990 poll by BVA (a French poll company) on biotechnology and genetic manipulation (8) showed that the latter term still evoked strong anxiety and fear about meddling with the sacred (see Tab. 1). Progress in genetic manipulation is viewed as a sorcerer's apprentice.

The CUBE study showed equally that the term genetic engineering, less familiar than the term biotechnology, had a more negative connotation. This does not necessarily mean that biotechnology is well known, however, as qualitative studies (9) carried out with groups of university level have shown, but rather that it benefits from a better image. The individuals interviewed by BVA perceived biotechnology as comprising traditional techniques, and as technologies which could improve their daily life. Brewing and fruit-tree grafting were mentioned, as was cutting-edge research in health and agrofood. Biotechnology was associated with a concept of tenderness, equilibrium with nature: the antithesis of chemistry. In short, biotechnology is viewed as advanced natural technology. These quirks of vocabulary raise the issue of the nature and content of the scientific discourse, no matter who the spokesmen, be they researchers, industrialists, or the media (notably journalists). How should the techniques being used be presented? Must we speak of the biological organisms used or limit ourselves to the applications expected? Simplification is often

Tab. 1. Some Concrete Examples of Knowledge and Acceptance of Biotechnology by the Public

Applications Proposed	Do you know? (% of YES)	Is it a good thing? (% of YES)
Insulin produced by bacteria	41	91
Bacteria clearing oil spills	54	83
Plant genetic engineering in order to create new varieties	57	55
Genetic engineering to improve livestock productivity	65	19

Source: BVA Environnement (1990), Le guide sociologique de l'environnement en France

misguided and carries false hopes. This is the difficulty of the dissemination, which must translate and portray without betraying.

Hands Off Our Animals!

Surveys have looked at whether people view research differently, depending on the animal under study. They do, but in a subjective way that is invariably ambiguous. Bacteria arouse ambivalence. They are synonymous with the world of medicine: the cure, but also illness and "microbes", thus giving rise to a fear of lack of their control. Plants, as in the case of tree-grafting mentioned earlier, arouse little fear, because they carry no notion of suffering. Nectarines and tomato-cherries are seen as proof that biotechnology equals diversity and abundance. But mention the idea of herbicide-resistant plants, and people call it perverse, associating it with increased use of chemicals in the field. Within the animal kingdom, any suggestion of manipulation provokes rejection and worry; words like "monster" and "eugenics" are often quoted.

Eurobarometer is one of the rare surveys which have attempted to find out the public's perception of each sector and the application of biotechnology within it: agriculture, breeding, food, medicine, the environment... Each question is framed instructively, but very positively, citing the potential progress (10); rare dissents are thus more striking. Health gets a massive vote of approval, thanks to the production of therapeutics such as insulin and vaccines, as does the environment, thanks to microbes capable of purifying "dirty water", recycling rubbish, and cleaning up oil spills (Fig. 2). Transgenic animals arouse deep suspicion, as to a lesser extent does food biotechnology. The French, like most Europeans, do not want genetic engineering of farm animals, but are less concerned about its applications in humans; gene therapy is generally accepted. Research which will solve health problems is supported (note the massive public support for Genethon), but the public finds it difficult to see any real need for genetic engineering in breeding when there are already agricultural surpluses (the BST example is well known), on top of large resistance to animal experiments in general. Once again the Frenchs' overall support for biotechnology falls slightly above the European average.

Even Existing is Extremely Dangerous!

French support does not amount to a blank check, however: although a great many are interested in research, this is mixed with a perception of risk (Fig. 3). According to *Eurobarometer*, the more people are interested in a research subject, the less they associate it with risks.

Risk assessment provokes dissension. "It's not a matter of obscurantism, when you reflect upon the risks of dissemination" exclaimed Daniel Chevallier, rapporteur of the Assemblée Nationale's Committee of Production at the session of 25th May 1992 on genetically modified organisms. Axel Kahn

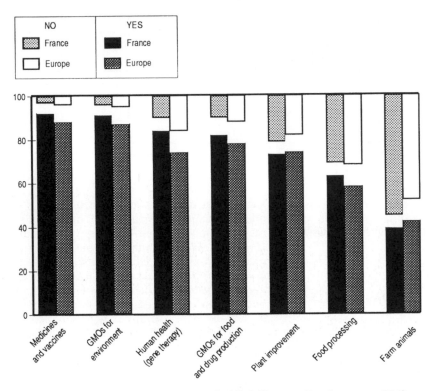

Fig. 2. Is research in biotechnology worthwhile? (Source: *Eurobarometer* 35.1)
For each application described, people had to indicate to what extent they agreed with the following statement: "Such research is worthwhile and should be encouraged".

sees it differently as a process of being "presumed guilty until proven innocent". Commenting on the decisions taken by the depu-
ties at the same session, KAHN says "these
measures are based on principles without
scientific foundation, based on dubious ideol-
ogies" (11).

Biotechnology is a particular case. Unlike
chemistry and nuclear power, controversy has
been generated before any actual industrial
development and in the absence of any incidents. Because of this unique situation those
in charge of development, the researchers
and industrialists, will have to deal with new
regulations at an early stage of their activities;
it also demands reconsideration of their entire communication strategy. With the exception of breeding, the utility of research does
not seem to be put in question. But people's
questions center on the ability of scientists
and politicians to identify the risks and to

control the applications of research. These
risks, still badly defined in certain areas such
as the release of GMOs (genetically modified
organisms) into the environment and gene
transfer (12), are all the worse perceived by
the public. The direct exchange of information at the local level should permit greater
transparency in the assessment of risks, identified or potential. It is without doubt at the
level of the company, the town, or the region
that concertation and information must be
developed, with respect to risk assessment
and control, while being honest about areas
of uncertainty.

Wise Judgement Wanted

Whatever the sector of application, the demand for control (governmental, in the questions of CUBE) is massive (Fig. 4). People

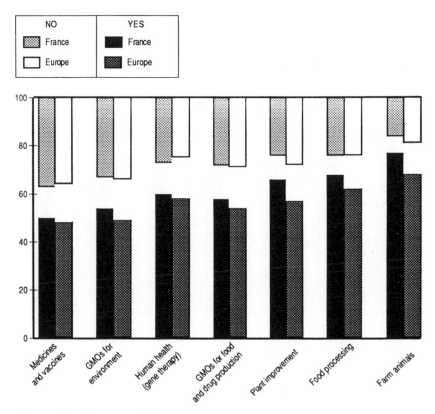

Fig. 3. Risk perception. (Source: *Eurobarometer* 35.1)
For each application described, people had to indicate to what extent they agreed with the following statement: "Such research may involve risk to human health or to the environment".

questioned by BVA expressed equally a need for regulation and ethics. But they do not want to leave this control in the hands of politicians or researchers, but with independent persons, sociologists and scientists, brought together in an expert committee. To our knowledge this is the conclusion of all the enquiries carried out over the past decade with respect to risk perception and the need for control. This preoccupation expressed consistently by the public raises the problem of the transparency of scientific work and political choices. They shed light on a desire for participation, democracy.

Is Conflict a Catalyst?

Opinions are based on values and symbols more than scientific and technological truths. This is all the more true in domains with an ethical dimension ("life patenting", for example). Moreover, it is often through conflict and controversies, often highlighted by the media that people first learn of such topics, and acquire opinions. Must we need conflicts? In reality when a conflict is presented by the media, it might put forward the conflict itself (who must we blame? what will the authorities do?), thus obscuring the main point, which is about risk and technological data. What about more public discussion? Attention by the national representatives is one im-

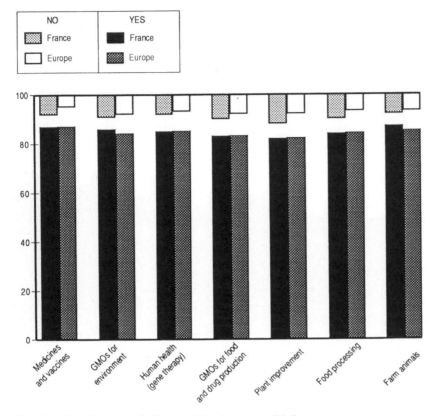

Fig. 4. Asking for control. (Source: *Eurobarometer* 35.1)
For each application described, people had to indicate to what extent they agreed with the following statement: "In any case, this research needs to be controlled by the government".

portant factor in the diffusion of information concerning advances related to science. The drawback is that years can pass before such bodies catch on to a problem in society, as in the case of the technology of assisted reproduction. With respect to genetic engineering, at the time of writing, the French parliament is discussing the implementation of the EC directives into national law: this has unfortunately turned into a debate of procedures, rather than a discussion of the fundamental issues involved.

On the face of it biotechnology holds all the winning cards needed to enter French daily life: a green label, a connotation with health. It has had a passage of fire, but the day will arrive where agrofood products will be found on supermarket shelves. But we should not declude ourselves: "We must not wait until the fears, worries and delirium disappear entirely from culture, even the most scientific" says sociologist DENIS DUCLOS. To which he adds "the social control of competence has become a central stake in the constantly rejuvenated face of science" (13).

Notes

(1) *La culture scientifique et technique des Français.* Poll carried out in January–February 1991 for the City of Science and Industry (1111 persons aged over 15, a representative sample of French population).

(2) *Eurobarometer* 35.1 "European opinions about biotechnology, in 1991". Poll carried out in July 1991 for the EEC (DG XII, CUBE) (12000 Europeans of which 1070 French, aged over 15).

(3) We can break these findings down further: in general, men are more knowledgeable than women, young people more than the elderly, higher social classes more than lower. To this can be added that those who pursue their studies longer are less ignorant about biotechnology. These sociodemographic trends support the findings of the City of Science and Industry's poll: the same erudite classes are consequently more interested in science.

(4) If 67% of those polled consider the early detection of cancer as biotechnology, only 53% are conscious that this is also true for the production of foodstuffs, e.g., the use of yeast in beer- and bread-making.

(5) Those in favor of "genetic manipulations" mostly had a high level of education (58%), were young people (18–24) (54%) or professionals (51%).

(6) LIMOGES, C., CAMBROSIO, A. (1991), Controverses publiques: les limites de l'information, *Biofutur* **100**, 87–90.

(7) The effect of socioeconomic factors on global optimism (toward the seven technologies proposed) parallels those described in (3).

(8) *Le guide sociologique de l'environnement en France; attitudes comparées du grand public et des agriculteurs*. BVA. Poll carried out in 1990 (1000 persons).

(9) In addition to the studies already mentioned – those of SOFRES (1980) and BVA (1990) – the work of a Spanish sociologist relates the results of several qualitative studies carried out throughout Europe: LEMKOW, L. (1991), *Public Attitudes to Genetic Engineering* (Workshop on Public Attitudes to Genetic Engineering, Madrid May 29–31, 1991). European Foundation for the Improvement of Living and Working Conditions, Dublin.

(10) *Eurobarometer.*

(11) KAHN, A. (1992), Une suspicion injustifiée. *Le Monde* 27th May, 1992. AXEL KAHN, director of the Unit of Genetic and Molecular Pathology (INSERM) is the president of the French Commission de Génie Biomoléculaire.

(12) MILLET, A., COËRS, P. (1989), Le frisson sécuritaire, *Biofutur* **81**, 4–7.

(13) DUCLOS, D. (1989), *Le savoir et la peur*, Éditions La Découverte. Sociologist at the CNRS, DENIS DUCLOS has been involved since 1980 in research on industrial risks and especially on the social forms of construction and perception of dangers.

16 Government, Researchers, and Activists: The Critical Public Policy Interface

SUSANNE L. HUTTNER

Los Angeles, CA 90024-1570, USA

"Although everybody with a political agenda routinely professes great respect for the wisdom of the American people, the actual behaviors reveal not respect but thinly veiled contempt." (ANDERSON, 1990)

1 Policy-Making at the Nexus of Converging Interests

Biotechnology policy-making embraces sophisticated science and superstition, government budgets and stock market prices, trade barriers, regulatory policies, risk assessment, and environmental and consumer activism. The new biotechnology's relatively brief and tumultuous policy history is not unique in the broader context of our scientific and industrial political culture. The course of events may, nevertheless, provide a window on the prospects for introduction of future technologies.

Biotechnology has been heralded as a major economic development vector for the 21st century. Throughout the industrialized world, public policy-making on biotechnology has been strongly influenced by action at the nexus of interests of government, industry, academia, and environmentalist groups. National policies have been and continue to be formulated within a climate of tension and competing goals. There is a single dominant issue at the interface that characterizes a nearly two decade long debate: whether government regulation should turn on the characteristics of products modified by recombinant DNA (rDNA) techniques or on the use of rDNA technology, per se. For those supporting product-based regulation, the new genetic techniques are regarded as an extension and refinement of more conventional breeding methods. They recommend that new biotechnology products be treated much the same as similar products created using conventional technologies. Those supporting regulation of the genetic modification process view the new biotechnology as an undeveloped science about which we have little knowledge or experience. They contend that biotechnology is sufficiently different and distinct from conventional technologies that its use warrants special regulatory rules and strictures.

The "product versus process" debate has continued for nearly 20 years with the same fundamental issues and claims reappearing at virtually every juncture of biotechnology's development. The debate has, at heart, conflicting views of what constitutes appropriate public policies on technology development and of how government should define the "public interest" to be served by those policies. How the "public interest" is characterized, who is characterizing it, and the impact it has on research and development form one theme of this chapter.

Another theme is focused on the efforts of individuals in all sectors represented at the public policy nexus to reform the organizational structures that fuel or restrain science and technology in our modern society. They bring to bear differing visions of appropriate criteria for targeting and managing technology and for judging economic and social benefit. The conceptions participants have of their efforts and of the place they would occupy in the broader context of history is as relevant as the impact they have. The biotechnology debate may prove to be a critical testing ground for continuing efforts to interject into governmental policies socioeconomic and sociocultural measures – the so-called "fourth criterion." Proponents argue that measures of efficacy, quality, and safety are insufficient to judge the potential risk associated with technologies and products. They would add social and moral claims. The manner in which these claims are injected, often indirectly and in the guise of more conventional regulatory concerns, deserves assessment. In biotechnology policy-making we have already witnessed a certain willingness by governments and science advisors to equate populist conceptions of potential risks with scientific measures of risk underlying historical regulatory criteria.

In the short term, the conflict is having a palpable effect on the pace of agricultural and environmental biotechnologies. Biomedical applications, on the other hand, have progressed relatively rapidly. In the long run, the ultimate resolution of the biotechnology debate on non-medical applications may prove to be pivotal in the history of governmental policy on science and technology, broadly, and on agriculture, in particular. Longstanding interest group campaigns against agricultural and environmental practices may have found fertile ground in the biotechnology debate, where public understanding of technical issues is extremely limited and public interest in agriculture and food production is, at best, ambivalent.

In this exploration of biotechnology policy-making and of the principal players involved, the history of events in the United States will be used as a case study. This is not because features of the U.S. story differ from events in other countries. Indeed, one can find both similar and conflicting experiences in Europe and Japan. The U.S. story is, however, especially rich in examples of conflict at the interface that is the subject of this chapter. The United States is the birthplace of rDNA technology. It supports a disproportionately large share of research and training in the life sciences, and it is home to more biotechnology companies than are found throughout the remainder of the world. The U.S. has also been host to the earliest and some of the most vocal controversies on biotechnology oversight policies, including the famous Asilomar Conference on laboratory research and the ice-minus/Frostban bacterium controversy on field research. A brief and selective overview of the history of U.S. biotechnology policy-making will be helpful to understanding the discussion of the participants in policy debates.

2 Biotechnology Development and Government

More than for most other technologies, the rate and directions of seminal biotechnology developments in the U.S. have been strongly affected by governmental actions. The relative magnitude and impact of conflicts at the policy-making interface have been markedly different for biomedical compared to agricultural or environmental applications. As discussed elsewhere (HUTTNER, 1993), the differences can be traced, in part, to two aspects of the federal science and technology policy

development process: biomedical research is both better funded by federal agencies and presented a more reasonable regulatory system [i.e., principally at the Food and Drug Administration (FDA) and the National Institutes of Health (NIH)]. Indeed the regulation of medical biotechnology is little different from that for medical research or products in which other technologies are used.

In contrast, basic research in plant and environmental biotechnology is significantly less well funded and has been yoked with new regulatory requirements and proposals. From a risk assessment perspective, these regulatory schemes are relatively vague, capturing both the risky and the benign for equally stringent governmental review. NINA FEDEROFF (1987), a plant scientist at the Carnegie Institution, expressed a concern common among researchers: "... instead of promoting the development of genetically modified organisms for environmental applications, the Government is creating legal obstacles. Instead of assembling scientists with relevant expertise to figure out what organisms and what genetic modifications do and do not need to be regulated, we're behaving as if we know nothing".

For example, for the first time in the history of crop breeding in the U.S., the federal government (i.e., the U.S. Department of Agriculture's Animal and Plant Health Inspection Service, USDA-APHIS) is regulating early, small-scale field trials of new plant cultivars and breeding lines. The new regulations (U.S. DEPARTMENT OF AGRICULTURE, 1987) and the environmental assessments, monitoring, and reports they require, were applied selectively only to plants modified with rDNA techniques, and virtually to all of them. At the Environmental Protection Agency (EPA), regulatory policies have been proposed for plants that resist pests or disease and for a wide range of microorganisms modified to enhance their usefulness in bioremediation or various industrial applications, but only where modern genetic techniques have been employed. While these proposals have yet to be finalized, on an interim basis every newly proposed field experiment (and some contained applications) is subject to environmental assessment by the agency. Both the

USDA and EPA policies represent new roles for the federal government in early stages of research that can influence the rate and pattern of discovery and technological innovation.

Plant pathologist ANNE VIDAVER (1991) has questioned the statutory bases of these new roles: "Based on perceptions of unknown risks to humans and the environment, both the U.S. EPA and USDA-APHIS regulatory agencies, responded to calls for oversight by some members of the public and some scientists. This new oversight was imposed even though neither agency's statutes were written by Congress with the intent to oversee research, including commercial research. The statutory authorities of the regulatory agencies in this regard are highly questionable, but have not been challenged. The statutes of TSCA and FIFRA of EPA and the Plant Pest Act of USDA-APHIS clearly relate to commercial and risk-based intent. Thus, it is interesting that a Vice President of Monsanto (CARPENTER, 1991) expressed alarm that scientific professionalism and academic freedom were at risk in the current regulatory environment."

On the recommendation of the Bush Administration's Office of Management and Budget, in January 1993 the EPA released for public comment a set of alternative proposals for delimiting the scope of regulation for certain biotechnology-derived biological pest control agents. Biological control agents promise more environmentally safe and effective replacements for chemical pesticides. Currently, small-scale field research with new biotechnology agents is subjected to case-by-case governmental assessment under the Federal Insecticide, Fungicide, and Rodenticide Act (EPA-FIFRA). Many of the new pest control approaches are derived directly from familiar, long-used biological control systems that are not subject to such regulation. Many represent genetic and physiological modifications analogous to crop improvements achieved by plant breeders through selective breeding and grafting that are also free from regulation. Some researchers have expressed concern that regulation under EPA-FIFRA effectively screens out biological control strategies that do not provide sufficient financial benefit to offset the cost of registration (see, e.g., COOK, 1992). Many, if not all, of these strategies target small niche markets with limited sales. The relatively high cost of biotechnology research makes this concern more acute.

One of the scope-definition proposals under consideration at the EPA offers an explicitly risk-based scheme that focuses regulatory scrutiny principally on those products with characteristics that may pose a moderate to high degree of risk in the environment. The remaining two proposals focus on the method of genetic modification, bringing greater regulatory scrutiny to products of the new biotechnology. The agency historically has favored approaches that capture most biotechnology products as a class.

The EPA proposals for small-scale field research involving other rDNA modified microorganisms have been somewhat similarly conceived, but provide even less focus on risk management. An EPA-TSCA proposal presented in the summer of 1992 would require stringent review of rDNA-modified organisms while exempting dangerous radiation-mutated *Clostridium botulinum* or *Bacillus anthracis,* simply because the first were modified by modern genetic methods while the second were either "naturally occurring" or similar to naturally occurring organisms and modified using older, cruder, and more familiar methods. As discussed in the following sections, this type of approach strains scientific logic on risk. It also gives governmental imprimatur to unsubstantiated claims that rDNA products as a class present a special set of new and, as yet, uncharacterized risks. Moreover, it directs limited governmental resources away from serious, well known risks and focuses them on new technologies simply because they are novel.

Policy proposals like these reflect the fact that biotechnology regulation has been, first and foremost, a policy issue to be resolved by political considerations. It has been driven only in part by science. Science, alone, would not justify USDA's and EPA's regulatory schemes that capture the majority of new biotechnology products any more than it would justify the regulation of all products of cross breeding, grafting, or chemical mutagenesis (see, e.g., NATIONAL RESEARCH COUNCIL, 1989). With the support of various sectors, agencies have developed regulatory schemes that move new biotechnology products forward with a governmental stamp of approval. For some agency decision makers, the process-based approaches may seem to simplify their jobs by clearly delineating what is to be regulated within a broad and fairly complex set of technologies and products.

The circumstances leading to these scientifically tenuous regulatory approaches and their associated costs are instructive of the kinds of governmental challenges that can confront new technologies. At the time USDA and EPA requirements were first being developed in the mid-1980s, there was concern and uncertainty about public acceptance of field research. The world had been introduced to agricultural biotechnology by a photograph (Fig. 1) of Dr. JULIEANN LINDEMANN, dressed in a "moonsuit," spraying the environmentally benign but genetically engineered Frostban bacteria on strawberry plants. To viewers her attire conveys a message that the bacterial spray is highly dangerous. Yet, a few yards away and just outside the photograph frame stands a group of journalists, some dressed in shorts and sandals and eating donuts. In the four years it took to progress from the first request for permission to conduct the experiment to the eventual field trial, the proposed experiments were subjected to four separate but essentially identical risk assessments and three lawsuits widely covered in the media. As an historical note, the Frostban bacterium has not been commercialized as a frost protection agent, despite demonstrations of its safety and efficacy.

The controversy surrounding those first field trials has had a palpable effect on members of the scientific research community. JEREMY RIFKIN and his Foundation on Economic Trends (FET), the litigants in the various lawsuits, took advantage of the environmental assessment process provided under the National Environmental Policy Act to use litigation to disrupt governmental approval of the proposed ice-minus and Frostban field trials. Researchers became apprehensive of having their own experiments subjected to

Fig. 1. DR. JULIEANN LINDEMANN of Advanced Genetic Sciences is dressed in protective clothing, including goggles and a respirator, as she sprays Frostban bacteria on strawberry plants in the first field trial of a genetically engineered organism. The bacterium had been modified by deleting a gene responsible for ice crystal formation and frost damage on crops. It is similar to a ubiquitous gene deletion mutant found in many parts of the world.

similar litigious actions. There was also concern that local governments would enact prohibitive restrictions on field research. As recorded in public dockets at government agencies, many scientists have supported scientifically flawed but workable regulatory schemes, like those of USDA and EPA, in an effort to create a buffer against possible public interference in research and to gain a more predictable regulatory process.

What has largely resulted, however, is governmental policy focused on imagined risk scenarios, often to the exclusion of empirical data on actual risk or conceivable benefits. After the ice-minus/Frostban litigation, both the EPA and USDA developed process-based schemes that selectively sequester new biotechnology products for special scrutiny and requirements. From a risk assessment perspective, these approaches present essentially a "black box" that research proposals enter and exit with little indication of which characteristics of the research or organisms posed risks warranting governmental intervention. Biotechnology research and products have been treated much as if they are unlike any others the agencies have dealt with before. This was underscored by a recent USDA-APHIS policy change (U.S. DEPARTMENT OF AGRICULTURE, 1993) intended to reduce reporting requirements for field research. The reduced requirements were limited to only those six crop species for which the agency had conducted multiple environmental assessments under the 1987 USDA-APHIS regulations. The new policy goes no further than the original in defining the specific plant pest risks subjected to USDA regulation that are associated with rDNA-modified organisms. Research with hundreds of other rDNA-modified crops will continue to be scrutinized by USDA on a case-by-case, every case basis.

The implication, in the absence of clear risk criteria, is that rDNA as a process and rDNA-modified products as a class are judged by the agencies to pose greater uncertainty and risk than similar products produced with conventional genetic methods. The agencies present strategies for gathering data and filling what are represented as major gaps in scientific understanding of the performance and risk of genetic modifications. In the absence of vigorous initiatives to support development of agricultural biotechnology through federal research funding, the agencies impress the interested public more with concern than enthusiasm for the prospects of applying modern genetic technologies for social benefit.

The new USDA-APHIS policy reflects the difficulty with which scientific consensus on risk (discussed in the next sections) is injected into regulatory policy on biotechnology. Notably, neither USDA nor EPA has undertaken the environmental impact statement, supported by that consensus, that would demonstrate that the use of rDNA methods, *per se*, does not confer inherent risk or unpredictability. Such an analysis would underscore the usefulness and adequacy of existing regulatory frameworks and many standard good laboratory, breeding, and field research practices. It would also likely obviate the need for most of the agencies' present regulations and procedures for the majority of new biotechnology products.

The market for new agricultural biotechnology products is expected by some analysts to double from US$ 236 million in 1991 to US$ 479 million in 1998 (BIOTECH FORUM EUROPE, 1992). The potential markets are virtually as large as those for conventional products, yet throughout the past decade private sector investment in agricultural biotechnology has lagged far behind that in the biopharmaceutical sector (see MILLER, Chapter 2 this volume). As a result, an array of potentially useful new pest control strategies, improved foods, and novel agricultural products sit on laboratory shelves awaiting private sector investment for delivery to commercial markets. Environmental applications of biotechnology remain largely on the drawing board. This is a result of many factors, including evolving and uncertain regulatory strictures. Taken together, these factors are conspiring to limit the application of the same tools that have so tangibly benefited the public through medicine.

Yet, the opportunities and needs for applying new technologies to agricultural and environmental problems are certainly no less great than in medicine. In some cases the public mandate is arguably as great or greater, as with serious health threats posed by toxic waste and with the dilemma of growing public intolerance of chemical pesticides versus a scarcity of effective alternatives that can sustain food production. As discussed in the next section, the slow pace of development in those sectors may ultimately determine the degree to which the technology is utilized in pursuit of social goals. How is it that potential social benefits and opportunities in agriculture and environment have not prevailed at the policy-making interface? What is the greater goal pursued by those who would limit or deflect entirely the application of new genetic tools to pressing problems?

3 Balance of Power

During the past decade, the balance of power has tipped in favor of caution in response to claims raised by representatives of certain special interest organizations and others who seek stringent regulation of modern genetic techniques and products, as a class. The more prominent of these organizations include the National Wildlife Federation, the Environmental Defense Fund, the National Audobon Society, the Committee for Responsible Genetics, the Consumers' Union, and the FET. Unsubstantiated claims of health or environmental risk have been met with an uneven response from academia and industry. The tangible policy effect has been seen in the ice-minus/Frostban situation and related policy decisions at the EPA and USDA. The effect is seen in more common form in the continuing attitudes of some government regulators who have come to consider field research and rDNA-modified organisms as hazardous until proven otherwise by governmental review.

Some federal regulators point to the more than 1500 field trials of new genetically engineered plants (and a few microorganisms) that have now taken place worldwide as a positive sign. The good news in the U.S. is that none was accompanied by significant public protest. The bad news is reflected in the view of the conventional U.S. plant breeder, who introduces into the environment thousands of new plant genotypes each year. In comparison, 1500 in more than a decade must seem insignificant.

The rate at which the agronomic performance of new biotechnology plants and microorganisms is being validated through small-scale field tests can affect commercialization. The current rate may prove too slow to generate the enthusiasm needed for farmers, food processors, and retailers to adopt the new technologies and products on a large scale. Indeed, a slow pace can actually foster disincentives to commercialization when a new technology or product faces public controversy. As products trickle out of the development pipeline, one at a time, they are easy targets for negative campaigns. For example, the anti-biotechnology PureFood campaign of JEREMY RIFKIN has been making thinly veiled threats of consumer boycotts against individual U.S. companies should they use products derived from the new biotechnology or even should they fail publicly to disavow the use of biotechnology products. The high competition and narrow profit margins that characterize the food production industry provide little buffer against even small potential losses in market share that may result from threatened consumer boycotts. Some companies have considered their best defense against such threats to lie in governmental regulation and approval of new products at all stages of development. The balance of power shifts perceptibly towards caution and excessive governmental strictures – the approach favored by the special interest groups.

As new agricultural biotechnology products move closer to the marketplace, the debate has shifted towards food safety and the policy outcomes have been markedly different, so far. Activist efforts to have FDA require pre-market testing and labeling (as "genetically engineered") for all new biotechnology-derived foods could affect commercialization. The campaign is based on the view that foods modified by new genetic methods, as a class, are inherently different and riskier than foods produced by conventional methods. The goal is to impose requirements beyond those applicable to similar food products. Working within a statutory mandate focused on food safety and wholesomeness, FDA historically has judged that information on genetic origin and production technologies is not substantively important to the consumer. The agency considers nutrients, allergens, and other fundamental food safety concerns in formulating requirements for pre-market testing and label information. Consistent with this approach, in May 1992 FDA announced that its existing statutes and procedures were fully applicable and adequate for new foods derived from genetically engineered plants (DEPARTMENT OF HEALTH AND HUMAN SERVICES, 1992). The statement emphasized that regulation and labeling would continue to be based on the characteristics and content of food products, and not on production technology.

A different government decision – that is, to require across the board labeling for all new genetically engineered foods – would undoubtedly impact the food production system. For producers, labeling would require special handling to keep biotechnology products separate from other similar products at all stages, from the field to the market. The added handling costs are vertically integrated from stage to stage, resulting ultimately in a more expensive consumer product. Only those new products will be judged commercially viable that present sufficiently valuable "improvements" to command market prices that offset the added production costs. Calgene has developed a comprehensive production system to deliver the new FlavrSavr tomato from the field to the fresh market. The success of this relatively novel strategy will depend on the tomato garnering premium prices.

The effect of mandatory labeling may be most acute in the processed food industry where fresh produce is commonly pooled before processing and canning, product prices are generally low and competitive, and profit margins are narrow. Moreover, most food processors serve consumers whose product choices are driven principally by cost. If the additional production costs cannot be balanced in the marketplace, the food processor will likely bypass even tangible nutritional or processing improvements offered by new biotechnology products. A plant breeding company notes that "(a) typical processing plant in California handles massive quantities of tomatoes each day from many growers and numerous fields. Processing plants are not designed to segregate tomatoes; rather substantial mixing occurs. Even with massive re-engineering and cost, opportunity for mixing would still be significant in the field and processing plant and would require enormous resources to monitor. ... Unwarranted labeling would create much additional cost and could have the effect of killing the technology" (STEVENS and GREEN, 1993).

DAVE ASHTON (1993) of Hunt Wesson, Inc., puts the situation in practical terms: "the routine labeling of foods produced through biotech modification would deter food com-

panies from commercializing the products. It is generally agreed that such a 'label' would alarm some consumers without providing information upon which they might make informed choices. For example, one share point of spaghetti sauce, a \$ 1.1 billion category, equals \$ 11 million in sales. There is little incentive to incur risk in the face of potential consumer alarm". Thus, in both fresh and processed foods sectors, labeling would likely drive product prices up or simply eliminate new biotechnology produce from the system, altogether. Those products that do reach the market will be relatively expensive. Ultimately, labeling could result in the economic sequestration of new biotechnology foods, keeping them out of reach of many low and middle income consumers.

In the May 1992 statement, FDA took a conservative approach under the Food Additive Provisions of the Food, Drug and Cosmetic Act to regulate certain foods containing substances that are new to the food supply, but rejected calls for general labeling. In this decision, the balance of power tilted towards scientific measures of risk and safety. Demands continue, however, for labeling on the basis of the consumer's "right to know", and the FDA recently called for public comment on safety issues related to the agency's statutory mandate for labeling (DEPARTMENT OF HEALTH AND HUMAN SERVICES, 1993). Incoming letters favored the May 1992 FDA approach, outnumbering critics by nearly 10:1.

Whether for early stage research or for premarket assessment, special regulatory requirements that are not justified by scientifically credible risk assessments put new biotechnology products in the spotlight for public scrutiny – for better or worse. "The bottom line is that regulated products such as pharmaceuticals, agricultural chemicals, food additives, and now genetically altered crops can be monitored and their market entry anticipated. The step that discloses the emergence of a new genetically modified crop is the requirement for an environmental release permit" (BECK and ULRICH, 1993). While this may be seen as an advantage for stock brokers and other industry analysts, it can also provide a mechanism for targeting products for negative campaigns at early stages of development, often before the agronomic or potential market performance has been demonstrated.

Viewed internationally, the situation has attained worst case scenarios in Germany and Denmark. These countries adopted restrictive national laws that affect virtually every aspect of research and commercial activity involving new genetic technologies. In response, German companies have moved entire research and production facilities (KAHN, 1992; MILLER, this volume Chapter 2). Other countries, like Japan, have very stringent regulations on agricultural and environmental applications of biotechnology. Not surprisingly, the pace of development and commercialization in such countries is slow, and in some cases imperceptible (see also Sect. 12). In each of these cases, the balance of power has swung decidedly away from scientific consensus on fundamental issues of risk.

Anywhere the policy debate occurs you find activists promoting unsubstantiated risk scenarios and a cadre of scientists who depart from strict scientific measures of risk to support the purported need for regulation on the basis of perceptions of risk. For many academic scientists, their actions are partly an expression of the legacy of the first field experiments with the ice-minus/Frostban bacteria and partly an effort to demonstrate that the scientific community is willing to accommodate public concerns about research. Companies have often supported unnecessarily stringent regulatory schemes for additional reasons, including the benefits they accrue from governmental imprimaturs of research and products, and the distance they may gain from competitors entering the product development pipeline (OFFICE OF TECHNOLOGY ASSESSMENT, 1992; TOWNSEND, 1993). Whatever the motivation, the research community has been sending government and the public mixed messages on the validity of risk claims.

4 The Science Connection – A Paradox in Policy-Making

Scientific advice based on empirical evidence and experience can be fairly characterized as a selectively used tool rather than a driving force in biotechnology policy-making. For example, there was never any significant scientific doubt about the safety of the ice-minus/Frostban field trials. In fact, field trials with two phenotypically similar strains (produced by spontaneous or chemically induced mutations) were quite appropriately free of any governmental review. Ironically, data from those field trials were submitted in support of the request for permission to field test the rDNA-modified strain. The ice-minus/Frostban experience is but one example of many in the history of biotechnology policy-making, beginning with the Asilomar Conference, where the use of objective and rigorous science advice was affected when confronted with highly controversial issues. In hindsight, perhaps one would say "gratuitously controversial." In such situations some scientists have supported scientifically flawed biosafety schemes they hoped would allow them to get on with their research, albeit more slowly and at greater cost. Hence, the promulgation of USDA and EPA regulatory procedures has been supported although they generally provide no incremental improvement in safety.

From a historical perspective, the recommendations of the Asilomar conferees and the biosafety process that was subsequently created by the National Institutes of Health (NIH) may be viewed as pivotal. They set the course for biotechnology policy-making by other agencies worldwide. Rather than building the biosafety scheme upon the broader relevant context of existing biosafety procedures for microbial research involving similar classes of research risk, the containment needs assessment proscribed in the NIH Guidelines (DEPARTMENT OF HEALTH AND HUMAN SERVICES, 1976) explicitly identified rDNA and rDNA-modified microorganisms as different and warranting special containment precautions. The NIH assessment process was built on the tacit assumption that some level of containment was necessary for most rDNA research. As early as 1978, however, NIH began to relax the rDNA Guidelines so as to bring greater focus to categories of research that pose serious risk. By 1980 nearly 99% of all rDNA research originally circumscribed by the NIH Guidelines was exempt. Nevertheless, the fundamental assumptions of the original NIH approach echo in biotechnology policies around the world. Those assumptions also provided an important part of the foundation on which RIFKIN's ice-minus/Frostban lawsuit against NIH was based.

Let us consider the basic paradox of policy-making for a technology that has been isolated and touted as something new and special, but for which scientific assessments would support no more regulation than is appropriate for older, more familiar technologies and products. On the one hand, several federal policies selectively regulate new biotechnology as a class of genetic modification processes. On the other hand, Tab. 1 presents a partial list of national and international scientific reports recommending that regulatory requirements should be limited to those that are justified and commensurate with scientific measures of product risk, regardless of the genetic modification technique. One might concede that the product versus process debate has been made more complex by biotechnology's broad sweep, affecting a major segment of the health care system, encompassing virtually the entire food production pipeline, and even reaching into environmental clean-up and marine-related industries. Yet, it has been argued compelling (NATIONAL ACADEMY OF SCIENCES, 1987; MILLER and YOUNG, 1988) that the new biotechnologies in each of these sectors lie on a continuum with older, more familiar technologies and that the new molecular methods are simply refinements of cruder techniques. Genetic modification to improve products has certainly been as broadly applied, historically.

The scientific arguments against biotechnology-specific regulations for agricultural field research and food safety stem from nearly twenty years experience with rDNA techniques in the laboratory and a decade in the greenhouse and field. This experience builds on more extensive knowledge of conventional plant breeding and agricultural practice. The case is well made in the findings of the U.S. NATIONAL ACADEMY OF SCIENCES (1987) and the U.S. NATIONAL RESEARCH COUNCIL (1989). The earlier report emphasized that there is no evidence that the introduction into the environment of organisms modified by rDNA techniques presents unique hazards, but rather that the risks are the same in kind as those incurred in the introduction of unmodified organisms or similar organisms modified by more conventional genetic methods. The Academy (1987) enunciated a fundamental principle that safety assessment and regulation *"should be based on the nature of the organism and the environment into which it will be introduced, not on the method by which it was modified"*.

This principle was developed further for practical application in a report by the NATIONAL RESEARCH COUNCIL (1989) that concluded: *"Crops modified by molecular methods should pose risks no different from those modified by classical genetic methods for similar traits. As the molecular methods are more specific, users of the methods will be more certain about the traits they introduce into the plants."* That is, the precision of rDNA techniques actually *enhances* determinations of safety and risk. The confidence expressed in these and other scientific reports is not reflected in federal regulations on agricultural and environmental field research.

Tab. 1. Major Reports Finding Risk is Not a Function of Genetic Modification Process

U.S. National Academy of Sciences (1987)
U.S. National Research Council (1989)
U.S. National Biotechnology Policy Board (Executive Office of the President, 1992)
Organisation for Economic Co-operation and Development (1993 a, c)
International Council of Scientific Unions, SCOPE and COGENE (1987)
NATO Advanced Research Workshop (1988)
UNIDO/WHO/UNEP Working Group on Biotechnology Safety (1987, 1992)

The National Research Council proposed a straight-forward framework for judging the risks and necessary mitigation precautions for testing new plants and microorganisms in the field. This framework is based on three questions:

(1) Are we familiar with the properties of the organism and the environment into which it may be introduced?
(2) Can we confine or control the organism effectively?
(3) What are the probable effects on the environment should the introduced organism or a genetic trait persist longer than intended or spread to nontarget environments?

The essential concepts of this framework were later built into a useful algorithm for oversight of field research by MILLER et al. (1990). By this scheme, the degree of regulatory oversight (and associated financial and administrative burdens to researchers) can be tailored to the degree of risk posed by a proposed experiment. The algorithm can be used to judge the potential risk of organisms to be tested in the field and to identify adequate risk-mitigation procedures (i.e., a variety of physical, spatial, genetic, or temporal containment and confinement practices), where needed. It possesses sufficient flexibility that it may be adapted to the needs of many agencies. It accommodates any organism, whether "wild-type," selected, or genetically manipulated in the laboratory using long-established or new methods.

Since 1986, a number of federal policy reports have been published that reiterate the findings of the scientific community on fundamental issues of risk evaluation and management. The preface of the Coordinated Framework (OFFICE OF SCIENCE AND TECHNOLOGY POLICY, 1986), a multi-agency federal report on biotechnology regulation, emphasized that oversight should focus on the product and not on the method by which that product was genetically modified. More recently, a broad U.S. policy statement (EXECUTIVE OFFICE OF THE PRESIDENT, 1992) was announced by the White House that described:

"a risk-based, scientifically sound approach to the oversight of planned introductions of biotechnology products into the environment that focuses on the characteristics of the biotechnology product and the environment into which it is being introduced, not the process by which the product is created. Exercise of oversight in the scope of discretion afforded by the statute should be based on the risk posed by the introduction and should not turn on the fact that an organism has been modified by a particular process or technique".

Later that year, the National Biotechnology Policy Board (1992) also recommended risk-based (i.e., product-based) regulatory approaches and further emphasized that "(t)he health and environmental risks of not pursuing biotechnology-based solutions to the nation's problems are likely to be greater than the risks of going forward." The report further stated that "regulation should provide net social benefits. Only when risk-reduction benefits exceed the cost of oversight, can everyone win. Then, resources will be allocated to productive uses, providing valuable consumer products and protecting health and safety. Gearing oversight efforts to the expected risks of products and their applications is essential".

Product-based approaches have also been recommended by international reports, including the Organisation for Economic Cooperation and Development (OECD) (ORGANISATION FOR ECONOMIC COOPERATION AND DEVELOPMENT, 1993a; see also 1993b) that recommends an approach to risk assessment and regulation of new biotechnology foods, consistent with the 1992 FDA policy statement:

"If the new or modified food or food component is determined to be substantially equivalent to an existing food, then it can be treated in the same manner with respect to safety. No additional safety concerns would be expected. ... Where substantial equivalence is more difficult to establish because the food or food component is either less well-known or totally new, then the identified differences, or the new characteristics should be the focus of further safety considerations."

At its July 1991 meeting in Rome, the Codex Alimentarius Commission endorsed a report of a Joint Food and Agriculture Organization/World Health Organization Consultation on Biotechnology that concluded that foods derived from modern biotechnology were not inherently less safe than those derived from traditional biotechnology. Taken together, scientific and policy reports in the U.S. and Europe repeatedly recommend against circumscribing new genetic techniques for special regulatory consideration, as a class.

These reports, their thoughtfulness and consistency notwithstanding, have played a remarkably peripheral role in the policy-making process. Where one would expect to see them emphasized and integrated into decision making, one instead sees a tendency to consider regulatory proposals apart from such scientific consensus and from within the bureaucratic perspective of the agency involved. A more significant role has often been played by agency science advisory panels, like the EPA's Biotechnology Science Advisory Committee (EPA-BSAC) and USDA's Agricultural Biotechnology Research Advisory Committee (USDA-ABRAC), and to some extent by individual scientists who respond to policy proposals during public comment periods and agency-sponsored workshops. Many of these scientists include in their advice and recommendations considerations of issues other than the strict scientific merits of regulatory proposals.

Their willingness to move beyond science respresents a role-shift, with the scientist-advisor stepping out of the traditional role of objective scientist and assuming a role of political participant. The science advice process, itself, is relevant here. In some situations, as when serving on certain agency advisory panels, the extramural scientist's role is directed by the agency. Commonly advisory panels are asked to deal only with parts of problems, rather than with broader aspects of an impending agency decision. It is the agency official overseeing the panel that has greatest control in determining what issues will be considered, by whom, and in what depth. This approach suits the agency decision making process by providing expert advice on identified issues needing clarification.

In some cases, however, the way in which an agency organizes and exercises control over advisory panels can seriously undermine the quality of science advice generated. An agency that does not appreciate the need or does not want to be advised can remain unaffected by any insitutional arrangements for advising them. EPA, for example, has been criticized (U. S. ENVIRONMENTAL PROTECTION AGENCY, 1992) for: (1) not having a coherent science agenda to support its focus on relatively high-risk environmental problems; (2) not defining or coherently organizing a science advice function to ensure that policy decisions are informed by a clear understanding of science; (3) not considering appropriate science advice in the decision making process; and (4) not evaluating the impact of its regulations on environmental improvement. This might help explain how the EPA-BSAC came to support process-based regulatory proposals under FIFRA and TSCA that are far afield of scientific indicators of risk.

5 Somersaulting Science Advice

The ambiguous behavior of some scientists is certainly not unique to the biotechnology community. EDITH EFRON commented on it in its extreme form in "The Apocalyptics" (EFRON, 1984), her study of the environmental politics of cancer and chemicals. Although thoughtful recommendations on scientific issues related to rDNA-modified organisms have been published in a variety of major reports and have been discussed by agency advisory panels, such advice has not always found its way into policy decisions. Some scientists have supported policy proposals built on assumptions contrary to those supported by broad scientific consensus. The result is policy built on the scientifically flawed assumption that the use of rDNA techniques and other modern molecular methods, in and of itself, provides adequate justification for governmental intervention in research and commercialization.

This disconnects the need for regulation from empirical measures of risk. It links it, instead, to imagined risk scenarios. Judging from letters responding to the most recent USDA-APHIS regulatory proposal (U.S. DEPARTMENT OF AGRICULTURE, 1993), some scientists suggest that the regulators are providing essential contributions to the scientific definition and prediction of risk. How they reconcile that with the general absence of heuristic risk criteria in the USDA policy is unclear.

As to the motivation of regulators, have they ever made a policy decision voluntarily that diminished the scope of their bureaucracies? As former FDA Commissioner FRANK YOUNG has said, "Cows moo, dogs bark, and regulators regulate". It is not necessary that a product or research activity actually poses a risk; it is sufficient that it be perceived to pose a risk. One former high level USDA official said that the Department's rationale is that if people think biotechnology is risky, that perception makes the risk real to those people, and if the risk is real then government must regulate it. This peculiar logic and the passive assent it receives from scientific advisors has produced regulatory policy based on governmental perceptions of public perceptions of risk, rather than on empirical evidence of risk.

Characterizing and responding to public perceptions is a fundamental part of risk communication and public education, both *bona fide* roles of federal government. However, where perceptions are unsubstantiated or lack heuristic strategies for evaluating perceived risks, they provide little support for efforts to identify tangible risk criteria and measures that warrant investment of limited governmental resources. Ultimately, approaches that emphasize public perceptions over scientific consensus will undermine government regulation insofar as they marginalize the role of empirical data. Looked at another way, however, regulators often make decisions that agrandize their administrative bureaucracies and, once new bureaucracies are built, it is not in their interest to change the status quo and roll them back in favor of more reasonable regulatory schemes.

The conflicting recommendations of scientists are, in some cases, undoubtedly a result of the careful manner in which scientific advisory committees are organized, meeting agendas are set, and discussions of policy proposals are controlled by agency staff. The potential for conflict of interest warrants that agency institutional interests in sustaining or expanding a regulatory domain be acknowledged and protected against.

In some cases, the scientifically flawed recommendations of scientific advisors may reflect an element of surrender to fear of public intervention in research. Many imagined scenarios where activists could destroy experiments or drag them into court. Unfortunately, they were not paying attention to the fact that since the ice-minus/Frostban experiments field research had been progressing, albeit slowly, in test sites around the U.S. without public controversy. Some who feared states and communities would create new legislative restrictions on research did not notice that of the states considering proposed legislation, the majority rejected new regulations (see HUTTNER and SMITH, 1991; HUTTNER, 1993).

Fear of public intervention was articulated in 1990 when the NIH held a series of regional meetings to explore the future of the NIH rDNA Research Advisory Committee (NIH-RAC) and to determine whether there was sufficient need to continue the RAC and the NIH Guidelines for Research Involving rDNA Molecules. As reported by a subcommittee to the RAC (NATIONAL INSTITUTES OF HEALTH, 1991):

> "(T)he great majority of speakers (in the regional meetings), however, were very supportive of the status quo: that is, continuation of the RAC's role in supervising the IBCs, retaining central review of some experiments, and keeping the guidelines current. It is striking, however, that this support exists despite an almost unanimous acknowledgment that the original rationale for establishing the RAC is no longer valid. Those who supported the status quo, almost to a person, gave two other rationales instead. One can be called the placebo theory: the public perceives extraordinary risks associated with recombinant-DNA and, however inaccurate the perception, the existence of the RAC is comfort-

ing and therefore beneficial. The second theory is related to the first and could be called the receptor-blocking theory. Recombinant-DNA is an arena that attracts political controversy, and if the RAC were not benignly occupying the site, it would be attacked by other, more pathogenic, agents in the form of new legislation and regulations at various levels of government. These theories are entirely plausible, although it is possible, as was pointed out, to take the opposite view. The existence of the RAC may have contributed to the public perception of risk more than it has assuaged it. Moreover, instead of blocking unreasoned opposition to recombinant-DNA, the RAC might serve as a catalyst – a point of attack for NEPA lawsuits, for example."

In other cases, recommendations for regulating unsubstantiated risks may stem from a misperception that an agency's proposals are "business as usual." How many of the scientists supporting the 1987 USDA-APHIS proposal to regulate genetically engineered plants realized that the proposal created *de novo* the authority for the Department to regulate small-scale field research? Before 1987, USDA did not regulate early field research with new plant cultivars or breeding lines under the Plant Pest Act (HUTTNER, 1993; HUTTNER et al., 1992). Moreover, the 1987 regulation created a new definition – the "regulated article" – that made it unnecessary for a plant to pose an actual plant pest risk to be subject to the Plant Pest Act and its regulatory requirements. As a result, hundreds of field experiments with plants posing little or no plant pest risk have been subjected to the administrative burdens and financial and time expenditures associated with environmental assessments, simply because the plants were products of genetic engineering. This includes virtually all of the more than 400 release permits issued by the USDA-APHIS, to date. None of the genetically engineered plants have displayed any unexpected or hazardous behaviors (see, e.g., LANDSMANN, 1992).

Have the scientists participating in the development of USDA and EPA policies understood their longer range implications? "Dr. DAVID KINGSBURY, who led the effort to slice up the biotech whale for President RONALD REAGAN during much of the 1980s,

looks back: "The period from late 1983 to 1986 really was one of hysteria. Everyone was feeling the pressure from the Hill (Congress) to *do* something, but a lot of us really didn't understand the ramifications of regulation. I could sense that a bureaucracy was brewing, but it was like a juggernaut – things were moving and taking on a life of their own. Congressional pressure and turf wars among agencies were growing acute and a lot of us did not understand the implications" (WITT, 1990).

The impact on research is not easily assessed, but there are some indications that the pace of research has been affected. The costs of regulatory compliance are not trivial for many academic researchers, who generally rely on funding from grants. It is estimated that the cost of a typical field trial of a genetically engineered plant is many times higher than that of a conventionally bred plant (TOWNSEND, 1993; see also FREIBERG, 1993). To date, only about 17% of all of the issued USDA-APHIS field test permits have gone to academic researchers, individually or in collaboration with industry (T. MEDLEY, personal communication). Of the 1000 or so field trials that have received permits, that amounts to fewer than 170 trials by academic researchers in the entire U.S. in the past six years. Yet, academic researchers have played a significant role in the development of germplasm and new cultivars used in commercial agriculture. The genetic histories of most major U.S. crops can be traced back to land grant universities. CHARLES ARNTZEN of Texas A & M University aptly observed that "(a) pipeline filled with commercial rDNA-modified plant products does not necessarily mean that consumers will rapidly and efficiently benefit from scientists' discoveries ... U.S. government regulations ... remain complicated by bureaucratic detail that stifles the research which provides the basis for new product development" (ARNTZEN, 1992). Have scientists supporting USDA's scientifically tenuous and burdensome regulatory scheme taken this into account?

Of course, the last-ditch defense of a controversial recommendation can always turn on the principle that every scientific measurement contains a margin of error. Regardless of the motivating factors, bad science makes bad public policy on risk management, and the course of public policy is difficult to change once it is set. Witness the very slow improvement of the USDA-APHIS regulations. Therefore, in pursuing anyone's characterizations of public concerns, it is worth remembering that public opinions change on a time frame of weeks and days, whereas public policy (laws and regulations) changes over decades.

6 Is the Public Really Demanding Governmental Intervention?

Within scientific circles it is often lamented that the public has become increasingly skeptical of science and technology. Just look, concerned scientists say, at the debates about animals in research, pesticides, and biotechnology – all expressions of the public's extravagant expectations of a risk-free world. True, the debates have been vigorous. But are they representative of public opinion?

Not according to national opinion polls in the U.S. – consider the biotechnology debate. Journalists have reported battles over lab research, field research, transgenic animals, and genetically engineered foods. Yet, national opinion polls over the same period have regularly found that while the public is generally poorly informed about biotechnology, they are nonetheless "cautiously optimistic" about likely benefits (NOVO INDUSTRI, 1987; OFFICE OF TECHNOLOGY ASSESSMENT, 1987; MILLER, 1985; HOBAN, 1989, 1990; HOBAN and KENDALL, 1992). Surveys of the public and identified stakeholders find that even though the majority may believe in stringent regulation, they do not want governmental intervention to restrict innovation and growth (SIDDHANTI, 1988; OFFICE OF TECHNOLOGY ASSESSMENT, 1987).

Notably, SIDDHANTI's (1988) survey of individuals actively involved in the biotechnology debate (government, industry, academic, and special interest groups) found that the majority of respondents (65.97%) supported product-based approaches that may require only minor alterations of existing regulatory frameworks, and only 3.2% expressly supported process-based regulation. She concluded that the "(p)rocess-based approach is the least opted for approach, this means that most people believe that the process of genetic engineering is not risky in itself, but it is the product that could have potentially risky impacts. This also confirms the general belief that process-based approach to regulation would unnecessarily single out the technology for regulatory scrutiny, instead of treating it as any other technology and concentrating on the risk of its products".

Public support for scientific research is stable, if not growing (Fig. 2), steadily nurturing the U.S. international preeminence in scientific research and training, and the nation's marked lead in biotechnology commercialization. Although Congress continues to demand evidence of social return from public investment in research in all fields, there is no sign that federal support for biotechnology research is in jeopardy. To the contrary, the FEDERAL COORDINATING COUNCIL ON SCIENCE, ENGINEERING, AND TECHNOLOGY (1993) recently proposed a broad cross-cutting initiative to enhance funding for biotechnology research at all relevant federal agencies.

Another reflection of public attitudes towards biotechnology and rDNA, specifically, can be found in the tens of thousands of Americans who have accepted and benefited from medical diagnostics and drugs provided by the biotechnology industry. These include human insulin for diabetics, erythropoietin for kidney dialysis patients, and sensitive diagnostics that are helping to keep the blood supply free of dangerous pathogens, like HIV and hepatitis virus. Public acceptance of these products did not depend upon the development of new federal regulations or biosafety procedures. They have been regulated under existing FDA regulations. Moreover, activists generally have not pressed their genetically-engineered-products-are-risky campaign against biomedical products. Taken together, the polls, taxpayers' support, and the rapid

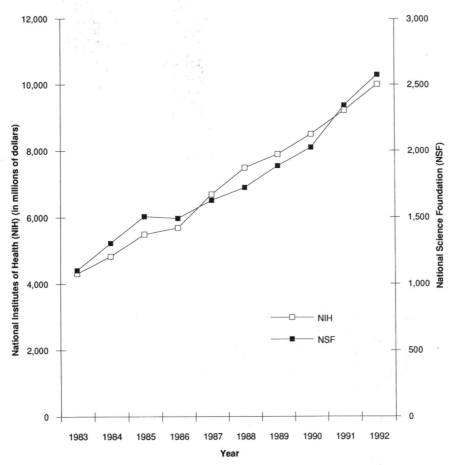

Fig. 2. Federal funding for life science research has steadily increased over the past decade at both the U.S. National Institutes of Health and the U.S. National Science Foundation.

acceptance of biomedical biotechnologies belie claims of wide-spread public concern.

7 Sources of Debate

This is not to minimize the often impassioned debate. While medical applications have met with a degree of acceptance typical of other medical technologies, agricultural applications have prompted vigorous objections. Who is objecting? According to news reports and government records, it is certain environmental organizations and, recently, a number of food professionals. What are their objections? Principal expressed concerns focus on two classes of issues: (1) Safety: the reliability of genetic engineering and the possible environmental and health consequences of genetically engineered plants, animals, and microorganisms; and (2) socioeconomic issues: particularly the possible impacts of new biotechnologies on already strained agricultural systems, such as small farms, the compatibility of biotechnology strategies to visions of sustainable agriculture, and the moral acceptability of genetic engineering.

On safety, as discussed above, numerous scientific reports (Tab. 1) have found that genetic engineering is in many ways more precise and reliable than traditional breeding techniques. Where plant breeders attempt to create improved varieties by combining in a new hybrid offspring thousands of genes from two parent plants, they routinely introduce deleterious traits as well as the sought improvements. The unwanted traits are incrementally eliminated by laborious "back-crossing" and by progressively focusing on desired traits (such as disease resistance or fruit yield). In an important refinement of this process, genetic engineering allows breeders to transfer or modify *single* traits, one at a time, by targeting *single*, well characterized genes.

Common sense dictates that using common crop plants and changing one or a few genes compared to thousands of genes means greater control and precision, and *more simplified risk assessment*. This realization, backed up by extensive laboratory and field experience, was the basis of the FDA decision (KESSLER et al., 1992; MILLER, 1993) to regulate certain new biotechnology foods by considering the product and the nature of the specific genetic modification but not the genetic method, itself. Outside of FDA, this common sense approach has largely eluded public policy debates in U.S. agencies.

Some critics are unmoved by the scientific evidence, our long history of safe crop breeding, and the prospects for accruing the kinds of improvements in agriculture and food that have already been achieved in medicine. In the skeptic's view, scientific evidence is often considered as unreliable and but one, secondary element within a complex of perceived environmental, social, and economic concerns. This inability, or in some cases, unwillingness to accept scientific evidence has led to a central anti-biotechnology mythology based on imagined risk scenarios involving the use of rDNA methods. Taken together with imagined wide-spread public outrage and conflicted responses from the scientific community, the mythology has been embraced by some in government as a compelling rationale for stringent regulation.

The anti-biotechnology mythology has several basic tenets: (1) that rDNA methods are inherently unreliable and risky; (2) that genetically engineered organisms are unpredicta-

ble and unsafe; (3) that current oversight systems are inadequate to control the "unique" and yet to be discovered risks of genetically engineered organisms or products; and (4) that the biotechnology research enterprise is driven by industrialist greed and socioeconomic elitism. Each of these concerns is presented by advocates of regulatory caution as an accurate and objective assessment of tangible risks. That they are actually laced with inaccuracies and faulty "scientific" reasoning has been discussed above and elsewhere (HUTTNER, 1993; MILLER, 1989, 1991). They have nonetheless been pursued over the past decade and a half and from issue to issue in agricultural biotechnology with a determination and consistency that is more characteristic of a crusade than of efforts to achieve rational policy.

Several years ago STEPHEN JAY GOULD, the noted evolutionary theorist, said of one of JEREMY RIFKIN's anti-biotechnology publications:

> "I regard Algeny as a cleverly constructed tract of anti-intellectual propaganda masquerading as scholarship. Among books promoted as serious intellectual statements by important thinkers, I don't think I have ever read a shoddier work. Damned shame, too, because the deep issue is troubling and I do not disagree with RIFKIN's basic plea for respecting the integrity of evolutionary lineages. But devious means compromise good ends, and we shall have to save RIFKIN's humane conclusion from his own lamentable tactics" (GOULD, 1987).

At base, the anti-biotechnology mythology and its proponents are principally driven by an idiosyncratic world view and deeply held personal values. That is not to discount the rights of the proponents to hold and express them. But as a foundation for public policy-making, one reasonably and fairly must ask, whose views and whose values? MICHAEL FUMENTO writes (FUMENTO, 1993):

> "When someone says we are outraged that the government allows Alar on the market when a certain dose of Alar's breakdown product in one study on one type of animal showed that it might be carcinogenic, but they

couldn't care less that our food is crawling with natural chemicals that have caused cancer in several species of animals at much lower doses in many tests, what they are saying is that they simply don't like Alar. When they demand that food irradiation be banned because it might someday be proved to cause cancer even though we know that animal testing has linked numerous other types of food preparation to cancer, they're saying that they simply don't like food irradiation. And likewise when they tell you that the negligible risk standard of as much as one cancer caused per million people using a synthetic pesticide is one cancer too many but that it's okay that one in five thousands of us die each year in automobile accidents, they're saying that to them automobiles are okay and synthetic pesticides are not. That reflects their values and/or their vested interests ... But they may not be your values."

8 Historical Echoes

Beyond individual views and values, one still must account for the vigor and momentum of the movement toward the common policy goal of governmental intervention in basic research – including publicly funded research at academic institutions. The anti-biotechnology effort recalls historian RICHARD HOFSTADTER's study of religious and political movements in American politics (HOFSTADTER, 1952):

> "Although American political life has rarely been touched by the most acute varieties of class conflict, it has served again and again as an arena for uncommonly angry minds. ... Behind such movements there is a style of mind, not aways right-wing in its affiliations, that has a long and varied history. I call it the paranoid style simply because no other word adequately evokes the qualities of heated exaggeration, suspiciousness, and conspiratorial fantasy that I have in mind."

HOFSTADTER described the basic elements of what he calls the "paranoid style": "The central image is that of a vast and sinister conspiracy, a gigantic and yet subtle machinery of influence set in motion to undermine and de-

stroy a way of life." He notes that there is a characteristic "curious leap in imagination that is always made at some critical point in the recital of events". Consider the following view of agricultural biotechnology presented by MARGARET MELLON (1991a) as a spokesperson for the National Wildlife Federation:

> "One area is the social and economic impact of the technology, especially in agriculture. In the United States and around the world, biotechnology products will affect the number of farmers, the choice of agricultural practices and the control of agriculture. In the United States a major concern is that genetic engineering will accelerate the trend towards the environmentally destructive practices of industrial agriculture, such as high chemical inputs that pollute land and ground water. In the third world, the displacement of local export crops, such as vanilla and coffee, by cell-culture substitutes engineered elsewhere would cause major social and agronomic damage."

Viewed from HOFSTADTER's model of the paranoid style, the "conspiracy" here lies in industrial agriculture and the "leap in imagination" lies in the assertion that biotechnology is at base bad for agriculture, farmers, and developing nations. All biotechnologies? What about plants resistant to disease, insects, or drought, veterinary diagnostics and vaccines, or grains with enhanced nutrient content?

REBECCA GOLDBURG of the Environmental Defense Fund uses a similar rhetorical approach:

> "The powers of genetic engineering contrast sharply with those of selective breeding, the traditional technique for genetically modifying crop plants and livestock. The genetic traits available to traditional breeders are limited by the ability of the organisms to interbreed; only traits from the same species, or sometimes closely related species, can be used in breeding programs. Thus, while traditional breeding can alter the characteristics of organisms used as food, breeders generally only introduce genetic traits from similar organisms that humans have long consumed."

Consider that statement beyond the central myth of rDNA's "powers" and riskiness.

HOFSTADTER's model would likely find the "conspiracy" in the presumption that scientists will be transferring genes into food without regard for health. The "leap in imagination" leads the reader to conclude that foods derived from traditional breeding are safe, while it is abundantly clear that virtually every common food contains naturally occurring substances that are seriously toxic when eaten in sufficient quantities (AMES et al., 1987). The statement shows limited understanding of crop breeding and of the fact that truly wide crosses are found in the breeding histories of some common crop plants (GOODMAN et al., 1987).

Tomatoes, a nutritional staple in many parts of the world, have been cultivated for more than five centuries and contain germplasm intentionally transferred from wild, inedible, and often toxic distant relatives (RICK, 1978). The speaker's implied confidence that genetic traits introduced from related species and even from common food sources presents little risk is not supported by the experience of plant breeders. In conventional plant breeding there is always the possibility of surprise, with unwanted characteristics appearing in new breeding lines. Naturally occurring plant toxins, many of which are quite dangerous at even low doses, have been affected by conventional breeding efforts (e.g., high solanine potatoes). Through a carefully developed breeding plan (Fig. 3), breeders routinely and effectively contend with problems arising from the relatively crude methods of conventional cross-breeding and grafting.

Sidestepping this history of safety in plant breeding and the relationship between breeding and new genetic methods, GOLDBURG recasts safety issues related to genetic engineering as akin to chemical food additives, such as

1. Identify source of germplasm for traits of interest

2. Introduce germplasm into a plant population

3. Select hybrid with desired traits combinations

4. Back-cross, if necessary, to eliminate unwanted characteristics

5. Evaluate in the greenhouse and the field the hybrid's agronomic

   performance, including among other factors:

   a) growth characteristics

   b) yield

   c) relative viability and longevity of

   superior traits

   d) susceptibility to insects and

   pathogens

   e) potential genetic exchange with

   sexually compatible relatives

6. Evaluate the safety of the selected plants, including:

   a) ecological effects

   b) food safety

7. Evaluate the selected plants for consumer acceptability

8. Release superior plants for general use. Destroy or return to research any unacceptable plants.

Fig. 3. A typical crop breeding plan.

artificial sweeteners. In an effort to gain strict regulation of all new biotech foods, she would invoke the relatively stringent testing requirements of the Food Additive Act:

> "We argue simply that substances added to foods by genetic engineering ought to be judged according to the same legal presumption as substances added to foods by any other means: they are guilty until proven innocent" (GOLDBURG and HOPKINS, 1992).

In some instances, particularly where the safety of introduced substances cannot be inferred from a history of safe use in the food supply, careful scrutiny is warranted. The majority of new biotech foods, however, will involve common crop plants and the kinds of traits and improvements traditionally sought by plant breeders. Many if not the majority of the newly introduced genes will be derived from organisms that are common to the food supply and have a history of safe use (see, for example, DEPARTMENT OF HEALTH AND HUMAN SERVICES, 1992, and Organisation for Economic Cooperation and Development, 1992a). Historically, the wide spectrum of genetic modifications presented by new crops developed by traditional breeding methods, with the exception of exotic varieties, generally have not been regulated by FDA. As genetic engineering becomes more commonplace in plant breeding, however, the regulatory scheme recommended by GOLDBURG would create a federal mechanism for regulating foods from more and more new crop varieties before they enter the marketplace. The scientific rationale for the associated reporting burdens and costs to producers and consumers is, at best, ambiguous.

Of course, activists' scenarios do not need to be scientifically rational to be effective. Each presents the specter of imposed risk to an American public known for its ready accommodation of voluntary risks and its utter fear of manmade hazards. As FUMENTO (1993) observes, "(p)eople tend to apply a grandfather clause to risks. Old risks are considered much better than new risks, even if the old risks are much greater. ... We spend much of our time pursuing trace amounts of modern chemicals like dioxin that, so far as we know, have killed no one, even while we

continue to gulp down alcohol, puff away at cigarettes, and clog our arteries with fat. Those are enemies, but they are old ones. We became used to them all long ago. But the presence of pesticides on vegetables in parts per quadrillion or dioxin in water at parts per quintillion – now that's considered terrifying". The anti-biotechnology scenarios create a vision of an entirely new class of involuntary risks with which, the public is told, we have had little or no experience and can exercise little or no control.

This approach also echoes the earlier activist-inspired debate on cancer and chemicals, described by EFRON (1984): "This vision of the (chemicals and cancer) problem and its cure clearly implies that carcinogens are, for the most part, artificial inventions and do not ordinarily appear in nature." Unnatural is completely unacceptable, according to a Natural Resources Defense Council (NRDC) spokesperson (FUMENTO, 1993): "Allowing the EPA to condone continued use of a chemical whenever the benefits outweigh the risks is absolutely anathema to the environmental community." Even when the benefits outweight the risks? This pro-natural, anti-pesticide philosophy has been extended to agricultural biotechnology by the National Wildlife Federation's MELLON (1991a):

> "It is time to articulate a vision of sustainable agriculture and ask how best to achieve it. Our goal should not be merely more new pesticides but a 75 percent reduction in pesticide use. By setting the proper goal, we will avoid the danger of spending millions trying to genetically engineer 'better' pesticides, when for far less we could have taken our agriculture systems off the pesticide treadmill forever."

By this view, using naturally occurring pest management systems, such as beneficial microorganisms, is acceptable so long as those organisms are intact and unmodified. On the other hand, it would be unacceptable to use other approaches to take advantage of the biochemical mechanisms the microorganisms employ in controlling insects. For instance, if the microorganism's gene responsible for insect control is identified and transferred directly into a crop plant where it is expressed and defends the plant against pest damage by the very same biochemical mechanism as in the microbe, it is not an acceptable pest control strategy. This conclusion strains logic.

It is commonly asserted by environmental groups that because genetically engineered organisms, particularly those presenting interspecies combinations of genes, are not identical to organisms found in nature, the "problem" they pose and the potential "cure" needed are both considered to warrant rigorous governmental intervention. From a scientific vantage point, such demands appear both unsound and inconsistent with public calls for environmentally safer pest control strategies.

That biotechnology opponents continue to adhere to scientifically refutable risk scenarios may signal a more central agenda. MARK SAGOFF (1988) observed that:

> "Agricultural biotechnology will also affect the environment indirectly by making production more independent of the functioning of natural ecosystems. Phosphate solubilizing and nitrogen fixing plants now in the experimental stage, for example, can compensate for poor soil conditions; why worry, then, about erosion? Likewise, drought-resistant crops, also being developed, substitute for water. Environmentalists, then, may lose important utilitarian or prudential arguments for protecting natural ecosystems. These arguments become harder and harder to defend as we find cheap technological substitutes for nature's gifts."

What is wrong with cheap technological substitutes for nature's gifts if they are safe and effective? The concern about the environmentalist's diminishing rhetorical armamentarium conflicts with what one might expect most environmentalists would be enthusiastic about – namely, that these kinds of crop improvements will enable farmers to utilize marginal lands, protect against further encroachment against primitive forests, and even help restore some farmland to more pristine natural systems.

The Environmental Defense Fund's GOLDBURG has described how safety issues are raised as a surrogate or straw man for other concerns: "I think a lot of people just don't feel right in their gut about recombinant DNA in agriculture – they feel on some level

it's not right to mix plant and animal genes. But unfortunately, health concerns are the only mechanism available to them to express their doubts. We have to talk about whether these products are safe, not whether they are necessary or desirable" (SEABROOK, 1993). This is echoed in an interview with RIFKIN in the same article:

> "I asked RIFKIN whether he thought that Calgene's tomato was safe. He leaned forward again, lowered his voice confidingly, and said that Calgene's tomato probably is safe. Then he gave me an argument, 'The tomato is the classic example of the old way of thinking: whatever increases productivity is good and will find a market,' he said. 'I call it World's Fair thinking. But now we have a new way of thinking. What we will see in food in the nineties is going to be a battle between the World's Fair view of the world and the new, ecologically based stewardship of the world. Food is an intimate issue. I'm telling you food is going to be the focus of all green-oriented politics. And this is only going to gather momentum – in a year, this movement is going to be so strong that no genetically engineered product will make it to the supermarket'."

The National Wildlife Federation's MELLON (1991a), a vocal supporter of nontechnological approaches to sustainable agriculture, points to another issue that drives opposition-limited federal funding:

> "The problem is that the hype surrounding biotechnology diverts our attention from those solutions by focusing attention on technologically dazzling new products. The bias towards products deprives the systems-based approaches of the research and extension resources that are required to achieve their full potential."

TED HOWARD (1993) of the Pure Food Campaign reveals the broadest spectrum of concerns: "Our opposition is for a large variety of reasons, from ethical and philosophical, to human health and nutrition, to environmental safety and our concern for the family farm and sustainable agriculture in this country." It is clear that it would be mistaken to conclude from the imagined risk scenarios that garner media attention that the goals and motivations of activists are driven by misunderstandings of scientific issues of risk. They are more likely fueled by an historical disaffection for industrial agriculture. This would conform with the fact that biomedical applications of biotechnology remain essentially unscathed by the anti-biotechnology mythology and are more marked by dramatic progress and the relative absence of activist campaigns.

It is interesting that the anti-biotechnology effort has been selectively targeted at certain applications and not others. The public's high regard for biomedical research and new therapeutics is well known, whereas public support for agricultural innovation is less vigorous undoubtedly because food is abundant and cheap. Even within agriculture, however, there are distinctions. For instance, where bovine somatotropin (BST) has been aggressively targeted, chymosin produced from genetically engineered bacteria has not. Chymosin (also known as rennet) is used to clot milk to make cheese. STEVE WITT (1993) provides an analysis:

> "Until chymosin, calves' stomachs were ground up in salt water. Ten calves' stomachs were required to produce one gallon of rennet – a very ugly image, especially for animal rights activists. Worse, only 5% of the (stomach extract) that was produced was actually rennet. Now enter biotech and Pfizer and what do you do? You splice the (chymosin) gene into your favorite bacteria, which you grow in a fermenter and what do you get? ... Ninety-five percent pure enzyme at one-half the cost of traditional rennet. The most amazing fact is that as much as 3 million gallons of (stomach extract) is used each year. At 10 calves' stomachs per gallon, that's 30 million calves worldwide that still have stomachs, thanks to (genetically engineered) chymosin. When I spoke to people at Pfizer, I was told that (RIFKIN's) PureFood Campaign and others talked about chymosin when it was first proposed, but they faced a tremendous dilemma. Some of the greatest, if not the greatest, proponents of chymosin are the animal rights activists throughout the world. (Chymosin is) a real home run for biotech with 40-45% of the market and revenues of about $ 50 million a year in the United States.

9 Values, Emotions, and Information

There is an emotional, even philosophical dimension that may for some underscore the appeal of the anti-biotechnology position espoused by activist groups. For example, SAGOFF (1988) recommends:

> "If we are concerned about the effects of biotechnology, we should not focus on the risks. We should think about the ways entrepreneurs and public officials will use the weapons of biotechnology purposefully to defeat and, one might say, humiliate nature."

This statement reveals a perhaps not so idiosyncratic, but deeply emotional view of nature anthropomorphized and victimized. In his analysis of individuals' psychological responses to environmental problems, psychiatrist GLENN SWOGGER, Jr., (1992) identified a common "wish to retreat to purity and simplicity in the face of complexity." Such a wish has been articulated by some participants in the anti-biotechnology effort; for example, MELLON (1991b) has said: "At some deep level I am disturbed by genetic engineering. Basically, I am conservative. I feel an affection for the natural world the way it is – the way four billion years of evolution have made it. I resist the notion of improving nature in the future just as I lament the loss of nature as it was in the past." The suggestion that the products of four billion years of evolution have been either environmentally benign, absent of genetic change, or unaffected by humans is simply wrong-headed.

Genes are continually transferred, mutated, and regulated in all species and throughout nature. In fact, genetic change has been a critical natural driving force on evolution and an essential factor in the development of the remarkable diversity of life we have on earth, today. For that matter, the practice of exchanging or controlling genes is not unique to the genetic engineers or the traditional breeders. There is also tremendous homogeneity in genes across broadly divergent species owing to nature's adherence to molecular systems that work. Geneticist FRANCISCO AYALA (1991) pointed this out for one common class of metabolically important molecules: "Cytochrome c molecules are slowly evolving proteins; that is, the rate of amino acid substitutions per unit of time is low. Therefore, organisms as different as humans, moths, and *Neurospora* molds have a large proportion of amino acids in their cytochrome c molecules in common." People have simply taken advantage of nature's conservative hereditary plan, identifying and refining the practices of nature in order to bring to social use the world's vast genetic resources. Modern genetic technologies bring any gene from any source into potential service, at least theoretically.

The lay person's general understanding of the genetic basis of evolution is sufficiently low, however, to support such naive expressions of evolutionary angst as that expressed by MELLON, AYALA continues:

> "The stark Darwinian picture of our origin still encounters enormous skepticism. Though the evidence for the evolutionary continuity of all organisms is widely accepted, the idea that it occurred through a natural, unguided process requiring billions of years is another matter. Virtually all biologists recognize this mechanism as fundamental for the unifying, central role of evolution in biology today, and they see it as enormously strengthened by the evidence from molecular genetics. But in a poll of a random sample of the population in this country, only 9 percent of respondents believed in an unguided process."

Returning, however, to the emotional and valuative content of the environmentalist's message, it would be foolhardy to presume that the interested lay person would recognize the scientific naivete. As DOROTHY NELKIN (1987) points out in her study of press coverage of science and technology, imagery often replaces content and the reader has grown used to incomplete and dramatic accounts of events. "Although there is more information about science in the press today than there was in the past, public understanding of science and technology is often distorted; this is an age of science journalism, but also science fantasy and scientistic cults."

While NELKIN's study is principally focused on the role of scientists in shaping press coverage of science events, it seems equally applicable to the role of other interest groups, as well.

Moreover, the "facts" supporting the anti-biotechnology mythology are often articulated by degreed scientists working in environmental organizations; those academic credentials confer credibility in the eyes of a lay person unaware of the speakers' diversion from scientific rigor and of their motivations based in personal world views. As EFRON (1984) noted:

> "There are scientists in existence today who are deeply distrustful of, or even hostile to, science as well as technology. They are guided by the belief that Western industrial civilization will, if not arrested and reversed, destroy life on earth. The idea of the apocalypse has greater power over such scientists than any attachment to the disciplines of science, and in the name of that idea they are entirely willing to reject those disciplines. Even when they are fully aware that the known data do not support that idea, they do not tolerate rational limits on their knowledge."

The anti-biotechnology visions of environmental apocalypse due, for example, to the rampage of rDNA-modified plants containing DNA from known plant pathogens (such as *Agrobacterium tumefaciens* Ti plasmid) are vivid and compelling to those who have no idea that the ability of an organism to cause disease is complex and tenuous, that relatively few genes from a pathogen are related to pathogenicity, and that vectors can be molecularly "disarmed." For the lay person, the environmentalist vision protrays serious involuntary risks.

Major environmental groups have considerable reach through their membership, newsletters, and media campaigns. Either intentionally or through ignorance, the scenarios they present are based upon scientific half-truths and inconsistencies. The result is either misinterpretation or disinformation that ultimately misleads on issues of risk. What may be at stake here was described by EFRON (1984) in the chemicals and cancer debate:

"(i)nevitably, the 'axiom' of nature's noncarcinogenicity left its mark on a large portion of the American public. In fact, it was the layman above all who was supremely conscious of that 'axiom.' His ignorance of the scientific issues was so profound that he could not be distracted by details. For millions, after years of 'environmentalist' instruction, all that was really graspable was the fact that 'nature' was good and 'chemicals' were evil."

We see the impact of this sort of axiom in the nature-is-good-but-genetic-engineering-is-dangerous approach pursued by the EPA's policies under TSCA and FIFRA (above). In contrast, while the impact of environmentalist efforts on government is clear, there is little sign that the public has yet grown generally antagonistic towards agricultural and environmental applications of biotechnology (see next). Government has actually outpaced public opinion in embracing the faulty axiom of nature's safety and biotechnology's inherent danger. Through the implementation of scientifically unwarranted regulatory systems, government has provided an imprimatur to the anti-biotechnology mythology. The public may not completely trust the government's efficiency and competence, but they are quick to deduce that a label announcing "product of genetic engineering" or special confinement requirements for field tests constitute at least a material and uncertain difference, and at worst a warning. When pseudoscientific arguments are embraced in policy, government becomes a channel for disinformation.

10 Who is Representing the American Public?

That the influence of the anti-biotechnology mythology (described in Sect. 7) is still relatively limited probably would not have surprised historian HOFSTADTER. He noted more than 40 years ago (1948) that: "(i)n material power and productivity the United States has been a flourishing success. Societies that are in such good working order have a kind of mute organic consistency. They do not foster ideas that are hostile to their fundamental working arrangements." Similarly for public attitudes towards biotechnology, a national poll conducted during one of the peak periods in the biotechnology policy skirmish by the U.S. Congress OFFICE OF TECHNOLOGY ASSESSMENT (1987) found that 82% of the American public believed that research in genetic engineering and biotechnology should be continued, and the majority expressed a certain willingness to accept relatively high rates of risks to the environment to gain the potential benefits of genetically engineered organisms. While the public surveys identify concerns, those concerns are usually not focused on genetic engineering, *per se*, but on specific applications and uses that can be managed.

Furthermore, GARY BEALL and JAMES HAYES (personal communication) conducted two studies of media newspaper coverage during the period January 1990 to April 1992 and found that editors of selected U.S. newspapers did not view agricultural biotechnology as a priority issue. Even though 20 percent of all agricultural biotechnology reports focused on the contemporaneous bovine somatotropin controversy, most were buried in news or back sections, such as business sections. Moreover, judging from editorial page coverage (either in editorials or letters to the editor), they found, public interest was scant. Editorials and letters tend to increase in frequency when people are riled up about an issue. BEALL and HAYES note, on media coverage generally, that "most of the stories pictured biotechnology in a favorable light and made liberal use of sources from the agricultural biotechnology industry."

Without regard for these and other indicators of either the public's ambivalence or even support for biotechnology, policy-making continues to favor governmental intervention into the majority of agricultural and environmental research. That the anti-biotechnology mythology has been embraced by government regulators should cause considerable unease among scientists striving to employ the best and most appropriate tools to solve environmental and agricultural problems. Instead, as we have seen, the anti-biotechnology mythology and scientifically dubious regulatory schemes gain at least tacit support again and again from some members of the academic research community and from industry. Thus, in the biotechnology controversy many players – environmentalists, industry, academic scientists, and government officials – have demonstrated sufficient motivation and interest in subordinating empirical evidence to populist conceptions of risk.

Some scientists have attempted to correct errors of scientific fact and provide support for more scientifically sound regulatory schemes. Their efforts have met with a weak and sometimes conflicting response from their peers on scientific advisory committees who variously endorse scientific principles and articulate compromises that mix science and non-scientific issues. As STEPHEN JAY GOULD (1981) cautions, "(i)f it is to have an enduring value, sound debunking must do more than replace one social prejudice with another. It must use more adequate biology to drive out fallacious ideas. (Social prejudices themselves may be refactory, but particular biological supports for them can be dislodged.)" This has clearly not yet been achieved in the biotechnology policy debate, where "more adequate biology" seems often to have been judged even by some as simply another "social prejudice".

On the other hand, the anti-biotechnology mythology and the regulatory structures it has spawned are built on real and well articulated social prejudices that deserve an airing – e.g., that genetic manipulation of nature is morally unacceptable, or that high technology has little place in "sustainable" agricultural systems. As GOLDBURG said (above), health claims are often a necessary entree to the policy debate for those seeking social and economic remedies. In order to expose those prejudices and socioeconomic issues to debate and analysis, the scientifically unverifiable claims of risk must be soundly and consistently debunked wherever sufficient scientific knowledge to the contrary exists; and that debunking must be accepted by policy-makers. Beyond that, science – an empirical system of posing questions, testing hypotheses, and making measurements on tangible issues of health and environmental risks – cannot help resolve issues of personal value or emotion

and is often outside the grasp of the layperson. It is nonetheless an essential part of rational policy-making, and is rejected at the risk both of excluding our very best safety assessment tools and of providing the public with little clarity or guidance.

The erratic and contradictory response of the scientific community to inaccurate characterizations of risk has allowed central social prejudices to remain veiled by a contrived scientific "controversy". Yet those very prejudices are perhaps a more appropriate substrate for public debate than the scientific and technical issues of risk, because they present issues about which the layperson has greater knowledge and personal experience. It would likely be wiser to let those societal issues rise to the surface for debate. As HOFSTADTER (1955) observed, "(t)he most prominent and pervasive failing (of the American political culture) is a certain proneness to fits of moral crusading that would be fatal if they were not sooner or later tempered with a measure of apathy and of common sense". Rather than distance the public from the debate by cloaking the issues in esoteric pseudo-scientific jargon, the scientific community would do a greater public service by stripping away the false issues and baring the central social concerns that would benefit from broader public discussion.

11 Biotechnology Policy as Public Relations

JERRY CAULDER (1988) of Mycogen Corporation has voiced an industry perspective on the need for public acceptance of biotechnology: "Biotechnology cannot proceed without public consent. And public consent is impossible in the face of public ignorance. Eliminating this ignorance does not require training the entire population in molecular biology: people may have a poor grasp of immunology, for instance, yet they don't fear vaccinations."

CAULDER is right. When highly vocal controversies occur, as they have for biotechnology, then a new technology and its products must gain public consent, for it is the public who will ultimately grant or withhold them in the marketplace. And one need not (and, in fact, cannot) educate the entire population in the subtleties of molecular and plant biology; however, vaccination is not an appropriate analogy to agricultural biotechnology. In fact, it falls into a "risk/benefit trap" which goes something like this: for biopharmaceuticals, most of which offer substantial tangible benefits (e.g., tPA for treating heart attacks, colony stimulating factors for bone marrow transplants and enhanced cancer chemotherapy, alpha-interferon for hairy cell leukemia, improved hepatitis B vaccine, and beta-interferon for multiple sclerosis), the public has judged even a moderate amount of risk to be acceptable. The products' entrance into the marketplace has not been controversial.

However, in applying the same paradigm to agricultural products, one finds that the critics question whether even small risk of an imagined "catastrophe" justifies the consumption of new tomatoes, potatoes, or soybeans. This is an argument that scientists cannot disprove, because they cannot prove a negative – that is, that the risk of a significant problem approaches zero. That the argument cannot be mounted is more frustrating when viewed from the perspective that the risk of using rDNA techniques to create new foods pales besides other known, common food risks.

The problem lies in the very different values people place on health versus agriculture and food production. In applying CAULDER's paradigm one must consider that most new agricultural biotechnology products will represent incremental improvements in aspects of food production that may appear to consumers to be at best indirect and likely esoteric. Rather than make analogies to high benefit technological innovations, it may better serve to put the new biotechnology into the broader context of how the products of "conventional biotechnology" – microbial fermentation, new crop varieties, and the rest – have been successfully and safely introduced into the marketplace. This is not an easy task. The public is repeatedly confronted with the anti-biotechnology mythology in media reports.

There is, at best, an inconsistent message coming from scientists in academia and industry. Government, itself, presents a culture of bureaucracy building and risk aversion that fosters imagined risk scenarios and subverts scientific evidence on actual risk. Taken together, it is not readily apparent that a new, more accurate context for judging biotechnology can be conveyed to the public.

Consider, however, a very different scenario lacking the stormy history of agricultural biotechnology. In that scenario, all things being equal between conventional and new biotechnology products, the government provides assurances focused on tangible issues of health and safety, and consumers seek only that additional information needed to identify other features they value, such as cost, appearance, nutritional quality, etc. With appropriate and accurate information at hand, consumers would find no cause for anxiety, and the new agricultural biotechnology products would join the ranks of the vast majority of other new food products.

Might agricultural and environmental biotechnology be reinvented to fit that scenario? That we will soon get beyond the agricultural biotechnology policy impasse is unlikely. Among pessimists, there is some concern that biotechnology may be more likely to join the ranks of food irradiation, where a technology was judged of sufficient perceived risk to warrant U.S. government-mandated product labels and industry interest in adopting the technology virtually disappeared. If that occurs, history may judge the scientific community more harshly than the anti-biotechnology environmentalist effort. Academic and industry scientists have repeatedly acquiesced to scientifically unsound policy debates and regulatory schemes in an effort aimed more towards gaining regulatory clarity than reasonableness, and more towards activist support than public understanding. Federally funded scientists who strayed from scientific assessments may be judged as having failed in their implied roles as stewards of the public's interest in obtaining social benefit from publicly supported research.

Public acceptance need not, and arguably should not, be derived from misleading governmental assurances based on unneeded assessments of imaginary risks. View the situation from the camp of the promoters of the anti-biotechnology mythology. Their current predominant policy approach depends upon an imperfectly refuted anti-biotechnology mythology and tactics intended to reach a broad audience. These tactics are less in step with efforts to serve the "public interest" than with the American advertising industry which, according to historian DANIEL BOORSTIN (1961), finds its roots in the antics of P.T. BARNUM, "... the first modern master of pseudoevents, of contrived occurrences which lent themselves to being widely and vividly reported".

The developmental course of the biotechnology debate is marked by many colorful "pseudoevents" orchestrated principally by JEREMY RIFKIN of the FET (often indirectly supported by press releases and statements by National Wildlife Federation and Environmental Defense Fund spokespeople). They include "public" campaigns against the harmless ice-minus/Frostban bacterium field tests, against milk from dairy herds treated with bovine somatotropin, and most recently against rDNA-modified foods, in general. Each campaign claimed dire consequences: an environment irrevocably damaged by uncontrolled mutant microbes, children exposed to dangerous growth-promoting chemicals, and genetically engineered foods that harbor invisible health risks. The ultimate impact of efforts like these depends in large part on the policymakers who can rally the authority to restrain product development by creating regulatory hurdles. Scientifically rational policy decisions depend, in turn, on consistent and sound science advice from the scientific community.

Each campaign claimed socioeconomic consequences, often related to the tenuous status of small farms or the consumer's "right to know." Each campaign professed to represent an outraged public, but these claims are belied by contemporaneous public opinion polls to the contrary (above). In their vigorous pursuit of government "protection" from imposed risks, the proselytizers and proselytes (including, recently, certain American chefs) represent a minority view. All the more need, perhaps, for extravagant claims and public relations efforts. In disseminating the anti-biotechnology mythology, these aggressive campaigners represent P.T. BARNUM's modern incarnation, seen darkly. As BARNUM rationalized (BOORSTIN, 1961):

> "I confess that I took no pains to set my enterprising fellow-citizens a better example. I fell in with the world's way; and if my 'puffing' was more persistent, my advertising more audacious, my posters more glaring, my pictures more exaggerated, my flags more patriotic and my transparencies more brilliant than they would have been under the management of my neighbors, it was not because I had less scruple than they, but more energy, far more ingenuity, and a better foundation for such promises."

With that he presented to his patrons a "mermaid": the preserved head of a monkey attached to the dried body of a fish. Contemporary American audiences are regaled with visions of "mutant bacteria," invisible food poisons, and threats to global biodiversity accompanied by outraged demands that the public be "protected" by governmental intervention. Replacing BARNUM's entrepreneurial entertainment today is a now familiar public policy agenda (described by EFRON, 1984) that appeared more than twenty years ago in response to environmental "crises." That agenda stems from perceptions of the tyranny of expertise and industrialism over "public good" and from a concerted effort to portray scientific assessments as socially flawed constructs. In both can be seen efforts to recast governmental policy on science and technology to consider scientific evidence as but one of several criteria needed to judge safety and risk. Others would include issues related to social justice, in economic and cultural terms framed by activists.

12 Approaching a Fourth Criterion: An International Perspective

The biotechnology debate has highlighted diverging interests and values extending beyond issues of safety and risk. Some believe that those interests and values deserve a place in policy decision-making, forming a "fourth criterion" to be added to safety, efficacy, and quality in governmental risk assessments. Others characterize them as presenting a "fourth hurdle" that applies vague and unfair standards to research and product development. What do policy-makers think? In the past year the eight-year struggle to commercialize BST for use in dairy farming culminated in the FDA's approval of the product for use and an almost immediate federally mandated ninety-day moratorium on sales. The European Community's European Commission proposed a seven-year ban on the use of BST. Governors in Minnesota and Wisconsin signed BST moratoria. That none of these moratoria was based on safety concerns signals a certain willingness by policy-makers to grant social and economic claims stature and even precedence over scientific indications on safety and risk.

Some observers discount the importance of the BST decisions in predicting an eventual fourth criterion:

> "The biotech lobby fears that the commission's opposition to BST will make large investment in European biotech a thing of the past. But some observers, such as MARK CANTLEY who follows biotech policy for the Organization for Economic Cooperation and Development, believe that the commission's action should not necessarily be viewed as a precedent. They point out that in 1990, commission agriculture and environment officials were pressing hard for socioeconomic criteria to be applied routinely in the evaluation of biotech products ... But that idea was thrown out in a high-level commission document published in 1991" (ALDHOUS, 1993).

In the U.S., the amendment to the federal budget reconciliation bill that established the ninety-day moratorium met with stiff bilateral opposition in both the Senate and House. Earlier draft legislation by the author had been unsuccessful, and it is unlikely that the moratorium effort would have succeeded at all if it had not been attached to critical federal budget legislation that could not be delayed for debate. For example, in a bipartisan letter to their colleague, Senator DAVID PRYOR, Senators EDWARD M. KENNEDY and NANCY LANDON KASSEBAUM (1993) argued:

> "If FDA determines that BST is safe and effective, then the marketplace, and not Congress, should determine the commercial fate of the product. Likewise, if FDA determines there is reason to label milk and milk products, we are confident FDA will require product labeling. If Congress second-guesses or preempts FDA on BST, it will establish a dangerous precedent that Congress knows better than FDA scientists about whether a product should be available or not. The pressure would intensify for political or socio-economic considerations to replace science in the regulatory arena."

There are, however, other indications that social and economic claims are gaining greater access to or circumventing the regulatory process. To cite a few examples, in the U.S. there are at least two instances in the states of North Carolina and Minnesota of special, parallel mechanisms established to provide public vetting of regulatory agency decisions on field research with rDNA-modified organisms. In Germany the Genetic Engineering Law enacted in 1990 requires formal review and approval of virtually all research involving genetic engineering, and even of exchange of recombinant organisms with other researchers. It has brought a stifling burden of bureaucratic oversight to basic research and prompted German scientists and companies to seek less restrictive environments (KAHN, 1992; DIXON, 1993). JÜRGEN DREWS (1993) of Hoffmann-La Roche observes:

> "... the European nations, especially the central European societies, are suffering from a growing backlog (of value-generating technologies and products). Scientifically unfounded and almost entirely politically motivated public opposition to biotechnology has

already led to the migration of many biotech investments, originally planned for Europe, to the United States or Asia. It is obvious from a recent study submitted by the SAGB, an advisory group of the European chemical industry, that this trend is continuing. According to this study, only 35% of all European chemical companies are still planning biotech investments in European countries. Forty-five percent have decided to invest in the United States and 20% will go to Japan. ... The societies of the European countries must come to understand that their attitude will not so much threaten the survival and well-being of multinational corporations based in Europe, but rather their own economic environment and their standards of living. International corporations can move – nations cannot."

Recently, the German Ministry of Health was prompted by a request from the European Commission and by growing concerns expressed by German industry and scientists to issue a draft proposal for revising the Gene Law and relaxing regulations on notification, waiting periods, and public hearings on low-risk genetically modified organisms. Notably, several German environmental and consumer organizations have softened their demands, agreeing to greater focus on human health and the environment.

Finland has been considering an extensive regulatory scheme that may be enacted early in 1995 and would:

> "have broad powers of supervision to make biotech organizations accountable for their actions. Each organization in Finland conducting biotechnology research would have to inform the Gene Technology Authority (GTA) of all ongoing and planned activities. The GTA would assess this research, as well as its impact on society and nature. ... The aim is to make it easier for organizations to conduct *'legitimate'* research projects, while curtailing them from conducting *'ambiguous'* projects" (emphasis added; O'DWYER, 1993).

Each of these regulatory schemes is based on fundamental risk assumptions that are more grounded in philosophical "any possible world" scenarios than in the rigor of scientific probability. Proponents gain a more transpar-

ent and accessible surveillance system, that at the same time holds the potential of constraining freedom of exploration if research is judged by some to be "ambiguous," inappropriate, or simply unnecessary. Especially for very early stage research, this places a difficult burden on scientists who must demonstrate to laypersons the social utility of experiments that may be intended only to broaden understanding of fundamental biochemical and physiological mechanisms or to characterize the nature and behavior of microbial ecosystems. The experience of regulators in North Carolina has demonstrated that extra-regulatory schemes can make efficient regulatory oversight vulnerable to unwarranted demands of single individuals who seek to delay or block research.

The impact of governmental actions based on extra-scientific concerns can be seen in the rate of development of agricultural biotechnology in different countries. A recent analysis of field research permits (issued through 1992; BECK and ULRICH, 1993), taken as an indicator of the pace of research and commercialization, shows that nations with rigid and burdensome regulatory schemes have had little or no field research activity. Denmark (3 permits), Germany (2), and Japan (1) had minimal activity compared to the U.S. (316), Canada (302), and France (77). "What underlies these enormous national differences? Cultural resistance to crop improvements by modern genetic methods and lack of research could both help explain low or declining activity. There may also be countries where regulation is patchily applied, or ignored, or nonexistent, so that some releases may simply be undocumented. This is rare, we think, if it happens at all. Clearly the variation in regulatory policies must be a factor in both the rates of submissions for, and approvals of, permits, although hard data is difficult to come by" (BECK and ULRICH, 1993). Normalized for the relative sizes of the industrial and academic research enterprises in each country, France's 77 permits and Belgium's 62 permits could be viewed as comparable to or perhaps exceeding activity in individual U.S. states or Canadian provinces.

There are also non-regulatory strategies for impeding research and development. Using a more tangential approach to circumvent the limitations of historical regulatory strictures, activists in Switzerland exploited constitutionally mandated legal venues for challenging building permits to interfere with the development of industrial research facilities and to force debate on the safety and appropriateness of genetic engineering (ALDHOUS, 1992; DIXON, 1993). At least partly stimulated by the hostile Swiss environment, the Swiss-based companies Hoffmann-La Roche, Ciba-Geigy, and Sandoz have developed research facilities elsewhere. In Germany the courts were used by activists to delay the development of a Hoechst production plant for nearly 10 years. In the U.S., RIFKIN used the courts to delay the first field trials of the ice-minus/Frostban bacterium by more than four years. Activists in the San Francisco Bay Area used federal and state environmental laws to delay and to increase substantially the costs of development of new research facilities at Stanford University, the University of California at San Francisco, and the University of California at Berkeley.

Pressure for considering socioeconomic issues has been considerably stronger in Europe than in the United States. In both regions and in other industrialized nations, the clearest signs of success of those striving for consideration of social and economic issues are the substantial delays and new regulatory requirements that have accompanied biotechnology policy-making in most countries. In his analysis of European opposition to biotechnology, BERNARD DIXON (1993) found that the focus and vehemence varies from country to country and from group to group. As described above for some U.S. activists and by DIXON, the debates in many instances are increasingly focusing on issues driven by personal values, such as views on sustainable agriculture, food production, and population growth. They are moving away from earlier claims of catastrophic biosafety risks.

To the extent that, in pursuit of their values and goals, the activists seek governmental limits or restrictions on the development and delivery of new products and technologies to society, their claims of social, cultural, or economic impact demand governmental scrutiny but not necessarily governmental imprimatur.

Where they would redirect the process of setting research funding priorities, redefine the criteria and participants involved in regulatory approval processes, and gain greater control over technological innovation and industrial development, their value judgments and motivations should be placed on the table. The current policy process in the U.S. and some other industrialized nations, however, shrouds socioeconomic claims and deflects dialogue and debate, essentially creating a back door to policy-making for special interests. Ethicist ROGER STRAUGHAN (1992) writes:

> "Deliberately to restrict consumers' freedom of choice by not allowing certain products to reach the market place is, in effect, to make value judgements paternalistically on behalf of those consumers. But the validity of a value judgement cannot be tested or established in the same way as a factual statement; the amount of salt in a tin of tuna, for example, can be measured and this factual information conveyed to consumers, but the judgement that tuna ought not to be offered to the consumer because certain methods of tuna fishing are morally wrong is not factual and cannot be proved or disproved by experts or authorities. ... That is not to say that deliberate restrictions of consumer choice may never be justified, but that any proposed restrictions must be scrutinised carefully and that the onus of justification rests upon those wishing to impose the restriction and thereby to make a judgement about what is right for other individuals."

13 Conclusion: A Call for a New Paradigm

Agricultural and environmental biotechnologies hold tremendous promise for providing effective new solutions for long-standing problems. They provide effective and precise new tools applicable broadly to agriculture and environmental reclamation. Yet, policymakers worldwide have focused special regulatory burdens on the use of these tools that discourage research and sustain a public mythology of risk.

As described in this chapter, policy-making in the U.S. has been marked by more fits than starts. The history of events demonstrates that conflicting interests, whether of environmentalists, industry, academia, or government bureaucrats, have converged and created gridlock. Lowest common denominator policies – those that are conservative enough to find some appeal in all interest groups – have been proposed or implemented that impose unnecessary and scientifically unwarranted regulatory burdens on research and innovation. Researcher perceptions of these burdens have created practical disincentives to field research needed to validate the effectiveness of new plant or microbial systems (RATNER, 1990; RABINO, 1991).

This course of events parallels fairly closely the tumultuous history of environmental policy-making in the U.S.: gridlock, belief that even trivial risk is intolerable (and zero risk is achievable), and a mythology of risk poorly refuted by a disorganized and variously motivated scientific community. In both biotechnology and past environmental debates, policy-making has been paralleled by sluggish private sector investment and slow innovation rates compared to other sectors, such as pharmaceuticals or microelectronics (see MILLER, Chapter 2, this volume). I believe that the inability or unwillingness to refute the anti-biotechnology mythology has been a singularly important factor in policy decisions affecting research and innovation. I have devoted a part of this essay to exploring the opportunities and impact of scientists seeking to clarify risk issues and to promote research-related policy goals. The scientific community apparently has learned little from the strife of earlier environmental controversies and, as a result, has been doomed to repeat the errors of the past.

This characterization may be somewhat unfair, however. The problem may lie in the mistaken conceptualization of American scientists as members of a "community" that can be called upon for action on policy matters. "Community," according to the American Heritage Dictionary, implies common interests, sharing, participation, and fellowship. Research scientists and indeed the U.S. research system are characterized foremost by the hallmarks of independence and rewards for individualistic creativity. Scientific societies, while providing a disciplinary tent for scientific interaction, generally provide few and imperfect mechanisms for organizing policy action. That is not to say that there is not substantial discussion within scientific organizations about the state of affairs of federal research policies; and it is not to trivialize the role of seminal reports on the nature of rDNA-related risks by the National Academy of Sciences, the National Research Council, and government sponsored policy panels. The problem lies in the tendency for such reports to gather dust on bookshelves while individual scientists fail to recognize or simply waste opportunities to bring scientific consensus to bear in the policy arena.

I have also explored what constitutes the "public" in public policy-making on biotechnology. Principal policy players have been limited to self-identified stakeholders, such as academic and industrial researchers, regulators, and environmentalists and other special interest groups. Policy-making gridlock over pseudoscientific risk issues has deflected attention from a critical regulatory policy issue – the public's interest in being protected from unreasonable risk and unjustified drains on their tax dollars. Government agencies that sidestep the public's mandate on risk at the cost of taxpayer dollars, place the public in double-jeopardy of having their tax dollars wasted both on research that is progressing too slowly towards social benefits and on expensive regulatory bureaucracies that, once built, are seldom recalled. The public also foots the bill for the inevitable concomitant costs of *not* having new and innovative approaches to serious unattended problems.

This situation illustrates the crisis in U.S. science policy-making that has marked agricultural and environmental regulations for three decades, and that may have reached a milestone in biotechnology policy. It demands a new paradigm that distinguishes the worthy historical policy goals of health and environmental safety from non-negotiable personal values, emotion-laced concerns, and scientifically indefensible risk scenarios. This paradigm would have several core characteristics. First and foremost, it would acknowledge the authenticity, rigor, and limitations of scientific knowledge. It would also separate the empirical from the emotional and, in so doing, it would end the confused collaboration between regulatory officials and special interests on fundamental scientific criteria of safety, efficacy, and quality. It would also recognize and respond to the growing demand for information on the nature and goals of basic scientific research, the process by which funding priorities are set, and the implications of science and technology developments for society. This would require new resolve to support rational decision-making – but who among the principal players has an interest in changing the status quo?

Perhaps the academic researcher can do so, called upon to provide science advice on agency advisory panels, in legislative testimony, in public comment periods, and in response to media queries. Once science advice becomes just that – factual scientific information unvarnished by any political second-guessing – then and only then will it be more difficult for activists and policy-makers to adhere to anti-scientific mythologies and to support contrived demands for protection against imaginary risks. Only then will social and economic issues come into clearer focus for dialogue and debate.

Is this a realistic expectation? Some scientific societies, notably the American Society for Biochemistry and Molecular Biology, the American Society of Microbiologists, the American Phytopathological Society, and the Federation of American Societies for Experimental Biology, have effective, integrated organizational mechanisms for developing and communicating scientific positions on research policy issues. They have to a degree successfully injected sound science information into the biotechnology debate. More importantly, over the course of their analyses they have developed an institutional "memory" of biotechnology policy issues. The societies can provide member scientists with an appropriate context within which to judge the merits of scientific questions posed by agencies.

This is not, however, a call merely for more science advice but a recommendation that the process of advising bears analysis. The con-

cerns are certainly broader than biotechnology and extend to the basic role of science advice in government and how agencies and legislators utilize it. As HARVEY BROOKS (1988) notes, "Cynics would argue that the nation today is using science advice more, but enjoying it less ... Scientists are listened to and consulted much more at all levels of government, but, at the same time, granted much less deference". Add to this the fact that scientists are not apolitical, insofar as they are also members of the greater society. No human enterprise, science included, is devoid of social biases. As the biotechnology debate became more polarized and scientific indicators of risk were routinely deflected, certain scientists and scientific organizations, such as the National Association of State Universities and Land Grant Colleges, became more aligned with identifiable political positions than with the scientific consensus.

Such social biases and political shortcuts devalue or make less relevant empirical evidence on issues of risk and safety much to the extent that the scientific community allows. In the biotechnology debate, there are many examples where scientists allowed considerable latitude in the face of public fears or anti-scientific challenges. As DAVID PACKARD (1988) wrote, "(s)cientists should recognize that their advice is indispensable on scientific issues, but their advice on other matters deserves no special credibility over other well-informed people". Thus, it would seem wise to stick to the science.

Where many scientists promote public science literacy as the key to sustaining the development of science and technology, I would argue that even a well informed public relies on experts on sophisticated technical issues. Indeed, as the public generously funds basic research in an effort to expand the frontiers of knowledge, they reasonably expect that scientists will use that knowledge to help forge social policies as well as to advance health and economic well-being. As part of the social contract attached to public funding for research, the scientific community serves as the public's proxy on technical issues related to policy-making. What is needed is a more coherent, sound, and consistent message from the scientific community to help policy-makers strike a balance between converging demands. Perhaps greater involvement by scientific societies will help communicate to the scientific community what is at stake when they allow science advice to take a back seat to extra-scientific considerations on fundamental issues of safety and risk. For the United States, early scientific and commercial prowess in agricultural biotechnology may have been critically underexploited. The governmental burdens and delays applied to research and commercialization are effectively leveling the playing field for worldwide competition and development.

14 References

ALDHOUS, P. (1992), Swiss drug giants seek antidote to activists, *Science* **256**, 608–609.

ALDHOUS, P. (1993), Thumbs down for cattle hormone, *Science* **261**, 418.

AMES, B.N., MAGAW, R., GOLD, L. SWIRSKY (1987), Ranking possible carcinogenic hazards, *Science* **236**, 271–280.

ANDERSON, W.T. (1990), *Reality Isn't What it Used to Be: Theatrical Politics, Ready-to-Wear Religion, Global Myths, Primitive Chic, and Other Wonders of the Postmodern World*, New York: Harper & Row.

ARNTZEN, C.J. (1992), Regulation of transgenic plants, *Science* **257**, 1327.

ASHTON, D. (1993), Presentation to the *Agricultural Biotechnology Roundtable*, September 24, Davis, California.

AYALA, F.J. (1991), Evolution, in: *The Genetic Revolution* (DAVIS, B.D., Ed.), pp. 178–195, Baltimore–London: The Johns Hopkins University Press.

BECK, C.I., ULRICH, T.H. (1993), Environmental release permits, *Bio/Technology* **11**, 1524–1528.

BIOTECH FORUM EUROPE (1992), Biotech food and nutrition market to double, *Biotech Forum Eur.* **9** (7/8), 426.

BOORSTIN, D.J. (1961), *The Image: A Guide to Pseudo-Events in America,* New York: Atheneum Macmillan Publishing Company.

BROOKS, H. (1988), Issues in high-level science advice, in: *Science and Technology Advice to the President, Congress, and the Judiciary* (GOLDEN, W., Ed.), pp. 51–64, New York: Pergamon Press.

CARPENTER, W. (1991), Image and stand the scientific community needs to take in the public eye, presented at the *Annual Meeting of the American Phytopathological Society*, August 17–21, St. Louis, MO.

CAULDER, G. (1988), Last Word, *Bio/Technology*, August.

COOK, R.J. (1992), *Letter* dated September 17 *to the Environmental Protection Agency Public Response Section*, Field Operations Division.

DEPARTMENT OF HEALTH AND HUMAN SERVICES (1976), National Institutes of Health: Guidelines for Research Involving Recombinant DNA Molecules; Notice, *Fed. Reg.* **51** (88), 16958–16985.

DEPARTMENT OF HEALTH AND HUMAN SERVICES (1992), Food and Drug Administration: Statement of Policy: Foods derived from new plant varieties; Notice, *Fed. Reg.* **57** (104), 22984–23005.

DEPARTMENT OF HEALTH AND HUMAN SERVICES (1993), Food and Drug Administration: Food labeling; Foods derived from new plant varieties, *Fed. Reg.* **58** (80), 25837–25841.

DIXON, B. (1993), Who's who in European antibiotech, *Bio/Technology* **11,** 44–48.

DREWS, J. (1993), Biotechnology and the generation of economic value, in: *The Biotechnology Report* (PRICE, H.S., Ed.), pp. 54–55, Hong Kong: Campden Publishers, Inc.

EFRON, E. (1984), *The Apocalyptics: How Environmental Politics Controls What we Know about Cancer,* New York: Simon & Schuster, Inc.

EXECUTIVE OFFICE OF THE PRESIDENT (1992), Office of Science and Technology Policy, *Fed. Reg.* **57** (39), 6753–6762.

FEDERAL COORDINATING COUNCIL FOR SCIENCE, ENGINEERING, AND TECHNOLOGY (1993), Biotechnology for the 21st Century: Realizing the promise, a *Report by the Committee on Life Sciences and Health,* Bethesda, MD: National Institutes of Health Office of Recombinant DNA Activities.

FEDEROFF, N. (1987), Impeding genetic engineering, *The New York Times*, September 2.

FREIBERG, B. (1993), A practical look at biotechnology, *Biotech Reporter*, April, p. 13.

FUMENTO, M. (1993), *Science Under Siege: Balancing Technology and the Environment*, New York: William Morrow & Company.

GOLDBURG, R.J., HOPKINS, D.D. (1992), What EDF wants, *Bio/Technology* **10**, 1384.

GOODMAN, R.M., HAUPTLI, H., CROSSWAY, A., KNAUF, V.C. (1987), Gene transfer in crop improvement, *Science* **236**, 48–54.

GOULD, S.J. (1981), *The Mismeasure of Man*, New York-London: W.W. Norton & Company.

GOULD, S.J. (1987), Integrity and Mr. Rifkin, in: *An Urchin in the Storm*, pp. 229–239, New York-London: W.W. Norton & Company.

HOBAN, T.J. (1989), Sociology and biotechnology: Challenges and opportunities, *Southern Rural Sociol.* **6**, 45–63.

HOBAN, T.J. (1990), An educational needs assessment of agricultural biotechnology, *Final Report to the North Carolina Biotechnology Center*, Raleigh, NC: North Carolina State University.

HOBAN, T.J., KENDALL, P.A. (1992), *Consumer attitudes about the use of biotechnology in agriculture and food production*, Raleigh, NC: North Carolina State University.

HOFSTADTER, R. (1948), *The American Political Tradition And The Men Who Made It*, New York: Vintage Books.

HOFSTADTER, R. (1952), *The Paranoid Style in American Politics*, Chicago: University of Chicago Press.

HOFSTADTER, R. (1955), *The Age of Reform*, New York: Vintage Books.

HOWARD, T. (1993), In, A garden of unearthly delights, R. Mather, D446,831, *Detroit News*, February 9.

HUTTNER, S.L. (1993), Risk and reason: An assessment of APHIS, in: *U.S. Agricultural Research: Strategic Challenges and Options* (WEAVER, R.D., Ed.), pp. 155–170, Bethesda, MD: Agricultural Research Institute.

HUTTNER, S.L., SMITH (1991), State case histories on biotechnology oversight, *UC Biotechnology Research and Education Program*, Los Angeles, CA: UCLA.

HUTTNER, S.L., ARNTZEN, C.J., BEACHY, R., BRUENING, G., DEFRANCESCO, L., NESTER, E., QUALSET, C., VIDAVER, A. (1992), Revising oversight of genetically modified plants, *Bio/Technology* **10** (9), 967–971.

INTERNATIONAL COUNCIL OF SCIENTIFIC UNIONS (1990), Joint SCOPE/COGENE Statement, in: *Introduction of Genetically Modified Organisms into the Environment* (MOONEY, H.A., BERNARDI, C., Eds.), p. XIX, London–New York: John Wiley & Sons.

KAHN, P. (1992), Germany's gene law begins to bite, *Science* **255**, 524–526.

KENNEDY, E.M., KASSEBAUM, N. LANDON (1993), *Letter* dated July 22 *to Senator David Pryor*.

KESSLER, D.A., TAYLOR, M.R., MARYANSKI, J.H., FLAMM, E.L., KAHL, L.S. (1992), The safety of foods developed by biotechnology, *Science* **256**, 1747–1749.

LANDSMANN, J. (1992), Rapporteur's *Report* on the *Second International Symposium on the Biosafety Results of Field Tests of Genetically Modified Plants and Microorganisms*, held in Goslar, Germany, May 11–14.

MELLON, M.G. (1991a), Biotechnology and the environmental vision, in: *Agricultural Biotechnology at the Crossroads: Biological, Social, and Institutional Concerns*, pp. 66–70, Ithaca, NY: National Agricultural Biotechnology Council Report 3.

MELLON, M. (1991b), *Biotechnology Views*, American Biotechnology Association.

MILLER, H.I. (1989), Debunking the silly myths that hobble biotechnology, p. 11A, *Minneapolis Star Tribune*, March 25.

MILLER, H.I. (1991), Regulation, in: *The Genetic Revolution* (DAVIS, B.D., Ed.), pp. 196–211, Baltimore-London: The Johns Hopkins University Press.

MILLER, H.I. (1993), Foods of the future: The new biotechnology and FDA regulation, *J. Am. Med. Assoc.* **269**, 910.

MILLER, H.I., YOUNG, F.E. (1988), "Old" Biotechnology to "New" Biotechnology: Continuum or disjunction?, in: *Safety Assurance for Environmental Introductions of Genetically Engineered Organisms* (FIKSEL, J., COVELLO, V., Eds.), *NATO ASI Series*, G18, New York: Springer Verlag.

MILLER, H.I., BURRIS, R.H., VIDAVER, A.K., WIVEL, N.A. (1990), Risk-based oversight of experiments in the environment, *Science* **250**, 490–491.

MILLER, J.D. (1985), *The Attitudes of Religious Environmental and Science Policy Leaders Toward Biotechnology: A Final Report*, DeKalb, IL: Northern Illinois University Public Opinion Laboratory.

NATIONAL ACADEMY OF SCIENCES (1987), *Introduction of Recombinant DNA-Engineered Organisms into the Environment: Key Issues*, prepared for the Council of the National Academy of Sciences by the Committee on the Introduction of Genetically Engineered Organisms into the Environment, Washington, DC: National Academy Press.

NATIONAL INSTITUTES OF HEALTH (1991), *Report of the Planning Subcommittee of the Recombinant DNA Advisory Committee*, Office of Recombinant DNA Activities, Bethesda, MD: National Institutes of Health.

NATIONAL RESEARCH COUNCIL (1989), *Field Testing Genetically Modified Organisms: Framework for Decisions*, Committee on Scientific Evaluation of the Introduction of Genetically Modified Microorganisms and Plants into the Environment, Board on Biology, Commission on Life Sciences, Washington, DC: National Academy Press.

NATO ADVANCED RESEARCH WORKSHOP (1988), Workshop summary, in: *Safety Assurance for Environmental Introductions of Genetically Engineered Organisms* (FIKSEL, J., COVEL-

LO, V.T., Eds.), *NATO ASI Series*, Berlin-Heidelberg-New York: Springer-Verlag.

NELKIN, D. (1987), *Selling Science: How the Press Covers Science and Technology*, New York: W.H. Freeman & Company.

NOVO INDUSTRI A/S (1987), *The Novo Report: American Attitudes and Beliefs about Genetic Engineering*, Wilton, CT: Novo Laboratories, Inc.

O'DWYER, G. (1993), Finland considering far-reaching biotech rules, *Bio/Technology* **11**, 1515.

OFFICE OF SCIENCE AND TECHNOLOGY POLICY (1986), Coordinated Framework for Regulation of Biotechnology; Notice, *Fed. Reg.* **51** (123), 233-2–23350.

OFFICE OF TECHNOLOGY ASSESSMENT (1987), *New Developments in Biotechnology, 2: Background Paper: Public Perceptions of Biotechnology*, Springfield, VA: National Technical Information Service, U.S. Department of Commerce.

OFFICE OF TECHNOLOGY ASSESSMENT (1992), *A New Technological Era for American Agriculture*, U.S. Government Printing Office: 192 O - 297–938.

ORGANISATION FOR ECONOMIC CO-OPERATION AND DEVELOPMENT (OECD) (1993a), *Safety Evaluation of Foods Derived by Modern Biotechnology: Concepts and Principles*, Paris: Organisation for Economic Co-operation and Development.

ORGANISATION FOR ECONOMIC CO-OPERATION AND DEVELOPMENT (OECD) (1993b), *Traditional Crop Breeding Practices: A Historical Review as a Baseline for Assessing the Role of Modern Biotechnology*, Paris: Organisation for Economic Co-operation and Development.

ORGANISATION FOR ECONOMIC CO-OPERATION AND DEVELOPMENT (OECD) (1993c), *Safety Considerations for Biotechnology: Scale-up of Crop Plants*, Paris: Organisation for Economic Co-operation an Development.

PACKARD, D. (1988), Science and the federal government, in: *Science and Technology Advice to the President, Congress, and the Judiciary* (GOLDEN, W., Ed.), pp. 245–250, New York: Pergamon Press.

RABINO, I. (1991), The impact of activist pressures on recombinant DNA research, *Sci. Tech. & Human Values* **16** (1), 70–87.

RATNER, M. (1990), Survey and opinions: Barriers to field-testing genetically modified organisms, *Bio/Technology* **8**, 196–198.

RICK, C.M. (1978), The tomato, *Sci. Am.* **239**, 76–87.

SAGOFF, M. (1988), Biotechnology and the environment: Ethical and cultural considerations, *Environ. Law Rep.* **19**, 10521–10526.

SEABROOK, J. (1993), Tremors in the hothouse, *The New Yorker*, July 19, pp. 32–41.

SIDDHANTI, S.K. (1988), Multiple perspectives on risk and regulation: The case of deliberate release of genetically engineered organisms into the environment, *Dissertation*, Ann Arbor, MI: University of Michigan Dissertation Services.

STEVENS, M.A., GREEN, C.E. (1993), *Letter* dated July 14 *to the Food and Drug Administration* docket No. 92N-0139.

STRAUGHAN, R. (1992), Freedom of choice: principles and practice, in: *Your Food: Whose Choice?* (NATIONAL CONSUMER COUNCIL, Ed.), pp. 135–156, London: HMSO Publications.

SWOGGER, G., JR. (1992), Why emotions eclipse rational thinking about the environment, *Priorities*, Fall, pp. 7–10.

TOWNSEND, R. (1993), APHIS and agricultural research, in: *U.S. Agricultural Research: Strategic Challenges and Options* (WEAVER, R.D., Ed.), pp. 141–154, Bethesda, MD: Agricultural Research Institute.

UNIDO/WHO/UNEP (1987), *Working Group Report on Biotechnology Safety*, Paris: UNIDO.

UNIDO/WHO/UNEP (1992), *Release of Organisms into the Environment: Voluntary Code of Conduct*, Biotech Forum Eur. **9** (4), 218–221.

U.S. DEPARTMENT OF AGRICULTURE (1987), Animal and Plant Health Inspection Service 7 CFR, Parts 330 and 340: Plants pests; introduction of genetically engineered organisms or products; Final Rule, *Fed. Reg.* **52** (115), 22892–22915.

U.S. DEPARTMENT OF AGRICULTURE (1993), Animal and Plant Health Inspection Service 7 CFR Part 340: Genetically engineered organisms and products; Notification procedures for the introduction of certain regulated articles; and Petition for nonregulated status, *Fed. Reg.* **58** (60), 17044–17059.

U.S. ENVIRONMENTAL PROTECTION AGENCY (1992), *Safeguarding the Future: Credible Science, Credible Decisions*, EPA/600/9-91/050.

VIDAVER, A.K. (1991), Oversight and regulation of biotechnology, presented at the *4th International Symposium on Biotechnology and Plant Protection*, October 21-23, University of Maryland at College Park.

WITT, S.C. (1990), *BriefBook: Biotechnology, Microbes, and the Environment*, San Francisco: Center for Science Information.

WITT, S.C. (1993), presentation to the *Science, Symbol, and Substance Conference*, June, Research Triangle Park, NC.

17 Press Coverage of Genetic Engineering in Germany: Facts, Faults and Causes

HANS MATHIAS KEPPLINGER
SIMONE CHRISTINE EHMIG

Mainz, Federal Republic of Germany

1 Introduction

Genetic engineering is widely known as one of the branches of science with the most promising future. This concerns its contribution to the diagnosis and cure of heriditary defects and serious diseases, to the production of vaccines and drugs, to growing plants and breeding animals as well as to its application in the area of environmental protection. Genetic engineering is at the same time, however, one of those branches of science, which are exposed to massive criticism. Alterations to the genetic make-up of microorganisms, plants, animals and humans make genetic engineering appear to be an especially serious interference in intact nature, since the alterations are invisible and their consequences seem to affect the whole population. People's fears range, therefore, from biological risks such as the uncontrolled spread of changes in organisms caused by genetic engineering to social risks such as decreasing acceptance of handicapped people. Here particular reference is made to long-term consequences, unknown consequences and irreversible consequences.

The opponents of genetic engineering in Germany group themselves in organizations such as Gen-ethisches Netzwerk (genetic-ethical network), Berlin, Öko-Institut (ecological institute), Freiburg, and Komitee für Grundrechte und Demokratie (committee for constitutional rights and democracy), Sensbachtal, who claim to represent public welfare, or rather the natural order of things. They appeal to the public with numerous publications and sometimes also offer services such as supplying speakers for the fight against genetic engineering. The magazine *natur* provides them with a forum which receives considerable attention. Examples for their activities are the years of struggle against the manufacture of human insulin using genetic engineering by the Hoechst AG company and against an outdoor experiment with petunias changed by genetic engineering by the Max-Planck-Institut für Züchtungsforschung (institute for plant breeding research) in Cologne.

Today, genetic engineering in Germany is at a cross-road. Its rapid development seems quite possible, but it may also be blocked in the foreseeable future. With this as a background, the investigation presented in this chapter intends to describe the representation of genetic engineering in the mass media and to analyze some essential reasons for its criticism and acceptance by the mass media (for details see KEPPLINGER et al., 1991b). Therefore, we have quantitatively analyzed the extent and the manner in which genetic engineering is portrayed in six national daily newspapers, four regional daily newspapers, two weekly papers, two weekly magazines and four popular science magazines. As a rule, the basis of the analysis is a random sample from one sixth of all editions in the years 1987 to 1989. The popular science magazines are an exception – from these we took account of half of all editions. In addition, we investigated 30 experts, 30 scientific journalists and 30 political editors, respectively. In the case of the experts, we were dealing with persons in leading positions at research institutions in universities, Max-Planck institutes as well as research enterprises. They may be regarded as distinguished specialists, whose judgement possesses considerable significance. This is also true in a similar way for the scientific journalists we interviewed, who were among the most respected in their field. The relevance of the answers is, therefore, based not on the number but on the special competence of those interviewed. In the case of the political editors, we were concerned with the employees of daily newspapers. They are not representative, statistically speaking, nor do they have specialist knowledge.

2 Representation of Genetic Engineering

In the editions investigated from 1987 to 1989, the 18 newspapers and magazines published a total of 524 articles with 4494 evaluating statements on genetic engineering. Projected to all editions in the period of investigation, this corresponds to 2688 articles with 22024 statements. We took into consideration here the varying publication frequency of the papers. Most of the articles and statements appeared in the national dailies. They contained a total of 302 articles with 2294 statements. Thus, it can be established that the journalistic discussion of genetic engineering in the period of investigation took place chiefly in the national dailies (Tab. 1).

Tab. 1. Number of Articles and Statements from 1987 to 1989. Basis: One sixth of all editions; half of all editions of the popular science magazines

	Coded:		Projection:	
	Articles n	Statements n	Articles n	Statements n
National dailies				
Frankfurter Rundschau	61	366	366	2196
Süddeutsche Zeitung	59	477	354	2862
Frankfurter Allg. Zeitung	83	516	498	3096
Die Welt	41	298	246	1788
die tageszeitung	47	543	282	3258
Handelsblatt	11	94	66	564
Total	302	2294	1812	13764
Average	50.33	382.33	302	2294
Regional dailies				
Westdeutsche Allgemeine	6	33	36	198
Kieler Nachrichten	2	35	12	210
Münchner Merkur	31	108	186	648
Rheinzeitung Koblenz	11	79	66	474
Total	50	255	300	1530
Average	12.5	63.75	75	382.5
Weekly papers				
Die Zeit	24	308	144	1848
Rheinischer Merkur	17	173	102	1038
Total	41	481	246	2886
Average	20.5	240.5	123	1443
Magazines				
Der Spiegel	12	167	72	1002
Stern	5	62	30	372
Total	17	229	102	1374
Average	8.5	114.5	51	687
Popular science magazines				
natur	51	539	102	1078
Spektrum der Wissenschaft	17	139	34	278
P.M. Magazin	8	187	16	374
bild der wissenschaft	38	370	76	740
Total	114	1235	228	2470
Average	28.5	308.75	57	617.5
All papers	524	4494	2688	22024
Average	29.11	249.67	149.33	1223.56

Tab. 2. Authors of the Articles on Genetic Engineering

	National Daily Newspapers ($n=302$) %	Regional Daily Newspapers ($n=50$) %	Weekly Newspapers ($n=41$) %	Magazines ($n=17$) %	Popular Science Magazines ($n=114$) %	Total ($n=524$) %
Scientific journalists	20	12	22	6	13	17
Other journalists	20	14	7	6	3	15
News agencies	18	16	—	—	—	12
Other identifiable authors	2	2	4	—	5	3
Non-identifiable sources	38	28	66	29	72	47
No information	3	28	—	59	7	8
Total	101	100	99	100	100	102

By far most articles on genetic engineering were not written by scientific journalists. Only 17% of the articles analyzed were written by authors who were clearly proven to be scientific journalists or who were identifiable as such. The source of a large number of articles could not be identified. Even if one supposed that half of these articles were written by scientific journalists – which is very improbable –, it can be established that the image of genetic engineering in the press is determined mainly by journalists without specific expert knowledge. Here the significant role of the news agencies is remarkable, above all for news coverage in the daily newspapers (Tab. 2).

In contrast to popular science magazines, daily newspapers, weekly papers and weekly magazines have clearly defined editorial sections, which generally include a special scientific section. However, most statements on genetic engineering do not appear – contrary to the widespread assumption – in the scientific sections of the daily and weekly papers, but in their political sections. The scientific sections took only second place here. All the other sections played no great part. In the political sections, genetic engineering was characterized on the whole rather negatively, in the scientific sections in contrast definitely positively. (The tendency of the statements was assessed on a scale of -3 to $+3$.) It should be remembered that the political sections are given attention by many more readers than the scientific sections. Thus, a central prob-

lem has been identified: genetic engineering is given press coverage particularly in those places where its presentation rather tends to be negative and where it is widely read (Fig. 1).

The articles concerning genetic engineering dealt with a wide range of topics, which can only briefly be summarized here. General

Tab. 3. Statements Concerning Various Areas of Application of Genetic Engineering in Newspapers and Magazines[a]

	Number of Statements ($n=4.494$) %	Tendency of Statements ($n=4.014$)[b] x
Genetic engineering in general	41	-0.18
Human genetics	22	-0.17
Human health	12	$+0.89$
Animal production	12	-0.26
Plant production	11	$+0.51$
Environmental protection	2	-0.20
Other areas of application	1	$+0.49$
Total	101	$+0.03$

[a] 6 national dailies, 4 regional dailies, 2 weekly papers, 2 magazines, 4 popular science magazines
[b] Mean values ($+/-3$). Without statements on the general political, economic and legal framework of genetic engineering

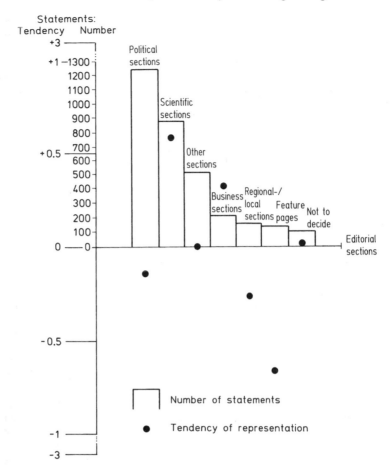

Fig. 1. Making genetic engineering a subject of discussion and the tendency of its representation in various editorial sections. Number of statements and mean values (+/−3; without statements concerning conditions).

questions concerning genetic engineering were definitely the center of interest. Human genetics was also reported relatively frequently. Both topics on the whole revealed a slightly negative tendency. This was also true for the application of genetic engineering in animal production. In contrast, the significance of genetic engineering for human health and for plant production was given definitely positive coverage (despite the discussion of an outdoor experiment with petunias in the period of the investigation). These two topics were only touched on in one third of all statements. Thus, a second problem has been identified: topics given a negative coverage were dealt with frequently, whereas topics given a positive coverage were discussed rarely (Tab. 3).

The papers investigated reported the opinions of various prominent persons concerning genetic engineering. Some of these persons belong to organizations which develop and apply genetic engineering, others belong to organizations which criticize both. Among the first-named organizations are academic research institutions and industrial companies, among the last-named organizations are the alternative research institutions, citizens' action groups and, for example, the Greens (the German ecological party). In the political sections of the daily and weekly papers especially those members of the latter organizations had a say, whose opinion of genetic engineering was very negative. In the scientific sections, especially those members of the first-mentioned organizations had a say, whose attitude was relatively positive. Thus, a third problem has been identified: Those who develop and make use of genetic engineering have no sufficient say in the places where the readers obtain their information (Tab. 4).

In theory, the tendency of the news coverage in the news items should be independent of the tendency of the opinions expressed in the comments. In reality, however, in most papers there was a clear connection between the tendency of the statements in the comments and the tendency of the statements in the news. Thus, a fourth problem has been identified: The journalists' views in the various editorial departments, expressed in the comments, had an influence on the tendency

Tab. 4. Statements by Various Authors in the Political and Scientific Sections of Newspapers and Magazines[a]

	Political Sections		Scientific Sections	
	Number of Statements n	Tendency of Statements[b] $\bar{x}$	Number of Statements n	Tendency of Statements[b] $\bar{x}$
Academic research institutions, industry	140	+0.90	184	+0.56
Alternative research institutes, interest groups, churches, The Greens	189	−1.64	28	−1.86

[a] 6 national dailies, 4 regional dailies, 2 weekly papers, 2 magazines
[b] Mean values (+/−3). Without statements on the general political, economic and legal framework of genetic engineering

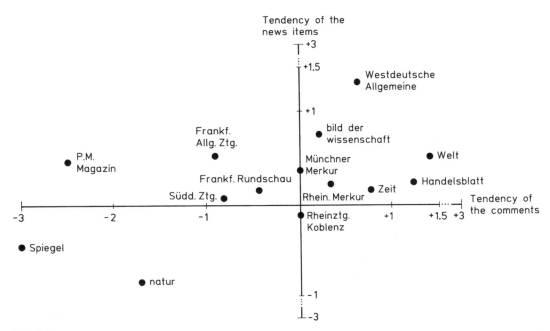

Fig. 2. Tendency of the representation of genetic engineering in comments and news items in individual newspapers and magazines. Basis: Evaluating statements in all editorial sections: mean values (+/−3); without statements concerning conditions.
Abbreviations stand for: *Frankfurter Allgemeine Zeitung, Frankfurter Rundschau, Süddeutsche Zeitung, bild der wissenschaft, Rheinischer Merkur, Rheinzeitung Koblenz.*

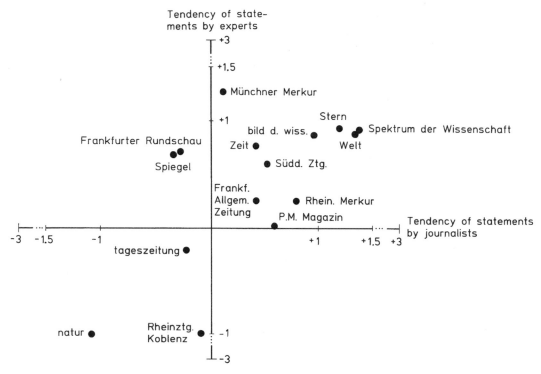

Fig. 3. "Instrumentalization" of experts. Basis: Statements by journalists and scientists: mean values (+/−3); without statements concerning conditions. Abbreviations see Fig. 2.

of the representation of genetic engineering in the news and reports. Thus, the news and reports on current events seemed to confirm the views of the commentators (Fig. 2). (In German newspapers, usually both the comments and the news are written by the same journalist.)

In many instances, experts differ in their opinions. In most cases, however, there is a clear majority for one point of view. In the present study, all or almost all of the experts interviewed judged genetic engineering positively. They assessed it as being very useful and considered its potential for doing harm quite low (cf. KEPPLINGER et al., 1991b, pp. 31–35). Most newspapers and magazines let those experts have their say, who characterized genetic engineering positively. They thus conveyed a thoroughly adequate picture of the experts' views. Some papers used the opposite technique. They let – corresponding to the tenor of the journalists' attitudes –

those experts have a say, who gave genetic engineering negative news coverage and thus portrayed a misleading picture of actual views among the experts. Thus, we can identify a fifth problem: The journalists' views have a decisive influence on the public visibility of experts (Fig. 3).

3 Causes for the Way of Representation

The representation of genetic engineering in the papers investigated can be attributed mainly to six causes:

1. A considerable number of journalists have a negative attitude towards genetic engineering (KEPPLINGER et al., 1991b, pp. 28–

30). This is especially true for political editors, but also for part of the scientific journalists. The journalists' attitudes were evident both in their general opinions concerning genetic engineering as well as in their views concerning its risks (Tab. 5).

2. A considerable number of journalists mistrust scientists, especially political editors; but this is also true, with reservations, for scientific journalists. The journalists' mistrust is shown both in their judgements concerning the credibility of the employees of the industrial companies (see KEPPLINGER et al., 1991b, pp. 61–65), as well as in judgements about the trustworthiness of employees in academic research institutions (Tab. 6).

3. The shortcomings in the representation of genetic engineering in the press to a considerable extent are due to flaws in the organization of the editorial departments: those who have specialist knowledge are often not responsible, and those who are responsible frequently do not have specialist knowledge. Specialist knowledge available is not used adequately. Scientific journalists competent in their special fields seldom write articles on genetic engineering.

Tab. 5. Opinions of Journalists and Scientists Concerning the Risks of Genetic Engineering. Question: "Various risks are being discussed in connection with genetic engineering. Please go through the following list: which of these risks are, in your estimation, to be taken very seriously, quite seriously, less seriously, or can be ignored completely?" Number of people questioned who were of the opinion that a risk was "to be taken very seriously" or was "to be taken quite seriously"

	Scientists $(n=30)$ n	Scientific Journalists $(n=30)$ n	Political Journalists $(n=30)$ n
"Discrimination of individuals with unusual genetic features on the job market"	15	23	18
"Decreasing willingness to accept the life of handicapped people"	12	21	19
"Military use of organisms altered by genetic engineering"	12	20	26
"No clear-cut transitions from the correction of defective genes to optimization of human hereditary factors"	11	19	26
"State-enforced abortion of unborn babies with genetic defects"	11	11	8
"Uncontrolled spread of manipulated organisms in outdoor experiments"	10	22	27
"Increased use of chemical pesticides as a result of the development of herbicide-resistant plants"	10	19	20
"Pressure for rationalization and accelerated structural change in agriculture"	10	18	16
"State-enforced information on genetically caused risks to one's own health"	10	16	18
"Impairment of health in laboratory staff through contacts with new combinations of organisms"	5	18	21
"Unintended creation of new pathogenic organisms from originally non-pathogenic material"	3	23	28
"Systematic 'breeding of humans' by private or state organizations"	3	13	16
"Environmental pollution due to accidents due to new combinations of organisms"	2	18	27
Total	114	241	270

Tab. 6. Confidence in Scientists in the Case of Controversial Science Projects. Question: "Please imagine the following case: Scientists at a university institute are planning to test plants altered by genetic engineering in an outdoor experiment. Scientists from an alternative scientific institution claim that there is danger of an uncontrolled spread of these plants. The scientists concerned dispute this and refer to appropriate preliminary investigations. Which side would you rather trust?"

	Scientists *n*	Scientific Journalists *n*	Political Journalists *n*
"The scientists concerned"	24	12	4
"The scientists of the alternative institution"	—	3	6
"Both to the same extent"	3	7	10
"Neither"	1	7	10
Other	2	1	—
Total	30	30	30

Tab. 7. Presumed Causes of Shortcomings in Press Coverage. Question: "The mass media's representation of science and technology is often criticized as being inaccurate. Regarding this I would like to read you an opinion. Please tell me whether you think that this opinion is correct, partly correct or incorrect. The opinion is as follows: The shortcomings in the representation of science and technology are not based so much on the fact that the scientific journalists have had poor training and are badly informed. It is due much more to the fact that their articles hardly get a chance in day-to-day press coverage, because the political editors take every topic of general interest out of their hands."

	Scientists *n*	Scientific Journalists *n*	Political Journalists *n*
"I agree completely with this opinion"	6	4	1
"I agree in part with this opinion"	12	14	11
"I do not agree with this opinion"	8	8	17
"I don't know"	1	—	—
No definite answer	3	4	1
Total	30	30	30

Moreover, they have no influence on the articles of their non-specialist colleagues in the political department (Tab. 7).

4. The attitudes of (German) journalists to the subjects of their news coverage have an effect on the way they represent them. A great many journalists approve of the conscious playing up of information which supports their own views. These views significantly influence the choice of news (see KEPPLINGER et al., 1989, 1991a). We call this "instrumental actualization" of information: information is instrumentally selected and published.

5. Numerous scientists recognize the necessity of actively informing the public about their own work. Only a few, however, take the initiative and publish articles about their activities in the press. Thus, for example, two thirds of the scientists interviewed were of the opinion that scientists "(must) on their own initiative inform (journalists) about the background knowledge which is vital for understanding (their research)". Only seven, however, had published an article in a daily or weekly newspaper. In doing so they have – to a large extent – left the field to the opponents of genetic engi-

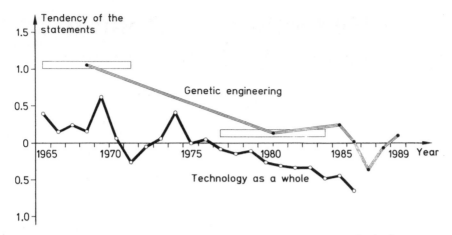

Fig. 4. Representation of genetic engineering seen in the long term. Basis: Statements concerning genetic engineering in the political sections of the *Frankfurter Rundschau, Süddeutsche Zeitung, Frankfurter Allgemeine Zeitung, Welt, Zeit, Spiegel* and *Stern* from 1965 to 1989 as well as statements concerning technology as a whole in the named sources from 1965 to 1986 (without statements on conditions). From KEPPLINGER (1989)

neering (see KEPPLINGER et al., 1991b, pp. 90–93).

6. Journalists' attitudes to genetic engineering, and linked to this the characterization of genetic engineering in the press, are the expression of a general "Zeitgeist", which the mass media were involved in bringing about (for details see KEPPLINGER, 1989, pp. 157–164). This is demonstrated by comparing the representation of genetic engineering with that of other technologies (energy, chemical industry, transport, etc.). Data from the recent past indicate that here a change in trend is taking place or has already taken place (Fig. 4).

4 References

KEPPLINGER, H. M. (1989), *Künstliche Horizonte. Folgen, Darstellung und Akzeptanz von Technik in der Bundesrepublik*, Frankfurt am Main/New York: Campus.

KEPPLINGER, H. M., BROSIUS, H.-B., STAAB, J. F., LINKE, G. (1989), Instrumentelle Aktualisierung. Grundlagen einer Theorie publizistischer Konflikte, in: *Massenkommunikation. Theorien, Methoden, Befunde* (KAASE, M., SCHULZ, W., Eds.), pp. 199–220, Opladen: Westdeutscher Verlag.

KEPPLINGER, H. M., BROSIUS, H.-B., STAAB, J. F. (1991a), Instrumental actualization: A theory of mediated conflicts, *Eur. J. Commun.* **6**, 263–290.

KEPPLINGER, H. M., EHMIG, S. C., AHLHEIM, C. (1991b), *Gentechnik im Widerstreit. Zum Verhältnis von Wissenschaft und Journalismus*, Frankfurt am Main/New York: Campus.

18 The Regulation of Modern Biotechnology: A Historical and European Perspective

A Case Study in How Societies Cope with New Knowledge in the Last Quarter of the Twentieth Century

Mark F. Cantley

Paris, France

Introduction

The events described below occurred over a period of two decades, from the mid-1970s, covering the birth and early years of modern biotechnology and the first practical and commercial applications of genetic engineering. The story is a complex one, for the several reasons discussed below; a story of many strands, but also of many interlinkages, many interactions, between the various strands. The danger for any narrator of such a story is that the more comprehensively and thoroughly he presents the complexity, and the mass of detail, the greater is the risk of concealing the salient lessons; and there are important lessons to be learned, by all the different players concerned, but especially for public policy in Europe. Therefore, selectivity, omission and compression have had to be applied, in the interest of greater significance and transparency.

The "story" is about how different societies have coped, are coping, with a sudden surge of knowledge and understanding about the basic structures and mechanisms of living things. In focusing on public policy and regulations, the intention is less to tell the story of the remarkable scientific discoveries and technological developments and inventions, well described elsewhere, than to study how societies learn to digest the new knowledge, and to manage its consequences; in the hope of showing how they may learn faster and manage better.

The various strands of the story are spun and interwoven through time, so time will serve as a common base and organizing principle; although various countries are at different points on their respective "learning curves" of experience, and the curves are not identical.

The learning process which can be perceived across the various strands is a multidimensional one, in several distinct respects:

– multi-disciplinarity of the scientific base
– multi-sectoral applications
– multi-constituency
– multi-national
– multi-international-institutional.

Each of these facets is here briefly introduced, and recurs in the narrative that follows.

The scientific base of biotechnology is multi-disciplinary, drawing upon elements from biochemistry, microbiology, molecular biology, genetics, process engineering (including especially, but not restricted to, fermentation); and no area of the life sciences and technologies has remained unaffected by the surge in knowledge and technique, particularly at the molecular level.

A similar multiplicity characterizes the sectors of application. In every area of human interaction with living entities and systems, there is scope for the application of greater knowledge, and consequently greater subtlety and efficacy of intervention – in agriculture and food production, in health care, and in the recycling of organic wastes, water purification, and protection and management of the natural environment. More "bio-rational" approaches may be seen as good in themselves, for values aesthetic, ecological or economic; or as essential responses to competitive pressures, from competitors to whom the global knowledge base is no less accessible; or as essential tools for maintaining (or achieving) decent standards of health and nutrition, for a human population approaching and passing ten billion, and while reversing the current degradation of the planet.

As issues relating to biotechnology and genetic engineering came into greater prominence, the number of interested "constituencies" increased. The scientists themselves, at Asilomar, initiated through systematic engagement with journalists, a wider debate engaging the political leaders and their staff, and the general public. The previous paragraph refers to the various economic sectors, agricultural, industrial and other, who were progressively drawn into the debate. The circle of debate was further widened by ethical considerations, to draw in philosophers and theologians. Within government, ripples spread from Research Ministries, concerned with both basic and applied science, to applications involving Ministries of Industry and Trade, Agriculture, Health, Environment, Education and others; and issues involving regulatory and legal aspects, engaging Patent

Offices and Ministries of Justice. The sequence in which they awoke or reacted varied from country to country, although within the European Community the processes of Community legislation tended to bring similar Ministries into synchrony in their preoccupations, with biotechnology as with other topics.

The debates about biotechnology have from the start been international, because of the natural internationalism of science. As the technologies moved into application, this internationalism was doubly reinforced: by the continuing progress towards an open world economy; and by the increasing significance within that economy of knowledge-intensive sectors (such as biotechnology), and consequently of intellectual property. The former development could be underlined by reference to the 7-year, "Uruguay Round" of GATT negotiations, culminating in the signature of the agreement at Marrakesh in April 1994; the latter aspect, by the significant attention and controversy which biotechnology attracted within the trade-related intellectual property element of the GATT agreement.

The multi-national policy implications and multiplicity of aspects mentioned above had their institutional counterparts, epitomized by the presence of biotechnology on the agenda of practically every agency of the United Nations. UNIDO, UNEP, WHO and FAO collaborated in the development of biotechnology guidelines. Common biotechnology-related policy questions face international agricultural research centers, or international conferences on harmonization of authorization/registration procedures for pharmaceuticals. At the "Earth Summit" in Rio de Janeiro, June 1992, biotechnology figured significantly in the debates on biodiversity, and in the articles of the resulting convention; within the "Agenda 21" development plans for the 21st century, there are many references to biotechnology, including the whole of Chapter 16.

This multidimensional character will be brought out in what follows, drawing particularly on developments within the European Community or Union; whose own constitution and institutional structures were undergoing rapid evolution during the same decades. This aspect interacted significantly, and sometimes adversely, with the process whose description is the central aim of this essay: how societies digest and learn to manage the surge of new knowledge and methods summarized by the word "biotechnology". Some points are commented on in the narrative; a final section draws together a synthesis and conclusions.

1 Origins and Beginnings: From Avery to Asilomar, and Capitol Hill

1.1 Slow Progress: The Decades Before Asilomar

1994 saw the 50th anniversary of the classic paper by OSWALD AVERY and colleagues (AVERY et al., 1944), in which they identified DNA as the molecule uniquely associated with the storage and transfer of genetic information between different strains of bacteria. His work on *Pneumococcus*, the bacterium responsible for pneumonia, had started three decades earlier during World War I. A decade after AVERY's paper, WATSON and CRICK used crystallographic data and biochemical reasoning to elucidate the structure of DNA, in the UK MRC's (Medical Research Council) Laboratory of Molecular Biology, Cambridge. In the same year and laboratory, SANGER published the amino-acid sequence of the protein insulin. X-ray crystallographic methods – developed in Cambridge's Cavendish Laboratory by the BRAGGS, father and son, in the 1920s – had there first been applied to biological molecules by BERNAL and PERUTZ in the 1930s; an initiative which led, two decades later, to the double helix.

Following the double helix discovery, the genetic code was elucidated; following AVERY, and through the work of LEDERBERG and others, the field of bacterial genetics was progressively developed. But more decades of work and progress elapsed before COHEN and BOYER at Stanford could (in 1974) publish (and subsequently patent) their use of restriction enzymes with bacterial plasmids for the fundamental "cut and stitch" activities which became known as genetic engineering. Also in the mid-seventies, SANGER, and GILBERT and MAXAM at Harvard, published their methods of reading, nucleotide-by-nucleotide, genetic sequences.

This history of some of the most significant discoveries of twentieth-century science has been often and more fully described, for their significance attracts the historians of science, and will long and rightly continue to do so (see, for example, the review by WITKOWSKI, 1988). As background to the policy and regulatory debates which developed around biotechnology, the purpose of recalling the history is to identify the salient factors which lead many observers to speak of the "Genetic Revolution" (e.g., DAVIS, 1991). A useful chronology of the two latest decades was assembled by RYSER and WEBER (1990).

1.2 The Genetic Revolution: Acceleration, and Irreversible Knowledge

As is indicated by the dates quoted, the gestation periods for these major scientific discoveries and developments were measured in decades, rather than years. But from these slow beginnings, a steep acceleration has followed. Understanding of the molecular mechanisms of all living systems was a progressive, interactive process. The interactions stimulated further insights, hypotheses, experiments and discoveries; the process was carried forward in an increasing number of centers around the developed world; the knowledge thus gained was cumulative, irreversible, and globally available. Subversive and pervasive, the discoveries could not be reversed, nor the powerful but simple techniques and methods disinvented. Could they, should they, be controlled and regulated? That was the original question at Asilomar; and was recurrent through the years that followed.

The subsequent developments in the life sciences and technologies, over the decades from the mid-70s, interacted increasingly with the "Information Revolution" of data storage, software sophistication, computing power, and global electronic networking. By the early '90s, DNA sequence data read by automated machinery was pouring into the 2 or 3 global databanks at a rate of millions of nucleotides per month. This flood of new knowledge, of which DNA sequence remains merely one aspect, albeit the most fundamental, has been generated and driven by massive increases in the financial, human and technical resources devoted to the R & D effort, by both governments and private sector. The technical resources themselves have become enormously more efficient and productive, both within the biology laboratory (e.g., in sequencing technology), and in the information handling within and beyond the laboratory: at all levels of scientific data, extending also to clinical, and bibliographic, and beyond science to the provision and use of commercial and legal information, including a related massive growth of patent applications and intellectual property rights.

This quantitative surge of knowledge, and the acceleration of its rate of further expansion, have had, and continue to produce, shock waves: qualitative effects rippling across scientific disciplines, government and international institutions and policies. The shock waves extend through agriculture and food production, health care, and environmental management, reaching into philosophy, theology and ethics. UNESCO in 1993 created an International Committee on Bioethics; over the previous two decades every UN agency had found itself involved in the implications of the new knowledge. An OECD study of long-term economic impacts of biotechnology in the mid-1980s, was obliged to appear with a modified title: "Economic and Wider Impacts of Biotechnology"; for the "Wider" impacts would be noticed first (OECD, 1989).

The changes within the life sciences are profound, as the traditional disciplines are flooded with illumination from the molecular level, from scientists who know not the name of LINNAEUS. Virologists who have split over the years into sub-groups focused on bacterial, insect, plant, animal, clinical or other viral sub-disciplines, now find a re-emergence of common interests, in projects such as the World Virus Databank. Molecular evolution re-examines and illuminates the legacy of DARWIN; taxonomy, systematics and nomenclature are reinvigorated and (particularly in the context of declining biodiversity) recognized as essential to the rational structuring and management of the new knowledge.

The implications for public policy of the biological revolution did not strike all departments of government simultaneously, nor similarly. If one uses an orchestral metaphor, the violins of Science had the privilege of initiating the new theme; other sections of government responded as in a fugue, with answers or variations on the theme; and with the increase in the number of sections of government participating, the need for a conductor of the orchestra became ever more apparent. There were to be many orchestras, and many conductors; harmony, and harmonization between traditions and places, proved elusive.

1.3 Asilomar

In agricultural or medical research, in plant or animal breeding, and in the production of fermentation antibiotics, the continuing development and application of the life sciences during the early post-war decades were routinely pursued, in familiar compartments with relatively little inter-sectoral interaction, beyond a perfunctory acknowledgement of common roots in biology. Basic science, the dramatic discoveries referred to above, were saluted with Nobel prizes, but impinged only sporadically or slowly on practical concerns. That has changed since the mid-70s.

It has become a convenience or a cliché to date histories of biotechnology from a conference held at Asilomar, California, in February 1975; whose origins were somewhat earlier. In February 1973, a conference on biohazards was held at Asilomar, California. It attracted little attention, but stimulated further thought. In June 1973, the annual session of the Gordon Conference on Nucleic Acids was held in New Hampton, New Hampshire, and was devoted to the problem of hazards in recombinant DNA research. The co-chairs of

the conference, MAXINE SINGER and DIETER SOLL, drafted a letter addressed to the National Academy of Sciences and the Institute of Medicine, requesting the formation of a study committee, to assess the biohazards posed by recombinant DNA research, and recommend appropriate action. The letter was published in *Science* (SINGER and SOLL, 1973).

As a result, the National Academy of Sciences announced in February 1974, that PAUL BERG would chair the study committee. The 11 members, all active in recombinant DNA research, were conscious of the quickening pace of research, and apprehensive about possible accidents. Their report was also published in *Science*, on 26 July 1974 (BERG et al., 1974), and almost simultaneously (slightly abridged) in *Nature*. The text of the "Berg letter" is reproduced below:

"Potential Biohazards of Recombinant DNA Molecules

Recent advances in techniques for the isolation and rejoining of segments of DNA now permit construction of biologically active recombinant DNA molecules in vitro. For example, DNA restriction endonucleases, which generate DNA fragments containing cohesive ends especially suitable for rejoining have been used to create new types of biologically functional bacterial plasmids carrying antibiotic resistance markers and to link *Xenopus laevis* ribosomal DNA to DNA from a bacterial plasmid. This latter recombinant plasmid has been shown to replicate stably in *Escherichia coli* where it synthesizes RNA that is complementary to *X. laevis* ribosomal DNA. Similarly, segments of *Drosophila* chromosomal DNA have been incorporated into both plasmid and bacteriophage DNAs to yield hybrid molecules that can infect and replicate in *E. coli*.

Several groups of scientists are now planning to use this technology to create recombinant DNAs from a variety of other viral, animal, and bacterial sources. Although such experiments are likely to facilitate the solution of important theoretical and practical biological problems, they would also result in the creation of novel types of infectious DNA elements whose biological properties cannot be completely predicted in advance.

There is serious concern that some of these artificial recombinant DNA molecules could prove biologically hazardous. One potential hazard in current experiments derives from the need to use a bacterium like *E. coli* to clone the recombinant

DNA molecules and to amplify their number. Strains of *E. coli* commonly reside in the human intestinal tract, and they are capable of exchanging genetic information with other types of bacteria, some of which are pathogenic to man. Thus, new DNA elements introduced into *E. coli* might possibly become widely disseminated among human, bacterial, plant, or animal populations with unpredictable effects.

Concern for these emerging capabilities was raised by scientists attending the 1973 Gordon Research Conference on Nucleic Acids, who requested that the National Academy of Sciences give consideration to these matters. The undersigned members of a committee, acting on behalf of and with the endorsement of the Assembly of Life Sciences of the National Research Council on this matter, propose the following recommendations.

First, and most important, that until the potential hazards of such recombinant DNA molecules have been better evaluated or until adequate methods are developed for preventing their spread, scientists throughout the world join with the members of this committee in voluntarily deferring the following types of experiments.

Type 1: Construction of new, autonomously replicating bacterial plasmids that might result in the introduction of genetic determinants for antibiotic resistance or bacterial toxin formation into bacterial strains that do not at present carry such determinants; or construction of new bacterial plasmids containing combinations of resistance to clinically useful antibiotics unless plasmids containing such combinations of antibiotic resistance determinants already exist in nature.

Type 2: Linkage of all or segments of the DNAs from oncogenic [cancer-inducing] or other animal viruses to autonomously replicating DNA elements such as bacterial plasmids or other viral DNAs. Such recombinant DNA molecules might be more easily disseminated to bacterial populations in humans and other species, and thus possibly increase the incidence of cancer or other diseases.

Second, plans to link fragments of animal DNAs to bacterial plasmid DNA or bacteriophage DNA should be carefully weighed in light of the fact that many types of animal cell DNAs contain sequences common to RNA tumour viruses. Since joining of any foreign DNA to a DNA replication system creates new recombinant DNA molecules whose biological properties cannot be predicted with certainty, such experiments should not be undertaken lightly.

Third, the director of the National Institutes of Health is requested to give immediate consideration to establishing an advisory committee charged with (i) overseeing an experimental program to evaluate the potential biological and ecological hazards of the above types of recombinant DNA molecules; (ii) developing procedures which will minimize the spread of such molecules within human and other populations; and (iii) devising guidelines to be followed by investigators working with potentially hazardous recombinant DNA molecules.

Fourth, an international meeting of involved scientists from all over the world should be convened early in the coming year to review scientific progress in this area and to further discuss appropriate ways to deal with the potential biohazards of recombinant DNA molecules.

The above recommendations are made with the realization (i) that our concern is based on judgements of potential rather than demonstrated risk since there are few available experimental data on the hazards of such DNA molecules and (ii) that adherence to our major recommendations will entail postponement or possibly abandonment of certain types of scientifically worthwhile experiments. Moreover, we are aware of many theoretical and practical difficulties involved in evaluating the human hazards of such recombinant DNA molecules. Nonetheless, our concern for the possible unfortunate consequences of indiscriminate application of these techniques motivates us to urge all scientists working in this area to join us in agreeing not to initiate experiments of types 1 and 2 above until attempts have been made to evaluate the hazards and some resolution of the outstanding questions has been achieved.

Paul Berg, Chairman
David Baltimore
Herbert W. Boyer
Stanley N. Cohen
Ronald W. Davis
David S. Hogness
Daniel Nathans
Richard Roblin
James D. Watson
Sherman Weissman
Norton D. Zinder
Committee on Recombinant DNA
Molecules Assembly of Life Sciences,
National Research Council,
National Academy of Sciences,
Washington, DC 20418"

The Asilomar conference was widely discussed in advance, and received worldwide press coverage. The many subsequently published descriptions and interpretations bring out several facets of what the conference was and what it achieved.

At the most factual level, the occasion was an invitation-only scientific meeting in which eminent specialists discussed the possible risks which might be associated with recombinant DNA techniques or experiments; and means for managing or reducing these conjectural risks. There was discussion of various levels of risk for classifying experiments; and corresponding levels of physical containment. Among the more constructive ideas, which British participants such as BRENNER emphasized at the conference, was the concept of biological containment: the use of strains of microorganism disabled in ways which would limit their ability to survive or reproduce outside the contained vessel and special conditions provided in the experiment. This area of the Asilomar debate was the starting-point for a great deal of risk assessment research over the following years – practically all of it reassuring, but always limited by the logical impossibility of "proving" a negative.

As an innovation in scientific communication, Asilomar could be seen both at the time and subsequently in various lights. Many commentators are inclined to congratulate the organizers on the integrity and transparency with which they were prepared to communicate their concerns to a broader public. Press representatives were invited to the conference, with the understanding that they would listen to the whole four-day conference before reporting. The resulting reportage was serious and competent, and generally acknowledged the obvious sincerity of the scientists themselves.

Some commentators nonetheless set Asilomar in the context of tradition they would describe as "elitist", characterized by the arrogant assumption that on complex matters, only those who understand the complexities should be involved in making decisions. Against such elitism it is argued that democratic procedures require the involvement of representatives of a broader constituency – of the taxpayers who have paid for publicly funded research, of the workers who might be the most immediate victims of a laboratory accident or infection, and by the same logic, of the general public who, on various conjectures, might also be victims either of an accident (such as an epidemic initiated by a recombinant organism), or be exposed to risks associated with products placed on the market.

The developments in genetic engineering to which Asilomar drew attention catalyzed a fundamental debate about the control of science and technology; or, insofar as such a debate was already in progress, extended and amplified it to all areas of the life sciences and technologies, their applications, and implications.

1.4 From Asilomar to Capitol Hill: A Dialogue of Scientist, Public and Regulator

The debate triggered by Asilomar was intense and widespread. The temptation to exaggerate and simplify, for journalists, cartoonists, or politicians, local and national, was great; and was not resisted. Scientists were often angered by the misrepresentations, and by the strident and hostile tone of the attacks they encountered – in some notable cases from major environmental movements. But a few eminent scientists supported the critics, in their calls for intense security provisions or a total moratorium on all rDNA research.

The result was a high profile, sometimes heated, public debate, with scientists often facing the ill-informed hostility of "public interest groups" or local politicians. The construction at Harvard University of a high security ("P3") laboratory for recombinant DNA research led to one such battle, in summer 1976, which was widely reported. This featured the colorful language of the local mayor of Cambridge, Massachusetts, ALFRED VELUCCI:

"It's about time the scientists began to throw all their god-damned shit right out on the table so that we can discuss it …. Who the hell do the scientists think they are that they can take federal tax dollars that are coming out of our tax returns and do research work that we then cannot come in and question?"

The period of angry debate lasted several years; a period summed up a decade later by NORTON ZINDER (1986) as "[divisible] into three periods:

Asilomar (1974–76),
the 'recombinant DNA wars' (1976–78),
and detente".

The American experience during 1974–1978 offered points of comparison for Europe, facing the same issues at approximately the same time; and for both the US and Europe during the second wave of concerns which arose during the latter half of the 1980s. Essentially similar issues were involved in each case.

Public concerns in the United States rose to a peak during 1976–1977, with the corresponding introduction in Congress (both House of Representatives and Senate) of bills to regulate recombinant DNA research. At the same time, following the Asilomar conference, the National Institutes of Health (NIH), under Director DONALD FREDRICKSON, had been active in developing guidelines for the conduct of such research, through the NIH Recombinant DNA Advisory Committee (NIH RAC). The first version of these guidelines was released by NIH on 23 June 1976, and published in the *Federal Register* on 7 July.

A major feature of the debates in the US was the progressive development of a well-organized, articulate and balanced response by the scientific community. The leading role was played by the American Society of Microbiology (ASM), but many other professional associations of biological and medical sciences joined with ASM in a broad alliance, through semiformal linkages via their executive officers, and widespread networks capable of providing rapid responses.

The recommendations of the ASM were summarized in a nine-point statement, approved in May 1977, and widely reported:

"1. That all responsibility for regulating action relative to the production and use of recombinant DNA molecules should be vested in HEW (the Department of Health, Education and Welfare).
 2. That to advise and assist the Secretary of HEW, an Advisory Committee should be established whose membership in addition to lay people should include representatives with appropriate technical expertise in this field.
 3. That institutions and not individuals should be licensed.

4. That in each institution engaged in DNA recombinant activities, to the maximum extent possible, direct regulatory responsibility should be delegated to a local biohazard committee. These committees should include both members with appropriate expertise conducted at that institution, and representatives of the public.
 5. That experiments requiring P1 [the lowest category of physical containment in the NIHRAC guidelines] requirements should be exempt from these regulations.
 6. That license removal is an effective and sufficient deterrent to obtain compliance. Further, that ASM is opposed to the bonding of scientists or to the establishing of strict individual liability clauses in the conduct of DNA recombinant activities.
 7. That ASM goes on record favoring uniform national standards governing DNA recombinant activities.
 8. That the Secretary of HEW should have the flexibility to modify the regulations as further information becomes available. Further, we support the inclusion of a sunset clause in the legislation, i.e., that legislation will be re-evaluated after a fixed period of time.
 9. That ASM expresses its concern that in establishing such important legislation governing research, and that this proceed only after due and careful deliberation." (From HALVORSON, 1986)

The ASM nine points emphasize relevant competence and technical expertise; delegation of responsibility to local committees, applying uniform national standards; the exemption from regulation of low-risk experiments (requiring P1 containment); and flexibility to adapt and re-evaluate legislation in the light of experience. All these points remained valid and important in the discussions during subsequent years and in other countries and legislatures.

During 1977, the scientific concerns were effectively communicated both in public, and to the staff of interested Congressmen. Amendments to earlier bills were prepared, progressively incorporating the scientific advice; informed scientists communicated their personal views on pending legislation to their respective senators or representatives. The information indicating the absence of harmful spread or effects of recombinant organisms was influential. By September, Senator AD-

LAI STEVENSON wrote (in a letter to the President's Science Advisor FRANK PRESS) that most of the legislation being considered was ill-designed for achieving its stated objective, namely, protection of the public without impeding research. He indicated his intention to explore the use of existing statutes to regulate recombinant DNA research.

At hearings in November 1977, the ASM expressed their concerns about "the apparent intemperate rush to establish legislation to regulate recombinant research without first consulting with the appropriately qualified scientific and medical experts, the need to understand that early allegations concerning recombinant DNA research were characterized by uncontrolled imagination and excessive claims by individuals who lacked knowledge of infectious disease, and the need for minimal interim legislation to extend appropriate guidelines to all recombinant DNA activities regardless of funding source."

During the fall of 1977 and in 1978, the ASM continued to work closely with Congressional committees, and the prospect of federal legislation declined.

1.5 Observations

The US experience in the post-Asilomar period was of significance, as a successful example of open dialogue between the scientific and political communities. The successes can be related to the flawless safety record of genetic engineering in the US over the following years; and to the position of scientific and economic leadership in biotechnology which the US maintained. More generally and importantly, the US experience provided, for scientists in all fields and legislatures everywhere, an object lesson in how to manage the interface between science and society in a way that was democratic and transparent; and that in consequence, was generally accepted and effective.

This successful experience was not inevitable. At a national conference in October 1980, on "recombinant DNA and the Federal Government", presentations were made by Federal officials responsible for agency concerns on the subject (17 Federal agencies participated), by former Congressmen and their key aides responsible for legislative activity in the field, and by Washington lawyers specializing in such issues. Lawyer STEPHAN LAWTON had been involved in drafting one of the main House Bills (for Congressman ROGERS) on regulating DNA research. Having narrated in detail the events during March 1977 to mid-1978, he concludes:

"Here ends the story, but not necessarily the lessons. I believe that all of us can learn from this experience with the recombinant DNA research legislation that did not pass.
Lesson number one is that Congress is very willing and quite able to act very quickly when the public health is at stake.
Lesson number two is that whoever dreamed up the legislative process was a genius to the extent that the legislative process is slow enough to prevent a stampede of unwise legislation.
Lesson number three is that Congress, in my judgement, had and continues to have an extremely healthy opinion of the scientific community. Congress is willing to listen to well-reasoned arguments and, believe it or not, Congressmen have the capacity to change their minds when confronted with new and well-reasoned arguments."

Presciently, he went on to ask:

"Is the legislation dead? Yes, for the time being. Is it dead forever? Maybe and maybe not. The issue itself will not be dead in Washington for a long time. It will be kept alive for several reasons. First, there are several committees of Congress that continue to hold hearings on the activities of the NIH and on recombinant DNA generally, principally the two science committees which don't have legislative jurisdiction over the recombinant DNA issues but continue in a very positive way to expose themselves and the public to the issue. Second, there is a very interested, sophisticated medical press in Washington, who will follow the issue. Third, I think the sheer excitement of the issue will keep the question of whether or not there should be governmental involvement with respect to recombinant DNA research alive. These three factors – the science committees' interest and their oversight hearings, the press, and the sheer excitement of the issues – will keep Congress interested in the issue." (LAWTON, 1981)

He concluded with the "old adage in Washington" (of Bismarckian vintage and origin), that "if you want to respect your laws like you love your sausage, do not watch too closely the process by which either is made".

2 Beginnings in Europe: From "GMAG" to the EC's 1982 Council Recommendation

2.1 The UK Working Parties, Health and Safety Legislation, and "GMAG"

When concern about the conjectural hazards of recombinant DNA research was expressed in 1974 (by the Berg letter), United Kingdom scientists were among those most affected in Europe, in view of the number of programmes and centers concerned. Such research was all funded through universities or research councils, and it was therefore straightforward for the Department of Education and Science to institute suspension of all work involving the recombinant DNA technique until government guidelines had been promulgated. A committee – the Ashby Working Party – was set up in July 1974 to advise on whether such work should proceed in the UK. Its report was published in December 1974, and recommended that such work should go ahead provided adequate safeguards were established (HMSO, 1974). The Ashby report especially focused upon the concept of "biological containment", by crippling the plasmid vector and its bacterial host. By reporting so promptly, the concepts developed by the Ashby Working Party could be used by UK or other scientists in the February 1975 Asilomar conference.

The government responded by setting up under Sir ROBERT WILLIAMS another working party in August 1975, to prepare a code of practice for work involving genetic manipulation. This reported promptly, in August 1976 (HMSO, 1976), like Ashby's working party completing its report within 12 months; but the scientific community was frustrated by the two-year delay in its work.

The Williams Working Party also recommended the establishment of an advisory body to maintain surveillance of work involving genetic manipulation, and advocated that statutory regulations be developed. As a result, the Genetic Manipulation Advisory Group (GMAG) was set up by the Department of Education and Science, and held its first meeting in January 1977. Its membership comprised eight scientific and medical experts; four "public interest" representatives; four trade union representatives, for worker interests; and two representatives of management, one nominated by the Confederation of British Industry, the other by the Committee of University Vice-Chancellors and Principals.

The concept of "public interest" representatives was an innovation in opening up the committee proceedings and allowed a greater flow of information to the public; although the GMAG meetings, unlike those of the NIH Recombinant DNA Advisory Committee, were held in private.

GMAG initially followed the advice of the Williams report regarding guidelines for the review of proposals for recombinant DNA work. The Williams guidelines emphasized good working practices, and four physical containment levels ranging from least to most stringent. GIBSON (1986) summarizes the differences from NIH as follows:

"Category I equated approximately to the NIH P2 facility requirements and Category IV is more demanding than the NIH requirements for its highest containment category. The categorization of experiments outlined in the Williams report is based on a scheme reflecting evolutionary relatedness, that is, the closer the animal source of the DNA insert was to that of man the higher the containment requirement would be for that particular experiment. This scheme was abandoned in 1979 in favor of a risk assessment scheme."

Publication of the Williams report and the creation of GMAG enabled scientific work to resume in the UK, where some ten Category III containment facilities were constructed. At GMAG's request, the Medical Research Council also sponsored training courses for biological safety officers at the government's Microbiological Research Establishment at Porton Down – now the Centre for Applied Microbiology and Research. Following the introduction in 1979 of the GMAG assessment scheme, most experiments were recategorized to the lower, "Category I", containment level, where the requirements were simply those of "good microbiological practice".

The 1970s saw a growing demand in many countries for improved health and safety at work, and the rising consciousness of risks influenced the rDNA debate. The UK had in 1974 adopted a comprehensive statute, the Health and Safety at Work Act, which gave wide-ranging powers to the government's Health and Safety Commission, implemented through the Health and Safety Executive (HSE) and the Factory Inspectorate. Under this Act, there followed more specific statutory regulations requiring all establishments (including ministries and research institutes) to set up local safety committees. "Regulations on Genetic Manipulation" (SI 1978 No. 752) were also introduced, operative from 1 August 1978. These provided that "persons should not carry on genetic manipulation unless they have previously notified the Health and Safety Executive and the Genetic Manipulation Advisory Group". Although local biological safety committees were not a statutory requirement, GMAG would only give its advice when proposals had been discussed with the local biological safety committee.

The UK genetic manipulation regulations introduced a definition that was used in subsequent European and many national legislative proposals:

"'genetic manipulation' means the formation of new combinations of heritable material by the insertion of nucleic acid molecules, produced by whatever means outside the cell, into any virus, bacterial plasmid, or other vector system so as to allow their incorporation into a host organism in which they do not naturally occur but in which they are capable of continued propagation".

GMAG's development of a risk assessment scheme in 1979, implemented from January 1980 onwards, diminished scientific concerns about overzealous safety committees, unnecessary delays, and excess disclosure of ideas

to competitors. Some rDNA work was exempted from regulations, and almost all work could be done at Category I level.

As the UK lacked a definitive description of "good microbiological practice", a guidance note on "Guidelines for Microbiological Safety" was developed by scientists in a "Joint Coordinating Committee for the Implementation of Safe Practices in Microbiology", and this was accepted by GMAG in July 1980.

GIBSON (1986) describes the Joint Coordinating Committee's action, by the scientific community, as "almost exceptional in the rDNA debate in the United Kingdom"; and emphasizes the role of individual scientific contributions. In particular, "SYDNEY BRENNER (of the Medical Research Council's Laboratory of Molecular Biology) first gave, in July 1978, the GMAG the initial concept for the risk assessment scheme that proved so successful. It was introduced in March 1979, and revised in January 1980. Local biological safety committees were able to operate with ease the risk assessment scheme, which was sufficiently flexible to allow scientific or medical information to be introduced when reaching a decision. The scheme resulted in most work being recategorized to either Category I or GMP. it was also Dr. BRENNER's development of the disabled Medical Research Council (MRC) *E. coli* strains (MRC 8 in particular) which allowed GMAG to incorporate the concept of biological containment in its risk assessment scheme with greater confidence."

Categorization of experiments was carried out by the local biological safety committee. From 1980 onwards, it was only in cases of uncertainty, or where work in Categories II, III or IV was envisaged, that GMAG was asked to advise on a case-by-case basis.

The UK Williams report and the GMAG Code of Practice appeared before the publication (June 1976) of the NIH guidelines, with the result that countries elsewhere in Europe involved in rDNA work initially decided to adopt the GMAG guidelines. In general, the absence of framework legislation such as the UK's Health and Safety at Work Act made it more difficult to introduce legislation to cover genetic engineering. As a result, when the NIH guidelines were introduced, with less demanding containment requirements and with a comprehensive and codified system for categorization of experiments, almost all mainland European countries decided to adopt these. Implementation methods and standards varied widely, and this was the background to the first legislative initiative by the European Commission, described below.

2.2 The European Commission, DG XII: Towards the First Biotech Research Programme, "BEP"

The European Commission's Directorate-General for Science, Research and Development ("DG XII") had since the mid-1970s advocated a Community R & D programme to address on a Europe-wide basis the challenges and opportunities of the advances in molecular biology. Study reports were commissioned from eminent academics to buttress the case: from DANIEL THOMAS of University of Compiegne, on enzymology (THOMAS, 1978); from ARTHUR RÖRSCH, then of University of Leiden, on genetic engineering (Rörsch, 1978); and from CHRISTIAN DE DUVE, of the International Institute of Cellular and Molecular Pathology, Brussels (DE DUVE, 1979).

At that time, and up to the implementation (in 1989) of the Single European Act, there was no specific legal basis under the founding treaties of the European Communities for R & D programmes (other than a reference to co-ordination of agricultural research in Article 41 of the Treaty of Rome, founding the European Economic Community; and the research aims pervading the Euratom Treaty). The R & D programmes proposed by the Commission during the 1970s and 1980s therefore had to use as their legal basis the very general provisions of Article 235:

"If any action by the Community appears necessary to achieve, in the functioning of the Common Market, one of the aims of the Community in cases where this Treaty has not provided for the requisite powers of action,

the Council, acting by means of an unanimous vote on a proposal of the Commission and after the Assembly has been consulted, shall enact the appropriate provisions."

This had the drawback of requiring a unanimous vote of the Member States. It was therefore considered politically essential to include in the proposed programme provision for research and development of risk assessment methods for rDNA work. A background study on risk and safety aspects was commissioned (SARGEANT and EVANS, 1979). The Biomolecular Engineering Programme, or "BEP", was finally adopted by Council on 7 December 1981, with a budget of 15 million ECU, to run from 1982 to 1986. Of its 103 projects, in fact only two addressed risk issues; but this aspect was in principle initiated by BEP, and expanded substantially in later programmes.

Expertise on genetic engineering matters within the Commission services was located in DG XII, the Directorate-General for Science, Research and Development; indeed, it was only by the chance presence of staff recruited for biological safety-related work under the Euratom Treaty that there was within the Commission a small but alert nucleus of staff cognizant of the subject, and capable of identifying, and interacting with, competent external scientific advisers such as the authors of the study reports cited above.

2.3 The 1978 Proposal for a Council Directive on rDNA Work

The DG XII biology staff, in consultation with the Scientific and Technical Research Committee (CREST), the European Science Foundation (see below), and other external sources of scientific advice, formulated in 1978 a "Proposal for a Council Directive establishing safety measures against the conjectural risks associated with recombinant DNA work" (EUROPEAN COMMISSION, 1978a). As for the research programme proposals, the legal basis was Article 235 (see above).

This proposal was submitted by the Commission to the Council on 5 December 1978. At all the ten places where it referred to "hazards" or "risks", the word was preceded by the adjective "conjectural". The preamble referred in positive terms to the value of basic and applied science, and the need to combine protection of man, his food supply and environment with the development of recombinant DNA work. It stressed the international and epidemiological character of the conjectural risks; and the risk that differing provisions in force or in preparation in Member States would affect their scientific and technological competitiveness. The rapid evolution of understanding, the need to consider local circumstances, and the need to safeguard scientific and industrial secrecy and intellectual property, were similarly acknowledged.

The definition of "recombinant DNA work" was identical to that of "genetic manipulation" in the UK regulations, quoted above.

The substance of the proposed Directive was to require prior notification to, and authorization by, national authorities, prior to all research or other work involving recombinant DNA. National authorities would develop categorization systems for rDNA work, keeping the Commission informed; and the Commission would publish them. Member States would submit to the Commission at the end of each year the list of authorizations during that year, and a general report on their experience and problems.

Of lasting significance was the final Article, No. 5, which read: "Because of the unceasing progress of knowledge and techniques in the field of basic and applied biology, this Directive and its continued applicability to production activities of industries shall be thoroughly reviewed, and revised if necessary, at regular intervals not exceeding two years."

As this proposed Directive moved into debate in the European Parliament, the Commission's staff, the scientists and administrators in DG XII, were alert to the evolution of opinion in the US during 1978, towards nonlegislation. The experience of the NIH RAC, and of the UK GMAG, was accumulating; the scientific debate was progressing; and a consensus was developing that some of the initial expressions of scientific (and other) concern had been exaggerated. NIH Director DON

FREDRICKSON visited DG XII Director-General GUNTHER SCHUSTER in 1978, to convey to him the lessons of the US experience; and emphasized the desirability of avoiding fixed statutory controls.

The Parliament had commenced its scrutiny, and was adding amendments more rigidly specifying containment requirements; but inspired by the US and UK experience, the Commission on the advice of SCHUSTER decided in 1980 to withdraw their proposed Directive, and replace it by a proposal for a Council Recommendation (EUROPEAN COMMISSION, 1980). This (non-binding) proposed Recommendation was that Member States adopt laws, regulations and administrative provisions requiring notification – not authorization – of recombinant DNA work.

2.4 The CREST Paper

At a meeting with CREST (the European Committee – of Member State officials – advising on matters of research, science and technology) in September 1978, the Commission already acknowledged (EUROPEAN COMMISSION, 1978b) that rDNA technology opened up new avenues for fundamental and applied work, which would lead to "an enormous improvement in our knowledge of genetic structures and genetic functions", which, "in the long run, could completely revolutionize certain production methods in agriculture and industry". The risks associated to rDNA work, deliberate malice apart, were "at the moment conjectural and, to a very large extent, controllable". Reference was made to experiments and other considerations leading to the view that "man and his environment, since they survived the continuous flow of information between species, may possibly be considered as relatively tolerant to any new form of recombinant DNA".

The NIH detailed guidelines for classification were cited, but with the comment that "At the moment only the laboratories sponsored by the NIH are compelled to follow these guidelines which are presently under revision and will probably be made less stringent". Regarding the (mid-1978) EC situation,

"Within the European Community, the United Kingdom has rendered compulsory advance notification of genetic manipulation work by all those who intend to carry out such work on the national territory. A code of practice, which differs from that of the NIH on essential points dealing with classification procedures and containment methods, is at the moment operated on a voluntary basis but with the understanding that the inspectors of the Health and Safety executive have extensive powers to enforce duties as well as precautions recommended by the British Advisory Group. The other Member States have also prepared or adopted guidelines for research with recombinant DNA which in some cases adhere to either the British or the American system and in others represent a compromise between the two sets of guidelines. In France, the Netherlands, Denmark and Belgium, the National Advisory Committees have been assigned the task to register the work at hand and to review research proposals. While a declaration of agreement has been drafted in France under which governmental, academic and industrial laboratories will submit to review and approval any project for recombinant DNA work, only one Member State (the Netherlands) has, in addition to the United Kingdom, clearly indicated an intention to introduce legislation for recombinant DNA research."

The Commission note acknowledged that the elaboration or adoption of guidelines in the Member States and in other European countries had been greatly facilitated by the critical reviews and recommendations issued by the Ad Hoc Committee on rDNA research of the European Science Foundation (ESF; see below), and by a Standing Committee on recombinant DNA of the European Molecular Biology Organisation (EMBO).

In spite of the declining assessment of the seriousness of the conjectural risks, the Commission put forward six reasons for national legislation:

(i)   gravity of the hazards – "conjectural" did not mean benign, and it was because of these heavy potential consequences that containment methods (physical and biological) had been developed;

(ii)   expansion of rDNA work – as rDNA use was becoming widespread, then the risk, if any, was "increasing with time in proportion with the total number of sites";

(iii) transnational nature of the risks – as viruses and bacteria do not respect fron-

tiers, this "reduces the liberty of individual nations to define and to follow independent policies in the field of genetic manipulations". Agreements and guarantees on the objectives and scale of protection systems in neighbouring countries could "best be generated through legal dispositions, taken in each country, which are based upon a core of principles adopted in common";

(iv)   research in laboratories from private enterprises – publicly funded rDNA might allow effective control over research activities in universities and public research institutes, but in the absence of legal dispositions it was difficult to envisage the laboratories of private industry following the same rules;

(v)   harmony between Member States – to avoid large variations in research potentialities and "the development of differing conditions of safety, work and access between Member States, and, subsequently, to the concentration of research activities at the most permissive sites";

(vi)   the exemplary value of legislation on recombinant DNA technology – the interesting argument was advanced of using rDNA as a prototype, "a choice material for establishing compatibilities between legislation and the development of modern technologies and for preparing a first basis to the dispositions which will undoubtedly have to be taken in the future to protect man against his own achievements. Provided that the legislation is tolerant, flexible, and associated to a stimulation of research through funding, the opportunity should not be missed".

The self-interest in the last sentence may be noted – DG XII had to argue for another three years and more before their research programme was adopted.

2.5 The Economic and Social Committee

The proposal for a Community Directive went forward to Council in December 1978, but the first official response was that of the Economic and Social Committee (ESC), a consultative body statutorily involved in all Community legislation, and representing the "social partners" – business, consumers, trade unions. Their report, delivered in July 1979, reviewed the history of the previous years, including US experience. It emphasized the declining assessments of risk, the absence of specific problems, and noted that the provision of costly safety measures was not in itself evidence of danger. Industry and agriculture would benefit from applications of the new knowledge, and in general, they questioned the urgency of a Directive if one considered only conjectural risks.

However, this assessment depended upon continued self-discipline by the scientific community, which could not be assured; and some countries – Netherlands was mentioned – felt that legislation would help to reduce the latent mistrust among the general population. They noted that if legislation were adopted, it would need to be adaptable to rapid scientific change.

The ESC report on balance supported the Commission proposal for a Directive; but proposed also that the ESC itself held a public hearing, jointly with the Commission, in view of the differences of scientific opinion, and in order to bring together opinion also of unions, industry, agriculture and public interest groups.

The Commission responded to the changing climate of opinion by withdrawing its proposal for a Directive, and putting forward in June 1980 the proposal for a Council Recommendation cited above (EUROPEAN COMMISSION, 1980). Its opening sentence summarized the reason for the change:

"An analysis of the current situation in the course of the last two years, in the United States as well as in Europe, to evaluate the importance of the dangers resulting from genetic engineering, has shown that the conjectural risks associated with the work involving the production or utilization of recombinant DNA are probably non-existent or small."

2.6 The European Science Foundation Adds Its Voice: "No Significant Novel Biohazards"

During the post-Asilomar years, an important role was played by the European Science Foundation. This body was set up in 1974, as a meeting-place for the research councils of Europe; with a mission for setting objectives and stimulating basic research in Europe. By the 1990s, it had over fifty members, from some 20 countries. From the start, it also had a close relationship with the European Community. In addition to its various activities in the field of basic science, it played a significant and influential role in articulating the voice of the European scientific community concerning the regulation of recombinant DNA research and biotechnology. Its annual assembly functions as a kind of "parliament" of 70–80 scholars delegated by the member countries to review its activities.

The ESF had in 1976 established an "Ad Hoc Committee on Recombinant DNA Research (Genetic Manipulation)". They brought together scientists involved in rDNA oversight and on national committees from many European countries, and one result of their work was to promote a progressive convergence of scientific principles for rDNA safety. The evolution of their views between 1976 and 1981 also illustrates the rapid change of scientific opinions regarding risks, which was everywhere a feature of those years, in the US, in the European Commission, and at national level in Europe – cf. further examples in Sect. 5.

Meeting in Amsterdam on 10 September 1976, the Committee discussed both the recently finalized first version of the NIH Guidelines; and the UK Report of the Working Party on the Practice of Genetic Manipulation (the "Williams report"). They noted the differences: the NIH system of safeguards relied to a greater extent on biological containment, the UK on physical. The American guidelines were written down in more detail than the UK recommendations and code of practice. The NIH administrative system was designed to cover research funded by the NIH, the UK system would address *all* laboratories, under the aegis of a national central advisory committee.

The ESF discussed the two systems, each coherent in itself, but decided against mixing them; and recommended the UK system. Its reasons mentioned the points above; and stressed also the importance of the statutory support under the British Health and Safety at Work Act. The Committee "emphasized strongly the importance of such legal support to ensure the effectiveness of an advisory service", and "assumed that similar support existed in comparable legislation in other European countries, or could be provided". Flexibility was also stressed, given the rapid development of the field.

EMBO's Advice Sought

The European Molecular Biology Organisation, EMBO, was also an active participant in the rDNA debates of the 1970s. It was founded informally by a group of molecular biologists meeting at Ravello, in Italy, in 1963, and the following year acquired official status as a non-profit association under Swiss law. Under pressure from the scientific community, and recognizing the significance of the new discipline, a number of European governments agreed to set up an organization, the European Molecular Biology Conference, in 1970. Among the activities of the EMBC was the donation of funds for fellowships, courses and workshops, managed by EMBO. The inter-governmental agreement that led to the EMBC also paved the way for the creation of the European Molecular Biology Laboratory, EMBL, which came into existence in July 1974.

EMBO, the original organization, acquired a high reputation for the quality of its activities – it also publishes the successful *EMBO Journal*. Commenting on EMBO on the occasion of its silver jubilee, *Nature* remarked that "the keys to its success are the select committees and reviewers, who decide which applications to support, and JOHN TOOZE, EMBO's executive secretary" (ANONYMOUS, 1989). TOOZE was an energetic participant in the rDNA debate – see in Sect. 8 the reference to his "Documentary History of the DNA Sto-

ry" – and it was a natural step for the ESF to seek a second opinion from the body which in Europe effectively represented the scientific discipline most concerned.

The ESF Committee sought further advice, inviting the EMBO Standing Advisory Committee on Recombinant DNA to compare the US and UK provisions. The EMBO committee's report was completed in October 1976, and circulated at the ESF Assembly meeting on the 26th of that month. The result, based on their Ad Hoc Committee's report, but including amendments adopted at the Assembly itself, was ESF's recommending that:

"1. subject to appropriate safeguards, research on Recombinant DNA molecules should be promoted and further developed in Europe.

2. adequate measures be taken to ensure the protection of the public and its members as well as of animals, plants and the environment. These measures are set out in Recommendations 3 to 6 below.

3. guidelines for such research, designating in sufficient detail the appropriate safeguards for the different kinds of experiments with Recombinant DNA, be rigorously followed by all researchers and all laboratories carrying out such experiments.

4. in the first phase, the recommendations and code of practice in the United Kingdom Report of the Working Party on Genetic Manipulation be adopted as the guidelines for Recombinant DNA research in Europe, *provided that the European Committee which is referred to in Recommendation 9 below is set up immediately in order to ensure that the same levels of containment are used in the different European countries for the same categories of experiments.**

5. HWATSON and TOOZE. Few ofthosewho for interpreting the recommendations and code of practice for Recombinant DNA research referred to above, for advising researchers in their use, and for supervising their implementation should be established forthwith in those European countries which have not already set up such a body.

6. in European countries national registers of laboratories carrying out Recombinant DNA research be established and that all laboratories, whatever their source of financial support, be legally required to declare relevant aspects of their programmes of Recombinant DNA research to a national register. We envisage such registers as being non-classified documents and we advise that they should be exchanged

among the national bodies referred to in Recommendation 5.

7. in each country the necessary administrative and supervisory measures be devised, in connection with appropriate exsting national legislation, to ensure that the guidelines adopted be strictly adhered to by all laboratories – governmental, industrial or private – or in universities or other research establishments.

8. the national bodies in Europe referred to in Recommendation 5 should keep in close contact with each other, with the EMBO Standing Advisory Committee on Recombinant DNA, with the appropriate committees of the NIH in America and with committees for Recombinant DNA research in other countries. We wish to emphasize the importance, both for the safety of the public and for the development of the science, of ensuring that the same experiments are not classified as differing in level of risk, and hence level of containment, in the different countries.

9. to give effect to Recommendation 8 above, a European Committee of representatives of those national bodies for Recombinant DNA research *referred to in Recommendation 5 above,** of the EMBO Standing Advisory Committee on Recombinant DNA research and of the European Medical Research Councils *and with representation of Agricultural research** should be established under the aegis of the ESF. It should come together at an early date and hold regular meetings, at frequent intervals in the first phase.** These meetings will provide an opportunity for mutual information, consultation and advice on general policy for Recombinant DA research and for the discussion of decisions on specific experiments, *in particular concerning biological containment.** This process should lead to the building up of common body of experience in all questions concerning Recombinant DNA research, at European level. A further task of this Committee will be to keep the guidelines under continuous review and to consider whether and, if so, how they should be revised in the light of the development of the science and of safety measures."

Footnotes to ESF recommendations:
* The words in italics are amendments which were proposed and agreed during the Assembly.
** It is proposed that this committee should meet first in January 1977 and at three-monthly intervals during the year.

The Assembly also agreed to maintain a "Liaison Committee on Recombinant DNA",

to continue the work of keeping the situation under review.

Four years later, by January 1981, scientific opinion regarding DNA risks had considerably evolved, and ESF's Liason Committee considered their work was done. They expressed this view vigorously in a brusque, one-page report, here reproduced in full:

"Statement from the ESF Liaison Committee on Recombinant DNA Research

The ESF Liaison Committee on Recombinant DNA, at its meeting on 14-15th January, 1981, unanimously decided that its work of promoting the necessary harmonization of national recombinant DNA guidelines is now sufficiently complete for the Liaison Committee to be disbanded. Although the national guidelines of some countries are still evolving towards the position already reached by others, the Committee believes that there is no further need for formal and regular liaison at the ESG between representatives of national recombinant DNA committees.

Extensive information supports the view that recombinant DNA work per se entails no significant novel biohazards. This is already accepted by some national recombinant DNA committees and in those countries special safety precautions beyond good microbiological practise together with the use of appropriate host organisms are no longer required for recombinant DNA work except when known pathogens or toxin producing organisms are involved. It has been established that recombinant DNA techniques offer safer ways of studying and using pathogens and for the production and study of toxins than conventional methods. The Liaison Committee endorses these national decisions.

In several countries certain microorganisms produced by recombinant DNA methods are now being grown on an industrial scale. Although some concern has been expressed about the biosafety of these large scale operations, the Liaison Committee points out that the fermentation industry already has long and extensive experience of large scale fermentation, including that of known natural and dangerous pathogens. Consistent with its statements in the preceding paragraph the Committee emphasizes that the large scale fermentation of microorganisms produced by recombinant DNA methods does not pose novel or special problems.

Finally, the Liaison Committee reaffirms its opinion that there is no scientific justification whatsoever for new legislation specific for recombinant DNA research and moreover, it sees no justification for further extensive recombinant DNA risk assessment programmes."

The report was approved by the assembly of ESF on 12th November 1981.

Their view that the debate was virtually over was widely shared – see the quotation in Sect. 8 from the 1981 "Documentary History" by WATSON and TOOZE. Few of those who participated in the ESF Assembly of November 1981 could foresee that thirteen years later, they would again be called upon to reiterate a similar message (see Sect. 7.4).

This view expressed at the ESF Committee in January 1981, and endorsed by the Assembly in November, was influential; many scientists gave similar opinions at the Commission-supported hearings in May 1981, held by the Economic and Social Committee. The ESC hearing focused more on social risks, such as the concentration of knowledge in the hands of industry, leading to commercial (or even military) applications not necessarily beneficial. *Nature*'s report (BECKER, 1981a) summarized views on safety and regulation as follows:

"Although many of the speakers argued that the DNA guidelines were safety regulations looking for a hazard, some members of ESC confirmed afterwards that they were in favour of legally enforced safety regulations, even if they needed to be frequently revised to take into account changes in risk assessment. It was suggested that the Community was merely going through the same process that had taken place in the United States a few years ago, but a little later because of the several countries involved."

2.7 Report of the North Atlantic Assembly

In parallel with the ESF and ESC deliberations, two Parliamentary bodies had also been considering the topic of rDNA safety: the North Atlantic Assembly, an inter-parliamentary assembly of the North Atlantic Alliance, which acts as a link between NATO authorities and member parliamentarians; and the Parliamentary Assembly of the Council of Europe, which brings together the democracies of Western Europe (and whose membership in the 1990s has expanded, almost *pari passu* with the democratization of Eastern Europe).

The North Atlantic Assembly (NAA) established in November 1978 a sub-committee on genetic manipulation, which in eighteen months completed a study on "the potential benefits of recombinant DNA research, and the postulated risks; on whether there was a need for regulations or legislation; and on aspects of commercialization". The group was chaired by ROBERT MCCRINDLE of the UK, and included French, German and US members.

The NAA group contacted many key scientists, including the ESF group, who drew to their attention the *de facto* emerging consensus on good laboratory safety practices and the similarity of national guidelines, whether modelled on the US NIH RAC, the UK GMAG, or a combination. NAA's group noted the rapid adaptation of the guidelines under the NIH RAC, and commended its (January 1979) expansion from 11 to 25 members, to increase non-scientific representation – similarly to GMAG.

NAA consulted also the World Health Organization in Geneva, in February 1980, on the question "whether, in the opinion of WHO, a universal set of guidelines regulating recombinant DNA research was necessary and if so, whether WHO could act as the regulatory body?

In reply it was felt that precedents exist for international guidelines (e.g., in the worldwide movements of biological and pharmaceutical products) and that WHO has the scientific and legal ability to draft such guidelines but that any action would of necessity have to emanate from national initiatives. Given the general feeling that risks from recombinant DNA research had been exaggerated it was considered that even if a resolution was passed by the World Health Assembly it would be difficult to envisage member states treating it as binding."

The conclusions of the NAA sub-committee, published in March 1981 (NORTH ATLANTIC ASSEMBLY, 1981), were "that the benefits of recombinant DNA research outweigh the risks and that maximum encouragement should be given to develop this research for the benefit of mankind. Nevertheless we felt that controls are still advisable on certain aspects of the research such as experiments using highly dangerous pathogens or on the germ cells of human beings. We argue however for a flexible set of control guidelines which both protect from any possible dangers that may arise – but at the same time – do not hamper research so that the public may benefit, as soon as possible, from all the possibilities offered by the implementation of this new technology."

2.8 Report of the Council of Europe

A similar consensus seemed to develop in the Council of Europe, in a report prepared by Mr. ELMQUIST and presented by their Legal Affairs Committee. They had organized a Parliamentary Public Hearing in Copenhagen in May 1981, under the title, "Genetic Engineering: risks and chances for human rights". Their report was based on that hearing, and took account of the work cited above – by the NAA, the European Commission, the Economic and Social Committee, and the ESF. It reviewed various models for regulation of genetic engineering, acknowledging that different countries would adopt different solutions; but focused particularly upon the (then distant) prospects for the engineering of human genes, and related issues for human rights.

The final Recommendation was adopted by the Parliamentary Assembly on 26 January 1982. It distinguished between concerns "arising from uncertainty as to the health, safety and environmental implications of experimental research"; and "those arising from the longer-term legal, social and ethical issues raised by the prospect of knowing and interfering with a person's heritable genetic pattern" (COUNCIL OF EUROPE, 1982).

Regarding the first concern, the resolution emphasizes the potential value of the new techniques and of the advances in understanding; refers to freedom of scientific enquiry as a basic value, but one carrying duties and responsibilities in regard to the health and safety of the public and the non-contamination of the environment; refers to earlier uncertainties, but notes that these "have in recent years been largely resolved – to the point of allowing substantial relaxation of the control and containment measures initially instituted or envisaged"; and advocates that "strict and comparable levels of protection should be provided in all countries for the general public and for laboratory workers against risks involved in the handling of pathogenic micro-organisms in general, irrespective of whether techniques of genetic engineering are used".

Regarding the legal, social and ethical issues, the Resolution refers to Articles 2 and 3 of the European Convention on Human Rights as implying "the right to inherit a genetic pattern which has not been artificially changed"; but goes on to add that "the explicit recognition of this right must not impede development of the therapeutic applications of genetic engineering (gene therapy), which holds great promise for the treatment and eradication of certain diseases which are genetically transmitted".

2.9 Division of Opinion in the European Parliament; Council Recommendation 82/472 Adopted

During the debate on the Commission's proposal for a Council Recommendation, rather than a Directive, on the control of rDNA work, opinion in the European Parliament became deeply divided. Following their May 1981 hearing, the Economic and Social

Committee had disagreed with the Commission proposal, and advocated a Directive. In the European Parliament, the rapporteur was DOMENICO CERAVOLO, an Italian Communist. His report argued that even if a risk was only due to a hypothetical chain of events, this was not a justification for thinking it any less valid or significant. The conjectural risks could not be dismissed, because no suitable criteria were available for assessing them.

However, the liberals and conservatives, at that time a majority in the Parliament, supported the Commission proposal, concerned that too much legislation would slow down the growth of Europe's biotechnology industry (BECKER, 1981b). The proposal for a Recommendation was approved by Parliament early in 1982, and adopted by Council in June that year.

In October 1984, the Committee of Ministers of the Council of Europe adopted almost the same text as a recommendation for their (then) 21 Member States; with slightly greater flexibility for Member States, who were left free to decide on the categories of risks requiring notification (because, it was said, the biological risks had previously been overestimated). It was also indicated that work would continue on studying other, ethical, questions.

3 From Research to Strategy, Co-ordination and Concertation, by Commission and Industry

3.1 Calm Before the Storms: The Early 1980s and the International Scientific Networks

With the adoption of Council Recommendation 82/472, recommending national systems of notification of recombinant DNA work, the regulatory debate in Europe seemed to quieten for some years. The recommendation was a rational response to uncertainties; it allowed the Member countries with significant research activities, and with correspondingly developed oversight structures, to develop further and to adapt these structures to evolving perceptions of need. Conscious of public sensitivities, the scientists involved co-operated readily with national authorities; international harmonization developed through the usual scientific networks, and through various international bodies, such as those referred to below.

European Science Foundation

The European Science Foundation had been active in the early years, as mentioned above, and in the January 1981 report of its rDNA group appeared to have concluded its work; but at the request of the Council of Europe, it conducted in 1982 a survey of national regulatory developments, referred to in Sect. 5.1 below.

The European Federation of Biotechnology was founded at Interlaken in 1978, through an initiative led by DECHEMA (the German Chemical Equipment Manufacturers Association), the (UK) Society of Chemical Industry, and France's Societé de Chimie Industrielle. The Federation expanded through the 1980s to include over 60 learned societies; from the various disciplines of the life sciences and technologies, and from all parts of Europe (mainly, but not exclusively, Western). DECHEMA had since the early 1970s within Germany been a leading advocate and source of information in relation to biotechnology. The subject was seen at that time mainly as the modern development of the fermentation industry, requiring an inter-disciplinary synthesis of biochemistry, microbiology and process engineering; and relatively slowly, DECHEMA and the EFB extended their concept of biotechnology to include the new genetics.

The EFB has various Working Parties, many of which interacted significantly with the European Commission, by conducting studies, organizing meetings, etc. Among the groups active in this way were and are those on Education, Environmental Biotechnology, and (during recent years), Public Perception; but of particular significance to the regulatory debate was that on Safety, which published a series of influential technical papers on various aspects of safe concepts and practices in biotechnology (KÜENZI et al., 1985, 1987; FROMMER et al., 1989, 1992, 1993). As key members were also active in national and OECD groups, and (after 1991) in CEN working groups (for European standards), duplication of activity was avoided, and consensus (e.g., on risk-based categorization of microorganisms) was readily spread between the various interested constituencies.

Within the ICSU (International Council of Scientific Unions) "family" of world-wide scientific unions, COGENE, the Scientific Committee on Genetic Experimentation, was established by the 16th General Assembly of ICSU in October 1976, to serve as a source of advice concerning recombinant DNA activities. Its purposes were defined as follows:

(i)   to review, evaluate and make available information on the practical and scientific benefits, safeguards, containment facilities and other technical matters;
(ii) to consider environmental, health-related and other consequences of any disposal of biological agents constructed by recombinant DNA techniques;
(iii) to foster opportunities for training and international exchange; and
(iv) to provide a forum through which interested national, regional and other international bodies may communicate.
(BERNARDI, 1991)

In March 1979, COGENE's Working Group on Recombinant DNA Guidelines presented a report, and the Committee organized the first COGENE Symposium, on Recombinant DNA and Genetic Experimentation, in Wye, UK. BERNARDI's view is that "These actions played an important role in downgrading the excessively stringent containment conditions which were imposed until then". Certainly the timing of these actions – three months after the European Commission had proposed its Directive (Dec. '78 – see Sect. 2.3 above) – and the international

character of ICSU, were favorable for influencing the European authorities and their advisers.

COGENE continued to work for rationalization of conditions imposed on rDNA work, their Working Group on Risk Assessment publishing in July 1991 a report which was circulated to other international agencies. More widely influential may have been a three-page statement prepared at a joint workshop by COGENE and SCOPE, ICSU's Standing Committee on Problems of the Environment, in 1987 (published in MOONEY and BERNARDI, 1990). For it addressed an issue which at that time was emerging into the political prominence it has since retained: the release of genetically engineered organisms into the environment. Again, the timing was significant in relation to regulatory proposals of the European Commission (see Sect. 4), although the relative weight of scientific advice was by then diminishing in competition with the influence of other actors. But the core of the SCOPE/COGENE statement was a succinct and balanced summary of scientific views on the risks of field release, and the appropriate means to approach these.

On risks:

"In view of the great potential of new technologies for addressing environmental and other problems, and because most introductions of modified organisms are likely to represent low or negligible ecological risk, generic arguments against the use of new genetic methodologies must be rejected. Indeed, the spectrum of available tools represents an evolving and expanding continuum, which includes conventional methods, rDNA techniques, and others. While much attention has been focused on the methods used to modify organisms, it is the products of these technologies and the uses to which they will be put that should be the objects of attention, rather than the particular techniques employed to achieve those ends.

Similarly, one must reject generic safety arguments based on the assertion that all introductions must have occurred some time during the course of evolution. Therefore, each introduction of an organism, whether modified or not, must be judged on its own merits, within the context of the scale of the application and the possible environmental costs and benefits.

Size, geographical scale, and frequency of introduction are among the factors that are important in determining whether a particular introduction will become established or spread. Therefore, small-scale field testing involves different considerations than does large-scale (e.g., commercial) application. This does not suggest that small-scale testing should be exempt from examination for regulation, but simply that any risks are likely to be much smaller and more easily managed than those for large-scale applications. Nor does this suggest that all large-scale applications will be problematical, since we have many examples to the contrary, including vaccines, biological control methods, and the use of rhizobia in agriculture."

On risk assessment:

"The properties of the introduced organism and its target environment are the key features in the assessment of risk. Such factors as the demographic characterization of the introduced organisms; genetic stability, including the potential for horizontal transfer or outcrossing with weedy species; and the fit of the species to the physical and biological environment. The scale and frequency of the introductions are important related factors. These considerations apply equally to both modified or unmodified organisms, and, in the case of modified organisms, they apply independently of the techniques used to achieve modification; that is, it is the organism itself, and not how it was constructed, that is important.

Each proposed introduction must be treated on its own merits, but this does not suggest that each needs to be considered *de novo*. As experience accumulates with particular kinds of introductions in particular environments, more generic approaches to these classes of introductions can be developed. The bases for classification should be refined continually, providing a set of criteria that will allow any proposed introductions judged to be innocuous to be carried out speedily and those judged to be problematic to be given the attention they deserve."

Statements such as those of SCOPE/CO-GENE quoted above, and similar opinions expressed by the EFB Working Group on Safety in Biotechnology, and the European Science Foundation, remained consistent and apparently valid throughout the 1980s and later years. But the situations in which they competed for influence were changing in key respects.

The continuing scientific successes, and the inescapably central role which the new knowledge and techniques had to play in any research concerned with understanding biological structures and functioning, led inexorably towards application. Application of the new biotechnology, in the production of molecules of interest by genetically modified host organisms, was normally in fermenters of large-scale, or at least, on a scale greater than laboratory. In agriculture, application would mean the release in progressively less confined conditions (greenhouse, test plot, etc.) of genetically modified plants and microorganisms. "Large-scale" and "Field Release" were terms that entered political discourse in the next round of legislative debate on biotechnology regulation in Europe. But before describing this, the debate on regulation should be set in a wider context of the debate on strategy, which also developed from the early 1980s.

3.2 Scientists Redeploy Effort, from Regulation to Research, Forecasting and Strategies

The Commission staff in Directorate-General XII (Science, Research and Development) who had been active in the "post-Asilomar" activities on regulation were in 1982 able to turn their energies to the management of the new research activity, the Biomolecular Engineering Programme ("BEP"; 15 million ECU, 1982–86), for which they had argued and negotiated during some six years. The phrase "Biomolecular Engineering" combined references to genetic engineering (reflecting the study by RÖRSCH, 1978) and enzymology (reflecting the study by THOMAS, 1978), as well as risk assessment work (drawing on the study by SARGEANT and EVANS, 1979).

The significance of research and technology for economic competitiveness acquired ever-increasing political visibility through the latter decades of the twentieth century; but apart from a reference to the co-ordination of agricultural research (Article 41), research did not feature in the original EEC Treaty of Rome (1957), founding the European Economic Community. Separate Treaties in 1951 and 1957 had founded the Coal and Steel

Community, and Euratom for atomic energy research; but it was only in 1974, after much study and debate, that the situation started to change; and a further decade-and-a-half elapsed before research had an explicit legal basis, via the coming into force (in July 1989) of the Single European Act.

At a Council meeting in Paris on 14 January 1974, four Resolutions were adopted on the development of a common policy on science and technology, including the co-ordination of national policies and the joint implementation of projects of interest to the Community (EUROPEAN COMMISSION, 1977a). RALF DAHRENDORF of Germany was the first Commissioner to have responsibility (1973–76) for Community R & D policy.

DAHRENDORF established in 1974 a "Europe +30" group of experts, under the chairmanship of MEP WAYLAND YOUNG (subsequently Lord KENNET), charged with considering how long-term (i.e., up to 30-year) forecasting could feed into the choice of Community R & D objectives. At that time, long-term studies and "futurism" tended to be dominated by US think-tanks – RAND Corporation, Hudson Institute and others – and the Club of Rome, famous for its "Limits to Growth" (MEADOWS et al., 1972), was not as close to public policy as the US Congressional Office of Technology Assessment, founded in 1972.

The "Europe +30" report was ambitious (KENNET, 1976), proposing the creation of a new Community institution for futures studies; the final implementation was modest, being the adoption (by a Council Decision in 1978) of a Community programme of "Forecasting and Assessment in Science and Technology", or FAST; with a budget of ECU 4.4 million, for the five years 1978–83. The new unit was housed within DG XII; and was the origin of a new initiative on biotechnology.

The mandate of FAST was broad. It included "highlighting prospects, problems and potential conflicts likely to affect the long-term development of the Community and defining alternative courses of Community R & D action to help solve or achieve them or to render concrete these possibilities" (EUROPEAN COUNCIL, 1978). With six graduate staff, it divided its work into three strands:

- Technology, Work and Employment
- "Information Society"
- "Bio-Society".

The third strand comprised a series of studies, workshops and widespread network consultations, conducted with a long-term perspective, and assisted in particular by the European Federation of Biotechnology. The results were used to assemble the "Bio-society" section of the first synthesis report of the FAST programme, in September 1982 (FAST, EUROPEAN COMMISSION, 1982). This became, with little adaptation, the first draft of a "Community strategy for biotechnology in Europe"; and reports with that title were published both by DECHEMA (on behalf of the EFB), in 1983, and by the Commission as a "FAST Occasional Paper" in March 1983 (FAST, EUROPEAN COMMISSION, 1983). The whole FAST report, including the "Bio-society" sections was subsequently published under various titles (FAST, COMMISSION DES COMMUNAUTÉS EUROPÉENNES, 1983; FAST, EUROPEAN COMMISSION, 1984; FAST-GRUPPE, KOMMISSION DER EG, 1987).

3.3 The Framework Programme for R & D, and the 1983 Communications: A First Community Strategy for Biotechnology

From 1981 to 1984, the Commissioner responsible for both Research and Development (mainly DG XII) and Industrial Affairs (half of DG III, the other half being Internal Market) was Vice-President ETIENNE DAVIGNON, of Belgium. He had during 1982 been politically responsible for the first "ESPRIT", a European Strategic Programme of R & D in Information Technology; with a budget of 750 million ECU, at that time a massive programme. It was, therefore, seen as significant when Commission President GASTON THORN, in his annual programme speech to Parliament in February 1983, announced that for biotechnology, the Commission would "follow the same approach as for ESPRIT".

The absence until mid-1974 of a formal commitment to Community level R & D policy has already been noted. Through the 1970s and early 1980s, the policy objectives for R & D tended to be unsystematic and *ad hoc*. The integration of Euratom's research activities, including those conducted at the Joint Research Centre's various facilities, and the political aftermath of the 1974 oil-shock, led to an initial predominance of energy-related research. Similarly, environmental research was emphasized in the years following RA-CHEL CARSON's "Silent Spring" (1962), the Stockholm conference of 1972, and in the context of the apocalyptic predictions of the Club of Rome's successive publications – "Limits to Growth" (MEADOWS et al., 1972), "Mankind at the Turning Point" (MESAROV-IC and PESTEL, 1974), and others. By the early 1980s, the relative weakness of European industrial performance vis-à-vis US and Japan was increasingly emphasized.

As Commissioner responsible for Industry and Research, DAVIGNON reacted to the previously unsystematic approach to Community R & D priorities by proposing the institution of a comprehensive, systematic, multi-year "Framework Programme", with explicit objectives, to be pursued through specific R & D programmes. To the international economic challenge, he responded by seeking to increase the share of expenditure devoted to industrial competitiveness, from under 10% in the early 1980s, towards 50% or more by the end of the first Framework Programme, 1983–86. This Framework Programme was prepared by a dozen background papers, each being a "Plan by Objective"; most of them written by outside experts, but that on biotechnology by FAST staff (CANTLEY, 1983). Like the FAST report, this paper followed the DAVIGNON philosophy of setting Community R & D activities in the context of a broader strategy and objectives. In a resolution of 25 July 1983, the Council endorsed the concept of the Framework Programme; and approved the development of biotechnology as part of the objective, "promoting industrial competitiveness".

President THORN's February 1983 speech was the Commission's first official mention of the word "biotechnology", and reflected a perspective much broader than just "rDNA research". Drawing upon the recommendations of the FAST report, and with some input from DG III (particularly its Food and Pharmaceuticals divisions), the Commission presented a first communication to the Council at the Stuttgart summit of June 1983: "Biotechnology: the Community's role" (EURO-PEAN COMMISSION, 1983a). A parallel "Background note" (same reference) provided an overview of national initiatives for the support of biotechnology, including not only EC but US and Japanese details (EUROPEAN COMMISSION, 1983b).

This communication, like the FAST report, emphasized the great progress in the life sciences in recent years; the broad range of application and likely scale of future markets; and the strengths of the US and Japan. This case was strengthened by selective quotation from contemporaneous US reports, by the Office of Technology Assessment and others (see Sect. 5.1). The relative weakness of biotechnology in the Member States, in spite of considerable scientific strengths, was attributed to "lack of coherence in R & D policies and the absence of structures on Community scale".

Although BEP, the Biomolecular Engineering Programme, had started only the previous year, it was said to be "regarded as a starting-point for the more ambitious action recommended by the 1984–86 framework programme for Community R & D." There should be training and mobility of scientists and technicians; efforts to put biotechnology across to the general public; a strengthening of basic biotechnology through projects halfway between research and applications; and improvement of the research context including data banks and cell collections.

Even with these support and training activities, biotechnology would be "unable to develop in the Community unless a favorable context is provided to encourage it"; and the three "necessary factors" identified were:

- access to raw materials of agricultural origin, on the same conditions as competitors outside the Community
- adaption of industrial, commercial and intellectual property systems
- rules and regulations.

On the last, the communication emphasized the need to harmonize the internal market, particularly for the health industries, and above all to prevent the appearance of specific national standards. "The advisability of introducing new regulations should be examined from the outset at Community level, as far as possible using the various instruments already available to the Community". The communication was favorably received by Council, and the Commission was encouraged to develop over the following months a fuller action programme.

The following communication in October that year (COM(83)672, EUROPEAN COMMISSION, 1983b) noted US reports dismissive of Europe's competitive strength in biotechnology, and outlined six action priorities to meet the challenges and opportunities of biotechnology:

(i) Research and training: horizontal actions "focused upon the removal of bottlenecks which prevent the application of modern genetic and biochemical methods to industry and agriculture"; and "information infrastructure and logistic support for the life sciences R & D through databanks, collections of biotic materials and related information/communication networks; and the sophisticated data capture technologies which are generating such information in ever-increasing quantities"; together with specific actions "intended to stimulate certain particular developments within well-defined sectors of biotechnology which can contribute to the solution of problems related to the common policy in agriculture and in the health care sector. (The Biotechnology Action Programme was proposed the following year – see EUROPEAN COMMISSION, 1984, and further references below).

(ii) Concertation of Biotechnology Policies: "a central activity of concertation: inter-service, international, and between Community and Member States", with "active monitoring and assessment of strengths, weaknesses and emerging opportunities and challenges"; the principal tool to be "an expanded series of networks, established in cooperation with the Member States to provide an *ad hoc* system of collaboration between individuals, specialized groups and institutions", coupled with "an information base, regularly updated by scanning, selecting, interpreting and sorting in an organized way the incoming flow of information". (A decade later the BIODOC documentation center would have some 60000 documents in 2000 files, and a steady flow of visitors; providing the basis for regular press reviews, an internal News-Sheet, and the European Biotechnology Information Service, EBIS.)

(iii) New Regimes on Agricultural Outputs for Industrial Use: the Commission announced its intention to propose to Council new regimes for sugar and starch for industrial use. (This was subsequently implemented, from 1986 onwards.)

(iv) A European Approach to Regulations affecting Biotechnology (discussed below).

(v) A European Approach to Intellectual Property Rights in Biotechnology: the communication noted that it was "indispensable that the industrial property laws and instruments available in the Common Market match the need of science and industry and of the Community's goals"; and that "Failure to provide such protection for intellectual property will drive firms to protect themselves by commercial secrecy. Such secrecy will inhibit precisely the collaborative patterns of activity which are needed in this complex inter-disciplinary field." The major unresolved issues noted were:
– "the patentability of biotechnological inventions as such;
– the implications and conditions associated with the rules of practical protection requirements and procedures (e.g. associated with the deposit of microorganisms and conditions for release to third parties);
– the additions to the existing problems concerning plant and animal variety protection, of complex new reactions with patent law".

It was suggested that Member States should be invited to share with the Commission information about the aims and context of ongoing and planned work in this area (as they might have submitted to the OECD 1982 survey – subsequently reported in BEIER et al., 1985). Member States were urged to ratify the European and Community Patent Conventions. The Commission proposal for a Directive on the protection of biotechnological inventions followed four years later (EUROPEAN COMMISSION, 1988) after extensive consultation. The final adoption became a matter for heated political debate and contentious amendments, in 1994; it was one of the first topics to require the "conciliation procedure" established by the Maastricht Treaty, for reconciling a deadlock between Council and Parliament, a deadlock still unresolved in December 1994.

(vi) Demonstration Projects: "designed to facilitate the transition between research developments and full scale exploitation on a commercial basis". (The first such projects formed part of the specific agro-industrial "RDD" – Research, Development and Demonstration – during the 3rd Framework Programme, 1990–1994.)

3.4 "A European Approach to Regulations Affecting Biotechnology", 1983

The October 1983 Commission communication, under the above title, treated this topic under three sub-headings:

● Biological Safety
● The Consumer and the bio-industry
● The Regulation of Products and their free circulation.

On the first topic, "Biological Safety", it noted that "Public and parliamentary opinion is divided between admiration of the new discoveries in biotechnology, and concern about some of the possible implications or conjectural risks of their use. This concern is reflected in the extensive discussion, studies and reports on the need for regulation and control of various aspects of the life sciences; a debate vigorously pursued throughout the developed industrial world".

Noting this widespread concern, and the similarly persistent concern with "issues such as animal welfare and their use in tests, which also impinge on biotechnology", the communication argued that "it is clearly a normal role for the Commission to ensure regulatory provision to maintain rational standards of public safety; to this end, monitoring the social dimensions of biotechnology and their interfaces with policy".

Regarding action, "in order to maintain awareness of evolving pressures for new policies or regulations, a monitoring function is needed, to collate at Community level the evolving views of national regulatory bodies and interested international organizations (OECD, WHO, Council of Europe, etc.), and hence to advise the Community on regulatory initiatives or international negotiations. This function can be appropriately combined with the concertation role described in item (ii) above [i.e. of the 1983 communication], and provision for it is included in the budget estimates".

Under the second topic, "The Consumer and the bio-industry", the communication noted that the roles of the public authorities, at both Community and Member State level, impinged at several points upon the "bio-industries" (pharmaceuticals, agro-food, etc.) and the consumer. The Commission sought to "encourage innovation, harmonize regulatory regimes, create a genuine common market and ensure that regulations were based on rational assessment and well-informed debate; while seeking always to maintain high standards of nutrition and safety". But regarding action, as with the previous heading, "the need is to maintain in the Commission the capacity to monitor the situation, and hence to concert necessary policy discussions and initiatives across the services, with Member States, and with other relevant groups (e.g. consumer associations)".

All these modest suggestions were in fact implemented over the following ten years, through the concertation action, supported under the R & D programmes BAP (1985–89, see below) and BRIDGE (1990–93), including the dialogue with consumer associations (see ROBERTS, 1980, KATZ and SATTELLE, 1991; ECAS, 1992). But the attitude towards biotechnology-specific regulatory initiatives implied by the 1983 communication could be summarized as attentive inaction.

Under the third heading, "The Regulation of Products and their free circulation", the Commission was addressing the central responsibility of DG III (Internal Market and Industrial Affairs). While emphasizing that the bio-industries, in view of their high entry costs and long time scales (for both R & D work, and regulatory approval) had great need of the full dimensions of European and world markets, the communication was prudent in this, its initial assessment:

"From a first review of the situation, it would appear that the application of current Community regulations in the various fields (pharmaceuticals, veterinary medicines, chemical substances, food additives, bioprotein feedstuffs) will meet current regulatory needs, provided that there is close cooperation between the competent authorities in the Member States and the Commission. Such cooperation can be achieved by greater recourse to the existing institutional or scientific committees and, as necessary, use of the new information procedure for technical standards and regulations adopted by Council in its directive 83/189/EEC of 28th March 1983".

Regarding possible future action, it continued:

"on the basis of its experience deriving from the use of these various instruments, the Commission will put forward general or specific proposals appropriate to create a regulatory framework suitable for the development of the activities of the bio-industries and for the free circulation of goods produced by biotechnology".

3.5 The Co-ordination Problem

The programme of activity which the Commission had thus published in its October 1983 communication reflected the multi-faceted character of biotechnology's challenges to policy. Within the Commission structure, it was clear that the interests of several Directorates-General would be involved.

This was a basic problem of public policy for biotechnology, with which the Commission, and governments, and international organizations, all had to wrestle increasingly over the following years: how to achieve a coordinated, or integrated, response to the wide-ranging but inter-connected challenges of biotechnology, through administrative structures which were essentially vertically divided. The problem was explicitly addressed by an OECD study report, on "Biotechnology and the Changing Role of Government" (OECD, 1988). An intense and continuous competition for attention characterizes the top of any administrative hierarchy or government, just as the front page of a newspaper has only a finite amount of space for one or two lead stories on any given day. This theoretical demand for simultaneous attention can be met only by chopping a problem into pieces treated in parallel by separate agencies. But the continual demands for conflict resolution and co-ordination require decisions to be pushed up to the higher levels, where they must compete for, or await, their turn for resolution.

The Commission through the early decades of modern biotechnology comprised a gradually increasing number of independently nominated Commissioners, two from each of the larger countries of the Community, one from each of the small; thus increasing from 9 at the start of the 1970s, to 13 (after UK, Danish and Irish accession in 1973), to 17 with the accession of Greece (1981), Spain and Portugal (1986), and reaching 20 with the accessions of Austria, Finland and Sweden in January 1995. Each Commissioner has responsibility, typically, for one or two Directorates-General, and is advised by a Cabinet selected for compatibility with his/her language, nationality and political orientation. The result is a body which, although formally committed to Community goals, has a continuing tendency to reflect the European reality of distinct and conflicting interests; and the Commissioners do not face the constraint of re-election, either personally or as a political group.

The consequent differences of outlook, interests and objectives are conveyed to the Directorates-General for which they are responsible. The main countervailing force to fragmentation is the strength of the center, i.e., the President and his staff, supported by the Secretary-General. These came significantly into play in biotechnology only from 1990 onwards; see Sect. 7.

The six-point strategy of October 1983, laid down in communication COM(83)672, recognized that several Directorates-General must be involved. Following an inter-service meeting in December 1983, and discussion at the level of several Commissioners' cabinets, DAVIGNON together with the Commissioners DALSAGER (for Agriculture, DG VI) and NARJES (for Internal Market, the other part of DG III) put forward to the Commission a paper on its internal co-ordination of policy for biotechnology. This was accepted on 2 February 1994.

3.6 The Biotechnology Steering Committee and Its Secretariat, "CUBE"; the Implementation of the 1983 Strategy, and the Different Interests of the Commission Services

The Commission agreed on the Davignon proposal to create a "Biotechnology Steering Committee" (BSC), to be chaired by the Director-General of DG XII (Science, Research and Development), and open to other services in function of their interests. A secretariat, which became known as the Concertation Unit for Biotechnology in Europe, or "CUBE", was established by redeploying the biotechnology staff from the FAST programme; with responsibilities focusing on monitoring developments in biotechnology, diffusing information about biotechnology to the services concerned, and supporting these services and the BSC in implementing the action priorities laid down in the October 1983 communication.

The Biotechnology Steering Committee provided in principle a forum for discussing biotechnology matters of common interest at a senior level – originally that of Directors-General. But it was not a decision-making body; and the time pressures on senior staff in the Commission made them reluctant to devote time to a mere "debating club". The consequence was dilution of the participation to a more junior level as the years went by; and a declining frequency of meetings: the numbers of meetings in the five years 1984 to 1988 were: 3, 3, 2, 1, 1.

The starting of the Biotechnology Steering Committee signalled a recognition by the Commission of the need for a co-ordinated strategic approach; the location of its Chair and Secretariat in DG XII was both a reflection of its background, and the seed of its subsequent failure; issues further treated in the following section.

The creation of the BSC was nonetheless a workable solution in the early 1980s. The issues seemed to be satisfactorily divisible; DG XII was promoting across the Commission a greater awareness of biotechnology, in particular drawing attention to the competitive challenge presented by the United States. DG XIII, responsible for Information Market Development, was in March 1982 persuaded to establish with DG XII a Task Force for Biotechnology Information – aided by the blessing of (then) Director-General RAY APPLEYARD, a Cambridge biologist by background. Reports by the US Congressional Office of Technology Assessment (1981) and the Department of Commerce (1982) were underlining the commercial potential, and extensive publicity accompanied the creation and expansion of the early biotechnology specialist companies such as Cetus (1971), Genentech (1976) and Biogen (1981). With similar perceptions growing in national administrations, the political will developed for a more substantial biotechnology R & D programme; and after extensive consultations with Member State experts, DG XII prepared and the Commission put forward in April 1984 a proposal for a "Biotechnology Action Programme" (BAP; EUROPEAN COMMISSION, 1984), discussed in Sect. 4.1 below.

Below the BSC, its Secretariat CUBE remained active in monitoring and information-diffusion activities; and acquired resources and an additional mandate with the adoption in March 1985 of the Biotechnology Action Programme; see Sect. 4.1.

Other parts of the 1983 strategy pursued a relatively separate development. For DG III, the main day-to-day business through the 1980s concerned the progressive development of harmonized Community-wide legislation in various sectors. Two of these – pharmaceuticals, and the food industry – might in theory become fields of application for the latest advances in the life-sciences; but the scientific discoveries, the technological inventions, and the journalistic enthusiasm for the word "biotechnology", did not yet perturb these sectorally-channelled activities. They were typically conducted as an intense dialogue between the Commission staff in the DG III division concerned, and the corresponding industrial trade association or lobby; interacting with corresponding debates at national level. The differences to be resolved usually centerd upon differences of national interest, with compromises being arrived at to offer something for all the major players, and no country's national champion(s) unacceptably penalized.

From the 1960s to the 1980s, extended debate and negotiation at national, Community, and wider international levels concerned the control of chemical products, and involved DGs III and DG XI (Environment). This impinged on biotechnology regulation only indirectly, in the late 1980s; and is described in Sect. 4.3 below.

A long-running argument required bilateral discussion between DGs III (Industry) and VI (Agriculture), concerning the pricing of the raw materials for the fermentation industries. The microbes needed a source of carbon – fermentable sugars, or starches (which could be enzymatically hydrolyzed to sugar). But whatever the raw material, it was agricultural in origin, and its pricing and control were matters for the Common Agricultural Policy. By 1986, new price regimes for non-food uses of sugar and starch were adopted by the Council of (Agriculture) Ministers, after long and contentious arguments. For in spite of a nominal commitment (in Article 39 of the EC Treaty) to the encouragement of technical innovation, the "managers" of the

Common Agricultural Policy – in practice, the Ministries of Agriculture rather than the Commission staff in DG VI – were at best ambivalent about the innovations offered by biotechnology – be it isoglucose and peptide hormones in the 1980s (increasing productivity in sectors plagued by production surpluses, and troubling consumer lobbies), or transgenic animals in the 1990s (bringing the wrath of the animal welfare movement to bear on the farmer).

3.7 Should European Bio-Industries Create a Lobby? The Davignon Meeting of December 1984

Such arguments in the mid-1980s seemed to be a world away from the "gene wars" of 1975–82, and from the controversies raging noisily in the United States between various federal agencies claiming oversight, or facing court challenges from JEREMY RIFKIN. To organize the defence and pursue the common interests of the nascent bio-industry in the USA, two industrial associations had been created. The Industrial Biotechnology Association, IBA, was founded in 1981, by a dozen or so large firms for whom a $ 10000/year subscription was negligible. For the more numerous start-ups, typically based on academic science, venture capital and entrepreneurial flair, there was the Association of Biotechnology Companies, subscription $ 600, formed in 1983; which by the end of the 1980s had several hundred member companies. IBA and ABC merged in 1994, to form the "Bio-Industry Organization, BIO".

The FAST report and the 1983 Commission communications had drawn attention to the US situation, and as indicated, there was a theoretical assent at political level by DAVIGNON and his cabinet, to the need for a "Community strategy". But the earnest arguments recorded in these communications received tolerant indifference at service level in the other DGs, who saw the mysteries of biotechnology as still playthings of DG XII and their scientific community, or abstractions in the unread scribblings of the FAST team. Real issues were matters such as the price of sugar.

European industry was invited to meet the Commission in 1984, to discuss the Community actions in biotechnology, the nature and mechanisms of the liaison between industry and the Commission, and the question of possible need for a specific bio-industry association. Key leaders were sought, from the sectors most concerned. President that year of CEFIC, the Council of the European Chemical Industry, was JOHN HARVEY-JONES, Chairman and Chief Executive Officer of ICI. To find a date convenient for both DAVIGNON and HARVEY-JONES was no easy matter, and it was finally on 12 December 1984, in the last month of the 1981–84 Commission, that the meeting took place.

Some 17 industrialists were present, to meet with Vice-President DAVIGNON, and Directors-General FERNAND BRAUN of DG III and PAOLO FASELLA of DG XII, accompanied by supporting staff. In addition to HARVEY-JONES were JACQUES SOLVAY of Solvay; HANS-GEORG GAREIS, head of Research at Hoechst; CALLEBAUT of Amylum, the starch refiners; A. LOUDON of Akzo, and other senior men from major companies; for the middle ranks, HILMER NIELSEN of Novo; and for the smaller firms, ALEXANDER STAVROPOULOS of Vioryl, Athens, and JOHN JACKSON of Celltech, (DENIS CREGAN of Kerry Creameries was fog-bound at London airport). Not present, to their irritation, were Monsanto Europe; CEFIC were still reticent about their "European-ness". DIETER BEHRENS of DECHEMA was present, a salute to his leadership in the European Federation of Biotechnology, and their work with FAST; but was treated with indifference, as a general with no troops.

The company was mixed, the agenda uncertain, and DAVIGNON was called away – with apologies – to a more demanding engagement after the first hour. The price of sugar was a tangible problem: on that, there could be a demand for action, and (two years later), a concrete result. But on abstractions such as "biotechnology", "rDNA" and "regulation", the discussion was desultory, the participants too high-level to know the detail,

and the agenda too vague for them to have been briefed.

A press release was issued next day, indicating that broad consensus had been reached on a list of topics: the need to avoid compartmentalization of the market, the need to modify for industrial use the price of raw materials of agricultural origin, the need for legislation at European level concerning patents, and a coherent action concerning the procedure for placing on the market of new products. Most of the press release referred to R & D and training actions, and the establishment of "centers of excellence" in biotechnology R & D.

Regarding the question of an industry association, there was a conspicuous silence. HARVEY-JONES recognized the Commission's need for an industrial interlocutor to give hard answers, and CEFIC had set up a group which could help; but it could not speak for the other two associations mentioned, EFPIA (the European Federation of Pharmaceutical Industry Associations), or CIAA (the Confederation of the Agro-Food Industry). He referred to the possibility of a loose grouping of the three federations, with the secretariat provided by CEFIC; and this was what subsequently developed.

3.8 The European Biotechnology Co-ordination Group, EBCG; ECRAB; and Their Extended Family

The European Biotechnology Co-ordination Group was established in June 1985, including the three Federations mentioned, together with GIFAP (the international association of the agrichemical industry), and AMFEP (the Association of Microbial Food Enzyme Producers), which included among its principal members Gist-Brocades and Novo. There was considerable reluctance in some of the sectoral trade associations – particularly EFPIA – to go beyond the loose structure of EBCG, which had neither Chairman nor budget, and simply met in rotation in the premises and under the chairmanship of its 5 founding associations.

EBCG was created as part of a coherent structure, in which a biotechnology R & D sub-group would be set up, under CEFIC leadership, as an element or extension of their quite active R & D group; and on regulatory matters, EFPIA would lead via a body known as "ECRAB": European Committee on Regulatory Aspects of Biotechnology. On patent matters, it was noted that UNICE, the general representative body for industry in the EC, already had a working group on patenting, which could handle any need for action or response by industry on that subject.

During the following years, additional European sectoral associations joined EBCG. FEDESA, for the Animal Health Industry, was created in March 1987, to give stronger representation for a sector troubled by the unhappy history of their steroid hormone products or analogues. Although cleared by the relevant scientific advisory bodies, yield or growth-enhancing products attracted public and political hostility, and were little welcomed by the Community authorities at a time when the managers of the Common Agricultural Policy were wrestling with growing problems of surpluses. The shadow of the steroid hormone battles of the mid-eighties – including a related trade dispute between the EC and the US, who had authorized use of the products – fell heavily across the subsequent discussion of the peptide hormones, starting with bovine somatotropin (BST) in the late eighties and nineties. Being produced by genetic engineering, just as the analogous human growth hormone (HGH) had been one of earliest molecules targeted (successfully) for industrial production by the bio-pharmaceutical industry, BST became a flagship molecule for the bio-industry. Many other ships in that large and ill-defined fleet would have preferred to salute other flags, and not be drawn into conflicts with agricultural policy or animal welfare concerns remote from their own preoccupations; but "misfortune acquaints a man with strange bedfellows".

The European Plant Breeders' Association, COMASSO, brought another broad dimension to EBCG; correctly acknowledging that the scientific success in plant cell genetic engineering and other breakthroughs would inevitably bring them into increasing involvement in biotechnology.

Parallel with the extension of its family in various sectoral directions, EBCG was faced with insistent requests for some form of membership, or associate relationship, from national bio-industry associations. Among these, most prominent and well-organized was the French association, Organibio (Organisation Nationale Inter-professionnelle des Bio-industries). Their founding president, GERARD NOMINÉ, subsequently described the alphabet soup of national associations and their aspirations, in an article in *Biofutur* (NOMINÉ, 1987).

In spite of this broadening of interest, and the policy challenges emerging at Community level in the 1980s; and in spite of various efforts to design and finance a more effective body; EBCG was finally abandoned by industry as ineffectual. A meeting was proposed for some date in 1991; a convenient date was difficult to find; and the conclusion, "Why bother?" was the *de facto* demise, the *nunc dimittis*, of EBCG. For by then, more effective industrial representation had emerged; largely in response to the failures of EBCG and ECRAB to anticipate or counter effectively the challenges to industry on the regulatory front in the late 1980s.

4 1985–1990: From Strategy to Legislation

4.1 Research and Concertation as Elements of a Strategy: The Biotechnology Action Programme, "BAP", 1985–1989

The October 1983 communication on strategy for biotechnology was favorably received at Council meetings of Ministers of Industry (November 1983) and Research (November 1983, February 1984). An ambitious proposal was drafted, in consultation with national and industrial representatives, thus enabling the content of the programme "to be better adapted to medium and long-term requirements of European industry and agriculture, and to make it fully compatible with national activities". The Commission's "Industrial R & D Advisory Committee" (IRDAC), a group of industrial advisers, created by Commission decision in February 1984, contributed their advice and opinion on the content of subsequent biotechnology R & D programmes. IRDAC held a Round Table meeting on biotechnology in October 1985, and subsequently created a specific Working Party ("WP-5"), which met from June 1986 onwards under the chairmanship of HILMER NIELSEN of Novo Industri. Its mandate being limited to "R & D", it was constrained to commenting on broader issues of Community policy; although several of its active participants were subsequently instrumental in the launch of the SAGB (see Sect. 7.1).

The proposal for the research action programme in biotechnology was published in April 1984 (EUROPEAN COMMISSION, 1984). It itemized the topics for a Research and Training Action in biotechnology, comprising:

- Sub-programme 1: Contextual measures (bio-informatics, and collections of biotic materials)
- Sub-programme 2: Basic biotechnology: pre-competitive topics in the following areas:
  - technology of bioreactors
  - genetic engineering
  - physiology and genetics of species important to industry and agriculture
  - technology of cells and tissues cultivated *in vitro*
  - screening methods for the evaluation of the toxicological effects and of the biological activity of molecules
  - assessment of risks
- Training grants.

Although the research and training action was the core of the proposal, and accounted for the bulk of the ECU 88.5 million requested, the April 1984 communication was also the occasion for a full re-statement of the Commission's strategy for biotechnology. The document opened with quotations from US reports dismissive of European capabilities, and re-stated the action priorities of the October 1983 communication. The proposal focused on two of the priorities, Research (see above) and Concertation (see below), but it reviewed action also on the other four.

On new regimes on agricultural outputs for industrial use, proposals for revisions to the sugar and starch regimes were announced as imminent for the former, and in preparation for the latter. On intellectual property rights in biotechnology, the Commission mentioned the establishment of an inter-service working group on biotechnology, identifying needs for improvement which could be used to prepare the Community position in the discussions and negotiations likely to follow the OECD enquiry, then in progress, on patenting in biotechnology. On demonstration projects, proposals would follow later, once R & D projects were in progress and targets could be better evaluated.

On "A European approach to regulations affecting biotechnology", the communication presented the Commission's views in some detail. Monitoring" of questions of biological safety, in particular the regulation of recombinant DNA ("genetic engineering") work and the "social dimensions" of biotechnology would be included in the "Concertation Action" (see below).

The document reiterated the general case for a common internal market, then referred to consultations the Commission had undertaken with committees of government experts, with industry, and with its own expert scientific committees in various sectors touched by biotechnology such as pharmaceuticals, chemicals, human nutritions, animal feedstuffs etc." The conclusion from these consultations and review was stated: "it would appear that the application of current Community regulations in the various fields will meet current regulatory needs, provided that there is close cooperation between the competent authorities in the Member States and the Commission. Such cooperation can be achieved by greater recourse to the existing institutional or scientific committees and, as necessary, use of the new information procedure for technical standards and regulations adopted by Council in its Directive 83/189/EEC of 28th March 1983. On the basis of its experience derived from the use of these various instruments, the Commission will put forward general or specific proposals appropriate to create a regulatory framework suitable for the development of the activities of the bio-industries and for the free circulation of goods produced by biotechnology."

The document then discussed the pharmaceutical sector, and announced the Commission's intention, arising from consultations with industry and the Pharmaceutical Committee, to submit legislative proposals to Council in July 1984. These would have as objective the implementation of concertation at Community level for "biotechnologically originating medicaments", and would include

- a proposal for a framework of consultation at community level prior to national decisions on the marketing of drugs produced by biotechnological or other high technology processes (this led in 1987 to the Directive 87/22, establishing the "prior expert concertation" procedure);
- proposals to facilitate the rapid adaptation to technical progress of the provisions of existing directives on proprietary medicinal products and veterinary medicinal products;
- a proposal to control the admission to the market of copies of innovative products, "to protect the R & D investment of the research-based manufacturer".

On "Concertation" (in the broader sense, not to be confused with the 87/22 procedure referred to above), the communication elaborated at length the logic of a network-based monitoring and information function, to be conducted in close liaison with "those responsible for biotechnology-related policies in the Member States", with industry (via industry associations) and other organized interests, and "in conjunction with the relevant services at Community and Member State level". The following list of nine tasks was proposed, and subsequently adopted, as the mandate of the concertation action; the only material amendment between the April 1984 proposal and the March 1985 Council Decision being the

inclusion in item 8 (at the request of Parliament) of the words, "and risks".

"Concertation activity will be implemented with the objectives of improving standards and capabilities in the life sciences, and enhancing the strategic effectiveness with which these are applied to the social and economic objectives of the Community and its Member States.

In conjunction with the relevant services in the community and the Member States, the following tasks will be executed:

1. monitoring the strategic implications of developments elsewhere in the world for biotechnology-based industry in Europe,
2. working with the services of the Community, Member States and other interested parties to identify ways in which the contextual conditions of operations for biotechnology in the Community may be further improved, to promote its development in all useful applications, and the supporting scientific capabilities,
3. responding to the needs for research and information in support of the specific actions of other services of the Commission,
4. identifying opportunities for enhancing through concertation and cooperation the effectiveness of biotechnology-related programmes in the Member States, and promoting collaborative initiatives in biotechnology with and between industry and universities,
5. consideration of how the safe and sustainable exploitation of the renewable natural resource systems in Europe may be enhanced by the application of biotechnology,
6. promoting in co-operation with developing countries and relevant institutions the pursuit of the same task (see preceding paragraph) within their respective regions,
7. monitoring and assessing developments in biotechnology bearing on safety and other aspects of the 'social dimension'.
8. disseminating knowledge and increasing public awareness of the nature and potential and risks of biotechnology and the life sciences,
9. establishing an *ad hoc* system of collaboration between groups and individuals with

interests and capabilities in the life sciences and biotechnology, so creating networks, as informal and flexible as possible, adapted to the particular problems under study: the networks to have the triple function of providing an active input into the programme, encouraging coordination through the exchange of information between the participants, and assisting the broader diffusion of information envisaged in the preceding task."

The "BAP" proposal was adopted by March 1985 with relatively little controversy, albeit with the budget initially reduced to ECU 55 million. Following the review of the programme, this was increased by ECU 20 million, to cover the accession of Spanish and Portuguese laboratories, and the allocation of increased resources for bio-informatics and risk assessment studies.

The programme supported the Community's first substantial programme of risk assessment research. More than ECU 6.0 million was allocated, to support the research of 58 laboratories across the Member States, in 16 transnational projects, which could be grouped under 5 headings:

1. Risk Assessment of Microorganisms under Physical Containment
2. Risk Assessment of Depollution Bacteria
3. Risk Assessment of Plant-interacting Bacteria
4. Risk Assessment of Transgenic Plants
5. Risk Assessment of Genetically Engineered Viruses.

This work contributed to the initial development of a scientific base of relevance to regulatory needs, although inevitably many open-ended questions remained. Further risk assessment research was conducted in the successor programme, BRIDGE (1990–93), on activities seen as complementary to the projects in BAP:

"*Monitoring and control techniques:* sampling and probes for engineered organisms and introduced segments of DNA; methods and instrumentation for high resolution automated microbial identification and the establishment of adequate data bases; creation of a bank of specific probes and chemical signatures for a large number of specific microorganisms; eradication methods.
Assessment techniques: biological containment; gene stability and gene transfer; development of microcosms and stimulating methods for impact analysis.
Acquisition of fundamental knowledge on gene behaviour (horizontal transfer between species, rearrangement of introduced genes in the host organisms), and on the survival and adaptation of released organisms, in particular soil bacteria, and including modification of host range and tissue range for engineered viruses.
Novel constructions: biologically contained organisms, suicide vectors or constructions which cannot develop outside the host organisms, engineered organisms which can be destroyed in the environment by known and specific techniques.

62 laboratories from the EC and EFTA countries are selected in BRIDGE to work in 14 projects aiming at the evaluation of possible interactions between GMO's and related species. In addition, a large, so-called 'targeted', project involving 10 laboratories is also underway for the development of methodologies rendering possible the automated identification of microorganisms in the soil. Finally six laboratories are working in identifying the state of the art on the identification and molecular taxonomy of fungi in agricultural applications." (VAN HOECK and DE NETTANCOURT, 1990)

4.2 The European Parliament Reviews Strategy: The Viehoff Hearings and Report

The debate on the BAP proposal triggered in the European Parliament not only discussion of the scope and content of the research programme, but a wider interest in the issues raised by biotechnology. The Parliament decided to prepare an "own initiative" report – i.e., not merely the formal opinion on, and proposed amendments to a Commission proposal, but an in-depth investigation of a topic initiated by the Parliament itself. The "rapporteur" nominated was Dutch Socialist MEP Ms PHILI VIEHOFF, who was assisted by Ms ANNEMIEKE ROOBEEK, of University of Amsterdam.

The Parliamentary investigation started with a public hearing on 20–21 November 1985. A background dossier was provided (VIEHOFF, 1985), with contributions from various invited experts, addressing a broad range of topics. In her Foreword, the rapporteur highlighted "four issues that need more attention in the years to come":

- First, the issue of access and distribution of biotechnology
  (the potential of biotechnology for the Third World, but their inability to get access to the technology; the possible effects of the US export control policy on high technology)
- Second, the issues of regulation and risks
  ("a pity … that the international efforts at OECD level to establish a common concept for the attainment of a harmonization system on international guidelines have been delayed, because of the fact that the American delegation could not accept the proposed text")
- Third, the restructuring of agriculture
  ("Biotechnology will enforce a restructuring of agricultural policy on a much broader scale and perhaps at a much faster speed than acknowledged until now")
- Fourth, the stimulation of socially useful products
  (malaria vaccine; orphan drugs; food production in Third World; toxic waste treatment; local, small scale energy production).

The Commission also submitted to the Hearing a paper closely based on the April 1985 BAP proposal; indicating that Commission services were "active in monitoring the evolution of the regulatory questions raised by developments in biotechnology … have participated in international discussions … A central issue in these debates concerns the adequacy of existing regulatory regimes to cover the new challenges posed by biotechnology.

In some areas, it is likely that they are already adequate: e.g. the Directive 82/471 on novel protein feedstuffs …"

In other areas, the Commission indicated that it was currently assessing the position; and referred to the recently-created Biotechnology Regulation Inter-service Committee (BRIC). Opinions and formal inputs to the Viehoff report over the following year (1986) involved no fewer than six of the Committees of the Parliament: covering Research and Technology, Agriculture and Food, Environment & Consumer Protection, Economic Affairs and Industrial Policy, Social Affairs and Employment, Legal Affairs & Citizens' Rights. The final report and resolution were adopted in Plenary Session of the Parliament in February 1987. There had been widespread consultation, debate and argument.

Controversy was not absent, for a parallel "own initiative" report on biotechnology and European agriculture was being conducted under a German, Green rapporteur, MEP FRIEDRICH WILHELM GRAF ZU BARINGDORF. His report and standpoint were severely critical of the "industrialisation" of agriculture in general, and of biotechnology and genetic engineering in particular. Inevitably, the Part A of his report – the formal Resolution – was heavily amended to the point where the final resolution, as voted in the same (February 1987) Plenary Session, largely contradicted the Rapporteur's original intentions.

The Resolution on the Viehoff report (EUROPEAN PARLIAMENT, 1987) was entitled: "On biotechnology in Europe and the need for an integrated policy", an expressive title, which summarized the comprehensive nature and central message of the report. The resolution opened with references to the review of the ongoing BAP research programme; to a 1986 Commission discussion paper on agro-industrial development (which led later to the ECLAIR Programme of Agro-Industrial Research); and to the opinions of the six Parliamentary Committees mentioned above. Next, eleven recitals were included, emphasizing the need to enlarge the research programme (not least, to enable the new Member States to participate), but also stressing five negative views or reasons for prudence:

"– determined that the social, economic, ecological, ethical, legal and health aspects of new developments in the field of biotechnology should be evaluated and assessed at the research and development stage,

– recognizing that biotechnology has up to now researched only a few aspects of the way in which DNA works, and convinced that it would be extremely dangerous and irresponsible to rush to transform these processes into marketable products, in view of the complex structures of the organisms and ecosystems being interfered with,

– aware of the predominantly negative effects of the use of biotechnology by the industrialized countries on the Third World countries,

– whereas to date there are no reliable scientific methods for assessing the medium and long-term effects of the irreversible release of genetically manipulated organisms into the environment,

– whereas the first experiments in release of genetically manipulated organisms are already underway and, for the first time in the Community, such an experiment was authorized and carried out this summer (i.e., 1986) in the United Kingdom, despite the fact that there are no comprehensive and binding provisions at national or Community level governing safety and liability in respect of such experiments."

Within the body of the Parliamentary Resolution's 29 numbered statements or demands, there was the usual mixture of positive statements (cf. item 1 below), and special interests; but also many cautious and risk-emphasizing statements and demands, such as those quoted below:

"1. Asks the Commission to review its biotechnology programme with a view to providing the Community with an effective strategic programme that affords Europe the means of increasing its competitiveness on world markets vis-a-vis Japan and the United States and enables the Community on the one hand to diversify its biotechnological research in such a way that resources are made available above all for basic research which takes a holistic view of life and work, gears its objectives to regional needs, safeguards, the bases of life and promotes forward-looking ecological research and, on the other hand, to investigate the medium and long-term effects of the use of genetic engineering in all its potential fields of application;

(...)

8. Considers that human medical research must exclude the manipulation of human genes, particularly human germ-cells, and that veterinary medical research must respect the integrity of individual animals and species and breeds of animals;

9. Stresses the importance of research in the areas of medical biotechnology and environmental biotechnology (e.g. degradation of toxic substances), which are under-represented in the current programme, but at the same time warns against the dangerous illusion that environmental protection can be replaced by 'biotechnological repairs' to the environment;

11. Expects the Commission to give priority in future to projects studying the problems posed by the intentional release into the environment of genetically engineered natural microorganisms ('deliberate release') and demands that such releases be banned until binding Community safety directives have been drawn up, possibly on the basis of the relevant OECD recommendations;

(...)

12. Asks the Commission to be aware in a general sense of the repercussions of each research project on the environment and suggests that for each research project a sum yet to be determined should be set aside in the overall financial package for an environmental impact study;

13. Calls for an assessment of the political and ecological repercussions of possible risks of epidemics or any restriction of gene resources and suggests a feasibility study for a European Institute for Ecology;

14. Demands that the principles and guidelines of good practice with regards to the safety of workers in laboratories, including university and research institutes, be strictly respected and broadened in scope in view of the special risks associated with genetic engineering methods;

15. Welcomes the recent inventory by the Commission's Biotechnology Regulation Interservice Committee as an important step towards creating a European biotechnology regulation system and calls for harmonization of Member States' provisions with regards to safety and the environment to provide for common procedures for risk assessment and imposition of conditions at each stage of the development of projects involving microorganisms carrying genetic material and suggests a step-by-step approach for regulating the various phases of biotechnology processes (laboratory, trials, limited production, mass production) and a case-by-case approach for approaching new biotechnology products; (...)

17. Calls on the Commission to submit a safety study examining the present state of legislation and desirable standards as regards liability and insurance cover, disaster prevention and protection against sabotage or terrorist attacks both in respect of research establishments and of production plants; (...)

28. Demands that existing provisions regarding the designation of products be adapted to take into account both the genetic production procedure employed and the composition of the products".

The Viehoff Resolution of February 1987 aptly summed up the political situation for biotechnology at European Community level. It showed a broad awareness of the potential of biotechnology, and the need for a coherent strategy, responding to the need for international competitiveness. But on regulation, it was unequivocally conservative, seeing "special risks associated with genetic engineering methods" (para 14), demanding a complete ban on field releases "until binding Community safety directives have been drawn up" (para 11), and with similarly restrictive views on gene therapy (somatic or germ-line) and animal transgenesis (para 8).

There seemed to be little evidence of US influence, and of the earlier US consensus on the absence of need for new, biotechnology-specific legislation; but that consensus related essentially to laboratory rDNA work. As large-scale field production and then field releases became current issues in the US, the

possibility of legislation remained very much alive. For example, a major conference – the "Second Annual Brookings conference on Biotechnology" – was held at the Brookings Institution in Washington, DC, in February 1986, and PHILI VIEHOFF attended this. The background paper (MELLON, 1986) reviewed the issues, defining the "elements of an adequate regulatory regime", and presenting the case for "A New Integrated Statute: A Hypothetical Statute for the Interim Regulation of Releases of Engineered Organisms".

4.3 BSC Creates BRIC, Debates Strategy, and Fades

The Commission communications on biotechnology, up to 1985, although presenting a strategic approach, were in practice largely drafted by the research service, DG XII, with marginal additions by other services. These additions, so far as concerned regulations, had on regulatory questions generally defended the adequacy of existing or planned sectoral measures. The services concerned were DG III (Internal Market and Industrial Affairs), and DG VI (Agriculture).

The development of large-scale industrial production, and slightly later, but foreseeably, the field release of genetically modified microorganisms and plants, fuelled increasing debate about the adequacy of existing regulatory structures and approaches. These debates took place in the US, in various European countries, in the OECD Group of National Experts on Safety in Biotechnology, and within the Commission services.

In all countries, such debates clearly required an inter-agency or inter-ministerial forum for policy debate; and in the Commission services, this was apparently provided by the Biotechnology Steering Committee. However, DG XII's main interest was in winning the resources for larger R & D programmes, and managing them efficiently, within a process of developing the more systematic approach represented by the "Framework Programme".

Certainly the FAST programme had produced broad and generally competent strategic analyses; but the programme was located within DG XII, not in an autonomous Community institution, as recommended by the "Europe +30" group; and its prescriptions were therefore unlikely to be implemented unless they coincided with other powerful interests, within the Commission services and beyond. The creation by the Commission President of a separate "Forward Studies Unit", and the creation by the European Parliament of a "Science and Technology Options Assessment" unit (sometimes described as the "poodle" of the Committee for Energy, Research & Technology Committee by MEPs active in other Committees) each underlined the non-centrality of FAST. None of these forecasting and policy assessment initiatives came close – at least, during the 1980s and early 90s – to the authority and central position of the US Congressional Office of Technology Assessment (OTA). The broad scientific networks of DG XII, and of its Concertation Unit for Biotechnology in Europe, CUBE, had little political weight in the developing conflicts over regulation.

The Directorate-General for the Environment, DG XI, started to attend the BSC from its 5th meeting (in July 1985), and were conscious of the discussions of regulation of field release, in which the US Environmental Protection Agency and Department of Agriculture were engaged. Technical arguments about regulatory details were not appropriate for the Directors-General at BSC level. Consequently in July 1985, the BSC agreed to establish BRIC, the Biotechnology Regulation Inter-service Committee.

The functions of BRIC were defined to be:

"a) to review the regulations applied to commercial applications of biotechnology;
 b) to identify existing laws and regulations that may govern commercial applications of biotechnology;
 c) to review the guidelines for rDNA research;
 d) to clarify the regulatory path that products must follow;
 e) to determine whether current regulations adequately deal with the risks that may be introduced by biotechnology and to initiate specific actions where additional

regulatory measures are deemed to be necessary;
 f) to ensure the coherence of the scientific data which will form the basis of risk assessment and in particular to avoid unnecessary duplication of testing between various sectors."

The Chair was divided between DGs III and XI; the Secretariat would be provided by DG XII's CUBE. DG VI (Agriculture) indicated that, while accepting this arrangement, they felt no less involved, having significant regulatory experience and responsibilities in fields which would surely be important areas of application for biotechnology. DG V (Employment and Social Affairs), having responsibilities for both worker safety and public health, were also active participants.

BSC remained in theory the parent of BRIC, to resolve disputes arising; but in practice, this was unworkable, for two reasons. Firstly, BRIC met almost every month, having 15 meetings from its first in September 1985, to its 15th in October 1986; BSC met three times during this period. Secondly, any basic inter-DG dispute could not with certainty be resolved at BSC, under its DG XII chairmanship; but might ultimately have to be resolved at Commissioner level between cabinets, or in the Commission itself. Although the Commission met weekly, its ability to handle technical disputes would require careful preparation and briefing, and could be used only infrequently, on major issues.

In consequence, BRIC was the active center within which the inter-service discussions on regulation of biotechnology were developed within the Commission, from 1985 to 1990.

A first and useful initiative by BRIC was the preparation of an inventory of existing Community regulations relevant to biotechnology (EUROPEAN COMMISSION, 1986a). The subsequent role and work of BRIC are presented in the following sections.

The Biotechnology Steering Committee faded out, for the reasons indicated above; meeting only once in 1987, and one final time in 1988. Yet its final meeting in July 1988 was in many respects its most significant. Occurring five years after the 1983 strategy papers, and a few weeks after publication (May 1988) of the Commission proposals on biotechnology regulation (Sect. 4.7 below), the BSC meeting of July 1988 sought to launch a comprehensive review of strategy for biotechnology. It might thus have developed and restated the "need for an integrated approach" (to quote the title of the European Parliament's Viehoff Report), and could have been a guiding influence on the subsequent evolution of the May 1988 regulatory proposals.

The basic papers for the July 1988 BSC meeting itemized the many changes in the strategic environment for biotechnology, since 1983:

- the scientific progress, and (consequent) pervasive significance of biotechnology;
- the industrial development;
- the GATT (Uruguay Round) negotiations, which would liberalize agricultural trade, and alter the attitudes of Agricultural Ministries to S & T innovation;
- the competitive challenge;
- the Single European Act, adopted July 1987, to become effective July 1989; facilitating (by majority voting on Article 100A) the progress towards a common internal market by 1992; and with an extensive new Section (Title VI) devoted to Research and Technological Development, and starting with the blunt language of Article 130F: "The Community's aim shall be to strengthen the scientific and technological basis of European industry and to encourage it to become more competitive at international level."

The meeting was attended by an unprecedented galaxy of Directors-General or other senior staff, from Directorates-General I ((External Relations), III (Internal Market and Industrial Affairs), V (Employment and Social Affairs), VI (Agriculture), VIII (Development), XI (Environment), and XII (Science, Research and Development). The arguments for a strategic review were accepted; but as the background papers had not been widely read or fully digested, it was agreed to invite all services to develop their views on the papers, and/or the review of strategy for biotechnology.

These views were received and assembled, in an internal "Interim Document: Towards a Redefinition of Community Strategy for Biotechnology". But an institutional/practical question was now pressing. The final trimester of the 4-year Commission, 1986–89, was approaching; and the thoughts of Commissioners and their cabinets were preoccupied with their personal and political futures. Few could expect to be in the same posts in January 1989 – if indeed they returned to the Commission at all after Christmas. Why waste something as important as the re-statement of Community policy for biotechnology on the dying days of an outgoing Commission? Surely better to await the new year, and enlist the enthusiasm of the incoming team?

In circulating the Interim Document in December 1988, the BSC Chairman referred to some salient communications that had appeared since July, such as the proposal (October 1988) for a Directive on the Protection of Biotechnological Inventions; to others in preparation, such as the DG VI proposal for a Community system of plant variety rights, and DG XII's proposal, already circulated in draft, for the next biotechnology R & D programme, BRIDGE (ECU 100 million, 1990–94); and to the outcome of the mid-term review of progress on the Uruguay Round. Also mentioned were the recent "Bio-Ethics" and "Bio-Safety" conferences (Mainz, 8–9 September 1988, and Berlin, 27–30 November 1988). He indicated the intention to continue preparing a draft communication on biotechnology for consideration by the new Commission, taking into account the Interim Document and other matters referred to, with a view to integrating these in a draft document for consideration at the next meeting of the Biotechnology Steering Committee.

But the BSC did not meet again. The new Commissioner responsible for research and technological development had other priorities, particularly for institutional re-organization of DGs XII and XIII, for implementing the second Framework Programme (1987–91) and preparing the third (1992–95), and for improving the management procedures for research contracts. Vice-President PANDOLFI and his cabinet took no part in either the regulatory debate on the two biotechnology Directives being negotiated through Parliament and Council, or the broader strategic policy issues of biotechnology; beyond dismantling the unit of DG XII which had been principally concerned, and delaying the proposed EC Human Genome R & D Programme for fuller consideration of its ethical dimensions.

4.4 1986: The "Hinge" Year: (1) BRIC Starts Work: The Background of Sectoral and Chemicals Legislation

After the first "Gene-Splicing Wars" of the post-Asilomar years, and the corresponding discussions in Europe which led to the 1982 Council Recommendation, there followed some years of relative calm, so far as concerned new initiatives for the regulation of biotechnology. As indicated in the preceding descriptions of Commission communications in 1983, 1984 and 1985, the general feeling in the Commission services responsible for various sectoral products was that the applications within these sectors of the new biotechnology would not pose insuperable problems, and could be handled within sectoral legislation. The research aspects had been addressed in the earlier debates, and resolved at Community level by the national notification requirements which followed (or, for the countries mainly concerned, preceded), the adoption of the 1982 Council Recommendation.

The situation was changing, as biotechnology moved towards applications in large-scale industrial production facilities, and field release of genetically modified organisms (GMOs) – microorganisms or plants. Public and political attention were being continually stimulated by high-profile press coverage, which could be summarized under three categories:

(i) the scientific discoveries, and the related progress of techniques, were generating a flood of new insights; the corresponding excitement diffused from the leading scientific journals into the popular press;

(ii) the economic potential and implications, although ultimately unquestionable, were often exaggerated by the promoters of new, start-up companies seeking to attract risk capital, or to sell equity to large companies; again, such "hype" was prone to exaggeration and distortion in the popular press, and readily merged into science fiction;

(iii) both in the US and in parts of Europe, opposition and hostility were continuing and in some areas increasing due to a mixture of motives, the mixture varying from country to country. There was a long-standing and justified concern by environmental groups that the track record of big industry and modern agriculture included disruption, destruction and persistent pollution; and the scientists financially linked to these interests had been as venal and prone to error and bias as any other human group. Anything which industry promoted with enthusiasm was *ipso facto* suspect.

Given that some of the technologies, and certainly the new knowledge, could be applied to matters involving life, reproduction and death, there was a sensitivity to ethical aspects which would be intensified by various current, or imminent and foreseeable, applications of the new knowledge. These would range from specific aspects such as genetic screening, or *in vitro* fertilization, to more generalized concerns about "interfering with nature", or "scientists playing God"; those who spoke, wrote and broadcast about such concerns did not lack an audience. The World Council of Churches prepared a rather hostile report (WCC, 1989), containing serious inaccuracies.

The entry of environmental interests into the policy debates on biotechnology was thus an obvious development in the mid-eighties, as the public authorities at national and Community levels interpreted their general responsibilities for the protection of the environment in relation to the challenges of the new processes and products resulting from biotechnology.

DG XI, the Directorate-General responsible within the European Commission for matters of environmental policy, took up with characteristic energy the responsibility of co-chairing the Biotechnology Regulation Inter-service Committee. They brought to the new issues their extensive regulatory experience, of which two areas of safety legislation were considered particularly relevant:

- chemicals (see below, the "Sixth Amendment" Directive);
- potentially hazardous industrial activities.

Although each BRIC-participant service had its own interests and experience, the co-chairs DGs III and XI were particularly influential, being responsible for drafting the subsequent biotechnology directives. The experience of DG XI with Chemicals legislation was of strong relevance as an influence on their thinking, and subsequently on their drafting, as a paradigm for regulating the products of biotechnology.

The rules for the control of chemicals were one of the first areas to receive extensive attention and efforts for the development of harmonized Community-wide procedures. The history was long and complex, and many of the issues discussed and contested at length from the mid-sixties through to the mid-eighties and beyond were relevant to other areas of legislation, at least in general terms. The difficulties and disagreements would lie in the details, of how far rules and procedures developed in relation to new (or existing) defined chemical entities could or should be applied to genetically modified organisms, in all their mystery and multi-molecular structural complexity. The central issue of principle was whether the fact of modification by certain techniques of DNA recombination – genetic engineering – defined an activity or a class of products *ipso facto* requiring regulatory oversight. If this point was conceded, whether for reasons of scientific uncertainty, expert disagreement, public and political concern, or some mixture of these (and in effect the 1982 Council Recommendation acknowledged the political legitimacy of focusing upon recombinant DNA organisms), then a series of practical questions would follow. To these practical questions, a collection of "off-the-shelf" answers were potentially available from the chemicals experience.

The early 1980s saw the coming into force of the provisions of Council Directive 79/831/EEC – a directive "amending for the sixth time Directive 67/568/EEC on the approximation of the laws, regulations and administrative provisions relating to the classification, packing and labelling of dangerous substances"; known for brevity as the "Sixth Amendment". Although five amendments to the Annexes and Articles of the original 1967 Directive were adopted between 1969 and 1983, the Sixth Amendment replaced the whole of the substantive parts of the parent Directive and the first five amending Directives, with the exception of the formal articles on the introduction of national laws, regulation and administrative provisions. It repeated in expanded form the provisions of the parent Directive dealing with classification, packaging and labelling of dangerous substances, and added a procedure for testing and notification of new chemicals. Certain product categories were excluded from the scope of the Directive – particularly medicinal products, foodstuffs and feedingstuffs; which in Commission terms were responsibilities of DG III and DG VI.

The Sixth Amendment defined or required discussion of many matters of subsequent relevance to the biotechnology Directives:

- *Notification:* Prior to being placed on the market, substances would be notified to the competent authority (see below) of the Member State in which the substance was first manufactured or imported; subject to a range of exceptions for most polymers, substances test marketed for research and analysis, small quantities, and substances already on the market before 18 September 1981 (an inventory of such would be prepared subsequently by the Commission). The notification had to include a technical dossier containing a "base set" of information and the results of tests, a declaration on unfavorable effects in relation to envisaged uses, a proposed classification and labelling, and proposals for recommended precautions for safe use.
- *Testing Requirements:* Annex VII defined the "base set" of information required with notification, essentially identical to the

"Minimum Pre-Market Set of Data" sufficient for an initial hazard assessment, as set out in an OECD Council Decision of December 1982 (see Sect. 6.2). Annex VIII defined additional test requirements ("step sequence testing") to be applied when successive quantity thresholds were exceeded. These additional tests concerned mainly long-term health and environmental effects, including toxicity to aquatic species.

– *Competent Authority:* Member States were required to appoint the competent authority responsible for receiving the notification, examining conformity with the Directive, possibly requesting further information, and transmitting a copy of the dossier or a summary to the Commission, who would then forward it to other Member States.

– *Confidentiality:* The notifier could indicate the information he considered to be commercially sensitive: the competent authority would decide.

– *Classification:* The Directive set out 14 danger categories, including "dangerous for the environment"; and Annex V defined procedures for determining the physico-chemical properties and toxicity and eco-toxicity of the substance. Annex VI set out the general principles of the classification requirements.

– *Committee for Adaptation:* A committee was established with power to adapt to technical progress the Directives concerning elimination of technical barriers to trade; the committee could take decisions by qualified majority. Annexes VI (Part 1), VII and VIII were excluded from this procedure.

The evolution of the Sixth Amendment was strongly influenced by the US legislative activity, and international discussions at OECD; particularly in the preparatory work for a 1974 OECD Council Recommendation on the Assessment of the Potential Environmental Effects of Chemicals. In 1975, the Commission had established a working group which linked discussions ongoing in various Member States; and as a result of a French draft law and these consultations, the Commission published a proposal for a Directive

in just 15 months. In 1976, the USA adopted their "Toxic Substances Control Act" (TOSCA), and from 1977, the Environmental Protection Agency published implementing rules. Thereafter, the US legislation became an increasingly important factor in the negotiations in the Community.

HAIGH (1984) comments critically on the procedure through which the Sixth Amendment proposal was developed between publication of the Commission's proposal (September 1976) and adoption of the Directive (October 1979). Although the basic structure remained as in the proposal, there were "very numerous changes of detail", so that "the ultimately adopted Directive must be considered as having been fundamentally changed". HAIGH remarks, with references to the UK situation, that:

"In general, it must be said that both the Economic and Social Committee and the European Parliament failed to appreciate the significance of the draft Directive or even to reflect what were to become the major issues of subsequent concern. In Britain only the House of Lords gave substantial attention to the draft, and even so only covered part of the major problems.
The Directive as finally adopted reflects very long, detailed and complex consideration in the Council working group. The nature of changes incorporated, sometimes involving subtle shifts from Article to Article, are a clear indication of the difficulties which were encountered. It is consequently one of the most difficult of Directives to understand. The most important change after the Directive was proposed was the introduction of step sequence testing at the initiative of the German government which was being prodded by its chemical industry. In actual fact, no public allusion to step sequence testing – let alone any text setting out its principles and provisions – can be found in any official document prior to publication of the Directive in the *Official Journal*. But quite apart from this major omission, by late 1977 it was already increasingly evident that the Directive as finally adopted would differ significantly from the proposal making the Sixth Amendment one of the most obvious cases where lack of intermediate public information makes the Community legislative process so difficult to reconstruct."

Both the content of the Sixth Amendment Directive, and the procedure leading to its adoption, offer illuminating parallels to the subsequent development of the biotechnology Directives.

Each service participating in BRIC brought to it its own background, precedents, and perceptions of need shaped by their specific experience. Reference has been made to the sectoral interests of Industry (DG III) and Agriculture (DG VI). The background of legislation for worker safety was contributed by DG V; like DG XI's responsibility for environmental protection, DG V's was a "horizontal" approach, applying across various sectors. The scientific culture of DG XII was similarly cross-cutting. Finally, in the later meetings of BRIC, the recently established Consumer Policy Service started to participate, having general interests in product safety.

Such were the diverse strands within the Commission which were brought together in BRIC.

4.5 1986: The "Hinge" Year: (2) National and International Developments

The year 1986 was the hinge, on which subsequent developments in biotechnology legislation turned. It was marked by a number of events of national, European, and global significance, which created an insistently influential background to the deliberations and activities of BRIC.

National developments in biotechnology regulation are further discussed in Sect. 5. Three had particular influence in 1986:

– Denmark became the first European country – probably, the first country in the world – to adopt legislation specifically on biotechnology, with the adoption in June 1986 of the Gene Technology Act.
– Germany's Bundestag had on 29 June 1984 established a "Commission of Enquiry on Prospects and Risks of Genetic Engineering", under the Chairmanship of Social Democrat member WOLF-MICHAEL CA-

TENHUSEN. Although the Commission's report was published in January 1987, its existence and the probability of resulting national legislation in Germany were widely recognized in 1986.
– The United States, through its Office of Science & Technology Policy (in the Executive Office of the President) published in the *Federal Register* of 26 June 1986, the announcement of the "Coordinated Framework for the Regulation of Biotechnology", comprising: the policy of the Federal agencies involved with the review of biotechnology research and products"; and indicating that existing statutes provided "a basic network of agency jurisdiction over both research and products".

Also in 1986, the OECD published its report, "Recombinant DNA Safety Considerations", including the Council Recommendation of 16 July 1986 (further discussed in Sect. 6.2). Although published later in the year, the content of this was already known and agreed in government circles in the early months of 1986. The OECD report was a major consideration, as could be illustrated by the case of chemicals. The influence of the OECD on worldwide developments in this sector was underlined by the formal agreement, in 1984, between the OECD and the International Programme on Chemical Safety (IPCS), regarding reciprocal use of guidelines, methodologies and evaluations developed by the two organizations. (IPCS itself was established by a memorandum of understanding between WHO (the World Health Organization), UNEP (the United Nations Environment Programme) and ILO (the International Labour Office) in April 1980). Thus the OECD work on the testing of chemical products and related concepts, such as Good Laboratory Practice, would be used by these UN agencies as the basis for worldwide rules in these areas. The discussion above of the negotiation of the Sixth Amendment Directive indicates the influence of OECD on the content of the Community Directive; a similar influence was exerted on the data requirements referred to in the US Toxic Substances Control Act.

The Commission was thus faced with conflicting arguments. On the one hand, the US administration, and the Commission services with sectoral responsibilities, were arguing for the adequacy of sectoral legislation for biotechnology products. Regarding the oversight of research, the NIHRAC guidelines and various national provisions in Europe, coherent with Council Recommendation 82/472, appeared to be meeting needs. The OECD report and Council Recommendation had explicitly recognized that "there is no scientific basis for specific legislation to regulate the use of recombinant organisms". On the other hand, the Danish and German initiatives made clear the threat to a unified common market presented by divergent national legislation; and the ongoing debates (on the Viehoff report) in the Committees of the European Parliament made clear their developing wish for legislation specific to biotechnology.

4.6 1986: The "Hinge" Year: (3) European Responses by Industry, Member States and the Commission

The Commission services through BRIC prepared and published in November 1986 a communication from the Commission to the Council, COM (86)573, entitled "A Community Framework for the Regulation of Biotechnology". This referred to a high-level meeting between Commission staff and Member State officials, that had taken place in April 1986; and a few days prior to that, ECRAB, industry's "European Committee on Regulatory Aspects of Biotechnology", had also met.

ECRAB (see Sect. 3.8 above) had prepared their position paper directly in response to the request of the Commission. Their paper stated bluntly that, in the view of their five participating organizations (AMFEP, CEFIC, CIAA, EFPIA and GIFAP), it was "neither necessary nor desirable to formulate a single set of guidelines and rules which cover every aspect of biotechnology". The paper quoted the OECD Report's statement (quoted above); and added:

"The industry associations concur with the OECD view that no new legislation is needed for the industrial use of genetically modified organisms for the manufacture of new products ... It is the industry view that any additional requirements for product specifications can be accommodated in the present legislation and that there is no basis for any discrimination against biotechnological products."

Where the paper acknowledged a need for caution, risk assessment (although "the risks involved are ... to a large extent conjectural") and further research was in the planned release of live GMOs. The paper reviewed the various beneficial applications in agriculture and the environment, again referring to the OECD report for further details; and put forward a "Proposal for a risk assessment scheme".

The recommendations here were for a stepwise development process, again quoting the OECD report: " in a stepwise fashion, moving from the laboratory to the growth chamber and greenhouse, to limited field testing and finally to large-scale field testing". The language of the paper was careful in its vocabulary, speaking of guidelines and rules; but acknowledging the obligation to document the risk assessments at each stage and the conclusions drawn, and to make this information available to the authorities in an agreed form. This should be on a commonly accepted basis in Europe, applying to all organizations whether industrial or non-industrial, private or publicly owned, and with mutual recognition by the various countries of evaluations and conclusions from the risk assessments.

The conclusion of the ECRAB paper was that the involvement of the authorities in the scale-up steps could be satisfied by a notification procedure. Authorities should be advised by expert committees, comprising competent scientific technical and public representatives, from government, academia and industry.

The ECRAB paper was clear and timely, delivered shortly before the meeting of 29-30 April between Commission staff and Member State officials, at which it was circulated.

The meeting was co-chaired by Mr. Tony Fairclough, Deputy Director-General of DG XI, and Mr. Tom Garvey, Director of the DG III Directorate (III-A) including Pharmaceuticals, Foods and Chemicals. There were twenty-six participants from Member States: 6 from the UK; 5 from Denmark; 3 each from France, Germany, and Italy; 2 each from Belgium and the Netherlands; 1 each from Ireland and Spain. Commission staff were in attendance from DGs III, V, VI, XI and XII – the "BRIC-participant" services.

A "Summary of Discussion" was subsequently prepared and circulated, and is the basis of the quotations below (European Commission, 1986b).

> "Extensive reference was made to the recently-completed OECD report, "Recombinant DNA Safety Considerations" (to be published later in 1986); several of those present (including Commission staff) had participated in this expert group, and all accepted its work as a good starting-point. Specific reference was made to its concept of "Good Industrial Large-Scale Practice" (GILSP). Also noted was its emphasis on the absence of rational basis for treating the regulation of rDNA organisms separately from the regulation of other organisms. This "non-discrimination" point was stressed by several participants – the basis for regulation should be risks of toxicity, pathogenicity, ecotoxicity, etc. rather than the fact of recombination. Also referred to was a report produced a few days earlier by "ECRAB" (European Committee on Regulatory Aspects of Biotechnology), representing the five major European industrial associations with biotechnology interests. [see above]
> …
> On Industrial containment: Many states felt that existing legislation was adequate, given notification, and observance of GILSP. A concern expressed particularly in relation to compulsory notification of industrial recombinant activities was that this could raise problems of commercial confidentiality which might inhibit compliance, or make such proposed legislation unacceptable.
> On Field release: It was accepted that the scientific basis for prediction of effects was inadequate, and research for this purpose should be reinforced; in the meantime, there

should be notification and case-by-case consideration prior to approval.
> While the desirability of a Community framework of regulation was generally agreed there was some reticence expressed. Some States, particularly the U.K., France and the Netherlands, seemed inclined to view existing legislation as a basic requirement to which countries might add further requirements relevant to their particular situation – geographical, climatic or regional; although Commission staff stated that such regional specificity could be and sometimes was included in Community Directives.
> In other Member States, legislation specifically relating to recombinant organisms was said to be imminent (Denmark's parliament passed such an Act a few weeks after the meeting, on 30 May), or under review, as in Germany.
> Other points raised in the debate:
>
> – the problems of definition can be severe – of biotechnology, or even of "recombinant": is a gene deletion a recombination? Or should all "modifications" be considered? Definition would be important in determining the coverage of any legislation;
> – there was acknowledgement that pressure for regulation was motivated by public concern, and a spillover effect from other areas of technology where accidents of unexpected nature or scale have occurred;
> – one participant emphasized the risks and loss of life resulting from delaying certain types of innovations (e.g. vaccines, therapies);
> – the speed of development of the science was so much greater than the speed of legislation that the latter should focus on principles and essentials, not on technical details;
> – conformity to a common level of safety does not have to imply detailed technical standardization;
> – international harmonization with major trading partners should be sought, on basis of OECD guidelines;
> – some speakers suggested an inventory – or at least an illustrative list of organisms; there was a feeling that there should be uniform containment standards throughout the Community for any given organism.

Regarding possible systems for Community legislation, reference was made to two existing systems of potential relevance to biotechnology; that for pharmaceuticals, and that for chemical products.

Towards the end of the meeting, Mr. GARVEY summarized the consensus as follows:

1. Agreement on the need for a Community approach to preserve a common market
2. Guidelines regarding physical and biological containment at production level, including waste management, could be a useful initiative for the Commission to propose
3. Non-release – to be approached on the basis of existing legislation
4. The ECRAB report was a constructive contribution; it called for action only on deliberate release, and that action should be Community action
5. On deliberate release – for most people the top priority for action – there was need for very close collaboration between Member States, based on a step-by-step process and notification, bearing in mind the need for flexibility to take account of special regional or climatic needs. The German delegation had emphasized the need to define our terms precisely.

The key question was what should follow this meeting; possibilities mentioned included a forum for regular information exchange on specific topics, particularly on containment and on deliberate release, research questions, the role of the Commission, discussion with industrial representatives and other interested groups.

Mr. GARVEY said that BRIC demonstrated the interest of the Commission. Its direction by the DGs for Environment and for Industry underlined the pervasive nature of biotechnology. This had been the first meeting with Member States, to explain roughly where we had got to, and to obtain Member State reactions.

The Commission's objective would be a Communication to Council, outlining a strategy and work programme in the area of biotechnology, and focusing on health and environmental protection, and on the internal market. This would be produced in the next month or two."

BRIC now had the inputs from industry and Member States; in fact no further meetings of such broad and general character as that of April '86 took place, and of the subsequent smaller consultative meetings, few if any records were kept or published.

The promised communication to Council, "A Community Framework for the Regulation of Biotechnology", was published in November 1986 (EUROPEAN COMMISSION, 1986c); it comprised just twelve paragraphs, summarized below:

(i)   introduced the new techniques known as "genetic engineering";
(ii)   raised the question of risks, for consumer/worker health and safety, or for the environment. More precise modification posed no *a priori* extra or new risks in enclosed manufacturing processes, although laboratory work had been subject to regulatory oversight in both the EC and the US; the prospect of deliberate release in agricultural and environmental applications had sparked further debate;
(iii) hence various national reports, and the OECD report – distinguishing *enclosed use* and *planned release*;
(iv)   OECD's GILSP – "no new or additional risks, either for the workers involved, the environment, or in respect of the resultant products";
(v)   OECD report concludes on planned release, that "while risks exist, they can be assessed to some extent by analogy with … existing organisms. However, there is insufficient experience at this stage to lay down a coherent set of regulations. Instead the report recommends a prior case-by-case evaluation of all planned release applications";
(vi)   reviewed Commission actions from the 1982 Council Recommendation to date, including the creation of BRIC, the review of regulations, the April 1986 meeting with Member States, and the ECRAB paper; and mentioned the Community and national risk assessment research programmes;
(vii) concluded that "the Commission believes the rapid elaboration of a Community framework for biotechnology regulation to be of crucial importance to the industrialization of this new

technology in the Community. Equally, citizens, industrial workers, and the environment, need to be provided with adequate protection throughout the Community from any potential hazards arising from the applications of these technologies. The internal market arguments for Community-wide regulation of biotechnology are clear. Microorganisms are no respecters of national frontiers, and nothing short of Community-wide regulation can offer the necessary consumer and environmental protection";

(viii) announced the intention of the Commission "to introduce proposals for Community regulation of biotechnology by Summer 1987 with a view to providing a high and common level of human and environmental protection throughout the Community, and so as to prevent market fragmentation by separate unilateral actions by Member States. The Commission's proposals will address two distinct aspects of the use of genetic engineering, viz:

A. Levels of physical and biological containment, accident control and waste management in industrial applications, and,

B. Authorization of planned release of genetically engineered organisms into the environment."

(ix) referred to the possibility that proposal (A) might cover other biological agents used in industry;

(x) stated "Because international experience of risk assessment in the field of 'planned release' is still limited, it is not possible to propose any general guidelines or testing requirements for the time being. The Commission will be proposing a Community case-by-case evaluation and authorization procedure based on mandatory phased notification by industry. This is in line with industry's own proposals and with the recommendation of the OECD report. The stages at which Community notification would be mandatory, the procedures for dealing with agricultural and environmental applications, and the general question of *a priori* exemptions, have yet to be agreed and will be a matter for further discussion with experts and with Member States officials in the light of the re-evalua-

tion of existing Community legislation referred to in par (vi)";

(xi) emphasized the international dimension and the need to maintain broad harmonization, in particular with principal trade partners;

(xii) announced that "The Commission is convinced that the development of a Community regulatory framework, which will both provide a clear, rational and evolving basis for the development of biotechnology and also ensure adequate protection of human health and the environment is an urgent necessity. To this end the Commission services, working together in the framework of BRIC, are launching the necessary work to draft proposals for legislation on genetically engineered organisms to be presented to the Council by Summer 1987. In the meantime, the Member States are requested to inform the Commission of their activities and intentions in the fields of biotechnology regulation and risk assessment research".

The language of this communication clearly went beyond the views and recommendation of the ECRAB paper, the Member State opinions of April '86, and the OECD report; it represented a determined thrust towards legislative proposals. At the same time, it may have anticipated accurately the probable pressure from the European Parliament – the views of the Viehoff report were noted in Sect. 4.2.

4.7 The Preparation of the 1988 Legislative Proposals, and Their Adoption

The division of responsibilities within BRIC was not controversial. DG V were preparing a specific "industrial Directive" (the seventh such) on worker safety vis-a-vis biological agents, within the context of a general Framework Directive on Worker Safety: originally Council Directive 80/1107/EEC on the protection of workers from the risks related to exposure to chemical, physical and biological agents at work. This framework was superseded by the adoption in June 1989 of Council Directive 89/391 "on the introduction of measures to encourage improvement in the safety and health of workers at work". The contained use of genetically modified microorganisms in industrial processes would be a joint responsibility between DGs III and XI. On the deliberate release to the environment of GMOs, chef-de-file would be DG XI. Through BRIC, mutual transparency, consistency and co-ordination would be assured, including scientific inputs from DG XII, and agricultural views from DG VI.

To services inexperienced in regulatory politics, this seemed a satisfactory arrangement; and where a common base of scientifically based risk assessment was present, so it proved.

The proposal for a Directive for the protection of workers against the risks of exposure to biological agents was prepared by DG V staff, from the division responsible for industrial medicine. Their networks included Member State authorities responsible for hospital pathology laboratories, and pharmaceutical companies familiar with handling dangerous pathogens, e.g., in connection with the development, testing and production of vaccines.

The DG V Directive addressed in principle all biological agents, defined as "microorganisms, including those which have been genetically modified, all cultures and human endoparasites, which may be able to provoke infection, allergy or toxicity" and classified these into four risk groups, on the basis of the risks they presented, and the availability of prophylaxis and therapy:

"1. group 1 biological agent means one that is unlikely to cause human disease;
 2. group 2 biological agent means one that can cause human disease and might be a hazard to workers; it is unlikely to spread to the community; there is usually effective prophylaxis or treatment available;
 3. group 3 biological agent means one that can cause severe human disease and present a serious hazard to workers; it may present a risk of spreading to the community, but there is usually effective prophylaxis or treatment available;

4. group 4 biological agent means one that causes severe human disease and is a serious hazard to workers; it may present a high risk of spreading to the community; there is usually no effective prophylaxis or treatment available."

Similar classifications had been developed and discussed by various national bodies, in Europe and elsewhere; by the "Safety in Biotechnology" working group of the European Federation of Biotechnology; and by the World Health Organization. The DG V proposal envisaged that any genetically modified microorganisms should for the purposes of the Worker Safety Directive be allocated to one of the four categories, depending on the characteristics of the host organism and genetic insert. The final Directive 90/679 (adopted in November 1990), as it required assessment of biological agents and their classification into the above four categories, did not require any specific references to genetic engineering (beyond a reference in Article 1 to the Directive's application being "without prejudice to the provisions of the biotechnology Directives 90/219 and 90/220", which had been adopted seven months previously).

Progress of the DG V directive was at BRIC largely a matter for information and report, and it provoked no significant conflict of views. It was valuable nonetheless to other services as an exemplar in handling certain general issues, of which the risk-based classification of biological agents was the most significant.

Member States were required to bring into force the laws, regulations and administrative provisions necessary to comply with the Directive within three years, i.e. by 26 November 1993; and by 26 May 1994, the Council was required to adopt a first list of group 2, group 3 and group 4 biological agents (Directive 93/88, providing this, was adopted on 12 October 1993; initially omitting coverage of genetically modified microorganisms). For exposure to group 1 biological agents, most of the Articles of the Directive (on risk reduction, information to competent authority and to workers, hygiene and individual protection, health surveillance, etc.) do not apply;

but the principles of good occupational safety and hygiene should be observed.

Technical adjustments to the Annexes in the light of technical progress, changes in international regulations or specifications and new findings in the field of biological agents can be adopted by a standard procedure of the Framework Directive 89/391.

BRIC met ten times between the November 1986 communication COM (86)573 (discussed in the preceding section), and the publication in May 1988 of the communication COM (88)160, comprising proposals for two Council Directives: one "on the contained use of genetically modified microorganisms"; and one "on the deliberate release to the environment of genetically modified organisms".

A number of significant points, some of them controversial, had to be resolved in the inter-service discussions. Scientific interests (approximately represented by DG XII, but as there are scientifically competent staff in all the services concerned, none could or did claim a monopoly on scientific expertise) were basically uneasy about the question of scope, and the stigmatization implied by focusing upon the rDNA techniques. Yet it was the progress in these techniques which had been trumpeted in the media; highlighted at Asilomar; financed by industry; and specifically addressed by the 1978–1982 Community deliberations leading to Council Recommendation 82/472. These techniques were the object of the OECD report "Recombinant DNA Safety Considerations" (OECD, 1986), and the ICSU-COGENE "Committee on Genetic Experimentation" (BERNARDI, 1991). The OECD Council Recommendation had recognized that "there is no scientific basis for specific legislation to regulate the use of recombinant DNA organisms", and this had equally been emphasized by ECRAB. But the European Parliament's resolution of February 1987 (on the Viehoff report) had called for a ban on deliberate release of GMOs until binding Directives were drawn up.

The compromise adopted was to emphasize, in the explanatory memoranda to each of the two proposed Directives, that some naturally occurring microorganisms might also be of concern, and that the Commission was "working towards the development of co-herent methods of risk assessment ... and will on this basis examine if and how the accompanying proposal could be modified or extended to cover non-genetically modified organisms". The field release proposed directive similarly referred to the rise of public concern as the reason for an approach "which focuses on the new techniques of genetic engineering" as " the first and most urgent step in the regulatory process; however, this will not impede evolution towards a more organism-related approach. Thus, the Commission will, as experience and knowledge on the matter build-up, undertake to regulate the release of certain categories of naturally-occurring organisms, such as known human, plant, or animal pathogens, and non-indigenous organisms. Moreover, different categories of organisms and/or techniques may be established, allowing different requirements for organisms of different levels of risk".

The definition of "genetically modified organism" in the contained use proposal was essentially still that of the 1978 UK regulations and the 1982 Council Recommendation; in the deliberate release proposal, the techniques were briefly and vaguely defined in a separate Annex 1, as "organisms which can be obtained by such techniques as recombinant DNA, microinjection, macroinjection microencapsulation, nuclear and organel transplantation or genetic manipulation of viruses." The proposal envisaged that the Commission would adapt the annexes of the Directive to technical progress, by amendments concerning "new techniques to be covered or deleted as appropriate".

There was disquiet at the essentially political reasoning for focusing on rDNA organisms, and DG XII continued to develop ideas for a "European Bio-Safety Science Board", similar in function to the US Biotechnology Science Coordinating Committee (see Sect. 5.1). The need for an independent, high-level scientific review mechanism to advise upon and to review regulatory aspects of biotechnology in Europe was discussed at the March 1985 meeting of the Biotechnology Steering Committee, and DG XII was invited to draft and circulate a proposal.

It was envisaged that the principal role of such a Board would be to consider and advise upon basic principles conducive to safe practice in biotechnology; and to promote exchange of information between regulatory authorities and advisory bodies at Community level, in the Member States, in other countries, and in discussion at international bodies. It would define needs for research, and provide a credible source of objective information for all persons engaged in the field, and for the general public.

Although such a mandate was drafted, there was indifference or outright opposition from other services in BRIC and BSC; and Vice-President NARJES (the Commissioner responsible for Research, and Industrial Affairs, 1985–88) did not press the proposal at Commission level. The debate on the drafting of the contained use and field release Directives dominated the time and attention available for inter-service attention to biotechnology, and DG XII itself was preoccupied with the management of the current R & D programmes, and the preparation of their ever-larger successors (at least, for biotechnology), in the following Framework Programme. (The BRIDGE programme proposal, requesting ECU 100 million, for 1990–93, was prepared during 1988; but published only following agreement at Council on the budget envelope for the third Framework Programme.)

Both proposals took as legal base Article 100A, the customary basis for common internal market legislation. Paragraph 2 of this Article indicated that "a high standard of protection for human health and the environment" was to be incorporated in any common market legislation; paragraph 4 correspondingly indicated a rather stringent opposition to national legislation with different standards.

Particularly for the deliberate release Directive, Article 130S, under the new Title VII ("Environment") of the Single European Act, might have seemed the more appropriate legal basis; but this had the tactical disadvantage of requiring unanimity at Council. Article 100A required only a qualified majority; and since the deliberate release Directive would contain a section relating to the placing on the (common internal) market of products, this could justify the corresponding legal basis.

The contained use Directive dealt with the safety in contained use of genetically modified microorganisms (GMMs), including questions of waste and accident prevention. Microorganisms were classified into Group I, on criteria basically identical to those of the OECD 1986 report, for "GILSP" microorganisms; Group II were "non-Group I". It should be noted in passing that the criteria for definition of GILSP drew largely upon the experience and standards of the pharmaceutical industry, and were not necessarily ideal for the definition of good practice or low-risk microorganisms in other sectors, or in basic research. There was nonetheless a tendency to view "Group II" microorganisms as a "higher risk" category, since it included higher risk organisms; but the definition was not precisely risk-based, and would soon give rise to needs for review.

The contained use Directive defined two categories of operation (we use here the language of the Directive as finally adopted; in the case of this Directive, there was little change from the original proposal):

> Type A operation: "shall mean any operation used for teaching, research, development or non-industrial or non-commercial purposes and which is of a small scale (e.g., 10 liters culture volume or less);
> Type B operation: "shall mean any operation other than a Type A operation".

The activities covered by the Directive could thus be summarized in a 2×2 table, depending on the microorganisms (Group I or Group II) and type of operation (Type A or Type B); see Fig. 1.

The Directive defines the role of the competent authority or authorities (to be designated by Member States), the various information requirements, and the role of the committee of representatives of Member States, defined by Article 21. The Article 21 Committee procedure could also adapt Annexes II to V to technical progress.

During the negotiations, the scope of both Directives was defined by Annex I, listing the techniques to be included in, or excluded

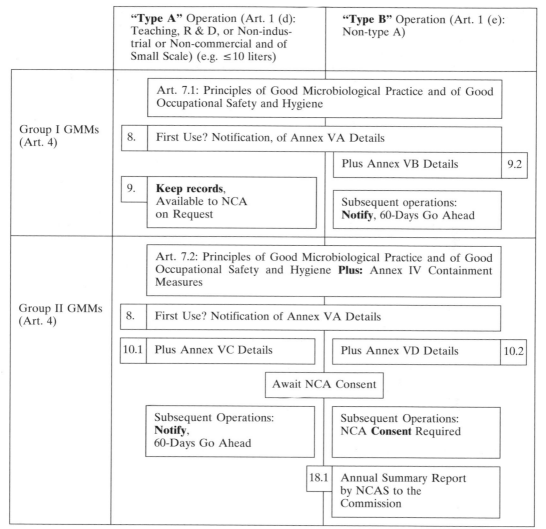

Fig. 1. Outline of Council Directive 90/219 on the Contained Use of Genetically Modified Microorganisms.

from, the definition of "genetically modified microorganism" (as envisaged in the original proposal for field release). But, also in both Directives, Annex I was during the negotiations at Council removed from the scope for amendment by Committee procedure. As the techniques listed were defined in some detail, in terms of inclusions and exclusions seen as relevant at the time of drafting or negotiation, the effect of this was to "freeze" the scope definitions; which elsewhere – particularly in the US context – were the most central topic for discussing the evolution of regulatory oversight requirements.

At a late stage in the negotiations, the legal basis of the contained use Directive was switched to Article 130S, as unanimity among the Environment Ministers appeared attainable, and this title indicated that it was essentially "floor" legislation, on top of which Member States might add national requirements (subject to a general requirement of compatibility with the aims of the EC Treaty).

The contained use and field release Directives were put forward in a single communication by the Commission, had a single rapporteur (Dr. GERHARD SCHMID, German Socialist) in the European Parliament, and generally followed a closely parallel track through the negotiations, to their adoption simultaneously as EEC Directives 90/219 and 90/220 at the meeting of the Council of Ministers (Environment) on 23 April 1990. Implementation in national legislation similarly was set for 18 months later (23 October 1991) for both Directives.

The deliberate release Directive remained based, as proposed, on Article 100A; and is summarized below in its form as adopted.

The preamble emphasizes preventive action; the self-reproducing character of organisms, hence possibility of irreversible effects; the need to ensure the safe development of industrial products using GMOs (i.e., the intent, however apparently restrictive in detail, is fundamentally positive); case-by-case evaluation; and a "step-by-step" approach from research to commercialization.

The Directive is in 4 parts, plus corresponding annexes, and can be summarized as follows:

> **Part A:** General (Articles 1–4): definitions, exclusions (Annex I), a general obligation laid on Member States (MS) to take "all appropriate measures to avoid adverse effects"; and the requirement to define a "national competent authority" (NCA).
>
> **Part B:** For R & D or other activities not involving placing on the market: (Articles 5–9): requires notification to NCA of details as specified in Annex II; NCA transfers to Commission within 30 days, who transmit to other MS within a further 30 days; the original NCA has to respond to the notifier within 90 days.
>
> Where sufficient experience is judged to have accumulated, an NCA may propose to the Commission a simplified procedure. NCAs may also require public consultation. The notifier has to

submit a post-release report on health and environment risks. There will be developed a system for the exchange of summary information.

> **Part C:** For placing on the market of products comprising or containing GMOs (Articles 10–18): consent to marketing of such products is to be given only if the products comply with an environmental risk assessment as specified in Annexes II and III. For products covered by EC legislation providing for a specific environmental risk assessment similar to the provisions laid down in this Directive, the remainder of Part C (Articles 11–18) need not apply. The Commission has to establish by 23 April 1991 a list of such EC legislation (by that date, there was none). Article 11 requires notification to NCA, which (Article 12) has within 90 days to provide a reasoned rejection, or forward the application to the Commission, who transmit immediately to the other Member States. Absence of objection within 60 days leads to automatic approval. If there are unresolved objections, there is a decision at Commission level in accordance with the Article 21 procedure.
>
> **Part D:** Final provisions (Articles 19–24): defines the Commission-level committee decision procedure, and includes provision for adaptation to technical progress of the requirements specified in Annexes II and III.

As was indicated in the previous section, the November 1986 Commission communication indicated a more restrictive approach than was advocated by the representatives of Member States with the greatest experience of biotechnology.

Some concessions to scientific sensibilities were made in the preambulatory and explanatory sections of the May 1988 proposal for the Directives, whose scope and content departed further from scientific and industrial advice.

As the Directives were discussed in the committees of the European Parliament, they acquired an increasing burden of more restrictive amendments. The rapporteur, Social Democrat Dr. GERHARD SCHMID, had a Ph.D. in chemistry, and had worked at the Institute for Biochemistry in Regensburg. But as a politician he was conscious of the growing support for the "Greens"; and therefore, conscious of the acute sensitivity of public opinion to "gene technology" issues, was inclined to a cautious and restrictive approach.

These developments were watched with growing alarm by scientific and industrial circles, in Europe and elsewhere in the world. In an editorial in *Gene* (YOUNG and MILLER, 1989), the Commissioner of the US Food and Drug Association and his assistant had explicitly criticized the Commission's proposed Directive on field release, on three grounds:

- the underlying premises on which it was based;
- the risk of a regulatory approach that would hinder research and development;
- the possible use of its provisions to erect non-tariff trade barriers to foreign products.

Developing the first point, they noted that:

> "The directive is focused on the regulation of "genetically modified organisms (GMOs)", which are defined as those manipulated with only certain recently developed techniques, including recombinant DNA. Thus the directive preferentially singles out for stringent regulation the newest techniques of genetic manipulation that enable the most precise and predictable genetic changes. This is at odds with the broad consensus that these newest techniques represent a clear refinement, an improvement on conventional techniques of genetic manipulation that enhances the precision and predictability of the effects of intervention. Such GMOs are clearly not a functional category, and most certainly not one correlated to risk."

On the second point, they described such a regulatory focus as "likely to become either too burdensome to allow progress, or too superficial to protect abuse"; as a departure from good science and the lessons of experi-

ence; and at odds with the published statements of the OECD and with US policy.

In a similar vein, the US Government through their Ambassador to the European Communities, made known their concerns about the Directives and the amendments adopted by the European Parliament, in a statement dated July 7, 1989, on "International Harmonization in the Biotechnology Field", from which the following quotations are taken:

> "It is in the mutual interest of the EC and the United States to resolve differences in the approaches used for the oversight of Biotechnology. To minimize adverse impacts on research and development, and to avoid the creation of barriers to international trade. In the spirit of achieving international harmonization, we offer the following comments on the proposed EC Directive on deliberate release of genetically modified organisms and the proposed amendments.
>
> The Directive is based on "genetically modified organisms" (GMOs) that are defined by the method of modification. By basing the Directive on the technique by which the organism is modified, the EC is regulating organisms produced by a given process. This is not a functional category directly correlated with the characteristics of the organism. As expressed in the US coordinated framework for the regulation of biotechnology, the US generally regulates products rather than the process by which they are obtained. We are concerned whether differences in approaches and their implementation may lead to difficulties in our attempts to achieve international harmonization. It is important to emphasize that whether an organism is "unmodified" or "genetically modified" is, in itself, not a useful determinant of safety or risk.
>
> We are concerned that the Directive may affect research and development and would ask how those sections of the Directive dealing with research will be implemented. The USG attempts to make the degree of regulatory review commensurate with the level of risk. Does the EC anticipate that it will provide guidance on the degree of review for negligible or trivial risk experiments and provisions for exemption?"

There followed various detailed criticisms or questions about the scientific expertise of the Article 21 Committees; the handling of data developed outside the EC; the handling of confidential business information; and other matters.

On the subject of definitions and scope, already mentioned:

"The definitions on "GMO" and "use" (Article 2) may present a trade barrier if companies do not have a clear understanding of the intended meanings.
The Directive does not appear to allow for the creation of exemptions as experience is acquired. This lack of flexibility in the regulatory structure may impede international harmonization with countries such as the United States where provision for exemption is made."

Finally, the US note comments on the amendments passed at first reading in the European Parliament:

"Several of the amendments passed by the European Parliament appear to be based in part on public concern and unsubstantiated fears about genetically modified organisms, rather than on accumulated scientific knowledge and experience with testing or organisms. If adopted, these amendments could have severe negative impact on research and development, innovation, product development and trade. We would, therefore, specifically urge the Council of Ministers to reject or modify the listed amendments for the following reasons." (There follows a list of various amendments, and arguments against them.)

It was not only from the US that such criticisms were received. A few months after the publication of the Commission's proposals, on 1st October 1988, at its 40th meeting, the Council of EMBO, the European Molecular Biology Organization discussed the regulation of rDNA research in Europe. Members of the EMBO Council explained the current situation in each of their countries. They were unanimously of the opinion that rDNA was a technique, and that any legislation should focus not on the technique but on the safety or otherwise of the products generated with it. Those present drew up and unanimously agreed the following statement for the attention of the EMBO Conference:

"Over the last 15 years, experience has shown that recombinant DNA methods, far from being inherently dangerous, are an important tool both for understanding properties of life and for developing applications valuable to humankind and the environment

(vaccines, diagnostics, pharmaceuticals etc.). EMBO strongly believes that there is no scientific justification for additional, special legislation regulating recombinant DNA research per se. Any rules or legislation should only apply to the safety of products according to their properties, rather than according to the methods used to generate them. Furthermore, EMBO believes that decisions concerning such issues should be taken at a European rather than a national level. At a time when European unification is accelerating, it would be incongruous for research to be illegal in one country and perfectly acceptable in a neighbouring state."

This statement, with a covering letter, was transmitted to members of the European Parliament on 16 May 1989, by Professors MAX BIRNSTIEL, Chairman of the EMBO Council, and LENNART PHILIPSON, Director-General of EMBL, the European Molecular Biology Laboratory. On 18 May, an open letter was addressed to the Presidents of the European Parliament, the EC Council, and the Commission, by the European Nobel Laureates in Medicine and Chemistry (16 in all):

"Dear Mr. President,
In the fourth week of May the European Parliament will debate in Plenary Session a subject which is of vital importance for the future development of science in Europe, namely the three Commission proposals for Council Directives on biotechnology. The Council will decide on this matter at the beginning of June.
Recombinant DNA is a method in biology, without which modern research in this field is not possible. It allows small and well defined changes to be introduced into the genomic set-up of an organism.
More than 90% of research and production use non-pathogenic and safe organisms.
There are well established and internationally accepted safety standards which have been followed by a community of about one hundred thousand researchers in the past 15 years.
The EC Commission has proposed three Council Directives, based on this experience. In principle there is no scientific justification to single out a technique for regulation instead of basing it on the properties of the generated organisms. Consequently, the proposal on "Worker Protection" relates to

"Exposure to Biological Agents" irrespective of the method by which these agents may have obtained their characteristics.

The proposals on "Contained Use" and "Deliberate Release" of genetically modified organisms are in line with already existing OECD recommendations and the guidelines of a number of major countries, such as the USA and Japan.

Amendments have been proposed which are based on unfounded fears rather than on scientific risk assessment. They are both impractical and widely inhibitory to the progress of knowledge and its responsible beneficial applications. We refer in particular to those tabled by the rapporteurs for the Environment, Public Health, and Consumer Protection Committee, as well as those accepted additionally in the same committee.

We would therefore appeal for your support for the Commission's proposals, unamended, in this important debate."

Following the adoption by Council of its "common position" in November 1989, the Directives returned to Parliament for second reading. On 8 February 1990, the Nobel Laureates wrote again to the three Presidents, shortly before the second reading in Plenary Session of the two proposed directives. Having recalled their earlier letter of May 1989, they continued:

"Now that the two directives "contained use" and "deliberate release" have been carried by EC Council and reviewed by the Environment Committee of the European Parliament, it appears to us that they contain a number of provisions relating to research which are both based on non-scientific criteria and so unduly burdensome as to be discouraging. Examples are, with regard to "contained use" measures based on volume used, and with regard to "deliberate release" the requirement that research projects be submitted for examination to authorities at three levels: member State concerned, Community, and all other member States.

If other amendments proposed by the Environment Committee pass the European Parliament at the second reading, this highly worrying situation may become even worse. We are therefore calling your attention to this problem in general, and we rely on your help to ensure that all due objectivity and calmness will be used in dealing with these two subjects."

This letter may have contributed to the narrow defeat of an amendment proposing a five-year moratorium on field releases.

It was during 1988–1989, in parallel with the debate on the Directives, that leading companies became convinced of the need to create a lobbying organization focused specifically on biotechnology; the Senior Advisory Group for Biotechnology (SAGB) was created in mid-1989. In spite of rapid efforts, it was too late to play a decisive role in the months leading up to adoption of the two Directives, on 23rd April 1990. Its efforts nonetheless became increasingly influential from early 1990 onwards, and are described in Sect. 7.1.

4.8 Horizontal or Sectoral Regulation, Process or Product? A Continuing Conflict

"Horizontal" and sector-specific legislation are not necessarily conflicting alternatives. There are many topics on which it is evident that the same rules or standards have to be observed, whatever the sector: for example, rules on financial reporting and taxation, or animal welfare, or worker safety (as in the framework Directive 89/391 and the specific Directive concerning exposure to biological agents, 90/679). Similarly there may be uniform rules for handling or disposing of radioactive materials.

However, the controversy arising in the context of biotechnology, or more specifically, relating to genetic engineering techniques, concerned whether these techniques posed or could – in some research activities or applications – pose risks of a character specifically attributable to these techniques; and if so, whether the assessment and management of those risks required expertise and regulatory oversight of a character essentially similar across diverse sectors. Scientific opinion, while admitting that genetic engineering could make possible gene transfers and functional consequences not otherwise attainable, held that the risk assessment which might in some sectoral applications be desirable or necessary did not differ in essentials from existing risk assessment requirements for the sector; e.g., the assessment of a live recombinant vaccine both for its safety and efficacy in intended use, and for side effects including environmental impacts, are not different in kind from the assessment of a non-recombinant live vaccine. The assessments for a biopesticide, a food product, and a vaccine are, however, very different, given the different circumstances of use and corresponding factors to be assessed.

These general points were central to arguments concerning, and leading to, the changes which occurred between the Commission's original proposal, and the version finally adopted, of the Directive 90/220 on field release. Of these changes, the most significant concerned the first Article of Part C: "Placing on the Market of Products Containing GMOs" [Article 8 in the COM(88)160 proposal, Article 10 in the Directive as finally adopted]. This issue is discussed in the following paragraphs.

Reference had been made repeatedly in the earlier Commission communications, and in the views expressed by industry, to the ability of existing or developing Community legislation specific to particular sectors to cope with products arising in these sectors that happened to have been made by biotechnology. This view was accepted by all Commission services; and consequently Part C of the Directive, whose inclusion was strongly contested by some services, was defined as a "stop-gap" measure for what remained after the opening article defined the exclusions, as follows:

"Articles 9 to 16 of this Directive, do not apply to:
– medicinal products,
– veterinary products,
– foodstuffs, feedingstuffs and their additives,
– plants and animals produced or used in agriculture, horticulture, forestry, husbandry and fisheries, the reproductive material thereof and the products containing these organisms,
– or to any products covered by Community legislation which includes a specific risk assessment."

In the Council Working Party on the Environment, and in the light of Parliamentary Amendments, this comprehensive exclusion from Part C was transformed into the version comprising Article 10:

"1. Consent may only be given for the placing on the market of products containing, or consisting of, GMOs, provided that:
   – written consent has been given to a notification under Part B or if a risk analysis has been carried out based on the elements outlined in that Part;
   – the products comply with the requirements of this Part of this Directive, concerning the environmental risk assessment.
2. Articles 11 to 18 shall not apply to any products covered by Community legislation which provides for a specific environmental risk assessment similar to that laid down in this Directive."

This change reflected a concern that existing legislation – Community or national – for the sectors which the Commission had proposed to exempt, did not in general contain explicit language specific to GMOs; still less the specific language in the Annexes of Directive 90/220. However, the clear preference of Commission and Council for sectoral legislation was expressed by the following declarations, recorded in the minutes of the Council Meeting of 23 April 1990:

Re Article 10 (2)

"The Commission undertakes, when preparing legislation on marketing authorization for products consisting of, containing or which could contain GMOs, to include in its proposals provisions for a specific environmental risk assessment of the product similar to that provided in this Directive. The Commission also undertakes, where appropriate, to propose modifications to existing product legislation in order to provide for such environmental risk assessment."

"The Council and the Commission agree that where any future legislation adopted by the Council provides for a specific environmental risk assessment similar to that provided in this Directive for any product covered by this Directive it will be indicated that

Articles 11 to 18 of this Directive shall cease to apply to that product."

This concession to sectoral legislation nonetheless implied a permanent connection, based on the Annexes of Directive 90/220, between that Directive and all future sectoral legislation for products containing or comprising living organisms.

The assumption thus expressed contrasted with the "sunset clause" approach which was written into US legislative proposals at the same period, as a recognition of the conjectural nature of the risks addressed; and underlined strongly the special character of genetic engineering techniques, which by virtue of their use would impose upon any products concerned, whatever their nature and whatever the genetic modification, a requirement for separate regulatory treatment. The Directive thus strongly contradicted the repeated demands by industry to avoid discrimination against the products of biotechnology, and the statements by the Commission that this would be avoided.

Plant Protection Products:
A Test Case

A specific product category in which this issue arose soon after adoption of the Directives concerned plant protection products (i.e., pesticides or similar). DG VI (Agriculture) had sought for many years to achieve agreement on a Directive concerning the placing on the market of these products. Bio-pesticides, although a small part of the pesticides market, were nonetheless seen as a significant sector, not least because of their potential "environmental-friendliness"; and it was not merely conceivable, but already clear, that improving the efficacy of bio-pesticides would be an early target of the genetic engineer. The world's first (experimental) field release of a genetically modified organism, in 1986, had been of a baculovirus used as a bio-pesticide.

DG XI made clear to DG VI, in September 1990, that as a result of the adoption of 90/220, although the Directive on plant protection products could apply to GMOs, in practice a clearance regarding the environmental impact of a GMO product, "because of its special characteristics", would be a necessary prerequisite for a marketing authorization under the Plant Protection Directive. It was pointed out by DG XI that all field tests of GMO products would permanently have to receive clearance under 90/220; and they argued that the reason for also including environmental risk assessment for GMO products was to ensure continuity between the evaluation at the field test stage and the product marketing stage, and "avoid the need to duplicate the highly expert Committees needed to review the particular environmental risks associated with GMOs".

This argument did not find favor with the Ministers of Agriculture in Council, who adopted the Plant Protection Products Directive 91/414/EEC on 26 June 1991, with a clause recognizing that for a transitional (three-year) period, recombinant bio-pesticides would be assessed under 90/220; but requiring the Commission to prepare for Council a proposal then assimilating the assessment of such products within the procedures for pesticides.

Similar arguments developed elsewhere; for example, in regard to medicinal and veterinary products. The Commission in November 1990 put forward a series of proposals relating to the future system for the free movement of medicinal products in the European Community, in communication [COM(90)283] (EUROPEAN COMMISSION, 1990b). Regarding environmental assessment, this stated:

"in view of the increasing awareness of the effects of medicinal products on the environment, the Commission is proposing that in appropriate cases an application for authorization for a medicinal product for human use include an assessment of the potential risks presented by the product for the environment. The details of the information required will be specified in the Annex to Directive 75/318/EEC on the testing requirement for medicinal products for human use.

So far as veterinary medicinal products are concerned, similar provisions have already been proposed by the Commission and are currently under consideration by the Council [COM(88)779 final]."

Thus the environmental risk assessment

would where appropriate form an integral element of the application for authorization for medicinal or veterinary medicinal products; a procedure consistent with the EC Treaty, and in particular Article 100A, paragraph 2 ("a high standard of protection of human health and the environment") and, in the Title VII (Environment), Article 130R, paragraph 2 ("consideration of environment shall be an integral element of the Community's other policies".

However, perhaps because the Pharmaceuticals division of DG III had been too preoccupied to follow the understandings, developed during the biotechnology debate, about the relation between the "horizontal" [90/220] and "vertical" (sectoral) legislation regarding GMOs, the communication continued:

"so far as live vaccines containing genetically modified organisms are concerned, Council Directive 90/220/EEC of 23 April 1990 shall apply".

This mistake was not repeated when the Food Products division of DG III drafted the proposed Regulation on Novel Foods; but clearly the continuing influence of 90/220 could not be avoided. In the original proposal [COM(92)295] of July 1992, the Article concerned included the following language:

1. Where the food or food ingredient falling under the scope of this Regulation contains or consists of a genetically modified organism within the meaning of Article 2 paragraphs 1 and 2 of Council Directive 90/220/EEC(10) on the deliberate release of genetically modified organisms, the information required in the request for authorization mentioned in Article 6 shall be accompanied by:
   – a copy of the written consent, from the competent authority, to the deliberate release of the genetically modified organisms for research and development purposes provided for in Article 6(4) of Directive 90/220/EEC, together with the results of the release(s) with respect to any risk to human health and the environment;
   – the complete technical dossier supplying the information requested in Annexes II and III of Directive 90/220/EEC and the environmental risk assessment resulting from this information.
   Articles 11 to 18 of Directive 90/220/EEC(10) shall not apply to food or food ingredients falling under the scope of Article 6 which contain or consist of a genetically modified organism.
2. In the case of food or food ingredients falling under the scope of this Regulation containing or consisting of a genetically modified organism, the decision mentioned in Article 6 paragraph 2 shall take account of the environmental safety requirements laid down by Directive 90/220/EEC.

However, the European Parliament made clear in its opinion on the proposal at first reading, adopted 27 October 1993, its continuing preoccupation with genetic engineering. Amendments were adopted demanding a separate authorization procedure under Part C of Directive 90/220 for foods or food ingredients containing or consisting of genetically modified organisms. The Commission could not accept the obvious contradiction with both its declaration to Council of 23 April (see above), and its reiteration of the "single procedure" policy in the April 1991 communication (see Sect. 7.2); but revised the text in terms coming close to a *de facto* separate procedure.

The inclusion of environmental risk assessments appropriate to a product category within the corresponding sectoral legislation had been the understanding inscribed in the Council Minutes of 23 April (see above), and was similarly understood by Member States. For example, a September 1990 UK Government list of "Priority tasks for the implementation of Directives 90/219/EEC and 90/220/EEC" notes under Number 1:

"The establishment of a list of Community legislation for the marketing of GMO products exempt from Part C of the Deliberate Release Directive, Article 10.
The response of the Commission at the July meeting on this aspect was not helpful. The consensus achieved in the negotiation and adoption of this Directive was, in part, dependent on rapid development of adequate product based clearance schemes. The UK believes that "dual" (i.e. both product and technology based) clearance systems should be

avoided and that Part C of the Deliberate Release Directive should operate only as a safety net while product Directives are developed."

Thus by the end of 1990, the new legislation at Community level had initiated two responses. On the one hand, the two new Directives would have to be transcribed into national legislation. Following this, there would be the corresponding establishment of national competent authorities, and the running in of their procedures; and, in parallel, the establishment of the necessary procedures at Community level. The whole process could be expected to take many years.

On the other hand, it was already becoming evident in 1990 that, although scientific and industrial opinion had lost the political battle regarding legislation, regulation and control of genetic engineering and GMO products, the consequences of this defeat would not diminish or disappear. Rather, the progress of the technology world-wide, and the central role of the new knowledge in all advanced areas of application of the life sciences, would inevitably amplify and highlight the consequences.

There was no reason for the speed of advance of knowledge and world-wide diffusion of the technology to be adapted to the rhythm of the legislative and administrative processes within the Community and Member States; the divergence between the two rapidly built up the pressure for review and adaptation. To this we return in Sect. 7, after reviewing some of the other actors and developments at global and national levels which impinged upon developments in Europe and at Community level.

5 Policy Evolution at National Level, in Different Continents, Countries and Cultures

Introduction

The story of modern biotechnology is one of the diffusion of new knowledge from an initially small number of point sources: from individuals or laboratories, and from the key publications describing landmark experiments (or experiments which in retrospect, after some delay – in the case of GREGOR MENDEL, several decades – come to be recognized as such). International scientific networks, and modern means of communication, broadcast and electronic, have in recent times shifted the shape of the diffusion process towards worldwide simultaneous awareness; but operational awareness depends still upon there being in any given laboratory, or in any given country, individuals and groups capable of understanding, digesting, replicating and adapting to their local circumstances and interests the new information and techniques. In the context of Third World development, the term currently in vogue is "capacity-building".

In the following sections, some highlights and main trends or events are described, mainly under country headings. Such a segmentation, natural enough in geographical or political terms, has defects for describing the diffusion of scientific knowledge. However, this chapter is not mainly about the diffusion of the scientific knowledge, but about the history of the regulatory response. Regulatory response has two contrasting characteristics. On the one hand, it has naturally to respond or react to the scientific developments, and its pattern in time and space must therefore reflect the characteristics of the scientific progress. (Ambitiously, it may even attempt to anticipate, as in the case of biotechnology.) On the other hand, regulatory processes are generally determined by national legislatures and executive authorities, driven by and reflecting national and local sensitivities and political priorities.

When scientists or regulators, of similar background training and awareness, are faced with similar challenges, they react in similar ways – all the more so because they are closely communicating with one another, and have objective reasons and effective means for seeking harmonization and efficient sharing of effort. There are two fundamental common factors: a single global scientific community, with a broadly shared understanding of a common scientific method; and our common humanity, with some shared concerns and almost universally acknowledged values, expressed through various solemn declarations, and given institutional form through bodies such as the agencies of the United Nations. Between these, or beyond them, lies everything that makes for individual, local and national diversity.

It would be tediously repetitive to describe the essentially similar ways in which all the successive national authorities responded to the perceived needs for regulatory oversight of biotechnology. In the following sections, the approach is therefore selective, perhaps arbitrarily so, because of the limitations in the materials readily available to the author. The aim has been to focus upon countries which by their scientific scale and eminence, or their unusual transparency or originality of approach, or particular values and sensitivities, illuminate the range of responses, the main historical trends, and the interplay between the local and the general.

5.1 USA: NIH RAC, Existing Agencies, and the Co-ordinated Framework

The dominance of America (i.e., the USA) in scientific, economic and linguistic terms is both a fact and an expository convenience, even when it bruises and irritates other national sensitivities. Sect. 1 has focused upon the first quinquennium after Asilomar, and was inevitably largely devoted to US developments: not because the US is the biggest, or the richest, or the most scientifically gest, or the richest, or the most scientifically

advanced country, but because it was in that country that the debate about the regulation of biotechnology was first played out in depth and in breadth.

In depth, in that it was in America that the leading edge scientists, both American and from other leading scientific countries, had first to participate in the debate, intensively, to the full limits of their abilities and understanding; including not only their scientific understanding, but their ability to communicate in various contexts and in unfamiliar ways. Their understanding and communication abilities had to address simultaneously multiple challenges: from scientific peers, from the general public, the media, and the activists. Science had to address the political community on Capitol Hill, and in State and even municipal authorities. Hence the debate was also "in breadth"; although in those early years there were absentees who would later become significant: the organized voices of industrial or agricultural interests were largely absent – the debate would ultimately have to become broader still. But in the early years, the absence of sectoral economic interests may have contributed to the integrity of the US debate, and consequent international acceptability of its conclusions.

It is easy in retrospect to select and ridicule some of the simplistic early fears and exaggerations; more constructive is to focus upon what emerged from the early American debate, for America and for others. We have sought to emphasize the constructive nature of the messy but open communication process between the scientific and political communities, and would count as regrettable the relative failure of Europe in subsequent years to profit from that debate: "He who does not learn from history is condemned to repeat it." But Parliaments and Congresses are not elected simply to endorse the lessons of history, or to import uncritically experience developed elsewhere. As children growing up or as countries facing contemporary challenges, we demand the right even to make our own "mistakes" – for who has the right to say they are "mistakes"?

With these caveats, we would summarize as follows the main "exportable" lessons emerging from the typically American noisy debate.

The early establishment by the Director of the National Institutes of Health of the NIH Recombinant DNA Advisory Committee (NIH RAC), its openness of debate, and the participation in its discussions of scientists, Federal officials from all the agencies concerned, and activists, built credibility, and confidence in the integrity of the process. This focal point provided not only an excuse, but good reason, to delay the legislative deliberations while the smoke of scientific uncertainties diminished.

The subsequent adaptability of the NIH RAC was no less important. As the successes of the basic science brought applications into prospect – large-scale industrial production, and field releases – the RAC initially addressed the challenges, creating appropriate advisory sub-groups, but then readily ceded oversight to the relevant regulatory agencies. As to which agencies were relevant, and as to whether existing statutes were adequate to the challenges posed by the new science and applications – that was the major US lesson of the early and mid-1980s, culminating in the "Co-ordinated Framework" of 1986, and the steadily growing confidence thereafter that "no new legislation" was the right decision – at least, for the US, given its existing Federal agencies, and their rule-making (and unmaking) possibilities within existing statutory frameworks.

The approach which thus evolved through the 1980s laid a sound basis for continued US scientific advances, and for commercial growth and expansion through the 1990s. There will as a result be progressive and relatively unconstrained assimilation of the new knowledge into the established channels and networks, specific to human needs and activities, in food production, health care, and the relationship with the biosphere.

NIH RAC

To underpin the above brief summary of the US experience, some of the main events and reference materials should be noted. The NIH RAC itself greatly contributed not only to contemporary debate but to the historical record, by the publication of the *Recombinant DNA Technical Bulletin*, from 1978 to March 1993; not only recording the formal business of the RAC, but serving as a focal point for much other information, debate and news relevant to the progress of recombinant DNA research, development and applications.

Formally, the RAC was (and remains) advisory to the Director of the National Institutes of Health. Observance of the Guidelines was an explicit obligation only on institutions in receipt of NIH grants; but such was its perceived authority in both scientific and regulatory circles that other parties conducting such research – in particular, industry – chose voluntarily to subject their proposed activities to RAC review. This may have been for various defensive motives, in the context of either the strategic threat of legislation, or the more specific threat of court action by activists criticizing a company's experiment. The NIH, with the endorsement of the Federal Interagency Committee on Recombinant DNA Research (through which the Federal agencies ensured mutual information and co-ordination of their relevant interests and actions), provided a mechanism for voluntary compliance.

From the start, it was recognized that the Guidelines must be dynamic. ELIZABETH MILEWSKI (1981) published in the *Bulletin* a review of the evolution of the guidelines. The original Guidelines were issued in 1976. The first major revision took place in 1978, after an extended process of public consultation; including the development of the adaptive mechanism itself, which MILEWSKI describes as follows:

> "A proposal to modify any section or to introduce new provisions into the Guidelines must be published for public comment in the *Federal Register* at least 30 days prior to to a RAC meeting. The proposal and any public comments are then discussed by the RAC in open session. A final decision on the proposal is rendered by the Director, NIH, after consideration of RAC recommendations."

Under these procedures, modifications were regularly made thereafter, typically every three months, with occasional re-publications of the Guidelines to take account of the accumulated changes; most recently, in June 1994. In July 1981, they were reissued to reflect a revision exempting from the Guidelines approximately eighty to ninety percent of all recombinant experiments.

The details of the progressive adaptation over the next decade are recorded in the *Bulletin* and in the *Federal Register*. The NIH RAC, after significant involvement in the first field release experiments and related controversies, was able to withdraw from that area as the interests of the Federal agencies with statutory powers were defined and developed. These interests were codified as formal rules under the various relevant statutes through rule-making procedures and public consultation via the *Federal Register* similar to those used for adapting the NIH Guidelines. By 1995, practically the whole of the RAC activity had shifted to the consideration of gene therapy protocols (the *Bulletin* was discontinued in 1993 as its functions could now be covered by the journal *Human Gene Therapy*). Other rDNA research was by that time either exempt; or delegated to local Institutional Biosafety Committees operating under NIH RAC Guidelines; or covered by the rules and statutes of other Federal agencies.

As gene therapy developed, and the number of protocols submitted started to grow rapidly in 1994, the NIH RAC changed gear again, withdrawing from the vetting of protocols for routine experiments with human gene therapy (for whose approval the FDA would remain responsible). From October 1994, only protocols raising novel scientific or ethical questions would be submitted to the Committee; which at the same time expanded its responsibility by acquiring an advisory role to the FDA.

The Coordinated Framework

The inter-agency debates about scope and boundaries of regulatory authority themselves developed in the early 1980s, increasingly requiring central co-ordination. It was therefore from the Executive Office of the President, Office of Science and Technology Policy (OSTP), that proposals for co-ordination were put forward, and ultimately led to executive decisions.

In April 1984, the Administration established an interagency working group, under the Cabinet Council on Natural Resources and the Environment, with the charge to study and coordinate the government's regulatory policy for biotechnology products. Specifically, the group was asked to:

1. Review the regulatory requirements which have been applied to commercialized biotechnologies.
2. Identify existing laws and regulations that may be applicable to biotechnology.
3. Review the function of the NIH Recombinant DNA Advisory Committee and its role in biotechnology commercialization and safety regulation.
4. Clarify the regulatory path that a company with a new product would follow to meet Federal health and safety requirements.
5. Determine whether current regulatory requirements and Federal review are adequate for new products.
6. Develop specific recommendations for administrative or legislative actions to provide additional regulatory review if warranted, while maintaining flexibility to accommodate new developments.
7. Review court rulings regarding the granting of patents for biotechnology.
8. Review other Federal actions such as support of basic research and training, US patents and trade laws, and other policy issues which affect commercialization and US position vis-à-vis international firms.

The results of the initial efforts by this interagency group were reflected in the *Federal Register* of December 31, 1984, where the OSTP published its proposal for a "Coordinated Framework for Regulation of Biotechnology", inviting comments by 1 April 1985. The purpose of the proposal was summarized as being:

> "to provide a concise index of US laws related to biotechnology, to clarify the policies of the major regulatory agencies that will be involved in reviewing

research and products of biotechnology, to describe a scientific advisory mechanism for assessment of biotechnology issues, and to explain how the activities of the Federal agencies in biotechnology will be coordinated."

These formal proposals were taking place against a background of scientific advances, increasing industrial investment activity and company start-ups, occasional high-profile Court disputes by activist groups challenging whether due process was being followed, and continuing Congressional interest. The Congressional Office of Technology Assessment (OTA) was continuing to build a good reputation for its competent and balanced reports, its Director JACK GIBBONS always prudently seeking to ensure that each investigation was based on bipartisan interest expressed by both houses of Congress: the Senate and the House of Representatives. OTA reports contributing significantly to the public and political debates at this period included:

- 1981: "The Impacts of Applied Genetics Applications to Microorganisms, Animals and Plants";
- 1984: "Commercial Biotechnology: An International Analysis".

Mention should also be made of a report (non-OTA) prepared for OSTP in 1983:

- "Competitive and Transfer Aspects of Biotechnology; Commercial Aspects of Biotechnology; Policy Option Papers", 27 May 1983;

and of the Department of Commerce's publication, July 1984, of a report, "Biotechnology", in its series, "High Technology Industries: Profiles and Outlooks".

These titles summarize that the balance of judgement in official circles and among influential advisers was now decisively shifting from concerns about risks, towards concerns about ensuring the exploitation of commercial opportunities, and capitalizing upon US scientific leadership. Flattering references would be made to the strength of science elsewhere, the lightness of their regulatory constraints, and the competitive capabilities of foreign companies, especially Japanese. (The European policy proposals at the same period were making similar references to the US situation – see Sect. 4.1; but were written in DG XII, Science, Research and Development, read only in research Ministries and scientific circles if at all, and did not achieve comparable political influence.)

The cited Department of Commerce report gives "a summary of the issues which emerged as being of the most immediate concern to the successful development of biotechnology products and consequent US international competitiveness"; the list commences with safety oversight, and a compliment to the NIH RAC:

"Industry representatives believe that the uncertainty over future oversight of safety issues related to commercial recombinant technology is creating delays in product testing, approval, realization of revenues, investment in products and in companies whose product lines may become subject to future regulation. Industry representatives would like to see the National Institutes of Health's Recombinant DNA Advisory Committee (RAC) expand its role as overseer of recombinant DNA research protocols into all areas that may be impacted by recombinant DNA applications, including the testing and production of recombinant DNA products. Whether or not the RAC continues/expands its role, a consensus emerged that additional regulatory committees should not be created within the government."

In June 1986, the OSTP announced again in the *Federal Register* (FR) the "Coordinated Framework for Regulation of Biotechnology", as refined in the light of comments received following the 1984 publication; which had also contributed to refinement of the interagency coordinating mechanism. The publication invited further public comment on certain new concepts which would be the subject of rule-making. Since the 1984 publication, the interagency group under the former Cabinet Council on Natural Resources and the Environment had been succeeded by a "Domestic Policy Council Working Group on Biotechnology". This Working Group was responsible for the Coordinated Framework, and also considered policy matters related to agency jurisdiction, commercialization, and international biotechnology matters.

The BSCC, and Interagency Coordination

At a more scientific level, there had been established in October 1985, under FCCSET, a "Biotechnology Science Coordinating Committee (BSCC)", charged with an advisory and coordinating role across Federal agencies undertaking research in biotechnology. FCCSET, the Federal Coordinating Council for Science, Engineering and Technology, was a statutory interagency coordinating mechanism managed by the OSTP. Chaired by DAVID KINGSBURY of National Science Foundation, the BSCC brought together senior staff from all the agencies principally concerned: the NIH, the FDA (Food and Drug Administration), the USDA (US Department of Agriculture) and the EPA (Environmental Protection Agency).

One of the primary activities of the BSCC had been the development of definitions, because a common scientific approach was seen as a fundamental requirement for a coordinated Federal regulatory framework. The June 1986 FR publication invited public comment upon BSCC-proposed definitions of "intergeneric organism", and "pathogen". These definitions would describe the combinations of genetic material that would cause a modified organism to come under review. Among the proposed exemptions from review, the BSCC proposed "genetically engineered organisms developed by transferring a well-characterized, non-coding regulatory region from a pathogenic donor to a non-pathogenic recipient". This particular point was a focus of later controversy, particularly in view of the importance to plant genetic engineering of the Ti (tumor-inducing) plasmid from the plant pathogen *Agrobacterium tumefaciens*.

The BSCC was also attempting to define what constituted "release into the environment", and was establishing a working group on greenhouse containment and small field trials in order to develop scientific recom-

mendations. It may be noted that, here as elsewhere, the work and preoccupations in the US regulatory debates interacted closely with the deliberations at the OECD (see Sect. 6.2); the key actors in the interagency debates in Washington also travelled regularly to Paris for meetings of the Group of National Experts on Safety in Biotechnology.

The FR publication describing the Coordinated Framework contained a section on "International Aspects". This emphasized US interest in the promotion of international scientific cooperation and understanding of scientific considerations in biotechnology; and in the reduction of barriers to international trade. It announced that the US was "seeking recognition among nations of the need to harmonize, to the maximum extent possible, national regulatory oversight activities concerning biotechnology". In this connection, the document referred to the OECD expert group responsible for the report, "Recombinant DNA Safety Considerations for Industrial, Agricultural and Environmental Applications of Organisms Derived by Recombinant DNA Techniques" (OECD, 1986); and quoted an extended summary of the main points of this report.

The Coordinated Framework announcement, a 93-page document (FR 51/21302), included policy statements from the Food and Drug Administration (FDA); the Environmental Protection Agency (EPA); the US Department of Agriculture (USDA); the Occupational Safety and Health Administration (OSHA); and the National Institutes of Health (NIH).

The 1986 FR publication did not end public or interagency controversy about the regulation of biotechnology. The Reagan Administration had made a fundamental policy choice, in determining that biotechnology could be safely regulated under existing regulatory authorities; and the "Coordinated Framework" was the justification and exposition of that approach. SHAPIRO (1990) in an extended and scholarly review of biotechnology regulation (including 420 references or footnotes in his 70-page study), summarizes as follows the potential problems of the Coordinated Framework:

"This structure poses three challenges to effective regulation. First, different agencies may have concurrent jurisdiction over the same experiment or product. While this problem is not unique to biotechnology, it deserves attention here because research and development in the still-infant biotechnology industry may be discouraged by the costs of duplicative regulation. Second, reliance on existing statutes has left gaps in the regulatory coverage of biotechnology: some products or types of research are not regulated by any agency. Finally, EPA has been forced to regulate biotechnology under authority given to it by Congress to regulate chemical substances. For this reason, EPA may lack sufficient regulatory authority to regulate some aspects of biotechnology."

In spite of these criticisms, SHAPIRO writing four years later, or six years after the original (1984) publication, comments that "Most observers to-day give the Coordinated Framework generally good marks for the road map it provides to agency responsibility for regulating biotechnology"; a judgement which many would still support in the mid-'90s. On the first criticism, of regulatory redundancy, agencies have through various procedures coordinated their interactions, and are meeting statutory deadlines for processing applications. Regarding regulatory gaps, there are some gaps over genetically engineered animals; but not large, e.g., there would be relevant statutes if the animal is a plant pest, an agent for animal or human diseases, or sold for food. On the third point, regarding adequacy of regulatory authority, EPA's use of TOSCA (Toxic Substances Control Act) to regulate microorganisms leaves on the agency the burden of proving that the "chemical" (for so microorganisms were defined, to allow the use of this Act) presents "an unreasonable risk of injury to health or the environment".

The 1986 Framework clarified many aspects of interagency coordination, but not all. The FDA had simply stated that they did not need to establish new procedures or requirements for the review of new biotechnology-derived products. However, the two other major regulatory agencies – USDA and EPA – announced policies for developing additional rules and guidelines.

In July 1987, after public notice and comment, USDA's Animal and Plant Health Inspection Service (APHIS) published final regulations addressing "the introduction of organisms and products altered or produced through genetic engineering which are or which there is reason to believe are plant pests". These regulations required case-by-case review of all organisms produced through recombinant-DNA techniques if the donor organism, recipient organism, or vector was listed by USDA.

EPA had proposed (in the Coordinated Framework) to develop new regulations that would give

> "… particular attention, under both FIFRA (Federal Insecticide, Fungicide and Rodenticide Act) and TSCA (Toxic Substances Control Act), to microorganisms that (1) are used in the environment, (2) are pathogenic or contain genetic material from pathogens, or (3) contain new combinations of traits."

From its inception (in 1985), the BSCC had been troubled by disputes concerning the scope of regulation of biotechnology. One of the most controversial concerned BSCC's proposed exemption of the case referred to above – organisms created by the addition of well-characterized non-coding sequences from a pathogen. On this, EPA's proposed rule was in conflict with the opinions of several other agencies at the BSCC; but in a court case brought against OSTP by the Foundation on Economic Trends, the BSCC had responded (in late 1986) that the definitions published in the Coordinated Framework were "not binding". The BSCC's standing was further damaged when the Chairman, DAVID KINGSBURY, was charged in October 1987 by a Congressional committee with a conflict of interest (having been a director of a biotechnology company at the time when he drafted the scope/exemption definitions), and subsequently resigned.

In response to EPA's insistence on the above point in its TSCA rule, the following Chairman of the BSCC (JAMES WYNGAARDEN, ex-Director of NIH, then at OSTP) wrote to the Office of Management and Budget (OMB) requesting them to withhold authorization of the EPA proposed rule, until the "other (BSCC) committee members could review the proposed rule from a scientific perspective". EPA remained obdurate, claiming a division of scientific opinion on the issue; and WYNGAARDEN again wrote (August 1988) to OMB, indicating that all other agencies opposed the EPA proposal, as "inconsistent with the conceptual basis of the Coordinated Framework", because they did not "use scientifically determined likelihood of risk bases for categories of regulated microorganisms". EPA refused to attend further meetings of the BSCC, thus blocking the achievement of consensus on the appropriate scope for Federal oversight.

Through 1989, the BSCC continued to wrestle with the scope issue, through a "Scope Subcommittee" chaired by the USDA Assistant Secretary for Science and Education, CHARLES HESS. Meanwhile, although industry complained about the ongoing dispute, and associated regulatory uncertainty, the interagency bickering was becoming seen as esoteric, philosophical, and in fact not holding up significantly a growing flow of field release research applications to both EPA and USDA. EPA's draft guidelines had been made public, and they continued to review all field releases of GMOs on a case-by-case basis.

The disagreement and polarization of opinion at BSCC also limited its ability to function as an advisory body to the State Department in connection with international activities – in particular at OECD, in connection with the March 1990 discussion document on "Good Developmental Practices (GDP)" for small-scale field trials (see Sect. 6.2).

Separating Research and Policy: The Council on Competitiveness

Analyzing the failure of BSCC, SHAPIRO emphasizes the failure to distinguish science from policy issues; and in effect, the response of the Administration was to shift the policy responsibility to the White House Council on Competitiveness, leaving research coordina-

tion to the Biotechnology Research Subcommittee of a reconstituted FCCSET Committee on Life Sciences and Health (CLSH). On the research side, the FCCSET CLSH produced in February 1992 an impressive report, "Biotechnology for the 21st Century", outlining how and why the US Federal Government through its various agencies would spend $ 4.03 billion on biotechnology R & D in Fiscal Year 1993. The preface, addressed to Members of Congress, was signed by OSTP Director ALLAN BROMLEY. Unperturbed by the change of Administration, the same subcommittee produced in June 1993 a similarly handsome report, "Biotechnology for the 21st Century: Realizing the Promise", with a budget of $ 4.3 billion, the preface now signed by President CLINTON's Science Advisor, JACK GIBBONS.

To resolve the interagency regulatory dilemmas, the Bush Administration instituted in 1990 a review of regulatory issues under the Council on Competitiveness. In August 1990, four "Principles of Regulatory Review of Biotechnology" were announced, as guidance for Federal agencies:

"1. Federal government regulatory oversight should focus on the characteristics and risks of the biotechnology product – not the process by which it is created.

2. For biotechnology products that require review, regulatory review should be designed to minimize regulatory burden while assuring protection of public health and welfare.

3. Regulatory programs should be designed to accommodate the rapid advances in biotechnology. Performance-based standards are, therefore, generally preferred over design standards.

4. In order to create opportunities for the application of innovative new biotechnology products, all regulations in environmental and health areas – whether or not they address biotechnology – should use performance standards rather than specifying rigid controls or specific designs for compliance."

Change of Administration, Continuity of Policy

The change from a Republican to a Democrat President with the CLINTON victory in the November 1992 election did not alter basic US national interests, nor the regulatory philosophy for biotechnology whose evolution has been outlined above. A specific illustration was provided by a USDA proposal published in the *Federal Register* on 6th November 1992.

Referring to the preamble published with the 1987 final regulations, the 1992 notice recalled that "APHIS stated its intention to modify or amend the regulations to ensure flexibility and to remove restrictions when warranted". The main point of the new notice was a proposal:

> "to provide for a notification process for the introduction of certain transgenic plants with which the (APHIS) has had considerable experience. ... This document also proposes to amend the regulations to allow for a petition process allowing for a determination that certain transgenic plants are no longer considered regulated articles. The proposed amendments would provide a procedure for filing a petition for determination of nonregulated status for those organisms which do not present a plant pest risk and therefore should no longer be regulated articles."

Subject to various conditions concerning the nature of the introduced genetic material, the new procedure proposed to introduce this simplified procedure for six major crop species, for which it was felt that sufficient experience had accumulated: corn, cotton, potato, soybean, tobacco, tomato. In March 1993, these rules were published in final form, with only minor modifications, in response to comments and the judgement of the incoming Administration (some details were revised, e.g., a 30-day notice period, rather than zero).

In fact such a simplification of procedure was becoming inescapable, given the acceleration of requests for field trials and the hiring freeze on additional staff at USDA. Release permits had roughly doubled every year from 1987 onwards, reaching 51 in 1990, 90 in 1991, 160 in 1992, 306 in 1993, and 521 in 1994 (to end-November). With the new system in place, 189 of the 1993 field trials and 454 of those in 1994 were notifications; thus reducing the authorizations to 67 of the 521 total. Given the multi-site nature of the requests, the 1994 figures corresponded to releases at 1708 sites, 1490 covered by notification.

A paper on renewal of the American economy through technology was published by the President and Vice-President in February 1993. This and subsequent speeches by GREG SIMON, Domestic Policy Adviser to the Vice-President (and a former Congressional staffer who had worked intensely on drafting an "Omnibus Biotechnology Act"), made clear the continuing determination of the US under the new Administration to maintain leadership in biotechnology. With the continued safe track record of the technology, the scale of Federal research expenditure, and the growing political perception in Congress of the significance of the technology for US international competitiveness (there is a bi-partisan biotechnology caucus of some one hundred Congressmen), the trend towards simplification and lightening of regulatory burden seems likely to persist.

5.2 UK: From GMAG to ACGM and ACRE; Statutory Developments in 1990–1992; the House of Lords Report and Government Response

The United Kingdom, birthplace of the double helix discovery and still a significant force in science, was a major player in the first decade of the biotechnology regulatory debate. Subsequently, its role had to be played in conformity with the rules of the European legislative frameworks.

GMAG: Creation, Mandate and Activities

Reference has been made in Sect. 1 to the establishment in December 1976 of

"GMAG", the "Genetic Manipulation Advisory Group"; whose "advice " regarding the conduct of rDNA research was in practice given statutory force by the requirement of the Health and Safety Executive (under the authority of the 1974 Act on Health and Safety in the Workplace), that research conform to the GMAG Guidelines. GMAG itself was created by the Department of Education and Science, as the Ministry then responsible for funding university research and the activities of the Research Councils; its secretariat was provided by the Medical Research Council.

Its membership comprised scientists, civil servants from potentially interested government departments, and "public interest" representatives, including trade union representation and lawyers; innovations in social policy for the oversight of science which attracted then and subsequently considerable academic interest, and general approval for its "transparency". Its meetings were nonetheless conducted in closed fashion, its participants required to sign the Official Secrets Act. The terms of reference were:

(1) to advise a) those undertaking activities in genetic manipulation, including activities relating to animals and plants, and b) other concerned individuals and organizations;
(2) to undertake a continuing assessment of risks and precautions (particularly any new means of physical or biological containment) and of any newly developed techniques for genetic manipulation, and to advise on appropriate action;
(3) to maintain contacts with relevant government departments, the Health and Safety Executive, and the Dangerous Pathogens Advisory Group;
(4) to maintain records of containment facilities and the qualifications of Biological Safety Officers; and
(5) to give advice on general matters connected with the safety of genetic manipulation, including health monitoring and the training of staff.

GMAG's activities were similar to those of the US NIH Recombinant DNA Advisory Committee. It initially advised on individual proposals, and on the establishment of guidelines for risk assessment of the conjectural hazards of genetic manipulation. It also provided guidance for the establishment of local Genetic Manipulation Safety Committees.

GMAG developed guidelines regarding appropriate levels of physical and biological containment within laboratories, for different types of organisms and experiments; and as experience accumulated without adverse incidents, the guidelines were progressively relaxed. There was close awareness of the evolution of scientific progress worldwide, and of the regular adaptation of the NIH RAC Guidelines.

However, with the successful progress of research and development, it was evident in the early 1980s that industry would wish to proceed to large-scale fermentation production, and use, of genetically-manipulated microorganisms. Progress in the genetic engineering of plants was similarly promising, raising the imminent prospect of field releases of genetically modified plants and microorganisms. But as genetic engineering developed towards more routine acceptance and applications, the position of GMAG under the DES began to look anomalous.

For example, by September 1982, in GMAG's third report, the Secretary of State for Education and Science, Sir KEITH JOSEPH was writing in the Preface:

"Since the Second Report [of GMAG] was published in December 1979, much scientific information on the biological applications of recombinant DNA technology has accumulated. Because the original fears about the work have so far not been substantiated and because of the very responsible attitude of those who carry out the work, this Report explains that the group were able, in early 1980, to introduce a risk assessment scheme for the categorization of experiments, and within a year to advise a further modification of the procedures governing the notification of individual experiments. ...
The Group have also revised their procedures for giving advice on the large scale use of genetically manipulated organisms. The Report records that the Group's advice on that area of work will in future be restricted to the biological properties of the organism being used; the Health and Safety Executive

will, as part of their responsibilities under the Health and Safety at Work Act 1974, continue to be concerned with the physical aspects of safety."

Such developments led the government to institute a review of GMAG by civil servants from DES and the Health and Safety Executive (HSE). The resulting report (April 1983) concluded that:

"It appears clear that the HSC (Health and Safety Commission) and the HSE (Health and Safety Executive) will continue to need advice in order to discharge their responsibilities, and that Health, Agriculture and Environment Ministers may well require advice on technical advice in their fields. There are good reasons for reconstituting the present GMAG as an HSC Advisory Committee with provision for advice to be available to Ministers of other Departments, and there is considerable support within government and outside for this solution. The likely structure and functions of an HSC Advisory Committee and the way in which it would work, are set out in the report. The conclusion of DES and HSE officials is that such a committee would be useful and workable and that, subject to the agreement of Ministers and the HSC and to the views of interested organisations, it should be established in place of GMAG as soon as possible."

ACGM: Terms of Reference

GMAG was in 1984 replaced by the Advisory Committee for Genetic Modification, with the following terms of reference:

"A. To Advise the Health and Safety Commission and the Health and Safety Executive, in Connection with Their Responsibilities Under the Health and Safety at Work Act 1974 on

(i)   the general standards of safe working to be observed by those under-

taking activities relating to genetic manipulation

(ii)   the categorization of experiments

(iii)   exemptions from the Health and Safety (Genetic Manipulation) Regulations 1978

(iv)   the assessment of risks and precautions (and in particular of any new methods of physical or biological containment) and of any newly developed techniques for genetic manipulation

(v)   at the request of HSE, the specific precautions necessary in individual cases of experimental work

(vi)   at the request of HSE, the biological aspects of individual cases of the non-experimental use of products of genetic manipulation

(vii) health monitoring and training of those undertaking genetic manipulation activities

(viii) the nature of any controls to be applied generally in laboratories and other workplaces engaged in genetic manipulation by way of regulations, codes of practice and guidance

(ix)   such other matters as may be referred to the committee by the HSC or HSE

B.   To Advise the Health, Agriculture, Environment, Industry and Northern Ireland Ministers

on such matters relating to genetic manipulation as may be referred to the committee by those ministers and offer comment on the technical or scientific aspects of any new developments in genetic manipulations which may have implications for their departments."

As sensitivity to the adverse public image of the word "manipulation" developed, the title was subsequently revised to "Advisory Committee on Genetic Modification". The ACGM continued over the following decade and currently to fulfil the advisory role thus defined. This mandate proved to be robust, and the operational performance was widely appreciated for its competence, including innovations and adaptability as necessary. As a result, the ACGM remained the central expert advisory body in the UK throughout a period of rapid expansion and innovation in biotechnology, which saw various major reports and legislative events at national and European Community levels.

The Committee focused largely on contained research and industrial work involving GMOs, its general approach including:

- the establishment of properly constituted genetic manipulation safety committees at each location where such work was conducted;
- the dissemination of guidance on procedures and risk assessment;
- the inspection of laboratories and advice on good practice.

ACGM established a working group to develop guidelines on the release of GMOs in the environment, including both ACGM members and additional experts and civil servants. The working group later became established as the "Intentional Introductions Sub-Committee of the ACGM", following the issue, with the approval of the HSC, of its guidelines. These guidelines, issued in April 1986, recommended that:

- the HSE should be notified of any proposal to release GEOs;
- the notifier, when making his initial assessment of the environmental consequences of a release, should be advised by an appropriately constituted local body including relevant scientific expertise and, where appropriate, a local environmental health officer; and
- a case by case examination of proposals should be carried out on behalf of the HSC on the basis of risk assessment material provided by the proposer in accordance with the guidelines.

These arrangements were implemented by the HSE largely on a voluntary basis, which in 1989 became statutory with the issue of the new Genetic Manipulation Regulations (see below).

Report of the Royal Commission on Environmental Pollution

The UK was (and remains) an active participant in the discussions of biotechnology safety and regulation in international fora, particularly in the European Community and in the OECD; often contributing to these from a basis of strong national expertise and interests. An important example was the role played by the long established Royal Commission on Environmental Pollution (RCEP). This Commission, under a new chairman at the start of 1986, Sir JACK LEWIS (Professor of Inorganic Chemistry in the University of Cambridge, and past President of the Royal Society of Chemistry), decided on its priority tasks over the following year or so, and included among these "the release of genetically engineered organisms into the environment". An invitation for evidence was issued in July 1986; their report was presented to Parliament three years later (RCEP, 1989).

The report gave a comprehensive overview of the subject of genetic engineering, its applications, and its regulations. It reviewed, albeit briefly, developments in many countries; noting Denmark's legislation; Germany's Catenhusen report (which had appeared in 1987); the US "co-ordinated framework"; and the related arguments. It summarized developments at the OECD, and in the European Community. Given the timing, the RCEP commented closely on the contemporaneous proposals for the EC Directives: both on contained use, and more particularly that on field release.

The RCEP report gave a strong and unequivocal recommendation in favor of statutory control of releases; the first five of its 67 recommendations were as follows:

"1. Statutory control of releases of genetically engineered organisms (GEOs) to the environment must be put in place.
2. Both the Secretary of State for the Environment and the Health and Safety Commission (HSC) (acting on behalf of the Sec-

retary of State for Employment) should be involved in decisions on release.

3. The Secretary of State for the Environment should take primary responsibility for control with respect to the environmental consequences of such releases.

4. The control of releases of genetically engineered organisms should be governed by a statute establishing controls in respect of environmental protection and providing a framework within which the Secretary of State would be empowered to make regulations including a system for licensing. The statute should, in addition, impose a duty of care obliging all those responsible for the release of a GEO, whether for experimental or commercial purposes, to take all reasonable steps for the protection not only of human health and safety but also of the environment.

5. A licence, which we refer to as a release licence, should be required before the release of a genetically engineered organism may also take place. It should be an offence, carrying a substantial penalty, to release a GEO without having first obtained a release licence or to fail to comply with any conditions attached to the licence."

The report strongly endorsed the EC's proposed Directive on field release, noting extensive similarities to its own conclusions and recommendations; its main criticisms of the Directives being their inadequacy in two respects:

(i) the proposed sectoral exemptions, in areas where Community-wide legislation did not exist – pesticides were particularly cited;

(ii) insufficient recognition, in the provisions for Community-wide market authorization, that the environmental risk assessment must be specific to local environments.

Environmental Protection Act, ACRE, EC Directives, the 1992 Regulations

The Government did not fully accept these recommendations, making clear in its positions at the EC Council Meetings its view favoring sectoral legislation. However, in response to the RCEP report, it included as Part VI of the Environmental Protection Act (1990) specific regulations relating to environmental protection from the possible risks of releasing GMOs. This Act required a safety assessment to be made and submitted to the Department of the Environment, in certain cases requiring a consent. The Advisory Committee on Releases to the Environment (ACRE) was created from the ACGM's Intentional Introductions Sub-Committee, as an advisory body to the Department of the Environment.

Given the RCEP report (1989), the adoption of the EC Directives on contained use (90/219) and field release (90/220), and the Environmental Protection Act (1990), the Government updated the HSE regulations (based on the 1974 Health and Safety at Work Act) and published new field release regulations under the Environmental Protection Act, thus giving guidance on the implementation of the latter and fulfilling its obligations under European legislation. The two sets of regulations, which are the current basis for the regulation of biotechnology in the UK, are:

● the Genetically Modified Organisms (Contained Use) Regulations 1992 (HMSO, 1992a);
● the Genetically Modified Organisms (Deliberate Release) Regulations 1992 (HMSO, 1992b).

The HSE and the DoE back these up with further clarificatory documentation, in the form of guidance (HMSO, 1993; DOE, 1993) and the "ACGM Newsletter" published by HSE.

The development of biotechnology in the UK has been rapid, and the advisory bodies ACGM and ACRE have developed a reputation for integrity, and scientific pragmatism. The UK has actively initiated in the EC Committee concerned, proposals for simplified procedures for various classes of field release, on the basis of ACRE's advice; simplifications for which the pressures have come from the UK scientific communities, academic and industrial; and from the consideration of the impact of regulations on international competitiveness.

House of Lords Report, 1993, and Government Response, 1994

The latter topic was the subject of particular scrutiny by the House of Lords Select Committee on Science and Technology, who established a subcommittee to conduct an enquiry into "Regulation of the UK Biotechnology Industry and Global Competitiveness". The sub-committee was chaired by Lord HOWIE OF TROON, and its report was published in October 1993, with an accompanying press release, volumes of evidence received, written and oral, and considerable publicity (HOUSE OF LORDS, 1993). The report was quite blunt, the Committee concluding that:

"– Early fears of scientists relating to genetically modified organisms (GMOs) in contained use turned out to be unfounded. As a general principle, except where pathogens are involved, separate regulation of GMOs in contained use is unnecessary over and above current good laboratory practice; and deliberate release of GMOs, except where bacterial or virus vectors, live vaccines or modification of the genome of animals are involved, is not inherently dangerous.

– United Kingdom regulations, which are in turn based on E.C. Directives, take an excessively precautionary line based on a view of the technology, which, in terms of scientific knowledge, was already obsolescent when the directives were being prepared in the late 1980s. In framing them the Commission appears to have been impervious to advice tendered to it by scientists, industry, and national experts. Not only does this place the UK at a competitive disadvantage *vis à vis* principal overseas competitors like the USA and Japan but even within the EC implementation is so uneven as to create serious inequalities."

By way of remedy, the Committee recommended that:

– "The Government must press for amendment of the EC contained use Directive so as to substitute a risk assessment system in place of the current classification of risk according to size of operation and pathogenicity; meanwhile, as interim measures, use of safe organisms should be subject only to a simple notification procedure whatever the scale of the operations; and the Health and Safety Executive should aim to give consent for use of unsafe organisms well within the 90 day maximum.

– The Byzantine structure of deliberate release regulation must be reformed so as to enable certain activities, as selected by a group of EC national experts, to be exempt from the present provisions; meanwhile, as interim measures the current number of questions to be addressed in the risk assessment questionnaire should be reduced by making them specific to the type of organism involved; applications should be processed in not more than 30 days; and universities and research councils should be exempt from paying fees on their applications.

– As a matter of principle, GMO-derived products should be regulated according to the same criteria as any another products. The present process-based system should be retained only for the limited areas where regulation is required – that is to say all work involving pathogenic organisms and for deliberate release of GMOs outside the low to negligible risk category; work on further process-based regulatory EC draft Directives on GMOs should cease forthwith.

– Promotion of public understanding of biotechnology is important but should not be carried out so far as to preclude evolution of regulation; because of its implications for competitiveness, DTI is ultimately responsible for ensuring that public perceptions are based on reason and knowledge."

In spite of the critical tone of the House of Lords report, the government's published response in a White Paper (DOE, 1994) was to express its broad agreement with most of the observations and recommendations, and to emphasize the ways in which it was moving in the direction advocated. This was in part because the publication of the report coincided with a period when the UK government was emphasizing on a broad front its commitment to a deregulatory philosophy regarding social legislation, the cost of which was claimed to be a significant factor in Europe's relative loss of competitiveness. The members of the House of Lords Committee were at pains to emphasize that there had been no deliberate intention on their part in this sense. Some quotations summarize the tone of the government response:

"The Government welcomes this timely and detailed report from the Select Committee on how regulation of biotechnology affects the competitiveness of the United Kingdom industry. The section of the enquiry dealing with the difficult issue of public understanding and acceptance of this fast moving technology is welcomed.
The Government believes that modern biotechnology has far-reaching implications for the UK economy and will have a major impact on products and processes across a wide range of sectors. It is a key enabling technology which has the potential to enhance significantly the competitiveness of many sectors of British industry. The Government recognizes the need to get both an appropriate regulatory framework and the right climate for investment to enable UK industry to remain competitive. Sensible regulation is crucial and, in line with Government policy, should be founded on the best scientific evidence available and be proportionate to any risk involved.
...
The Government considers that the regulatory system now existing in the UK leaves industry well placed to compete in global markets but recognizes that industry has not always shared this perception. However, rapid evolution has taken place in biotechnology regulation in the UK and elsewhere over the past few years, resulting in the introduction of fast track procedures for clearance of some deliberate releases of genetically modified organisms (GMOs) and simplification of notification procedures governing the contained use of GMOs.
...
Similarly, the evolution from process- to product-based regulation of GMOs, which the Committee supports, was foreseen during the drafting of the two relevant EC directives. The Government has been at the

forefront of discussions in Brussels to ensure that GMO products are regulated together with equivalent products derived by conventional means. The safety criteria applied to these products should not be affected by the way they are subjected to regulation. The Government will continue to press in the European Community for the evolution from GMO regulation to a product-based approach where this is practicable.
...
Because the need for flexible approach to regulation in this field was anticipated from the outset, the Government is able not only to endorse but also to implement the great majority of the recommendations in the report ..."

Similar remarks referred to the evolution of EC-level legislation and the Biotechnology Co-ordination Committee. Responding to the specific accusation by the House of Lords Select Committee that:

"In framing the Directives on which the United Kingdom regulation are based the European Commission took an excessively precautionary line which, in terms of scientific knowledge, was already out of date when they were being prepared in the late 1980's. Advice to that effect appears to have been ignored",

the Government replied that:

"Biotechnology is a fast-moving technology and it is therefore appropriate that legislative and administrative controls regulating the use of GMOs are in a state of continuing evolution. As scientific experience accumulates, regulations will be reviewed and tightened or loosened as is appropriate to risk. It is therefore no surprise that the Select Committee views some of the detailed provisions of the two EC directives governing the use of GMOs now to be dated and the Government agrees with this view. However, notable progress has been made in revising aspects of each directive in recent months, and the Government is committed to continuing to pursue the development of European legislation in line with the best scientific advice available."

The Government endorsed the view of the Committee regarding the importance of the work of the European Community's Biotechnology Co-ordination Committee.

5.3 Germany: The Catenhusen Enquiry, the 1990 Gentechnikgesetz, and the 1993 Revision

Germany, the largest and economically most powerful country of Western Europe and of the European Community or Union, has acted as a leading scientific and industrial force in biotechnology, and also as its sternest and most conscientious or concerned critic; but not simultaneously, nor from the same internal constituencies. In the early 1970s, it was a leader. Through the 1980s, German sensibilities, and the pressure of German public and political opinion expressed at three levels – the Land (or, at least, certain Länder); the federal Parliament, in both chambers; and the European Parliament – were the most conservative and obstructive factors affecting biotechnology in Europe, contributing to the construction of a restrictive regulatory framework. Finally in the 1990s, with all the zeal of a reformed sinner, German industry and government have been again in the leading position, pressing for reform and simplification of biotechnology regulation, at national and European levels; particularly during the German Presidency of the Council of Ministers, July to December 1994.

Such a brusque summary runs the risk of caricature. The particular intensity of the public debate in Germany reflected a close fusion of two strands of discussion which in other countries were for most of the time more or less separate: on the one hand, the usual and necessary deliberations and discussions on technical aspects of risk assessment in scientific and industrial contexts, including the ecological uncertainties associated with field release; on the other, the wide-ranging and troubling analysis and debate concerning ethical aspects arising in the use or possible abuse of the new knowledge and techniques, in areas such as pre-natal diagnosis, *in vitro* fertilization, and "human genetic engineering", i.e., gene therapy, somatic and germ-line; and extending to fundamental philosophical debates about the relation between man and "Nature".

Reaction to the disastrous and tragic experience of the nazi period led in the period of postwar reconstruction to the creation in Germany of a federal constitution and a judiciary particularly attentive to separation of powers and to propriety of procedure. Throughout the developed world, as KEVLES has particularly clearly expounded ("In the Name of Eugenics", 1991), and from FRANCIS GALTON ("Hereditary Genius", 1869) through the first half of the twentieth century, respected members of the scientific community repeatedly displayed a combination of hubris, naivety and indifference to moral standards and ethical values (at least as judged *a posteriori* by the standards of later generations). This was exemplified by the close working partnerships between scientific experts and public authorities in more and more fields of public policy as the century progressed; from genetics to eugenics was a short and (in the context of those decades) a natural step. The culmination of these trends, in the specific German experience leading up to 1945, cast a long shadow.

To this traumatic experience, Germany in common with other Northern European countries added a centuries-long romantic tradition, whether rooted in an instinctive animism, or arising in the last two centuries from a natural revulsion at the destructive, disruptive and ugly characteristics of the industrial revolution. From an anti-industrial sentiment to an anti-intellectual suspicion of science and scientists was a short step. Such matters require deeper historical and sociological analysis than can be offered in the context of this paper. Their current influence is in some measure illustrated by the results of opinion surveys referred to in Sect. 7.5. The practical consequences for biotechnology and its regulation are summarized in the following paragraphs.

Industrial Fermentation – DECHEMA

The German chemical and pharmaceutical industry has been and remains one of the largest and strongest in the world, since the beginnings of these industries. In the postwar years, its strengths were reflected both in the traditional food and drink fermentation sectors, and in a rapid build-up of fermentation pharmaceuticals in giants such as Bayer, BASF, and Hoechst.

A leading role in education and innovation was played by the German Chemical Equipment Manufacturers' Association, DECHEMA, under the leadership of DIETER BEHRENS for three decades, from 1960 to 1990. Among DECHEMA staff and the academic experts in their circle of advisers, there was in the 1960s and early 70s already a lively awareness of progress and potential in applied microbiology; and a far-sighted report was published in 1974 (DECHEMA, 1974). A developed version of this report was subsequently financed by the Federal Ministry for Research and Technology, BMFT. From that time, BMFT has played a key leading role in strengthening the scientific base of biotechnology in Germany, and in trying to stimulate the more effective transfer of scientific and technical knowledge from academic into industrial contexts. It was therefore natural for BMFT to encourage the role of DECHEMA, placed precisely at the interface between the academic and industrial communities.

Mention should also be made of the build-up of the national research capability through BMFT support for the GBF ("Gesellschaft für Biotechnologische Forschung") at Braunschweig; but this is less relevant to the story of biotechnology regulation.

The long history underlying the industrial tradition in biotechnology – in which German scientists played no small part – has been researched and narrated by ROBERT BUD (1993). The strength of the German industrial fermentation tradition and the alertness of DECHEMA gave an early and vigorous impulse to biotechnology. This was from a base which had little connection with the scientific innovations in recombinant DNA research; there seemed for several years to be a deliberate maintenance of distance between the industrial fermentation industry and the new genetics. In the BMFT-financed DECHEMA study published in 1976, the definition of biotechnology was as follows:

"Biotechnology is concerned with the use of biological activities in the context of technical processes and industrial production. It involves the application of microbiology and biochemistry in conjunction with technical chemistry and process engineering."

The separation was similarly reflected in the early history of the European Federation of Biotechnology (see Sect. 6.1). But from the mid-70s, another strand was developing which would come to dominate the regulatory debate, and profoundly shock and shake the leaders of German industry; and not simply because the successes of rDNA research were beginning to indicate the need to prepare for the move to large-scale production of a few particular molecules of pharmaceutical interest.

rDNA Research – The Guidelines

The Federal Ministry for Research and Technology (BMFT), in response to the US and international debate launched by the Asilomar Conferences (1973 and 1975 – see Sect. 1.3), appointed a committee of experts which drafted a proposal for safety guidelines in 1976, modelled after the NIH Guidelines, and relating to laboratory work with recombinant organisms. Following public discussion, the Guidelines were formally enforced in 1978. These Guidelines were legally binding only on research funded by the Federal Government, but in practice were followed by industry on a voluntary basis, as in the US. Again similarly to the NIH Recombinant DNA Advisory Committee in the US, the Guidelines were repeatedly amended – five times, up to 1986, before the adoption in 1990 of the Gene Technology Law (see below).

The Guidelines allowed genetic manipulation work to be undertaken only in registered laboratories or institutions, of which by 1990 there were about 1000. The Guidelines covered a range of topics – definitions, registration, categorization of experiments, supervision of rDNA work, health monitoring of workers, transport of GMOs, and a detailed appendix containing recommendations on the handling and containment of pathogenic microorganisms. Four laboratory safety levels were designated, ranging from low-risk, L1, to the riskiest, L4. Two biological categories were also defined, B1 and B2, in terms of lists of the microorganisms and strains corresponding to each. For production of microorganisms in volumes greater than 10 liters, safety measures LP1 to LP3 were also defined in the fifth edition, in accordance with the OECD "Blue Book" which was appearing at that time.

The degree of scrutiny of recombinant research was defined as a function of the degree of risk. In a registered and approved laboratory, L1 experiments could take place without further registration or approval. (This was an innovation in the fifth edition of the guidelines.) After 28 May 1986, only experiments at levels L2 to L4 required individual approval from ZKBS, the Central Commission for Biological Safety. Established in 1975 within the Federal Health Board (BGA: Bundesgesundheitsamt), ZKBS had 12 members: four experts in rDNA technology, four experts in biological research, and four persons from other fields – e.g., trade unions or industry. Prior to the fifth edition, every experiment of greater than 10-liter volume required individual authorization from the BGA. In accordance with the OECD Guidelines, the BGA could reduce safety requirements for working with low-risk organisms (such as can be handled under GILSP: Good Industrial Large Scale Practice).

In 1979, the BMFT conducted a closed hearing, involving both foreign and domestic experts, on "The Chances and Risks of Genetics Research". This was described by FRITZ GLOEDE (1992), of the Technology Assessment Unit of the Bundestag (TAB) as an occasion "in which scientific-technical, ecological and social risks were discussed controversially for the first time".

Ethical Aspects and Human Genetics – The Benda Report

The nineteen-seventies debate interested primarily the scientific community and industry. Widespread public debate emerged only in 1984, and this mainly in the area of the ethics of reproductive medicine and human genetics. Such ethical debate led to a series of commissions, in particular the establishment in May 1984, jointly by the Ministers for Research (HEINZ RIESENHUBER) and Justice (HANS ENGELHARD) of a Working Group on "*In Vitro* Fertilization, Genome Analysis and Gene Therapy", under the chairmanship of Professor ERNST BENDA. Their report (BENDA, 1986) published the following year, containing for each of the areas mentioned in the title carefully reasoned recommendations for legislation, or pointing to possible future needs. In conjunction with other debates, this led eventually to the adoption of the Law on the Protection of Embryos, in 1991.

The Catenhusen Commission of Enquiry

Also in 1984, the Bundestag launched a Commission of Enquiry, on "Prospects and Risks of Gene Technology" ("Chancen und Risiken der Gentechnologie"; the official English translation renders the last word as "Genetic Engineering"). Its terms of reference were as follows:

"I. The Commission is required to describe the prospects and risks of genetic engineering and concomitant new biotechnological research in their main current applications, particularly in the fields of health, nutrition, raw materials, energy production and environmental protection. Prime consideration should here be given to economic, ecological, legal and social effects and safety aspects.

Particular attention shall be given to the border area of genetic engineering as applied to humans, including its ethical aspects.

When preparing to make recommendations and laying the groundwork for political decisions the Commission shall be at particular pains:

1. to investigate any possible conflict of aims between the constitutionally guaranteed freedom of research and other basic rights,
2. to elaborate criteria for limits on the use of new genetic engineering and cell

biology methods on human cells and on humans as a whole,

3. to produce criteria and recommendations for directives and safety standards in the industrial application of genetic engineering methods,
4. to promote measures for promoting genetic engineering research in important fields of use."

The 17-person Commission comprised 9 members of the Bundestag (in the proportions 4:3:1:1 between the Christian Democrats, Social Democrats, Liberals (FDP), and Greens), and 8 experts. Elected as Chairman was WOLF-MICHAEL CATENHUSEN, Social Democrat.

The Commission worked for over two years, and produced in January 1987 a major, 400-page report (DEUTSCHER BUNDESTAG, 1987). In view of the Bundestag's explicit aim to contribute also to the international and European debate on genetic engineering, and to promote corresponding harmonization, it was translated and published in English later that year.

The Catenhusen Commission gave careful attention to the legal basis and constitutionality of the existing safety guidelines which had been promulgated under the executive authority (as a research funding agency) of the Research Ministry. The Commission's conclusion was that an explicit legal basis from Parliament was essential. Their specific recommendation – after considering the scope and applicability of various existing statutes – was that the existing Federal Law on Epidemics should be appropriately extended and renamed, as the "Law on the Regulation of Biological Safety".

The Commission also considered the constitutionality of a general ban on genetic engineering; but rejected this, remarking that "An obligation on the legislature to impose a general prohibition on genetic engineering could only be held to apply if any use whatever of genetic engineering were to involve a substantial danger to the legal position of citizens protected by the Basic Law, so that it would clearly fall in the socially harmful category. In the Commission's view, however, this is not the case".

The Green participants in the Commission (initially ERIKA HICKEL, from March 1985 HEIDEMARIE DANN) did not agree the final report and recommendations, but sought to make political capital by publicizing their disagreement. Contrary to the unanimously agreed rules of procedure for minority reports, a lengthy document was submitted to the Commission on 17 December 1986, one day before the final session, "and was presented to the public on the same date at a press conference as the comments of the Greens group on the Commission's report". (The quotation is from the Commission's own comments on the text submitted by Ms DANN). The Commission comments add, after reviewing other procedurally unacceptable aspects, that "Nevertheless the Commission has decided to include the submitted comments in the Report, [seeing] no reason to afford Mrs DANN and her group any pretext for representing themselves as a persecuted minority."

These matters might seem relatively trivial, but they foreshadowed and exemplify the contentious and polarized debate of the following years: in which the efforts and views of the Greens group were at least highly influential, whether or not they accurately represented popular sentiment. The Greens' comment ended with a demand that the development of research and technology be oriented to the needs of society, and a recommendation to the Bundestag "to call upon the Federal Government:

– to stop any application of genetic engineering, as this technology is not supported by a broad consensus of society;
– to create opportunities for a broad discussion on the ethical principles, aims, usefulness, the social and ecological tolerability of biomedical research and its practical application, the participants to come from society at large and not merely a small elite of parliamentarians, representatives of industry and scientists;
– to set up for this purpose a wide-ranging system of commissions at local, regional and national levels with the following characteristics, duties and authorities:
– equal representation of men and women;

- at least 50% participation by people who have neither any involvement whatsoever in the research, or in its industrial or any other use nor profit therefrom;
- representation of relevant civil initiatives, organizations, consumer associations, environmental and women's groups, and the like;
- monitoring of adherence to safety regulations and ethical, social or ecological restrictions;
- examining proposed research plans and the right to have a saying the allocation of monies for research;
- initiating research in neglected and socially useful areas or matters;
- the work of these commissions, and also of other relevant State organizations and authorities in research and technology policy to be open and public."

The agreed recommendations of the Commission itself ran to some 50 pages, the principal recommendations and findings being the following:

- The Safety Guidelines should apply to all genetic research establishments and production plants, and they should register with a central body. Violations of the Safety Guidelines should be punished.
- In general, the Commission agreed with the existing Safety Guidelines and UVVs (see below). In some areas they should be further developed, including more precise descriptions of safety measures in laboratories, pilot plants and production plants. Animal cell cultures and hybridomas should also be covered, and a review of safety guidelines for retroviruses and oncogenes should be undertaken.
- Biotechnology should be encouraged by the government and research institutions, in the usual range of application areas. In human genetics, genome analysis of embryos, newborns and adults should be allowed, subject to restrictions to prevent misuse of information by employers and insurance agencies.
- Somatic gene therapy should be allowed, germ-line therapy should be forbidden.

- Genetically engineered viruses should be used only for vaccines. The potential risks of releasing such viruses as pesticides should be very carefully evaluated.
- The release of GMOs should be banned for 5 years, whereupon the Parliament should assess whether to lift the ban. During this period, research should develop risk assessment techniques.
- Large-scale release of microorganisms might be allowed if they had not been genetically altered, or if they had been altered using conventional techniques only, or had only a single gene deleted. Such release must be authorized by the Federal Ministry of Health and the ZKBS (Central Commission for Biological Safety). The release of organisms pathogenic for humans and farm animals should not be allowed.
- The deliberate release of plants genetically engineered through biotechnology should be allowed, following risk evaluation, all requests to be reviewed by the ZKBS. There should be evaluation of toxic effects, and of whether the plant's genes could transfer to other plants. Genetically altered animals might be released, providing that they could be "recovered".
- There should be strict liability for any injuries caused through activities requiring approval under the Safety Guidelines.

From Catenhusen to the Gene Technology Law, via Hessen

The Catenhusen Commission's recommendation for legally binding safety guidelines was initially opposed by science and industry. DECHEMA, for example, included among its many activities in biotechnology a Working Group on Safety, which had produced in 1982 a collection of the various sectoral regulations concerning pharmaceuticals, agrichemicals, food, feed and chemicals, which were relevant to biotechnology (DECHEMA, 1982). FROMMER (1989), a leading member of the this Working Group on Safety (and on behalf of Germany, also of the OECD Group of National Experts on Safety in Biotechnology) commented that "In most cases, the existing regulations provide an adequate framework for ensuring the safety of products developed through biotechnology. There are, however, some ongoing efforts to adapt the current regulatory elements to biotechnology without having to change the present regulatory structure". At the time when FROMMER was writing this, there was ongoing in Germany a major debate about the need for legislation, following the Catenhusen report, and in the period leading up to the debate and adoption of the 1990 Gene Law.

FROMMER cited also the worker safety rules promulgated by the Berufsgenossenschaften (the Insurance Association for Occupational Safety and Health), which have statutory force, and are stringent. These rules ("Unfallverhütungsvorschriften" – UVV) had been issued for work in areas/activities such as chemical laboratories and medical laboratories. A UVV for biotechnology was issued in 1990. Other existing environmental protection laws could also potentially apply to biotechnology – e.g., those relating to human pathogens, animal pathogens, plant pathogens, and harmful emissions.

However, a general change in attitudes was triggered by problems with licensing practice under the Federal Emission Protection Act, and partly in response to the negative decision reached by the Administrative Supreme Court in Hessen relating to the commissioning of the Hoechst insulin factory at Frankfurt (November 1989). Again, the central issue was the lack of legal basis for genetic engineering and the Safety Guidelines. At the same period, the Greens movements were mounting an effective political campaign against "Gene technology". Moreover, the European Commission's proposed Community legislation was published in May 1988, and finally adopted in April 1990, making it inevitable that some legislation must take place at national level, in Germany as elsewhere.

The legislative process in Germany was initiated in 1989, a timing which thus enabled its progress and provisions to be closely co-ordinated with the progress of the Community Directives. In the European Parliament, both the Social Democrat rapporteur GERHARD SCHMID, and the Greens' biotechnology spokesman BENEDIKT HARLIN, were clearly applying in the European context ideas similarly in vogue in their national debate in Germany – e.g., the Catenhusen Commission proposal for a five-year moratorium on field release, against which the scientific community had so strongly lobbied in May 1989 and February 1990 - see Sect. 4.7. A series of parliamentary hearings took place in the Bundestag in 1989, and the Gene Technology Law ("Gentechnikgesetz") was adopted on 11 May 1990.

TAB, the Technology Assessment Unit of the Bundestag, merits mention as an illustration of the efforts of the Bundestag to inform itself effectively on complex matters, in a role similar to that fulfilled by the US Congressional Office of Technology Assessment (OTA). In May 1990, the same month as the new Gene Law was adopted, the Parliamentary Committee for Research, Technology and Technology Assessment asked the TAB to conduct a study on "Biological Safety in the Use of Genetic Engineering". This was to address uncertainties which had been highlighted in the Catenhusen deliberations. The new law also had provisions referring to the "current state of science and technology", thus demanding consideration of the question whether new knowledge with relevance for biological safety had emerged since the Commission's considerations some four years previously.

However, there was a change in the climate of Bundestag debate after the December 1990 elections (when the Greens, falling below the 5% threshold, were no longer represented); with the growing availability of international experience on the basic safety issues, by the time TAB gave an interim report in March 1992, their work was criticized as peripheral within the context of international discussion and too narrow in focus. By early 1992, complaints about the Gene Law were of more interest to the politicians (see GLOEDE, 1992).

The 1990 Gene Law imposed heavy costs and restrictions on academic and industrial biotechnology, in both research and commercialization. From early 1989, major German companies, contrary to their traditionally "strong, silent" style, began to express their criticisms publicly and vociferously. Hoechst was a founder member of the the Senior Advisory Group for Biotechnology (SAGB), in 1989 (see Sect. 7.1). Similar criticisms were being expressed no less publicly by academic leaders from the Max Planck Institute, and the Basic Research Institute (DFG).

A public hearing was organized in February 1992, under the auspices of the Parliamentary Committee for Research, Technology and Technology Assessment, at which a wide range of individuals and representative associations expressed their grievances. The Association of German Chambers of Industry and Commerce (DIHT) was especially outspoken in its criticisms, emphasizing that research and production were being driven out of Germany, and small firms inhibited from starting up.

The Chemical Industry Association (VCI) submitted a joint statement with the Federal Pharmaceutical Industry Association (BPI), pointing out that the Gene Law had meant a more than fivefold increase in the cost to companies of preparing applications and obtaining approval, without in any way raising safety standards in genetic engineering. The hearing was informed by eminent academic professors that level 1 or "no risk" experiments constituted over 80% of all biotechnology experiments, and similar figures were given for industry's use of low-risk organisms. Formal authorization had to be obtained, even for level 1 work; and the volume of documentation required differed little between level 1 and level 4 experiments.

The BGA and ZKBS supported the simplification of procedure, the BGA representative pointing out that the immense workload created by the authorization procedures for level one projects would have adverse effects on the quality of the assessment of projects falling within higher safety levels. Restrictive provisions on the movement of genetically engineered organisms would gravely impede German participation in international collaborative research.

Public hearings on deliberate release were another contentious matter. These had on occasion degenerated into media circuses, giving protest groups ample opportunity to hinder the approval process. In Spring 1994, for example, some 20 campaigners turned up at a hearing in the northern town of Einbeck,

dressed as giant sugar-beets; and occupied test sites to protest against the planned release of recombinant plants. The protesters were greatly outnumbered by journalists, guaranteeing extensive media coverage.

In more sinister vein, it had been found necessary to provide a 24-hour police guard for the Director and his family of a Max Planck Institute research station conducting the first field release trials in Germany, in 1990. That such direct action and the regulatory framework had an inhibiting effect was evident when in 1993 the OECD published an analysis of worldwide field releases (OECD, 1993d): of a world (OECD) total of 878 experimental release authorizations by end of 1992, Germany accounted for only 2 – against totals of 62 for Belgium, 77 for France, 45 for the UK, and 330 for the USA. A further three were authorized in 1993. Vandalism against trial plots continued; for example, in Summer 1994, AgrEvo, an agrochemicals joint venture between Hoechst and Schering, suffered the destruction of an outdoor test of genetically modified plants, the reported value of the damage being over DM 10 million.

Large bio-industrial companies expanding their R & D and their pilot plants in countries with less restrictive regulatory measures, small and medium sized enterprises and academic research institutes unable to afford the burden: the message was not lost on the Parliamentarians, at a time of soaring unemployment. The absence of the Greens from the Bundestag, since the December 1990 Bundestag election, may have reduced certain pressures on the two large parties. The Christian Democrats proposed a revised Gene Law in May 1992. Although agreeing with the majority of revisions, the Social Democratic Party (SDP) threatened to block the measure in the lower house (Bundesrat), over amendments relating to the elimination of public hearings at field releases, and shortening of the required period for authorizations; but the would-be-blocking group within the SPD was over-ruled, compromises were finally agreed, and the changes to the Gene Law were finally enacted at the beginning of January 1994.

As a reaction to the general discussion of the Gene Law and its enforcement, the Bundestag Research Committee enlarged the terms of reference of the TAB study (see above), asking for a comparative international review of regulations in other countries (in particular the USA and Japan), and experience with its practical enforcement; requesting delivery of the report in mid-1993.

Since the change of position in Germany – which could be defended as a natural evolution in response to accumulating experience, national and international – the German government has in the context of the European Union been a strong advocate of a more positive and less restrictive stance vis-à-vis biotechnology, including both simplification and review of the existing Contained Use and Deliberate Release Directives (see Sect. 7.4).

Not entirely by chance, the senior German Commissioner, Vice-President MARTIN BANGEMANN, an FDP politician of liberal economic views, was responsible for industrial policy and Directorate-General III, both in the 1989–92 and 1993–94 Commissions. He was thus well-placed to promote a renewed thrust for biotechnology from the Commission, as was successively exemplified by the April 1991 communication (see Sect. 7.2), the White Paper of December 1993 on "Growth, Competitiveness and Employment", and the follow-up communication on biotechnology at the Corfu European Council (see Sect. 7.4). Further action on the EU regulatory framework for biotechnology was therefore a natural expectation during the German Council Presidency, July-December 1994; and the outcome of the election on 16 October 1994, although bringing the Greens back into the Bundestag, does not seem likely to perturb the now broadly-based consensus behind the "de-regulatory" thrust.

5.4 France: The Two Commissions, Gallic Pragmatism

France has many reasons to be a major player in biotechnology: as the major agricultural producer in Europe, number 2 to the US in terms of world exports of agro-food products; and as a country with a proud scientific tradition going back to LAVOISIER (for chemistry), but above all focused on LOUIS PAS-

TEUR, the founder of modern microbiology. The 1995 centennial of his death provided an occasion for ostentatious celebration of the national tradition in this discipline.

French scientists from Pasteur Institute and other laboratories were to be found in the front ranks of molecular biology in the 1970s, and were fully au fait with the challenges and concerns voiced at the Asilomar conferences.

In March 1975, a month after the Asilomar conference and in parallel with the development of the NIH RAC and its guidelines, the DGRST (Délégation Générale à la Recherche Scientifique et Technique) established a "Commission Nationale de Classement des Recombinaisons Génétiques *in vitro*", charged with formulating safety rules to be applied to each case of genetic recombination which it received. At the same period, there was widespread interest in the ethical issues arising, an interest particularly strong in France.

The Royer Report, "La Sécurité des Applications Industrielles des Biotechnologies": One Element of a Strategic Approach to Biotechnology

In 1980, at the request of the Ministry of Industry, PIERRE ROYER of the DGRST prepared a report on the safety of industrial applications of biotechnology (ROYER, 1981); including a review of the regulations existing at that period. Before summarizing these, it is illuminating to set the specifics of regulation in the context of a broader strategic view; for throughout the period we are discussing, French public policy vis-à-vis biotechnology has been characterized by a consistency and determination independent of individuals, Ministries and administrations.

In agricultural policy or in GATT negotiations, it is evident that there are tensions between the historically dirigiste traditions of French government (a brief obeisance to NAPOLEON, COLBERT, and "managed trade" is customary), and the "Anglo-Saxon", liberal model of an open world trade system, and non-interventionist government. As France, host of the OECD and signatory of the

GATT agreement, progressively privatizes great state-owned companies, promising the European Commission that the next subvention to support restructuring of another of its major nationalized industries (Air France, Bull, Crédit Lyonnais, …) is positively the last, it is evident that the differences may be more of style than of substance. Nonetheless, in spite of the common and international constraints which tend increasingly to unify the policies pursued by all governments of the "Western" world (the adjective itself is an anachronism), France retains much more strongly than, say, the UK, an assumption of the responsibility of the state to identify strategic trends and to develop appropriate public policy responses to the challenges thus identified.

This emphasis on long-term forecasting, analysis and formulation of strategy – reminiscent of Japan's MITI or Science and Technology Agency, and perhaps returning to fashion in the UK with the "Foresight" exercizes of the Office of Science and Technology – was strong in France in the 1970s. The Nora-Minc report on "L'Informatisation de la Société" (1978) was a government-commissioned report, and paperback best-seller. Faced with the dramatic developments in the life sciences, it was therefore wholly unsurprising that the French government commissioned from three "éminences grises" of the life sciences the report which appeared in 1979, "Les Sciences de la Vie et la Société". This report, by FRANÇOIS GROS, FRANÇOIS JACOB and PIERRE ROYER, became the foundation stone for the elaboration of successive policy commitments by French governments to ensure that France corrected its current weaknesses, and did whatever was necessary to ensure that she became a major player in biotechnology on the world scene.

The Gros, Jacob and Royer report was general, strategic and inspirational; it therefore had to be accompanied, followed and complemented by more detailed, practical and operational works, such as the inventory of industrial activities and capabilities assembled by JOEL DE ROSNAY (1980), the Pelissolo report (1981), and the report by ROYER already cited. The Prime Minister (RAYMOND BARRE) charged JEAN-CLAUDE PELISSOLO

with preparing proposals to ensure, without calling into question the competences of the various Ministries concerned, the coherence of the different programmes, and to bring about the conditions enabling France to occupy its proper place in the field of life science applications, particularly biotechnology. It was in a similar perspective that the Minister for Industry, ANDRÉ GIRAUD, charged PIERRE ROYER with the leadership of a study group on the safety of the industrial applications of biotechnology. The mission was to:

> "enumerate, if there are any, the accidents or incidents which have occurred in the past through the industrial use of microorganisms; and try to extrapolate to the future the results of this enumeration, with an appreciation of the degree to which novel risks may appear. In the light of this analysis, the group should finally reflect upon the safety measures which may be required in the industrial sector."

The Royer report (in a chapter prepared by CHARLES MERIEUX) reviewed the existing "bio-industry" sectors, noting that beyond general risks of industrial accidents, there were few specific to these sectors, whose relatively "ordinary" conditions of temperature and pressure rendered their general risks less than those in sectors using physical and chemical means of transformation. The report did not address risks in hospitals, pathology laboratories or animal rearing establishments, in which it acknowledged that greater risks would be found.

The following chapter (by JEAN-PIERRE ZALTA) reviewed the potential dangers of the new biotechnologies, including recombinant DNA, discussing in detail the concept of "potential" danger, whose significance was in inverse relation to the progress of knowledge. Such potential dangers should therefore be continually re-assessed. An objectively identified danger could be addressed depending on its nature, e.g., by experimental protocols for the confinement of pathogenic bacteria; the nature of a potential danger similarly allowed one to conceive without difficulty the means for handling it.

Further chapters reviewed existing regulations, in France and abroad; methods of avoiding accidents in biotechnology; and public opinion aspects. The report summed up its conclusions in five areas:

1. Regarding existing dangers, of infection and allergic reaction, the vast majority were of agricultural or hospital origin; for the remainder, the occasional cases of infection had led to the development of refined safety systems in certain French industrial companies, and were often requirements for international trade. The proposal was to consider the generalization of the application of best industrial safety practice.
2. Regarding the potential dangers of biotechnology, there would be continuing rapid evolution of the state of knowledge. The proposal was therefore that safety rules should not be too fixed; there should be in operation a continuing system for the revision of the rules, at both national and international levels, to ensure that they correspond to the current state of knowledge.
3. Regarding existing regulations, many already applied to the bio-industries. For new activities arising from molecular and cellular biology, existing legislation for worker, environmental protection and product authorization largely provided the necessary means. The proposal emphasized the need to launch fuller studies, reviewing the evolution of legislation in other countries and at Community level, and to maintain industrial confidentiality, speedy procedure, and the possibility for slackening confinement rules if the progress of knowledge justified it.
4. On measures for the prevention of accidents, the instruments were technical measures, methods of organization, training and information, for workers, the public, and users. Various proposals could be made, on risk measurement, surveillance measures as necessary, the acceptance of reference data in the form of standards, and codes of good practice or of professional standards.

5. Regarding public perception in a field little understood but which touched the sources of life, concern could lead to opposition to industrial scale work – the study group had conducted an opinion survey – and their proposals related to four points:
   - create a climate of awareness;
   - increase the press services of scientific organizations;
   - establish press agencies devoted to the problems of scientific research and technological innovation
   - update the teaching of biology in schools and colleges.

The final conclusion of the study group was that faced with the rapid changes, and the need to avoid a rigid regulation, its hope was that their work should serve as a point of departure for a continuing assessment of the knowledge, dangers, regulations and techniques for preventing these, training, and public information on these problems. Such a Study Commission on Safety in Biotechnology would certainly be of value to science, industry and the French public.

The Committee on Genetic Engineering

The philosophy of the Royer report remained that of the French authorities during the following years. The work of the DGRST Commission was maintained, subsequently under the authority of the Ministry of Research, and under the chairmanship of Professor ZALTA. By an inter-ministerial decree, No 89-306, its role was enlarged and reinforced, and it was renamed the "Commission du Génie Génétique", the Committee on Genetic Engineering. This body has sole competence to draw up a scientific classification of existing or new organisms in all biotechnological applications, on the basis of the real or potential hazards they present.

Up to 1989, the system in operation was in effect one of self-control by the scientific community for rDNA research. The Committee comprised entirely scientists. The research institutions signed an agreement of standard form with the DGRST, and undertook to establish a local oversight body ("instance locale de surveillance"), including representatives of the personnel, responsible for ensuring respect of the conditions imposed by the Committee for the conduct of experiments. The applicant for a project would propose the level of risk classification, and complete a questionnaire, on the basis of which the Committee assesses the adequacy of the training of the researcher, the existence of the local oversight body, and decides upon the confinement category required. A standard letter of notification is then transmitted to the applicant, the Director of the laboratory, and the local oversight body.

The Biomolecular Engineering Committee ("Commission du Génie Biomoléculaire")

As the prospect of field release developed, the Ministry of Agriculture and Forestry created on 4 November 1986 the Biomolecular Engineering Committee. In the agricultural and agro-food sectors, the Committee is charged with the formulation of an "opinion on the risks associated with the use of products resulting from biomolecular engineering, in particular the risks of dissemination of living organisms resulting from these techniques". For each dossier examined, the Committee formulates recommendations on the precautions to be taken, and the conditions of use considered necessary. It pronounces on the usage of the modified organisms, and not on the host–vector systems, whose classification belongs to the Committee on Genetic Engineering (see above).

The Committee's opinion is sought by researchers and practitioners of the sector, whether from public or industrial organizations. The submissions may occur at any stage from research and development, through production, to placing on the market of a product. It can even advise on the design of a research programme.

Currently (and since its inception) chaired by Professor AXEL KAHN of INSERM (l'Institut National pour l'Enseignement et la Recherche Médicale), the Committee originally comprised fifteen members drawn from agronomic, medical or veterinary research, medicine, industry, and law, and representatives of consumers and professional and trade union organizations.

The Committee has a significant role in encouraging technical progress, with strict respect for consumer and public safety, and protection of the environment. It can request research and study as necessary for the evaluation of the potential risk from new techniques, and has on several occasions emphasized to the relevant Ministries the importance of national and Community programmes in this respect.

Implementation of the Community Directives 90/219 and 90/220

Following the adoption in April 1990 of the European Community Directives, legislation was prepared which resulted in the adoption, on 13 July 1992, of Law No 92-654 concerning the control of the use and dissemination of genetically modified organisms, and modifying Law No 76-663 of 19 July 1976 concerning establishments classified for the protection of the environment. This law covered confined use, field release, and placing on the market of genetically modified products.

For confined use, in research or at industrial scale, the Genetic Engineering Committee remains the authority responsible for classification of genetically modified organisms, and for prescribing the conditions for their production and use.

For evaluating the risks associated with field release, the Biomolecular Engineering Committee was replaced, or renamed, as the Committee for the Study of the Dissemination of Products resulting from Genetic Engineering; including also the risk assessment for placing on the market of products composed wholly or in part of genetically modified organisms.

Further details relating to the practical implementation of the new law were fixed by decrees of the Council of State; for example, that of 23 February 1993, Decree 93-325, fixed the composition of the new or renewed "Biomolecular Engineering Committee", now placed under the twin authorities of the Ministry of Agriculture and Forestry, and the Ministry of the Environment. Its composition was expanded to eighteen members, including (among other new members) a representative of an environmental protection association, a consumer representative, and a Deputy from the OPCST (Office Parlementaire des Choix Scientifiques et Techniques).

The role of the Committee was enlarged to include human vaccines, and gene therapy. (The new membership includes several medical specialists, and a representative of the Ministry of Health). The first gene therapy proposals were received (and approved) in 1993.

France was an active participant in the OECD Group of National Experts on Safety in Biotechnology, particularly in the work on "Good Industrial Large Scale Practice" as defined in the 1986 Blue Book. France also took the initiative to host in April 1992 an OECD seminar in Jouy-en-Josas (near Paris), at which the scientific advisers to the field release authorities in various countries were invited to exchange views on how they handled dossiers relating to three of the most commonly transformed crops: potato, oilseed rape, and maize (OECD, 1993a). Areas of disagreement and consensus were identified.

At this workshop, the degree of molecular characterization necessary for assessing safety in field tests of transformed plants was a major subject of discussion. Although all countries require molecular analysis of transformants, two fundamentally different points of view were expressed. The French authorities prefer extensive characterization of transformants before the first field trials, including determination of the limits of the DNA being transferred, whereas representatives of other countries appeared more willing to allow small-scale tests for establishing gene efficacy with transformants that are less well characterized, or that contain genes of unknown function.

The success of the French regulatory system for field releases may be judged, first of all by the absence of any reported adverse effects; and by the fact that by the end of 1992, 77 authorizations had been granted – the largest figure for any European country (OECD, 1993b). Moreover, the French authorities were the first to propose, in January 1993, at the "Article 21" Committee for the field release Directive 90/220, the institution of simplified procedures at national and Community levels, for experimental release applications which fulfilled the following conditions:

1. "plants whose interaction with the environment has been studied during a number of field trials with transgenic plants following an authorisation from the Competent Authorities. Tobacco, sugar beet and rape seed could fall in this category.
2. precisely described inserted sequences (constructs) whose behaviour in the transgenic plant species has already been studied. For example: herbicide resistance (genes) in constructions transferred with the *Agrobacterium tumefaciens* system, which (inserted sequences) contain only known marker genes, e.g., kanamycin or hygromycin resistance gene, or the herbicide resistance gene itself."

The acceptance of this proposal was an important step forward, in maintaining some convergence between European and US practice – it will be recalled (see Sect. 5.1) that the US was at precisely the same period (November 1992 through spring 1993) proposing and approving a substantial lightening of oversight for six of the main US crop plants.

Role of Standards and AFNOR, and of Organibio

France attaches more than average importance to the role of industrial standards, the national body being AFNOR (Association Française pour la Normalisation). It was also France which earliest and most effectively created a multi-sectoral representative body, "Organibio" (Organisation Nationale Interprofessionnelle des Bio-industries) to represent the spectrum of economic actors interested in biotechnology. Organibio provided and continues to provide an effective focal point for biotechnology interests, and has often interacted with the four government Ministries most involved in biotechnology: Industry; Research and Higher Education; Agriculture; and Environment.

In 1985, the Chairman of the Inter-Ministerial Group for Chemical Products (GIPC) , a body reporting to the Prime Minister, and the Commissioner for Standards at the Ministry of Industry, gave a joint task to AFNOR and Organibio. They were charged with studying the conditions for the establishment of coherent provisions for ensuring safety in biotechnology, but sufficiently flexible to allow for progressive adaptation to scientific and technological progress and taking into account the "potential risk" as defined in the 1981 Royer report (CADAS, 1993).

AFNOR has developed a standards programme proposed by Organibio with the support of the many other parties, public and private, involved in the vast field addressed by the totality of the biotechnologies. These cover sound research and production practices. Reference has been made to the role of AFNOR as secretariat of the CEN Technical Committee 233 on biotechnology – see Sect. 7.2; a role in which the French national work on biotechnology standards could immediately make an appreciated contribution to the advance of the work at the European level.

5.5 Japan: The Attentive Spectator

Strong Background in Fermentation

Japan has a long tradition of fermented foods and drinks. With the modernization that started in the late 19th century, the rapid industrial and technological developments were accompanied also by basic scientific development. Among the areas of success were the identification of key enzymes and organic acids, and an enhanced understanding of the traditional fermentation organisms and processes.

During the 1920s, the dependence of the growing chemical industry upon imported oil was recognized as a strategic weakness. SAKAGUCHI (1972) summarized the period as follows, in his opening address to the 4th International Fermentation Symposium, 1972, which for the first time was held in Japan, at Kyoto:

"The most serious problem in Japan before World War II was how to supply aviation fuel, synthetic rubber and synthetic resins with inadequate petroleum resources. It was decided as a matter of fundamental national policy that plant resources which were derived from solar energy be converted into basic raw materials for the synthetic chemical industry through biochemical means. Plans included ethanol, butanol, acetone, isobutanol and 2,3-butanediol. This policy was a major factor stimulating studies of the Japanese fermentation industry to change from a food-industry to a non-food industry. It also resulted in mechanization and an increase in the scale of the fermentation industry, and a great stimulus to research, which led to the studies on fermentation engineering by Terui and Aiba after the war."

After the Second World War, there was thus already a strong industrial and scientific base to assimilate the developments in fermentation production of antibiotics – by 1948, some 70 companies were engaged in penicillin production. A similar phenomenon took place four decades later, with the widespread development of capability to manufacture interferon by rDNA: although few companies would ultimately develop profitable sales from these specific molecules (penicillin, interferon), there was a widespread consensus in recognizing the importance of the field, and these chosen target molecules represented a sort of standard entry test and demonstration of capability. Again, both cases recognized the essential need to import knowledge, to develop international scientific links, and to build industrial collaborations, initially in R & D. In the field of molecular biology, a similar emphasis on international links is highlighted in a historical paper by HISAO UCHIDA (1993), for many years Japan's active leader in OECD biotechnology.

The Regulatory Context: Caution, Commitment, and the International Dimension

The title, "attentive spectator", would be unfair as a summary of Japan's strengths in biotechnology, when one considers its strength in fermentation. For example, the new fermentation technology for the production of amino acids was established in Japan (by Kyowa Hakko), and opened a new era in the Japanese fermentation industry. Ajinomoto has long been world leader, e.g. in glutamic acid and lysine production. And the pioneering (by Dianippon Ink etc.) of a hydrocarbon-based single-cell protein (SCP) process preceded and provided the technology for the subsequent European developments.

However, in the context of regulation of biotechnology, Japan has proceeded cautiously, and with close attention to international developments – particularly at the OECD. This caution reflects a strong sensitivity to public opinion, which cannot be, and is not, taken for granted. Simplistic Western descriptions of Japan as a "consensus" society do not sufficiently distinguish between habits, practices and policies aimed at achieving consensus – sometimes through lengthy arguments stretching over years, and with the search for consensus more evident than its achievement – and the reality of the usual differences in interests and opinions between different sectors of society, and between different Ministries and agencies of government.

The "Minamata" poisonings of the 1960s (caused by the discharge of mercury-containing industrial wastes, which accumulated in the food chain and hence poisoned fishermen and other consumers of fish), severely shook public trust in industry and in the authorities, and emphasized to the Ministries their responsibilities as defenders of the public interest and public safety.

This climate of responsible caution is illustrated by the story of the development of *n*-paraffin-based SCP for human consumption. Dossiers submitted to the Ministry of Agriculture, Fisheries and Food by the companies concerned – Dianippon, Kaneka, Kyowa Hakko – gave satisfactory evidence of safety, and the initial decision was favorable. But following protests from the Consumers' Union, a political decision was made to impose (in 1973) a 5-year moratorium. Ajinomoto abandoned the commercialization of a monosodium glutamate made from the same substrate.

The Mitsui Petrochemicals production of shikonin by fermentation culture of plant cells, and the marketing of a "Bio-lipstick" based on this, is sometimes cited as an illustration of the readiness of the Japanese consumer to accept new technology. (Shikonin, a red coloring substance with supposed beneficial properties, extracted from the dried roots of a plant, has long been used in cosmetics.) But there was considerable concern in Tsukuba, the "Science City", about the construction of a high security ("P4") containment facility for RIKEN, the Institute for Physical and Chemical Research of STA, the Science and Technology Agency. The decision of the public authorities was challenged by local citizens (albeit unsuccessfully) in the courts. In practice, although construction of the facility was completed, it is rarely used. There is also some doubt among industrialists as to whether "bio" still has positive connotations for the consumer.

In recent years, the fragmentation and change in political parties and the impact of global economic developments (trade, recession, arguments about protectionism and the opening up of agricultural markets) have shaken long-standing relationships and assumptions within Japanese society, and reinforced the recognition of the need for (bio-) technological innovation to proceed with caution. At the same time, the leaders of Japanese society, in government, industry and academia, recognized clearly from an early stage the significance of biotechnology for their economy and society. In an intensively developed industrial society, short of indigenous natural resources, the pursuit of continued economic growth and the constraints of space, environment and quality of life, all point to the need for economic development to be increasingly based upon information and knowledge.

Thus both the traditional strength in fermentation, and the clear perception of long-term economic interest and necessity, have led Japan to a very clear-cut identification of biotechnology as a key area of technology for future development. The public authorities are therefore deeply committed to a policy context which will encourage its development; but at the same time they recognize the strategic necessity of achieving a broadly-based consensus throughout the society. A comprehensive picture of public policy on biotechnology would therefore emphasize:

- the major life science developments which figure significantly in the regular (quinquennial) 30-year forecasts produced by the Science and Technology Agency (STA);
- the significant and long-term priority accorded to various areas of biotechnology in publicly-funded R & D, in collaboration with industry, and with the promotion of industrial collaborative research groupings;
- the commitment by the public authorities to safety and responsible regulation, on lines consistent with international standards (in particular, as defined by the OECD);
- the energetic efforts devoted to public information, through networks of teachers to reach schools in every prefecture, through an annual highly-publicized STA seminar (with strong television coverage), through publications, and through the exhibitions organized by the Japan Bioindustry Association, in Tokyo and Osaka in alternate years.

In the context of this paper, the remainder of this section summarizes the regulatory developments, which should be seen as parts of the broader national pattern and policy commitment emphasized above.

Regulatory Philosophy: Administrative Guidelines

The regulation of biotechnology in Japan has been achieved, and is maintained, through administrative guidelines. These are in principle voluntary, and in practice, followed. Although no laws define sanctions for non-compliance, the issue has not arisen; and the consequences of non-compliance in terms of public opprobrium, and the breach of relationships with government and other companies or laboratories which deliberate non-

compliance would create, constitute an effective constraint.

The administrative guidelines can be, and are, regularly adapted and updated to take account of accumulated experience and scientific progress, within Japan and worldwide. The process of adaptation involves extensive consultation and discussion; and the consultation is particularly attentive to international developments – at the OECD, in the US, and in the European Union.

The various guidelines administered under the different Ministries are briefly presented below. In summary, one could say that Japan has followed more closely the American reasoning and practice, in that the guidelines can be regularly adapted, just as the US agencies can make and modify their regulations; but with the difference that the process in the US is conducted on a defined statutory basis.

Intellectually, the Japanese leadership accepts and emphasizes the point, expressed in both OECD and US statements, that there is no scientific basis for discriminating against rDNA products. In practice, and bearing in mind what has been said about attentiveness to public perceptions, the guidelines have specifically addressed rDNA activities and products; and the progress to field release experiments has been relatively slow. Only five releases of recombinant plants or microorganisms had occurred by the end of 1993 – a figure similar to Germany; and contrasting with the several hundred (research) releases by then authorized in Belgium, Canada, France, the UK and the US.

The Agencies and Their Administrative Guidelines

Following Asilomar, the initial regulatory debates concerned only rDNA research – in both academic and industrial laboratories, and public research institutes. Two guidelines were created, essentially similar in content, concepts and definitions: that relating to university laboratories is administered by the Ministry of Education; that for all other laboratories, by the Science and Technology Agency.

rDNA Experiments

The guidelines define basic conditions to be observed by researchers and research managers, to ensure the safety of rDNA experiments. They include definitions, safety standards for containment, and organizational matters. The guidelines categorize rDNA experiments into three classes:

- institution-notified (requiring prior notification to the Director of the institution);
- institution-authorized (requiring the prior approval of the Director of the institution);
- non-standard experiment (requiring conduct under governmental guidance) following approval of the experimental plan by the Director of the institution.

For non-standard experiments, prior to starting the experiment, the "planning sheet" is referred to the government, for review by the national Committee for Recombinant DNA Technology, under the Council for Science and Technology.

Containment requirements are specified in terms of:

- physical containment
  - below 20 liters (P1, P2, P3, P4)
  - above 20 liters (LS-C, LS-1, LS-2)
- biological containment
  - B1 level (defined in terms of *E. coli* K12, *Saccharomyces cerevisiae*, *Bacillus subtilis*, and conditions on the introduced plasmid or phage)
  - B2 level.

Further details are tabulated in KATAYAMA (1993). Some 3000 "standard" experiments took place in STA's institutes in 1991; non-standard, 282. The inclusion of statistics for universities (from the Ministry of Education) would bring the total number of experiments to some 50000.

The STA Guidelines give extensive details on experimental systems, laboratory design, and laboratory practices. There have been since 1979 at least nine modifications, all in the direction of lightening and simplification.

The Ministry of Agriculture, Fisheries and Food (MAFF)

MAFF is the administrative guidance authority for the application of rDNA organisms in agriculture, forestry, fisheries, the food and related industries. MAFF has three GMO advisory committees, corresponding to microorganisms, plants and animals.

Plants are seen as potentially eliciting strong public reactions, but the procedure is established, full information is made publicly available, and by mid-1994 four plant releases had taken place, and eleven more were at greenhouse stage – mainly focusing on rice and tomato.

MAFF established in 1989 its "Guidelines for the Application of Recombinant DNA Organisms in Agriculture, Forestry, Fisheries, the Food Industry and Other Related Industries", with the aim of promoting the safe progress of agro-industries. The principles of the Guidelines are:

(1) safety assessment of rDNA plants on the basis of greenhouse experiments and limited small-scale field trials, conducted in a step-by-step procedure, and aiming at preparation for wider use under appropriate cultural practices, without specific confinement requirements;
(2) for rDNA microorganisms, appropriate confinement conditions (GILSP, category 1,2,3); for GMMs to be released in the environment, further studies are needed;
(3) for rDNA small laboratory animals, appropriate containment conditions have to be applied, enabling them to be put into commercial use. (Transgenic mice, such as "onco-mice", have been approved since 1992.)

These details are from HASEBE (1993).

The Ministry of Health and Welfare (MHW)

MHW, conscious of Japan becoming the country with the longest life expectancy in the world, has defined as a national objective to build "an animated society of longevity". In this context, the importance of advanced technologies including biotechnology is recognized for pharmaceuticals and medical devices.

The private sector has created in April 1986 the "Japan Health Sciences Foundation" (JHSF), a non-profit organization for the development of high technology including biotechnology; the contributing companies including pharmaceuticals, chemicals, foods, fiber and other areas.

The Ministry itself, also in 1986, established the "Basic Research Project on Human Sciences", to promote government–private sector joint research projects in medical and pharmaceutical sciences. The JHSF has been implementing this research programme in four areas:

- application of biotechnology,
- development of glycotechnology,
- development of medical materials, and
- elucidation of immune response mechanisms.

Again in 1986, the MHW established guidelines concerning the manufacture of rDNA products by biotechnology, closely based upon the OECD 1986 "Blue Book" (above MHW information based on HOJO, 1993).

MHW also has responsibilities in the area of food additives, and in December 1991 issued a policy statement called "Basic Principles on Safety Assurance for Foods and Food Additives Produced by Biotechnology". This was accompanied by two guidelines on rDNA techniques in food production:

a) a Manufacturing Guideline ("GMP" – Good Manufacturing Practice), and
b) a Safety Assessment Guideline.

These were enacted in April 1992. Drawing upon the concept of "substantial equivalence" developed in the OECD context, these apply to foods and food additives, "which are substantially equivalent to traditional counterparts, and in which recombinants themselves are not to be consumed" (SUZUKI, 1992; for the Guidelines, see MINISTRY OF HEALTH AND WELFARE, 1992).

Further work is continuing on guidelines for foods and food additives produced by oth-er biotechnology; discussion includes such topics as labelling, novel foods, selectable antibiotic resistance markers, etc.

The Ministry of International Trade and Industry (MITI)

MITI is responsible for application of the guidelines on industrial application of rDNA technology. It began its study of safety measures in the Chemical Product Council in 1985, based on the OECD work which led to the 1986 "Blue Book". In parallel with the OECD publication, the Chemical Product Council in May 1986 submitted a report on "Ensuring Safety for Industrial Application of Recombinant DNA Technology" to the Minister of MITI. Based on this report, the Minister announced the "Guideline for Industrial Application of Recombinant DNA Technology", which completely follows the OECD recommendation.

MITI has examined almost 300 applications under the guidelines by mid-1994, and so far no problems have been encountered.

The Environment Agency

Within the Environment Agency, an Expert Group on Biotechnology and Environmental Protection published in February 1989 an interim report on "Basic Points of View of Environmental Protection with Outdoor Use of rDNA Organisms". This outlined a procedure for evaluating the indirect impacts of releases of rDNA organisms (it being assumed that such releases would only be of organisms whose direct impacts were already considered highly safe). In the following year (March 1990), the Expert Group produced two further reports:

- "Technical Items on Ecosystem Impact Assessment for the Field Utilization of GMOs", and
- "Procedures of Risk Assessment for the Utilization of GMOs" (draft).

Administrative guidance on GMO releases (for research) is based on these reports. In

1991, the Agency, encouraged by contacts with the European Commission's Directorate-General XI (Environment), sought to have the guideline procedures translated into law; but this was strongly resisted by industry and other Ministries, and the proposal has remained "frozen".

Consideration of environmental risks was still continuing in 1994, with a final view not expected for a year or two. Existing guidelines cover research, and will be retained for this stage; the future situation for development and commercialization remains unclear, and new guidelines are under discussion. The Japanese government, the Diet, has requested a review of guidelines.

The Environment Agency is examining bioremediation, both as a potential user of the technology, and as a defender of the environment; and is developing guidelines for this application; which are also likely to involve prefectural governments.

5.6 Other Countries

To give a comprehensive history of the evolution of the regulatory debate in each and every country where it occurred would be time-demanding and tedious, because of the inevitable similarities: the same scientific challenges, the same value-laden fears about tampering with the great fundamentals of life, reproduction and the natural environment. Yet in spite of these basic similarities, and because they deal with values intimately bound up with national cultures, with each national psyche, the debates in each country were and continue to be intensely national or even local. On basic matters, people wish to control their own destiny; not to submit to a technocratic logic of economic efficiency, nor to import uncritically the conclusions of arguments and evidence accepted elsewhere. "Mutual acceptance of data" seems a desirable and logical goal to the international bureaucrat or the rational economist; but in sensitive areas, it must be recognized also as an act of trust. The choice of what constitutes a "sensitive" area is itself clearly a political one.

In the following sub-sections, a few notes and observations are recorded about other countries, to counter the impression that the debate in a few of the world's major developed countries sufficiently summarizes the global debate, and as basis for some generalizations in Sect. 5.7.

A legislature is a legislature, whether the nation numbers one, ten, one hundred or one thousand million inhabitants; the length of the debates, and of the resulting texts, are not in proportion to population. Moreover, the take-up of the debate by multi-agencies at many levels – the European Community, the OECD, the UN agencies – forced all countries, in as much as they were members of some of these bodies, to give thought to their national positions, whether or not they had previously been active in biotechnology. While initially, a quick response might be offered by a "follow-my-leader" stance, vis-à-vis the USA or other, the progressive development of biotechnology would sooner or later throw up in every country some specific case, some specific challenge, which it was not acceptable to treat in such simplistic terms. These points will be illustrated by some examples.

5.6.1 The Netherlands

The Dutch have played major roles in the public policy debates about biotechnology, including in particular its regulation. Their domestic situation combines several factors which have underlined their interest in biotechnology:

(i) a strong and broad scientific base in the universities: the Agricultural University at Wageningen and the nearby public agricultural research institutes; and the many other centers of excellence in the universities. Groningen (Protein Crystallography), Amsterdam (Microbial Physiology), Leiden (Pharmacology, and Plant Molecular Genetics – the first demonstrated transformation of monocotyledonous plants), Delft (Enzymology and Fermentation Science), are a partial list of some of their internationally known strengths;

(ii) significant (bio-)industrial activities – Akzo in pharmaceuticals (their subsidiary Organon launched the first genetically engineered product to be commercialized, a scours vaccine for piglets), Unilever's food research center at Vlaardingen, Gist-Brocades's pre-eminence in the production of penicillin and of enzymes;

(iii) an agricultural (and horticultural/ornamental) tradition of intensive, science-based, export-oriented production (lying behind only the US and France in value of agricultural exports, from a population base of some 15 million);

(iv) a public opinion attentive to nature and conservation (cf. Tab. 2 in Sect. 7.5: Netherlands lies third after Denmark and Germany in its risk perception of biotechnology), in spite of the massive and centuries-long interventions in – or against – nature, on which depend the very creation and survival of the "Low Countries" as a habitable and prosperous location for human settlement and activity;

(v) related to (iv), a strong public expectation of transparency and commitment to open dialogue, reflected in systematic public information provision and strong consumer movements (whose research arm, SWOKA, has published several reports on public attitudes to biotechnology, especially in foods; e.g., "Impacts of New Biotechnology in Food Production on Consumers", SMINK and HAMSTRA, 1994).

The CCGM

Given the strength of life sciences research in the Netherlands, the Dutch responded to the international debate at and around Asilomar with alacrity. A delegation of three Dutch scientists had been at Asilomar, and reported back to the Ministry of Science and Education. Following exchanges of letters between the scientific community, through KNAW (the Royal Netherlands Academy of Arts and Sciences), and the Ministry, a "Committee in Charge of the Control over Genetic Manipulation" (CCGM) was established in 1976 (reporting to the Ministry of Science and Education, but largely overlapping in its membership a similar Committee on Genetic Engineering set up to advise the Health Council).

The CCGM was strongly internationalist in outlook, acknowledging its use of the work and reports of the US NIH RAC, the UK Ashby and Williams reports, and the work of the European Molecular Biology Organisation (EMBO) and the European Science Foundation (ESF).

Its first report (March 1977) covered the introductory and background materials, from the 1973 Gordon Conference to date; reported an inventory of all ongoing and planned rDNA research in the Netherlands; referred to ethical questions (but set them aside for evaluation elsewhere); and gave clear and conservative conclusions, including an explicit demand for legally based regulation. The Committee recommended that:

"1. In consideration of the great scientific importance of recombinant DNA research, this also be developed in the Netherlands.

2. This research be conducted with strict precautions, according to clear guidelines.

3. Legal regulation of this research be laid down shortly in the form of a *lex specialis*. Such regulation must provide for a required registration of research projects in this field and must make binding the guidelines and supervision of their observance.

4. Within the framework of this legal regulation a Supervisory Commission for recombinant DNA research composed of the various scientific disciplines and branches of society, which will continue the work of the committee of the KNAW in charge of the Control of Genetic Manipulation be initiated.

5. For experiments at the C IV level the facilities which will be established outside this country in European connection be used.

6. In Holland, at least one laboratory be established at C III level according to the "Report of the Working Party on the Practice of Genetic Manipulation" in England.

7. Financial and material support be given in the training and education of research workers in this field, by participation in training courses and probationary periods in this country and outside it as well as in scientific congresses concerning the safety techniques for working with pathogenic organisms".

The CCGM argued for specific legislation on rDNA research, because much haste is needed"; and its language in this first report repeatedly implied an assumption of inherent danger.

The year 1977 was acrimonious for those interested in rDNA research. The Minister, the State Secretary of Science and Education, following some exchanges in Parliament, sent to the Governing Boards of Institutes of Scientific Education a letter referring to "utmost restraint in concluding contracts", a phrase which caused confusion, and appeared to ignore the risk differentiation advocated by the CCGM. The Committee felt (and said so in its next report), that "the majority of its recommendations had been disregarded. The Government had, for example, not retained the first recommendation that in view of its great scientific importance, recombinant DNA research should be developed in the Netherlands".

Public arguments were given extensive media coverage, with the issue of a counter-proposal" to the CCGM's first report, by the Union of Scientific Workers (BWA) and the General Association of Amsterdam Students (ASVA), in May 1977. At the request of the BWA/ASVA, the CCGM studied the "counter-proposal", and concluded that the Amsterdam project group "had given a tendentious and in many instances distorted picture of the CCGM and its proposals". A similar exchange of arguments took place with the Union of Scientific Research Workers (VWO).

The CCGM was also irritated by the failure of the government to act promptly on their recommendation for early legislation, and had therefore found itself obliged to proceed with a system of "voluntary" registration of rDNA research based on "gentlemen's agreements".

Full details of these arguments were published in the second annual report which the CCGM published in June 1978 (English translation, March 1979). Summing up the history of the preceding year, the report noted that: "a number of important developments have taken place in the field of recombinant DNA research:

1. The application of recombinant DNA technology in fundamental molecular-biological research yielded many new examples of the scientific importance of this technique.
2. Its application in the production of biologically important compounds was given a realistic foundation in experiments.
3. The appreciation of the security aspects of research increased and its foundation was improved.
4. In the growing public debate on recombinant DNA research political aspects of decision-making regarding science policy have played an important role."

The 110-page report, the Second Report of the CCGM, provided a detailed review of the state of the arguments, both national and international, about rDNA risks and their management during 1977–78. This full picture was the basis against which the Committee expressed their concern that Dutch research was lagging behind, as a result of the uncertain situation. Their Conclusions and Recommendations had a note of asperity:

"1. The recombinant DNA technology is of essential importance for fundamental molecular-biological research. As a result of the absence of clear policy intentions on the part of the government, confusion has arisen as to decision-making about recombinant DNA research. Another result is that the Netherlands lags behind several other countries where this process of decision-making is concerned. This in turn has caused the Netherlands to drop behind in molecular-biological research in comparison to the rest of the world. The CCGM considers this to be a sign that molecular-biological research is undervalued and feels this to be contrary to the priority given by ZWO (the Dutch Organization for Pure Scientific Research) to the development of molecular biology in the Netherlands.
2. The CCGM stands by the recommendation it made to the government last year that a legislative framework for this kind of research be set up as soon as possible. The CCGM considers such legal regulation of essential importance for the regulation of social control on the development of recombinant DNA research and the applications that may result from it.
3. The CCGM feels it is of the utmost importance that uniform international guidelines be

created for recombinant DNA research and recommends that this position be brought forward in international discussions.

4. The CCGM advocates co-operation in an international framework to evaluate the conjectural risks of recombinant DNA research. In a European framework, the EMBO is designated to co-ordinate such research.

5. The CCGM has come to the conclusion that the Dutch guidelines for recombinant DNA research are too strict on a number of points and intends to review these guidelines in the short term in the light of new insights into the hypothetical risks of recombinant DNA research, and to modify them wherever necessary."

These recommendations and the report encapsulate a paradox or dilemma which has persisted throughout the regulatory debate on biotechnology, in several European countries. Faced with scientific uncertainties and public fears, a legal framework is advocated: to provide authorization for researchers to continue their work, legal security for their institutions, and public reassurance. The public and political debate inevitable to the legislative process then tends to strengthen both the general demand for control, and the weight of the detailed safety requirements which are then imposed. Meanwhile, the objective evidence on the seriousness of the risks themselves indicates that they have been overestimated, and the regulatory requirements exaggerated.

From CCGM to VCOGEM

In the Netherlands, the blunt and critical language of the CCGM may have helped to accelerate some resolution of their demands. As the CCGM noted in their third and final report (KNAW, 1980), the Minister of Public Health and Environment on 21 June 1979 announced the establishment of an "Ad Hoc Committee for Recombinant DNA Activities" which would take over the work of the Academy Committee. In autumn 1979, the Minister of Science Policy established also a "broad" committee to study the social and ethical aspects of activities dealing with hereditary material.

As to CCGM's first point, concerning support for research, the government responded

in 1981 with the creation of an "Innovation-oriented Research Programme for Biotechnology", and the creation of a high-powered "Programme Commission for Biotechnology" to oversee its implementation, spending 70 million guilders over the following five years.

The new Ad Hoc Committee for Recombinant DNA Activities (known as "VCO-GEM") was charged with:

a. an inventory of recombinant DNA techniques and their classification in risk categories;
b. drawing up guidelines, based on the classification in risk categories, for safety measures for activities with recombinant DNA, and the level of expertise of the persons involved in these activities;
c. assessing the potential dangers of concrete activities with recombinant DNA and their classification in risk categories.

Although its function was advisory, the committee was connected to the statutory responsibilities of the authorities by a further article in the Ministerial decree:

"On the request of the person doing the research or other activities with recombinant DNA, the government institution authorized to issue a permit regarding such activities, the regional inspector of the State Inspection of public health in charge of the control of the environment or the district manager of the Labour Inspection in whose district such activities are carried out, or the Minister of Social Affairs, the committee classifies a project involving such activities in a risk category it finds suitable and advises the applicant with regard to the safety measures that must be taken during this project."

Thus the control of research, large-scale production, or environmental release involving recombinant organisms was effectively connected to the statutory authority relating to existing responsibilities such as those implied by the references in the preceding paragraph (in particular, the Public Nuisance Act). This solution had the drawback that because of the regional structure of these other authorities, additional information might sometimes be demanded by a local authority, adding an element of uncertainty to the regulatory burden (a more centralized procedure was instituted in 1993).

However, this solution was pragmatic and effective. It enabled the VCOGEM to provide national oversight, without requiring a new statute; and it enabled the Netherlands to present itself as a country welcoming foreign investment in biotechnology, and not stigmatizing the new techniques. Thus in an attractive two-volume brochure on "Biotechnology in the Netherlands", prepared by the Committee on Foreign Investment, and published in 1986, the subject was addressed as follows under the heading of:

"Additional Advantages"
The Netherlands offers a number of favorable factors ... of special importance for biotechnological firms.
...
Attainable regulations in the field of health, safety and environment
...
As for recombinant DNA activities, a DNA commission reported in August 1983 to the Dutch parliament that genetic engineering is controllable and acceptable. The parliament has used this advice to decide that there will be no special law or regulation promulgated for genetic engineering. The existing regulation will be maintained, but centralized "advice" pertaining to the risks of genetic engineering will take place."

The development of European Community legislation obliged the Dutch to give further consideration to the regulation of biotechnology, and to modify the above view for production and field release. An order based on the Chemical Substances Act came into effect on 1st March 1990, regulating all genetically modified organisms, and anticipating the requirements of the Community Directives adopted in April that year.

In a publicity brochure for "Biotechnology in the Netherlands: Vigorous and Varied", issued later in 1990, the section on "Rules and Regulations " stated:

"A realistic set of rules and regulations covering biotechnological research or industrial applications significantly contribute to the favorable biotechnological climate in The Netherlands.

Safety legislation: R & D phase

In the mid-eighties, the Dutch government adopted recommendations, formulated by a recombinant-DNA committee, regarding work involving genetic engineering. These recommendations stated that:
– work on genetic engineering is controllable;
– existing regulations guarantee safety satisfactorily;
– the introduction of genetically-engineered organisms into the environment should be covered by general rules and regulations addressing environmentally hazardous substances;
– safety aspects for personnel should be covered within the framework of the already existing Workers Safety Act.

Thus the need for new rules and regulations is eliminated, while at the same time optimal safety is guaranteed without hampering the progress of biotechnological innovations. This was clearly evidenced when The Netherlands became one of the world's first countries where field tests with genetically transformed plants (virus-resistant potatoes) were carried out.

Safety legislation: production phase

In 1990, rules for both contained use and environmental release of genetically modified organisms have been published, creating a good formal framework for the development of biotechnology."

The situation of biotechnology in the Netherlands remains tense in the mid-1990s. For in spite of the clearly positive attitude of government, and the successful establishment of several biotechnology companies, Dutch and of foreign origin, there remains considerable public suspicion and hostility for some applications. Test plots of transformed plants have been vandalized. The *Eurobarometer* surveys of 1991 and 1993 indicate in regard to the creation and use of transgenic animals a high level of opposition.

On the positive side, government Ministries such as Environment, Economic Affairs and Agriculture, all continue to address with intelligent transparency and determination these challenges; conscious of the essential role which biotechnology must increasingly play in many of the major sectors of the national economy. Legislation proposed in late 1994 by the Minister for Agriculture is intended to provide a clear basis for the continued progress of work with transgenic animals, under appropriate safeguards.

Dutch internationalism expresses itself actively in biotechnology, particularly through the support for ISNAR (International Service for National Agricultural Research) in The Hague, from the Directorate General for International Cooperation (DGIS) in the Ministry of Foreign Affairs and for the former "BIOTASK" Task Force of the International Agricultural Research Centres funded through the Consultative Group for International Agricultural Research (see Sect. 6.3); through initiatives such as the Anglo-Dutch proposal on safety guidelines, described in Sect. 6.4; and through a vigorous participation in international fora such as the European Union, the OECD, and UN activities such as those deriving from the UNCED (Rio, Earth Summit). In the OECD, VCOGEM Chairman PETER DE HAAN was for many years Chairman of the Safety Working Group responsible for drafting several of the OECD key reports on safety in biotechnology.

5.6.2 The Scandinavians

There are strong objections to grouping and summarizing the behavior of groups of countries, given the diversity of opinions and cultures which even a single small country may contain: the days of *cuius princeps, huius religio* are unlikely to return in the era of the global electronic village. Yet for various compelling reasons, there have developed regional political-economic groupings such as the European Union, or the Nordic Council. To some degree, their geographical proximity, shared historical experience, common or interacting cultures, and joint multi-national institutions appear to display common elements distinctive to that grouping, and inspire similar policy responses to similar challenges. We describe briefly some of the Scandinavian experience of the biotech regulatory debate, before advancing to some more ambitious and broader generalizations.

In the national, European, OECD and world-level debates about the regulation of biotechnology, the Scandinavian countries have expended energies and exerted influence to a degree out of proportion to their size in economic or population terms; reflecting the strong public and political interest which the subject has elicited in these countries.

The subject of biotechnology risks was usually linked with environmental and with ethical concerns. Nor are these subjects in two water-tight compartments – rather, they are the two halves of the permanent discussion of the relation of Man and Nature; a relation which (in Scandinavian eyes) should concern all of mankind, Scandinavian and other. Thus to take a recent example from Norway, two reports which have recently been published (in English) are:

- "Ecological risks of releasing genetically modified organisms into the natural environment", Directorate for Nature Management, 1992;
- "Biotechnology related to Human Beings", Ministry of Health and Social Affairs, 1993.

Both are competent and well-written reports, treating subjects already much treated elsewhere, but with a seriousness of purpose, respect for "Nature", and intensity of expressed concern typically Nordic; one would not expect identical reports from Brazil, or China, or Nigeria.

It is not difficult to cite examples illustrating the readiness of Nordic Council countries to offer a world lead in environmental matters – the Stockholm Conference on the Human Environment in 1972, the continuing work of the Stockholm Environment Institute, or the chairing by Norwegian Prime Minister GRO HARLEM BRUNDTLAND of the World Commission on Environment and Development (established in 1983 at the request of the UN General Assembly, and charged with the preparation of long-term environmental strategies for sustainable development – a significant precursor of the 1992 Earth Summit at Rio).

Denmark

As is discussed in Sect. 7.5 on public perception, the Danes have long displayed in comparison to the other European Community nations by far the highest level of risk perception in relation to "gene splicing" or biotechnology. At the same time, the authorities were never blind to the importance of biotechnology for companies of great significance to their economy – Novo Nordisk (until the late 80s, two separate companies), Danisco, Carlsberg and others – and for the long-term competitiveness of their agro-food industry, a highly successful, science-based and export-oriented sector.

Conscious of the risks of exotic "invasions" (such as the cow parsnip, introduced as an ornamental but now out of control), the Danish government in October 1983 set up a committee to consider the need for regulatory controls. The committee was described by Environment Ministry staff as pursuing a "balancing act" between the promotion of biotechnology and the prevention of adverse consequences.

In response to the argument that legislation should be pursued on an international basis, Denmark argued that, here as in many other areas (e.g. acid rain), individual countries should take the first steps in setting standards. The "Environment and Gene Technology Act" was passed on 4th June

1986, and widely reported as the world's first such legislation. The Act was in six parts, and its general goal was defined in Article 1(1):

> "The purpose of this Act is to protect the environment, nature and health, including considerations of nutrition in connection with the applications of gene technology."

The Act was comprehensive in wording and restrictive in tone; but Article 3(2) allowed for exemptions:

> "The Minister for the Environment may lay down rules to the effect that certain genetically engineered organisms and cells and certain applications thereof shall be exempt from the provisions of this Act on such conditions as may be specified."

All deliberate release was formally banned, under Article 11(1); unless exempted (under Article 11(3)) as one of "certain cases" which the Minister could approve; the Minister undertook to debate in Parliament the first field release to be so authorized. This took place three years later – a herbicide-resistant sugar-beet developed by Danisco, in a test plot; the parliamentary debate in fact focused mainly on the characteristics of the herbicide concerned.

During the three years following the 1986 Act, there was an intense and widespread debate, and three amendments were introduced to remove some of its most problematic features. One allowed pilot plants to be treated as research rather than production facilities, and so be subject to less stringent regulations. A second relaxation removed the provision by which a company automatically had to stop work whenever a complaint was lodged with the Environment Appeal Board. The paperwork nonetheless remained heavy, although civil servants from the Environment Ministry emphasized that there had been a fast learning curve: the time required for approvals had diminished sharply (and the first two approvals had been delayed by appeals).

What was emphasized throughout the Danish experience was the absolute need for public debate and transparency – particularly through "consensus conferences", and through the development of special initiatives for the preparation of educational materials for schools. Again from the *Eurobarometer* surveys, it may be noted that the Danes have within the EC context the highest level of trust in government as a source of information on new technologies. Although these commitments to debate and transparency weighed heavy upon the scientific communities, academic and industrial, continuing substantial research funding for biotechnology confirmed the government's basically positive view of the technology.

Sweden

In Sweden, "scientists tried to interest the government in regulation during the early 1970s, but when that failed, implemented regulation through research councils, following the NIH guidelines" (MCKELVEY, 1994). The Asilomar debate rekindled interest, and around 1977, a heated public debate occurred in Sweden, with calls even from industry to regulate genetic engineering; thus pushing the government into investigating legislative changes. As a result, special regulations were introduced in 1980. At the same time, the Swedish Recombinant-DNA Committee was established.

MCKELVEY in her thesis on "Evolutionary Innovation" selected as case material the production by genetic engineering of human growth hormone; a story involving a partnership from the mid-70s between the recently started San Francisco based Genentech, and the Swedish pharmaceutical company Kabi. Although her central interest is in the process of innovation, her account casts interesting side-lights on the early policy debates in that country. From the start of the rDNA regulatory debates, the discussions touched upon matters of specific significance for some local players, thus providing some counter-balance to the natural (Nordic?) tendency to debate philosophical generalities.

Sweden also had a strong academic scientific tradition in key areas such as protein crystallography and molecular biology, and

their international connections were underlined by LENNART PHILIPSON's decade as Director-General of the European Molecular Biology Laboratory, EMBL – an institution deriving from the scientific community represented by EMBO, which played a major part in the European debates on regulatory policy for biotechnology. On the industrial side, Pharmacia should also be mentioned as a Swedish based multi-national of outstanding importance for the technology, instrumentation and separation/filtration/purification materials which it supplies to research and production facilities in the bio-sectors, and for genome sequencing. The Swedish debates have reflected tensions between the internationalist and scientific traditions, academic and industrial, and the more introspective tendencies expressed by intense and interminable debate on ethical and philosophical issues.

The history of official developments has been conveniently summarized in the 1992 report of a Parliamentary Committee of Enquiry, set up in 1990, "The Committee on Genetic Engineering". The following quotation is from the (English-language) summary of their report (SWEDISH GOVERNMENT COMMITTEE ON GENETIC ENGINEERING, 1992), starting with the 1980 regulations:

"In the Labour Environment Act a demand was introduced for preliminary approval of the use of a working method employing recombinant-DNA technology in a way as to constitute either an untried approach to research into recombinant-DNA technology or its industrial application. The authorizing authority was the Swedish National Institute of Occupational Health.

Changes to the Environment Protection Legislation were also introduced in 1980. After further changes, advance approval was required with regard to plants for activities involving recombinant-DNA technology, albeit not for plants for such research considered by the Swedish Recombinant-DNA Advisory Committee to belong to the lower risk classes P1 and P2. The authorizing authority was the Swedish National Licensing Board for Environment Protection.

At the time when these special rules were first introduced, the Swedish Parliament took the view that the administrative and legal consequences of establishing a special authority for the control of genetic engineering should be the subject of an enquiry. According to the report of the subsequently appointed Committee, the use of recombinant-DNA technology hardly gave rise to any reason for concern provided that accepted protection and security measures in the area of microbiology were adopted. With this risk assessment, the arguments for an authorization procedure had weakened and the Committee proposed that the regulations which had been introduced for advance approval should be waived with regard to both the labour environment and the external environment. These regulations were waived with effect from July 1, 1987.

Another Committee appointed by the Swedish government in 1984 made proposals mainly concerning ethical standards for the use of genetic engineering on human beings. The Swedish Ministry of Agriculture, in a special report issued in 1990, has surveyed genetic engineering research on plants and animals."

Subsequent legislation addressed ethical aspects of human genetics. The (1990-92) Committee on Genetic Engineering addressed in depth a wide range of issues, attentively studying a diversity of international opinions. Their report clearly acknowledged the point that risk lay in the nature of the organism, modified or not by recombinant-DNA techniques; but noted also that there was no accepted classification specifying which plants and animals are to be regarded as dangerous.

More pragmatically, the report acknowledged the general consideration, since the country was preparing for accession to the European Community, that "Swedish control of GMOs should *not be incompatible with the EC's directives*". This consideration clearly influenced many of the subsequent recommendations; and Swedish legislation based on these recommendations was developed over the following years.

A framework, Gene Technology Act, was passed by the Swedish parliament in June 1994. Together with a governmental ordinance and more detailed regulations written by different agencies with responsibility in the field, EC Directives 90/219/EEC and 90/220/EEC were accepted to be fully implemented in Sweden in the fall of 1994 (SCREEN, 1994), and thus just prior to accession to the European Union in January 1995.

5.7 Generalizations: From Concrete to Abstract, from National to Global

If time and space permitted, there are many interesting national experiences which should be added to a fuller history of biotechnology and its regulation, from countries large and small:

- **Belgium**, against a noisy background of constitutional argument and national/regional restructuring, quietly pioneered not only the science but the commercialization of vast areas of modern biotechnology – hepatitis-B vaccine made in a recombinant yeast; the elimination of rabies in the wild population of foxes by the first large-scale field release of a live recombinant viral vaccine; pioneering the use of the Ti plasmid for genetic engineering of crop plants; developing reliable diagnostic test kits for the AIDS virus ...
- **China**, one hundred times greater in population, and certainly not indifferent to the potential of modern biotechnology, boldly testing the effects of growth hormone in animal husbandry, conducting large-scale field releases of transformed plants (CHEN, 1992, 1995), and microorganisms, developing the production and exploration of the interferons and their activities ... and in the 1990s, gradually increasing participation in the international debates;
- **India**, from an early start based on the 1979 version of the NIH Guidelines, recognizing the potential and adopting scientific priorities for a biotechnology adapted to India's vast needs;
- **Russia**, emerging from the wreckage of the USSR, and an inheritance of environmental irresponsibility, in SCP plants as elsewhere, seeking to rejoin the "Western" world, accepting US FDA drug approvals, drafting in 1994 legislation crudely modelled on earlier European laws, themselves in course of recycling through the crucible of revision and review, but asking (through

UNESCO) to have their draft law internationally reviewed;

- **Switzerland,** home base to several of the world's leading pharmaceutical companies, nonetheless witnessing some of the most fundamentalist opposition to genetic engineering (RYSER and WEBER, 1990), sufficiently widespread to threaten (by a public referendum on genetic engineering, to be held in 1996 or 1997) the whole future of teaching, research and investment in modern biotechnology.

- in **Latin America**, in **Africa**, and in **South-East Asia**, from the early 1990s, conferences of countries – or a few individuals trying to speak for their countries – seeking to "jump-start" the creation of regionally harmonized bio-safety regulations, importing selectively some of the concepts and principles developed elsewhere …(cf. VAN DER MEER et al., 1993).

The preceding sections, on the major early players and on the Scandinavians, offer some basis for cautious generalizations. Cautious, since the countries discussed in detail – fewer than ten in number, some 10% in total of the world's population, and all of them OECD Member countries – are far from being a "random sample" of the world's peoples. In historical, scientific, industrial and other terms, they cover many of the most salient features of the early years of the "recombinant DNA story"; but it is precisely the universality of DNA, and the universality of biotechnology's fields of application – food, health and environment – that forces policy advisers to consider which aspects of the early experience are transient or locally specific, and which aspects are appropriate for more permanent and universal application. That is the challenge particularly to the agencies of the United Nations, whose actions relevant to the regulation of biotechnology we review in the following section.

The big, rich and scientifically advanced countries discussed at the start of this section had a natural monopoly in the early years of the debate. The recommendations of the US NIH RAC, and the UK GMAG, therefore

had enormous initial influence. Because they were discussed principally in scientific circles, the natural internationalism of science and its culture of critical scrutiny limited the extent to which these early guidelines were mere expressions of national self-interest. They represented an "honest start" to a debate about conjectural risks.

From that start, the debate developed in three directions, which might for brevity be summarized as objective/scientific; national self-interest; and global-idealist.

1. objective/scientific
   the accumulation of practical experience, and continued scientific discussion and exchange, enabled the uncertainties to be diminished, and the risks more clearly characterized – to the point where the focus of concern shifted away from the recombinant techniques, to the organisms, inserted materials, and the nature and context of intended uses;

2. national self-interest
   in national contexts, as practical implications and applications became apparent to local constituencies – scientific, industrial, agricultural, environmental, political, religious or whatsoever – special interests and local preferences made themselves felt, and influenced both the agenda and the outcomes of the national debates;

3. global-idealist
   given the continuing conjectural nature of the risks, those individuals, agencies or institutions not having a direct or concrete interest, what the French might call a "déformation professionnelle", were free to indulge both in unfettered speculations, and in globalist and universal prescriptions about the better ordering of the world.

It would be offensive and inaccurate to pretend that some countries are more honest than others, or that scientists have greater integrity than "laymen". Nor is it our intention to imply that any of these three tendencies is morally superior: the intention is descriptive, and we use these three terms to summarize what has been observed.

The objective/scientific approach is universally valued, and because of the universality of its approach has particular value in providing common ground for international relations and agreements. The US and European experience has contributed much of international validity and value, but the US Congress and the European Union institutions are not the government of the world; the OECD (see Sect. 6.2) may have greater authority in some respects, because it has to find the common ground between 25 diverse governments, and because it collaborates with the UN agencies with their wider responsibilities.

The pursuit of self-interest, individual or national, is inevitable, (whether or not one sanctifies it, like ADAM SMITH, as the working of the "invisible hand" of divine Providence), and is the business of politics and government. The interest of the Scandinavian experience often lies at the points of potential or actual collision between ideals and interests. A "Hands off" approach to Nature may be difficult to reconcile with the development of a productive and competitive salmon-farming industry, or the maintenance of a long-established tradition of whaling.

The definition and pursuit of broader multi-national aims, on a regional or global basis, is in some areas clearly the most effective, or the only and essential way, to defend all our local interests: the defence of the "global commons" makes co-operation a non-zero-sum game, and it can be formulated in terms of ideals: the rights of man, sustainable development. But implementation depends on local institutions and interests. Thus good policy-making for biotechnology has to recognize all three strands.

6 International Actors: European, the OECD, the UN Agencies and Rio

Introduction

Responses to the policy challenges posed by biotechnology were initially national, and the previous section has described developments in several countries, including those who have been major or significant players in the debates on the regulation of biotechnology. In this section, we consider the major spheres of action at international or supranational level, starting with Europe; enlarging the scope to the OECD area; and finally considering the agencies and actions at the global level represented by the UN agencies (which themselves have regional agencies, some of them significant in the biotechnology debates).

A tension which has been touched upon in the preceding section, is between processes which are rooted in local, national experience, and work up from there; and the more "global-idealist" attempts to develop and if necessary impose, "top down", a global agreement or convention. Given that some problems must be addressed in the latter mode, international action becomes necessary, to seek consensus, and to establish and use international mechanisms and fora for information, debate, decision and ultimately enforcement. Difficult though the latter might appear, there are precedents, some of them potentially relevant to biotechnology. Reference has already been made to at least the following four:

(i) the scientific consensus regarding the two-year moratorium, 1984–86, in response to the (July 1984) Berg letter which led to the Asilomar conference (see Sect. 1);

(ii) the Directives adopted through the Treaties and institutional machinery of the European Community/Union (discussed extensively throughout this chapter, see Sects. 2, 3, 4, 7 and 8);

(iii) Decisions and Recommendations adopted by the Council of the OECD (see Sect. 6.2 below, and in the history of chemicals control);

(iv) the history of international agreements and Conventions adopted through the actions of UN agencies (see remarks concerning chemicals control, in Sects. 4.4 and 6.3 below).

In the following sections, we consider more generally the historic and continuing activities of the international bodies – scientific and intergovernmental – which have been significantly involved in discussion and/or decisions affecting the regulation of biotechnology.

The first of the four examples referred to above was, and was widely seen and remarked upon as, a "first" and "once off": a voluntary moratorium, proposed by the scientists themselves, and subsequently endorsed where necessary by national research funding bodies in the US, the UK and in other countries. Asilomar was not only a "first", but a starting point: the scientific communities have since been obliged, often reluctantly, to maintain their engagement in political and public dialogue. This section therefore includes international action of both scientific and political (or intergovernmental) character.

The progressive shift to international or global bodies can also be viewed as a third re-run, now at global scale and in developing countries, of the surge and ebb of bio-safety concerns, the learning cycle, which has taken place successively in the US and Europe. Rather than a "cycle", the "learning curve" is a better metaphor – in which those furthest down the curve should surely (and in their own commercial interests) help those starting (or stuck) on the upper slopes.

6.1 Other Europes, Scientific (ESF, EMBO, EFB) and Political (EC/U, CoE, ECE)

Scientific: (1) The European Science Foundation, and (2) The European Molecular Biology Organization

The ESF, founded 1974, representing the research councils of Europe, and the EMBO, founded 1964, representing individual scientists in this discipline, have both been introduced in Sect. 2. In their respective roles, they played from the earliest years a significant part in the European area of the post-Asilomar debates. Both were still doing so two decades later in the 1990s, in the debate on the reform of the European Community legislation – see Sect. 7.4. Their Resolutions are quoted in the sections cited, and reference to these opinions was made by many others, including political authorities at national (cf. Sect. 5.6.1, The Netherlands), and at European Community level.

Scientific: (3) The European Federation of Biotechnology

The European Federation of Biotechnology was also introduced in Sect. 3.1. Founded in 1978, it rapidly expanded to become a federation of over 60 learned societies, reflecting the multi-disciplinary character of biotechnology (albeit with some delay in recognizing its links with molecular biology and genetic engineering). Its greatest strength and vitality arose from its various working groups; but after its early collaboration with the FAST programme (see Sects. 3.1 and 3.2), its work had little direct political impact upon the regulatory debate in Europe.

The Safety Working Group of the EFB maintained over many years, and continues to maintain close links with the European Commission, particularly with those responsible for defining the elements of the research programme relating to risk assessment. The group itself produced a series of technical papers, on matters such as the definition of safety categories for biological agents (see references to KÜENZI et. al., and FROMMER et al. the successive Chairman; MARTIN KÜENZI also chaired for several years the OECD Group of National Experts on Safety in Biotechnology). By the overlapping networks which were created, through individuals active in several contexts (e.g. national, European, sectoral, professional, standards bodies, OECD, ...), close concertation was achieved between scientists in these different contexts, and consequently between the regulations based on such scientific advice.

Where the regulations departed from scientific advice, under the pressure of other political forces, tensions arose (see Sect. 7.3, the Frommer letter), and the EFB Safety Group had difficulty in making its advice influential. In the long-term, the continued good quality work by the EFB Safety Group contributed significantly to the underpinning work. But as regulatory developments became more politicized, its participants were not of a culture or disposition to descend into the political arena, and the burden of lobbying for national and science-based regulation fell on other shoulders.

In September 1988, the EFB held a 10th-anniversary International Workshop at Interlaken, and produced "Strategy Papers" from its Working Parties, under the heading "10 Years of Biotechnology in Europe and Strategic Planning for a Second Decade"; but these had little public or political impact, as attention focused on the progress of the then recently proposed "Biotechnology Directives".

Political: (1) The European Community/Union

Reference has been made in earlier sections (2, 3, 4) to the actions taken by the European Community/Union institutions, first in 1978–82, and again from 1986 onwards. The further evolution is described in Sect. 7, and much of the final section, 8, addresses specifically this dimension. In effect, it is difficult to separate the history of biotechnology regulation in Europe from the evolution of the Community institutions themselves. In wider political arenas, such as those further discussed below, the European Commission on behalf of the Community also became an increasingly significant player – again, reference has been made to the influential example of the development of regulations and agreements for the control of chemicals.

Political: (2) The Council of Europe

The Council of Europe was created by a Statute signed in London on 5th May, 1949, as the association of the democracies of Western Europe. By mid-1994, with the recent accession of ex-communist states from Eastern Europe, it had 32 members, including all the Member States of the European Union and the European Free Trade Area. In the preamble to the founding statute, the members affirm their devotion to various commonly held values, the common heritage of

their peoples and the true source of individual freedom, political liberty and the rule of law; principles which constitute the basis of all genuine democracy. They also express their belief that for the furtherance of these ideals and in the interests of economic and social progress, there is a need for greater unity among like-minded countries of Europe.

The stated aim is:

"to achieve greater unity between its Members for the purpose of safeguarding and realising the ideals and principles which are its common heritage and facilitating their economic and social progress".

Although the pursuit of these aims has been partly overshadowed by the creation and progressive enlargement of the European Community, the Council of Europe has nonetheless maintained its distinct and important activities. It pursues its aim through the organs of the Council by discussions of questions of common concern and by agreements and common actions on economic, social, cultural, scientific, legal and administrative matters; as well as the maintenance and further realization of human rights and fundamental freedoms.

Its organs are the Committee of Ministers, which is the decision-making body, and the Consultative Assembly, composed of members of national parliaments. The two legal instruments are Conventions, and Recommendations. A Convention is a multilateral treaty representing a Europe-wide consensus; once approved by the Committee of Ministers, it is open for signature by Member States, who on ratifying it commit themselves to respect the obligations defined in the Convention. On topics not suited to Conventions, the Committee can adopt Recommendations, addressed to governments of Member States, providing guidelines for national legislation or administrative practice.

The Council of Europe has been a significant player in the development of common agreements on the labelling and classification of chemicals. It has also been responsible for the Convention on the European Pharmacopoeia; for work on Flavoring Substances; and on Pesticides; among various other topics. Some of its Recommendations have been taken over by the Council of the European Community in Community Directives. On other occasions, Community Recommendations – including the 1982 recommendation concerning the registration of rDNA work – have been taken over by the Council of Europe.

The Council of Europe on several occasions addressed aspects of biotechnology. The Parliamentary Assembly, at Helsinki in 1981, had debated the challenges of new technologies, including biotechnology. In 1984, the Council adopted a Recommendation concerning national registration of recombinant DNA work, closely based on the 1982 Council Recommendation of the European Community (see Sect. 2.8).

The Council's report and resolution of 1982 (see Sect. 2.8) emphasized not only safety aspects, but the ethical issues raised by the possible applications of modern genetics directly to human beings. The subsequent work of the Council of Europe continued to address ethical questions, particularly through the CAHBI group, described below.

In 1983, the Council decided to set up an Ad Hoc Committee of Experts on Genetic Engineering, "CAHGE". Its task was to study the ethical and legal problems raised by new techniques of artificial (or medically assisted) procreation, and by the new DNA-based technologies. The ultimate objective was the harmonization of the policies of Member States. After some initial confusion about the relationship between *in vitro* fertilization and genetic engineering, it was decided at the first meeting (December 1983) to start with a detailed study of the ethical and legal aspects of artificial procreation. The group subsequently changed its title to "Ad Hoc Committee of Experts on Progress in the Biomedical Sciences", known as "CAHBI". A succession of plenary meetings over the following years addressed a range of related subjects, and greatly clarified differing views regarding the status of the human embryo; but without achieving consensus on this fundamental question.

The Council published in June 1994 the draft of a proposed "European Convention on Bioethics", announced as the first international treaty explicitly addressing the potential risks of progress in biology and medicine to the rights and freedom of human individuals. The Convention itself is described (e.g., BUTLER, 1994) as "little more than a lowest common denominator of member states' wishes"; but supporters indicate that it provides a basis of broad common principles, to which member states in their legislation could add more detail or stricter terms.

The Council's activities underline the continuing sensitivity of Europeans to ethical questions raised (or potentially raised) by modern biology, and the readiness of both politicians and public to see genetic engineering as part of a single spectrum of activities including genetic screening, *in vitro* fertilization, gene therapy and other biomedical applications of modern genetics.

The UN Economic Commission for Europe

The UN bodies of global scope are discussed in Sects. 6.3 and 6.4; it will be appropriate to mention here one of their regional offices, the UN Economic Commission for Europe, or ECE.

The ECE played a significant role in the post-war years in Europe in the response to problems of chemicals safety and control which were widely shared across Europe, East and West. It became the global office for certain activities and Committees, including those relating to the international transport of hazardous substances.

The ECE was alert in the 1980s to the positive potentials of biotechnology, for example organizing in September 1986 a seminar in Bulgaria on "Biotechnology in the Chemical Industry". Also in that year the Conference on Security and Cooperation in Europe held a meeting in Vienna, and one item from its Concluding Document stated:

> "18. Taking note of the progress made in, and the new opportunities offered by, research and development in biotechnology, the participating States consider it desirable to enhance the exchange of information on laws and

regulations relating to the safety aspects of genetic engineering. They will therefore facilitate consultation and exchange of information on safety guidelines. In this context, they emphasize the importance of ethical principles when dealing with genetic engineering and its application."

ECE therefore recommended to its "SAST" (Senior Advisers to ECE Governments in Science and Technology) that it assume responsibility for the inventory on existing safety guidelines in biotechnology; a recommendation subsequently agreed. Since 1988, on a regular basis, ECE has invited governments to transmit such materials to them, which it has issued as monographs. It acts as a depository for national reports, regulations and legislation related to safety in biotechnology, and by 1994 had materials from 29 governments, as well as from UNIDO, the European Commission, and the OECD. The Senior Advisers at a meeting in September 1994 underlined the importance attached to this work, and decided to publish the following year a compilation of the most recent summaries of national submissions.

The UN ECE's Committee on the Transport of Hazardous Substances was also the venue for discussion of the treatment of GMOs in international trade. Following the April 1990 adoption of the European Community Directives on Contained Use and Field Release, DG XI with DG VII (Transport) initiated proposals for the categorization of GMOs into existing hazard categories for the purposes of transport. The proposals did not lead to any new Community legislation, as it was recognized that existing rules regarding the transport of cell lines, medical samples etc. would apply, and it was the characteristics of the biological materials rather than the fact of genetic engineering that was relevant to the risk classification.

6.2 The OECD Role in Biotechnology and Safety

Introduction: An Authoritative Position, Based on Four Strands

In May 1992, the World Bank and ISNAR (the International Service for National Agricultural Research – see Sect. 6.3) published for widespread diffusion throughout their client countries (effectively, the whole of the developing world) a 40-page report entitled, "Biosafety: The Safe Application of Biotechnology in Agriculture and the Environment". In its Executive Summary, the report outlined its purpose; its main recommendation; and a remarkable reference to the work of the Organisation for Economic Co-operation and Development (OECD):

> "The purpose of this document is to provide a practical guide for policymakers and managers with limited time and restricted access to the extensive documentation that is accumulating on the safe use of biotechnology …
> This document recommends that a national biosafety system should be established within the existing regulatory framework and draw on existing institutions, personnel, and current legislation to the greatest extent possible. It recommends that the focus in the regulatory process should be on the nature of the product itself, not on the techniques used to produce it …
> …
> All these applications of modern biotechnology are being developed under safeguards for good laboratory practices and recombinant-DNA safety considerations, as described by the Organisation for Economic Co-operation and Development (OECD) in 1986 for laboratory-based experimentation and extended in 1992 to guidelines for small-scale field trials. They remain the most authoritative set of internationally agreed-upon guidelines presently available and provide a sound basis for national policymakers in all countries." (PERSLEY et al., 1992).

In the following section, the history is reviewed of the activities through which the OECD's international standing was gradually achieved. The history can be summarized in terms of four strands, treated further below or elsewhere as indicated:

1. the OECD from the 1960s onwards played a major role, in conjunction with the UN agencies, in developing the systems and procedures for ensuring the safety of chemical products (being placed on the market, entering into world trade, etc.): this work is referred to in Sect. 6.3, along with the roles of the UN agencies concerned;
2. starting in 1981, the OECD published a series of authoritative study reports on various aspects of biotechnology of interest to public policy-makers (and many others);
3. also starting in the early 1980s, the OECD worked intensively on, and published a series of reports on, generic safety matters concerning recombinant DNA research and products (see below, the work of the "GNE");
4. in the 1990s, continuing work on safety led to reports which increasingly discussed this and other aspects of biotechnology in terms specific to particular sectors.

The general policy reports and the series of safety-related studies both document a progressive evolution of perspective, from generic biotechnology issues, to essentially sectoral aspects.

The Historical Background

The Organisation for Economic Co-operation and Development was created out of the earlier Organisation for European Economic Co-operation, an *ad hoc* instrument to assist with postwar reconstruction in Europe and the implementation of the Marshall Plan. The value was recognized by the Member countries, of having such a forum for discussing policy matters of common interest, and for developing instruments of value to all. Focusing originally on the collection of economic statistics and harmonization of terms and definitions, the scope of the organization's activities slowly expanded, as did its membership (to 25 Member countries by mid-1994) – comprising the countries of the European Community/Union, the European Free Trade Area, North America (all three NAFTAns), Australia, New Zealand, Japan; and Turkey.

The Committee for Scientific and Technological Policy, CSTP, had been created in 1972; and soon after the Asilomar meeting (1975), the Secretariat raised the possibility of recombinant DNA and genetic research as one example among others, of areas which might be studied,

"not to describe new scientific advances nor to analyse in depth the technological or other consequences of scientific research. Rather, the aim is to determine what mechanisms, procedures and other capabilities governmental or paragovernmental bodies presently have or may require for anticipating, monitoring and controlling the consequences stemming from advances in certain areas of scientific research, and to assess their ability to deal effectively with the attendant policy implications. Thus, the emphasis would be not so much on what the specific content of such science-related issues are, but on how the necessary capabilities can be marshalled so as to ensure that the implications of research and the procedures for its conduct are integrated with major science policy concerns." (OECD, 1976)

Such pre-occupations remain fully current two decades later, perhaps tinged with greater cynicism about the ability to forecast long-term technology trends, and still less long-term economic and political developments, normally of greater significance than most technologies. But in the 1970s, such needs were widely felt: reference has been made to the European Community's FAST Programme of Forecasting and Assessment in Science and Technology. This itself was partly stimulated by the existence and perceived usefulness of the US Congressional Office of Technology Assessment (created 1972). In the mid-80s, the European Parliament established its "Science & Technology Options Assessment" group, STOA. Further examples could be multiplied around the world – POST in the UK, OPCST in Paris, NOTA in the Netherlands, TAB in Bonn, etc.

Returning to OECD, it was only in 1980 that the CSTP agreed to include biotechnology in its work programme for the following year. CSTP, supported by the corresponding Directorate for Science, Technology and Industry, was the first, and for a number of years, the only body of the OECD to deal extensively with biotechnology, due to the science-based origins of this technology. More recently, the increasing "pervasiveness" of biotechnology applications has led other Committees of the OECD to develop interests and relevant activities – particularly the Environment Committee and Directorate; still more recently the Committee for Agriculture; and most recently, the Industry Committee.

CSTP work on biotechnology started in 1981. The subject has remained on the agenda of the Committee ever since, thus becoming a continuous feature of OECD work. Content and direction of the Committee's biotechnology work followed a logical plan, but one which significantly reflected biotechnology developments in the world at large; and acquired some influence upon them.

The first activity reviewed the dominant scientific and technological trends and pointed to the main policy problems, leading to the publication BULL et al., 1982.

The GNE, and the Work on Generic Bio-safety Issues

Based on the 1982 broad report, the Committee decided to carry out four follow-up activities: two on urgent and specific policy issues (safety and patent protection), and two on more general policy areas (government policies in R & D, and long-term economic impacts). Of these four, the safety issue was recognized to be of overriding importance and thus become a continuous activity, with a series of major publications; major both in the scale of resources invested in their preparation and negotiation, and in their influence. For although consensus was difficult to achieve, given wide divergences both in the perceptions of risk in biotechnology, and in the state of scientific development in different

Member countries, there remained a persistent and common interest in mutual learning, and a willingness to compromise.

Work on safety started with the completion of the preceding activity on trends and perspectives. An initial meeting of Secretariat experts in December 1982 defined a number of principles, the first of which was about scientific rationale: "Guidelines, rules and regulations have to be based on the best available scientific knowledge, and they have to be sufficiently flexible to adapt to new knowledge." This simple postulate, uncontroversial for the scientific community and basic to all rational discussion, has been much argued about, and has politically not always been accepted.

As safety assurance came to dominate government policies with regard to biotechnology, it consequently also became an ongoing OECD activity, the centerpiece of OECD's biotechnology work. The organization thus responded to the public and government concern about genetic modification, in a political atmosphere marked by environmental movements which mirrored or reinforced those concerns.

The work on biotechnology safety acquired a significance for Member countries, for reasons other than those which had led to the organization's long-term involvement in nuclear safety and in the control of chemicals. These two OECD activities responded to a history of serious risk problems and accidents, some of which endangered the environment or human health. In contrast, the biotechnology safety activity accompanied the very first developments of the new technology and even preceded them, and was maintained with high interest in spite of a continuing excellent safety record.

Interest in the safety activity strengthened in 1983, as the applications of biotechnology began to take place outside contained laboratory conditions, and the first products were moving towards commercialization. In that year, the Committee created an "Ad Hoc Group of National Experts on Safety and Regulations in Biotechnology". This group was charged with establishing scientific criteria for the safe use of genetically engineered organisms in industry, agriculture and the environment.

Approximately eighty experts, including representatives of scientific research, industry, government and regulatory bodies, as well as members of national rDNA Committees, worked for three years under the chairmanship of Dr. ROGER NOURISH from the UK Health and Safety Executive, to draft the report *rDNA Safety Considerations* (OECD, 1986). This was subsequently widely cited, worldwide; and became popularly known as the "Blue Book". The European Commission, through DG XII as chef de file, was an active member of the expert group; practically all members were also active in their national contexts, concerning biotechnology regulation.

The report devised general guidelines for the evaluation of large-scale use of rDNA organisms, and represented a major step forward in the history of biotechnology since Asilomar. The earlier NIH RAC and GMAG guidelines related to small-scale research. Agreement in the expert group constituted a first step in the harmonization process of safety principles and practices among the Member countries of the OECD.

The OECD Council adopted in 1986 the recommendations of the report, which, though not formally binding, expressed a high degree of commitment by Member countries to adopt the common scientific framework set out in the report.

The government representatives in the group came from a number of different government ministries and agencies, all having some direct interest in biotechnology safety: science and technology, environment, public health, agriculture and others. Thus, the main task, but also the chief difficulty, of the group was to reconcile the different perspectives of these agencies and to promote an interdepartmental and interdisciplinary approach – as well, obviously, as an international one. The discussions of the group often reflected these inter-departmental differences, particularly between agencies for science and technology and agencies for the environment. To facilitate dialogue, the OECD Environment Directorate, and through it the Environment Committee, actively participated in these discussions and used the Group for help with policy analysis and programme co-ordination.

The 1986 "Blue Book" included the Council Recommendation, and made three fundamental points which convey the general approach of the experts:

– any risks raised by rDNA organisms are expected to be of the same nature as those associated with conventional organisms. Such risks may, therefore, be assessed in generally the same way as non-recombinant DNA organisms;
– although rDNA techniques may result in organisms with a combination of traits not observed in nature, they will often have inherently greater predictability than organisms modified by conventional methods;
– there is no scientific basis for specific legislation to regulate the use of recombinant DNA organisms.

On the basis of these general assumptions, a new concept was defined for the safe handling at large scale of industrial applications of low-risk rDNA organisms. This concept advocates a minimum level of control, "Good Industrial Large Scale Practice" (GILSP), based on existing good industrial practices. A number of criteria were also set out which rDNA organisms should meet in order to be assigned "GILSP" status, and to be handled under GILSP: these were set out in Appendix F of the "Blue Book" (reproduced below). They were also subsequently used as the defining criteria for "Type I" microorganisms in the European Community legislation on "contained use". It was nonetheless later argued (THORLEY, 1994) that these conditions reflected too closely the experience and requirements of the pharmaceutical industry; but at the time, no other sector of the "bio-industries" was present in the debate.

Appendix F of "Recombinant DNA Safety Considerations":

Suggested Criteria for rDNA GILSP

(Good Industrial Large Scale Practice) Micro-organisms

Host Organism:
● Non-pathogenic;
● No adventitious agents;
● Extended history of safe use; OR

- Built-in environmental limitations permitting optimal growth in industrial setting but limited survival without adverse consequences in environment.

rDNA Engineered Organism:
- Non-Pathogenic
- As safe in industrial setting as host organism, but with limited surviuval without adverse consequences in the environment.

Vector/Insert:
- Well characterized and free from known harmful sequences
- Limited in size as much as possible to the DNA required to perform the intended function; should not increase the stability of the construct in the environment (unless that is a requirement of the intended function)
- Should not transfer resistance markers to microorganisms not known to acquire them naturally (if such acquisition could compromise use of drug to control disease agents).

Industrial input to these deliberations was relatively limited, and restricted to individuals with backgrounds in the pharmaceutical industry. Again, THORLEY (1994) pointed out (THORLEY, 1994) that criteria such as the precise characterization of the vector insert were not necessarily appropriate to all the research and industrial contexts to which they were subsequently applied.

The importance of the GILSP concept was nonetheless great, given that the vast majority of industrial applications typically used intrinsically low-risk organisms. A specific recommendation was made for industry to utilize, wherever possible, such low-risk organisms in industrial applications of rDNA techniques.

The approach of the 1986 report to the safety of agricultural and environmental applications was, of necessity at that time, conservative. The OECD experts felt that the safety assessment of organisms for agricultural and environmental applications was less developed than for industrial applications. General safety guidelines or criteria were, therefore, premature and a provisional case-by-case approach was recommended. They acknowledged, however, that considerable data were available on the environmental and human health effects of living organisms and that this should be used to guide risk assessment.

Thus, a largely encouraging expert view had replaced the concerns of the 1970s: the risks long associated with biotechnology remained purely conjectural.

OECD Member countries, as well as India and Latin American countries, adopted in their national safety guidelines or legislation the general OECD safety principles. These principles often became a guide to the ministries and government departments sharing responsibility for biotechnology, and thus contributed to building up a common national, as well as international, approach.

As the flexible approach adopted in the "Blue Book" called for the adaptation of safety assessments to new knowledge, a revision was undertaken in 1988 by a follow-up "Group of National Experts on Safety in Biotechnology", widely referred to thereafter as "the GNE".

This revision aimed at the elaboration of GILSP criteria and the identification of general safety principles for agricultural and environmental applications of plants and microorganisms.

As the GILSP concept was relatively new, the GILSP criteria were elaborated to assist countries in their correct interpretation. The revision provided, for each criterion, an illustration of the nature of the different requirements and of the way these should be met.

The second major area of revision concerned the safety of the introduction of genetically modified organisms into the environment for agricultural or environmental purposes. The increasing number of field tests performed since 1986, and the increasing number of those planned, led the OECD experts to define a set of "Good Developmental Principles" (GDP). These principles were to guide researchers in the design of small-scale field experiments with genetically modified plants and microorganisms. By early 1993, more than 800 experiments, at more than 1100 sites, had been carried out in the world. New knowledge and this experience enabled the development of GDP, judged to be premature in 1986. But simplification for the oversight of low-risk releases was becoming essential: by end of 1984, there had been over 3000 site-releases in the US alone, numbers having doubled annually for several years.

GDP identifies three key safety factors: the characteristics of the organisms, the characteristics of the research site, and the use of appropriate experimental conditions. It also defines the different ways in which GDP can be met. While existing national or international codes of good practice for the safe conduct of research address primarily human health and worker safety, GDP also takes into account environmental safety.

The revision was published with the title *Safety Considerations for Biotechnology – 1992* (OECD, 1992).

Large-scale Release, and Sectoral Applications of Microorganisms

After 1991, the GNE broadened its activity to cover a number of areas, some of these new. Safety work in the Secretariat had been carried out in close co-operation between the Directorate for Science, Technology and Industry (DSTI) and the Environment Directorate, which shared the tasks. These areas were:

(i) Guiding principles for large scale releases of genetically modified organisms, extending to plants, microorganisms, and animals.

This work included the completion of general statements of safety principles for all modified organisms (the "Preamble" document; OECD, 1993c), as well as specific work on crop plants, leading in particular to the production of the reports: *Safety Considerations for Biotechnology – Scale-up of Crop Plants, Traditional Crop Breeding Practices: A Historical Review* and *Field Release of Transgenic Plants, 1986–92* (OECD, 1993d, e, f). It also includes programmes of work on various categories of microorganisms: biofertilizers, live vaccines, bioremediation/biomining, biopesticides, biofeeds, all leading towards corresponding reports. This shift towards sectoral safety studies reflected the general development of biotechnology and of related risk assessment work and regulatory developments.

The report on the Langen workshop of November 1993 (OECD & IABS, 1994) was an important example. This considered the non-target safety aspects of live recombinant vaccines, addressing both human and animal vaccines, viral and bacterial, with leading experts in each field drawn from many parts of the world. Speakers reviewed the past successes and problems of vaccines in relation to various diseases, demonstrating the great increase in understanding which modern molecular genetics was bringing to established vaccines, as well as opening the way to new and safer ones. Various potential problems specific to recombinant vaccines were reviewed, and the means of addressing these. In conclusion, it was reported that, "While recombinant should therefore be investigated with caution, it does not seem justified to treat them differently from other live vaccines".

(ii) Safety assessment of food produced by biotechnology, from terrestrial and aquatic organisms (by the Environment Directorate in co-operation with DSTI) was again a sectoral example with sectorally specific aspects.

Work focusing on terrestrial organisms, including a workshop in Copenhagen in 1991, led to the publication *Safety Evaluation of Foods Derived by Modern Biotechnology – Concepts and Principles, 1993* (OECD, 1993g). This book was timely. It came at a critical moment, responding to public concerns and discussions in some Member countries about new foods based on biotechnology. The book elaborates scientific principles to be considered in making evaluations of new foods or food components based on a comparison with foods that have a safe history of use. The most practical approach to determine the safety of foods derived by modern biotechnology is to consider whether they are "substantially equivalent" to analogous traditional food products. The case studies in this report illustrate the application of the concept of substantial equivalence. For new foods to which this concept was not applicable, further work continued.

An OECD workshop in Oxford, UK, in September 1994 opened the consideration of a broader class of questions on the safety evaluation of foods in general, particularly novel foods; and the close interaction with the UN WHO/FAO activities in this field was ensured by the presence of key experts also working in these other contexts (particularly in relation to the WHO/FAO expert consultation on food safety assessment methods of autumn 1995). This allowed the participating individuals, and hence their organizations, to define the respective roles of the different organizations in a coherent manner, thus facilitating the widespread acceptance of common scientific principles as the bases for regulations in this vast and fundamental sector.

(iii) Reviews of monitoring methods for genetically modified organisms in the environment (OECD, 1994a, b), based on the 1992 workshop in Ottawa; and the computerized pointer system for releases, leading to the annual publication on diskettes of the "BIOTRACK" pointer system (Environment Directorate).

The mandate of the GNE ended in March 1994. There was widespread consultation and debate over what should succeed it. General agreement on the need to address a wider range of issues than safety carried the discussion into areas of interest to several Committes and Directorates, underlining a greater need for co-ordination. The solution adopted at the end of 1993 was to establish an Internal Co-ordination Group for Biotechnology, chaired by a Deputy-Secretary-General, as a Secretariat responsibility, with involvement of the various interested Directorates (see OECD, 1994d). The parallel with the European Commission's "Biotechnology Coordination Committee" (see Sect. 7.2) is evident.

The CSTP itself agreed to establish a new "Working Party on Biotechnology", with the broader mandate reproduced below:

Working Party on Biotechnology

Of the Committee for Scientific and Technological Policy

1. The general objective of the Group shall be to keep under review and advise upon science, technology and innovation issues in biotechnology, with a view to assisting the development of its safe and effective use, by *inter alia*, encouraging the international harmonization of science-based principles and practices, and facilitating international scientific and technological collaboration and exchange.
2. The relevant elements of the Work Programme of the Committee for Scientific and Technological Policy (CSTP) shall determine the scope and content of the Group's activities. The Group's functions shall include:
– providing a forum for Member countries to make known and to discuss their needs and priorities;
– hence enabling the identification of significant issues for the Work Programmes of relevant OECD Committees;
– advising the CSTP upon the prioritization of suggested activities for future Work Programmes;
– managing the implementation of the biotechnology elements of the current CSTP Work Programme, including project activities.
3. The Group will report to the Committee for Scientific and Technological Policy, which shall keep other Committees – in particular the Environmental Policy Committee, the Industry Committee and the Agriculture Committee – informed on the progress of work relevant to their interests; in particular, through the Internal Co-ordination Group for Biotechnology. This should lead to avoidance of duplication, and the promotion of joint activities, where appropriate.
4. The Mandate shall run from 1 April 1994 to 31 August 1998, subject to modification by decision of the CSTP; with a mid-term review to assess the value, impact and effectiveness of the work of the Group.

The new Working Party first met in June 1994, within a Work Programme specification set by their parent Committee and with specific projects in all the three major areas of application: agro-food; health care; and environment. As in the earlier work, the OECD was not directly involving itself in the drafting of regulations; but continues through various Directorates and Committees to make two fundamental contributions:

– developing common scientific concepts and principles, to underpin regulation where this necessary;
– conducting enquiries in co-operation with Member governments, and in relation to various sectors of relevance to biotechnology (e.g., in 1994, on bio-pesticides; in 1995, on intellectual property), to encourage mutual understanding of the data requirements of regulators in different countries, and hence facilitate movement towards common approaches and ultimately mutual acceptance of data, and of product authorization decisions.

6.3 The UN Agencies (1): WHO, FAO, ILO, IPCS, UNIDO, the World Bank and CGIAR, ICGEB, UNESCO and ICSU

The pervasive character of biotechnology has brought it onto the agenda of practically every agency of the United Nations (UN). The perspectives would differ from agency to agency, but usually involved some aspects of promotion and/or regulation. Some of the salient developments are presented below. Although many of these UN initiatives were slow to develop or late in starting, they continue to demand political attention, and to promote or elicit calls for regulatory action at global level. Such calls have been encouraged by parallels with other topics in which global agreements or conventions have been adopted and are implemented, or moving towards implementation.

A list may serve to summarize and introduce the following sections.

1. World Health Organization (WHO) – experience of developing and diffusing safe working practices, interested in the potential of biotechnology for vaccines and other biologicals, and responsible for the publication of many Technical Reports on guidelines and requirements for such products (cf. WHO, 1983, for an early example). In later years, WHO became an outspoken advocate of the importance of biotechnology and genetic engineering for world health (WHO, 1994).
2. Food and Agriculture Organization (FAO) – interested in the potential of biotechnology for enhancing the quantity and quality of food production; involved in issues concerning conservation and use of genetic resources, particularly relating to crop plants.
3. UN Educational, Scientific and Cultural Organization (UNESCO) – longstanding programmes in applied microbiology, particularly the worldwide network of Microbial Resource Centres (MIRCENs), the international GIAM conferences (Global Impacts of Applied Microbiology), support for ICSU, the International Council of Scientific Unions.

4. UN Industrial Development Organization (UNIDO) – interested to promote the safe development and application of biotechnology in the interests of developing countries; launching of ICGEB, the International Centre for Genetic Engineering and Biotechnology; related service, BINAS (Biotechnology Information Network and Advisory Service); leadership of the inter-agency (UNIDO/ UNEP/WHO/FAO) Working Group on Biotechnology Safety (Voluntary Guidelines on GMO Release: UNIDO, 1991); co-ordinating role for Chapter 16 ("Biotechnology") in AGENDA 21 (see Sect. 6.4).
5. The World Bank – not only financing a growing number of projects involving biotechnology, but chairing the CGIAR (Consultative Group for International Agricultural Research), and thus involved in advising all the 18 IARCs (International Agricultural Research Centres) on policy questions of common interest – including those relating to intellectual property rights, and bio-safety and regulation. (cf. PERSLEY et al., 1992).
6. UN Environment Programme (UNEP) – see Sect. 6.4. Involved in the inter-agency Working Group formulating Guidelines for Deliberate Introductions.
7. Advanced Technology Assessment System (ATAS) – preparation and diffusion of informative reports on biotech applications of interest to developing countries, e.g., "Biotechnology and Development: Expanding the Capacity to Produce Food" (UN, 1992).

From the Control of Chemicals to the Control of Biotechnology – A False but Influential Analogy

In describing the history of the biotechnology regulatory initiatives in the European Community (Sect. 4.4), extensive reference was made to the prior and continuing international developments concerning the regulation and control of chemicals. The history of the international development of regulations and control for chemicals has been written elsewhere, in documents ranging from the details of regulatory texts and technical annexes, to partly autobiographical overviews by some of the individuals involved; the review by LÖNNGREN (1992) is particularly valuable.

As the chemical industry developed and expanded – and especially in the postwar decades – the control of chemicals was recognized to be essential, because of their effects upon human health – either directly, or via residues in the food supply; and (a later development in regulatory terms), because of their effects upon the environment. Thus the challenge of chemical safety engaged the responsibilities of several of the UN agencies, created in 1944 and subsequent years: particularly those concerned with public health, worker safety, food and agriculture, and the environment. All these areas of regulatory responsibility would also be later considered in relation to the applications of biotechnology, and its (or their) regulation.

Although the problems are in principle universal, the responsibilities for establishing and operating systems for regulation and control are necessarily local and national. Thus the development of the present systems for controlling the safety of chemicals involved several "levels", political or geographical, including the following:

- national (the US TOSCA being of particular significance also in international terms)
- European (see Sect. 4.4, concerning the evolution of the "Sixth Amendment" Directive)
- OECD (since most new chemicals, and the bulk of trade, originate in the OECD area)
- UN agencies, particularly WHO, FAO, ILO and UNEP.

The relative success of those efforts, at all the four levels and in the various contexts cited above, has strongly underpinned the assumptions and efforts of the same agencies – often the same individuals – when they sought to repeat in the context of biotechnology and/or genetically modified organisms the successful story of chemicals control.

The history of biotechnology regulation, and the evolution of the debates scientific and (later) political, indicate that attempts to apply to biotechnology the concepts, approaches and regulatory instruments developed for chemicals may be based upon a false

analogy; but the attempts to make such a transfer were of major influence. It is therefore important to an understanding of the initiatives of UN agencies vis-à-vis biotechnology to include also some elements summarizing their roles in the earlier (and continuing) success story.

More recently, as biotechnology moves into routine application in different sectors, the influence of the generic approach adopted in regard to chemicals might be expected to diminish in relevance. However, the three major sectors of application themselves coincide with the sectoral responsibilities of three of the UN agencies most active in the earlier work on chemicals:

- agriculture and food (FAO)
- health care and pharmaceutical products (WHO)
- environment (UNEP).

Thus the role of these and other UN agencies vis-à-vis biotechnology remains of considerable, or even increasing, importance.

Historical Developments – WHO and FAO – Additives and Residues in Food

The signing of the Charter of the United Nations at San Francisco, in 1945, provided a fresh starting-point and basis for international machinery to address common challenges such as health and food. The near-parallel founding of the "Bretton Woods" institutions, particularly GATT, reflected a similar willingness to design a world system more open than that of the pre-war period, and in which the growth of world trade would require common machinery to address such issues as the safety of chemicals, medicines and foods, and to avoid obstacles to trade in these products.

The Food and Agriculture Organization of the United Nations (FAO) was also created in 1945, and the constitution of the World Health Organization (WHO) was signed the following year. In the immediate post-war years, basic food supply and nutrition were obvious priorities, and the role of chemicals as fertilizers and to reduce losses to pests was obvious. But from the 1950s, concerns were shifting from assuring the quantity of food, to considerations of its quality and safety.

Concerns about chemical (particularly pesticide) residues in foods, and synthetic food additives, led WHO and FAO to collaborate in joint activities to bring the best scientific expertise in the world to bear upon these problems. The World Health Assembly – the governing body of the WHO – noted in 1953 that the increasing use of various chemicals in the food industry had in recent decades created a new public health problem. In response, WHO and FAO initiated two series of annual meetings: on food additives (Joint FAO/WHO Expert Committee on Food Additives, JECFA), and on pesticide residues. The first meeting on food additives was held in 1956, and that on pesticide residues in 1963. Joint Meetings of the FAO Panel of Experts on Pesticide Residues in Food and the Environment, and the WHO Expert Group on Pesticide Residues, are usually referred to as the Joint FAO/WHO Meeting on Pesticide Residues, or JMPR.

In 1980, the WHO activities concerned with the safety assessment of food chemicals were incorporated into the International Programme on Chemical Safety (IPCS), whose genesis and activities are described below.

In 1962, the Codex Alimentarius Commission was established, to implement the Joint FAO/WHO Food Standards Programme. The Commission was created as an intergovernmental body, which now has more than 120 Member nations, whose delegates represent their countries. The Commission's work is carried out through various Committees, such as the Codex Committee on Food Additives (CCFA). JECFA serves as the advisory body to the Codex Alimentarius Commission on all matters concerning food additives.

Since the early 1960s, the JMPR has evaluated a large number of pesticides. The WHO component of these Joint Meetings relied upon procedures developed by other expert groups – particularly JECFA – and developed specific principles for evaluating the various classes of pesticide that are used on food crops and may leave residues on them. The WHO issued a series of publications on "Environmental Health Criteria", of which number 70 on "Principles for the safety assess-

ment of food additives and contaminants in food" (WHO, 1987) summarized the assessment procedures used by JECFA.

A later (WHO, 1990) publication, on "Principles for the Toxicological Assessment of Pesticide Residues in Food" gathered together and updated the assessment principles relevant to this field. By that date, it seemed appropriate to include a page specifically on biotechnology, because of three areas of emerging concern:

– the production of chemicals of biological origin with pest control activity (e.g., pyrethrin and derivatives)
– the use of microbial pest control agents
– the development and use of genetically altered (bioengineered) organisms for specific purposes.

Biotechnology in Foods – Current FAO/WHO Initiatives

The shared interests in food safety of FAO and WHO over several decades have been indicated by reference to the work of the joint bodies JECFA and JMPR, both conducted since 1980 under the aegis of the IPCS. It became evident in the 1980s that modern biotechnology would find applications in the organisms used by the food industry (for fermentation, enzymatic processing, etc.) or in the foods themselves (either by improvement of traditionally used strains, or as novel foods).

A "Joint FAO/WHO Consultation on the Assessment of Biotechnology in Food Production and Processing as Related to Food Safety" was held in Geneva in November 1990. The conclusions of the consultation emphasized the long historical experience with various biotechnologies in food production, and its tone was positive.

> "The newer biotechnological techniques, in particular, open up very great possibilities of rapidly improving the quantity and quality of food available. The use of these techniques does not result in food which is inherently less safe than that produced by conventional ones."

After citing examples, the report (WHO, 1991) stated that "whenever changes are made in the process by which a food is made or a new process introduced, the implications for the safety of the product should be examined. The scope of the evaluation will depend on the nature of the perceived concerns."

The conclusions of the report continued:

"5. The evaluation of a new food should cover both safety and nutritional value. Similar conventional food products should be used as a standard and account will need to be taken of any processing that the food will undergo, as well as the intended use of the food.

6. Comparative data on the closest conventional counterpart are critically important in the evaluation of a new food, including data on chemical composition and nutritional value. The Consultation believed that such data are not widely available at the present time.

7. A new, multidisciplinary approach to the safety evaluation of new foods is desirable, based on an understanding of the mechanisms underlying changes in composition. Detailed knowledge of the chemical composition of the food, together with information on the genetic make-up of the organisms involved, should form the basis of the evaluation and will indicate the necessity for toxicity testing in animals. The approach will be facilitated by the integration of molecular biology into the evaluation process.

8. The Consultation agreed a set of scientific principles to be applied to the evaluation of the safety of foods produced by biotechnology, although at present they would need to be applied on a case-by-case basis.

9. In due course it will be possible to develop a framework for the evaluation of all new foods, including those produced by biotechnology. This will need to be flexible, the data needed depending on the nature and use of the product. There is at present little experience from which to develop general criteria for such a framework and, until such time as these criteria can be developed, a case-by-case approach is required.

10. As far as the products of the newer biotechnologies are concerned, detailed knowledge of their molecular biological properties will facilitate the evaluation process. It is already possible to identify many of the categories of data that will be necessary. In due course it will be possible to identify the genetic elements that are likely to be acceptable for use in food-producing organisms.

11. To facilitate the safety evaluation of foods produced by means of biotechnology, action at the international level will be necessary to provide timely expert advice in this matter to Member States of FAO and WHO, the Codex Alimentarius Commission, the Joint FAO/WHO Expert Committee on Food Additives and the Joint FAO/WHO Meeting on Pesticide Residues.

12. The Consultation concluded that, because of the rapidity of technological advances in this area, further consultations on the safety implications of the application of biotechnology to food production and processing will be advisable in the near future."

This report was in part consistent with the contemporaneous OECD work on food safety, Point 6 on comparison with the "closest conventional counterpart" being closely consistent with the concept of "substantial equivalence" advocated in the OECD report (OECD, 1993g); but much of the language of Points 6 to 10 was in fact hugely ambitious, or could be so read.

The FAO/WHO report's Recommendations went on to set out what was in effect an outline programme for work over the following years, or simply a broad statement of intent:

"1. Comprehensive, well enforced food regulations are important in protecting consumer health, and all national governments should ensure that such regulations keep pace with developing technology.

2. National regulatory agencies should adopt the strategies identified in this report for evaluating the safety of foods derived from biotechnology.

3. To facilitate the evaluation of foods produced by biotechnology, data bases should be established on:
   – the nutrient and toxicant content of foods;
   – the molecular analysis of organisms used in food production;
   – the molecular, nutritional and toxicant content of genetically modified organisms intended for use in food production.

4. Consumers should be provided with sound, scientifically based information on the ap-

plication of biotechnology in food production and processing and on the safety issues.

5. FAO and WHO, in cooperation with other international organizations, should take the initiative in ensuring a harmonized approach on the part of national governments to the safety assessment of foods produced by biotechnology.

6. FAO and WHO should ensure that timely expert advice on the impact of biotechnology on the safety assessment of foods is provided to Member States, the Codex Alimentarius Commission, the Joint FAO/WHO Expert Committee on Food Additives and the Joint FAO/WHO Meeting on Pesticide Residues.

7. FAO and WHO should convene further consultations at an appropriate time to review the Consultation's advice in the light of scientific and technical progress."

As biotechnology advanced from making marginal, well-characterized changes, towards its use (and/or with other modern technologies) in producing novel foods, the scope of discussion had to broaden, as illustrated by four events or activities in the 1990s:

(i)   The European Commission's draft Council Regulation on Novel Foods included genetically modified organisms in its scope; but the Commission sought to avoid any stigmatization of the techniques of biotechnology by setting them in the broader context indicated by the title of the regulation, and maintained a risk-based approach (subject to the obligatory reference to the environmental risk assessment requirements of the Field Release Directive 90/220 – obligatory if GMO food products were in future to fall under the new Regulation).

(ii) The OECD work on food safety was advanced by a workshop held in Oxford in September 1994, under the aegis of the former Group of National Experts on Safety in Biotechnology, (which in March 1994 was replaced by a body of broader mandate – see Sect. 6.2) but addressing the more general issue of "safety assessment in food". As biotechnology-re-

lated aspects moved from minor constituents or marginal changes, to providing foods which would constitute major elements of the whole diet, it was recognized that different problems were posed, different methods required: techniques, toxicological and other, designed to appraise minor or potentially toxic constituents, could not be scaled up by a multiplying factor as safety margin, if they were envisaged as major components of the whole diet, of man or of a test animal. Principal scientific issues were rather nutritional, including problems of extrapolating from animals to humans in the knowledge that the nutritional requirements of different species are far from identical.

(iii) The Danish Government in November 1994 hosted a WHO workshop to develop in more operational detail the concept of "substantial equivalence".

(iv) WHO and FAO prepared for late 1995 an expert consultation on the safety assessment of foods produced by biotechnology, addressing these broader issues.

In the context of the European Union, the debate on the Novel Foods Regulation became for a time the focus of political demands for GMO labelling, on the grounds of consumer information and transparency. While this seemed plausible in the context of products such the Calgene "Flavr Savr" tomato, it was clearly unrealistic to envisage for fungible products such as grains, where distinct labelling would imply a separation and duplication of the whole storage, transport and distribution systems, from the farm to the final bakery outlet or animal feed-lot.

Against this background of politicization and controversy, the continuing international and science-based approach of the UN Agencies FAO and WHO, and of the OECD, provided and continues to offer an important source of objective and internationally acceptable advice and reference materials.

Chemicals Control: ILO, EC, IPCS and OECD

A major role in the control of chemicals was played by the International Labour Organization, ILO; a surviving element from the 1919 Peace Treaty of Versailles. One of ILO's primary concerns was and remains occupational safety. In 1950, the ILO Chemical Industries Committee called for a classification and labelling system for hazardous chemicals. In 1956, it published a "List of Dangerous Substances Presenting an Occupational Hazard".

The evolution of an integrated approach to the risk assessment, regulation and control of chemicals had several strands; including both UN agencies, and the OECD. In addition, after the signing in 1957 of the Treaty of Rome, establishing the European Economic Community, the EEC also became a significant player in the international arenas concerned. The OECD itself, although based on the earlier (1948) Organisation for European Economic Co-operation, was created in its current form only in 1960.

Through the 1960s, concerns about the environmental impacts of pesticides and other chemicals continued to rise. The publication of RACHEL CARSON's "Silent Spring" in 1962 is often cited as a landmark in the recognition of the need for more effective control of chemicals, particularly pesticides. In the context of biotechnology, we note in passing the plea, less known but no less worth citing, which CARSON made for a more biologically-based technology:

"A truly extraordinary variety of alternatives to the chemical control of insects is available. Some are already in use and have achieved brilliant success. Others are in the stage of laboratory testing. Still others are little more than ideas in the minds of imaginative scientists, waiting for the opportunity to put them to the test. All have this in common: they are biological solutions, based on understanding of the living organisms they seek to control, and of the whole fabric of life to which these organisms belong. Specialists representing various areas of the vast field of biology are contributing – entomologists, pathologists, geneticists, physiologists, biochemists, ecologists – all pouring their knowledge and their creative inspirations into the formation of a new science of biotic controls."

This plea, preceding the arrival of modern biotechnology, may yet be answered by it; but we return to the history.

LÖNNGREN's (1992) "Historical Overview" includes a rich chronology of events, which summarize some of the defining moments in the evolution of international approaches to chemicals control. From LÖNNGREN and other sources can be compiled the following:

1961: Joint FAO/WHO expert meeting on the use of pesticides in agriculture

1962: Establishment of the Codex Alimentarius Commission
Publication by the Council of Europe of the first edition of "Yellow Books" with lists of some 500 chemicals and recommendations for labelling

1963: First Joint FAO/WHO meeting on Pesticide Residues

1966: OECD meeting, Jouy-en-Josas, France, on "Unintended Occurrence of Pesticides in the Environment"

1967: Adoption by the Council of the European Communities of Directive 67/548, on "the approximation of the laws, regulations and administrative provisions relating to the classification, packaging and labelling of dangerous substances"

1968: UN General Assembly decision to convene in 1972 a "Conference on the Human Environment"

1969: Establishment by ICSU (the International Council of Scientific Unions) of SCOPE, the Standing Committee on Problems of the Environment
OECD convenes a Study Group on the Unintended Occurrence of Pesticides

1970: OECD Council establishment of Environment Committee
World Health Assembly request to WHO Director-General to develop a long-term programme for environmental health (launched in 1973, as the Environmental Health Criteria Programme)

1971: OECD establishment of a Sector Group on the Unintended Occurrence of Chemicals in the Environment (which from 1975 becomes the Chemicals Group)
UNESCO launch of the programme Man and Biosphere

1972: UN Conference on the Human Environment, Stockholm
Establishment of the UN Environment Programme, UNEP

1974: OECD Council adoption of a Recommendation on Premarketing Assessment of the Potential Effects of Chemicals on Man and his Environment

1977: WHO Resolution 30.47, expressing concern about the toxic effects of exposure to chemicals, and requesting the Director-General of WHO to study the problem and propose long-term strategies
OECD Council adoption of a Recommendation concerning Guidelines in respect of the Procedures and Requirements for Anticipating the Effects of Chemicals on Man

1978: OECD Council decision to set up a Special Programme on the Control of Chemicals

1979: EC Council adoption of the "Sixth Amendment" of Directive 67/548

1980: The Heads of UNEP, ILO and WHO sign a Memorandum of Understanding establishing the International Programme on Chemical Safety (IPCS)

1981: OECD Council Decision concerning the Mutual Acceptance of Data in Assessment of Chemicals
Publication of OECD Test Guidelines
Publication of OECD Principles of Good Laboratory Practice (GLP)

1982: OECD Council Decision concerning the Minimum Pre-marketing Set of Data (MPD) in the Assessment of Chemicals
WHO expert meeting on "Health Impact of Biotechnology", in Dublin

1983: OECD Council adopts three Recommendations concerning problems on confidentiality of chemicals data and proprietary rights
OECD Council adopts a Recommendation concerning Mutual Recognition of Compliance with GLP

1984: Formal agreement between OECD Chemicals Programme and IPCS, calling for reciprocal use of guidelines, methodologies and evaluations developed by the two organizations

1985: IPCS Programme Advisory Committee supports the development of Health and Safety Guides and International Chemical Safety Cards, as complements to Environmental Health Criteria documents
FAO Conference adopts the International Code of Conduct on the Distribution and Use of Pesticides

1986: OECD publishes "Recombinant DNA Safety Considerations"; OECD Council adopts related Recommendation.

1987: UNEP Governing Council adopts the London Guidelines for the Exchange of Information on Chemicals in International Trade

1989: Incorporation of "Prior Informed Consent" (PIC) procedure in the FAO Code of Conduct on the Distribution and Use of Pesticides

1990: ILO General Conference adopts a Convention concerning Safety in the Use of Chemicals at Work

1992: The UN Conference on Environment and Development in Rio de Janeiro, Brazil; including adoption of the Convention on Biological Diversity, and of the report, "AGENDA 21" (see Sect. 6.3)
OECD publishes "Safety Considerations for Biotechnology – 1992".

1993: OECD workshop, Langen, Germany, on "Non-Target Effects of Live Recombinant Vaccines"

1994: WHO Conference, Geneva, on "Biotechnology and World Health: vaccines and other biologicals produced by genetic engineering, review of risks and benefits"
OECD Workshop, Tokyo, on "Bioremediation": coinciding with publication of "Biotechnology for a Clean Environment"
First Conference of the Parties to the Convention on Biological Diversity, launches a one-year consideration of the need "for a Protocol on Bio-safety.

These specific decisions and events are a shorthand representation, the tips of icebergs, of a great and ever-increasing volume of international activity; the events and the activity reflecting a political will, and/or a perceived necessity to address the problems of chemicals control, and to develop the corresponding machinery on an international basis. Much of the activity could be characterized as reactive, an attempt to repair situations of already severe damage; but the same perceptions and experience argued for attempting to shift to a more anticipatory, "pro-active" mode.

LÖNNGREN notes that "by the early 1970s, officials in many advanced industrialized countries realized that the burden of carrying out the necessary research and development for effective chemicals control was too great for any single country. Member countries of the OECD, for example, began to explore ways in which they could effectively share necessary technical, manpower and financial resources. Given the significant trade in chemicals, another incentive to international co-operation was the recognition that independent or isolated national action could lead to international confusion and the view of such actions as technical barriers to trade".

Significant roles in the evolution of international co-operation in chemicals control were played also by the UN Economic Commission for Europe; and by the Council of Europe (see Sect. 6.1).

Thus for all of the bodies mentioned above, there developed around chemicals control an awareness and perception of need, and a corresponding political will; with the consequent resource commitments; and the gratifying experience of successful development and adoption of many international instruments. This experience and its timing influenced in various ways the responses of the various bodies involved to the diverse challenges of biotechnology.

As indicated already, the WHO was interested in both the positive potential, and the safety aspects, of biotechnology for health care applications. In 1982, an expert meeting was convened in Dublin, Ireland, by the European office. Its conclusions were cautiously positive, including the following:

"– The conjectural risks of the application of recombinant DNA and other techniques of biotechnology can be assessed and managed with current risk assessment strategies and control methods.
– Biotechnology in general is regarded as a safe industry. To ensure that future developments take place in an equally safe manner, the implementation of new biotechnological techniques and processes should be monitored for long-term, adverse effects.
– Biotechnology contributes in several ways to the improvement of human health, and the extent of this contribution is expected to increase significantly in the near future. To support this development further, the free exchange of information in this area between all parts of society, as well as between nations, should be strongly encouraged.

– A general responsibility of all those concerned with biotechnology is to inform the public of its impact on health and the environment. Pamphlets and other educational materials on biotechnology should be developed and distributed to those involved in the education of the public." (WHO, 1984)

The promise of biotechnology for health care applications took some years to be confirmed, but the 1980s provided sufficient concrete successes of biological production methods (interferons, interleukins, growth hormones, monoclonal antibodies), diagnostics (monoclonals, enzymes, PCR), and safe and effective new vaccines through genetic engineering.

By the end of the 1980s, the methods of biotechnology were figuring ever more significantly in the toolbox of the WHO's development programmes. For vaccines especially, the goal of a multi-valent, single-shot live vaccine for the main childhood diseases could be considered attainable. In November 1994, the WHO Conference in Geneva on "Biotechnology and World Health" made clear the value of modern biotechnology to the aims of the Organization (WHO, 1994).

UNIDO

The UN Industrial Development Organization (UNIDO), WHO and UNEP in 1985 organized an informal working group to consider biosafety in relation to research institutions, industry and the environment. The Working Group took into account the fact that UNIDO was then in course of establishing the International Centre for Genetic Engineering and Biotechnology (ICGEB) and pressed for its active role in the study of actual and conjectural hazards, in developing a risk assessment methodology, in conducting assessments, and in developing biosafety guidelines for its Member countries.

Of the ambitious objectives defined in 1985, many have since been addressed by other bodies, or changed in significance; but the working group gradually focused upon the drafting of a Voluntary International "Code of Conduct" for Biosafety, with particular reference to the release of GMOs into the environment. It was envisaged that the code could provide a minimum framework or guideline to help governments in developing their own regulatory infrastructure and in establishing standards, or in obtaining appropriate advice and support in cases where a regulatory infrastructure did not yet exist.

The key elements of the Code of Conduct could be summarized as follows:

- *Scope:* comprehensive, in principle covering all types of GMOs (plants, animals, microbes) and their products;
- *General Principles:* focus on the product, rather than the process; work with well-characterized sequences; safety precautions should be appropriate to the level of risk; monitoring; public information; reporting of adverse incidents; science-based risk assessment; review systems adaptable in the light of scientific information;
- *Actions for Government and Regulatory Authority:* designation of a national bio-safety authority; examination of existing regulatory mechanisms to consider their application to GMOs, if necessary with revision; independent risk assessment of proposals, possibly using experts from another country; respect for confidential business information; exchange of information;
- *Responsibilities of the Researcher/Proposer:* risk evaluation, record keeping, notification, obtaining approval, disclosure of all relevant information, suggestion of independent review mechanisms if the country does not yet have a designated national authority.

Where regulatory review within a country was not feasible, it was suggested that those wishing to conduct such GMO releases might call upon a supranational "Biotechnology Information Network and Advisory Service", BINAS, based on UNIDO/ICGEB.

The draft code of conduct (UNIDO, 1991) met with widespread agreement, and was completed in the early 1990s. The Preparatory Committee for the UNCED conference (see Sect. 6.4) acknowledged the work of the UNIDO/UNEP/WHO/FAO group (FAO having now joined it), and commended it to the Conference Secretariat.

World Bank/CGIAR

Since the early 1970s, donor funding (from national governments and private foundations) for International Agricultural Research Centres (IARCs) has been channelled through the Consultative Group for International Agricultural Research (CGIAR). This ensures transparency, and provides a basis for information exchange and policy development on matters of common interest – such as biosafety, and intellectual property rights. The CGIAR Chair and Secretariat are provided by the World Bank.

In addition to the IARCs in the various regions – e.g., the International Rice Research Institute (IRRI) in the Philippines, CIMYTT in Mexico for wheat and maize, etc., there is in the Netherlands, at The Hague, the International Service for National Agricultural Research (ISNAR). With the support of the Dutch Government (through DGIS, the Directorate for International Development Cooperation), a BIOTASK Task Force was established, based at ISNAR, to provide for all the IARCs some common guidance on the policy challenges of biotechnology.

Secretary of BIOTASK was GABRIELLE PERSLEY, editor of a collection of World Bank Technical Papers on Biotechnology and Agricultural Development, and author of a summary report based on these (PERSLEY, 1990a, b). PERSLEY with the collaboration of VAL GIDDINGS of USDA and CALESTOUS JUMA of ACTS (the African Centre for Technology Studies, Nairobi), wrote for World Bank/ISNAR a short report on Biosafety. The report commends existing work by OECD (see Sect. 6.2) and the UNIDO/UNEP/WHO/FAO group (see above); advocated the creation of National Biosafety Committees; but emphasizes primarily the use so far as possible of existing national regulatory structures for the product categories concerned (PERSLEY et al., 1992).

6.4 The UN Agencies (2): UNEP, Rio, AGENDA 21, CSD and the Convention on Biological Diversity

UNEP, the Rio "Earth Summit", and the Convention on Biological Diversity

The UN Environment Programme, launched at the Stockholm Conference in 1972 has a much shorter history than most of the UN agencies mentioned above; and therefore has not had the opportunity to build up the long and widespread track record of expert activity such as WHO and FAO have achieved. Moreover, although the scientific problems of food production, nutrition, toxicology and human health are complex, their complexity is relatively well explored and understood as compared with the vast uncertainties and complexities of the whole planetary ecosystem, and human interactions with it.

Thus in terms of scientific mastery of its domain of responsibility, UNEP is at a much earlier point on its learning curve than WHO and FAO. Yet the urgency of the environmental challenges has demanded an acceleration of efforts and action, not only scientific, but political and regulatory/operational. Current industrial practices and consumption habits, in both the developed world, and the developing world, are environmentally polluting and unsustainable in many obvious respects. The most optimistic rebuttal of the Club of Rome's "Limits to Growth" scenario (MEADOWS et al., 1972) would argue that resource shortages, price rises, market responses and human ingenuity may yet evolve an adaptive trajectory to continue economic growth and development in a sustainable manner. But the current evidence of pollution and damage; the population forecasts; and the forecasts and expectations of rapid economic growth in the developing world; combine to indicate that the costs of the transition to sustainability will be high; and higher still, the longer necessary changes are deferred.

Such was the widely accepted reasoning articulated by MAURICE STRONG, Secretary of the preparatory group for the June 1992 "Earth Summit", the UN Conference on Environment and Development, held in Rio de Janeiro two decades after the Stockholm Conference of '72. Rio was highly publicized in the world's media, and served to highlight environmental challenges; and its various activities inevitably included extensive consideration of biotechnology matters.

The treatment of biotechnology at Rio was nonetheless ambivalent, for several reasons: acknowledging its importance to the aims of sustainable development, but tending to reiterate in the global forum the long-rehearsed arguments about conjectural risks and uncertainties associated in particular with the field release of genetically modified organisms. Four reasons may have contributed to this less than balanced picture:

(i) The conference itself had a strong presence of non-governmental organizations (NGOs), a presence emphasized by their ability to attract media attention; and many of the NGOs were hostile to biotechnology, not least because they distrusted the developed countries and industrial interests seen as promoting biotechnology; although there was industrial representation and communication (e.g., a written statement by the International Biotechnology Forum – IBF, 1992), this made relatively little impact.

(ii) UNEP's lack of historical scientific tradition (e.g., as compared with WHO and FAO), and an apparent tendency to favor political visibility over scientific reasoning if the two were in conflict.

(iii) In the government delegations at Rio and in its preparatory activities, there was a predominance of Environment Ministry representation, naturally enough; but given the breadth of the topics addressed, the exclusion of other parts of government weakened the quality and credibility of some of the resulting outputs. For example, the European Commission was represented by DG XI, who were strongly resistant, e.g., to a DG XII presence at the Rio conference to cover bio-technology policy aspects.

(iv) Although the scientific dimensions of the challenges of environment and development were explicitly addressed by a special ICSU conference in Vienna, in November 1991 – ASCEND 21 – both in the Rio conference and in the subsequent follow-up and instrumental developments, the scientific aspects were not strongly sustained; a complaint strongly voiced by the President of the International Union of Biological Societies, in their journal *Biology International*, and at the UNESCO-IUBS-Diversitas Conference in Paris, September 1994 (DI CASTRI, 1994).

Preparatory work, including the substantive negotiations on the texts of UNCED documents and agreements, took place over the preceding three years; there were many meetings, in New York, Nairobi and elsewhere.

"Agenda 21", a 700-page, 40-chapter report, endorsed at the Rio Summit, outlined the trajectory to a sustainable development path in the 21st century. It sought "to provide a blueprint for a global partnership aimed at reconciling the twin requirements of a high quality environment and a healthy economy for all peoples of the world", and to act "as a guide for business and government policies and for personal choices into the next century". It "lays out what needs to be done to reduce wasteful and inefficient consumption patterns in some parts of the world, while encouraging increased but sustainable development in others"; "offers policies and programmes to achieve a sustainable balance between consumption population and the Earth's life-supporting capacity", and "describes some of the technologies and techniques that need to be developed to provide for human needs while carefully managing natural resources". (Quotations from KEATING, 1993).

Rio produced five documents, comprising two international agreements, two statements of principles, and "Agenda 21" as an agenda for action on world-wide sustainable development. The statements of principle were:

- the Rio Declaration on Environment and Development; 27 principles defining the rights and responsibilities of nations as they pursue human development and well-being;
- a statement of principles to guide the management, conservation and sustainable development of all types of forests, which were emphasized as essential to economic development and the maintenance of all forms of life.

Two international conventions were negotiated separately from – in parallel with – preparations for the Earth Summit, and were signed by most governments meeting at Rio:

- the UN Framework Convention on Climate Change, aiming to stabilize greenhouse gases in the atmosphere at levels that will not dangerously upset the global climate system, requiring a reduction in emission of "greenhouse gases" such as carbon dioxide;
- the Convention on Biological Diversity (CBD), requiring that countries take action to conserve the variety of living species, and to ensure that the benefits from the exploitation of biological diversity would be equitably shared.

UNEP participated in the UNIDO-led interagency group that produced the Code of Conduct on Releases (UNIDO, 1991), and recognized the potential of biotechnology, for environmental as for other aims. But the natural constituency of Environment Agencies to which it looked for support and by whom its priorities and actions were governed, did not display a strong commitment to scientific methods, and the rhetoric of "conjectural risks of GMOs" tended to be endlessly repeated, irrespective of whether microbes, palm trees or elephants were the organisms concerned.

Agenda 21, Chapter 16; the CBD; and the Problems with Biotechnology

Chapter 16 of AGENDA 21 made clear the promise of biotechnology to contribute significantly to health, increased food production, better reforestation, more efficient industrial processes, decontamination of water and the cleanup of hazardous wastes.

Noting that most of the developments in biotechnology had been in the industrialized world, Chapter 16 referred to the "new opportunities for global partnerships between these countries – rich in technological expertise – and developing countries, which are rich in biological resources but lacking in funds and expertise to use them" (KEATING, op.cit). This concept of bargaining access to developing country germplasm for access to developed country biotechnology was written into the CBD, particularly its Article 16; and tended to foster several illusions. Firstly, the illusion that biotechnology applied to biodiversity would generate economic benefits sufficient to pay for conservation of the latter; secondly, the illusion that the barrier to such exploitation was the intellectual property held by the developed world; and thirdly, a confusion between the powers of governments in the developed world, and the property rights of companies. Such weaknesses caused US President BUSH to refuse signature of the CBD; and although President CLINTON on Earth Day in April 1993 indicated his intention to sign, Republican opposition made clear that US ratification, requiring a two-thirds majority in the Senate, would require bipartisan consensus. Certainly it proved unattainable in the summer of 1994, in time for the deadline after which non-signatories could not be full participants in the first "Conference of the Parties". This, the decision-making body under the Convention, held its first session in the Bahamas, in November-December 1994. The Republican victories in the November 1994 Congressional elections did not appear likely to soften the US opposition.

Chapter 16 of Agenda 21 also spoke of safety in biotechnology: "Care must be taken that new techniques do not damage environmental integrity or pose threats to health. People need to be aware of both the benefits and the risks of biotechnology. There is a need for internationally agreed principles on risk assessment and management of all aspects of biotechnology." These Chapter 16 references to safety were brought up in 1994; the more intense discussions of biotechnology safety and regulation took place in the context of the CBD.

It was a general problem of the CBD that the subject attracted more passion than precision. A scientifically vast and complex topic became the object of debates which drifted ever further from the original objective of conservation. Reference has been made already to the illusory "bargain" of Article 16.

In the ASCEND 21 scientific conference in Vienna, December 1991, and the resulting report (ICSU, 1992), many references had been made to the need for biotechnology; nowhere was it suggested as a threat to biodiversity. But as the negotiations towards Rio progressed, biotech safety and regulation was another field of argument imported into the preparation of the CBD.

Bio-safety in the CBD

During 1989 to 1992, the European controversy over the biotechnology Directives adopted in April 1990 had become intense. DG XI's chef-de-file position in the Rio preparations, aided by Environment Ministry representatives from other European governments (particularly Scandinavian), enabled them to press for similar elements at Rio – if the Community Directives were sound and beneficial, why indeed should their benefits be restricted to Europe, and why should not the Europeans offer their benefits worldwide?

However, as indicated elsewhere (Sects. 2 and 5.1), the United States had taken a strongly different view of the need for biotechnology-specific regulation, as had the OECD (Sect. 6.2). The discussions on biotech safety references within the CBD text therefore became contentious. A paper on "Biotechnology and Biodiversity" was commissioned by UNEP, from GIDDINGS of USDA, and PERSLEY of the World Bank (GIDDINGS and PERSLEY, 1990).

Their paper detailed the various techniques of biotechnology which could contribute directly to biodiversity conservation, and noted the indirect benefit attributable to productivity-enhancing technology which by increasing output of currently cultivated areas could take the pressure off marginal lands and ecologically sensitive areas; but did not identify any significant risks to biodiversity specifically attributable to biotechnology.

The outcome of the negotiations, the final text of the CBD, was a compromise. In Article 8(g), there was the warning of biotechnology risks:

"Each Contracting Party shall, as far as possible and as appropriate:
...
(g) Establish or maintain means to regulate, manage or control the risks associated with the use of release of living modified organisms resulting from biotechnology which are likely to have adverse environmental impacts that could affect the conservation and sustainable use of biological diversity, taking also into account the risks to human health."

More significantly, under Article 19, "Handling of Biotechnology and Distribution of its Benefits", in Article 19.3, the argument about the need for a binding international measure was during the negotiations with difficulty deferred, by including instead the commitment to "consider the need":

"The Parties shall consider the need for and modalities of a protocol setting out appropriate procedures, including, in particular, advance informed agreement, in the field of the safe transfer, handling and use of any living modified organism resulting from biotechnology that may have adverse effect on the conservation and sustainable use of biological diversity."

The reference to "prior informed consent" (PIC) drew concept and language from the long discussions of PIC procedure which had taken place in the late 1980s in the context to the London Guidelines for the Exchange of Information on Chemicals in International Trade, and the FAO Code of Conduct on the Distribution and Use of Pesticides. Article

19.4 added emphasis to the obligations in terms of an information requirement:

"4. Each Contracting Party shall, directly or by requiring any natural or legal person under its jurisdiction providing the organisms referred to in paragraph 3 above, provide any available information about the use and safety regulations required by that Contracting Party in handling such organisms, as well as any available information on the potential adverse impact of the specific organisms concerned to the Contracting Party into which those organisms are to be introduced."

Following the Rio Conference and in relation to the steps towards implementation of the CBD, UNEP Executive Director MOSTAFA KEMAL TOLBA set up Expert Panels to advise on 4 aspects:

1. Priorities for action for conservation and sustainable use of biological diversity and agenda for scientific and technological research
2. Evaluation of potential economic implications of conservation of biological diversity and its sustainable use and evaluation of biological and genetic resources
3. Technology transfer and financial issues: Issues and options
4. Consideration of the need for and modalities of a protocol setting out appropriate procedures including, in particular, advanced informed agreement in the field of the safe transfer, handling and use of any living modified organism resulting from biotechnology that may have adverse effect on the conservation and sustainable use of biological diversity.

The panels, each of some 15–20 people were constituted by UNEP's, invitation, most countries participating in at most one panel; but the US, on all four. The European Commission, via DG XI's representative, participated in panel 4, on bio-safety. They met three times in 1992–93, the panel reports being finalized at Montreal in March '93; and presented at a conference in Trondheim, May '93; hosted by the Norwegian Government. The report of expert panel 4 did not achieve consensus. A minority of the panel, including the US representative, expressed themselves unconvinced of the need for a protocol; a majority were in favor; and the report recorded the arguments of the two sides. Given the arbitrary choice and small total number of panel participants, the terms "majority" and "minority" were of little significance.

The panels were advisory to the Executive Director of UNEP (TOLBA was in January 1993 succeeded by ELIZABETH DOWDES-WELL); for the preparation of the agenda of the Conference of the Parties, an Intergovernmental Committee on the Convention on Biological Diversity (ICCBD) was convened, supported by the creation of a small Interim Secretariat.

The ICCBD first met for a week, 11–15 October, 1993, in Geneva. Procedural arguments dominated most of the week, particularly over the financial mechanism envisaged in Article 20 of the CBD. The discussion on the need for a bio-safety protocol was compressed into the final half-day, and indicated the continuing division of opinion. The US and Japanese delegations expressed their clear opposition to a protocol; the European Community gave a more ambiguous signal, the Belgian representative (speaking on behalf of the Community, because his country held the Presidency of the Council of Ministers in the second half of 1983) having received conflicting briefings.

Various developing countries emphasized the need for a bio-safety protocol, a demand also vociferously supported by some of the non-governmental organizations.

Given the abridged nature of this and other debates, it was agreed that the ICCBD should meet a second time, for a longer period; and a two-week meeting took place in Nairobi from 20th June to 2nd July 1994. The debate on a bio-safety protocol again proved divisive and controversial; the following elements are taken from Earth Negotiations Bulletin (ENB), an objective reporting service funded by Canadian Government and other sources.

"Germany, on behalf of the European Union, stated that in the short-term the development of technical guidelines on biosafety was favored without prejudice to the medium-term development of international legal instruments on biosafety, while assessing the need for and modalities of a protocol. The Netherlands stated that an international agreement on safety in biotechnology should eventually be given the form of a legally-binding agreement and that such an agreement should pay attention to both capacity building and timing ...

The UK referred to joint work with the Netherlands and announced its plan [to] co-host a meeting on biosafety in 1995 in Asia. The US stated that a protocol on biosafety is not warranted, but it did recognize the needs specified in Article 19 and stated that guidelines were not a substitute for scientific evaluations and did not replace needs-based requirements. Japan stated that decisions on this issue should be made on the basis of accumulated scientific knowledge and on-going examinations, such as those conduced by the OECD.

...

[Regarding] a Protocol: Many [developing] countries ... supported the need to have a legally-binding agreement on biosafety. Malaysia was concerned about the issue of transnational corporations using developing countries as a place to transfer and test living modified organisms (LMOs). Supported by Sweden, Malaysia called for the establishment of a subsidiary body to consider the issue and report to the COP. Sri Lanka also noted that LMOs are subject to mutations. Sweden, supported by Norway and India, pointed out that concerns expressed in the report of UNEP Panel 4, regarding potential threats to biological diversity of LMOs from biotechnology, were not adequately reflected in the initial document prepared by the Interim Secretariat. China wanted guidelines based on regional initiatives that would then lead up to a legally-binding international instrument.

Greenpeace, Third World Network and Genetic Resources Action International (GRAIN) stressed the need for a legally-binding international regulatory mechanism. The Third World Network proposed an addition supported by all three of the NGOs who had intervened: 'They stressed the urgent need for a protocol because of the serious risks posed by the transboundary nature of the export of LMOs and examples were given that northern companies had already started carrying out hazardous genetic engineering experiments in the South. They also asked that the destabilising socio-economic aspects of biosafety form part of such a protocol.'

There was heated debate on the report of the discussion, [leading a small drafting group to summarise it by the sentence:] 'A significant number of representatives expressed support for immediate work on a protocol, while others expressed support for the Conference of the Parties establishing a step-by-step process to consider the need for and modalities of, a protocol.' [But this was not accepted by the Plenary session.] The only area where there was general agreement was that the Committee should recommend that the issue of biosafety should be on the agenda of the first meeting of the COP."

The matter of biotechnology regulation at global level, by an international binding protocol, thus remained open and unresolved, as the first Conference of the Parties (COP) took place in the Bahamas, 28 November–9 December 1994. Preoccupied with formal business of rules of procedure, methods of working, and the more central issues such as the financial mechanism, and the institutional and geographical location of the CBD Secretariat, the COP might have deferred further substantive discussion of matters such as the need for a bio-safety protocol to its medium-term work programme; but there was an insistent demand from both the Nordic countries and the "G77" group of developing countries to initiate work more rapidly on building a protocol. The compromise was an agreement to give the matter intensive consideration over the following year, including a week-long open meeting (in Madrid), and gathering all relevant information and experience of risk assessment, guidelines and legislation.

The Anglo-Dutch, "Chapter 16" Initiative

Following the endorsement at Rio of Agenda 21, a Commission on Sustainable Development (CSD), located at UN Headquarters in New York, was charged with responsibility for its follow-up.

Within the Ministries of the Environment in the Hague and London, staff were convinced of the importance of a framework of safety guidelines for biotechnology as a basis for its responsible and acceptable application world-wide. Based on the safety references in Chapter 16 of Agenda 21, it could be said that there had been international agreement on the development of such guidelines; and a draft was prepared by the two Ministries. This was discussed at working meetings in the UK (March 1994), and in the Netherlands, (May 1994); and at international conferences in Harare, Zimbabwe (African Regional Conference on Safety in Biotechnology, October 1993 – see VAN DER MEER et al., 1993) and in Colombia (June 1994), and Indonesia (October 1994).

The draft was put forward in Brussels at the April 1994 meeting of the "Article 21" Committee of National Competent Authorities responsible for Directive 90/220 (see Sect. 7.3), where it was favorable received; although it was not discussed at the Commission's Biotechnology Co-ordination Committee, nor adopted as Commission policy.

The two Ministries (or governments) also put forward the draft guidelines at the meeting in New York, June 1994, of the Commission on Sustainable Development. The proposal was also debated at the second ICCBD meeting in Nairobi, June 1994; but it did not attract significant interest, nor divert attention from the more noisy and polarized debate on CBD Article 19.3.

Rather belatedly, the UK-NL draft guidelines incorporated reference also to the UNIDO/UNEP/WHO/FAO guidelines on "introductions" (see Sect. 6.3); it being claimed that they were more precise and operational than the latter. However, as UNIDO was the UN agency responsible for co-ordination of activity on Chapter 16 of Agenda 21, its staff tended to prefer the existing multi-agency guidelines.

The text of the UK-NL guidelines also gave rise to ambiguity, some industrial participants in the working meetings being led to believe that they were not about "genetically modified organisms", and therefore could not be said to stigmatize them. In fact, this was playing with words, since the draft guidelines refer repeatedly to "novel organisms" – having defined them as "organisms produced by modern genetic modification techniques whose genetic make-up does not occur naturally".

The basic political case for the Anglo-Dutch initiative was that they could be rapidly adopted, like the UNIDO guidelines, on a voluntary basis; and their existence would not preclude the development of a binding protocol on the lines envisaged under the CBD Article 19.3.

In a somewhat defensive article in the Dutch-government-supported Biotechnology Development Monitor of September 1994, BERT VISSER of DGIS summed up the uncertain situation of the initiative at that stage:

> "Assuming that the guidelines appear to be a valuable practical tool in introducing safety mechanisms and fostering growing international co-operation and harmonization, three major questions remain.
>
> The *first* question is which organization can function as the institutional basis for these guidelines. Without an institutional home to promote and guide their implementation, the impact of the guidelines may remain limited. Although the *United Nations Environment Programme* (UNEP) has been mentioned as a potential candidate, it is unclear whether the organization is able to give sufficient priority to the initiative. However, the issue has been taken up by the organization and will be thoroughly discussed.
>
> The *second* question, which is related to the first one, is how developing countries, which might most benefit from the guidelines, perceive the initiative. The organizers hope that many countries will realize the need for safety mechanisms soon, and that they may benefit from a readily available, technical document. Many delegations present at the meeting of the *Intergovernmental Committee on the Convention on Biological Diversity*, which took place in Nairobi in June 1994, supported the initiative, and this warrants optimism. The *third* question is whether sufficient resources will become available to allow developing countries to implement the guidelines. International organizations and donors may have an important role to play here."

In their November 1994 statement, timed for the first Conference of the Parties to the CBD, the bioindustry emphasized the positive potential of biotechnology for biodiversity, and the use of existing guidelines for safety, possibly through a Task Force led by relevant UN agencies such as UNIDO and UNEP (IBF, 1994).

7 Re-thinking and Review: 1990–1995

7.1 The Bio-Industries in Europe Find Their Voice

The demise of the Biotechnology Steering Committee after its final meeting of July 1988 appeared to carry with it as it sank the prospects of renewing and strengthening the co-ordinated strategic view of biotechnology, which had been initiated by DAVIGNON in 1983. Did this matter?

The significance of the loss was not immediately evident. It has been said (CYERT and MARCH, 1963) of companies, that they do not have objectives; individuals have objectives. Analogously, one could say that industry sectors and associations do not have interests; their member companies do. The development of biotechnology-based industry in Europe may well be in the public interest; but the joint-stock corporation's primary formal responsibility, in the open economies of the western world, has been to its shareholders, and to securing a profitable return on their investment. The managers of European "bio-industry" companies had indicated at the December 1984 meeting with Vice-President DAVIGNON (see Sect. 3.7), and subsequently by their behavior, their reluctance to devote effort and resources to lobbying or representational activities on behalf of biotechnology in Europe. A vacuum had developed, which others filled. Speaking at the Toxicology Forum in Berlin in October 1987, Sir GEOFFREY PATTIE (the former UK Science Minister) had warned the bio-industries that if they failed to "occupy the high ground" of public opinion in advance of their critics, it would be difficult to recapture it later; his warning was accurate, and ignored.

The situation changed at the end of the 1980s. Several major companies were not only recognizing the significance of biotechnology for their own capabilities and competitiveness; but were recognizing that the actions being undertaken by the Community institutions and/or national authorities in relation to biotechnology would impinge directly, and perhaps adversely, upon their own activities. Three examples illustrate this.

ICI in 1984 were still at an early stage in their strategic decision to enter the seeds sector. By the end of the 1980s, it was clear that the whole basis of protection of agricultural crop plants would ultimately be transformed by the new knowledge. Plant protection would shift away from simply spraying pesticidal chemicals on fields, towards the incorporation of self-protective capacity in the plant genome, and to subtler effects depending on interactions between the protection products administered and the characteristics of the plant (and target pests). The whole public policy framework vis-à-vis transgenic plants – affecting research, development, testing, patenting, safety rules, and commercialization – would be of crucial significance to their business.

Hoechst in Germany had encountered costly opposition and legal delays to operating in Hessen their plant for producing insulin by a recombinant bacterium, and had the frustration of seeing competitors Lilly and Novo Nordisk develop their businesses in this area in less hostile policy climates. Their immediate costs, and in the longer term the location of their research, development, pilot plants, and full-scale production facilities, would be increasingly influenced by decisions in Frankfurt (for Hessen), Bonn, and Brussels.

Monsanto had from the early 1980s invested heavily in checking the safety, demonstrating the efficacy, and developing and testing the production technologies and systems for bovine growth hormone, and for its administration to cows. By 1989, saw the dark shadow of the steroid hormone debate falling over the topic; the regulatory approach adopted by the European Community would be similarly crucial to the prospects for commercialization in Europe of their peptide hormone. They had visited the Commission to discuss the project in 1984; a decade later, the product would be on the market in three continents, including the US; but the decision in the European Union would still be deferred. Debate on hormones had been widened in the European Parliament – in particular by Environment Committee Chairman KEN COLLINS – into a broader demand for a fourth criterion, of "socio-economic assessment" to be added to the established criteria of quality, safety and efficacy – familiar in the context of pharmaceuticals. The vague menace of this "4th hurdle" – a menace to which the ban on steroid hormones gave credibility – was seen by industry as a threat and disincentive to research and development in any area where innovation might affect established and influential interests.

The Senior Advisory Group for Biotechnology (SAGB) was created, under the aegis of CEFIC (the Council of the European Chemical Industry) in 1989. The three companies mentioned above joined with a French (Rhône-Poulenc), an Italian (Montedison, by then part of the Ferruzzi Group), a food company (Unilever) completing the range of sectors represented, and Sandoz from Basel, bringing a link to the Swiss pharmaceutical majors. GÜNTER METZ of Hoechst, President of CEFIC, wrote in July 1989 to Commission President JACQUES DELORS, to inform him of the creation of SAGB and its aims; and in January 1990, their first publication was delivered to various Commissioners in their respective languages (SAGB, 1990a).

The January 1990 booklet, "Community Policy for Biotechnology: Priorities and Actions" introduced SAGB and its purpose: "to promote a supportive climate for biotechnology in Europe". It advocated a "coherent and supportive Community policy", noting that the USA and Japan were ahead of Europe in this respect. Among the specific proposals, under "Community Safety Regulation" SAGB endorsed five guidelines for Community regulation:

"– Clearly define and assign responsibility for regulation;

– Apply existing, non-discriminatory approaches for safety in research and industrial processes;
– Regulate products on the basis of their inherent characteristics and intended use;
– Observe common principles for product sector regulation;
– Develop regulatory approaches in common with major competitors."

Gently though these points were stated, they were clearly opposing in principle the two rDNA-specific draft Directives by that stage moving towards second reading in the European Parliament.

Other statements in the January 1990 booklet emphasized the need for a common market for biotechnology processes and products, the need to strengthen the Community role in pre-competitive R & D, the need for patent protection for inventions of biotechnology, and the need for market authorization based only on objective scientific criteria for safety, quality and efficacy – i.e., by implication opposing the "fourth hurdle".

A section on social and ethical issues stressed that broad debate on these aspects of biotechnology was essential; and explicitly noted that the communication took no policy position on genetic modification of the human germ-line, an activity in which none of the member companies was engaged.

Vice-President BANGEMANN, responsible for Industrial Affairs, reacted briskly to the SAGB and other pressures relating to biotechnology, asking his DG III staff to prepare a Commission communication for the Council meeting of Industry Ministers in March. A DG III working paper was indeed produced for that deadline; but to complete a new Commission communication embracing the whole range of major issues raised by biotechnology was to take not three, but sixteen busy and contentious months, engaging the energies of staff in several DGs, and demanding the resolution of several major questions.

February and March 1990 saw the delivery to Parliament of the Nobel prize-winners' letter (see Sect. 4.7), no doubt with some influence on the voting. But the common posi-

tion of the Environment Ministers was little altered in the final compromises, and the Directives were adopted on 23 April 1990; in spite of last-minute uncertainties by Belgium, France and the UK, any of whose Ministers might have opposed, had one of the others given a lead, and who together could have been a blocking minority.

The concern in scientific circles, academic and industrial, was immediately sharpened. A second SAGB communication (SAGB, 1990b), on "Community Policy for Biotechnology: Economic Benefits and European Competitiveness", described in blunt language, supported by unfavorable statistics on patents and investment, industry's answer to the question, "Why can't Europe compete for commercial biotechnology investment?":

"– The United States has a strong climate of support for biotechnology. The European climate for biotechnology is viewed as negative.

– Europe's hostile political attitude toward biotechnology, reflected in incoherent and adversarial regulatory systems, creates unacceptable risk and cost for all biotechnology investors.

– Small start-up companies suffer disproportionately from inefficient regulation because they cannot afford to move their operations elsewhere to escape the impacts.

– American investment culture and incentive schemes remain innately more attractive for risk capital, especially start-up venture capital."

In parallel with the development of the SAGB, the national bioindustry associations of Belgium (BBA), Denmark (FBID), France (ORGANIBIO), Italy (ASSOBIOTEC), Spain (Asociacion de Bioindustrias), The Netherlands (NIABA) and UK (BIA) developed within their countries, and established (in 1991), a "European Secretariat of National Bioindustry Associations" (ESNBA). This brought a useful additional voice into the policy debates in Brussels; and in spite of the different scales of their member companies or organizations, the SAGB and ESNBA were generally concordant in the aims pursued and arguments advanced.

7.2 The Birth of the BCC, and the April 1991 Communication

The growing controversy over biotechnology did not escape the attention of the President's office, and in summer 1990 the President's Chef de Cabinet PASCAL LAMY called on DGs III, XI and XII to produce short papers covering respectively the international competitiveness, regulatory, and ethical issues. Following these, the Secretary-General DAVID WILLIAMSON convened on 15 November a meeting of the Directors-General concerned, to consider the policy issues of biotechnology and related problems of coordination.

Thus was born the "Biotechnology Co-ordination Committee", or BCC, although some further months and meetings were required to finalize its terms of reference, which were then accorded official standing by a Commission decision, in March 1991. Since transparency was one of its watchwords, a press release was then issued, announcing its terms of reference, which included these statements:

"The European Commission recognizes the importance of biotechnology for Europe's future. Clear signals are needed to interested parties of the Commission's intentions in this field.

Biotechnology is an increasingly important element in many areas of Community activity. It is imperative that interdepartmental co-operation in this field be reinforced ...

The Commission therefore has created a new high-level interservice group to develop a well-balanced Community policy in biotechnology. The Biotechnology Coordination Committee (BCC) is chaired by the Commission's Secretary General, Mr. DAVID F. WILLIAMSON.

The BCC covers all sectors and activities of the Commission in the field of biotechnology, with the participation of all relevant services. Its main tasks are three-fold. First, it is to examine new initiatives made by the Commission's services and to prepare the Commission's final decisions. Second, it will create, if need be, a system of Round Tables involving the Commission, industry and other interested parties. Third, it will evaluate the existing Community policy on biotechnology."

The BCC was born in the middle of the inter-service battlefield over the biotechnology communication being prepared under DG III leadership, at the request of Vice-President BANGEMANN. Given the mounting demands on the Secretary-General, in the months leading up to the "Maastricht" Treaty on European Union, and with the accession negotiations developing for several candidate countries seeking membership of the Community, it seemed clear that he could not long afford to devote personal attention to the BCC. Indeed at its first meeting, in November 1990, he indicated his intention to transfer the responsibility to "the service most concerned". However, some tens of meetings and four years later, DAVID WILLIAMSON was still in the chair; *res ipsa loquitur.*

The BCC provided the essential central machinery to resolve the inescapable interservice conflicts over biotechnology, which neither the Biotechnology Steering Committee (1984–88), nor its creation, the Biotechnology Regulation Interservice Committee (1985–90), had had the authority to do. Although the BCC had no formally delegated authority from the "college" of the 17 Commissioners, it was clear that a service, or services, isolated or in a small minority at BCC, would be unlikely to obtain a different result by taking the issue to the Commission. The Commissioners, chaired by President DELORS and with the Secretary-General as secretary, would normally ask for the BCC view. It was thus potentially, and sometimes in practice, more powerful in its inter-DG co-ordination role than the US interagency Biotechnology Science Coordination Committee (see Sect. 5.1)

The communication was finally published in April 1991 (EUROPEAN COMMISSION, 1991a), under the title "Promoting the competitive environment for the industrial activities based on biotechnology within the Community" – a precise and carefully negotiated, committee-crafted title. It was easier to refer simply to "the April 1991 communication". Since the formal establishment of the BCC by Commission decision had taken place the previous month, and was advertised in the April communication as the means of ensuring a cohesive approach to biotechnology, the birth of the BCC and the April 1991 "Bible" were in their origins closely associated.

The significance of the April 1991 communication was underlined over the following years. It was the spur to the preparation of a fresh "own initiative" report by the European Parliament; an appropriate response, since almost six years had elapsed since the 1995 Viehoff hearings (see Sect. 4.2). But the responsibility of rapporteur for drafting this was given to a German Green MEP, HILTRUD BREYER; whose very entry into politics had been triggered by her concerns about "gene technology". The resulting working papers (BREYER, 1992, 1994) were not a balanced reflection of the diverse opinions in the parliamentary committees concerned; and although a relatively balanced public hearing was held finally in March 1994, the initiative was overtaken by the June 1994 elections; it would be left to the new Parliament to pick up the threads, or start afresh.

BREYER's critique – developed in various ways by her advisers (e.g., WHEALE and MCNALLY, 1993) – was that the Commission was effectively being dictated to by industry, in particular through the SAGB; was ignoring the special risks of genetic engineering; and was (via the April 1991 communication and its re-emphasis on sectoral legislation) seeking to dilute or dismantle the April 1990 Directives (219 and 220) through which control of these risks was to be achieved. It was a critique inviting and receiving several strong responses.

Legally, the Commission was on clear ground, for the Single European Act had strongly emphasized industrial competitiveness as an aim of Community policy – for example, by adding to the founding EEC Treaty articles such as 130F, within the new Title VI on Research and Technological Development:

"The Community's aim shall be to strengthen the scientific and technological basis of European and to encourage it to become more competitive at international level."

Politically, against a background of soaring unemployment and documented charges that over-regulation was rendering companies uncompetitive and driving investment out of the Community, the Commission could hardly be reproached for being attentive to industrial demands for a policy environment less hostile to a key technology. Indeed leading socialist MEPs were among those in close dialogue with SAGB.

Factually, it was ironic to see the bio-industry accused of over-effective or even counter-productive lobbying (e.g., see TAIT and LEVIDOW, 1992), when the problems of industry in the 1990s could in part be attributed to their inadequate responses to the Davignon invitation of December 1984 (see Sect. 3.7) and over the following years.

Within the Commission, Vice-President MARTIN BANGEMANN as Commissioner responsible for industry, presented the April 1991 communication to the Council of (Industry) Ministers in November that year, where it was favorably received. The sensitivity of BANGEMANN and his cabinet to the needs and problems of the bio-industry was heightened by the intense arguments over gene technology in Germany itself (see Sect. 5.3), leading eventually to the revision of the 1990 Gene Law; arguments in which both the major companies (Hoechst and others) and the research organizations (e.g., the Max Planck Gesellschaft) played leading parts.

BANGEMANN gave the communication further prominence by combining it with two other commission communications, under the heading "European Industrial Policy for the 1990s", and having it published in an attractive format as a special supplement to the *Bulletin* of the EC (EUROPEAN COMMISSION, 1991a). The other papers in this booklet were a general communication on "Industrial policy in an open and competitive environment: guidelines for a Community approach"; and a communication on the information technology and electronics industries, which like biotechnology, were of strategic significance, in rapid evolution, and generating challenges to public policy. This 3-piece "Bangemann Communication" foreshadowed in its content and policies the "Delors White Paper" of December 1993 (see Sect. 7.4); and signalled clearly that biotechnology had now achieved a prominence in the context of industrial policy which it had lacked in the mid-80s. Similarly it signalled that DG III was now a leader for Commission policy in biotechnology, and would be a major player within the Biotechnology Coordination Committee.

The follow-up to the April 1991 communication was intimately linked with the character of the BCC and the content of its discussions. Increasingly and significantly, it was referred to as "the Williamson Committee"; and it was becoming evident why the Secretary-General could not easily escape or delegate the chairmanship, but had to remain, like a medieval monarch wrestling with his fractious barons. Four major baronies were now contending in the biotech policy domain; and while none sought or claimed the whole territory, there were boundary disputes to be resolved, and with the whole kingdom expanding, new opportunities to be seized and disputed:

– DG III (Industry), the natural interlocutor for SAGB and the national bio-industry associations, was the general advocate of international industrial competitiveness, as well as specific defender of sectoral legislation, particularly in pharmaceuticals and foods; (in chemicals, the main authority lay with DG XI);

– DG VI (Agriculture), generally preoccupied with the management responsibilities and political problems of the common agricultural policy, and especially the GATT negotiations in 1990–93, discovered that their 'in'-trays were full of biotechnology, in both the research and the legislation Directorates within DG VI:

● the modest and neglected agricultural research activity had to collaborate with DG XII and join (under the common "Framework Programmes") in fast expanding, integrated agricultural and agro-industrial programmes of research, development and demonstration, intimately involving biotechnology;

● the "hormone wars" had spilled over to the BST (bovine somatotropin) battle, with transgenic animals in the not-so-distant future, these and other specific challenges becoming generalized into the "Fourth Hurdle" argument for socio-economic assessment;

● other agricultural legislative responsibilities included those relating to plant breeders' rights (interacting with patents, and changed by the international revisions from the March 1991 meeting of UPOV, the International Convention on Plant Varieties); and specific

sectoral legislation on topics such as pesticides (see Sect. 4.8), transgenic animals (DG VI being chef de file for the welfare of agricultural animals), and transgenic plants (both as inputs to agriculture, via seeds, or as outputs, for placing primary produce – e.g., fresh fruit – on the market);

– DG XI (Environment; and, until 1991, Consumer Protection), as advocate and guardian of the two April 1990 "biotechnology" Directives (DG III's co-interest in 90/219 was seldom obtrusive before 1993), had the uncomfortable position of having simultaneously to pursue the implementation of these Directives through national legislation; to manage (after 23 October 1991) the Commission's executive responsibilities under the Directives; and to defend the horizontal Directives against a rising volume of criticisms, internal and external, before they had even been implemented in national legislation to become operational;

– DG XII (Science, Research & Development), although at times seeing themselves primarily as a research funding agency, with broad responsibilities and a fast-expanding budget, had also to recognize that they were part of the Commission, and were expected by the scientific communities outside (upon whom the burdens of biotechnology-specific regulations were falling) to act as the voice of scientific reasoning and interest inside the Community institutions.

In addition to these four "baronies", the BCC had to include many lesser but not insignificant voices. DG I (External Relations) had to cope with the fall-out of trade wars over technical standards (cf. the EC–US dispute over hormones), with international consequences of conventions such as those at Rio (see Sect. 6.4), and their repercussions in other international contexts (e.g., GATT and the new World Trade Organization, WTO). DG V's role in worker safety legislation has been mentioned (Sect. 4.7). DG VIII (Development) had several potentially significant reasons to participate, but in practice rarely did so. Other services such as DG VII (Transport) and Consumer Policy Service would similarly have occasional points where biotechnology interacted with their interests. After January 1993, in the Third Delors Commission, responsibility for the much-debated Directive on the Protection of Biotechnological Inventions, the "Biotech Patents" Directive, shifted from DG III to DG XV (Internal Market and Financial Services), but co-operation on this dossier, with DG III and through BCC, remained close.

The April 1991 communication provided, in its preparation and for several years subsequently, a permanent agenda, a Vade Mecum, for the Biotechnology Co-ordinating Committee. At each BCC meeting, an updated "State of Play" document would summarize the various dossiers, particularly those concerning legislation. A review of progress on the commitments in the communication was published in October 1992 (EUROPEAN COMMISSION, 1992b), but did not significantly alter the agenda, whose main headings had emerged during (and before) the 1990–91 months of drafting. The communication obviously included statements about the importance of biotechnology, of European competitiveness, and of Community R & D programmes; but in the context of a history of biotechnology regulation, five topics merit special mention:

- intellectual property rights
- ethics
- socio-economic impact assessment: the "Fourth Hurdle"
- the relationship between "horizontal" legislation (the April 1990 Directives) and sectoral, or "vertical" legislation
- standards.

On intellectual property issues in biotechnology, the need to clarify and harmonize many questions of interpretation and principle in patent law had been addressed by the Commission's October 1988 proposal (EUROPEAN COMMISSION, 1988). Points where the inventions of genetic engineers interacted with plant genomes required that the discussion await the proposal for a harmonized Community system of plant breeders' right, which was put forward (drafted by DG VI) in 1990 (EUROPEAN COMMISSION, 1990a). The international UPOV meeting of March 1991 introduced changes requiring amendments to the Commission proposal, and causing further

delay. Parliamentary concerns about ethical aspects of the patenting of life, particularly animals, and about patenting parts of the human body, deferred agreement throughout 1994. The differences between Parliament and Council brought the matter up, on 28 November 1994, as one of the first examples for the new "conciliation procedure" established by the Maastricht Treaty; illustrating again the arbitrary but inevitable interplay between the evolution of Community policies for biotechnology, and of the constitution and institutional procedures of the Community itself.

In response to the ethical issues raised repeatedly in Parliamentary debates on biotechnology proposals of all kinds (whether dealing with R & D programmes, patent law, or safety regulation), the Commission acknowledged their significance in the April 1991 communication, itemizing several of them, and indicating the need for "an advisory structure on ethics and biotechnology". Soon afterwards, the Commission established the "Group of Advisers on Ethical Implications of Biotechnology". This group of 6 eminent persons held their first meeting in March 1992, with an opening welcome by President DELORS. The group was the subsequent source of balanced and thoughtful opinions on a range of topics addressed to it by the Commission, or chosen on their own initiative; it was subsequently (1994) increased to 9 persons. A report was published on their first two years' activities (EUROPEAN COMMISSION, 1994e).

The preparation of the April 1991 communication coincided and interacted with the ongoing debate on socio-economic assessment, summarized as the "Fourth Hurdle" (i.e., additional to those of quality, safety and efficacy, long used in pharmaceutical legislation), the controversy and subsequent Community ban on steroid hormones and analogues spilling over onto the biotechnology-produced peptide hormones, particularly bovine growth hormone or somatropin, known as BGH or BST. The expressions of consumer concern, the opposition of some farming interests, the inconvenience of productivity increases for the managers (at Community and national levels) of a Common Agricultural Policy already wrestling with surpluses, the concerns of the animal welfare movements, and the resulting opposition in the European Parliament, were all given political expression and focus by demands for a socio-economic impact assessment as an element of product approval procedures.

To these demands, the Commission in its April 1991 communication gave a carefully reasoned but clear-cut negative; at least so far as concerned the routine systems for product authorization. The Commission was not indifferent to the socio-economic dimensions, and had indeed its means of assessing such aspects; but these dimensions had to be addressed in ways other than the standard authorization criteria for products. The communication referred to the FAST programme and the work of the European Foundation for the Improvement of Living and Working Condition ("the Dublin Foundation") which had undertaken and published extensive work on the social assessment of biotechnology (YOXEN and DI MARTINO, 1989). The Commission reserved the right on exceptional occasions to over-rule the three science-based criteria for acceptance, "in the light of its general obligation to take into account other Community policies and objectives".

Regarding the Regulatory Framework, the April 1991 communication, published just one year after the Council Decisions on the Biotechnology Directives 90/219 and 90/220, and six months before the required deadline for implementing national legislation by Member countries, considered these regulations and announced its intention to review them.

The description of the "Regulatory Framework" in the April 1991 communication started by noting that "not all products derived through biotechnological methods will require a specific assessment and/or authorization procedures." The vast majority of biotechnology products, it continued, were produced through traditional methods; and "As far as new biotechnology products are concerned which involved gene manipulation, each product will have to be considered on a case-by-case basis and assessed as necessary".

The communication described the "Regulatory Framework" in terms carefully bal-

anced between the "horizontal" (DG XI and DG V) Directives, adopted the previous year, and product legislation (DG III and DG VI), with its diverse strands and different histories for the various sectors:

"Those products which do require governmental activity may be assessed and authorized under the regulatory framework for biotechnology which has been developed by the Community. This regulatory framework, which is based upon scientific analysis and evaluation, covers horizontal (environmental and worker protection) and product legislation. This latter is based on the three criteria of safety, quality and efficacy,* which are also applied when assessing whether a product can be authorized for distribution on the open market. The horizontal framework ensures that all stages of pre-industrial development and environmental aspects are covered.
* It should be noted that these three criteria are nowadays considered to include impact on nature and safety for the environment."

That was the current picture; the communication went on to outline the Commission's intended actions, and clearly acknowledged the need for adaptation and change. The horizontal directives had provisions relating to adaptation to technical progress, which would be used. The Commission would examine whether existing product legislation was appropriate, or could be slightly amended, to take account of action specific to biotechnology. The horizontal legislation would cover the gaps between sectoral legislation.

The dynamic and innovatory nature of biotechnology posed a challenge for legislators, and this meant "a constant assessment of the appropriateness of existing and proposed legislation". Duplication of testing and authorization procedures would be avoided; and the Commission would ensure that "one integrated assessment and notification procedure covers all that is required for product authorization". Thus was stated emphatically what became known as the "one door, one key" policy.

Standards Programme in Biotechnology, Through CEN

The April 1991 communication also announced the intention of the Commission to institute a programme of development of standards for biotechnology. This suggestion had been raised in the context of BRIC, the Biotechnology Regulation Inter-service Committee, some years before, by DGs III and XII. Encouraging the development of European standards had long been one of the most obvious means of creating a harmonized single European market, just as the use of national rules about technical standards was one of the easiest ways in which national authorities could practice an effectively protectionist policy under respectable cover.

The complete harmonization of national standards proved exceedingly time-consuming in cases of even moderate complexity, and after 1991, the Community switched to the "new approach", emphasizing mutual acceptance of products which satisfied national standards in any EC Member State. However, in a new area such as biotechnology in which few or no standards existed, the case for a direct drive for European standards was easy to make.

The procedure for the development and adoption of standards is a well-developed procedure, operated through the CEN: Comité Européen de Normalisation, with defined voting rules, working in three languages (English, French and German), and covering not only the EC countries, but the European Free Trade Area (EFTA). The working groups which prepare the draft standards generally comprise experts, nominated by national standards bodies, and drawn from relevant expert circles, industrial or other. It has therefore been quite common for Community legislation to be kept short and flexible by a reference to "accepted international standards".

The original suggestion in BRIC was not well-received by DG XI, who were not Chef-de-file for standards (it was a DG III responsibility), and feared that such a step might open their drafting of the biotechnology Directives to technical criticisms from independent sources. The Directives were consequently drafted in considerable technical detail – inevitably reflecting technical perceptions of Environment Ministry and DG XI advisers as of the mid-1980s; a drawback underlined by the Ministers' decision in Council to remove from the scope of adaptation to

technical progress the Annexes defining (in some technical detail) the scope of the Directives.

However, the April 1991 communication re-opened this issue with a clear commitment to a programme for the development of standards for biotechnology. A mandate was prepared and (after some inter-service argument) agreed for CEN, Vice-President BANGEMANN emphasizing that "standards are not an alternative but a useful complement to legislation". The CEN programme referred to the two Biotechnology Directives 90/219 and 220, and the Worker Safety Directive, 90/679.

CEN established a Technical Committee, TC 233 (Biotechnology), and clarified some points of potential overlap with other areas of ongoing work on standards development. Details of the 54 standards under development have been summarily described by KIRSOP (1993a,b), and full details are available from the TC 233 Secretariat; a function which for this Committee was awarded to AFNOR, the French standards organization.

Elected as first Chairman of TC 233 was SAGB Director BRIAN AGER, a former expert adviser to the European Commission (DG XII) and to OECD, previously secretary of the UK Advisory Committee on Genetic Modification. The standards development work was subdivided between four Working Groups (the examples quoted are not a comprehensive list):

1. Laboratories for research, development and analysis: 10 standards (e.g., categorization of laboratories, reporting on existing lists of animal and plant pathogens, examining classification work in support of the classification work required under the Worker Safety Directive 90/679, codes of good practice)
2. Large-scale process and production: 7 standards (e.g., building and equipment specification, according to degree of hazard, control procedures for materials and energy, codes of practice, waste handling, inactivation and testing)
3. Standards relating to modified organisms for application in the environment: 13 standards (e.g., insert characterization and sequencing, methods for detection and determi-

nation, sampling methods, molecular markers, quality control of diagnostic kits)
4. Equipment: 24 standards (testing procedures for cleanability, sterilization, leak-tightness, performance criiteria for the various classes of components – pumps, valves, shafts, filtration, etc).

This extensive work on standards was another illustration of the operational and practical work resulting directly from the April 1991 communication.

7.3 Implementing the 1990 Directives: DG XI, the Committee of Competent Authorities, and National Developments

The Directives 90/219 and 90/220, on contained use and field release, were adopted on 23 April 1990; the final Article in each case requiring Member State governments to have legislation in place to give effect to them within 18 months, by 23 October 1991.

At the start of the 1990s, these regulatory developments were the most prominent feature of public policy in Europe, concerning biotechnology. The leading role was played by the Environment Directorate-General in the European Commission (DG XI), chairing on behalf of the Commission the Committee of National Competent Authorities (NCA) set up, generally under a corresponding Ministry or agency in each national capital, but with significant variations from country to country. Many other developments were occurring in parallel, as indicated in the preceding section; interacting with and complicating the implementation under national law of the Directives 90/219 and 90/220.

The Directives themselves would not come into effect until 23 October 1991, eighteen months after their adoption. During that period, the Commission through DG XI energetically organized a series of "National Experts Meetings on Environmental Aspects of the Use of GMOs", the first on 2–3 April 1990, three weeks before adoption of the Directives. Seven more such meetings took place

before 23 October 1991, in Brussels or in pleasant locations in host countries. The meetings were efficiently organized, and contributed to the development of a certain *esprit de corps* among members of the nascent authorities.

The practical business of the meetings was substantial, as many details had to be addressed in preparing for the implementation of the two Directives. Reporting on the progress of corresponding national legislation in each Member State became a routine item. The meetings were an opportunity to discuss questions of definition, interpretation, clarification. Combining formal business with opportunities for informal interaction, they enabled problems to be identified and discussed, various solutions considered. National authorities – particularly those with significant experience of national legislation, or who had addressed in detail some specific issue – could offer presentations; whose merits as models for Community-wide application could then be debated.

The agenda was often broadened beyond the formal business, to address other matters of interest; for example, interaction with sectoral legislation, in sectors such as pharmaceuticals, pesticides, seeds, and novel foods.

After 23 October 1991, the meetings became formal meetings of the competent authorities set up under each Directive: typically three meetings being held adjacently, one in relation to Directive 90/219, one in relation to 90/220, and a joint meeting where matters of wider common interest could be raised. Decisions under Article 21 of either directive required formal convocation of the "Article 21 Committee", or a Commission decision by written procedure.

BRIC, the former Biotechnology Regulation Inter-service Committee, had held its final meeting on 19 November 1990, and had been formally dissolved on the creation of the Biotechnology Co-ordination Committee. Although many topics were in progress or under discussion in various services, it was clear that the BCC would be the future forum for their discussion. Indeed, there were some early references to it as "Super-BRIC"; but it was made clear that its mandate would include determining the overall objectives for Community action in biotechnology, thus resuming the concept of strategy that had fallen into abeyance after 1988, but had been restated by Vice-President BANGEMANN.

Few tears were shed for the demise of BRIC. Although the April 1991 communication had indicated the need to review constantly the regulatory framework, this was merely a statement of Commission policy and intentions, and did not have the legal force and authority of the 219 and 220 Directives. Thus the meetings of the national competent authorities, under the chairmanship of DG XI, were for them a more satisfactory forum than BRIC. These meetings could become a quasi-autonomous policy-making center for biotechnology; provided that the topics discussed were connected to regulatory aspects, and avoided direct conflict with other services, e.g. responsible for sectoral legislation. Other Commission services would in any case be reluctant to dispute with DG XI in front of the Member States; and the representatives of Member States present would in most cases be from agencies sympathetic to DG XI and the aims of the Directives.

From this "power-base", a global policy might conceivably be launched; and indeed, if the European Community had undertaken some pioneering and successful work, why should they not seek to spread widely its benefits and the Community's influence and intellectual authority? But the basis of such influence building remained contentious, not to say flawed: for it had always to be related to the control of genetically modified organisms (GMOs) as defined in the Directives. Thus, under "international activities", the meeting of competent authorities held in Denmark on 10 July 1992 discussed seven topics:

(i)   European Free Trade Area, EFTA
(ii)   EC/US bilateral discussions
(iii)   CEN, the European Standards Committee
(iv)   OECD (see Sect. 6.2)
(v)   UN Conference on Environment and Development; (the "Earth Summit", or "Rio Conference) (see Sect. 6.4)
(vi)   Council of Europe
(vii)   UNEP Biodiversity Convention (see Sect. 6.4).

In parallel with this expansionist development of the Committee of Competent Authorities, the Biotechnology Co-ordination Committee was seeking to enhance policy coherence, and ensure greater transparency between the Commission services. During 1993–94, a series of Commission decisions and communications reflected a general commitment to greater openness (see EUROPEAN COMMISION, 1993 c, d; 1994 g).

DG VII (Transport) collaborated with DG XI in drafting a Directive relating to the transport of GMOs; and in participating at Geneva in the UN Economic Commission for Europe's Committee on the Transport of Dangerous Substances. On that topic, industrial representatives at the Geneva Committee made known to other Commission services (e.g., Industry, DG III, and Science, DG XII) their concerns, thus enabling other DGs to attempt to join in the discussion; such information did not flow freely within the Commission until the BCC was involved. DG XII, for example, in conjunction with the curators of the microbial culture collections in Europe, and in the MINE project (Microbial Information Network for Europe), had extensive networks of experts familiar with the packaging and labelling of dangerous organisms, but their advice was not invited, because their opinions were critical. The OECD conclusion about the absence of basis for legislation specific to rDNA organisms was similarly not evident in the transport discussions, and a DG XII offer to co-finance with DG VII an expert workshop received no reply.

Item (vi) above referred to the inclusion of specific reference to GMOs in the Council of Europe Convention on Civil Liability arising from Dangerous Activities; again, there was no internal consultation of scientific advice within the Commission services.

There was a coherence in these actions, if the assumption was maintained that there were special risks inherent in GMOs. However, this was not being borne out by scientific opinion, or from reports being published in the US. The basis remained conjectural; regulatory initiatives including the Community Directives increasingly justified themselves by reference to the need for public reassurance; and yet the suspicion was voiced, that the emphasis on regulatory activities focused on GMOs was itself reinforcing the message of inherent riskiness, and stigmatizing the techniques.

DG XI encouraged the widespread international imitation of the Community legislation, for example, encouraging by direct visits and discussions the Japan Environment Agency to introduce similar proposed measures in the Diet in 1991; an initiative which, having short-circuited the usual diplomatic channels and other Ministries interested, once discovered found little favor with the Ministries principally involved in biotechnology, and was quickly withdrawn.

In the preparations for the Earth Summit at Rio, DG XI were chef de file, biodiversity conservation being an "environment" matter; and dicouraged attempts by other services to follow closely the evolution of the debates. The resulting Convention on Biological Diversity is discussed in Sect. 6.4, including reference to the patenting and bio-safety articles which might have benefited from broader perspectives in their drafting; the latter again stigmatizing GMOs. (An exception was the Prieels report, 1991, on the New Delhi meeting of 23–25 October that year. Her report highlighted the curious dialogue, between developing country scientists asking for access to biotechnology and its benefits; and environmental experts from the developed world insistently offering regulations.)

The UK House of Lords Report (see Sect. 5.2), published in 1993, was critical of the drafting of the Directives, as having been "based on out-of-date science", and "impervious to scientific advice". There was some substance to this reproach, particularly prior to adoption of the Directives, when scientific criticism was feared as threatening the success of the political process. For example, when the draft directives were about to be adopted in April 1990, copies of draft technical guidelines relating to their implementation which had been sent to Member States were forwarded by several Member State experts to the Working Party on Safety in Biotechnology of the European Federation of Biotechnology (see Sect. 6.1), with requests for their opinion. This was the most broadly representative expert group of biotech safety specialists

in Europe. In response to their having circulated the Commission drafts, LAURENS JAN BRINKHORST, Director-General of DG XI, wrote to the Safety Group Chairman in angry and threatening terms.

The Chairman, Professor FROMMER, responded (2 October 1990):

"I have to say that we were most taken aback by both the content and the tone of your letter which is critical of attempts to arrange up to date scientific input to the production of technical guidelines in biotechnology.

We understand that DG XI is engaged in producing a series of guidelines under the GMO Directives and yet you appear to indicate that the scientific community who, along with industry, will be most directly affected by your activities, has no place in commenting on your proposals.

As far as access to the documents is concerned, given that they had been distributed by your staff at a "national experts" meeting in April and, given also that several EC countries, quite properly and democratically, seek the views of interested parties (scientists, industry, trade unions and so on), it can surely be no surprise that they found their way to the EFB Working Party from several different sources.

We suggest that you contrast the transparent, consultative approach of various Member States with that of your services. A secretive approach is the last way to generate confidence either with the practitioners of biotechnology or the public.

You will recall that the Working Party has in the past offered to advise and provide scientific guidance and, notwithstanding the above comments, this offer remains open."

Similar sentiments internally made clear the reluctance of DG XI to accept advice from the Directorate-General for Science, Research and Development, DG XII. The Environment DG preferred to consult its own experts, for example in a study report on recombinant vaccines (MCNALLY and WHEALE, 1991), which it refused (in spite of repeated requests) to share with DG XII, although DG XII was managing research projects in this field, within the Community R & D programme; whose results were of course routinely published.

These matters were not simply differences of house style; they contrasted with a central feature of the US regulatory approach, which was its openness to criticism and debate, and the consequently better scientific basis and adaptability of the US regulations. Draft technical guidelines were routinely published in the *Federal Register*, with a 60 or 90 day period for public comment; a similar openness characterized the meetings of the NIH Recombinant DNA Advisory Committee.

However, it was politically desirable to be seen to be receiving scientific advice; and provided that it was not overtly subversive of the basis of the Biotechnology Directives, this was increasingly acceptable to and sought by, DG XI. The (closed) meetings of national competent authorities were ideal for this purpose, and DG XII were invited to give presentations of their risk assessment research. Similarly in the context of the bilateral environmental meetings between EC and US, a permanent technical working group on biotechnology was set up between DG XI and staff of the US Environmental Protection Agency, and through exchange of views at this group, convergence on various practical aspects of regulation could be encouraged. DG XI's longstanding contacts with the Community's Joint Research Centre's facilities at Ispra could similarly be used to strengthen the appearance of scientific basis.

In October 1993, the UK competent authority organized a workshop on environmental risk assessment for releases of GMOs; and in May 1994, DG XI organized an EC/US workshop on release of transgenic plants. It was becoming increasingly clear that microorganisms, plants and animals required distinct treatment. Moreover, a "tiered" approach of successive questions was inevitable encouraging the development of the concept of categories meriting exemption from regulatory oversight; as was the "competitive pressure" from the simplification of the US regulations (e.g., on the main crop plants, in 1993) but Annex I, defining scope, could not be amended by the committee procedures.

A "Risk Assessment Group" was set up in June 1994 under the Committee of Competent Authorities, to build up in a controllable way the committee's scientific capability; and enable them more effectively to use or resist independent scientific advice or criticism, from DG XII or elsewhere. DG III and DG XII's attempts to strengthen independent

scientific advice – reflected in the December 1993 White Paper and subsequent communication to the Corfu Summit (see Sect. 7.4) – were strongly resisted by DG XI, using in particular the Committee of Competent Authorities.

Regarding the implementation of the Directives in national legislation, the Commission was required to produce triennial reports, of which the first was due in 1993, but publication delayed by translation and other difficulties until the end of 1994. Details are presented in Sects. 5.1 to 5.5 of historical developments in some of the major countries; the overall situation regarding implementing legislation in the European Union for the Contained Use Directive, 90/219, could by end of 1994 be summarized as follows.

In the *United Kingdom* the Genetically Modified Organisms (Contained Use) Regulations (adopted under the Health and Safety at Work Act 1974) came into force on the 1st February 1993, replacing the Genetic Manipulation Regulations of 1989. The principles of notification, risk assessment, application of containment, etc. found in Directive 90/219/EEC were already features of the UK's existing legislation.

In *Denmark*, the 1986 Gene Technology Act was reviewed following the adoption of the EC Directives and revised in 1991. The new law (Law No 356 of 6 June 1991 on the Environment and Genetic Engineering) entered into force on 23 October 1991, with six associated Orders. For the working environment, an amended Order (Labour Inspectorate Order No 684 of 11 October 1991) entered into force on the same date.

In *Germany*, the Genetic Engineering Law (Gentechnikgesetz) and its implementing Regulations were adopted in July 1990 in order to transpose Directive 90/219/EEC of 23 April 1990 on the contained use of genetically modified microorganisms into national law; a revised form of the law was formally adopted in January 1994 (see Sect. 5.3).

In *Ireland*, the Environmental Protection Agency Act, 1992, enacted on 23 April 1992, includes a Sect. 111 on Genetically Modified Organisms which enables the Minister to give full effect to the Directive by means of regulations, after consultation with the Minister for Industry and Commerce and any other Minister of the Government concerned. These Regulations under Sections 6 and 111 were issued in November 1994, by Statutory Instrument 345/94.

In *France*, the Directive was transposed by the adoption by the French Parliament of Law no. 92.654 of 13 July 1992 concerning the control of use and release of genetically modified organisms and modifying Law no. 76.663 of 19 July 1976 concerning installations controlled for environmental protection. A number of detailed regulations have been adopted (Decrée nr 93-774 of 27.3.93, Decrée nr 93-773 of 27.3.93).

In *Spain*, a framework legislative text was presented to Parliament in 1993, and adopted in 1994.

In *Portugal*, framework legislation implementing Directive 90/219/EEC was adopted by the Council of Ministers on 28 January 1993. Detailed regulations have since been adopted.

In *Luxembourg*, a draft legislative text was presented to Parliament in 1993.

In *Greece*, no legislation had yet been presented to the decision making bodies by mid-1994. A draft text was being discussed at inter-ministerial level.

In *Italy*, the Framework Law enabling the transposition of a number of EEC Directives including 90/219/EEC was adopted in February 1992. The Legislative Decree laying down the details of implementation was adopted in March 1993 (Decreto Legislativo nr 91, 3.3.93, *Gazzetta Ufficiale* nr 34, 3.4.93).

In *Belgium*, implementing legislation for Directive 90/219/EEC was adopted in the region of Flanders in December 1992 by the adoption of VLAREM II.

In the Brussels Region, a draft of regulation transposing Directive 90/219/EEC into regional competences as regards environment was submitted to the regional Council of Environment for a first reading on July, 12th 1993 and for a second reading on September 13th 1993.

As a first step, Directive 90/219/EEC had been transposed into the framework of the General Law on Work Protection. As a second step, the same decree would be adopted and integrated into the framework of the new

regional law for environmental licensing when it came into force. The draft regulation extends the field of application of the Directive to all GMOs; and to human, animal and plant pathogens. A list of 1765 pathogens was established, with a class of risk assigned to each pathogen.

In the Wallonia region, a draft text was under discussion in 1993 and 1994.

In the *Netherlands*, the Directive 90/219/EEC had been largely implemented by the Nuisance Act. To fulfill all the obligations enforced by the Directive 90/219/EEC a paragraph was added to the "Genetically Modified Organisms Decree pursuant to the Chemical Substances Act" (GMOD, enforcing Directive 90/220/EEC, *Bulletin of Acts and Decrees*, 25 January 1990, 53). In this paragraph procedures and classification for contained use were described. Technical details are regulated by a Ministerial Regulation which gives the possibility of quick response to new developments. Both the modified GMOD and the Ministerial Regulation came into force on 1st October, 1993.

The Commission is expected (indeed, under Article 18.3 of the Directive, is required) to produce a report giving details of the appointments of competent authorities, measures taken for practical implementation, and a short summary overview of activities and installations in each country and experience with the implementation of the Directive. Late submission of national data delayed this beyond its scheduled appearance in 1993 (EUROPEAN COMMISSION, 1994d).

Other simplifications and streamlining of procedures – formats for the summary notification of information, revisions of the technical annexes – advanced steadily through the Committees of NCAs, as recorded by successive decisions of the EUROPEAN COMMISSION (1994b, c, f).

Given this extended and complex process of implementation of the directives, involving the legislative processes in both national and (for federal countries such as Belgium and Germany) regional parliaments, followed by the development of new administrative procedures to give effect to these laws, it was understandable that there was little enthusiasm in the Committee of National Competent Authorities for reviewing the legislation. Indeed, there was considerable opposition.

But while the legislative and administrative processes, debates, and accompanying uncertainties persisted, there was a sort of "planning blight" on the investment in research, development and innovation which the new opportunities were eliciting in the US, and (to a lesser degree) in the more pragmatically governed parts of Europe. This hostile climate, or perceived hostile climate, led the Commission to recognize the need to revise and simplify procedures so far as possible within the existing Directives; but beyond that, to consider revising them more fundamentally in revised proposals to the European Parliament and Council, even before the 1990 Directives were fully implemented. This revisionist drive, pressed by industrial lobbies such as SAGB with strongly argued and well presented documentation, brought the need to review strategy for biotechnology into prominence in the December 1993 "Delors White Paper".

7.4 The White Paper and the Next Stage: From Corfu to Essen, and the 1995 Review

Political Priority Shifts to Employment

By mid-1994, unemployment in the developed world economies of the OECD area had reached 35 million. In his successful presidential campaign up to the November 1992 election, candidate BILL CLINTON defeated incumbent President GEORGE BUSH in a campaign featuring the slogan, "It's the economy, stupid". In February 1993, within a few weeks of inauguration, the new President and Vice-President issued in their own names the policy document, "Technology, Growth and Employment: renewing the American economy through technology" (CLINTON and GORE, 1993).

This policy statement reflected Vice-President AL GORE's well-known enthusiasm for the concept of "information highways"; but biotechnology also featured prominently in the report. The following quotations illustrate

the flavor; including the very explicit references to the impact of the regulatory framework:

"Technology is the engine of economic growth. ... Breakthroughs such as the transistor, computers, recombinant DNA and synthetic materials have created entire new industries and millions of high-paying jobs.

...

We can promote technology as a catalyst for economic growth by:

- directly supporting the development, commercialization, and deployment of new technology;
- fiscal and regulatory policies that indirectly promote these activities;
- investment in education and training; and,
- support for critical transportation and communication infrastructures.

...

In many technology areas, missions of the agencies coincide with commercial interest or can be accomplished better through close cooperation with industry."

Under the heading, "A World-Class Business Environment For Private Sector Investment", item 7 stated:

"Ensure that federal regulatory policy encourages investment in innovation and technology development that achieve the purposes of the regulation at the lowest possible cost: Regulatory policy can have a significant impact on the rate of technology development in energy, biotechnology, pharmaceuticals, telecommunications, and many other areas. The caliber of the regulatory agencies can affect the international competitiveness of the agencies they oversee. At the same time, skillful support of the new technologies can help business reduce costs while complying with ambitious environmental regulations. A well designed regulatory program can stimulate rather than frustrate attractive directions for innovation. We will review the nation's regulatory "infrastructure" to ensure that unnecessary obstacles to technical innovation are removed and that priorities are attached to programs introducing technology to help reduce the cost of regulatory compliance."

The US job situation was bad, but in terms of job creation over the 1980s, the European performance was significantly worse. Within the European Community, or European Union as it became known after the November 1992 ("Maastricht") Treaty on European Union, continuing high unemployment in all Member countries made incumbent governments of whatever political position unpopular; improving economic performance to regenerate growth and expansion of employment became the top priority of national governments and of the European Commission and Council. The global character of the challenge was underlined at the Detroit Summit of the G7 group in April 1994. Similarly at the OECD Ministerial meeting of Economics Ministers in June 1994, the basic document was the OECD report on measures to combat unemployment, a report commissioned two years previously.

The OECD report's diagnosis emphasized human resources, and the need for governments to facilitate and to enhance the flexibility of operation of labor markets. Within Europe, and particularly under the influence of UK criticisms and the British decision to opt out of the "Social Chapter" of the Maastricht Treaty, the accusation gained political currency that the weight of regulation – particularly social legislation, adding to the costs of employment, and rendering it difficult to dismiss labour – was rendering Europe inadequately competitive, and thus contributing to and maintaining the high level of unemployment. To the slight political embarrassment of Lord HOWIE (a Labour Party member of the House of Lords), his Committee's report (October 1993 – see Sect. 5.2), criticizing the inappropriate or excessive regulation of biotechnology, tended to be seen as part of the same "deregulatory" drive. It was a natural confusion.

In spite of the high unemployment, the associated rise of political demands for protectionism did not prevail; the North American Free Trade Area was approved in the US Congress, with bipartisan support. The long-drawn-out Uruguay Round of GATT negotiations was finally agreed in December 1993, and signed at Marrakesh in May 1994; in December 1994, ratification by both houses of the US Congress confirmed this success. The new World Trade Organization was established, starting in January 1995. The economies of the Western world remained committed to the open world trade system on which the Bretton Woods institutions and the OECD were founded.

This general background, including particularly the accusation that excessive European regulation was damaging competitiveness, interacted strongly with the ongoing arguments about the regulation of biotechnology. There was consensus that economic growth and new jobs might derive from new-technology-based industries; but that in any case the pervasive impact of new technology, particularly of new information technologies, made them impossible to ignore if competitiveness was to be maintained. Biotechnology was also demonstrating a similar tendency to pervasiveness, its supposed risks remained conjectural, and it continued to offer no evidence to give concrete justification for technology-specific regulation.

The "Delors" White Paper: "Review the Regulatory Framework for Biotechnology"

Through 1993, the Commission services prepared a major policy document, which President DELORS presented to the European Council in December: on "Growth, Competitiveness and Employment"; also known as the "Delors White Paper" (EUROPEAN COMMISSION, 1993b). Given the above background, it was not surprising that biotechnology featured prominently as one of the three technologies explicitly addressed in this document, along with information technology and the audio-visual sector(s).

On biotechnology, the White Paper referred to its emergence "as one of the most promising and crucial technologies for sustainable development in the next century"; and noted that the Community was highly competitive in the sectors to which biotechnology was particularly relevant; "sectors which cover chemicals, pharmaceuticals, health care, agriculture and agricultural processing, bulk and specialized plant protection products as well as decontamination, waste treatment and disposal. These sectors where biotechnology has a direct impact currently account for 9% of the Community's gross value added (+/- 450 billion Ecu) and 8% of its employment(+/- 9 mio)".

Having compared the growth rate of these sectors in Europe with growth rates for the US and Japan and estimates of the growth of the total market, the paper concluded that "the Community growth rate will have to be substantially higher than at present to ensure that the Community will become a major producer of such products, thereby reaping the output and employment advantages while at the same time remaining a key player in the related research area".

Going on to review the adverse factors, however, it emphasized:

– publicly financed research and development expenditure lagging behind US levels;
– privately financed R & D insufficient to compensate for the shortfall in public; rather, "available indicators identify a delocalization – an investment outflow";
– "regulation concerning the safety of applications of the new biotechnology is necessary to ensure harmonization, safety and public acceptance. However, the current horizontal approach is unfavorably perceived by scientists and industry as introducing constraints on basic and applied research and its diffusion and hence having unfavorable effects on UK competitiveness;
– technology hostility and social inertia in respect of biotechnology have been more pronounced in the Community in general than in the US or Japan ..."

In its conclusions and recommendations, the paper stressed:

"overcoming existing constraints by creating appropriate channels for biotechnology policy development and co-ordination and by acting on the following recommendations:

a) Given the importance of regulations for a stable and predictable environment for industry and given that they influence locali-

zation factors such as field trials and scientific experimentation, the Community should be open to review its regulatory framework with a view to ensuring that advances in scientific knowledge are constantly taken into account and that regulatory control is based on potential risks. A greater recourse, where appropriate, to mutual recognition, is warranted to stimulate research activities across Member States. Furthermore, if the Community is to avoid becoming simply a market rather than a producer of biotechnologically-derived products then it is vital that Community regulations are harmonized with international practice. The development of standards will supplement regulatory efforts.

b) The Commission intends to make full use of the possibilities which exist in the present regulatory framework on flexibility and simplification of procedures as well as for technical adaptation. To sustain a high level of environmental protection and to underpin public acceptance it is important to reinforce and pool the scientific support for regulations. An advisory scientific body at Community level for biotechnology diffusion drawing on the scientific expertise within and at the disposal of the existing committees could play a crucial role in intensifying scientific collaboration and in providing the support needed for a harmonized approach to the development of risk assessments underlying product approval. This body could also advise on the development of a further Community strategy for biotechnology.

The remaining recommendations stressed strengthening R & D, with improved co-ordination between Community and Member State efforts; the creation of "a network of existing and new biotechnology science parks"; and additional incentives to "improve the investment climate for biotechnology and facilitate commercialization."

Follow-up at Corfu: The "Next Stage" and the "Two-track Approach"

Under the Biotechnology Coordination Committee, the recommendations of the Delors White Paper were developed in greater detail by the services concerned, and presented to the European Council (of Prime Ministers) at the Corfu Summit in June 1994, near the end of the Greek presidency (EUROPEAN COMMISSION, 1994a). The communication, entitled "Biotechnology and the White Paper on Growth, Competitiveness and Employment: Preparing the Next Stage", was addressed to the Council, the European Parliament, and the Economic and Social Committee, and had chapters on:

- Regulatory Framework
- Strengthening of Scientific Advice
- Research and Development
- Biotechnology and SMEs
- The Investment Climate
- Public Understanding
- Ethics.

An Annex presented the "State of Play of the Biotechnological Regulatory Framework", thus further underlining the major role which regulatory issues now played in the Commission's proposed Community strategy for biotechnology.

In the chapter on the Regulatory Framework, the Committee referred to the growth of knowledge and experience, from which "it may be concluded that the risks involved in the contained use of GMMs are substantially less than were once foreseen ... the potential for horizontal gene transfer resulting in novel and harmful properties being acquired by microorganisms has not been shown to present hazards to human health and the environment ... evidence is accumulating to the effect that genetically modified plants do not differ from non-modified plants other than in the specific character conferred by the introduced gene".

After further references to the impact of the regulatory framework on industrial competitiveness, but also the results of surveys indicating its important role in building public confidence in biotechnology, the document

announced the Commission's intention to apply "the following two-track approach for the future development of the biotechnological regulatory framework:

- the exploitation of existing possibilities for revising measures/procedures/degree of oversight/requirements, through use of the "light" procedure of adaptation to technical progress (regulatory Committee procedure); (internal amendment)
- the bringing forward of amendments to existing legislation in order to incorporate changes which cannot be achieved by technical adaptation while leaving the basic structure of the framework intact (external amendment).

The immediate intentions focused on the "contained use" Directive, 90/219, where four objectives for action were identified:

(i)   streamlining and easing of the administrative/notification/consent requirements where this does not compromise safety;
(ii) ensuring that the classification of the genetically modified microorganisms and of the activities in which they are used is appropriate to the risks involved;
(iii) ensuring that the conditions of use are appropriate to the risks involved;
(iv) extension of the flexibility of the Directive so that it can be more easily adapted to technical progress by regulatory Committee procedures.

Some of these aims could be achieved by using the Committee procedure of the Directive itself; in particular the Annex II definition of "Type I" microorganisms could be simplified from the original requirements, based on the 1986 OECD "Blue Book", to simpler statements about microorganisms unlikely to endanger human or animal health, or the environment. This was subsequently proposed to the "Article 21" Committee for 90/219, and adopted in November 1994 (EUROPEAN COMMISSION, 1994f).

Coincidentally, the Commission published

a report (THORLEY, 1994) by a consultant who had studied sector-by-sector the interpretation in 12 sectors of the OECD concept of "Good Industrial Large Scale Practice". His conclusion was that, insofar as industrial experience had been taken into account during the definition of "GILSP", it had been exclusively that of the pharmaceutical industry, with the high standards for characterization and purification corresponding to the needs of pharmaceutical products. As biotechnology moved into application in other sectors, the inappropriateness of the horizontal legislation became more apparent.

The Commission saw the need to lighten the Directive in other respects, but as this could not be achieved within the current terms, it would require specific amendments to the Directive:

- replacing the consent requirements by record-keeping, or notification for information purposes, for certain low-risk activities;
- replacing the explicit consent requirements by implicit consent for certain higher-risk activities;
- reduction of time periods involved in implicit/explicit consent procedures;
- adapting the present risk classification system for GMMs, in accordance with new safety considerations;
- removal of the differentiation between activities in research laboratories and production plants.

The Commission stated its intention to consult widely, and place its proposals for amendment before the Essen Summit at the end of 1994.

Regarding field release and Directive 90/220, the Corfu communication pointed out the changes which the Commission had already been making (following proposals from several Member States), through Commission decision procedures. These concerned in particular simplification of the notification requirements for releases of plants (EUROPEAN COMMISSION, 1994b, c). (It may be noted that simplifications of procedures for major crop plants had been introduced in the United States in 1993: see Sect. 5.1.) However, the

document announced that "further experience is necessary in order to determine the right balance between the need for safety, public reassurance and the minimum restraint on industry and research work". The Commission undertook to review the Directive during the first half of 1995, under the French presidency, to assess the need for proposals in relation to:

- extending the flexibility of Directive 90/220/EEC, so that its scope and the procedures to be followed are always appropriate to the risks involved, and are easily adaptable;
- strengthening more uniform decision-taking between Member States in the case of research and development releases;
- introducing further opportunities for notifiers (industry and researchers), so that they can benefit more from the existence of a uniform Community system;
- facilitating the link between this Directive and product legislation.

The Annex to the Corfu communication reviewed progress on the implementation of the horizontal Directives 90/219 (contained use) and 220 (field release), and the worker safety Directive 90/679, whose transposition into national law was proceeding more slowly. On specific product legislation, a similar review noted Directives (or amendments to Directives) adopted (on additives in feedingstuffs, and on high technology/biotechnology-based medicinal products), or in preparation (on plant protection products, amending 90/414); as well as those under discussion, in particular the draft Regulation concerning novel foods and food ingredients.

The Annex also presented a succinct summary statement of the underlying principles of the Community's biotechnology regulatory framework:

"– Necessity: the Commission will propose legislation in this area only if it is shown to be necessary by a thorough examination, on a case-by-case basis, of the characteristics inherent in specific biotechnological applications.

– Efficient interaction: biotechnologically-derived products will be subject to only one authorization and assessment procedure before being placed on the market.
– Evaluation criteria: product evaluation will take place in accordance with the three established criteria of safety, quality and efficacy. The Commission will normally follow scientific advice. In exceptional cases, however, it reserves the right to take a different view in the light of its general obligation to take into account other Community policies and objectives.
– Adaptation to progress: the regulatory framework will be kept up to date with scientific and technical progress. This is of particular importance in a rapidly developing field such as biotechnology.
– Standards: the development and existence of standards may be used to complement legislation, particularly on technical details of good practice and safety procedures.
– International obligations: the Commission will ensure that all decisions in the field of biotechnology will be in conformity with international obligations, in particular with the provisions resulting from the Uruguay Round negotiations."

Although the Commission's communication attracted widespread coverage in the specialist biotechnology press, its timing virtually coincided with the election for the European Parliament; with the result that in the immediate pre- and post-election period, there was no immediate political reaction.

The Corfu communication was welcomed by the Council of (Industry) Ministers, on 29 September 1994, who called upon the Commission to present by the end of the year proposals on modifying Directive 90/219, and to envisage changes to 90/220 "in the short term". They called upon the Environment Council – which met on 4 October – to examine the subject "without delay"; but the subject was not placed on the agenda, and was barely discussed. Doubts therefore persisted about the prospects for change.

Industrial Reaction – the SAGB "Yellow Book"

Following the Commission's December 1993 White Paper, the Senior Advisory Group for Biotechnology (see Sect. 7.1) pre-

pared, and in April 1994 published, a 20-page yellow booklet entitled: "Biotechnology Policy in the European Union: Prescriptions for Growth, Competitiveness and Employment: a response to the Union's 1993 White Paper on Growth, Competitiveness and Employment" (SAGB, 1994a). Listing its 29 members, it noted that their combined turnover was $ 305 billion; their annual R & D expenditure $ 17 billion, and investment $ 24 billion; and that they provided employment for nearly 2 million people world-wide. A parallel publication analyzed more specifically the policy needs to stimulate innovation, and the development of start-up businesses (SAGB, 1994b).

Responding in positive tone to the Commission's White Paper, the SAGB yellow booklet noted that the White Paper identified the causes of increased challenges for the European Union as including "lack of adaptation to new technologies, in particular biotechnology"; and went on to propose a five-point action plan for facing the challenges:

● Regulatory Reform
● Lowered Barriers to Investment
● Biotechnology-Supporting Fiscal Measures
● Increased Labor Flexibility
● Provision of Leadership in Wealth Creation, Competitiveness and Prosperity.

It further proposed the creation of a "Special Economic and Competitiveness Task Force for Biotechnology"; the management to comprise appropriate expertise in the investment and economic domains, with representation from "Ministers of Industrial and Economic Affairs, Industry, Agriculture, Academia, the Investment Community, Employment Representatives, and the Commission's Representatives for Industrial and Economic Competitiveness".

On the first Priority Action, Regulatory Reform, its message was blunt. It emphasized that

"A viable regulatory structure is essential for a stable and predictable environment in which to operate, develop and maintain competitiveness and which can cope with rapid technological advances. In the Union, this means using the best experience of individual Member States, recourse to mutual recognition where possible and regulatory structures based on scientific evaluation of real and measurable risk."

Citing the UK House of Lords report (see Sect. 5.2), the German Ministry of Health and their own earlier reports (see Sect. 7.1), the SAGB paper stated that a viable regulatory structure should encompass:

> "Rapid construction of an agreed legal framework within which regulations on biotechnology issues should be encompassed.
> Evaluation of existing regulations (including those currently under development) to determine fit and compatibility with the framework. Due to the multi-disciplinary nature of the application of biotechnology, the evaluation process must involve all those involved in using and regulating biotechnology.
> Where necessary, replacement of existing regulatory structures by new structures which are both consistent with the framework and with scientific progress.
> Institution of transparent consultation processes with the prime users of regulation to ensure viability of regulatory proposals before any new proposals are put forward."

It concluded that fundamental changes were needed, in view of the recognition that "the horizontal biotechnology process-based European Directives which touch on many other regulations are basically flawed"; and therefore called for "a halt to the development of new regulatory structures until the [SAGB proposed] action plan has been put in place and the basic framework has been developed and agreed".

However, in several countries, national bio-industries that had laboriously established working arrangements with their national competent authorities began to lose heart for continued struggle, or for perturbing public opinion by even the appearance of "deregulation". If their local appeasement

was successful, why should they pursue quixotic battles for broader or abstract concepts of European public interest? No lobby argued for the non-existent firms, products or services deterred from coming into being: "les absents ont toujours tort".

Scientific Reaction – the ESF Resolution of July 1994

In parallel to the industrial initiatives, the European Science Foundation (see Sects. 2.6 and 6.1) had been brought back into the recombinant DNA debate, because of the impact which genetic engineering legislation was having on their members. A survey of EMBO scientists had indicated considerable concern (RABINO, 1991, 1992). Similarly, among the researchers involved in the Community's own biotechnology R & D programme, an investigation (offering strict anonymity, to elicit frank responses) confirmed that the legislation – in some countries more than others – was felt to be onerous, time-consuming, and irrational in several respects.

The ESF in 1994 responded to the Delors White Paper of December 1993, with a Resolution "welcoming ... the opportunity to comment on the existing legal framework for biotechnology and for genetic engineering", and agreeing that "the present regulations go beyond what is necessary to control the risks (real or potential) ... and may therefore act as a disincentive to the development of the European biotechnology research and industry". It urged the Commission "to correct and improve the conditions for effective research in Europe by providing for an amendment of the legal framework and for an adaptation of its regulations to the state of the art in research and technology ... Such changes are possible without increasing the risk and can thus be easily realized while fully upholding the present standards for the safety of human health and the environment".

A "Resolution on the Legal Framework for Genetic Engineering" was adopted by ESF in mid-1994, culminating in a statement of nine principles to be observed in changing the legal framework for genetic engineering and biotechnology:

"1. Genetic engineering is a method which has led to important progress in many areas of biological and medical research (including new methods of diagnosis and therapy) and of ecological research. Given its scientific potential, Europe should play a great role in the industrial application of this method. Otherwise Europe may face the danger of becoming a market rather than a site of innovation and production. In this respect legal frameworks can exhibit different effects: whereas the United States of America has not passed laws on the use of genetic engineering techniques, the European framework restricts research and the application of new research results in industry by exaggerated administrative requirements. If this development were to continue, it could soon lead to reduction in activities and in educational and research incentives, to a deterioration in the quality of university education and of research quality in the respective fields in Europe.

2. Over 20 years of experience with recombinant DNA and modern biotechnology in general has established that the risks involved can be assessed and minimized and controlled by adequate safety measures drawn up according to the state of the art, especially in biological research. As with regulations concerning the use of unmodified organisms, the legal framework for the contained use of genetically modified microorganisms should only be concerned with organisms with a risk potential and should differentiate the relevant procedures more effectively in accordance with this potential. Therefore:

– for the contained use of genetically modified microorganisms classified as safe or with no risk, no notification to the competent authority should be required;
– for the contained use of genetically modified microorganisms classified as with low risk potential, only information to the competent authority should be required, so that such use may commence immediately without any further consent by the authority;
– for the contained use of genetically modified microorganisms classified as with high risk potential, the requirement should be for the notification to the competent authority before such use commences, as prescribed in the existing Council Directive 90/219/EEC.

The classification of genetically modified organisms (GMOs) used by the US National Institutes of Health (Guidelines for research involving recombinant DNA molecules) might be a good example to be followed. To adapt those regulations might also ease communication and exchange between scientists in Europe and the United States and even beyond.

3. For the definition of operations (at present "Typ A" for teaching, research, development, or non-industrial or non-commerical purposes, and "Typ B" for any other operation), neither the scale (culture volume) nor the purpose seems to be appropriate as the sole and exclusive criterion. In consultation with academic and industrial research scientists, a workable risk base approach should therefore be developed.

4. For a better implementation of the legal provisions, the Council Directive should clearly differentiate between the active use, i.e. operations directly resulting in a genetic modification of (micro)organisms, and the more passive handling of GMOs, i.e. operations connected with storing and inactivation of GMOs and waste treatment. For the latter operations, the existing regulations for hazard goods should be applied as far as possible.

5. Considering the fact that risks for human health and the environment in the case of microorganisms are limited to operations with microorganisms with a high risk potential, the Directive should provide for the possible consultation of specific groups or the general public only on aspects of the proposed contained use of microorganisms with such a high risk potential.

6. Possibility to exchange material is a key issue for research, since it permits to achieve verification of results prior to and after publication and for subsequent work at other laboratories. In the case of GMOs, there should be no doubt possible that the international exchange of sam-

ples for purposes of research should not depend upon prior authorization and should not be subject to regulations governing the "placing on the market" of GMOs.

7. World-wide, the majority of field trials in which GMOs are deliberately released into the environment are carried out with plants. Nevertheless, very few such trials have been carried out up until now in most European countries compared with, for example, other OECD countries. This is mainly due to existing European regulations. Simplification of these regulations should be possible without creating an unacceptable increase in risk to human health and the environment, and should follow the "one stop, one shop" principle:

  – Once the criteria for a simplified consultation procedure have been established as sufficient in one case, all subsequent applications concerning that GMO or GMO-combination should automatically be dealt with in the same way.
  – When the risks caused by a release are restricted to a limited area (location), and thus human health and the environment in other countries are not in danger, the necessary decision should be a purely national one; a prior consultation procedure within the EU should therefore no longer be required, being replaced, if deemed necessary, by notification of the other competent authorities.

8. In order to implement research achievements for the benefit of medicine and public health, genetic engineering in the field of human and veterinary medicine aims to develop new diagnostic and therapeutic methods and new pharmaceuticals. To promote such research and to further its implementation, for the sake of mankind:

  – applications of diagnostic or therapeutic methods using genetical modification techniques should be explicitly excluded from the provisions of "deliberate release" or "placing on the market" of GMOs;

  – a human being or an animal to whom substances containing GMOs are applied for diagnostic or therapeutic purposes or to whom individual cells are reintroduced after genetic modification should not be regarded as GMOs themselves;
  – procedures for the permission to introduce new pharmaceuticals developed in a European laboratory into clinical trials or onto the market should ensure that the chances for success of an application are not dependent on the country in which the laboratory or the company making the application has its legal or effective base.

9. Gene therapy, in general, is one method among others which aims at maintaining and restoring health, especially of human beings. The use of such therapy in a specific case must be the sole responsibility of the physician and be subject to the same regulatory framework as other therapeutic methods. Such therapy should, however, be applied in close consultation with and under the scientific control of persons competent in the particular field of genetic engineering. Furthermore, it is recommended to establish national registers to collect data on cases where gene therapy has been applied. No special legal framework should therefore be promulgated for gene therapy at European or national levels."

This was transmitted to President JACQUES DELORS with an accompanying letter on 11 June 1994, on behalf of the "Board and Executive Council of the European Science Foundation, representing 55 member research councils, academies and institutions devoted to basic scientific research in 20 European countries". The letter urged the Commission to amend the legal framework on genetic engineering, and in doing so to adhere to the following guidelines:

"– The legal framework for the use of genetically modified microorganisms should differentiate the relevant procedures more effectively in accordance with their risk potential.

- The definition of criteria for operations with such microorganisms would benefit from a new and workable risk assessment based approach which would be developed in consultation with academic and industrial research scientists.
- The regulations concerning field trials with genetically modified organisms should be simplified.

The European Science Foundation underlines that the amendments proposed in the enclosed Resolution do not create any unacceptable increase in risk to human health and the environment."

German Industry and Science Visits the President

On 26 July 1994, in the first month of the German Presidency of Council, Dr. ERNST-LUDWIG WINNACKER of the Max Planck Institute, and Dr. HANS-JÜRGEN QUADBECK-SEEGER of BASF, President of the GDCh, the Society of German (industrial) Chemists, visited Commission President JACQUES DELORS, to discuss improvements in the EU regulatory framework for biotechnology. They left with him a 4-page statement, referring to the reasons why reform was needed, its scope and objectives, details of the changes required to 90/219 and 90/220, and proposed accompanying measures to ensure the continued success of biotechnology in the European Union.

Such a formidable combination of industrial, scientific and (some) governmental pressure for change of the 1990 Directives might seem irresistible; but beyond the individual or institutional sensibilities defending the Directives, loomed vaguely a much larger conservative force. At a time when new biotechnology products were starting to come to market, in such sensitive areas as consumer foods and animal productivity enhancers, the major uncertainty for investors and governments alike was "public opinion".

7.5 Public Opinion: The Joker in the Pack – or the King?

A Tense and Dynamic Trilogue

It may seem illogical to have an opinion about a topic of which one is largely ignorant; and where strong opinions co-exist with ignorance, the knowledgeable may reasonably describe it as prejudice, and therefore – less reasonably – dismiss it. This is not an option for elected politicians, who must respect, or at least take some account of, the views of their electors – even when these are dismissed as ignorant prejudice by the experts. The experts are vulnerable to charges of elitism if they dismiss the opinions of the general public, who are taxed to finance publicly supported research; and who may resent the imposition of risks (real or conjectural) that may flow from such research.

This triangle of tensions and communications between the general public, the scientist and the politician is shifting and ill-defined, but has in recent decades become increasingly significant in many areas of new technology, from irradiated food to nuclear power. Inevitably, it has been and remains a major influence on the evolution of biotechnology regulation; changing over time, and varying widely from country to country and between different areas of application, and different techniques.

In the USA, the post-Asilomar debate in the late 1970s has already been described (Sect. 1.4) as a dialogue – more correctly, a trilogue – of scientist, public and regulator. Yet in spite of the public ignorance and fears about conjectural risks, in the US conditions at that time, the pressures on Congress did not lead to legislation, and the counter-arguments for using existing statutes prevailed. A major factor in that outcome was the level of trust, by both public and Congress, in some of the scientific leaders, and particularly in the leadership of the science-based agencies such as FDA and the National Institutes of Health. The role of the NIH RAC (Recombinant DNA Advisory Committee) has been referred to (Sects. 1.4 and 5.1), and the transparency of its public meetings has been frequently cited as a factor in its credibility. Although the transparency in Europe has been less, the need for the involvement of lay participants and representatives of public interest has been acknowledged in the constitution of national committees such as the UK's GMAG and ACGM (see Sect. 5.2).

Reports Link Perceptions, Regulations and Trust

The importance of public perception for the progress of biotechnology was evident, from the Asilomar debate of the late 1970s and subsequently. Many authors or analysts made the parallel with nuclear power, or more generally with chemicals. Such parallels would emphasize the initial overselling in the early years of the technology; the naivety, hubris or dishonesty of scientists or engineers; the sacrifice of safety to personal, political and economic interests; and ultimately the trust-demolishing discovery or disclosure of side-effects and consequences previously unforeseen or deliberately concealed.

The first FAST report (FAST, 1982) had for these reasons emphasized the need for public information and education:

"education for the Bio-Society has a wider dimension. The strategic projects to be pursued through the key centres should respond to (or anticipate) the needs expressed, through market-place or political decision, by a democratic society; and must in that context be capable of winning the political, financial and social support necessary for their implementation. Such support depends upon a degree of public acceptability and comprehension. Obtaining such support can be more difficult than solving the technical problems, and the consequences of failing to do so can be more costly than the development of the technology itself. General education, through the school system and the public media, should therefore aim to provide widespread understanding, sufficient to permit informed discussion and appraisal of the acceptability of proposed developments before expensive commitments are made."

On the other hand, it was already evident that excessive caution could strangle innovation, and might already be doing so; PERUTZ (1981) was quoted in the FAST report:

"The time taken from patenting a new compound to its marketing averaged three years in the early 1960's, it rose to 7 years and a half in the early 1970's and to nine years in 1978–79, largely as a result of the demand for ever more elaborate trials and safety tests ... The number of chemically new drugs put on the market is falling and the fraction being spent on development continues to rise at the expense of research."

The FAST report emphasized that:

"There is a growing evidence that we are mismanaging the processes of societal learning and self-improvement of which technological progress is an element. Lead-times for innovation are increasing, with associated greater costs. The balance between broader societal interests and defence of the existing (products, jobs, institutions, ideas) is typically counter-productive in the medium- or long-term."

The FAST report, among its "Contextual Recommendations", emphasized both attention to the public education in the life sciences throughout the school system; and

"the creation of clear and consistent regulatory frameworks, applicable throughout the Community, on all aspects of laboratory development, factory manufacture, testing, and marketing of products and services; with particular reference to the novel products and services likely to arise in the fuel, food, chemical and pharmaceutical industries. These activities are already being pursued within the Commission Services through committees."

Similar concerns arose, somewhat later, in the United States. By 1985, in a National Academy of Sciences conference on "Biotechnology: An Industry Comes of Age", a recurrent concern at the meeting was voiced by McGINTY (OLSON, 1986):

"In the end, whether or not these new biotechnologies really get off the ground in this country is going to depend upon whether we can erect a regulatory regime that can secure public trust."

Thus on both sides of the Atlantic (and indeed in Japan: see Sect. 5.5), there was recognition of the linkage between public perceptions of risk, regulatory regimes, and public trust in these. The linkage seems wholly proper in democratic societies; although the existence of the linkage does not give any obvious conclusion as to how to manage a situation in which public perceptions of risks (of biotechnology, or of black magic) diverge significantly from scientific perceptions. In any event, the recognition of this linkage led to the need to measure public understanding and opinion, so far as possible objectively; rather than depend upon self-appointed activists claiming to represent "the public interest".

A paper in Swiss Biotech (CANTLEY, 1987) reviewed the situation and the available US and European data, noting in the European context the heterogeneity of national opinions, at least as suggested by general surveys (in 1977 and 1979) of public attitudes towards science and technology (EUROPEAN COMMISSION, 1977b, 1979). Tab. 1 was reproduced from the 1979 Community-wide survey.

The OTA Poll, 1986/87

The first quantitative survey work on public opinion specifically on biotechnology was a nation-wide (US) survey commissioned by the Congressional Office of Technology Assessment (OTA), conducted in October–November 1986. The results were published in May 1987 (OTA, 1987). The poll was based on a national sample of 1273 adults. The survey, using an extended questionnaire, sought information about knowledge and opinion on science and technology issues in general, and on genetic engineering and biotechnology in particular.

Summarizing the results, the OTA noted that "nearly half (47%) of the adult population of the United States describe themselves as very interested, very concerned, or very knowledgeable about science and technology"; and defined this population as "the science observant public".

80% of all Americans expected developments in science and technology to benefit them and their families; 71% expected at least some risks; but when directly faced with the choice between the risks and the benefits to society from continued technological and scientific innovation, 62% felt that the benefits outweighed the risks, as against 28% holding the opposite view.

The report made clear the range of views, and the high uncertainties; but on balance gave fairly clear support for continued but cautious progress. Its tone is conveyed by these quotations from the Executive Summary of the (OTA, 1987) report:

"… 52% believes that genetically engineered products are at least somewhat likely to represent a serious danger to people or the

Tab. 1. Biotechnology-Relevant Questions and Responses, from the Survey of Attitudes to Eight Research Areas (All Figures Represent % of Respondents), 1979

Country/Question	EC	B	DK	D	F	IRL	I	L	NL	UK
Genetic Research										
Worthwhile	33	38	13	22	29	41	49	37	36	32
Of no particular interest	19	20	10	16	22	20	19	31	17	21
Unacceptable risks	35	22	61	45	37	22	22	18	41	36
Don't know	13	20	16	17	12	17	10	14	6	11
	100	100	100	100	100	100	100	100	100	100
Development of Research on Synthetic Food										
Worthwhile	23	16	13	34	10	23	11	25	23	34
Of no particular interest	21	26	21	16	20	29	20	39	30	25
Unacceptable risks	49	44	50	36	66	38	65	25	42	36
Don't know	7	14	16	14	4	10	4	11	5	5
	100	100	100	100	100	100	100	100	100	100

environment. Nonetheless, a two-thirds majority of the public (66%) says that it thinks that genetic engineering will make life better for all people.

When all other factors are equal, the public says it is more favorably disposed toward genetic alteration of plants, animals and bacteria than manipulation of human cells. ...

A large majority of the public (82%) favors environmental applications of genetically altered organisms on a small-scale, experimental basis. In fact, 53% would favor and 14% say they would not care if their community were selected as a site to test a genetically altered organism. However, only 42% of the public think commercial firms should be permitted to apply genetically altered organisms on a large-scale basis. ...

The issue of human cell manipulation is more sensitive than other forms of genetic engineering. While a majority of the public (52%) believes it is not morally wrong to change the genetic make-up of human cells, a significant minority (42%) says that it is. When confronted with specific applications of human cell manipulation, however, many Americans relax their position ... a majority of the public says it favors the correction of potentially fatal genetic defects in germ line cells ... as well as somatic cells. A majority of those who feel human genetic manipulation in general is morally wrong nonetheless says it would approve its use in specific therapeutic applications. ...

A large majority of the American public (82%) believes that research in genetic engineering and biotechnology should be continued. Support for this research appears in all segments of the population. ...

The public believes that Federal agencies are distinctly less able than university scientists to assess potential risks. Moreover, in disputes between Federal agencies and environmental groups over risk statements, the majority of the public says it is inclined to believe the environmental groups. ...

The survey finds that while the public expresses concern about genetic engineering in the abstract, it approves nearly every specific environmental or therapeutic application. And, while Americans find the end-products of biotechnology attractive, they are suffi-

ciently concerned about potential risks that a majority believes strict regulation is necessary. Moreover, the majority of Americans believes a government agency or an external scientific body should be responsible for deciding about environmental use of genetically altered organisms. At the same time, a majority (55%) believes that the risks of genetic engineering have been greatly exaggerated, and 58% feel that unjustified fears of genetic engineering have seriously impeded the development of valuable new drugs and therapies.

... people favor the continued development and application of biotechnology and genetic engineering because they believe the benefits will outweigh the risks. And, while the public expects strict regulation to avoid unnecessary risks, obstruction of technological development is not a popular cause in the United States in the mid-1980s."

The results of the OTA poll were seen as relatively reassuring for the progress of biotechnology, and the fact that there have been few, if any, subsequent nation-wide polls in the US is itself an expression of that confidence. There were major challenges to biotechnology activities, academic and industrial or agricultural – especially concerning the early field release experiments; and to Federal agencies such as EPA and NIH. But these challenges were mainly through the courts, by isolated or numerically insignificant activists. Such actions did not appear to catalyze major or lasting public or Congressional concern; and neither did isolated events, either of vandalism to test plots, or of widespread reports, alleging irresponsible behavior by individual scientists or companies.

Federal agencies undertook efforts at public communication – particularly USDA, through various public information initiatives, such as widely advertized regional conferences, open alike to the press and the critics as well as the interested public. In general, a continuing openness to questions and transparency to public and press, together with a continuing absence of adverse incident, proved sufficient to maintain, if not wholehearted acceptance, at least an absence of serious opposition to continuing research and innovation in the US.

The OTA study foreshadowed similarly divided views in other countries and times. The relative rankings, e.g., of different applications, are very similar from country to country, although the numbers vary widely between countries. But the methodological problems of conducting public opinion surveys are great in general, and particularly so in establishing meaningful, unbiased and consistent measures of something as nebulous as attitudes to topics about which the general level of awareness and technical understanding are naturally low. Slight differences of wording can make substantial differences to the results. For example, the principal conclusion of the OTA study (see highlighting in final paragraph of the quotation above) was contradicted in a study the following year conducted on behalf of the firm Novo Industri:

"... the same mixture of ignorance, optimism and worry as previous surveys. Nearly 40% of about 1000 respondents had never heard of genetic engineering. Of those who had, most expect it will yield significant benefits.

On the other hand, a full 70% do not believe that the benefits will outweigh the perceived risks of genetic engineering. And an equal number favor 'strict limitations on the kinds of genetic engineering research that scientists can do'. Regulations should be formulated primarily by scientists in collaboration with the government, two-thirds of the respondents say; only one-third favor involvement by private firms. Also, the poll finds broad support for international controls on biotech research." (NOVO INDUSTRI, 1987)

Danish Sensibilities: First to Legislate, Inventors of "Consensus Conferences"

This poll may have been one of the factors which in subsequent years led Novo Nordisk (their name changed following the merger with Nordisk Gentofte) to adopt a particularly energetic and high profile effort to publicize themselves as a "green", environmentally-friendly and biotech-based company. It may also be because they are Danish; and of all countries significantly involved in biotechnology, none has more energetically than Denmark pioneered and promoted widespread debate by the entire citizenry about the nature, merits and possible risks of genetic engineering and biotechnology.

The Danish experience is of special interest in several respects. Even in the 1970s, general public opinion surveys about attitudes to various areas of science in different European countries indicated in Denmark an exceptional sensitivity to the possible risks of genetic research (see Tab. 1 above). Germany was in second place (EUROPEAN COMMISSION, 1977, 1979). These two countries were respectively the first and second in the world to introduce legislation specific to "gene technology". Denmark also pioneered carefully structured "consensus conferences", to bring popular or lay opinion into dialogue with the biotechnology experts, and to force the experts to explain in comprehensible language what they were doing or attempting.

National opinion surveys in Denmark over the years following the 1986 Gene Law indicated quite significant shifts in opinion (BORRE, 1990), with at least a suggestion that, while the perception of risks remained high, there was some shift towards greater acceptance (see Tab. 2). Danes are in the European context also exceptional in the level of trust in their government as a source of information on new technologies – twice the European average, and approaching 50%; so that the national policy to support and promote biotechnology (in spite of the fears) may have been significant in influencing the shift towards acceptance. But this observation anticipates the results of the *Eurobarometer* surveys in the early 1990s, described below. BORRE himself (1990) discusses the effects of other variables – educational, political, etc., noting considerable politicization of the gene technology debate on left/right lines.

Calls for Measurement and Research

In a paper on public perception of biotechnology, prepared for the International Biotechnology Symposium at Hannover, TAIT (1988b) paid tribute to the "pro-active risk regulation" being developed in the field (in

Tab. 2. Agreement and Disagreement in Denmark with Opinions on Gene Technology, Percent, 1987–89

	Sept. 1987	May 1988	May 1989
1. It is important that we do not get behind when the potential of gene technology is to be exploited			
Agree	59	67	69
Disagree	32	24	20
Don't know	8	9	11
Total	100	100	100
2. It is morally wrong to interfere with the natural genetic materials of animals and human beings			
Agree	77	70	77
Disagree	18	22	15
Don't know	5	8	8
Total	100	100	100
3. The risk that dangerous new microbes escape during experiments with gene splicing, is strongly exaggerated			
Agree	35	43	35
Disagree	38	32	40
Don't know	27	25	25
Total	100	100	100
4. An international can should be imposed on any kind of gene splicing			
Agree	39	29	25
Disagree	51	60	61
Don't know	10	11	14
Total	100	100	100
5. Only experts and settle questions on gene splicing – ordinary people know too little about it			
Agree	72	76	—
Disagree	24	20	—
Don't know	4	4	—
Total	100	100	—
6. I would protest if a factory in my neighbourhood were permitted to work with gene-spliced bacteria			
Agree	61	52	46
Disagree	28	36	37
Don't know	11	12	17
Total	100	100	100

Source: BORRE, 1990

the EC); but commented that such regulation "reinforces the need for more detailed information on public perception of the industry". Although some industry managers and government were optimistic, feeling that "the thoroughness of these regulatory initiatives is out of all proportion to the real extent of the risks", TAIT stressed that

"Serious outbreaks of public resistance to new biotechnology developments, for example in the United States, West Germany and Denmark, should, however, warn against undue complacency. The industry does have the potential to arouse strong passions, not least because it raises genuine moral and ethical issues for society. A small minority, if sufficiently committed and vocal, can change the attitudes of the majority."

The article summarized the results of the polls referred to above, and of a similar UK poll conducted for the Department of Trade and Industry; but she was critical of opinion polling as a 'broad-brush' technique:

"Opinion polling is inevitably a 'broad-brush' technique. It does not provide detailed information on the nature of perceptions about an issue, why they are favorable or adverse, whether these perceptions will remain latent or are likely to be translated into actions and for which groups this is most likely to be the case. Most important, it does not indicate the extent to which any potential conflict is motivated by concern for the interests of protagonists or by ethical and value-based considerations. The former is epitomized by the well-known NIMBY syndrome ('not in my backyard') (GERVERS, 1987) but the distinction between this and the alternative NIAMBY ('not in anybody's backyard') is rarely recognized. NIMBY tends to be used as a blanket term to describe all conflicts over potentially hazardous or otherwise unwanted activities."

She argued for methodological development, calling (TAIT, 1988a) for a complete programme of research on perceptions to include the following elements:

– an opinion poll carried out on the general public, as for the OTA survey described here;
– identification of specific sections of the public likely to have an important influence on the biotechnology debate; unstructured, tape-recorded interviews with representatives of such groups to provide material for a structured questionnaire to be answered by a random sample of group members; data analysis to indicate the contributions of values and of interests to various perceptions, the relation-

ships between perceptions and relevant behaviors, and the policy implications of the survey results;
– research on the customers for new biotechnology products – how are they likely to use new biotechnology products in relation to what regulators will expect of them; what will be the pressures on them to violate regulations; how can monitoring systems be made more effective (see, for example, Ref. 8);
– research on interactions and relationships among scientists, regulators and the business community; what is the range of perceptions among them of the hazards presented by biotechnology developments; how could the effectiveness of their communication with one another and with the general public be improved;
– broad-ranging research on the social, environmental and political impact of biotechnology developments, individually and in combination with the existing systems of, for example, land use and health care.

Similar sentiments were being expressed, needs articulated, elsewhere in Europe; a thoughtful review was provided by VAN DEN BROCK (1988), in which both industry (through NIABA, the Netherlands Industrial and Agricultural Biotechnology Association) and the Parliament's NOTA (Netherlands Organization for Technology Assessment) stressed the need for research and measurement. LYDI STERRENBERG, co-ordinator of the NOTA biotechnology programme, was quoted:

"With all scientific and technological innovations that are expected to have a dramatic impact, the public opinion, knowledge and attitude towards the new developments should in fact always be investigated. The public should also be informed more systematically about the developments at an early stage."

Opinion research and survey measurements were undertaken in many European countries from 1990 onwards. A critical resume of the surveys was presented by LEMKOW (1992) as the background to research which he undertook in several European countries for the European Community's "Dublin" Foundation (see below):

"it can be argued that divergent results are a product of markedly different societies, values and systems of access to information. Another conclusion of an analysis of the quantitative studies of public opinion might be that the methodology provides an incomplete, even misleading picture of public attitudes and acceptance of biotechnology. Other (qualitative) approaches can be used, such as discussion/focus groups, workshops and in-depth interviews, to sharpen the picture. The quantitative studies have provided some tantalizing clues and have shown the ambivalence of public opinion to new biotechnology and its applications. The data which they have provided have been used as the basis or the design of much of the most recent qualitative research and vice versa.

It was precisely because of the contradictory and somewhat confusing results of earlier survey research that The European Foundation for the Improvement of Living and Working Conditions and Directorate General XI of the European Commission opted to support some qualitative research to gain a better understanding of the complex set of attitudes to biotechnology identified in earlier quantitative studies. This has involved research in Britain, France, Germany and Spain and was based on focus/discussion groups with members of the "informed public" in each of the four countries as well as workshops involving representatives of pressure groups and interested parties (pharmaceutical and food processing industries, agricultural interests, medical profession, trade unions, environmental groups, public administrations)."

He described at some length the results of the "focus group" discussions, their strengths and weaknesses, and the difficulties in addressing a little-known and complex subject of widespread significance. All groups focused at length on risks; they "felt that research could get out of hand, and adequate controls were lacking". On "Regulation and Control", LEMKOW summarized the outcome as follows:

"Regulation was another of the major themes raised in all the groups. There was an evident "lack of faith" in the competence and will of governments to control developments in genetic engineering or other areas of research where there are possible safety and public health risks. A feeling of vulnerability was manifest in all the groups – being quite independent of the ideological positions of the respondents.

– It was repeatedly argued that regulation should be based on sound, honest and objective informa-

tion about the usefulness of biotechnology and the possible risks attached to some of its applications.
– Regulation should be free from the interference of commercial or industrial interests.
– Some groups proposed that supra-national organizations play a more active role in the regulation and control of new biotechnologies."

Clearly such results could be cited in support of a clear-cut regulatory framework for biotechnology, including international dimensions. LEMKOW also stressed information":

"information issue was the "star" theme in the "focus" groups among the interested public. The workshops were no exception and many of the interventions of articulate professionals, experts in the field of biotechnology, paralleled those of the non-specialist public. It was understood by all the participants that the general public was eager for information on biotechnology but with certain conditions attached:

– Information should be provided in clear, accessible and understandable language, and not couched in lofty specialist terminology.
– It should be objective and truthful.
– It should come from a reliable source which should be identified.

The above were seen as prerequisites for strengthening public confidence. This is as far as consensus went. Representatives from industry tended to argue that the facts spoke for themselves. Information could not always be provided because of the need for confidentiality in the face of competition in the development of new products for the market. Interest groups in this context often gave the impression that industry regarded the principle of secrecy as more important than the provision of information. Consumer organizations and environmental groups were highly critical of current information policy in biotechnology which they considered to be inadequate. Access to information was a right and there should be mechanisms for guaranteeing this where risks for the environment, health and safety may be involved. The withholding of information would only reinforce the existing lack of confidence in commercial activities in the scientific and technological field."

In fact the European Commission under the BRIDGE programme grant supported a growing number of publications and initiatives for public information – the various language versions of "Biotechnology for All"

(KATZ and SATTELLE, 1991), consumer dialogue workshops and related publications, an EC-US bilateral workshop in Dublin on "Communicating with the Public about Biotechnology" (US-EC TASK FORCE, 1993), a network of teachers in the European Initiative for Biotechnology Education, and support for the EFB Task Group on the Public Perception of Biotechnology.

The *Eurobarometer* Surveys

LEMKOW, TAIT and others recommended more "depth" in the research – for example, taking account of the socio-economic circumstances of the consumer, clarification of the relationship between attitudes to science in general and perceptions of biotechnology, the need to distinguish more clearly attitudes to different applications of biotechnology, the need to analyze the (in)stability of attitudes over time, and the need to take into account ethical questions.

Although LEMKOW saw surveys as providing "at best, only a superficial impression of the state of opinion on scientific applications in such areas as genetic engineering", in fact all the points noted in the preceding paragraph could be addressed through surveys, given large enough sample sizes, adequately structured questions, full analysis of the results, and repetition on two or more occasions.

These conditions describe the *Eurobarometer* biotechnology surveys conducted in March/April 1991 and March/April 1993. The twice yearly *Eurobarometer* exercise, conducted since the early 1970s, is a very professional operation. Survey firms are competitively selected, to cover each of the countries of the European Community or Union, in its national language(s). A sample of approximately 1000 adults is interviewed in each country; including (since the 1990 reunification) an additional 1000 in (former) East Germany; only 500 in Luxembourg; and 300 in Northern Ireland, in addition to 1000 in Great Britain. In addition to standard questions about Europe, it is open to any department of the Community institutions to "purchase" additional questions.

With the BRIDGE programme, the resources devoted to the concertation action enabled the Commission to undertake a comprehensive Community-wide survey, of public attitudes to biotechnology – or genetic engineering. It was typical of the care devoted to the study that the samples were split – half being asked about "biotechnology" (a word now existing, give or take spelling differences, in all 9 languages), and half about "genetic engineering" (or similar terms), or (for German and Dutch samples), " gene-technology", and for Danish, "gene-splicing". (In some countries, the choice of words made major differences to responses, "genetic engineering" being more negative; elsewhere, differences were slight.)

The questions addressed essentially four issues:

- knowledge (for subsequent questions, each was preceded by a short explanation)
- opinion about regulation and applications
- sources of information
- trust in different sources.

The results of the survey were published in Europe-wide and national reports by the EUROPEAN COMMISSION (1991b), and reported elsewhere (e.g., by MARLIER, in DURANT, 1992); and the data tapes placed in public data banks in Europe and America.

A similar exercise, using some of the same questions, and some additional questions on ethics, was conducted in the spring of 1993; and ERIC MARLIER again prepared the report, (EUROPEAN COMMISSION, 1993a), including extensive comparisons with the 1991 data.

The "key results" are reproduced below. The full results are available from the Commission, and the depository data banks.

Marlier Summary

For the purposes of the present paper, on the evolution of regulations, *Eurobarometer* results offer several points of interest:

– they confirm the earlier (1979) data indicating wide differences of attitude between different Community countries;

– the reported attitudes, e.g., to risk, appear well correlated with political behavior – e.g., the highest and second-highest risk perceptions are found in the countries who were first (Denmark) and second (Germany) to legislate on biotechnology;
– the rankings of countries on various topics appear to be consistent over time, even with the 1979 survey (when it touched genetic research).

Some Key Results of the Spring 1993 Survey (EUROPEAN COMMISSION, 1993a; Report Prepared by ERIC MARLIER)

– A large number of persons interviewed – particularly in Greece, Spain, Ireland and Portugal – were unable or unwilling to answer certain questions. Compared to the previous survey carried out on the same subject (spring 1991), this proportion has dropped however.
– As in the 1991 poll, the two main sources of information used by Europeans for what concerns "new developments that affect our way of life" are, in ranking order, television (the supremacy of which has yet again been confirmed) and newspapers.
– In ranking order, the most reliable sources of information on biotechnology/genetic engineering are considered to be environmental organizations, consumer organizations and schools/universities.

In 1991, consumer organizations slightly supplanted environmental organizations.

If consumer organizations have lost their predominance as "the most reliable source" it is not because they have become less popular than in 1991 but because environmental organizations have themselves made considerable progress.
– Less than one respondent in five believes that Public authorities provide a reliable source of information regarding biotechnology/genetic engineering. In Denmark, however, this percentage is nearer one in two.

In 1991, the situation was similar but not as pronounced: the Danish result was weaker and the European average slightly higher.
– Each of the seven new technologies analyzed is perceived by a large majority of persons interviewed as "improving our way of life in the next 20 years".

The only two technologies for which this majority is not absolute but relative are genetic engineering (as opposed to biotechnology) and space exploration. As in 1991 these find, overall less favor.

The level or "optimism" regarding genetic engineering has lessened considerably since the last survey. This drop is very pronounced in Germany and particularly in the five new Länder.
– 48% of interviewees believe that biotechnology/genetic engineering "will improve our way of life in the next 20 years"; 15% think the opposite. In 1991, "optimism" was at 50% and "pessimism" at 11%.
– In general, when there exists a significant difference, the term "genetic engineering" is less well known and has a more negative connotation than the term "biotechnology". This was already the case in 1991.
– Support for biotechnology/genetic engineering, as well as "optimism" regarding it, is a positive function of what is known on the subject. As in the survey two years ago it depends to a great extent on the type of application and is linked to the risk associated with it; a risk which is considered to be neither negligible nor dramatic, regardless of the application analyzed.
– Except for research on farm animals and, to a lesser extent, food research, where opinions are mixed, those interviewed "tend to agree" that the various kinds of research into biotechnology/genetic engineering discussed in the questionnaire are "worthwhile and should be encouraged". It was already the case in 1991.
– Regardless of the nationality and the application of biotechnology/genetic engineering in question, demand for governmental control of the various applications is massive. This was even clearer in 1991.
– The classification of the different types of research according to the degree of support given to them is identical in 1991 and 1993. It is the same for the classifications linked to the associated risk or related to the level of "demand for control".
– Since the last survey, support for the different applications analyzed has, overall, slightly dropped. In Germany and especially in the five new Länder this drop in "global support" is particularly pronounced.

The "global risk" associated with these applications has remained stationary whereas the level of "global demand for control" has somewhat dropped.

– Whereas the perception of risk is particularly high in Denmark (it is the highest in the Twelve), the support recorded here is around the European average.

Although weaker than that registered in Denmark, the perception of risk is also very high in West Germany (it is the second highest in the Community). On the other hand, support here is a great deal lower than the Community average (it is the weakest in the Twelve).

This divergence in attitudes has increased in comparison to 1991. One plausible explanation of this result is that the Danes (see above), even more now than two years ago, are proportionally many more than the West Germans to trust Public Authorities "to tell the truth about biotechnology/genetic engineering".

– In Luxembourg, global support, perception of risk as well as global demand for control have noticeably increased since the previous survey. In Portugal, on the other hand, we observe a considerable rise in the global perception of risk, accompanied by a significant drop in global support and global demand for control.

– As for research into biotechnology/genetic engineering involving human beings as well as animals and plants, at least three out of four interviewees declare that "there should be clear ethical rules" indicating when research "may not in any way" be undertaken.

– There is the usual ranking of seven application areas, similar across countries, from medical applications (most acceptable) down to the modification of farm animals (least);

– for all application areas, there is a high and uniform demand for control: regardless of applications 83% to 87% (1991: 82% to 89%) of EC citizens agree that it "should be controlled by the government";

– on the key question of trust, the public interest groups (for environment, or for consumers) are seen as more credible by far than government, and still more so than industry, sources.

On the last point, a disturbing trend between the 1991 and 1993 results was summarized by the headline in BBN (*Biotechnology Business News*, 1993), "*Eurobarometer* results discouraging". They presented the results as follows:

"The results of the second *Eurobarometer* survey on awareness and attitudes about biotechnology have been published in a full report. A major goal of the second EC-wide public opinion poll was to identify trends with respect to the first biotechnology survey of this type conducted in March of 1991. A total of 12500 persons were interviewed.

A first analysis is not encouraging, considering the results summarized below:

– The number of people that believe biotechnology will contribute to life and living conditions has decreased.

– There is speculation that public attitudes are being based on feelings rather than factual knowledge.

– There is an increase in risk perception in almost all applications of biotechnology, combined with a decrease in readiness to support research in each area, except in the case of pharmaceuticals.

– There is a generally reduced demand for public control which, in light of a higher risk perception, indicates a reduced trust in those who do control.

– This is supported by the unchanged ranking order of various organizations that act as a source of information."

Tab. 3. Comparison of 1993 and 1991 Polls

Who do you trust most to tell you the truth about biotechnology?	(% mentioning)	
	1991	1993
Environmental organizations	52.6	60.8
Consumer organizations	52.3	55.5
Schools, universities	37.1	38.5
Animal welfare groups	29.1	32.2
Public authorities	20.4	16.8
Religious organizations	9.7	8.2
Industry	6.0	5.6
Trade unions	5.2	5.2
Political organizations	4.9	4.0

The outcome of these surveys and the analysis of public opinion can be summarized as follows:

- Public ignorance and uncertainty about biotechnology or genetic engineering gives rise to a high level of apparent demand for regulation.
- Depth interviews with "focus groups" suggest a strong demand for transparency, and for comprehensible and trustworthy information.
- Government is little trusted, industry still less, as a source of information, so that either statutory or voluntary regulation may still not build trust; (though clearly this would not justify any government's abandoning regulatory activity when they believe it to be objectively necessary).
- In these circumstances, there is both an opportunity for environmental and consumer organizations, and a responsibility on them, to provide objective information and reasoned opinions about developments in biotechnology.
- There is little evidence for the view that government regulation itself builds or would build trust in biotechnology.

A comprehensive comparative summary of polls was presented by ZECHENDORF (1994). Activities on research and communication continued to be promoted in the mid-90s by the Task Group on Public Perception of the European Federation of Biotechnology, under the chairmanship of JOHN DURANT, the UK, and perhaps the world's first professor of the "public understanding of science". The Commission, through the Biotechnology Coordination Committee, and with continuing encouragement from the European Parliament to accompany R & D programmes by studies on social, ethical and other dimensions, seemed likely to continue through the 1990s the *Eurobarometer* measurement process launched in 1991 and 1993.

8 Synthesis and Conclusions: Learning from History

In 1981, the Nobel prize-winning American, JAMES WATSON, and the Secretary of the European Molecular Biology Organization, JOHN TOOZE, published "The DNA Story: A Documentary History of Gene Cloning". With their narrative, they interspersed the principal documents associated with the pre- and post-Asilomar discussions, from 1973 through to the end of the 1970s, documents which illustrate the rise of the once-threatening tide of Congressional legislation, and of the widespread public concern and criticism which drove it. A few of the same elements are briefly reviewed in the first section of this review; the WATSON and TOOZE compendium fills over six hundred pages. In the closing paragraph of their final section, "Epilogue", the authors conclude with evident relief:

> "Politics and politicking preoccupied the first years of the recombinant DNA story, but that phase, fortunately, is fast becoming history. This book is our epitaph to that extraordinary episode in the story of modern biology."

More than a decade later, no such facile conclusion can be offered in a history of biotechnology regulation; one thinks rather of a contemporary historian in Europe's Thirty Years' War, or in the Anglo-French 100-year conflict, invited at year 10 or year 20 to give an overview and prediction of outcomes ...

For the "politicking", although it paused in the early 1980s, picked up momentum thereafter and has increased ever since, *pari passu* with the progress of the science and the diffusion of biotechnology. In Europe, the politicking was more intense, and the initial outcome less happy than in the US; for the surge of knowledge and innovations coincided with two other historic processes. The mid-80s saw a surge of political support for environmental movements, which in parts of Europe tapped into older romantic traditions, containing strong anti-intellectual and anti-technological elements. LÖNNGREN (1992) speaks of "the politicization of chemicals control". At the same time, the political will and leadership in Europe, at both Community and national levels, was ready to drive forward the processes of constitutional change and development. The potential for such development had always been present in the founding EC Treaty, but the drive was accelerated from the mid-1980s by an impatience with slow progress, and by a will to "build Europe". These were given concrete expression by the 1992 target date for completing the common internal market, and by the Single European Act (adopted 1987, effective 1989) as the instrument for its completion. Majority voting for proposals under Article 100A (the legal basis for harmonizing legislation), and for specific R & D programmes within a (still unanimity-requiring) multi-year Framework Programme, were among the several significant innovations of this Act.

The momentum was maintained, through the three successive Commissions during the ten-year presidency of JACQUES DELORS – at least to its penultimate year, 1993, which saw the ratification of the Treaty on European Union, signed at Maastricht in December 1992. By 1993, however, it was "a damn close-run thing" (as WELLINGTON remarked at an earlier defining moment in European history), with a second plebiscite required in Denmark, a wafer-thin assent even in France, and the ruling British Conservative party almost mortally split. The continued decline in numbers voting in European Parliamentary elections (in June 1994), and the divisive political arguments accompanying the 1994 plebiscites in Austria, Finland, Sweden and Norway, on accession to the European Union, underlined the slackening of political will.

This political backdrop interacted repeatedly, and often unnecessarily and unhelpfully, with the development of biotechnology in Europe. The politicking hindered, where it should have facilitated, the effective integration of the new knowledge into the activities and sectors that needed it. Conversely, the history of Community strategy for biotechnology in Europe, and the history of biotechnol-

ogy regulation to which for some years it seemed to be reduced, illuminated structural weaknesses within the Commission, and within the Community's institutional structure. These structures were ill-adapted to managing the challenges and complexities of biotechnology; for even when these were clearly identified and described, in good time, the communication to the political level was generally ineffectual; and political action was blocked or diverted into irrelevant and unhelpful actions by the weight of other interests.

Many factors render obscure the legislative and other actions of the Community institutions, shielding them from effective democratic scrutiny, and limiting their transparency: the multi-institutional complexity (Commission, Parliament, Council, etc.) of the machinery; the distance from national politics (where "Brussels bureaucracy" is a convenient scapegoat for nationally unpopular measures), and from citizens and local communities; and the inescapable diversity of Europe's languages and cultures, at once its glory and a permanent political constraint.

This lack of transparency means that on complex subjects, only a sustained and determined effort of communication can ensure that all parties with relevant interests and knowledge have the opportunity to participate in preparing proposals and decisions; and when the mass of information and opinions is effectively elicited, there has to be a radical condensation and filtering to summarize the debate into the drafting or amending of a legislative text, or to enable the elected parliamentarian to cast his vote. Both in the communication, and in the condensation, the opportunities for distortion, accidental or willful, are legion.

Complexity without transparency allows, even encourages, the pursuit of individual and institutional self-interest. Key individuals involved in the biotechnology regulatory agenda differed widely in their interests, their style of operation, and their attitudes to science, innovation and industry.

Within the Commission, each Directorate-General has its "déformation professionelle", and the linguistic barriers are trivial in comparison to those between DGs. The Commission as a whole is by constitution naturally activist, and that constitution reflects the political aims of the founding Treaties: there was much to be done. This maps down to the level of the individual, particularly in the Directorates-General concerned with legislation: success tends to be equated with the adoption of a new Directive or Regulation, however flawed.

Thus in DG III, legislation was essential to creating a common market – for food products, pharmaceuticals, etc., and ultimately to achieve new structures such as the European Medicines Evaluation Agency. Similarly for DG XI, the control of chemical products for the protection of human health and the environment, was a major challenge, the legislation a major achievement, and the basic need for such control – whatever the disputes about details – essentially an unquestionable imperative, world-wide. DG V's responsibilities for promoting uniform high standards of worker safety similarly demanded and brought forth a constructive and successful framework of Community law.

The Commission embraces other aims and their corresponding cultures. The Common Agricultural Policy was the creation of the Community institutions, its management and defence the burden of DG VI; who had simultaneously to respond to world-wide pressures for change – for the liberalization of agricultural trade under GATT, for protecting rural interests under the pain of "rationalization", and for reconciling the diverse European interests represented by the Ministers of Agriculture. Biotechnology, uninvited, came insistently onto the DG VI agenda via agricultural research and agricultural legislation, offering productivity increases in sectors plagued by excess production.

The culture of DG XII, especially in its earlier decades, was scientific in its sympathies and roots. They were reluctant legislators in 1978, glad to retire from such matters in the mid-80s. Global trends – the move towards knowledge-based economies, the natural internationalism of science, its perceived relevance to economic competitiveness, the increasingly expensive and specialized character of research – led to rapidly expanding biotechnology R & D programmes at European

level in the late 1980s and 1990s. The pressures of managing these increasing resources with a static or declining complement of staff forced DG XII to focus on the politics of winning these heavier research budgets, and on managing efficiently the selection and administration of vast numbers of projects. These pressures further diminished the appetite for inter-DG arguments over legislation; but paradoxically increased the need for such interaction, as the expanding R & D activities, and the global trend to more knowledge-based economies, were inexorably increasing the scientific content in the agenda of other DGs. DG XIII – responsible for the large R & D programmes in information technology and telecommunications – was from the mid-80s closely involved in the full range of research, industrial policy, and related legislative activities. For biotechnology, no such monopoly was conceivable, as the pervasive significance of the new knowledge obtruded across the range of DG interests, and no single DG could pretend to a monopoly of scientific wisdom, even within the life sciences and technologies.

The first FAST programme, and the Commission's 1983/84 responses, establishing the Biotechnology Steering Committee, reflected an adequate perception and analysis of the challenges, followed by apparently appropriate action. The "need for an integrated approach" was similarly endorsed by Parliament, in the 1987 Viehoff report and the resolution adopted.

The subsequent fading of the Biotechnology Steering Committee has been described. As the new techniques of genetic engineering were emerging from the laboratory to cross the road to the market-place, the bus of environmentalism was accelerating; and although the new techniques could fairly claim a place on the bus, as "Clean Technologies", the interaction in Europe was more a collision than an accommodation or a welcome.

The self-confidence of success led to an uncritical and inappropriate transfer of the culture of chemicals control to legislation focused on, and by inescapable implication stigmatizing, a technology. Many factors reinforced this strategic blunder: widespread scientific illiteracy, sensationalism in the media, bureaucratic and political opportunism, agricultural protectionism, mistrust of industry, an anti-industrial, anti-intellectual populism – and the usual scientific uncertainties and caution.

Oversight in the form of notification and monitoring is a rational response to uncertainty, and enables uncertainties to be diminished by the accumulation of experience, and resources for risk assessment and management to be rationally deployed. This was the approach adopted by the Community in the 1982 Council Recommendation; and it worked satisfactorily, not least because there was no attempt by DG XII, the service chef-de-file, to exploit the opportunity to build up a permanent bureaucracy.

In the United States, the effective dialogue between scientific and political communities headed off the threat of technology-specific legislation; and even those who (unsuccessfully) advocated and prepared such legislation in the 1970s and 1980s, typically incorporated in their Bills a "sunset clause", which would automatically terminate the legislation after a set period, if there was no further Congressional action taken to renew or amend it.

Such a "provisional" or "learning" approach was a rational and scientific response to uncertainties about a new phenomenon, such as a new technology. But for new chemical substances, or pharmaceutical products, there is the practical certainty of a continuing stream of new entities requiring testing and oversight; and the corresponding administrative structures are therefore conceived on permanent lines, give or take some future adaptation.

The imposition of this "permanent" character on novel technologies both stigmatized them, and built a bureaucratic structure at Community and national levels with an in-built tendency to justify and defend its continued existence.

Within the European Parliament, the active members, coping with a flood of documentation, and a complex and exhausting life-style (between home, committee work in Brussels, and plenary sessions in Strasbourg), could in general devote little time to understanding complex dossiers such as biotechnology.

While there could be real concerns about ethical aspects of the use and abuse of new technologies (e.g., in relation to human genetics or animal welfare), and popular suspicions of "mad scientists" and mistrust of industry, in general the esoteric character of genetic engineering meant that practically all MEPs would leave such a dossier to the rapporteur – or, if the rapporteur was not of their political group, would designate a member of their group to follow the dossier. The basis for formulating the parliamentary opinion on legislation relating to biotechnology was therefore typically a very narrow one; in an area which shared (with nuclear energy) the most concentrated attention of the "Greens" fraction in the Parliament. Moreover, even MEPs not of this fraction, were in many countries acutely conscious in the late 1980s that the major political parties were losing ground to the Green movements; and to recapture these votes, were anxious to demonstrate their own "Green" credentials. A severely restrictive approach to the highly publicized new gene technology appeared to be a painless and popular way of doing so.

Against this coincidence of popular fears, political self-interest and bureaucratic opportunism, the voices of scientific protest were few, feeble and disregarded. DG XII lost the arguments inside the Commission, and had at the critical moments no interested allies. The protests to Parliament by Nobel prize-winners did not represent a politically significant constituency. The OECD report on rDNA safety, indicating no scientific basis for legislation specific to recombinant DNA, was quoted for its prestige and authority, in support of precisely such legislation. The advice of the safety specialists of the European Federation of Biotechnology was aggressively rejected by the Director-General of DG XI. The House of Lords Committee's report noted that in drafting the legislation, the Commission had been "impervious to scientific advice"; in fact the efforts of DG XII to offer such advice, as they were (by the Council Decisions on BAP and BRIDGE programmes) required to do, were vigorously repulsed and successfully counter-attacked.

A similar "knee-jerk" reaction greeted the suggestion (in the Biotechnology Regulation

Inter-service Committee, around 1987) that the details of a fast-changing and complex field might best be addressed by technical experts in standards committees. DG XI was chef-de-file for biotechnology legislation, but not for standards. As a result, technical details of scope – a central issue in the US debates – were defined in Annex I of each "Biotechnology" Directive, 90/219 (contained use), and 90/220 (field release), in terms specific to the legislators' understanding of the science of the 1980s, as modified by the experts chosen by the Environment Ministers; who then removed these defining Annexes from the scope of the committee procedure for adaptation to technical progress. The consequences in costs, delays and controversies would dominate the regulatory debate throughout the 1990s.

The silencing of GALILEO no doubt seemed to contemporaries a matter of limited significance, beyond the scientific and theological communities; but by 1990, biotechnology was beginning to matter, and countervailing forces were coming into play, to correct the strategic error. Industry in Europe, following the widely publicized meeting with DAVIGNON of December 1984, had established a communication network for the expression of bio-industrial interests, but failed to endow this with muscle. By continuing to devote their main energies to sectoral channels, they confirmed a similar conservatism within the Commission.

The change of perspective from 1989 was attributable to the significance accorded to biotechnology in less constrained environments (such as the USA), or in those where long-term strategic vision was taken seriously (as in Japan). Multi-national companies operating in several continents could most readily compare the differences of approach, and their implications for regulation. Although they could to some extent re-locate their activities and investments to adapt to circumstances, this had costs and discomforts, particularly for those whose base operations and major investments were in Europe; and for all firms, wherever based, the European market was a major element of the global total.

The loss of investment (actual and threatened), and the loss of R & D activities and personnel, the seed corn for future industries, inevitably attracted political concern, particularly once linked with the rising political concerns about employment. The constitution of the SAGB at European level, the various national bio-industry associations, and the US examples, ensured an attentive hearing for industry once it started to express itself vigorously at political level, from 1990 onwards; but their intervention was late, and did not have enough momentum to divert the legislative juggernaut in that year.

Within the European Commission, the consequences of the failure of inter-service co-ordination were gradually recognized at the highest level; and in 1990 the Secretary-General at the request of President DELORS initiated the Biotechnology Co-ordination Committee. More importantly, he maintained and developed the central role of the BCC within the Commission services; thus acting as a brake on the autonomous behavior of individual DGs, and enforcing a greater degree of horizontal transparency within the house. Also during the early 1990s, the Commission was responding to the need and political demands for greater external transparency (EUROPEAN COMMISSION, 1993); and within the BCC framework, "Round Tables" with industry and with a wide range of ngo (non-governmental organization) interests became a regular feature of its activities. The 1991 communication similarly announced that CEN (the European Standards Committee) would be charged with a mandate to develop standards in biotechnology.

These developments were neither trivial nor obvious: the suspicions and hostility vis-à-vis biotechnology which had driven, and been reinforced by, the 1990 legislation were far from being dissipated. If a Directorate-General was disgruntled at BCC, a 'phone call or a fax could quickly trigger a forceful letter from a sympathetic MEP to the Secretary-General, and there would not be lacking groups and activist organizations to carry the argument to the public domain, *mutatis mutandis* – and the mutations could be remarkable.

Moreover, the "public domain" for argument was dramatically enlarged as the UN agencies progressively recognized the need or opportunity for each of them to engage with biotechnology. Particularly damaging were the renewed and amplified opportunities for stigmatization offered by Article 19.3 of the Convention Biological Diversity, with its invitation to consider the need for an international "bio-safety" protocol. Using these international fora to reinforce one's local position was an instinct as natural to the conflicts in Brussels as in Bosnia.

The prominence of biotechnology regulatory matters in the Commission's December 1993 White Paper on "Growth, Competitiveness and Employment" has been noted in the previous section, along with the follow-up action in the communication at the Corfu Summit, and in the regulatory proposals submitted during the German and French Council presidencies of 1994–95.

These developments clearly display the capacity of the European Commission, of industry, and of national political leaders to be responsive, and to limit and reverse the past mistakes. But as HERACLITUS observed, one cannot step twice in the same river. The waters of public opinion have been muddied by misrepresentation, and there remains enough continuing uncertainty and concern to slow the work of reorienting policies and of adapting or dismantling the legal and administrative structures whose foundations are now questioned.

The European Parliament has yet to re-address the central issues of biotechnology regulation. As the renewed European Parliament (after the June 1994 election) struggles for increased power in the Union's inter-institutional debate of the mid-90s, it is difficult for it to acknowledge that it goofed in earlier years. Institutional face-saving is no less endemic in the national Ministries concerned, and within the Directorates-General of the Commission. However, bureaucratic drafting skills, changes of government, and internal reorganization are all instruments through which such changes can be respectably managed, and all will have their role.

Parliamentary debates – and votes – on specific challenges such as the Directive on the Protection of Biotechnological Inventions, or the Novel Foods Regulation, continue to give cause for concern to those focusing on Europe's economic competitiveness. Bio-

technology is not yet recognized as integral to the future competitiveness of agriculture and of major sectors of industry, as well as to the effective improvement of public health and the protection of the environment. Ethical issues, such as those highlighted in the Council of Europe's draft Convention on Bioethics (1994), will continue to attract greater prominence in Europe; with the risk of consequent relative neglect and damage to the bases of Europe's economic (and consequent political) weight in the 21st century.

To paraphrase the Watson and Tooze "Epilogue" quoted at the start of this section: politics and politicking preoccupied the first years of the recombinant DNA story, and that phase, in Europe and more than a decade later, became, unfortunately, not "history", but a story of arrested development. The internal conflicts within the Commission are for the moment better controlled, but much energy in Brussels is still devoted to inter-institutional and Community–national conflicts, on constitutional matters which the USA settled thirteen decades ago; and to the geographical expansion of the Community.

Insofar as wider international relations and activities come into play – for example, through EC–US bilateral, OECD, or UN agencies – the tendency is for the contending interests, within the Community institutions and at national level, to use such wider dimensions to reinforce their position in domestic conflicts.

As Europe's political leaders and public servants battle for control on the bridge of their Ship of State, and prepare for the Inter-Governmental Conference of 1996–97, they must remember there's ocean out there (and rocks) – not just more and more ship. On the swelling and stormy oceans of knowledge, not least, of the life sciences and technologies, forecasting and navigational skills, and institutional and political structures capable of using them intelligently, will be more than ever essential.

Personal Statement

Opinions expressed engage only the author, and not his current or previous employers, or their Member Governments. Mr. CANTLEY currently (1994) heads the Biotechnology Unit in the Directorate for Science, Technology and Industry of the Organization for Economic Co-operation and Development. He previously (1979–92) worked in Directorate-General XII (Science, Research and Development) of the Commission of the European Communities – originally in the FAST team (Forecasting and Assessment in Science and Technology); then from 1984 to 1992 as Head of "CUBE: the Concertation Unit for Biotechnology in Europe", established as Secretariat of the Biotechnology Steering Committee, and the Biotechnology Regulation Interservice Committee; and responsible for the "Concertation Action" within successive Community R & D programmes in biotechnology, from 1985 to 1992.

Acknowledgement

The author has benefitted in the preparation of this article from the help of colleagues and friends too numerous to mention individually. He would, however, particularly wish to set on record the invaluable assistance provided by BERNHARD ZECHENDORF, documentalist in DG XII of the European Commission, and by the BIODOC centre which he maintains; whose tens of thousands of carefully organized documents are now the indispensable resource alike for the biotechnology policy-makers of to-day, and the policy-researchers of to-morrow.

9 References

ACGM (Advisory Committee on Genetic Manipulation) (1986), *The Planned Release of Genetically Manipulated Organisms for Agricultural and Environmental Purposes – Guidelines for Risk Assessment and for the Proposals for Such Work*. Health and Safety Executive, London.

ACGM (1987), *Guidelines for the Large Scale Use of Genetically Manipulated Organisms*. Health and Safety Executive, London.

ACGM (1993), *The Genetically Modified Organisms (Contained Use) Regulations 1992*. Health and Safety Executive, London.

ANONYMOUS (1989), EMBO: Workshops to postdocs and publishing, *Nature* **338,** 725, 27 April.

ATAS (Advanced Technology Assessment System) (1992), *Biotechnology and Development: Expanding the Capacity to Produce Food*. Department of Economic and Social Development, United Nations, New York.

AVERY, T., MACLEOD, C., MC CARTHY M. (1944), Studies on the chemical nature of the substance inducing transformation of pneumococcal types. I. Induction of transformation by a deoxyribonucleic acid fraction isolated from *Pneumococcus* type III, *J. Exp. Med.* **79,** 1373–158.

BECKER, J. (1981a), Bioengineering hazards: European doubts, *Nature* **291,** 181, 21 May.

BECKER, J. (1981b), Recombinant DNA research: EEC safety dispute, *Nature* **294** (24) 31 December.

BEIER, F. K., CRESPI, R. S., STRAUS, J. (1985), *Biotechnology and Patent Protection: An International Review*. OECD, Paris.

BENDA, E. (1986), *Bericht der gemeinsamen Arbeitsgruppe des Bundesministeriums für Forschung und Technologie und des Bundesministeriums der Justiz: In-vitro-Fertilisation, Genomanalyse und Gentherapie* (Report of the Benda Commission).

BENNETT, D. J., GLASNER, P. E., TRAVIS, L. (1986), *The Politics of Uncertainty: Regulating Recombinant DNA Research in Britain*. Routledge & Kegan Paul, Henley-on-Thames.

BERG, P., et al. (1974), Potential biohazards of recombinant DNA molecules, *Science* **185,** 303.

BERNARDI, G. (1991), The Scientific Committee on Genetic Experimentation (COGENE), *Sci. Int.*, Special Issue, September. ICSU, Paris.

BERTAZZONI, U., FASELLA, P., KLEPSCH, A., LANGE, P. (1990), Human embryos and research, in: *Proceedings of the European Bioethics Conference* in Mainz, 7–9 November 1988. Campus-Verlag, Frankfurt–New York.

BORRE, O. (1990), Public opinion on gene technology in Denmark 1987 to 1989, *Biotech Forum Europe* **7,** 471–477, 6 December.

BREYER, H. (1992), *Draft Report on the Commission communication to the Parliament and the Council promoting the competitive environment for the industrial activities based on biotechnology within the Community*. European Parliament, PE 203.456/B.

BREYER, H. (1994), *Draft Report on biotechnology within the Community, for the Committee on Energy, Research and Technology of the European Parliament*. European Parliament, PE 208.602/A, 7 March.

BUD, R. (1993), *The Uses of Life: A History of Biotechnology*. Cambridge University Press, Cambridge.

BULL, A., HOLT, G., LILLY, M. (1982), *Biotechnology: International Trends and Perspectives.* OECD, Paris.

BUTLER, D. (1994), Europe plans convention on social impacts of biomedical technologies, *Nature* **370**, 3, 7 July.

CADAS (Comité des Applications de l'Académie des Sciences) (1993), *Les techniques de transgénèse en agriculture: Applications aux animaux et aux végétaux.* Technique & Documentation, Lavoisier, Paris.

CANTLEY, M. F. (1983), *Plan by Objective: Biotechnology.* XII-37/83/EN, Commission of the European Communities.

CANTLEY, M. F. (1987), Democracy and Biotechnology: Popular attitudes, information, trust and the public interest, *Swiss Biotech* **5** (5), 5–15.

CARSON, R. (1962), *Silent Spring.* Houghton Mifflin, Boston, MA.

CHEN, Z. L. (1992), Field releases of recombinant bacteria and transgenic plants in China, in: *Proceedings of the 2nd International Symposium on the Biosafety Results of Field Tests of Genetically Modified Plants and Microorganisms,* 11–14 May, 1992, Goslar, Germany (CASPER, R., LANDSMANN, J., Eds.). Pub. Biologische Bundesanstalt für Land- und Forstwirtschaft, Braunschweig, Germany.

CHEN, Z. L. (1995), Field tests of transgenic plants against viral and insect infections, in: USDA, op. cit.

CLINTON, W. J., GORE, A. (1993), *Technology for America's Economic Growth, A New Direction to Build Economic Strength.* Office of Science and Technology Policy of the White House, February 22.

COHEN, S. N., CHANG, A. C. Y., BOYER, H. W., HELLING, R. B. (1973), Construction of biologically functional bacterial plasmids *in vitro, Proc. Natl. Acad. Sci. USA* **70**, 3240–3244.

COUNCIL OF EUROPE (1982), *Thirty-Third Ordinary Session of the Parliamentary Assembly,* Recommendation 934 (1982), on Genetic Engineering. Text adopted by the Assembly on 26 January 1982 (22nd sitting).

COUNCIL OF EUROPE (1986), *Towards a European Regulation of Genetic Manipulations.* Press Release I (84) 71.

COUNCIL OF EUROPE (1993), *Draft European Convention on Civil Liability for Damage Resulting from Activities Dangerous to the Environment.*

COUNCIL OF EUROPE (1994), *Draft European Bioethics Convention* (for the protection of human rights and dignity of the human being with regard to the application of biology and medicine).

CYERT, R. M., MARCH, J. G. (1963), *A Behavioral Theory of the Firm.* Prentice-Hall, Inc., Englewood Cliffs, NJ.

DAVIS, D. (Ed.) (1991), *The Genetic Revolution: Scientific Prospects and Public Perceptions.* Johns Hopkins University Press, Baltimore–London.

DECHEMA (Deutsche Gesellschaft für chemisches Apparatewesen e.V.) (1974), *Biotechnologie: Eine Studie über Forschung und Entwicklung – Möglichkeiten, Aufgaben und Schwerpunkte der Förderung.* DECHEMA, Frankfurt.

DECHEMA (1982), Sichere Biotechnologie, in: *Arbeitsmethoden für die Biotechnologie.* DECHEMA, Frankfurt.

DECHEMA (1983), *Biotechnology in Europe – A Community Strategy for European Biotechnology.* Report to the FAST-Bio-Society Project of the Commission of the European Communities. DECHEMA, Frankfurt, on behalf of the European Federation of Biotechnology.

DE DUVE, C. (1979), *Cellular and Molecular Biology of the Pathological State: A Proposal for a Community Programme in Biopathology.* Document XII/112/79 of the European Commission.

DOE (Department of the Environment) (1993), *DOE/ACRE Guidance Note 1: The Regulation and Control of the Deliberate Release of Genetically Modified Organisms.* DOE, London.

DOE (1994), *Regulation of the United Kingdom Biotechnology Industry and Global Competitiveness.* Government Response to the Seventh Report of the House of Lords Select Committee on Science and Technology, 1992–93 Session. Cm 2528, Her Majesty's Stationary Office, London.

DE ROSNAY, J. (1980), *Biotechnologie et Bio-industrie.* Seuil, Brussels.

DEUTSCHER BUNDESTAG (1987), *Bericht der Enquete-Kommission „Chancen und Risiken der Gentechnologie"* des 10. Deutschen Bundestages (Vorsitzender: W.-M. CATENHUSEN). – *Report of the Commission of Enquiry on „Prospects and Risks of Genetic Engineering"* (official English translation of the report of the "Catenhusen Commission").

DI CASTRI, F. (1994), Rigor and openness in biology. *Biology International: The News Magazine of the International Union of Biological Sciences (IUBS)* **29**, July.

DIRECTORATE FOR NATURE MANAGEMENT (1992), *Ecological Risks of Releasing Genetically Modified Organisms into the Natural Environment.* DN Report 1991–7b. Directorate for Nature Management, Trondheim.

DURANT, J. (1992), *Biotechnology in Public: A Review of Recent Research.* Science Museum, London, for the European Federation of Biotechnology.

ECAS (Euro-Citizen Action Service) (1992), *The European Citizen:* Biotechnology. 17, Special Issue, June.

ECONOMIC AND SOCIAL COMMITTEE OF THE EUROPEAN COMMUNITIES (1981), *Genetic Engineering: Safety-Related Aspects of Recombinant DNA.* Colloquium organized by the ESC, 14–15 May.

ECONOMIDIS, I. (Ed.) (1990), *Biotechnology R&D in the EC: Biotechnology Action Programme (BAP),* Part I: Catalogue of BAP Achievements on Risk Assessment for the Period 1985–1990. European Commission.

ECONOMIDIS, I. (1994), Research activities of the European Communities in biosafety, in: *Scientific-Technical Backgrounds for Biotechnology Regulations* (CAMPAGNARI, F., et al., Eds.), pp. 31–36. ECSC, EEC, EAEC, Brussels – Luxembourg, printed in the Netherlands.

ECRAB (European Committee on Regulatory Aspects of Biotechnology) (1986), *Safety and Regulation in Biotechnology.*

ERNST & YOUNG and SAGB (1994), *Biotechnology's Economic Impact in Europe: A Survey of its Future Role in Competitiveness.* Ernst & Young, London, and SAGB, Brussels.

EUROPEAN COMMISSION (1977a), Common policy for science and technology, *Bulletin of the European Communities,* Supplement 3/77.

EUROPEAN COMMISSION (1977b), *Science and European Public Opinion.* Document XII/922/77.

EUROPEAN COMMISSION (1978a), Proposal for a Council Directive establishing safety measures against the conjectural risks associated with recombinant DNA work, *Official Journal of the European Communities,* C301/5–7.

EUROPEAN COMMISSION (1978b), *Note from the Commission Services to Members of the CREST, Concerning a Proposal for a Council Directive Establishing Safety Measures Against the Conjectural Risks Associated to Recombinant DNA Work.* Document XII/698/78.

EUROPEAN COMMISSION (1979), *The European Public's Attitudes to Scientific and Technological Development.* Document XII/201/79.

EUROPEAN COMMISSION (1980), *Draft Council Recommendation, Concerning the Registration of Recombinant DNA (Deoxyribonucleic Acid) Work.* COM (80) 467, 28 July.

EUROPEAN COMMISSION (1983a), *Biotechnology: The Community's Role, Communication from the Commission to the Council.* COM (83) 328, 8 June; a separate "background note" with the same reference number is sub-titled "National Initiatives".

EUROPEAN COMMISSION (1983b), *Biotechnology in the Community, Communication from the Commission to the Council.* COM (83) 672; a separate annex presents the action priorities in fuller detail.

EUROPEAN COMMISSION (1984), *Proposal for a Council Decision Adopting a Multinational Research Programme of the European Economic Community in the Field of Biotechnology.* COM (84) 230.

EUROPEAN COMMISSION (1986a), *The European Community, and the Regulation of Biotechnology: An Inventory.* BRIC/1/86.

EUROPEAN COMMISSION (1986b), *Biotechnology Regulation: Meeting of Commission Staff with Member State Officials,* Brussels, 29–30 April 1986: Summary of Discussion. BRIC/2/86.

EUROPEAN COMMISSION (1986c), *A Community Framework for the Regulation of Biotechnology, Communication from the Commission to the Council.* COM (86) 573, 4 November.

EUROPEAN COMMISSION (1988), *Proposal for a Council Directive on the Protection of Biotechnological Inventions, Communication from the Commission to the Council.* COM (88) 496, October.

EUROPEAN COMMISSION (1990a), *Proposal for a Council Regulation on Community Plant Variety Rights, Communication from Commission to Council.* COM (90) 347. *Off. J. Eur. Commun.* No. C244/1–27.

EUROPEAN COMMISSION (1990b), *Future System for the Free Movement of Medicinal Products in the European Community* (including a proposal for a Council Regulation laying down Community procedures for the authorization and supervision of medical products for human and veterinary use and establishing a European Agency for the Evaluation of Medical Products; and three proposals for Council Directives amending or repealing existing Directives). COM (90) 283.

EUROPEAN COMMISSION (1991a), *Promoting the Competitive Environment for the Industrial Activities Based on Biotechnology within the Community.* SEC (91) 629. Subsequently included in *Bull. Eur. Commun.* Suppl. 3/91, "Industrial Policy in the 1990s".

EUROPEAN COMMISSION (1991b), *Opinions of Europeans on Biotechnology in 1991.* Report undertaken on behalf of the Directorate-General for Science, Research and Development of the Commission of the European Communities, "CUBE" Unit (Concertation Unit for Biotechnology in Europe), by INRA (Europe) sa/nv, Brussels.

EUROPEAN COMMISSION (1992a), *Proposal for a Council Regulation (EEC) on Novel Foods and Novel Food Ingredients.* COM (92) 295 Final – SYN 426.

EUROPEAN COMMISSION (1992b), *Biotechnology After the 1991 Communication: A Stock-Taking.* BCC Report, October.

EUROPEAN COMMISSION (1993a), *Biotechnology and Genetic Engineering: What Europeans Think About It in 1993.* Survey conducted in the context of *Eurobarometer* 39.1.

EUROPEAN COMMISSION (1993b), *Growth, Competitiveness, Employment: The Challenges and Ways Forward into the 21st Century.* White Paper. COM (93) 700, 5 December. (The "Delors" White Paper)

EUROPEAN COMMISSION (1993c), *Increased Transparency in the Work of the Commission, Off. J. Eur. Commun.* No. C63/8, 5 March.

EUROPEAN COMMISSION (1993d), *Public Access to the Institutions' Documents, Off. J. Eur. Commun.* No. C156/5, 8 June.

EUROPEAN COMMISSION (1994a), *Biotechnology and the White Paper on Growth, Competitiveness and Employment: Preparing the Next Stage.*

EUROPEAN COMMISSION (1994b), *Commission Decision of 15 April 1994,* amending Council Decision 91/596/EEC concerning the summary notification information format referred to in Article 9 of Council Directive 90/220/EEC, *Off. J. Eur. Commun.* No. L 105/26–44.

EUROPEAN COMMISSION (1994c), *Commission Directive 94/15/EC of 15 April 1994,* adapting to technical progress for the first time Council Directive 90/220/EEC on the deliberate release into the environment of genetically modified organisms, *Off. J. Eur. Commun.* No. L 103/20–26.

EUROPEAN COMMISSION (1994d), *Summary Report from the Commission on the Implementation by Member States of Council Directive 90/219/EEC on the Contained Use of Genetically Modified Micro-organisms.* (Publication expected late 1994/early 1995)

EUROPEAN COMMISSION (1994e), *European Commission's Group of Advisers on the Ethical Implications of Biotechnology.* Activity Report 1991–93.

EUROPEAN COMMISSION (1994f), *Commission Directive 94/51/EC of 7 November 1994,* adapting to technical progress Council Directive 90/219/EEC on the contained use of genetically modified micro-organisms, *Off. J. Eur. Commun.* No. L297, 29–30, 18 November.

EUROPEAN COMMISSION (1994g), *Commission Decision on Public Access to Documents, Off. J. Eur. Commun.* No. L46/58, 18 February.

EUROPEAN COUNCIL (1978), *Council Decision of 25 July 1978 on a Research Programme of the European Economic Community on Forecasting and Assessment in the field of Science and tech-nology, Off. J. Eur. Commun.* No. L225/40 of 16 August.

EUROPEAN COUNCIL (1982), *Council Recommendation of 30 June 1982 Concerning the Registration of Work Involving Recombinant Deoxyribonucleic Acid (DNA)* (82/472/EEC).

EUROPEAN COUNCIL (1990), *Council Directive of 23 April 1990 on the Contained Use of Genetically Modified Micro-Organisms; and Council Directive of 23 April 1990 on the Deliberate Release into the Environment of Genetically Modified Organisms, Off. J. Eur. Commun.* No. L117/1–27.

EUROPEAN PARLIAMENT (1987), *Resolution on Biotechnology in Europe and the Need for an Integrated Policy.* Doc. A2-134/86. *Off. J. Eur. Commun.* C76/25–29, 23 March.

EUROPEAN SCIENCE FOUNDATION (1982), Report on the Current State of Regulations concerning Recombinant DNA Work (result of an inquiry conducted among the members of the former Liaison Committee for Recombinant DNA Research). ESF, Strasbourg.

FAST, COMMISSION DES COMMUNAUTÉS EUROPÉENNES (1983), *EUROPE 1995: Mutations Technologiques et Enjeux Sociaux:* Rapport FAST. futuribles, Paris.

FAST, EUROPEAN COMMISSION (1982), *The FAST Programme,* Vol. I: Results and Recommendations.

FAST, EUROPEAN COMMISSION (1983), *A Community Strategy for Biotechnology in Europe.* FAST Occasional Papers No. 62.

FAST, EUROPEAN COMMISSION (1984), *EUROFUTURES: The Challenges of Innovation* (the FAST Report), Butterworths, London, in association with the journal *Futures.*

FAST-GRUPPE, KOMMISSION DER EUROPÄISCHEN GEMEINSCHAFTEN (1987), *Die Zukunft Europas: Gestaltung durch Innovationen.* Springer-Verlag, Berlin.

FEDERAL REGISTER (1986), *Office of Science and Technology Policy, Coordinated Framework for Regulation of Biotechnology;* Announcement of Policy and Notice for Public Comment, pp. 23302–23393, June 26.

FROMMER, W. (1987), in GIBBS et al. (1987), Chapter 19, *Federal Republic of Germany.*

FROMMER, W., and 18 other authors (1989), Safe biotechnology III: Safety precautions for handling microorganisms of different risk classes, *Appl. Microbiol. Biotechnol.* **30**, 541–552.

FROMMER, W., and 27 other authors (1992), Safe biotechnology (4): Recommendations for safety levels for biotechnological operations with microorganisms that cause disease in plants, *Appl. Microbiol. Biotechnol.* **38**, 139–140.

FROMMER, W., and 27 other authors (1993), Safe biotechnology (5): Recommendation for safe work with animal and human cell cultures concerning potential human pathogens, *Appl. Microbiol. Biotechnol.* **39**, 141–147

GALTON, F. (1869), *Hereditary Genius.* Macmillan. (Republished in 1892 and in various modern editions, e.g., Fontana, 1962)

GERVERS, J. H. (1987), The NIMBY syndrome: is it inevitable? *Environment* **29** (8), 18–20.

GIBBS, J. N., COOPER, I. P., MACKLER, B. F. (1987), *Biotechnology and the Environment: International Regulation.* Stockton Press and Macmillan, London.

GIBSON, K. (1986), European aspects of the recombinant DNA debate, in: ZILINSKAS and ZIMMERMAN, pp. 55–71.

GIDDINGS, L. V., PERSLEY, G. (1990), *Biotechnology and Biodiversity.* UNEP/Bio. Div/SWGB. 1/3, Ad Hoc Working Group on Biotechnology, of the Ad Hoc Group of Experts on Biological Diversity. UNEP, Geneva and Nairobi.

GLOEDE, F. (1992), Biological safety in the use of genetic engineering, in: *Technology & Democracy: The Use and Impact of Technology Assessment in Europe, Proceedings of the 3rd European Congress on Technology Assessment,* Copenhagen, 4–7 November. Danish Board of Technology; printed by Kandrup, Copenhagen.

GROS, F., JACOB, F., ROYER, P. (1979), *Sciences de la vie et société.* Documentation française.

HAIGH, N. (1984), *EEC Environmental Policy and Britain.* ENDS (Environmental Data Services); subsequent editions published by Longmans.

HALVORSON, H. O. (1986), The impact of the recombinant DNA controversy on a professional scientific society, in: ZILINSKAS and ZIMMERMAN, pp. 73–91.

HASEBE, A. (1993), Regulatory system and application of rDNA organisms in Ministry of Agriculture, Forestry and Fisheries (MAFF), in: *Scientific Political and Social Aspects of Recombinant DNA,* 3–4 March, Tokyo, organized by Science and Technology Agency.

HMSO (1974), *Report of the Working Party on the Experimental Manipulation of the Genetic Composition of Micro-organisms* (Ashby Working Party), Cmnd. 5880, Her Majesty's Stationary Office, London.

HMSO (1976), *Report of the Working Party on the Practice of Genetic Manipulation* (Williams Working Party), Cmnd. 6600, Her Majesty's Stationary Office, London.

HMSO (1978), *Health and Safety at Work: Genetic Manipulation. Regulations and Guidance Notes.* Health and Safety Executive.

HMSO (1992a), *The Genetically Modified Organisms (Contained Use) Regulations 1992.* Statutory Instrument No. 3217/1992.

HMSO (1992b), *The Genetically Modified Organisms (Deliberate Release) Regulations 1992.*

HMSO (1993), *A Guide to the Genetically Modified Organisms (Contained Use) Regulations 1992.*

HOJO, T. (1993), Promotion of biotechnology in pharmaceutical industry, in: *Japan-U.S.A. Workshop on Scientific Political and Social Aspects of Recombinant DNA,* 3–4 March, Tokyo, organized by Science and Technology Agency.

HOUSE OF LORDS, Select Committee on Science and Technology (1993), *Report on Regulation of the United Kingdom Biotechnology Industry and Global Competitiveness.* HL Paper 80. Her Majesty's Stationary Office, London.

IBF (International Biotechnology Forum) (1992), *UNCED '92: Policies for Sustainable Development: The Role of Biotechnology.* IBF, Brussels–Tokyo–Washington, DC–Ottawa.

IBF (1994), *UNCED Biological Diversity Convention: A Statement of Principle by the International Bioindustry Forum.* IBF, Brussels–Tokyo–Washington, DC–Ottawa.

ICSU (International Council of Scientific Unions) (1992), *An Agenda of Science for Environment and Development into the 21st Century* (ASCEND 21), based on a Conference held in Vienna, November 1991. Cambridge University Press for ICSU, Paris.

KATAYAMA, S. (1993), Summary of guidelines for recombinant DNA experiments, in: *Japan-U.S.A. Workshop Scientific Political and Social Aspects of Recombinant DNA,* 3–4 March, Tokyo, organized by Science and Technology Agency.

KATZ, J., SATTELLE, D. B. (1991), *Biotechnology for All.* Hobsons Scientific, Cambridge. (Editions produced also in French, German, Italian, Spanish, Norwegian, Dutch)

KEATING, M. (1993), *The Earth Summit's Agenda for Change:* A plain language version of Agenda 21 and the other Rio Agreements. Centre for Our Common Future, Geneva.

KENNET, W. (1976), *The Futures of Europe.* Cambridge University Press, Cambridge.

KEVLES, D. (1992), *In the Name of Eugenics.* Pelican Books, El Toro, CA.

KIRSOP, B. (1993a), Development of European standards in biotechnology, *Pharm. Technol. Int.,* September, 36–44.

KIRSOP, B. (1993b), European standardization in biotechnology, *Trends Biotechnol.* **11**, 375–378.

KNAW (Royal Netherlands Academy of Arts and Sciences) (1977), *Report of the Committee in Charge of the Control on Genetic Manipulation.*

KNAW (1979), *Second Report of the Committee in Charge of the Control on Genetic Manipulation,* 1977/78.

KNAW (1980), *Third Report (and Final Report) of the Committee in Charge of the Control on Genetic Manipulation,* 1979–1980.

KÜENZI, M., and 17 other authors (1985), Safe biotechnology: General considerations, *Appl. Microbiol. Biotechnol.* **21,** 1–6.

KÜENZI, M., and 20 other authors (1987), Safe biotechnology 2: The classification of microorganisms causing diseases in plants, *Appl. Microbiol. Biotechnol.* **27,** 105.

LABORATORY OF THE GOVERNMENT CHEMIST (1986), *Biotechnology: A Plain Man's Guide to the Support and Regulations in the UK* (2nd Ed.). Department of Trade and Industry, London.

LAWTON, S. (1981), U.S. Congress: Congressional attitudes on recombinant DNA research, *Rec. DNA Tech. Bull.* **4,** 51–55.

LEMKOW, L. (1993), *Public Attitudes to Genetic Engineering: Some European Perspectives.* European Foundation for the Improvement of Living and Working Conditions, Dublin.

LÖNNGREN, R. (1992), *International Approaches to Chemicals Control: A Historical Overview.* KemI, Stockholm.

MCKELVEY, M. (1994), *Evolutionary Innovation: Early Industrial Uses of Genetic Engineering.* Kanaltryckeriet i Motala AB, Stockholm.

MCNALLY, R., WHEALE, P. (1991), *The Environmental Consequences of Genetically Engineered Live Viral Vaccines.* Report to DG-XI, Contract No. B6614/89/91.

MARLIER, E. (1992), *Eurobarometer 35.1: Opinions of Europeans on biotechnology in 1991,* in: DURANT, J., pp. 52–108.

MEADOWS, D. H., MEADOWS, D. L., RANDERS, J., BEJRENS, W. W. III (1972), *The Limits to Growth: A Report for the Club of Rome's Project on the Predicament of Mankind.* Universe Books, New York, and Potomac Associates, Washington, DC.

MELLON, M. (co-ordinator) (1986), *Biotechnology: A Legal/Regulatory Analysis.* A Paper Prepared in Conjunction with the Second Annual Brookings Conference on Biotechnology, February 18–19. Brookings Institution, Washington, DC.

MESAROVIC, M., PESTEL, E. (1974), *Mankind at the Turning Point: The Second Report to the Club of Rome.* Hutchinson, London.

MILEWSKI, A. (1981), Evolution of the NIH Guidelines, *Rec. DNA Tech. Bull.* **4** (4) 160–165.

MINISTRY OF HEALTH AND SOCIAL AFFAIRS, Oslo (1993), *Biotechnology Related to Human Beings.* (English language summary of) Parliamentary Report to the Storting No. 25 (1992–93), presented by the Norwegian Government on 12 March.

MINISTRY OF HEALTH AND WELFARE, Tokyo (1992), *Guidelines for Foods and Food Additives Produced by Recombinant DNA Techniques.* (In English and Japanese)

MOONEY, H. A., BERNARDI, G. (Eds.) (1990), *Introduction of Genetically Modified Organisms into the Environment.* Published on behalf of the Scientific Committee on Problems (SCOPE) and the Scientific Committee on Genetic Experimentation (COGENE) of the International Council of Scientific Unions (ICSU), by Wiley, New York.

NOMINÉ, G. (1989), La Bio-industrie s'organise, *Biofutur,* Nov., 41–42.

NORA, S., MINC, A. (1978), *L'informatisation de la Société.* La documentation française, Paris.

NORTH ATLANTIC ASSEMBLY (1981), *Genetic Manipulation* (Recombinant DNA research, application and regulation).

NOVO INDUSTRI (1987), *The Novo Report: American Attitudes and Beliefs about Genetic Engineering.* Prepared by Research & Forecasts, Inc., 110 East 59th Street, New York, NY 10022.

OECD (1976), (Restricted), Secretariat Note SPT (79) 1, 16 January, *Promises and Threats of Scientific Research.* OECD, Paris.

OECD (1986), *Recombinant DNA Safety Considerations.* OECD, Paris.

OECD (1988), *Biotechnology and the Changing Role of Government.* OECD, Paris.

OECD (1989), *Biotechnology: Economic and Wider Impacts.* OECD, Paris.

OECD (1990), *International Survey on Biotechnology Use and Regulations.* Environment Monographs No. 39. OECD, Paris.

OECD (1992), *Safety Considerations for Biotechnology – 1992.* OECD, Paris.

OECD (1993a), *Report on the Seminar on Scientific Approaches for the Assessment of Research Trials with Genetically Modified Plants,* 6–7 April 1992, Jouy-en-Josas. OECD, Paris.

OECD (1993b), *Biotechnology in the OECD Committee for Scientific and Technological Policy.* OECD, Paris.

OECD (1993c), *Preamble to Reports on Scientific Considerations Pertaining to the Environmental Safety of the Scale-up of Organisms Developed by Biotechnology.* OECD, Paris.

OECD (1993d), *Safety Considerations for Biotechnology: Scale-up of Crop Plants.* OECD, Paris.

OECD (1993e), *Traditional Crop Breeding Practices: A Historical Review to Serve as Baseline for Assessing the Role of Modern Biotechnology.* OECD, Paris.

OECD (1993f), *Field Releases of Transgenic Plants, 1986–1992: An Analysis.* OECD, Paris.

OECD (1993g), *Safety Evaluation of Foods Derived by Modern Biotechnology: Concepts and Principles.* OECD, Paris.

OECD and IABS (International Association for Biological Standardisation) (1994), *Proceedings of the OECD Workshop on Non-target Effects of Live Recombinant Vaccines,* Langen, 2–3 November 1993. Pub. IABS, Geneva.

OECD (1994a), *Ottawa '92: The OECD Workshop on Methods for Monitoring Organisms in the Environment.* Environment Monographs No 90. OECD, Paris.

OECD (1994b), *Compendium of Methods for Monitoring Organisms in the Environment.* Environment Monographs No. 91. OECD, Paris.

OECD (1994c), *Biotechnology for a Clean Environment: Prevention, Detection, Remediation.* OECD, Paris.

OECD (1994d), *Biotechnology Policies Given Closer Co-ordination by OECD. OECD Letter* 3/10, December.

OLSON, S. (1986), *Biotechnology: An Industry Comes of Age.* (US) National Academy Press Washington, DC. (Based on an academic conference on genetic engineering, 27–28 February, 1985)

OTA (US Congressional Office of Technology Assessment) (1981), *The Impacts of Applied Genetics: Applications to Microorganisms, Animals and Plants.* OTA, Washington, DC.

OTA (1987), *New Developments in Biotechnology: Background Paper 2: Public Perceptions of Biotechnology.* Congressional OTA, Washington, DC.

PELISSOLO, J. C. (1981), *La biotechnologie, demain?* Documentation française.

PERSLEY, G. J. (1990a), *Agricultural Biotechnology: Opportunities for International Development.* CAB International.

PERSLEY, G. J. (Ed.) (1990b), *Beyond Mendel's Garden: Biotechnology in the Service of World Agriculture.* CAB International.

PERSLEY, G. J., GIDDINGS, L. V., JUMA, C. (1992), *Biosafety: The Safe Application of Biotechnology in Agriculture and the Environment.* The World Bank/International Service for National Agricultural Research, The Hague.

PERUTZ, M. (1981), Why we need science, *New Sci.* **92**, 530–536.

PRIEELS, A.-M. (1991), *United Nations Conference on Environment and Development. Senior Experts Workshop on Environmentally Sound Applications of Biotechnology,* New Delhi, 23–25 October. Report to the Commission of the European Communities.

RABINO, I. (1992), A study of attitudes and concerns of genetic engineering scientists in Western Europe, *Biotech Forum Europe* **9** (10), 636–640.

RABINO, I. (1991), The impact of activist pressures on recombinant DNA research, *Sci. Technol. Human Values* **16** (1) Winter; 70–87.

RABINO, I. (1994), German genetic engineering scientists and the German public: Complementary perceptions in a changing European context, *Publ. Understand. Sci.* **3**, 365–384.

ROBERTS, E. (1989), *The Public and Biotechnology.* A discussion document, based upon the Workshop on Consumers and Biotechnology, Brussels, February 1989. European Foundation for the Improvement of Living and Working Conditions, Dublin.

RÖRSCH, A. (1978), *Genetic Manipulations in Applied Biology: A Study of the Necessity, Content and Management Principles of a Possible Community Action.* Document EUR6078, European Commission.

ROYAL COMMISSION ON ENVIRONMENTAL POLLUTION (1989), *Thirteenth Report: The Release of Genetically Engineered Organisms to the Environment.* Cm 720, Her Majesty's Stationary Office, London.

POYER, P. (1981), *La Sécurité des applications industrielles des biotechnologies.* Rapport du groupe de réflexion, présenté à M. le ministre de l'Industrie. Documentation française, Paris.

RYSER, S., WEBER, M. (1990), *Genetic Engineering – A Chronology* (Second Revised Edition). Editiones (Hoffmann-La Roche), Basel.

SAGB (Senior Advisory Group for Biotechnology) (1990a), *Community Policy for Biotechnology: Priorities and Actions.* SAGB, Brussels.

SAGB (1990b), *Community Policy for Biotechnology: Economic Benefits and European Competitiveness.* SAGB, Brussels.

SAGB (1994a), *Biotechnology Policy in the European Union: Prescriptions for Growth, Competitiveness and Employment.* SAGB, Brussels.

SAGB (1994b), *Biotechnology Policy in the European Union: Competitiveness, Investment and the "Cycle of Innovation".* SAGB, Brussels.

SAKAGUCHI, K. (1972), *Historical Background of Industrial Fermentation in Japan;* in: *Fermentation Technology Today, Proceedings of the IVth International Fermentation Symposium,* Kyoto (TERUI, G., Ed.).

SARGEANT, K., EVANS, C. G. T. (1979), *Hazards Involved in the Industrial Use of Micro-organisms: A Study of the Necessity, Content and Management Principles of a Possible Community Action.* European Commission EUR 6349 EN.

SCREEN (Swift Community Risk Evaluation Effort Network) (1994), *Swedish Legislation and Procedures, Screen* **2**, September, pp. 12–14.

SHAPIRO (1990), Biotechnology and the design of regulation, *Ecol. Law Quart.* **17**, 1.

SINGER, M. F., SOLL, D. (1973), DNA hybrid molecules, *Science* **181**, 1114.

SMINK, G. C. J., HAMSTRA, A. M. (1994), *Impacts of New Biotechnology in Food Production on Consumers.* SWOKA Research report no. 170, October. SWOKA, The Hague.

SUZUKI, Y. (1993), Assuring safety of foods and food additives produced by recombinant DNA techniques, in: *Japan-U.S.A. Workshop on Scientific Political and Social Aspects of Recombinant DNA*, 3–4 March, Tokyo, organized by Science and Technology Agency.

SWEDISH GOVERNMENT COMMITTEE ON GENETIC ENGINEERING (1992), *Genetic Engineering – A Challenge.* Allmänna Förlaget, Stockholm.

TAIT, J. (1988a), Public perception of biotechnology hazards, *J. Chem. Tech. Biotechnol.* **43**, 363–372.

TAIT, J. (1988b), NIMBY and NIABY: Public perception of biotechnology, *Int. Ind. Biotechnol.* **8** (6), 5–9.

TAIT, J., LEVIDOW, L. (1992), Proactive and reactive approaches to risk regulation, *Futures*, April, 219–231.

THOMAS, D. (1978), *Production of Biological Catalysts, Stabilisation and Exploitation.* Document EUR6079, European Commission.

THORLEY, J. F. (1994), *The Development of Good Industrial Large Scale Practice in Biotechnology.* European Commission.

TOOZE, J. (1977), Genetic engineering in Europe, *New Sci.*, 10 March, 592–594.

UCHIDA, H. (1993), Building a science in Japan: The formative decades of molecular biology, *J. Hist. Biol.* **26** (3), 499–517.

UNIDO (United Nations Industrial Development Organization) (1991), *Voluntary Code of Conduct for the Release of Organisms into the Environment.* Prepared by the UNIDO Secretariat for the Informal UNIDO/UNEP/WHO/FAO Working Group on Biosafety. UNIDO, Vienna.

USDA (United States Department of Agriculture) (1995), *Proceedings of the 3rd International Symposium on the Biosafety Results of Field Tests of Genetically Modified Plants and Microorganisms*, 13–16 November, 1994, Monterey, California. To appear.

US-EC TASK FORCE FOR BIOTECHNOLOGY RESEARCH (1992), *Methods of Communicating Biotechnology with the Public.* Final Report of the United States – Commission of the European Communities Workshop, 22–25 March.

Pub. by Office of Agricultural Biotechnology, US Dept. of Agriculture, Washington, DC.

VAN DEN BROEK, J. M. (1988), Public opinion on biotechnology: (un)known, (un)loved? *Biotechnologie in Nederland* **5**, 15/259–25/269.

VAN DER MEER, P. J., SCHENKELAARS, P., VISSER, B., ZWANGOBANI, E. (Eds.) (1993), *Proceedings of the African Regional Conference for International Cooperation on Safety in Biotechnology*, 11–14 October, Harare, Zimbabwe.

VAN HOECK, F., DE NETTANCOURT, J. D. (1990), Introduction (re biotechnology, safety and risk assessment), in: ECONOMIDIS (Ed.).

VIEHOFF, P. (rapporteur) (1985), *Biotechnology Hearing.* Outline, papers assembled as background for the Hearing, 20–21 November. PE 98.227/rev, European Parliament.

WATSON, J. D., TOOZE, J. (1981), *The DNA Story: A Documentary History of Gene Cloning.* W. H. Freeman and Company, San Francisco.

WHEALE, P., MCNALLY, R. (1993), Biotechnology policy in Europe: a critical evaluation, *Sci. Publ. Pol.*, August, 261–279.

WHO (World Health Organization) (1983), Quality control of biologicals produced by recombinant DNA techniques: WHO consultation, *Bull. WHO* **61** (6), 897–911.

WHO (1984), *Health Impact of Biotechnology.* Report on a WHO Working Group, Dublin, 9–13 November 1982. WHO Regional Office for Europe, Copenhagen; Health Aspects of Chemical Safety, Interim Document 16.

WHO (1987), *Principles for the Safety Assessment of Food Additives and Contaminants in Food, Environ. Health Crit.* **70**; published under the joint sponsorship of the UN Environment Programme, the International Labour Organization, and the World Health Organization, in collaboration with the Food and Agriculture Organization of the United Nations, Geneva.

WHO (1990), *Principles for the Toxicological Assessment of Pesticide Residues in Food, Environ. Health Crit.* **104**; published under the joint sponsorship of the UN Environment Programme, the International Labour Organization, and the World Health Organization, Geneva.

WHO (1991), *Strategies for Assessing the Safety of Foods Produced by Biotechnology.* Report of a Joint FAO/WHO Consultation.

WHO (1994), *Conclusions and Recommendations of the Working Group on Biotechnology and World Health: Vaccines and Other Medical Products Produced by Genetic Engineering. Review of Risks and Benefits.* WHO, Geneva, 6–8 November.

WITKOWSKI, J. (1988), Fifty years on: molecular biology's hall of fame, *Trends Biotechnol.* **6**, 234–243.

WORLD COUNCIL OF CHURCHES (1989), *Biotechnology: Its Challenges to the Churches and to the World.* Report by WCC Subunit on Church and Society. WCC, Geneva.

YOUNG, F. E., MILLER, H. I. (1989), "Deliberate releases" in Europe: over-regulation may be the biggest threat of all, *Gene* **75,** 1–2.

YOXEN, E., DI MARTINO, V. (Eds.) (1989), *Biotechnology in Future Society: Scenarios and Options for Europe.* Publication No. EF/88/08/EN of the European Foundation for the Improvement of Living and Working Conditions. Published by Dartmouth, Aldershot, England, for the Office for Official Publication of the European Communities, Luxembourg.

ZECHENDORF, B. (1994), What the public thinks about biotechnology, *Bio/Technology* **12,** 9 September, 870–875.

ZILINSKAS, R. A., ZIMMERMAN, B. K. (Eds.) (1986), *The Gene-Splicing Wars: Reflections on the Recombinant DNA Controversy.* Macmillan, New York, and Collier Macmillan, London.

ZINDER, N. D. (1986), A personal view of the media's role in the recombinant DNA war, in: ZILINSKAS and ZIMMERMAN, pp. 109–118.

Index